Circle

r is the radius   d is the diameter
$$d = 2r$$
Area $A = \pi r^2$
Circumference $C = 2\pi r$ or $C = \pi d$

Rectangular Solid

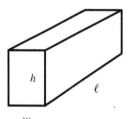

Volume $V = w\ell h$

Sphere

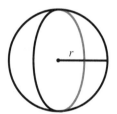

Volume $V = \dfrac{4}{3}\pi r^3$

Right Circular Cylinder

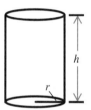

Volume $V = \pi r^2 h$
Surface Area $S = 2\pi r^2 + 2\pi rh$

Right Circular Cone

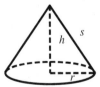

Volume $V = \dfrac{1}{3}\pi r^2 h$
Surface Area $A = \pi r^2 + \pi rs$

Right Pyramid

Volume $V = \dfrac{1}{3}Bh$
where **B** is the area of the base

INTERMEDIATE

Algebra

Larry R. Mugridge *Kutztown University*

INTERMEDIATE

Algebra

SAUNDERS COLLEGE PUBLISHING

Philadelphia Fort Worth Chicago San Francisco Montreal Toronto London Sydney Tokyo

Text Typeface: 10/12 Times Roman
Compositor: York Graphic Services
Acquisitions Editor: Robert Stern
Developmental Editor: Ellen Newman
Managing Editor: Carol Field
Project Editor: Maureen Iannuzzi
Copy Editor: Charlotte Nelson
Manager of Art and Design: Carol Bleistine
Art and Design Coordinator: Doris Bruey
Text Designer: Tracy Baldwin
Cover Designer: Lawrence R. Didona
Text Artwork: York Graphic Services
Director of EDP: Tim Frelick
Production Manager: Charlene Squibb
Marketing Manager: Denise Watrobsky

Cover Credit: Kandinsky: Circles in a Circle; Philadelphia Museum of Art: Louise and Walter
                       Arensberg Collection.

Printed in the United States of America

INTERMEDIATE ALGEBRA

ISBN: 0-03-009477-1

Library of Congress Catalog Card Number: 90-053232

123   032   987654321

This book is dedicated to my children,
Lyn and Jason

Preface

This book is intended for students who have a knowledge of elementary algebra and need to cover intermediate algebra before taking other college mathematics courses. Each concept is introduced by example before the definition is given, and the student is encouraged to become an active participant. A complete pedagogical system that is designed to motivate the student and make mathematics more accessible includes the following features:

Key features

Chapter Overviews   Each chapter begins with an Overview to introduce the student to the material and demonstrate its relevance to the real world.

Objectives   Each section contains Objectives to help the student focus on skills to be learned in the chapter.

Examples with Solutions   Approximately 550 examples with complete, worked-out solutions are provided; explanations are highlighted in blue print. Some of the examples contain steps that can be done mentally once the student has achieved a certain level of proficiency (although it may not be possible at first); this logo calls the student's attention to the highlighted parts of the equation. A blue triangle indicates the end of the solution.

Common Error Boxes   Wherever appropriate, students are shown errors that are commonly made. The correct approach is then illustrated and explained.

Strategies   After developing a mathematical technique, such as factoring a trinomial, a summary of the technique is given as a strategy. This will provide the student with a reference when doing problems from the exercise set.

Calculator Crossovers   Many of the examples are followed by problems that are specifically designed to be solved with a calculator. These problems are similar to the preceding examples and illustrate the role of the calculator in solving problems. Calculator Crossovers are set off by a calculator logo for easy reference. Calculator Crossovers may be used at the instructor's discretion.

Learning Advantages   These hints occur in Chapters 1 through 6 to give students additional help; they immediately precede the end-of-section exercises. For more detailed explanations on studying algebra, I recommend the pamphlet *How to Study Mathematics* by James Margenau and Michael Sentlowitz (The National Council of Teachers of Mathematics, Inc., Reston, Virginia, 1977) and *Mathematics at Work in Society: Opening Career Doors* by James R. Choike (Mathematical Society of America, Washington, D.C., 1988), from which some of the Learning Advantages are adapted.

Exercises   Graded by level of difficulty, the exercises (approximately 6500) are carefully constructed so that there is odd-even pairing. In addition, each exercise problem is related to an example from the section. The numbers used in the examples and exercises have been carefully selected so that the student becomes confident in using fractions and decimals in addition to the integers. The numbers, however, do not become so complicated as to ''turn off'' the student or teacher. All computations can be done without a calculator.

Beginning with Chapter 2, each exercise set contains **Review Exercises** to help the student understand and remember concepts studied in earlier sections. Each review exercise set covers background material essential to the following section. The **Enrichment Exercises** at the end of each section are challenging problems that the more capable students will be able to solve. In the student's edition, answers to the odd-numbered exercises, Review Exercises, and all of the Enrichment Exercises are given in the Answer Section following the text.

Applications   Whenever appropriate, word problems have been included so that the student constantly practices translating word phrases and statements into mathematical equations. The word problems are realistic and are culled from applications in geometry, physics, business, economics, and psychology. The structure of problem solving is emphasized throughout the text, beginning with the simple conversion of word phrases into mathematical expressions. Many concepts introduced in early chapters are repeated using different applications.

Chapter Summary and Review   Each chapter concludes with a comprehensive review which includes definitions and strategies learned in the chapter. All terms are keyed to the section number for easy reference. Marginal examples are included to further illustrate the concepts reviewed.

Review Exercise Set   These exercises are included at the end of the chapter and represent all types of problems to ensure that the student has attained a level of proficiency and is comfortable proceeding to the next chapter. All exercises are keyed to sections so the student can refer to the text for assistance. Answers to all of these exercises can be found in the back of the book.

Chapter Test   Each chapter concludes with a Chapter Test that contains representative problems from each section.

Pedagogical Use of Color   This text uses color in the figures and to highlight the various pedagogical features throughout the text. Multiple colors are useful, for example, to distinguish between two lines that are graphed simultaneously, or to highlight important statements. The complete color system is described in more detail on page xv.

An overview of the book

The main thrust of this book is to enable students to develop algebraic skills to be used to solve word problems. This is the primary aim of each chapter; we start with developing a skill and then use it to solve word problems. In addition, many geometric problems are included. A chapter-by-chapter overview follows:

Chapter 1
Operations and variables

Since the foundation of algebra is arithmetic, it is reviewed comprehensively in this chapter, starting with addition, subtraction, multiplication, and division of fractions. Next, we review basic symbols and order of operations. The set of real numbers is examined along with the subsets of natural, whole, rational, and irrational numbers. Then, we review the four basic operations on real numbers: addition, subtraction, multiplication, and division. In Section 1.4, we deal with the properties of real numbers. We study variables and variable expressions in the final section.

Chapter 2
Linear equations and inequalities

In this chapter, algebraic expressions are used to construct linear equations and inequalities. In Section 2.1, we investigate properties of equality that will be used to solve linear equations, which constitutes Section 2.2. In Section 2.3, we solve word problems that require solving equations. Next, we develop formulas with associated applications from geometry and trigonometry (area and volume formulas of regions and solids). We also develop formulas from business, such as simple interest and cost-revenue-profit. Section 2.5 deals with solving inequalities, and in the final section we examine equations and inequalities with absolute value.

Chapter 3
Polynomials and factoring

In the first section, we study multiplication properties with exponents and simplify expressions using properties of exponents. In the next two sections, we introduce the algebra of polynomials, which is followed by four sections on factoring. Coverage includes factoring of monomials; general trinomials; and special polynomials such as the difference of two squares, perfect square trinomials, and the sum of the difference of two cubes. We also develop factoring by grouping and state a strategy for general factoring. These techniques are then applied to solving quadratic equations that are factorable, which, in turn, leads to applications of quadratic equations. The applications include using the Pythagorean Theorem, solving geometric problems, and understanding applications to business that involve quadratic equations.

Chapter 4
Rational expressions

In the first section, we define rational expressions, evaluate them, and determine where a rational expression is defined. In Section 4.2, we develop the techniques of dividing polynomials, which leads to the algebra of rational expressions in Sections 4.3 and 4.4. Next, we discuss complex fractions, where we develop methods to simplify complex fractions into rational expressions. We then solve equations containing rational expressions and introduce formulas that involve rational expressions such as the total resistance formula.

Chapter 5
Roots and radicals

We introduce square roots and higher roots in Section 5.1 and connect these concepts with rational exponents in Section 5.2. In the next four sections, we deal with the algebra of radical expressions and simplifying radical expressions. Radical equations and word problems using radical equations are solved in Section 5.7. We conclude this chapter by defining a complex number in terms of i and covering the algebra of complex numbers.

Chapter 6
Quadratic equations

One of the main goals of this book is developing techniques for solving equations. In this chapter, we return to the problem of solving the general quadratic equation $ax^2 + bx + c = 0$, where the trinomial may not be factorable. We begin with a discussion of the square root method to solve quadratic equations and then develop the completing of the square method. The quadratic formula is discussed in Section 6.3 and then used in applications of quadratic equations. Next, we investigate quadratic and rational inequalities. In the final section, we solve quadratic equations using complex numbers and solve equations that are quadratic in form.

Chapter 7
Linear equations in two variables and their graphs

In this chapter, we investigate the connection between analytic geometry and its applications, beginning by plotting points in a plane and then developing linear equations and their graphs. In Section 7.2, we find the slope of a line given two points on the line. We also find the slope of a line from its graph. The difference between zero slope and undefined slope is explained, and the geometric meaning of positive and negative slope is discussed. In the next section, we find the equation of a line given (a) the slope and y-intercept, (b) the slope and a point on the line, or (c) two points on the line. Section 7.4 discusses graphing linear inequalities in two variables. Section 7.5 includes many business and economic applications of linear equations, such as linear depreciation, linear cost-revenue-profit, and linear supply and demand equations. The final section deals with variation and applications in such fields as physics.

Chapter 8
Systems of equations

In this chapter, we continue to examine the connection between analytic geometry and its applications. We show how a system of two equations in two variables enables us to study more complex applications. Our study of systems of equations begins by finding solutions using three methods: the graphing method, the elimination method, and the substitution method. In Section 8.3, we develop techniques to solve a system of linear equations in three variables, which is followed by the development of Cramer's rule and matrix methods for solving systems of equations in Sections 8.4 and 8.5. In the final section, we solve systems of linear inequalities.

Chapter 9
Conic sections

This chapter focuses on four conic sections in Sections 9.1 through 9.4: parabolas, circles, ellipses, and hyperbolas. In these discussions, our aim is to obtain the standard form equations for these conic sections. Nonlinear systems of equations and inequalities are covered in the last two sections of this chapter.

Chapter 10
Functions

There are many real-world examples of dependence of one quantity upon another; here we develop the mathematical basis to describe the concept of dependency. Beginning with the definition of relation and function in Section 10.1, we show how to determine when a relation is also a function. Function notation and the algebra of functions is introduced in Section 10.2, where the student learns to form the quotient $[f(x + h) - f(x)]/h$ and to simplify the result. Section 10.3 develops linear, quadratic, polynomial, and exponential functions, and in the last section of the chapter the student learns the concept of one-to-one and how to find an equation for the inverse function when it exists.

Chapter 11
Exponential and logarithmic functions

We introduce the exponential function to the base e and logarithmic functions that lead to an ever greater variety of applications. In Section 11.1, the exponential function $f(x) = e^x$ is introduced along with a plethora of applications such as exponential growth, compound interest, and growth and decay equations. The logarithmic functions are then developed along with the technique of solving the equation $y = \log_b x$ for y, x, or b. The next section deals with the properties of logarithms, which are used to solve logarithmic equations. The chapter concludes with a section that involves solving exponential equations by using either a calculator or by using the tables in the appendix.

Chapter 12
Sequences and series

This chapter focuses on a special kind of function—the sequence. The first section introduces the sequence notation as well as the technique for finding the nth term of a sequence. Section 12.2 then introduces the concept of series along with the sigma notation to write a series. Arithmetic and geometric sequences are discussed in the next section, showing the student how to find the sum of the first n terms of an arithmetic series, the general term of a geometric sequence, and the sum of the first n terms of a geometric series. Section 12.4 develops the binomial theorem along with Pascal's triangle to find the binomial coefficients.

State requirements

Particular attention has been paid to the testing requirements for various states (e.g., TASP, ELM, CLAST, etc.) Please see pages H1–H4 for additional information on how this text meets the requirements for your state.

Ancillary package

The following supplements are available to the student to accompany this text:*

Student Solutions Manual and Study Guide (Linda Holden, Indiana University) Contains step-by-step solutions to one fourth of the problems in the exercise sets (every other odd-numbered problem) in addition to providing the student with a short summary of the important concepts in each chapter.

Videotapes A complete set of videotapes (18 hours), free to adopters, has been created and scripted by the author. Keyed to the text, the videotapes explain,

---

*For the instructor we have an Instructor's Manual with solutions to all the exercises not included in the Student Solutions Manual; Prepared Tests with six tests for each chapter, as well as midterm and final examinations and a Diagnostic Test; a Computerized Test Bank for the Apple II, IBM, and Macintosh computers; and a Printed Test Bank containing tests generated from the Computerized Test Bank.

using computer graphics, examples with corresponding practice problems. The student can participate by stopping the tape to work the practice problems on his or her own and then checking the solutions by continuing the tape. The tapes are available in VHS format and include a list of topics and the amount of time spent on each section.

One-to-One Interactive Software (George W. Bergeman, Northern Virginia Community College)   Available for both Apple and IBM, this software allows students to test their skills and to pinpoint and correct weak areas. One-to-One presents worked examples and displays annotated, step-by-step solutions to problems answered incorrectly. Students may also choose to see the solutions to problems answered correctly and to view a partial solution if they need help to begin solving a problem. The software includes two useful review capabilities: Missed Problems Review and Disk Review. As the student works, One-to-One keeps track of problems that are answered incorrectly. When the work on a topic is completed, the students are given the option of reviewing all problems answered incorrectly. In addition, the Disk Review feature provides a quick and efficient review of all the topics on the disk.

One-to-One Solution Finder (George W. Bergeman, Northern Virginia Community College)   Available for IBM, this software allows students to input their own questions through the use of an expert system, a branch of artificial intelligence. Students may check their answers or receive help as if they were working with a tutor. The software will refer the student to the appropriate section of the text and will record the number of problems entered and correct answers given. Featuring a function grapher, the students can zoom in and out, evaluate a function at a point, graph up to four functions simultaneously, and save and retrieve function setups via disk files.

Acknowledgments

This text was prepared with the assistance of many professors who reviewed the manuscript throughout the course of its development. I wish to acknowledge the following people and express my appreciation for their suggestions, criticisms, and encouragement:

Robert Billups, Citrus College
Mary Chabot, Mt. San Antonio College
Arthur P. Dull, Diablo Valley College
Lois Higbie, Brookdale Community College
Linda Holden, Indiana University
Herbert Kasube, Bradley University
Susan McClory, San Jose State University
Jane Morrison, Thornton Community College
Karla Neal, Louisiana State University
Bob Olsen, Jefferson Community College
Harvey E. Reynolds, Golden West College
Howard Sorkin, Broward Community College
Jan Vandever, South Dakota State University

I would also like to thank the following reviewers who were invaluable in ensuring the accuracy of the text: June Oliverio-Wallace and Sue Landers.

Special thanks to the following people for their work on the various ancillary items that accompany *Intermediate Algebra:*

George W. Bergeman, Northern Virginia Community College (One-to-One Interactive Software and One-to-One Solution Finder)

Howard Sorkin and Mitchel Levy, Broward Community College (Prepared Tests)

Linda Holden, Indiana University (Student Solutions Manual and Study Guide)

Nicholas Kernene, Indiana University (Instructor's Resource Manual)

Mary Chabot, Mt. San Antonio College (Test Questions for Computerized Test Bank)

I want to thank the dedicated staff at Saunders College Publishing for their tireless energy throughout the past three years: Bob Stern, Senior Mathematics Editor, with his impeccable sense of timing; Ellen Newman, Developmental Editor, who spent long hours, that frequently shortened the evening, on the book; and Maureen Iannuzzi, Project Editor, who guided the book to a successful conclusion with her expertise in converting the manuscript to a polished full-color book.

Larry R. Mugridge
Kutztown University
December 1990

Pedagogical Use of Color

The various colors in the text figures are used to improve clarity and understanding. Many figures with three-dimensional representations are shown in various colors to make them as realistic as possible. Color is used in those graphs where different lines are being plotted simultaneously and need to be distinguished.

In addition to the use of color in the figures, the pedagogical system in the text has been enhanced with color as well. We have used the following colors to distinguish the various pedagogical features:

P R O P E R T Y

S T R A T E G Y

C A L C U L A T O R   C R O S S O V E R

D E F I N I T I O N

C O M M O N   E R R O R

R U L E

Table No.   Table title

Col. head   *Col. head*

Introducing the Book

The **Overview** introduces the material to the student and demonstrates its relevance to the real world.

Overview

The foundation of algebra is arithmetic. In order to perform *algebraic* operations, a thorough understanding of performing *arithmetic* operations is necessary. Therefore, in this first chapter, we review arithmetic starting with a review of fractions. Next, we study the structure of the set of real numbers together with operations on real numbers. Properties of real numbers are followed by a review of exponents, roots, and order of operations. In the final section, we study variables and variable expressions. At this time, we introduce the important concept of writing verbal statements as algebraic expressions. Converting verbal statements into algebraic expressions is the basis for using algebra to solve word problems and will be practiced throughout the text.

1.1

Fractions–A Review

OBJECTIVES

Objectives help the student focus on skills to be learned in the chapter.

▶ *To write fractions in lowest terms*

▶ *To multiply and divide fractions*

▶ *To add and subtract fractions with common denominators*

▶ *To add and subtract fractions with unlike denominators*

In our daily living we encounter the **whole numbers**

$$0, 1, 2, 3, 4, \ldots$$

as well as **fractions** such as $\frac{1}{2}$, $\frac{4}{5}$, and $\frac{3}{9}$. In this section we review some basic concepts involving fractions. In particular, we will review the techniques used to add, subtract, multiply, and divide fractions. These four operations will be discussed in greater detail in later sections. The purpose here is to strengthen your ability to work with fractions, since these same techniques will be used in

2

Strategies are highlighted with a gold and orange screen.

algebra. In the fraction $\frac{2}{3}$, the number 2 is called the **numerator** and the number 3 is the **denominator.** The fraction is in **lowest terms** if the numerator and denominator have no common factors. For example, the fraction $\frac{2}{3}$ is in lowest terms, while $\frac{4}{6}$ is not in lowest terms, since the numerator and the denominator have a **common factor** of 2.

The following observation is used when we write a fraction in lowest terms.

The value of a fraction remains the same, when the numerator and denominator are divided by the same nonzero number.

To write a fraction in lowest terms, we can use the following steps.

STRATEGY

To write a fraction in lowest terms

Step 1   Find the largest nonzero number c that is a common factor of the numerator and the denominator. That is, the fraction can be written as

$$\frac{\text{numerator}}{\text{denominator}} = \frac{a \cdot c}{b \cdot c}$$

Step 2   Then,

$$\frac{a \cdot c}{b \cdot c} = \frac{a}{b}$$

That is, we divide numerator and denominator by the common factor c, and the resulting fraction is now in lowest terms.

For example, to write $\frac{4}{6}$ in lowest terms, write the numerator as $2 \cdot 2$ and the denominator as $3 \cdot 2$. Therefore,

$$\frac{4}{6} = \frac{2 \cdot 2}{3 \cdot 2} \qquad \text{The number } c \text{ is 2 for this fraction.}$$

$$= \frac{2}{3}$$

Therefore, $\dfrac{60}{105} = \dfrac{4}{7}$.

The shorthand method to reduce this fraction looks like this:

$$\frac{60}{105} = \frac{\overset{\overset{4}{\cancel{12}}}{\cancel{60}}}{\underset{\underset{7}{\cancel{21}}}{\cancel{105}}} = \frac{4}{7}$$

COMMON ERROR

When reducing a fraction to lowest terms, be sure that the numerator and denominator are written as products and not as sums. Compare the following two sequences of steps.

Sequence 1: $\dfrac{12}{15} = \dfrac{4 \cdot 3}{5 \cdot 3} = \dfrac{4 \cdot \cancel{3}}{5 \cdot \cancel{3}} = \dfrac{4}{5}$ Correct.

Sequence 2: $\dfrac{9}{10} = \dfrac{1 + 8}{2 + 8} = \dfrac{1 + \cancel{8}}{2 + \cancel{8}} = \dfrac{1}{2}$ Incorrect.

Sequence 1 is correct, since 3 is a common factor. Sequence 2 is incorrect, since 8 is not a common *factor*.

We next consider combining fractions using the operations of addition, subtraction, multiplication, and division. Since the methods for multiplying and dividing fractions are simpler than for addition and subtraction, we start with multiplication. To multiply two fractions, we use the following rule.

RULE

Multiplication of two fractions

Given two fractions $\dfrac{a}{b}$ and $\dfrac{c}{d}$, then

$$\frac{a}{b} \cdot \frac{c}{d} = \frac{a \cdot c}{b \cdot d}$$

That is, the answer from multiplying two fractions is a new fraction obtained by multiplying the numerators and then multiplying the denominators.

Common errors are highlighted with a blue and yellow screen.

Rules are highlighted with a light and medium brown screen.

(b) A common denominator of the three fractions is 6. Therefore,

$$\frac{3}{2} + \frac{5}{3} - \frac{5}{6} = \frac{3 \cdot 3}{2 \cdot 3} + \frac{5 \cdot 2}{3 \cdot 2} - \frac{5}{6}$$

$$= \frac{9}{6} + \frac{10}{6} - \frac{5}{6}$$

$$= \frac{9 + 10 - 5}{6}$$

$$= \frac{\overset{7}{\cancel{14}}}{\underset{3}{\cancel{6}}}$$

$$= \frac{7}{3} \qquad\qquad\qquad 2 \text{ is a common factor.} \quad ▲$$

Examples provide complete worked-out solutions; each example begins with a light brown bar and ends with a blue triangle.

Example 9   Add: $4\frac{1}{2} + 1\frac{3}{4}$.

Solution   We first change each mixed number into a fraction.

$$4\frac{1}{2} = 4 + \frac{1}{2} = \frac{4}{1} + \frac{1}{2} = \frac{8}{2} + \frac{1}{2}$$

$$= \frac{8 + 1}{2} = \frac{9}{2}$$

$$1\frac{3}{4} = 1 + \frac{3}{4} = \frac{4}{4} + \frac{3}{4} = \frac{4 + 3}{4} = \frac{7}{4}$$

Next, we add the two fractions.

$$4\frac{1}{2} + 1\frac{3}{4} = \frac{9}{2} + \frac{7}{4}$$

$$= \frac{18}{4} + \frac{7}{4} \qquad \text{Replace } \frac{9}{2} \text{ by } \frac{9 \cdot 2}{2 \cdot 2} \text{ or } \frac{18}{4}.$$

$$= \frac{25}{4} \text{ or } 6\frac{1}{4} \qquad\qquad\qquad ▲$$

Learning Advantages appear in Chapters 1 to 8 to provide additional hints for students.

L E A R N I N G   A D V A N T A G E   *Before starting on each homework assignment, first study the notes you took in class and then read the appropriate section(s) from the book. If you have trouble on a particular exercise problem, find the example that most closely pertains to it. Remember that the problems in the exercise set are paired so that each even-numbered problem matches the previous odd-numbered problem.*

This logo alerts students to operations or steps that can be performed mentally.

Example 4   Find each sum.

(a) $\dfrac{1}{2} + \dfrac{3}{4}$ (b) $-3 + (-21)$

(c) $-4 + 12$ (d) $15 + (-35)$

Solution   (a) We find the sum of two positive fractions using the technique reviewed in Section 1.1.

$$\frac{1}{2} + \frac{3}{4} = \frac{2}{4} + \frac{3}{4}$$

$$= \frac{2 + 3}{4} \qquad \text{This step can be done mentally.}$$

$$= \frac{5}{4}$$

(b) Here the two numbers are negative. Therefore, we first add their absolute values, then attach a negative sign to the answer.

$$-3 + (-21) = -(3 + 21)$$

$$= -24$$

(c) Since the two numbers are of opposite sign, we subtract the smaller absolute value (4) from the larger absolute value (12). The sign of the answer is the sign of the *original* number with the larger absolute value.

$$-4 + 12 = (12 - 4)$$

$$= 8$$

(d) The two numbers are of opposite sign, so we subtract the smaller absolute value (15) from the larger absolute value (35). The answer is given the ''$-$'' sign, since the larger absolute value came from -35.

$$15 + (-35) = -(35 - 15)$$

$$= -20 \qquad \blacktriangle$$

R U L E

Subtraction of real numbers

For any two real numbers a and b,

$$a - b = a + (-b)$$

That is, to subtract b from a, add the opposite of b to a.

Introducing the Book

Definitions are highlighted with a light and medium tan screen.

Properties are highlighted with a gray and light blue screen.

1.3 Operations on Real Numbers

25

They are each the same distance from the origin. We say that 3 and -3 are *opposites* of each other. The number -3 is the opposite of 3, and 3 is the opposite of -3.

DEFINITION

Let a be a real number. The **opposite** of a is $-a$.

Therefore, to find the opposite of a, change the sign of a. For example, if a is $\frac{4}{5}$, the opposite of a is $-\frac{4}{5}$. If a is -20, the opposite of a is $-(-20)$ or 20. This illustrates the double negation property.

PROPERTY

The double negation property

For any real number a,

$$-(-a) = a$$

Example 1   **(a)** $-(-8) = 8$ **(b)** $-(-9.3) = 9.3$

(c) $-\left[-\left(-\frac{5}{2}\right)\right] = -\left[\frac{5}{2}\right] = -\frac{5}{2}$ ▲

To indicate distance but not direction, we use the *absolute value* of a number. Geometrically, the **absolute value** of a number, x, is the distance that the given number is from the origin on a number line. The absolute value of the number x is written as $|x|$ and is read "the absolute value of x."

We have given the geometric meaning of absolute value. We now state the formal definition of absolute value.

DEFINITION

Let x be a real number. The **absolute value** of x is defined as

$$|x| = \begin{cases} -x \text{ if } x < 0 \\ x \text{ if } x \geq 0 \end{cases}$$

These problems are specifically designed to be used with a calculator and are similar to preceding examples. These problems can be used or omitted at the discretion of the instructor. Calculator Crossovers are highlighted with a lavender screen.

Rational numbers $\left\{ \dfrac{p}{q} \,\middle|\, p \text{ and } q \text{ are integers and } q \neq 0 \right\}$

Irrational numbers $\{x \mid x \text{ is a number that is not rational}\}$

Real numbers $\{x \mid \text{either } x \text{ is a rational or an irrational number}\}$

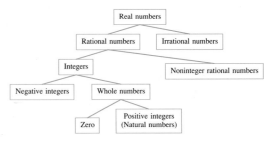

Example 4   (a) $\sqrt{2}$, π, $-\sqrt{21}$, and $1 + \sqrt{3}$ are irrational numbers.

(b) 0, 3.33, $\dfrac{4}{5}$, $-\dfrac{6}{48}$, $\dfrac{13}{1}$, and $\dfrac{4}{2}$ are rational numbers.

(c) -3, $\dfrac{6}{2}$, 4, $-\dfrac{560}{10}$, and $\dfrac{45}{9}$ are integers.

(d) All of the numbers in (a), (b), and (c) are real numbers.   ▲

C A L C U L A T O R   C R O S S O V E R

Using a calculator, write the decimal representations of the following numbers. If the number is rational, try to find the repeating block of digits. If the number is irrational, write as many digits in the decimal representation that your calculator displays.

1. $\dfrac{2}{3}$ **2.** $\dfrac{56}{99}$ **3.** $\sqrt{2}$ **4.** $-\dfrac{23}{6}$ **5.** $\dfrac{79}{7}$

Answers   **1.** $0.66666\dots$; 6 is the repeating digit.   **2.** $0.565656\dots$; 56 is the repeating block of digits.   **3.** $1.4142136\dots$. Since $\sqrt{2}$ is irrational, there is no repeating block of digits.   **4.** $-3.833333\dots$; 3 is the repeating digit.   **5.** $11.285714\dots$. There is a repeating block of digits, but it is not evident.

In order to use mathematics in applications, word phrases must be translated into algebraic expressions. We start by translating word phrases that involve the four basic operations of addition, subtraction, multiplication, and division. The following table lists some common word phrases and the equivalent algebraic expression. Let x represent the unknown number.

Word phrase	Algebraic expression
Addition	
The sum of a number and five	$x + 5$
Eight more than a number	$x + 8$
A number increased by 2.3	$x + 2.3$
Subtraction	
A number minus three	$x - 3$
Five and one half less than a number	$x - 5\frac{1}{2}$
A number decreased by 4.7	$x - 4.7$
A number subtracted from three	$3 - x$
Ten fewer than a number	$x - 10$
The difference of a number and nine	$x - 9$
The difference of nine and a number	$9 - x$
Multiplication	
Ten times a number	$10x$
Twice a number	$2x$
6.92 multiplied by a number	$6.92x$
Three fourths of a number	$\frac{3}{4}x$
Twenty-eight percent of a number	$0.28x$
Division	
A number divided by 8.1	$\frac{x}{8.1}$
The quotient of 12 and a number	$\frac{12}{x}$
The quotient of a number and 12	$\frac{x}{12}$
The reciprocal of a number	$\frac{1}{x}$

Tables are highlighted with light blue screens.

End-of-section exercises are graded in difficulty. Each exercise problem is related to an example from the section. Answers to the odd-numbered exercises are in the back of the book.

---

12   C H A P T E R **1**   **Operations and Variables**

EXERCISE SET 1.1

For Exercises 1–10, write each fraction in lowest terms.

1. $\dfrac{4}{6}$ **2.** $\dfrac{9}{6}$ **3.** $\dfrac{15}{18}$ **4.** $\dfrac{14}{4}$

5. $\dfrac{6}{21}$ **6.** $\dfrac{5}{15}$ **7.** $\dfrac{7}{49}$ **8.** $\dfrac{10}{50}$

9. $\dfrac{16}{20}$ **10.** $\dfrac{28}{4}$

For Exercises 11–30, find the product and write the answer in lowest terms.

11. $\dfrac{3}{5} \cdot \dfrac{2}{7}$ **12.** $\dfrac{11}{4} \cdot \dfrac{3}{5}$ **13.** $\dfrac{2}{5} \cdot \dfrac{1}{3} \cdot \dfrac{4}{5}$ **14.** $\dfrac{7}{5} \cdot \dfrac{1}{2} \cdot \dfrac{3}{4}$

15. $\left(\dfrac{2}{7}\right)\left(\dfrac{4}{3}\right)\left(\dfrac{4}{5}\right)$ **16.** $\left(\dfrac{4}{5}\right)\left(\dfrac{2}{3}\right)\left(\dfrac{6}{5}\right)$ **17.** $\left(\dfrac{7}{3}\right)\left(\dfrac{42}{10}\right)$ **18.** $\left(\dfrac{9}{15}\right)\left(\dfrac{4}{5}\right)$

19. $\left(\dfrac{12}{2}\right)\left(\dfrac{15}{25}\right)$ **20.** $\left(\dfrac{45}{35}\right)\left(\dfrac{18}{9}\right)$ **21.** $\left(\dfrac{4}{120}\right)\left(\dfrac{180}{3}\right)$ **22.** $\left(\dfrac{7}{140}\right)\left(\dfrac{270}{9}\right)$

23. $\left(\dfrac{4}{10}\right)\left(\dfrac{18}{9}\right)\left(\dfrac{15}{12}\right)$ **24.** $\left(\dfrac{28}{21}\right)\left(\dfrac{40}{5}\right)\left(\dfrac{2}{10}\right)$ **25.** $5\left(\dfrac{3}{2}\right)$ **26.** $6 \cdot \dfrac{4}{5}$

27. $\dfrac{4}{3} \cdot 18$ **28.** $\left(\dfrac{25}{8}\right)(32)$ **29.** $\left(\dfrac{1}{5}\right)\left(\dfrac{18}{3}\right)(5)$ **30.** $\left(\dfrac{2}{12}\right)\left(\dfrac{22}{40}\right)(10)$

For Exercises 31–40, find the quotient and write the answer in lowest terms.

31. $\dfrac{3}{2} \div \dfrac{5}{7}$ **32.** $\dfrac{5}{9} \div \dfrac{7}{4}$ **33.** $\dfrac{11}{2} \div \dfrac{5}{3}$ **34.** $\dfrac{1}{5} \div \dfrac{4}{13}$

35. $\dfrac{4}{18} \div \dfrac{2}{9}$ **36.** $\dfrac{16}{8} \div \dfrac{5}{15}$ **37.** $\dfrac{4}{5} \div \dfrac{12}{10}$ **38.** $\dfrac{25}{35} \div \dfrac{16}{12}$

39. $\dfrac{20}{12} \div \dfrac{32}{34}$ **40.** $\dfrac{38}{24} \div \dfrac{12}{56}$

For Exercises 41–46, write the expression as a single fraction in lowest terms.

41. $\left(\dfrac{15}{22}\right)\left(\dfrac{1}{7}\right) \div \dfrac{3}{14}$ **42.** $\dfrac{18}{8} \cdot \dfrac{2}{3} \div \dfrac{5}{6}$ **43.** $\dfrac{3}{4} \cdot 5 \div \dfrac{10}{12}$

44. $\left(\dfrac{6}{5}\right)(4) \div \dfrac{3}{2}$ **45.** $\dfrac{3}{4} \div 3 \div \dfrac{5}{16}$ **46.** $\dfrac{4}{5} \div 8 \div \dfrac{3}{2}$

Many word problems have been included to help students translate word phrases into mathematical equations.

Beginning with Chapter 2, each exercise set contains **Review Exercises** to help ensure that the student understands concepts studied in earlier sections. Answers to odd-numbered exercises are in the back of the book.

59. The sum of $\frac{2}{3}$ times a number and $\frac{5}{2}$ is the difference of 10 and $\frac{1}{3}$ of the number.

60. The sum of three halves times a number and four is the sum of one half of the number and five.

REVIEW PROBLEMS

The following exercises review parts of Section 1.5. Doing these problems will help prepare you for the next section.

Simplify.

61. $5x - 3x + 2 - 9$

62. $\frac{2}{3}y + \frac{5}{6}y - \frac{3}{4} - 1$

63. $4(1 - 3z) + 2$

64. $5x - 6(1 - 2x)$

65. $10(0.2 - 0.1a) + 2(3a + 1) + 6a - 12$

66. Find the value of $2 - 4t$ when $t = -\frac{3}{2}$.

67. Find the value of $3(2x - 4) + 1$ when $x = -2$.

For Exercises 68–72, convert the following word phrases into algebraic expressions. Use x to represent a number.

68. Twice a number increased by two.

69. The difference of ten and a number.

70. The reciprocal of the sum of a number and three.

71. Forty-five percent of a number.

72. The quotient of a number and -32.

These are challenging problems for the more advanced students. Answers to all of the **Enrichment Exercises** are in the back of the book.

E N R I C H M E N T   E X E R C I S E S

For Exercises 1–3, solve each equation.

1. $3(4 - 2x) = 1 - (1 - x)$

2. $\frac{2}{7} - \frac{5y}{14} = \frac{3}{4} - \frac{y}{7}$

3. $\frac{a + 1}{a} = -2$

For Exercises 4–7, write an equation for the following sentence. Then solve the equation.

4. A number divided by the sum of three times the number and four is negative five.

5. Four times the sum of five and twice a number is three times the difference of three times the number and two.

6. Three times the reciprocal of a number is negative five sevenths.

7. The reciprocal of the sum of twice a number and three is one sixth.

Answers to Enrichment Exercises are on page A.5.

91. The square of the sum of two consecutive odd integers.

92. The sum of the squares of two consecutive even integers.

93. The sum of the squares of three consecutive even integers.

94. The square of the sum of three consecutive odd integers.

E N R I C H M E N T　　E X E R C I S E S

For Exercises 1–4, simplify.

1. $2 - \{3 - [x - (4y - 2x) + 1] - 1 + 4(1 - 8y)\}$

2. $x[5 - 6(4 - 3x)]$

3. $ab[a^2b - ab(b - a)]$

4. $8p^2q^3 + 4pq\{1 - 2[pq^2 + p^2q]\} + 8p^3q^2$

5. A pipe of length 12 feet is cut into three pieces. If the first piece is x feet long and the middle piece is $\dfrac{1}{2}$ foot less than twice the first piece, express the length of the third piece in terms of x.

Answers to Enrichment Exercises are on page A.3.

The **Summary and Review** section includes definitions and strategies learned in the chapter.

C H A P T E R 1　　**S**ummary and review

Examples

$$\frac{12}{18} = \frac{2 \cdot 6}{3 \cdot 6} = \frac{2}{3}$$

$$\frac{12}{18} = \frac{\overset{2}{\cancel{12}}}{\underset{3}{\cancel{18}}} = \frac{2}{3}$$

Fractions (1.1)

A fraction is in **lowest terms** if the numerator and denominator have no common factors other than one. To write a fraction in lowest terms, we make use of the fact

$$\frac{a \cdot c}{b \cdot c} = \frac{a}{b}, \qquad c \neq 0$$

A common practice for writing a fraction in lowest terms is to use the **cancellation method.**

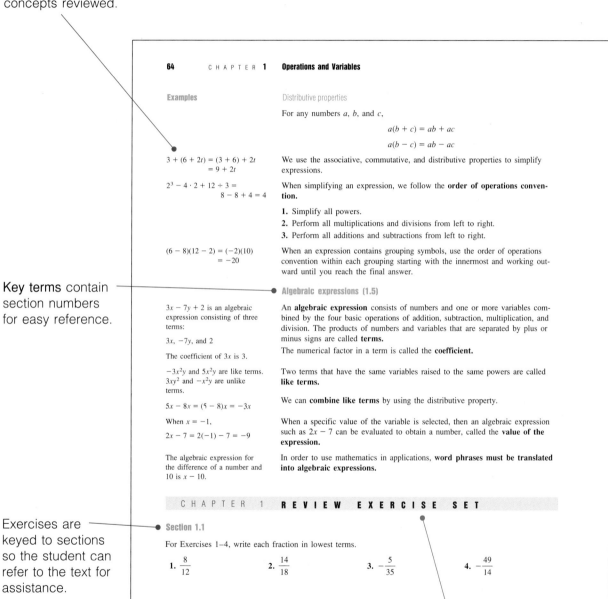

Marginal examples further illustrate concepts reviewed.

64   C H A P T E R **1**   **Operations and Variables**

Examples

Distributive properties

For any numbers a, b, and c,

$$a(b + c) = ab + ac$$

$$a(b - c) = ab - ac$$

$3 + (6 + 2t) = (3 + 6) + 2t$
$\qquad = 9 + 2t$

We use the associative, commutative, and distributive properties to simplify expressions.

$2^3 - 4 \cdot 2 + 12 \div 3 =$
$\qquad 8 - 8 + 4 = 4$

When simplifying an expression, we follow the **order of operations convention.**

1. Simplify all powers.
2. Perform all multiplications and divisions from left to right.
3. Perform all additions and subtractions from left to right.

$(6 - 8)(12 - 2) = (-2)(10)$
$\qquad = -20$

When an expression contains grouping symbols, use the order of operations convention within each grouping starting with the innermost and working outward until you reach the final answer.

Key terms contain section numbers for easy reference.

Algebraic expressions (1.5)

$3x - 7y + 2$ is an algebraic expression consisting of three terms:

$3x$, $-7y$, and 2

The coefficient of $3x$ is 3.

An **algebraic expression** consists of numbers and one or more variables combined by the four basic operations of addition, subtraction, multiplication, and division. The products of numbers and variables that are separated by plus or minus signs are called **terms.**
The numerical factor in a term is called the **coefficient.**

$-3x^2y$ and $5x^2y$ are like terms. $3xy^2$ and $-x^2y$ are unlike terms.

Two terms that have the same variables raised to the same powers are called **like terms.**

$5x - 8x = (5 - 8)x = -3x$

We can **combine like terms** by using the distributive property.

When $x = -1$,

$2x - 7 = 2(-1) - 7 = -9$

When a specific value of the variable is selected, then an algebraic expression such as $2x - 7$ can be evaluated to obtain a number, called the **value of the expression.**

The algebraic expression for the difference of a number and 10 is $x - 10$.

In order to use mathematics in applications, **word phrases must be translated into algebraic expressions.**

C H A P T E R   1   **R E V I E W   E X E R C I S E   S E T**

Exercises are keyed to sections so the student can refer to the text for assistance.

Section 1.1

For Exercises 1–4, write each fraction in lowest terms.

1. $\dfrac{8}{12}$ **2.** $\dfrac{14}{18}$ **3.** $-\dfrac{5}{35}$ **4.** $-\dfrac{49}{14}$

The **Review Exercise Set** allows students to pinpoint any difficulties that remain before proceeding to the next chapter. Answers to all of these exercises are in the back of the book.

The **Chapter Test** contains representative problems from every section of the text. Answers to all of these exercises are in the back of the book.

Exercises are keyed to sections so the student can refer to the text for assistance.

---

68   C H A P T E R **1**   **Operations and Variables**

C H A P T E R 1   T E S T

Section 1.1

1. Write the fraction $\dfrac{12}{21}$ in lowest terms.

For Problems 2 and 3, combine as indicated and write the answer in lowest terms.

2. $\dfrac{6}{5} \cdot \dfrac{35}{9}$

3. $\dfrac{6}{5} - \dfrac{1}{5}$

Section 1.2

For Problems 4–7, let

$$S = \left\{ -5, -\sqrt{2}, 0, \frac{1}{2}, 3 \right\}$$

Determine which members of S belong to the following sets.

4. The natural numbers

5. The integers

6. The rational numbers

7. The irrational numbers

8. Write either $<$ or $>$ between the two numbers to make a true statement.

$$-3\frac{1}{2} \qquad -4\frac{1}{2}$$

Section 1.3

For Problems 9–11, simplify

9. $-|-3|$

10. $7.4 - 3.2$

11. $\dfrac{|4-6|}{|13-5|}$

Section 1.4

For Problems 12 and 13, use the stated property to fill in the blank with the correct expression.

12. Associative property of addition: $3 + (2 + x) =$ _____

13. Associative property of multiplication: $4(3y) =$ _____

For Problems 14 and 15, use the appropriate associative property to rewrite each expression and then simplify.

14. $5 + (2 + w)$

15. $\dfrac{3}{2}\left(\dfrac{2}{3}x\right)$

16. Use the distributive property and then simplify. $3 - 2(1 - 4a)$

Contents

Chapter 4   **Rational expressions**   *226*

Chapter 5   **Roots and Radicals**   *305*

Chapter 6   **Quadratic equations**   *384*

Operations and Variables

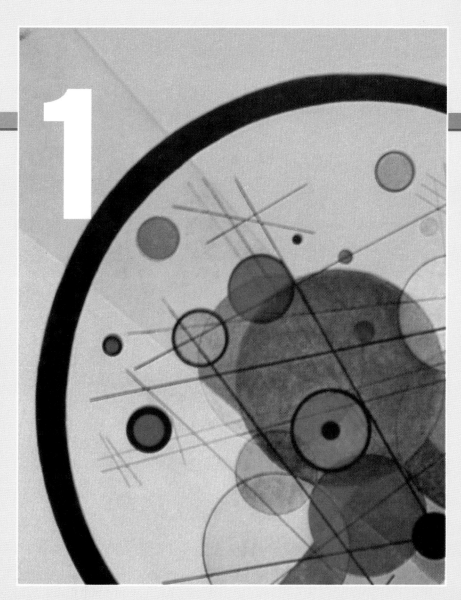

Overview

The foundation of algebra is arithmetic. In order to perform *algebraic* operations, a thorough understanding of performing *arithmetic* operations is necessary. Therefore, in this first chapter, we review arithmetic starting with a review of fractions. Next, we study the structure of the set of real numbers together with operations on real numbers. Properties of real numbers are followed by a review of exponents, roots, and order of operations. In the final section, we study variables and variable expressions. At this time, we introduce the important concept of writing verbal statements as algebraic expressions. Converting verbal statements into algebraic expressions is the basis for using algebra to solve word problems and will be practiced throughout the text.

1.1

Fractions–A Review

OBJECTIVES

▶ *To write fractions in lowest terms*

▶ *To multiply and divide fractions*

▶ *To add and subtract fractions with common denominators*

▶ *To add and subtract fractions with unlike denominators*

In our daily living we encounter the **whole numbers**

$$0, 1, 2, 3, 4, \ldots$$

as well as **fractions** such as $\dfrac{1}{2}, \dfrac{4}{5}$, and $\dfrac{3}{9}$. In this section we review some basic concepts involving fractions. In particular, we will review the techniques used to add, subtract, multiply, and divide fractions. These four operations will be discussed in greater detail in later sections. The purpose here is to strengthen your ability to work with fractions, since these same techniques will be used in algebra. In the fraction $\dfrac{2}{3}$, the number 2 is called the **numerator** and the number 3 is the **denominator.** The fraction is in **lowest terms** if the numerator and denominator have no common factors. For example, the fraction $\dfrac{2}{3}$ is in lowest terms, while $\dfrac{4}{6}$ is not in lowest terms, since the numerator and the denominator have a **common factor** of 2.

The following observation is used when we write a fraction in lowest terms.

> **The value of a fraction remains the same, when the numerator and denominator are divided by the same nonzero number.**

To write a fraction in lowest terms, we can use the following steps.

S T R A T E G Y

To write a fraction in lowest terms

Step 1   Find the largest nonzero number c that is a common factor of the numerator and the denominator. That is, the fraction can be written as

$$\frac{\text{numerator}}{\text{denominator}} = \frac{a \cdot c}{b \cdot c}$$

Step 2   Then,

$$\frac{a \cdot c}{b \cdot c} = \frac{a}{b}$$

That is, we divide numerator and denominator by the common factor c, and the resulting fraction is now in lowest terms.

For example, to write $\dfrac{4}{6}$ in lowest terms, write the numerator as $2 \cdot 2$ and the denominator as $3 \cdot 2$. Therefore,

$$\frac{4}{6} = \frac{2 \cdot 2}{3 \cdot 2} \qquad \text{The number } c \text{ is 2 for this fraction.}$$

$$= \frac{2}{3}$$

Example 1   Write $\dfrac{6}{21}$ in lowest terms.

Solution   We start by writing both numerator and denominator as products, then dividing by the common factor.

$$\frac{6}{21} = \frac{2 \cdot \cancel{3}}{7 \cdot \cancel{3}}$$

$$= \frac{2}{7} \qquad \text{Divide the numerator and the} \\ \text{denominator by 3.} \qquad \blacktriangle$$

In Example 1, notice that slashes were used to indicate that we divided both numerator and denominator by the common factor 3. Dividing the numerator and denominator by the common factor 3 can be written in a more compact form:

$$\frac{6}{21} = \frac{\overset{2}{\cancel{6}}}{\underset{7}{\cancel{21}}} = \frac{2}{7}$$

This shorter process, however, does not directly indicate that 3 is a common factor.

It is sometimes difficult to immediately find the *largest* number c that is a common factor of the numerator and denominator of a fraction. In the next example, we show how to slightly change the two-step process to reduce a fraction to lowest terms.

Example 2   Write $\dfrac{60}{105}$ in lowest terms.

Solution   Since the last digit of the numerator is 0 and the last digit of the denominator is 5, they have a common factor of 5.

Therefore,

$$\frac{60}{105} = \frac{12 \cdot \cancel{5}}{21 \cdot \cancel{5}}$$

$$= \frac{12}{21} \qquad \text{Divide the numerator and} \\ \text{the denominator by 5.}$$

While $\dfrac{12}{21}$ is a partially reduced form of $\dfrac{60}{105}$, it is not the final answer, because 12 and 21 have a common factor of 3.

$$\frac{12}{21} = \frac{4 \cdot \cancel{3}}{7 \cdot \cancel{3}}$$

$$= \frac{4}{7} \qquad \text{Divide the numerator and} \\ \text{the denominator by 3.}$$

Therefore, $\dfrac{60}{105} = \dfrac{4}{7}$.

The shorthand method to reduce this fraction looks like this:

$$\frac{60}{105} = \frac{\overset{\overset{4}{\cancel{12}}}{\cancel{60}}}{\underset{\underset{7}{\cancel{21}}}{\cancel{105}}} = \frac{4}{7}$$

▲

COMMON ERROR

When reducing a fraction to lowest terms, be sure that the numerator and denominator are written as products and not as sums. Compare the following two sequences of steps.

Sequence 1: $\dfrac{12}{15} = \dfrac{4 \cdot 3}{5 \cdot 3} = \dfrac{4 \cdot \cancel{3}}{5 \cdot \cancel{3}} = \dfrac{4}{5}$ Correct.

Sequence 2: $\dfrac{9}{10} = \dfrac{1 + 8}{2 + 8} = \dfrac{1 + \cancel{8}}{2 + \cancel{8}} = \dfrac{1}{2}$ Incorrect.

Sequence 1 is correct, since 3 is a common factor. Sequence 2 is incorrect, since 8 is not a common *factor*.

We next consider combining fractions using the operations of addition, subtraction, multiplication, and division. Since the methods for multiplying and dividing fractions are simpler than for addition and subtraction, we start with multiplication. To multiply two fractions, we use the following rule.

RULE

Multiplication of two fractions

Given two fractions $\dfrac{a}{b}$ and $\dfrac{c}{d}$, then

$$\frac{a}{b} \cdot \frac{c}{d} = \frac{a \cdot c}{b \cdot d}$$

That is, the answer from multiplying two fractions is a new fraction obtained by multiplying the numerators and then multiplying the denominators.

The answer to a multiplication of numbers is called a **product.** For example, the product of 4 multiplied by 3 is 12, that is, $3 \cdot 4 = 12$. The numbers 3 and 4 are called **factors.**

Example 3   Find the product and write the answer in lowest terms.

(a) $\dfrac{3}{4} \cdot \dfrac{5}{7}$ (b) $\dfrac{2}{3} \cdot \dfrac{1}{5} \cdot \dfrac{4}{5}$ (c) $\dfrac{4}{5} \cdot \dfrac{10}{3}$

Solution   (a) $\dfrac{3}{4} \cdot \dfrac{5}{7} = \dfrac{3 \cdot 5}{4 \cdot 7}$

$= \dfrac{15}{28}$ Multiply numerators together and multiply denominators together.

(b) To find the product of three (or more) fractions, simply multiply all numerators and then all denominators.

$$\dfrac{2}{3} \cdot \dfrac{1}{5} \cdot \dfrac{4}{5} = \dfrac{2 \cdot 1 \cdot 4}{3 \cdot 5 \cdot 5}$$

$$= \dfrac{8}{75}$$

(c) We write the numerator and denominator as products, then look for common factors.

$$\dfrac{4}{5} \cdot \dfrac{10}{3} = \dfrac{4 \cdot 10}{5 \cdot 3}$$

$$= \dfrac{4 \cdot 2 \cdot \overset{1}{\cancel{5}}}{\underset{1}{\cancel{5}} \cdot 3}$$

$$= \dfrac{4 \cdot 2 \cdot 1}{1 \cdot 3}$$ Divide the numerator and denominator by 5.

$$= \dfrac{8}{3}$$ ▲

Consider again the multiplication problem $\dfrac{4}{5} \cdot \dfrac{10}{3}$ of Example 3(c). Since any factor of a numerator ends up as a factor of the numerator in the product and any factor of a denominator ends up as a factor in the denominator of the product, we can "divide out" common factors *before* multiplying the two fractions. Therefore, Example 3(c) can be done in the following way:

$$\frac{4}{5} \cdot \frac{10}{3} = \frac{4}{\overset{}{\underset{1}{\cancel{5}}}} \cdot \frac{\overset{2}{\cancel{10}}}{3} \qquad \text{Divide out 5.}$$

$$= \frac{4 \cdot 2}{1 \cdot 3}$$

$$= \frac{8}{3}$$

Example 4   Multiply.

(a) $3\left(\dfrac{4}{5}\right)$
　　　　　　　　　　　　　　　　　　　(b) $\dfrac{25}{45} \cdot 12$

Solution   (a) First write 3 as $\dfrac{3}{1}$, then multiply.

$$3\left(\frac{4}{5}\right) = \frac{3}{1} \cdot \frac{4}{5}$$

$$= \frac{3 \cdot 4}{1 \cdot 5}$$

$$= \frac{12}{5}$$

(b) $\dfrac{25}{45} \cdot 12 = \dfrac{25}{45} \cdot \dfrac{12}{1} \qquad 12 = \dfrac{12}{1}.$

$$= \frac{\overset{5}{\cancel{25}}}{\underset{3}{\underset{\cancel{9}}{\cancel{45}}}} \cdot \frac{\overset{4}{\cancel{12}}}{1} \qquad \begin{array}{l}\text{Divide 25 and 45 by 5,} \\ \text{then divide 9 and 12 by 3.}\end{array}$$

$$= \frac{5 \cdot 4}{3 \cdot 1}$$

$$= \frac{20}{3}$$

▲

Two fractions are **reciprocals** of each other if their product is one. For example, $\dfrac{2}{5}$ and $\dfrac{5}{2}$ are reciprocals since

$$\frac{2}{5} \cdot \frac{5}{2} = 1$$

In general, the two fractions $\dfrac{a}{b}$ and $\dfrac{b}{a}$, where neither a nor b is zero, are reciprocals, since

$$\frac{a}{b} \cdot \frac{b}{a} = 1$$

The reciprocal is used to divide two fractions.

RULE

Division of two fractions

For two fractions $\dfrac{a}{b}$ and $\dfrac{c}{d}$, where b, c, and d are nonzero,

$$\frac{a}{b} \div \frac{c}{d} = \frac{a}{b} \cdot \frac{d}{c}$$

That is, to divide two fractions, multiply the first fraction by the reciprocal of the second fraction, called the **divisor.** The answer to a division problem is called the **quotient.** For example, the quotient of 6 and 3 is 2.

Example 5   Find the quotient and write the answer in lowest terms.

(a) $\dfrac{4}{3} \div \dfrac{5}{2}$ (b) $\dfrac{2}{3} \div \dfrac{4}{7}$

Solution   (a) $\dfrac{4}{3} \div \dfrac{5}{2} = \dfrac{4}{3} \cdot \dfrac{2}{5}$ Multiply $\dfrac{4}{3}$ by the reciprocal of $\dfrac{5}{2}$.

$$= \frac{4 \cdot 2}{3 \cdot 5}$$

$$= \frac{8}{15}$$

(b) $\dfrac{2}{3} \div \dfrac{4}{7} = \dfrac{2}{3} \cdot \dfrac{7}{4}$

$$= \frac{\overset{1}{\cancel{2}}}{3} \cdot \frac{7}{\underset{2}{\cancel{4}}}$$ 2 is a common factor.

$$= \frac{1 \cdot 7}{3 \cdot 2}$$

$$= \frac{7}{6} \qquad \blacktriangle$$

In the next example, we show how to simplify expressions that involve the two operations of multiplication and division.

Example 6   Write $\dfrac{15}{8} \cdot \dfrac{3}{2} \div \dfrac{3}{8}$ as a single fraction in lowest terms.

Solution   We use the definition of division and multiply the first two fractions by the reciprocal of $\dfrac{3}{8}$. Then, we look for common factors to simplify the triple product.

$$\frac{15}{8} \cdot \frac{3}{2} \div \frac{3}{8} = \frac{15}{8} \cdot \frac{3}{2} \cdot \frac{8}{3}$$

$$= \frac{15}{\overset{}{\underset{1}{8}}} \cdot \frac{\overset{1}{3}}{2} \cdot \frac{\overset{1}{8}}{\underset{1}{3}}$$

$$= \frac{15 \cdot 1 \cdot 1}{1 \cdot 2 \cdot 1}$$

$$= \frac{15}{2} \qquad \blacktriangle$$

When adding or subtracting fractions, recall that the fractions must have the same denominator.

R U L E

Addition and subtraction of fractions

$$\frac{a}{b} + \frac{c}{b} = \frac{a+c}{b}$$

$$\frac{a}{b} - \frac{c}{b} = \frac{a-c}{b}$$

That is, to add or subtract two fractions with the same denominator, add or subtract the numerators while keeping the common denominator.

Example 7   Add or subtract as indicated.

(a) $\dfrac{3}{5} + \dfrac{6}{5}$

(b) $\dfrac{7}{2} - \dfrac{5}{2} + \dfrac{11}{2}$

Solution

(a) $\dfrac{3}{5} + \dfrac{6}{5} = \dfrac{3 + 6}{5}$ Add the numerators.

$= \dfrac{9}{5}$

(b) $\dfrac{7}{2} - \dfrac{5}{2} + \dfrac{11}{2} = \dfrac{7 - 5 + 11}{2}$

$= \dfrac{13}{2}$ ▲

To add or subtract fractions with unlike denominators, first rewrite each fraction so that they all have a common denominator. To do this, we use the following property. Let $\dfrac{a}{b}$, $b \ne 0$, be a fraction. Then for any nonzero number c, we have

$$\dfrac{a}{b} = \dfrac{a \cdot c}{b \cdot c}, \qquad c \ne 0$$

That is, the value of a fraction remains the same when both numerator and denominator are multiplied by the same nonzero number c.

We use this property to change the denominator of a fraction while keeping its value.

Example 8   Add or subtract as indicated. Write the answer in lowest terms.

(a) $\dfrac{3}{4} - \dfrac{1}{8}$

(b) $\dfrac{3}{2} + \dfrac{5}{3} - \dfrac{5}{6}$

Solution

(a) A common denominator of the two fractions is 8, since both 4 and 8 divide into 8 with a remainder of zero. Therefore, multiply the numerator and denominator of $\dfrac{3}{4}$ by 2.

$$\dfrac{3}{4} - \dfrac{1}{8} = \dfrac{3 \cdot 2}{4 \cdot 2} - \dfrac{1}{8}$$

$$= \dfrac{6}{8} - \dfrac{1}{8}$$

$$= \dfrac{6 - 1}{8}$$

$$= \dfrac{5}{8} \qquad \text{Subtract numerators.}$$

(b) A common denominator of the three fractions is 6. Therefore,

$$\frac{3}{2} + \frac{5}{3} - \frac{5}{6} = \frac{3 \cdot 3}{2 \cdot 3} + \frac{5 \cdot 2}{3 \cdot 2} - \frac{5}{6}$$

$$= \frac{9}{6} + \frac{10}{6} - \frac{5}{6}$$

$$= \frac{9 + 10 - 5}{6}$$

$$= \frac{\overset{7}{\cancel{14}}}{\underset{3}{\cancel{6}}}$$

$$= \frac{7}{3} \qquad \text{2 is a common factor.} \quad \blacktriangle$$

Example 9   Add: $4\frac{1}{2} + 1\frac{3}{4}$.

Solution   We first change each mixed number into a fraction.

$$4\frac{1}{2} = 4 + \frac{1}{2} = \frac{4}{1} + \frac{1}{2} = \frac{8}{2} + \frac{1}{2}$$

$$= \frac{8 + 1}{2} = \frac{9}{2}$$

$$1\frac{3}{4} = 1 + \frac{3}{4} = \frac{4}{4} + \frac{3}{4} = \frac{4 + 3}{4} = \frac{7}{4}$$

Next, we add the two fractions.

$$4\frac{1}{2} + 1\frac{3}{4} = \frac{9}{2} + \frac{7}{4}$$

$$= \frac{18}{4} + \frac{7}{4} \qquad \text{Replace } \frac{9}{2} \text{ by } \frac{9 \cdot 2}{2 \cdot 2} \text{ or } \frac{18}{4}.$$

$$= \frac{25}{4} \text{ or } 6\frac{1}{4} \qquad\qquad \blacktriangle$$

L E A R N I N G   A D V A N T A G E   *Before starting on each homework assignment, first study the notes you took in class and then read the appropriate section(s) from the book. If you have trouble on a particular exercise problem, find the example that most closely pertains to it. Remember that the problems in the exercise set are paired so that each even-numbered problem matches the previous odd-numbered problem.*

For Exercises 1–10, write each fraction in lowest terms.

1. $\dfrac{4}{6}$ **2.** $\dfrac{9}{6}$ **3.** $\dfrac{15}{18}$ **4.** $\dfrac{14}{4}$

5. $\dfrac{6}{21}$ **6.** $\dfrac{5}{15}$ **7.** $\dfrac{7}{49}$ **8.** $\dfrac{10}{50}$

9. $\dfrac{16}{20}$ **10.** $\dfrac{28}{4}$

For Exercises 11–30, find the product and write the answer in lowest terms.

11. $\dfrac{3}{5}\cdot\dfrac{2}{7}$ **12.** $\dfrac{11}{4}\cdot\dfrac{3}{5}$ **13.** $\dfrac{2}{5}\cdot\dfrac{1}{3}\cdot\dfrac{4}{5}$ **14.** $\dfrac{7}{5}\cdot\dfrac{1}{2}\cdot\dfrac{3}{4}$

15. $\left(\dfrac{2}{7}\right)\left(\dfrac{4}{3}\right)\left(\dfrac{4}{5}\right)$ **16.** $\left(\dfrac{4}{5}\right)\left(\dfrac{2}{3}\right)\left(\dfrac{6}{5}\right)$ **17.** $\left(\dfrac{7}{3}\right)\left(\dfrac{42}{10}\right)$ **18.** $\left(\dfrac{9}{15}\right)\left(\dfrac{4}{5}\right)$

19. $\left(\dfrac{12}{2}\right)\left(\dfrac{15}{25}\right)$ **20.** $\left(\dfrac{45}{35}\right)\left(\dfrac{18}{9}\right)$ **21.** $\left(\dfrac{4}{120}\right)\left(\dfrac{180}{3}\right)$ **22.** $\left(\dfrac{7}{140}\right)\left(\dfrac{270}{9}\right)$

23. $\left(\dfrac{4}{10}\right)\left(\dfrac{18}{9}\right)\left(\dfrac{15}{12}\right)$ **24.** $\left(\dfrac{28}{21}\right)\left(\dfrac{40}{5}\right)\left(\dfrac{2}{10}\right)$ **25.** $5\left(\dfrac{3}{2}\right)$ **26.** $6\cdot\dfrac{4}{5}$

27. $\dfrac{4}{3}\cdot18$ **28.** $\left(\dfrac{25}{8}\right)(32)$ **29.** $\left(\dfrac{1}{5}\right)\left(\dfrac{18}{3}\right)(5)$ **30.** $\left(\dfrac{2}{12}\right)\left(\dfrac{22}{40}\right)(10)$

For Exercises 31–40, find the quotient and write the answer in lowest terms.

31. $\dfrac{3}{2}\div\dfrac{5}{7}$ **32.** $\dfrac{5}{9}\div\dfrac{7}{4}$ **33.** $\dfrac{11}{2}\div\dfrac{5}{3}$ **34.** $\dfrac{1}{5}\div\dfrac{4}{13}$

35. $\dfrac{4}{18}\div\dfrac{2}{9}$ **36.** $\dfrac{16}{8}\div\dfrac{5}{15}$ **37.** $\dfrac{4}{5}\div\dfrac{12}{10}$ **38.** $\dfrac{25}{35}\div\dfrac{16}{12}$

39. $\dfrac{20}{12}\div\dfrac{32}{34}$ **40.** $\dfrac{38}{24}\div\dfrac{12}{56}$

For Exercises 41–46, write the expression as a single fraction in lowest terms.

41. $\left(\dfrac{15}{22}\right)\left(\dfrac{1}{7}\right)\div\dfrac{3}{14}$ **42.** $\dfrac{18}{8}\cdot\dfrac{2}{3}\div\dfrac{5}{6}$ **43.** $\dfrac{3}{4}\cdot5\div\dfrac{10}{12}$

44. $\left(\dfrac{6}{5}\right)(4)\div\dfrac{3}{2}$ **45.** $\dfrac{3}{4}\div3\div\dfrac{5}{16}$ **46.** $\dfrac{4}{5}\div8\div\dfrac{3}{2}$

For Exercises 47–62, add or subtract as indicated. Write the answer in lowest terms.

47. $\dfrac{5}{7} + \dfrac{2}{7}$

48. $\dfrac{3}{5} + \dfrac{7}{5}$

49. $\dfrac{2}{3} - \dfrac{1}{3}$

50. $\dfrac{5}{6} - \dfrac{1}{6}$

51. $\dfrac{2}{5} + \dfrac{3}{5} - \dfrac{4}{5}$

52. $\dfrac{12}{7} - \dfrac{4}{7} + \dfrac{6}{7}$

53. $\dfrac{1}{3} + \dfrac{1}{6}$

54. $\dfrac{3}{14} + \dfrac{1}{7}$

55. $\dfrac{4}{3} - \dfrac{2}{9}$

56. $\dfrac{5}{12} - \dfrac{1}{24}$

57. $\dfrac{3}{2} + \dfrac{2}{3} - \dfrac{7}{6}$

58. $\dfrac{5}{3} - \dfrac{1}{4} - \dfrac{7}{12}$

59. $\dfrac{5}{3} - \dfrac{7}{15} - \dfrac{2}{5}$

60. $\dfrac{7}{4} - \dfrac{11}{20} - \dfrac{2}{5}$

61. $\dfrac{7}{6} - \dfrac{13}{24} + \dfrac{1}{4}$

62. $\dfrac{4}{5} + \dfrac{2}{3} - \dfrac{1}{15}$

63. A rectangle has two sides that are each $1\dfrac{1}{3}$ feet long and the other two sides are each $\dfrac{5}{2}$ feet long. Find the total distance around the rectangle.

64. Joan mowed $\dfrac{1}{3}$ of the lawn and her brother mowed $\dfrac{1}{4}$ of the lawn. How much of the lawn is mowed?

65. If a pizza is cut into eight pieces and you eat three pieces, how much pizza is left?

66. Sarah drank $\dfrac{1}{3}$ of a pitcher of cola and Frank drank $\dfrac{1}{4}$ of the pitcher. How much cola is left?

67. A recipe for oatmeal cookies calls for $1\dfrac{2}{3}$ teaspoons of cinnamon. If Jay plans to triple the recipe, how much cinnamon should he use?

68. A heating oil company is contracted to deliver $3\dfrac{2}{5}$ tanks of oil each month for 4 months. How many tanks of oil will be delivered during the 4 months?

69. What is wrong with the following sequence of steps?

$$3 = \frac{6}{2} = \frac{5+1}{1+1} = \frac{5+\cancel{1}}{1+\cancel{1}} = \frac{5}{1} = 5$$

70. What is wrong with the following sequence of steps?

$$\frac{7}{21} = \frac{7}{7 \cdot 3} = \frac{\cancel{7}}{\cancel{7} \cdot 3} = \frac{0}{3} = 0$$

ENRICHMENT EXERCISES

Simplify.

1. $\dfrac{10}{24} + \dfrac{8}{36}$

2. $-\dfrac{3}{92} - \dfrac{5}{23}$

3. $3\dfrac{1}{7} - 1\dfrac{3}{14}$

4. $\dfrac{\dfrac{7}{3}}{\dfrac{5}{8} + \dfrac{1}{4}}$

5. $\dfrac{\dfrac{1}{3} + \dfrac{1}{2}}{\dfrac{2}{5} - \dfrac{1}{10}}$

6. $\dfrac{4}{3}\left(\dfrac{1}{2} - \dfrac{2}{7}\right)$

7. $\dfrac{1}{4} \cdot \dfrac{12}{5} - \dfrac{2}{15} \cdot \dfrac{7}{2}$

8. $\left(\dfrac{12}{5} - \dfrac{2}{3}\right)\left(\dfrac{2}{7} + \dfrac{1}{4}\right)$

9. $\left(\dfrac{3}{8} + \dfrac{1}{4}\right) \div \left(\dfrac{5}{12} - \dfrac{7}{6}\right)$

10. A taxicab company charges 80 cents for the first $\dfrac{1}{4}$ mile and 50 cents for each additional $\dfrac{1}{4}$ mile. How much is the cab fare for a 3-mile trip?

11. Division is defined in terms of multiplication. For example, $12 \div 3 = 4$, since $4 \cdot 3 = 12$. Use this concept to justify division of two fractions $\dfrac{a}{b}$ and $\dfrac{c}{d}$, where b, c, and d are nonzero:

$$\frac{a}{b} \div \frac{c}{d} = \frac{ad}{bc}.$$

Answers to Enrichment Exercises are on page A.1.

1.2

The Real Numbers

O B J E C T I V E S

▶ *To investigate the structure of the real number system*

▶ *To understand the concept of variable*

Fundamental to mathematics is the concept of set. A **set** is a collection of objects. In particular, we will be interested in sets of numbers. It is common to use capital letters to denote sets. For example, $A = \{4, 6, 8\}$ is the set of the even numbers between 3 and 9. Each object in a set is called an **element** or **member** of the set. For example, 4, 6, and 8 are each members of the set A. To show membership, we use the following notation.

$a \in A$ means "*a* is a member of *A*."

$a \notin A$ means "*a* is *not* a member of *A*."

For example, $4 \in \{4, 6, 8\}$ whereas $3 \notin \{4, 6, 8\}$.

The set $A = \{4, 6, 8\}$ is an example of a **finite** set. The set $N = \{1, 2, 3, 4, 5, \dots\}$ is an example of an **infinite** set. The three dots mean that the pattern continues without end. The set N is called the set of **natural numbers** or **counting numbers.**

We often make statements about some or all elements of a set. We could list these elements, but if the number of elements is large, it would not be practical. Another way is to use a letter to represent a typical element. A letter used in this way is called a **variable.** The French mathematician Francois Viete (1540–1603) is given credit for introducing the concept of variable.

Using a variable, a set may be specified by describing its elements. For example, we specify the set A in *both* ways as follows.

Listing method: $A = \{4, 6, 8\}$.

Set-builder method:

$$A = \{x \mid x \text{ is an even number between 3 and 9}\}$$

Read ''the set of all x such that x is an even number between 3 and 9.'' In this case, x is the variable that represents an even number between 3 and 9.

Example 1

Let B be the set of all odd integers between 10 and 22. Specify B by the **(a)** set-builder method and **(b)** listing method.

Solution

(a) $B = \{x \mid x \text{ is an odd integer between 10 and 22}\}$.
(b) $B = \{11, 13, 15, 17, 19, 21\}$.   ▲

Two sets are **equal** if they each have the same elements. For example, $\{1, 2, 3\} = \{3, 1, 2\}$.

A is a **subset** of B if each element of A is an element of B. We write

$$A \subseteq B$$

For example, $\{1, 2, 3\} \subseteq \{1, 2, 3, 4\}$.

The **empty** set $\varnothing$ is the set with no elements. For example, $\{x \mid x \text{ is a number that is both positive and negative}\}$ is the empty set. The empty set is a subset of any set A. That is,

$$\varnothing \subseteq A$$

N O T E   *Keep in mind the distinction between the symbols $\in$ and $\subseteq$. For example, $3 \in \{2, 3, 4\}$ is a true statement, whereas $\{3\} \in \{2, 3, 4\}$ is a false statement. Instead, we should write $\{3\} \subseteq \{2, 3, 4\}$.*

Example 2   Let $A = \{2, 5, 12\}$ and

$B = \{x \mid x$ is an even integer between 1 and 11$\}$.

Answer each of the following as true (T) or false (F).

(a) $12 \in A$ **(b)** $A \subseteq B$ **(c)** $\varnothing \subseteq A$

(d) $A \in B$ **(e)** $\{4\} \subseteq B$ **(f)** $6 \notin B$

(g) $\{10, 12\} \subseteq A$

Solution   **(a)** T   **(b)** F   **(c)** T   **(d)** F   **(e)** T   **(f)** F   **(g)** F   ▲

In arithmetic, we deal with numerical expressions such as

$$3 \cdot 1 + 1, \qquad 3 \cdot 2 + 1, \qquad 3 \cdot 3 + 1, \qquad \text{and} \qquad 3 \cdot 4 + 1$$

In algebra, we can write all four of these expressions by using a variable. For example, the four numerical expressions above can be concisely written as the **variable expression** $3 \cdot x + 1$ or simply $3x + 1$, where x represents any number from the set $\{1, 2, 3, 4\}$. Notice that the multiplication dot may be omitted and $3 \cdot x$ can be written as $3x$. Any single number is called a **constant.** For example, $\frac{2}{3}$ and π are each constants.

Example 3   Replace the four numerical expressions by a single variable expression involving z, where z represents any member from the set $\{2, 4, 6, 8\}$.

$$1 - 3 \cdot 2, \qquad 1 - 3 \cdot 4, \qquad 1 - 3 \cdot 6, \qquad 1 - 3 \cdot 8$$

Solution   The appropriate variable expression is $1 - 3z$, since replacing z, in turn, by the numbers from the set $\{2, 4, 6, 8\}$ will generate the four numerical expressions.   ▲

In our study of algebra, we will use basic sets of numbers. As stated before, the **set of natural numbers** is given by $N = \{1, 2, 3, \ldots\}$.

When 0 is included with the natural numbers, we have the set of **whole numbers:**

$$W = \{0, 1, 2, 3, \ldots\}$$

The set of **integers** I is

$$I = \{\ldots, -3, -2, -1, 0, 1, 2, 3, \ldots\}$$

Notice that $N \subseteq I$, that is, every natural number is an integer.

A **rational number** is any number of the form $\frac{p}{q}$, $q \neq 0$, where p and q are integers. Examples of rational numbers are

$$\frac{1}{3} - \frac{9}{5}, \qquad 5, \qquad 0, \qquad 4.38, \qquad \text{and} \qquad \frac{15}{93}$$

Notice that $5 = \dfrac{5}{1}$, so the integer 5 is also a rational number. In general, any integer n is a rational number, since $n = \dfrac{n}{1}$. The decimal number 4.38 is also a rational number, since $4.38 = \dfrac{438}{100}$. In general, any terminating or repeating decimal is a rational number.

A number that is not rational is an **irrational number.** Examples of irrational numbers are

$$\pi, \qquad \sqrt{2}, \qquad -4\sqrt{3}, \qquad \text{and} \qquad 5.616116111611116\ldots$$

The rational numbers together with the irrational numbers form the set of **real numbers.**

Any real number can be expressed in decimal form. For example, the two rational numbers $\dfrac{5}{4}$ and $\dfrac{7}{11}$ have the following decimal representations:

$$\frac{5}{4} = 1.25 \qquad \text{and} \qquad \frac{7}{11} = 0.636363\ldots$$

The number $\dfrac{5}{4}$ has a terminating decimal representation, whereas, $\dfrac{7}{11}$ has a nonterminating decimal representation. Notice that the nonterminating decimal is formed by the two digits 63 repeating. In general,

> **Any rational number has a decimal representation that either is terminating or nonterminating and is formed by a repeating block of digits.**

The decimal representations of irrational numbers behave in a different way. For example, the decimal representation of the irrational number pi (π) for the first 20 decimal places is the following.

$$\pi = 3.14159265358979323846\ldots$$

The number pi is the ratio of the circumference of a circle to its diameter. The symbol π was popularized by Leonhard Euler in the 1700s.

Computers have been used to find π accurate to thousands of decimal places with no pattern being evident. In general, we have the following statement concerning the decimal representations of irrational numbers.

> **Irrational numbers have decimal representations that are nonterminating with no repeating block of digits.**

We summarize the types of numbers introduced and illustrate them in a family tree of numbers.

Natural numbers $\{1, 2, 3, \ldots\}$
Whole numbers $\{0, 1, 2, 3, \ldots\}$
Integers $\{\ldots -2, -1, 0, 1, 2, \ldots\}$

Rational numbers $\left\{\dfrac{p}{q}\,\middle|\,p \text{ and } q \text{ are integers and } q \neq 0\right\}$

Irrational numbers $\{x\,|\,x \text{ is a number that is not rational}\}$

Real numbers $\{x\,|\,\text{either } x \text{ is a rational or an irrational number}\}$

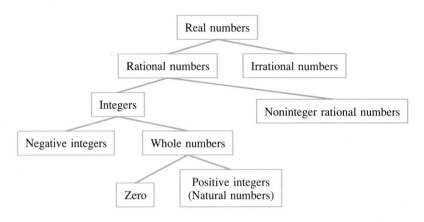

Example 4   **(a)** $\sqrt{2}$, π, $-\sqrt{21}$, and $1 + \sqrt{3}$ are irrational numbers.

(b) 0, 3.33, $\dfrac{4}{5}$, $-\dfrac{6}{48}$, $\dfrac{13}{1}$, and $\dfrac{4}{2}$ are rational numbers.

(c) -3, $\dfrac{6}{2}$, 4, $-\dfrac{560}{10}$, and $\dfrac{45}{9}$ are integers.

(d) All of the numbers in (a), (b), and (c) are real numbers.   ▲

CALCULATOR   CROSSOVER

Using a calculator, write the decimal representations of the following numbers. If the number is rational, try to find the repeating block of digits. If the number is irrational, write as many digits in the decimal representation that your calculator displays.

1. $\dfrac{2}{3}$ **2.** $\dfrac{56}{99}$ **3.** $\sqrt{2}$ **4.** $-\dfrac{23}{6}$ **5.** $\dfrac{79}{7}$

Answers   **1.** $0.66666\ldots$; 6 is the repeating digit.   **2.** $0.565656\ldots$; 56 is the repeating block of digits.   **3.** $1.4142136\ldots$. Since $\sqrt{2}$ is irrational, there is no repeating block of digits.   **4.** $-3.833333\ldots$; 3 is the repeating digit.   **5.** $11.285714\ldots$. There is a repeating block of digits, but it is not evident.

We can visualize numbers by graphing or plotting them on a number line. In the figure below, the natural numbers between 0 and 5 are graphed. We say that the number is the **coordinate** of the associated point. The point is called the **graph** of the number.

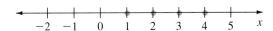

Each real number is the coordinate of a point on the number line. Some real numbers are plotted on the number line in the following figure.

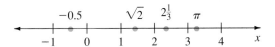

Example 5 Graph the numbers of the set $\left\{1, 3, -\dfrac{1}{2}, \dfrac{4}{3}\right\}$

Solution We draw a number line and plot the numbers as shown in the figure.

Example 6 Name the coordinates of the points graphed in the figure below.

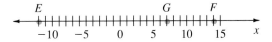

Solution -12 is the coordinate of E.
14 is the coordinate of F.
7 is the coordinate of G.

Numbers can be used to describe changes in direction (either positive or negative) that occur in real-life situations.

Example 7 Represent each quantity as either a positive number or a negative number.

(a) Fred loses $250 in a poker game.
(b) The fullback gained 23 yards on the play.
(c) The company's earnings were down 14%.

Solution (a) $-$250   (b) 23 yards   (c) -14%.

Inequalities

The number line not only displays the positions of the real numbers, but also compares numbers. The numbers are increasing from left to right. To compare two numbers we use **inequality symbols:**

< means **"is less than"**   > means **"is greater than"**

≤ means **"is less than or equal to"**

≥ means **"is greater than or equal to"**

For example, on the number line below, 2 is to the left of the number 4. We compare 2 and 4 by saying

2 is less than 4   or   4 is greater than 2

$$2 < 4 \qquad\qquad 4 > 2$$

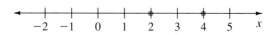

On a number line, the smaller of two numbers lies to the left of the larger number.

Example 8

Write either < or > between the numbers to make a true statement.

(a) 5   3   **(b)** −3   −2   **(c)** $-\dfrac{1}{2}$ $-\dfrac{3}{4}$

Solution   We use a number line to compare the two numbers.

(a) 5 is to the right of 3, so $5 > 3$.

(b) −3 is to the left of −2, so $-3 < -2$.

(c) $-\dfrac{1}{2}$ is to the right of $-\dfrac{3}{4}$, so $-\dfrac{1}{2} > -\dfrac{3}{4}$.   ▲

L E A R N I N G   A D V A N T A G E   *An important factor in successfully completing an algebra course is to* **attend class regularly.** *Furthermore, be an active member of the class. Do not hesitate to ask questions, even ones you think might be too elementary. Your teacher is understanding and there are probably other students who have the very same question.*

EXERCISE SET 1.2

1. Let A be the set of all odd natural numbers between 8 and 16. Specify A by the listing method.

2. Let B be the set of all even natural numbers between 7 and 13. Specify B by the listing method.

3. Let $B = \{4, 5, 6, 7, 8, 9, 10\}$. Specify B by using set-builder notation.

4. Let $A = \{5, 7, 9, 11, 13, 15\}$. Specify A by using set-builder notation.

For Exercises 5–12, let $A = \{3, 5, 8, 10\}$ and $B = \{x \mid x$ is an odd number between 2 and 12$\}$. Answer each of the following questions as true (T) or false (F).

5. $5 \in A$

6. $11 \in A$

7. $\varnothing \subseteq B$

8. $A \subseteq \varnothing$

9. $B \subseteq A$

10. $\{1, 7\} \subseteq B$

11. $\{3, 9\} \subseteq B$

12. $\{1\}$ is not a subset of B.

13. Replace the four numerical expressions by a single variable expression involving x, where x represents any member of the set $\{3, 6, 9, 12\}$.

$$-1 + 2 \cdot 3, \qquad -1 + 2 \cdot 6,$$
$$-1 + 2 \cdot 9, \qquad -1 + 2 \cdot 12$$

14. Replace the four numerical expressions by a single variable expression involving y, where y represents any member from the set $\{-1, -2, -4, -5\}$.

$$\frac{2}{3} - 9(-1), \qquad \frac{2}{3} - 9(-2),$$
$$\frac{2}{3} - 9(-4), \qquad \frac{2}{3} - 9(-5)$$

15. Replace the three numerical expressions by a single variable expression involving z, where z represents any member from the set $\left\{0, \frac{1}{2}, \frac{5}{4}\right\}$.

$$(4)\sqrt{3}, \qquad \left(4 + \frac{1}{2}\right)\sqrt{3}, \qquad \left(4 + \frac{5}{4}\right)\sqrt{3}$$

16. Replace the three numerical expressions by a single expression involving a, where a represents any member from the set $\{-0.1, 0, 0.2\}$

$$3(-0.1 + 2), \qquad 3(2), \qquad 3(0.2 + 2)$$

For Exercises 17–22, let

$$T = \left\{-\frac{4}{2}, \frac{5}{2}, -0.333\ldots, \frac{7}{14}, 5 - \sqrt{2}, 14\pi, 0, 0.1212112111211112\ldots, \frac{20}{10}\right\}$$

Determine which members of T belong to the following sets.

17. The set of natural numbers.

18. The set of whole numbers.

19. The set of negative integers.

20. The set of integers.

21. The set of rational numbers.

22. The set of irrational numbers.

For Exercises 23–26, answer true (T) or false (F).

23. All natural numbers are rational numbers.

24. All rational numbers are integers.

25. Some integers are irrational numbers.

26. All real numbers are either positive or negative.

27. Draw a number line and graph the numbers of the set $\{0, 1, 3, 5, 10\}$.

28. Draw a number line and graph the numbers of the set $\{-2, -1, 1, 2\}$.

29. Draw a number line and graph the numbers of the set $\left\{-\dfrac{3}{2}, -0.75, \dfrac{1}{2}, 1.25\right\}$.

30. Draw a number line and graph the numbers of the set $\left\{-2, -\dfrac{3}{4}, -0.5, 0, 1\dfrac{1}{2}\right\}$.

31. Name the coordinates of the points graphed in the figure below.

32. Name the coordinates of the points graphed in the figure below.

For Exercises 33–38, name the number described.

33. Jamie deposited $45 in his savings account.

34. In football, the quarterback was tackled for a loss of 18 yards.

35. The card store had a loss of $10,450 last year.

36. During the night, the temperature dropped 15°.

37. The population of Waupun, WI, decreased by 221 people last year.

38. The baseball team won by three runs.

For Exercises 39–46, write either $<$ or $>$ between the numbers to make a true statement.

39. $-2 \quad -4$

40. $5 \quad -6$

41. $\dfrac{2}{5} \quad \dfrac{1}{2}$

42. $0.33 \quad 0.333$

43. $-2\dfrac{5}{6} \quad -2\dfrac{6}{7}$

44. $-3\dfrac{1}{4} \quad -3\dfrac{3}{8}$

45. $0.654 \quad 0.645$

46. $-0.76 \quad -0.77$

For Exercises 47–56, determine which number is larger.

47. $\dfrac{3}{4}$ and $\dfrac{7}{8}$

48. $\dfrac{3}{5}$ and $\dfrac{2}{3}$

49. $-\dfrac{6}{5}$ and $-\dfrac{5}{4}$

50. $-\dfrac{5}{6}$ and $-\dfrac{7}{8}$

51. $-1\dfrac{2}{7}$ and $-1\dfrac{3}{8}$

52. $2\dfrac{1}{7}$ and $2\dfrac{3}{14}$

53. 4.2971 and -4.2972

54. -0.00101 and -0.000101

55. -10.1298 and -10.1289

56. 7.1521 and 7.1530

57. Any real number is either an integer or lies between two successive integers. For each of the following noninteger numbers, determine the two successive integers between which it lies.

   (a) $\dfrac{5}{3}$ **(b)** $-\dfrac{18}{4}$ **(c)** $\sqrt{13}$

58. Each of the following real numbers lie between two successive integers. Find the two successive integers.

   (a) $\dfrac{9}{4}$ **(b)** -12.82 **(c)** $\sqrt{22}$

For Exercises 59–68, if the set is nonempty, list the elements of the set. If the set is empty, write $\varnothing$.

59. The set of positive even integers less than 12.

60. The set of negative odd integers greater than -9.

61. $\{x \mid x$ is a natural number between -1 and 4$\}$.

62. $\{x \mid x$ is a whole number between -1 and 5$\}$.

63. The set of whole numbers less than 0.

64. The set of natural numbers that are not positive.

65. The set of integers that are greater than -3.

66. The set of integers that are negative.

67. The set of numbers of the form $3n$, where n is an integer.

68. The set of numbers of the form $4n$, where n is an integer.

69. In medicine, drugs used to treat a disease can have unwanted side effects. If the risk in taking a drug is rated on a scale of 1 to 10, where 10 is the highest risk, and if the benefit is also rated on a 1 to 10 scale, where 10 is the highest benefit, then the risk-benefit ratio is defined to be

$$\frac{\text{risk rating}}{\text{benefit rating}}$$

Two experimental drugs, A and B, are being developed to treat the acquired immune deficiency syndrome. Drug A has a risk rating of 5 and a benefit rating of 4, while drug B has a risk rating of 7 and a benefit rating of 6. Which drug has a lower risk-benefit ratio?

70. Refer to Exercise 69. Two drugs, A and B, are used to lower cholesterol. Drug A has a risk rating of 4 and a benefit rating of 7, whereas drug B has a risk rating of 3 and a benefit rating of 5. Which drug has the higher risk-benefit ratio?

E N R I C H M E N T   E X E R C I S E S

For Exercises 1–5, determine if the number is rational or irrational. If the number is rational, write it in the form $\dfrac{a}{b}$, where a and b are integers.

1. $\dfrac{\sqrt{3}}{\sqrt{48}}$

2. $4.141414\ldots$

3. $\sqrt{150}$

4. $\dfrac{\sqrt{20}}{3\sqrt{5}}$

5. $\dfrac{(\sqrt{3})^4}{2}$

6. On a number line, graph the two numbers -1 and 5. By looking at the line, find the number between -1 and 5 that is twice as far from -1 as it is from 5.

7. How many two-element subsets can be formed from the set $\{a, b, c, d, e\}$?

8. How many three-element subsets can be formed from the set $\{a, b, c, d, e\}$?

Answers to Enrichment Exercises are on page A.2.

1.3

Operations on Real Numbers

O B J E C T I V E S

▶ *To find the opposite of a number*

▶ *To find the absolute value of a number*

▶ *To review addition and subtraction*

▶ *To review multiplication and division*

In this section, we review the four basic operations on real numbers—addition, subtraction, multiplication, and division. These operations will be used in the study of algebra, so it is important to check your ability to use them.

These four operations rely on the concepts of opposites and absolute value. Consider the numbers 3 and -3 on the number line below.

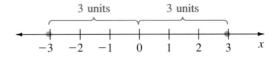

They are each the same distance from the origin. We say that 3 and -3 are *opposites* of each other. The number -3 is the opposite of 3, and 3 is the opposite of -3.

D E F I N I T I O N

Let a be a real number. The **opposite** of a is $-a$.

Therefore, to find the opposite of a, change the sign of a. For example, if a is $\frac{4}{5}$, the opposite of a is $-\frac{4}{5}$. If a is -20, the opposite of a is $-(-20)$ or 20. This illustrates the double negation property.

P R O P E R T Y

The double negation property

For any real number a,

$$-(-a) = a$$

Example 1   (a) $-(-8) = 8$ (b) $-(-9.3) = 9.3$

(c) $-\left[-\left(-\frac{5}{2}\right)\right] = -\left[\frac{5}{2}\right] = -\frac{5}{2}$ ▲

To indicate distance but not direction, we use the *absolute value* of a number. Geometrically, the **absolute value** of a number, x, is the distance that the given number is from the origin on a number line. The absolute value of the number x is written as $|x|$ and is read "the absolute value of x."

We have given the geometric meaning of absolute value. We now state the formal definition of absolute value.

D E F I N I T I O N

Let x be a real number. The **absolute value** of x is defined as

$$|x| = \begin{cases} -x \text{ if } x < 0 \\ x \text{ if } x \geq 0 \end{cases}$$

For example, if $x = -7$ then $|-7| = -(-7) = 7$ by the double negation property. Keep in mind that *the absolute value of a number is never negative.*

The absolute value of a nonzero number is always positive and the absolute value of zero is zero.

Example 2　(a) The absolute value of $\dfrac{56}{3}$ is $\dfrac{56}{3}$.

(b) The absolute value of -6.29 is 6.29.

(c) The absolute value of $-\sqrt{3}$ is $\sqrt{3}$.　　　▲

Expressions involving absolute values can be simplified as shown in the next example.

Example 3　(a) $-|28| = -28$　　　　　(b) $\left|-\left(-\dfrac{4}{7}\right)\right| = \left|\dfrac{4}{7}\right|$

$$= \dfrac{4}{7}$$

(c) $|-3 - 9| = |-12|$　　Simplify the numerical expression inside the absolute value first.

　　　$= 12$　　　　　　　　　　　　　　　　　　▲

C O M M O N　E R R O R

As indicated in Example 3(c), $|-3 - 9| \neq |-3| - |9|$. Notice that

$$|-3 - 9| = |-12| = 12$$

whereas,

$$|-3| - |9| = 3 - 9 = -6$$

We are now ready to review the four basic operations of real numbers.

R U L E

Addition of real numbers

1. To add two real numbers having the same sign, add the absolute values. The sum has the same sign as the original numbers.
2. To add two real numbers of opposite signs, subtract the smaller absolute value from the larger absolute value. The sign of the answer is the sign of the original number having the larger absolute value.

Example 4   Find each sum.

(a) $\dfrac{1}{2} + \dfrac{3}{4}$ (b) $-3 + (-21)$

(c) $-4 + 12$ (d) $15 + (-35)$

Solution   (a) We find the sum of two positive fractions using the technique reviewed in Section 1.1.

$$\frac{1}{2} + \frac{3}{4} = \frac{2}{4} + \frac{3}{4}$$

$$= \frac{2 + 3}{4} \qquad \text{This step can be done mentally.}$$

$$= \frac{5}{4}$$

(b) Here the two numbers are negative. Therefore, we first add their absolute values, then attach a negative sign to the answer.

$$-3 + (-21) = -(3 + 21)$$

$$= -24$$

(c) Since the two numbers are of opposite sign, we subtract the smaller absolute value (4) from the larger absolute value (12). The sign of the answer is the sign of the *original* number with the larger absolute value.

$$-4 + 12 = (12 - 4)$$

$$= 8$$

(d) The two numbers are of opposite sign, so we subtract the smaller absolute value (15) from the larger absolute value (35). The answer is given the "$-$" sign, since the larger absolute value came from -35.

$$15 + (-35) = -(35 - 15)$$

$$= -20 \qquad\qquad \blacktriangle$$

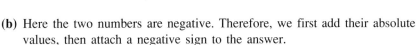

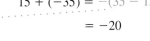

R U L E

Subtraction of real numbers

For any two real numbers a and b,

$$a - b = a + (-b)$$

That is, to subtract b from a, add the opposite of b to a.

Example 5

(a) $5 - 3 = 5 + (-3)$

$= 2$

(b) $40 - (-33) = 40 + 33$ The opposite of -33 is 33.

$= 73$

(c) $-15 - (-6) = -15 + 6$

$= -9$

(d) $8.47 - 8.47 = 0$

(e) $\dfrac{7}{2} - \dfrac{1}{6} = \dfrac{21}{6} - \dfrac{1}{6}$ Use 6 as a common denominator.

$= \dfrac{21 - 1}{6}$

$= \dfrac{20}{6}$ Subtract numerators.

$= \dfrac{10}{3}$ Write in lowest terms.   ▲

To find the sums and differences of several numbers, start with the first two numbers on the left and work to the right until you reach a final answer. This technique is illustrated in the next example.

Example 6   Find: $4 - 6 + 8 - 15$.

Solution   First find the difference, $4 - 6$, which is -2, then add 8, and so on, until you have the final answer.

$$4 - 6 + 8 - 15 = -2 + 8 - 15 \qquad 4 - 6 \text{ is } \ \ 2.$$
$$= 6 - 15 \qquad\qquad -2 + 8 \text{ is } 6.$$
$$= -9 \qquad\qquad 6 - 15 \text{ is } -9. \qquad ▲$$

Example 7   Evaluate each expression for $x = 6$ and $y = -4$.

(a) $x + y$ (b) $x - y$ (c) $|y - x|$

Solution   First, we replace x by 6 and y by -4, to obtain a numerical expression, then simplify.

(a) $x + y = 6 + (-4)$

$= 2$

(b) $x - y = 6 - (-4)$

$\qquad\quad = 6 + 4$

$\qquad\quad = 10$

(c) $|y - x| = |(-4) - 6|$

$\qquad\qquad = |-10|$

$\qquad\qquad = 10$

The multiplication of two numbers can be represented in various ways:

$$5 \times 9, \qquad 5 \cdot 9, \qquad 5(9), \qquad (5)9, \qquad (5)(9)$$

These are all notations for 5 times 9, or 45. In this textbook, however, we will not use $\times$ for multiplication since the letter x will be used as a variable.

The number 45 is called the **product** of 5 and 9. The numbers 5 and 9 are called **factors** of 45.

We now review the rules for multiplication.

R U L E

Multiplication of real numbers

To multiply two numbers a and b, first multiply their absolute values.

1. If the numbers have the same signs, the product is positive. That is,

$$(\text{positive})(\text{positive}) = \text{positive}$$

$$(\text{negative})(\text{negative}) = \text{positive}$$

2. If the numbers have opposite signs, the product is negative. That is,

$$(\text{positive})(\text{negative}) = \text{negative}$$

$$(\text{negative})(\text{positive}) = \text{negative}$$

3. If one number is zero, the product is zero. That is,

$$a \cdot 0 = 0 \text{ and } 0 \cdot a = 0, \quad \text{for any number } a$$

We can write the rules for multiplication of signed numbers in the following way.

For real numbers a and b,

$$(-a)(b) = -ab, \qquad (a)(-b) = -ab, \qquad (-a)(-b) = ab$$

Example 8   (a) $(-6)(-2) = 12$

(b) $(-3)(8) = -24$

(c) $\left(\dfrac{4}{3}\right)\left(-\dfrac{7}{12}\right) = -\left(\dfrac{4}{3}\right)\left(\dfrac{7}{12}\right)$

$\qquad\qquad\qquad = -\left(\dfrac{\overset{1}{\cancel{4}}}{3}\right)\left(\dfrac{7}{\underset{3}{\cancel{12}}}\right)$

$\qquad\qquad\qquad = -\dfrac{7}{9}$

(d) $0 \cdot (-2.5) = 0$

(e) $(-2)(3)(-4)(2) = (-6)(-4)(2)$ $\qquad$ -2 times 3 is -6.

$\qquad\qquad\qquad\quad = (24)(2)$ $\qquad\qquad$ -6 times -4 is 24.

$\qquad\qquad\qquad\quad = 48$ $\qquad\qquad\qquad\qquad\qquad$ ▲

Example 9   Jan bought 200 shares of Apex stock at $26 per share. How much did he pay for the stock?

Solution   The amount paid for the stock is given by

$$(\text{number of shares})(\text{the price per share})$$

Therefore, the amount is $200 \cdot 26$ or $5,200. ▲

Consider the product $4 \cdot 4 \cdot 4$ that contains the factor 4 used three times. A compact way of writing this triple product is 4^3. The number 4 is called the *base* and 3 is called the *exponent*. The expression 4^3 can be read as "four cubed," "four raised to the third power," or "the third power of four." In general, the product formed by *n* factors of a certain number is called the *nth power of the number*. The *n*th power of a number is symbolized using exponential notation.

D E F I N I T I O N

Exponential notation

Let *a* be any real number and *n* a positive integer. Then,

$$a^n = a \cdot a \cdot a \cdot a \cdots a$$

where *a* is a factor *n* times. The number *a* is called the **base** and *n* is called the **exponent,** and a^n is called the **nth power of a.**

Notice that the exponent tells how many times *a* is used as a factor in the product.

Example 10   Write without exponents.

(a) $x^2 = x \cdot x$ (b) $w^3 = w \cdot w \cdot w$ (c) $t^1 = t$ ▲

Example 11   Write each expression using the exponential notation.

(a) $b \cdot b \cdot b \cdot b = b^4$ (b) $5z \cdot z \cdot z \cdot y \cdot y = 5z^3 y^2$

If a numerical expression contains a power, then it can be simplified as shown in the next example.

Example 12   (a) $2^5 = 2 \cdot 2 \cdot 2 \cdot 2 \cdot 2 = 32$

(b) $(0.8)^2 = (0.8)(0.8) = 0.64$

(c) $\left(\dfrac{4}{5}\right)^3 = \dfrac{4}{5} \cdot \dfrac{4}{5} \cdot \dfrac{4}{5} = \dfrac{4 \cdot 4 \cdot 4}{5 \cdot 5 \cdot 5} = \dfrac{64}{125}$

(d) $(-9)^2 = (-9)(-9) = 81$

(e) $-9^2 = -9 \cdot 9 = -81$ ▲

N O T E   *Notice the difference between parts (d) and (e) of Example 12. The base in $(-9)^2$ is -9, whereas, the base in -9^2 is 9. The exponent applies only to the number or variable to its immediate left, unless grouping symbols are used to indicate the base.*

As indicated in the note,

$(-a)^n$ **is not necessarily the same as** $-a^n$.

C A L C U L A T O R   C R O S S O V E R

The value of a number raised to a power can be found by using the $\boxed{y^x}$ key on a calculator. For example, 47^3 can be found as shown.

Enter   47 $\boxed{y^x}$ 3 $\boxed{=}$
Display   47 47 3 103,823

Therefore, $47^3 = 103,823$.

Evaluate each of the following.

1. 7^5 **2.** 11^3 **3.** $(-64)^4$ **4.** -3^{14} **5.** 2^{25}
6. $(0.43)^{10}$ **7.** $(-289)^3$

Answers   **1.** 16,807   **2.** 1331   **3.** 16,777,216   **4.** $-4,782,969$
5. 33,554,432   **6.** 0.0002161   **7.** $-24,137,569$

Division of real numbers is defined in terms of multiplication.

DEFINITION

Let a and b be real numbers with $b \neq 0$. Then a divided by b, called the **quotient** of a and b, is the real number q such that $a = bq$. Using fraction notation,

$$\frac{a}{b} = q, \quad \text{provided that } a = bq$$

The quotient of a and b can be written as $\dfrac{a}{b}$ or as $a \div b$.

Here are some examples of division:

$\dfrac{10}{2}$ is 5, since $10 = 2 \cdot 5$.

$\dfrac{-10}{2}$ is -5, since $-10 = (2)(-5)$.

$\dfrac{10}{-2} = -5$, since $10 = (-2)(-5)$.

$\dfrac{-10}{-2} = 5$, since $-10 = (-2)(5)$.

These results can be generalized.

RULE

Observations about dividing two real numbers

To divide two real numbers, divide their absolute values.

1. If the real numbers have the same signs, then the quotient is positive.

$$\frac{\text{positive}}{\text{positive}} = \text{positive}$$

$$\frac{\text{negative}}{\text{negative}} = \text{positive}$$

2. If the real numbers have opposite signs, the quotient is negative.

$$\frac{\text{positive}}{\text{negative}} = \text{negative}$$

$$\frac{\text{negative}}{\text{positive}} = \text{negative}$$

What happens when division involves zero? Consider the following example:

$$\frac{0}{3} = ?, \quad \text{where } 3 \cdot ? = 0$$

In this case, we see that $\frac{0}{3} = 0$, since 0 is the unique real number so that $3 \cdot 0 = 0$.

In general,

If a is any nonzero number, $\dfrac{0}{a} = 0$.

Now, what about division *by* zero? Here are two examples:

$$\frac{3}{0} = ?, \quad \text{provided that } 0 \cdot ? = 3$$

There is no real number, however, that when multiplied by 0 yields 3. We conclude that $\frac{3}{0}$ is not defined.

$$\frac{0}{0} = ?, \quad \text{provided that } 0 \cdot ? = 0$$

In this case, *all* real numbers when multiplied by 0 result in 0, and so the quotient $\frac{0}{0}$ is certainly not unique. Therefore, $\frac{0}{0}$ is not defined because this quotient must be a *unique* real number. We conclude that

Division by zero is not defined.

We summarize the above discussion as a theorem.

T H E O R E M

If a is a nonzero real number, then

1. $\dfrac{0}{a} = 0$.

2. $\dfrac{a}{0}$ is not defined.

3. $\dfrac{0}{0}$ is not defined.

Consider the following examples involving division of signed numbers.

$$\frac{-21}{7} = -3, \quad \text{since } -21 = (7)(-3)$$

On the other hand, notice that $-\dfrac{21}{7} = -3$, and so $\dfrac{-21}{7} = -\dfrac{21}{7}$.

In a similar fashion, it can be shown that

$$\dfrac{21}{-7} = -\dfrac{21}{7} \qquad \text{and} \qquad \dfrac{-21}{-7} = \dfrac{21}{7}.$$

The following rules for division of signed numbers are suggested by the preceding examples:

For numbers a and b, with $b \neq 0$,

$$\dfrac{-a}{b} = -\dfrac{a}{b}, \qquad \dfrac{a}{-b} = -\dfrac{a}{b}, \qquad \dfrac{-a}{-b} = \dfrac{a}{b}$$

Example 13   (a) $6 \div 2 = \dfrac{6}{2}$

$\qquad = 3$

(b) $(-88) \div 4 = \dfrac{-88}{4}$

$\qquad = -22$

(c) $\left(-\dfrac{3}{4}\right) \div \left(-\dfrac{15}{28}\right) = \left(-\dfrac{3}{4}\right)\left(-\dfrac{28}{15}\right)$

$\qquad = \dfrac{3}{4} \cdot \dfrac{28}{15}$

$\qquad = \dfrac{7}{5}$

(d) $\dfrac{4.72}{0}$ is undefined.

(e) $0 \div \dfrac{4}{3} = 0$ ▲

Example 14   Karen made \$36,240 last year paid in 24 biweekly checks. What was her pay each check?

Solution   We divide 36,240 by 24: $\dfrac{36{,}240}{24} = \$1{,}510$ each check.   ▲

Example 15   Evaluate each of the following expressions for $x = -2$ and $y = -6$.

(a) xy (b) $(-x)y$ (c) $\dfrac{x}{y}$

Solution   Replace x by -2 and y by -6, then simplify.

(a) $xy = (-2)(-6) = 12$

(b) $(-x)y = [-(-2)](-6) = [2](-6) = -12$

(c) $\dfrac{x}{y} = \dfrac{-2}{-6} = \dfrac{2}{6} = \dfrac{1}{3}$ ▲

C A L C U L A T O R   C R O S S O V E R

Perform the indicated operations.

1. $94.58 + 32.71$ **2.** $27.661 - 32.807$

3. $(52.1)(-83.9)$ **4.** $328.9 \div 71.5$

5. $\dfrac{83.916}{-9.628}$ Round to 3 decimal places.

Answers   **1.** 127.29 **2.** -5.146 **3.** -4371.19 **4.** 4.6
5. -8.716

We finish this section with an important observation about reading an algebra book.

L E A R N I N G   A D V A N T A G E   *When reading this book, be sure to concentrate on what is being expressed. In mathematics, the English language is used in a very precise manner. For example, consider the two phrases:*

"4 less than 5"   and   "4 is less than 5"

Although they look similar, mathematically they each mean entirely different concepts:

"4 less than 5" means $5 - 4$

"4 is less than 5" means $4 < 5$

E X E R C I S E   S E T   1.3

For Exercises 1–8, simplify.

1. $-(-4)$ **2.** $-(-12)$ **3.** $-\left(-\dfrac{3}{4}\right)$ **4.** $-(-0.36)$

5. $-(-0)$ **6.** $-\left(-\dfrac{7}{4}\right)$ **7.** $-[-(-8)]$ **8.** $-[-(-7)]$

For Exercises 9–12, fill in the blank with the correct number.

9. The absolute value of $\dfrac{4}{11}$ is _____.

10. The absolute value of 0.29 is _____.

11. The absolute value of -3.82 is _____.

12. The absolute value of $-\dfrac{5}{6}$ is _____.

For Exercises 13–18, simplify.

13. $-|32|$

14. $-|81|$

15. $|-4.2|$

16. $\left|-\dfrac{7}{8}\right|$

17. $-|-90|$

18. $-|-834|$

For Exercises 19–46, perform the indicated operations.

19. $5 - 3$

20. $11 - 5$

21. $30 - (-20)$

22. $95 - (-30)$

23. $\dfrac{7}{8} - \dfrac{7}{8}$

24. $3.8 - 3.8$

25. $\dfrac{7}{5} - \dfrac{2}{5}$

26. $\dfrac{9}{4} - \dfrac{3}{4}$

27. $\dfrac{4}{3} - \dfrac{5}{6}$

28. $\dfrac{17}{8} - \dfrac{5}{4}$

29. $-\dfrac{1}{3} - \left(-\dfrac{1}{3}\right)$

30. $-0.4 - (-0.4)$

31. $14 - 18 + 12 + 10$

32. $4 - 5 - 7 + 9$

33. $\dfrac{1}{4} - \dfrac{3}{2} - \dfrac{3}{4} + \dfrac{1}{2}$

34. $\dfrac{2}{3} + \dfrac{1}{6} - \dfrac{4}{3} - \dfrac{7}{6}$

35. $(-3)(-4)$

36. $(-9)(-5)$

37. $(-2)(-3)(3)(-4)$

38. $(-1)(-2)(-3)(-4)(-5)$

39. $\left(\dfrac{6}{5}\right)\left(-\dfrac{1}{2}\right)$

40. $12\left(-\dfrac{1}{3}\right)$

41. $(3.77)(0)$

42. $(0)\left(\dfrac{2}{3}\right)$

43. $(6.3)(-0.1)$

44. $(0.1)(-2.8)$

45. $23\left(-\dfrac{1}{10}\right)$

46. $\left(\dfrac{1}{100}\right)(-45)$

47. Evaluate each expression for $x = 5$ and $y = -3$.
   (a) $x + y$
   (b) $x - y$
   (c) $|y - x|$

48. Evaluate each expression for $a = -3$ and $b = -2$.
   (a) $a + b$
   (b) $b - a$
   (c) $|a + b|$

49. Evaluate each expression for $s = 2$ and $t = -4$.
   (a) st
   (b) $s(-t)$
   (c) $\dfrac{t}{s}$
   (d) t^2
   (f) $-|-s^3|$

50. Evaluate each expression for $u = -3$ and $v = -6$.
   (a) uv
   (b) u^2
   (c) $\dfrac{u}{-v}$
   (d) $|(-u)v|$

For Exercises 51–56, write each expression without exponents.

51. s^2

52. y^5

53. c^1

54. $3z^2$

55. $5a^3b^4$

56. $2u^2v^3$

For Exercises 57–60, write each expression using exponential notation.

57. $s \cdot s \cdot s$

58. $b \cdot b \cdot b \cdot b$

59. $2r \cdot r \cdot t \cdot t \cdot t$

60. $7x \cdot x \cdot y \cdot z \cdot z \cdot z$

For Exercises 61–70, evaluate each expression.

61. 3^2

62. 4^3

63. $\left(\dfrac{2}{3}\right)^3$

64. $\left(\dfrac{4}{3}\right)^2$

65. $(0.7)^2$

66. $(0.4)^2$

67. $(-10)^2$

68. $(-7)^2$

69. -3^4

70. -11^2

For Exercises 71–86, perform the indicated division, when it is defined. When the division is meaningless, write "not defined."

71. $8 \div 4$

72. $18 \div 3$

73. $-34 \div -2$

74. $80 \div -5$

75. $0 \div 2.3$

76. $0 \div \dfrac{7}{2}$

77. $8 \div -0.1$

78. $0.54 \div 0.1$

79. $52 \div 0$

80. $6.43 \div 0$

81. $\dfrac{45}{8} \div \dfrac{9}{4}$

82. $\dfrac{35}{6} \div \dfrac{5}{6}$

83. $\dfrac{18}{25} \div -\dfrac{9}{35}$

84. $\dfrac{22}{15} \div -\dfrac{44}{9}$

85. $\dfrac{5.78}{0}$

86. $\dfrac{88.2}{0}$

For Exercises 87–92, simplify.

87. $|2 - 6|$

88. $2 + |-6|$

89. $|-4| - |-2|$

90. $|-4 - (-2)|$

91. $|-3 + (-5)|$

92. $|-3| + |-5|$

93. Is $|a + b| = |a| + |b|$ for all numbers a and b? To answer this question, evaluate $|a + b|$ and $|a| + |b|$ for $a = -4$ and $b = 3$.

94. Is $|a - b| = |a| - |b|$ for all numbers a and b? To answer this question, evaluate $|a - b|$ and $|a| - |b|$ for $a = -9$ and $b = 5$.

For Exercises 95–98, fill in the blank with either ''>'' or ''<'' to make a true statement.

95. The product of two negative numbers is _____ 0.

96. The sum of two negative numbers is _____ 0.

97. The quotient of two numbers of opposite signs is _____ 0.

98. The quotient of two negative numbers is _____ 0.

99. Sheryl bought 300 shares of Maridian stock at $15 per share. How much did she pay for the stock?

100. Jay bought 12 baseball cards at $3.50 each. How much did he pay for the cards?

101. At Ridgemont High School, there are 65 classes each with 30 students. How many students attend the school?

102. Denise bought eight cases of cola. How many cans of cola did she buy if each case contains 24 cans?

103. Joe buys a compact disc player from a store for $299. The store collects a down payment and then Joe pays $42 per month for 6 months. What was his down payment?

104. During the first half of the basketball game, the Lakers had out-rebounded the Hawks 14 to 6. If each rebound resulted in 2 points, how many more rebound points did the Lakers score over the Hawks?

105. In the banana trade, bananas are called *fingers*. A *hand* consists of fingers, and a *bunch* is a cluster of hands. If there are 12 fingers in a hand and 7 hands in a bunch, how many bananas are there in 10 bunches?

106. Each truck in an eight-truck convoy is carrying 1,500 pounds of oranges and 2,500 pounds of grapefruits. How many pounds of produce are being transported by the convoy?

107. Doris earned $128 in 4 days. How much did she earn per day?

108. Eight ice cream cones cost $9.60. How much did each cone cost?

E N R I C H M E N T   E X E R C I S E S

1. Which number is 10,000 more than 999,999?
   (a) 9,999,999   (b) 1,099,999   (c) 1,000,999   (d) 1,009,999

Find the missing number in the sequence.

2. 53, 43, 34, 26, _____, 13, 8, 4

3. 0.1, 0.09, 0.088, _____, 0.08766, 0.087655

4. In the human body, there are 23 chromosomes and 4,000 known genes attached to each chromosome. How many chromosome-gene pairs are possible?

5. Melody is planning to buy a car. She has a choice of six models, three engine sizes, and 14 colors. How many different choices (model, engine size, color) must she consider?

6. A set S is **closed** with respect to an operation such as addition, if for any two members a and b of S, then $a + b$ is a member of S. For example, the set of real numbers is closed with respect to addition and also with respect to multiplication. Is the set

$$\left\{ \ldots, -\frac{1}{81}, -\frac{1}{27}, -\frac{1}{9}, -\frac{1}{3}, -1, 1, \frac{1}{3}, \frac{1}{9}, \frac{1}{27}, \frac{1}{81}, \ldots \right\}$$

closed with respect to multiplication? division?

Let n be a positive integer. Then $n!$, read "n factorial," is defined as

$$n! = n(n - 1)(n - 2) \cdots 2 \cdot 1$$

For example, $3! = 3 \cdot 2 \cdot 1 = 6$.

For Exercises 7–10, find each of the following.

7. $4!$

8. $5!$

9. $\dfrac{4!}{2!2!}$

10. $\dfrac{7!}{3!4!}$

Answers to Enrichment Exercises are on page A.2.

1.4

Properties of Real Numbers

O B J E C T I V E S

▶ *To understand the basic properties of the real numbers*

▶ *To use the basic properties to reduce numerical expressions*

In this section, we state and discuss some of the basic properties of the real number system. These properties are vital in finding the value of a numerical expression. They will also be used in simplifying algebraic expressions in future chapters. You are already familiar with many of these properties from past experience. For example, you know that $2 + 3$ is the same as $3 + 2$. The answer is 5 either way. This fact of numbers is called the *commutative property of addition.*

We now state these properties.

Addition properties

Closure

For real numbers a and b, $a + b$ is a real number.

Example. $2 + 15 = 17$, and 17 is a real number.

Associative

For real numbers a, b, and c,

$$a + (b + c) = (a + b) + c$$

Examples. $53 + (7 + 94) = (53 + 7) + 94$.
$2x + (4x + 1) = (2x + 4x) + 1$.
$(at + bv) + bw = at + (bv + bw)$.

Commutative

For real numbers a and b,

$$a + b = b + a$$

Examples. $\dfrac{3}{5} + \dfrac{2}{7} = \dfrac{2}{7} + \dfrac{3}{5}$.
$6 + y = y + 6$.

Additive identity

The number 0 is the additive identity. That is, 0 is the only number with the following property. For any real number a,

$$0 + a = a + 0 = a$$

Examples. $0 + \dfrac{2}{3} = \dfrac{2}{3}$, and $r + 0 = r$.

Additive inverse

For any real number a, there is a number $-a$ such that

$$a + (-a) = (-a) + a = 0$$

The $-a$ is called the additive inverse of a. Any real number a has only one additive inverse.

Examples.   The additive inverse of $\dfrac{4}{3}$ is $-\dfrac{4}{3}$ and

$$\frac{4}{3} + \left(-\frac{4}{3}\right) = -\frac{4}{3} + \frac{4}{3} = 0$$

The additive inverse of $-2x$ is $-(-2x)$, or simply $2x$. Furthermore,

$$-2x + 2x = 2x + (-2x) = 0$$

Notice that the additive inverse of a, $-a$, is also called the opposite of a. Furthermore, the "negative sign" can be used in three different ways.

1. It is used to denote a negative number such as $-\dfrac{3}{2}$ $\left(\text{negative } \dfrac{3}{2}\right)$.

2. It is used to denote the additive inverse or opposite of a number. For example, the additive inverse of $\dfrac{3}{2}$ is $-\dfrac{3}{2}$ $\left(\text{the opposite of } \dfrac{3}{2}\right)$.

3. It is used to denote subtraction, as in $\dfrac{4}{5} - \dfrac{3}{2}$ $\left(\text{minus } \dfrac{3}{2}\right)$.

Multiplication properties

Closure

For any two real numbers a and b, ab is a real number.

Examples. $21 \cdot 100 = 2{,}100$, and $2{,}100$ is a real number.
$\dfrac{5}{2} \cdot \dfrac{4}{35} = \dfrac{2}{7}$, and $\dfrac{2}{7}$ is a real number.

Associative

For any three numbers a, b, and c,

$$a(bc) = (ab)c$$

Examples. $2(4y) = (2 \cdot 4)y$, and $\dfrac{3}{2}\left[\dfrac{4}{5}z\right] = \left[\left(\dfrac{3}{2}\right)\left(\dfrac{4}{5}\right)\right]z$.

Commutative

For any two numbers a and b,

$$ab = ba$$

Examples. $w \cdot 5 = 5w$, and $x(4y) = (4y)x$.

Multiplicative identity

The number 1 has the property that for any number a,

$$1 \cdot a = a \cdot 1 = a$$

Furthermore, 1 is the only number with this property.

Examples.　$1 \cdot x = x$, and $bv \cdot 1 = bv$.

The number 1 has many forms, since any *nonzero* number divided by itself is 1. For example,

$$\frac{2}{2}, \quad \frac{-38}{-38}, \quad \frac{t}{t}, \quad \frac{x-1}{x-1}$$

are all forms of the number 1, where $t \neq 0$ and $x - 1 \neq 0$.

Multiplicative inverse　**For any nonzero number a, the number $\dfrac{1}{a}$ has the property that**

$$a\left(\frac{1}{a}\right) = \left(\frac{1}{a}\right)a = 1$$

The number $\dfrac{1}{a}$ is called the **multiplicative inverse** or **reciprocal** of a.

Examples.　The multiplicative inverse of 3 is $\dfrac{1}{3}$, since

$$3\left(\frac{1}{3}\right) = \left(\frac{1}{3}\right)3 = 1$$

The multiplicative inverse of $\dfrac{5}{4}$ is $\dfrac{4}{5}$, since

$$\frac{5}{4} \cdot \frac{4}{5} = \frac{4}{5} \cdot \frac{5}{4} = 1$$

The multiplicative inverse of 0.3 is $\dfrac{1}{0.3}$ or $\dfrac{10}{3}$, since

$$(0.3)\left(\frac{10}{3}\right) = \left(\frac{10}{3}\right)(0.3) = 1$$

The multiplicative inverse of $x - y$ is $\dfrac{1}{x-y}$, since

$$(x - y)\left(\frac{1}{x-y}\right) = \left(\frac{1}{x-y}\right)(x - y) = 1$$

where $x - y \neq 0$.

The distributive properties　**For any three numbers a, b, and c,**

$$a(b + c) = ab + ac$$

$$a(b - c) = ab - ac$$

Examples. $5(s + t) = 5s + 5t.$
            $12(w - v) = 12w - 12v.$

Many of these properties are probably familiar to you. They are used in simplifying numerical expressions. These properties will also be used in the study of algebra.

The associative properties   Notice that there are two associative properties, one for addition and one for multiplication. When simplifying an expression such as $9 + 3 + 11$ or $5 \cdot 2 \cdot 6$, the associative properties state that the grouping of terms makes no difference. The numerical expression $(9 + 3) + 11$ will simplify to the same answer as the numerical expression $9 + (3 + 11)$ and $(5 \cdot 2) \cdot 6$ is the same as $5 \cdot (2 \cdot 6)$. That is, the answer is the same no matter how the numbers are grouped.

The associative properties of addition and multiplication are used to simplify expressions.

Example 1   Let x and y represent real numbers. Use an associative property, then simplify.

(a) $1 + (3 + x)$ **(b)** $10(3y)$

Solution   **(a)** We use the associative property for addition.

$$1 + (3 + x) = (1 + 3) + x = 4 + x$$

(b) We use the associative property for multiplication.

$$10(3y) = (10 \cdot 3)y = 30y \qquad \blacktriangle$$

The commutative properties   Notice that there are two commutative properties, one for addition and one for multiplication. These properties tell us that the order in adding or multiplying two numbers is immaterial. For example, $4 + 5$ and $5 + 4$ are both equal to 9. Similarly, $3 \cdot 7$ and $7 \cdot 3$ are both equal to 21.

The two commutative properties are illustrated in the next example.

Example 2   Let u, v, and w represent real numbers. Then,

(a) $w\left(\dfrac{2}{3}\right) = \left(\dfrac{2}{3}\right)w$ Commutative property of multiplication.

(b) $v + u = u + v$ Commutative property of addition.

(c) $7v + 3wu = 3wu + 7v$ Commutative property of addition.

(d) $v(2w) = (2w)v$ Commutative property of multiplication. $\blacktriangle$

Replacing a numerical expression such as $2 + 4 \cdot 9$ by a number that it represents is called **simplifying** the expression. For example,

$$14 - (2 + 5) = 14 - 7 = 7$$

$$\frac{2 + 18}{6} = \frac{20}{6} = \frac{10}{3}$$

When simplifying an expression, it is important to obey rules that govern the order of operations. For example, the expression $7 + 4 \cdot 9$ might be evaluated in two different ways. Only one of them is correct.

$$7 + 4 \cdot 9 = 11 \cdot 9 = 99 \qquad \text{Incorrect.}$$

$$7 + 4 \cdot 9 = 7 + 36 = 43 \qquad \text{Correct.}$$

Since a numerical expression should simplify to a single number, the following order of operations has been established.

R U L E

Order of operations convention

1. Simplify all powers.
2. Perform all multiplications and divisions from left to right.
3. Perform all additions and subtractions from left to right.

In the next example, we show how to observe the order of operations convention when simplifying a numerical expression.

Example 3   Simplify each numerical expression.

(a) $14 - 5 + 8 = 9 + 8$

$= 17$

(b) $15 \div 3 + 2 \cdot 4 - 1 = 5 + 8 - 1$ Perform division and multiplication first.

$= 12$

(c) $3^2 - 2^3 + \dfrac{5}{3} = 9 - 8 + \dfrac{5}{3}$ Simplify the powers first.

$= 1 + \dfrac{5}{3}$

$= \dfrac{8}{3}$

(d) $4\left(\dfrac{3}{8}\right) - \left(\dfrac{5}{2}\right)^2 + \dfrac{7}{2} = \dfrac{3}{2} - \dfrac{25}{4} + \dfrac{7}{2}$ Perform the multiplication and simplify the power.

$$= \dfrac{6}{4} - \dfrac{25}{4} + \dfrac{14}{4}$$

$$= \dfrac{6 - 25 + 14}{4}$$

$$= -\dfrac{5}{4}$$ ▲

C A L C U L A T O R   C R O S S O V E R

Does your calculator follow the order of operations convention? Try this experiment. Use your calculator to find the value of $12 + 4 \div 2$ by entering the numbers and operations as they appear from left to right.

If your calculator displays the correct answer 14, then it does obey the rule that division is performed before addition. Your calculator has an algebraic operations system.

If your calculator displays the incorrect answer 8, then it performs the operations in the order in which they occur. A way to obtain the correct answer is to divide 4 by 2 first, then add 12.

In more complicated numerical expressions, grouping symbols such as parentheses are used to change the order of operations. For example,

$$4 \cdot 7 - 5 \qquad \text{means} \qquad 28 - 5 \text{ or } 23$$

whereas

$$4(7 - 5) \qquad \text{means} \qquad 4 \cdot 2 \text{ or } 8$$

In the second expression, the grouping of the difference of 7 and 5 means that the subtraction of 5 from 7 is performed before multiplying by 4.

There are three basic grouping symbols: parentheses, (   ), brackets, [   ], and braces, {   }.

R U L E

The rule for simplifying an expression containing grouping symbols

Use the order of operations convention within each grouping, starting with the innermost and working outward until you reach the final answer.

Example 4   Simplify each numerical expression.

(a) $(3 + 2)(8 - 4) = 5 \cdot 4$

> Do the addition and subtraction within each parentheses first.

$$= 20$$

(b) We first start within the two parentheses.

$$4 + [(2^3 + 1)(8 - 2 \cdot 3)] = 4 + [(8 + 1)(8 - 6)]$$
$$= 4 + 9 \cdot 2$$
$$= 4 + 18$$
$$= 22$$

(c) $[(2 + 1)^2 - 5] \cdot 2 = [3^2 - 5] \cdot 2$

> Add 2 and 1 before squaring.

$$= [9 - 5] \cdot 2$$
$$= 4 \cdot 2$$
$$= 8$$ ▲

In an expression like $\dfrac{2 + 18}{1 + 3}$, the fraction bar is a symbol that indicates that the numerator and denominator are each considered as a single expression. Therefore,

$$\frac{2 + 18}{1 + 3} \quad \text{means} \quad \frac{20}{4}$$

Example 5   Simplify the following expressions.

(a) $\dfrac{3^3 - 2 \cdot 3}{3 \cdot 4} = \dfrac{27 - 6}{12}$

> Simplify the numerator and the denominator.

$$= \frac{21}{12}$$

$$= \frac{7}{4}$$

(b) $\dfrac{(15 - 7) \cdot 7}{(1 + 1) \cdot 4} = \dfrac{8 \cdot 7}{2 \cdot 4}$

> Work within each pair of parentheses first.

$$= 7$$ ▲

E X E R C I S E　　S E T　　1.4

For Exercises 1–24, use the stated property to fill in the blank with the correct expression.

1. Associative property of addition: $33 + (2 + 51) =$ _____

2. Associative property of addition: $(6 + 2) + 1 =$ _____

3. Associative property of multiplication: $3(5 \cdot 2) =$ _____

4. Associative property of multiplication: $(4 \cdot 8)6 =$ _____

5. Commutative property of addition: $7 + c =$ _____

6. Commutative property of addition: $(-3) + x =$ _____

7. Commutative property of multiplication: $t \cdot (-5) =$ _____

8. Commutative property of multiplication: $w(6) =$ _____

9. Associative property of multiplication: $(xy)z =$ _____

10. Associative property of multiplication: $a(2c) =$ _____

11. Commutative property of multiplication: $(ab)3 =$ _____

12. Commutative property of multiplication: $r(wv) =$ _____

13. Associative property of addition: $(34 + 2d) + (-r) =$ _____

14. Associative property of addition: $4f + (g + 2h) =$ _____

15. Associative property of multiplication: $(2r)t =$ _____

16. Associative property of multiplication: $p(qr) =$ _____

17. Additive identity: $4u + 0 =$ _____

18. Additive identity: $2a + 0 =$ _____

19. Multiplicative identity: $(x + y) \cdot 1 =$ _____

20. Multiplicative identity: $1 \cdot (2s - t) =$ _____

21. Distributive property: $2(r + v) =$ _____

22. Distributive property: $4(a + c) =$ _____

23. Additive inverse: $8c + (-8c) =$ _____

24. Multiplicative inverse: $\left(\dfrac{1}{3z}\right)(3z) =$ _____

For Exercises 25–36, state the property that each statement illustrates.

25. $6 + y = y + 6$

26. $(4 + c) + x = 4 + (c + x)$

27. $1 \cdot nm = nm$

28. $(w - r) + 0 = w - r$

29. $r(tk) = (tk)r$

30. $(3z)(z - 1) = 3[z(z - 1)]$

31. $(s + t) + w = w + (s + t)$

32. $3(0.2y) = [3(0.2)]y$

33. $4f + (-4f) = 0$

34. $(4G)\left(\dfrac{1}{4G}\right) = 1$

35. $3 + (-12)$ is a real number.

36. $4 \cdot (-2)$ is a real number.

For Exercises 37–46, use the appropriate associative property to rewrite each expression and then simplify.

37. $2 + (15 + 2w)$

38. $40 + (25 + a)$

39. $3(4r)$

40. $5(2x)$

41. $(4z - 1) + 5$

42. $(7t - 3) + 6$

43. $\dfrac{4}{3}\left[\left(\dfrac{3}{4}\right)r\right]$

44. $\dfrac{5}{6}\left[\left(\dfrac{6}{5}\right)u\right]$

45. $\left(-\dfrac{4}{9}\right)\left[\left(-\dfrac{21}{8}\right)st\right]$

46. $\left(-\dfrac{16}{35}\right)\left[\left(-\dfrac{45}{24}\right)w\right]$

For Exercises 47–60, use the distributive property and then simplify.

47. $4(s + 12)$

48. $5(3 + t)$

49. $3(r - 2)$

50. $6(2 - v)$

51. $\dfrac{3}{2}(s + 2)$

52. $\dfrac{4}{5}(f + 5)$

53. $\dfrac{2}{3}(c - 6)$

54. $\dfrac{4}{5}(R - 15)$

55. $10(2.5 - a)$

56. $100(y + 0.02)$

57. $\dfrac{3}{5}\left(\dfrac{5}{3} + b\right)$

58. $\dfrac{7}{4}\left(t + \dfrac{4}{7}\right)$

59. $\dfrac{6}{11}\left(y - \dfrac{11}{6}\right)$

60. $\dfrac{2}{5}\left(\dfrac{5}{2} - R\right)$

For Exercises 61–82, simplify each numerical expression.

61. $35 - 24 + 10$

62. $6 - 3 - 18$

63. $3^2 + 2^2 - (-5)^2$

64. $5^3 - (-10)^2 - 20$

65. $5\left(\dfrac{3}{25}\right) - \dfrac{4}{5} + \dfrac{1}{10}$

66. $\dfrac{7}{3} - 4\left(\dfrac{5}{6}\right) + 1$

67. $(-6) \div 12 + \dfrac{4}{3} - 2$

68. $48 \div (-8) + 6^2$

69. $(4 - 8)(4 - 3)$

70. $(8 - 19)(7 - 4)$

71. $5 + [(9 - 5)(22 - 16)]$

72. $[(9 - 3)(21 - 3^2)] - 4^2$

73. $\dfrac{3^2 - 4^2 + 1}{8 - (-3)^2 - 4}$

74. $\dfrac{6 - 3^2 - 5}{8 - (-2)^3 - 10}$

75. $|5 - 9| - |2^3 - 3^2|$

76. $4 - |4^2 - 21|$

77. $\dfrac{-5(7-5)+14-4}{6(-3)+1}$

78. $\dfrac{-3(4-7)+(-2)}{4(-5)-(-22+2)}$

79. $\dfrac{16}{21} \div \left(-\dfrac{2}{3}\right)^3$

80. $\dfrac{35}{-12} \div \left(\dfrac{5}{4}\right)^2$

81. $\dfrac{7}{4} \div \left[\left(\dfrac{49}{2}\right)\left(\dfrac{-10}{15}\right)\right]$

82. $\dfrac{3}{2} \div \left(\dfrac{21}{-4}\right) \div \left(-\dfrac{12}{7}\right)$

E N R I C H M E N T   E X E R C I S E S

1. The memory capacity of personal computers is usually measured in kilobytes. Furthermore, this memory is frequently a power of 2. For example, some computers have 64 K (kilobytes) of memory. Notice that $64 = 2^6$. Express the following memory capacities as powers of 2.
  (a) 128 K   **(b)** 256 K   **(c)** 512 K

Write each of the following as a power of x.

2. $x^2 \cdot x^3$

3. $x^5 \cdot x^{12}$

4. $x^n \cdot x^m$

5. $x^2 \cdot x^2 \cdot x^2$

6. $(x^2)^3$

7. $(x^n)^3$

8. $(x^n)^m$

Answers to Enrichment Exercises are on page A.2.

1.5

Algebraic Expressions

O B J E C T I V E S

▶ *To evaluate variable expressions and simplify them*

▶ *To convert verbal statements to algebraic expressions*

Recall that a *variable* is a letter that represents an unspecified number from a given set. The elements of the given set are called the values of the variable. An **algebraic expression** consists of numbers and one or more variables combined by the four basic operations $+, -, \times, \div$. Here are some examples of algebraic expressions.

$$7x + 2, \quad 3v^2, \quad \frac{bv(1+a)}{4a}, \quad \text{and} \quad -x^2yz^3$$

In an algebraic expression, the products of numbers and variables that are separated by plus or minus signs are called **terms.** For example, $4x - 5y^2 + 1$ has three terms: $4x$, $-5y^2$, and 1. The numbers and variables in a term are called the **factors** of the term. For example, 4 and x are the two factors of the term $4x$. The numerical factor is called the **coefficient** of the term. Examples of terms and coefficients are given in the following table.

Term	Coefficient
$3y$	3
x^3	1
$-0.6stw$	-0.6
$-a^5b^3c$	-1
$\sqrt{5}at^2$	$\sqrt{5}$

In the expression $3x + 7x$, there are two terms, $3x$ and $7x$. Both terms have the same variable part and differ only in their coefficients. They are examples of *like terms*. Two terms that have the same variables raised to the same powers are called **like terms**. Examples of like terms are the following.

$$2x \quad \text{and} \quad -8x \qquad 0.1s^2 \quad \text{and} \quad 2.4s^2 \qquad ab^2c, \quad -ab^2c, \quad \text{and} \quad 5ab^2c$$

Examples of pairs of **unlike terms** are the following.

$$2x \quad \text{and} \quad 2x^2 \qquad 3a \quad \text{and} \quad a^2c \qquad qp^3 \quad \text{and} \quad q^3p$$

Like terms can be combined into a single term under the operations of addition and subtraction. Consider for example the sum of $3x$ and $7x$. Using the distributive property,

$$3x + 7x = (3 + 7)x \qquad ba + ca = (b + c)a$$

$$= 10x$$

To combine like terms, we need not write the distributive property each time. Simply combine the coefficients of the terms using addition and subtraction.

Example 1

Simplify each algebraic expression by combining like terms.

(a) $16y + 9y$

(b) $0.6a^2bc - 0.7a^2bc$

(c) $x^2 - 3x^2 + 5 - 14 + 8x^2$

(d) $9(x + y) - 2(x + y)$

Solution

(a) $16y + 9y = (16 + 9)y$

$$= 25y$$

(b) $0.6a^2bc - 0.7a^2bc = (0.6 - 0.7)a^2bc$

$$= -0.1a^2bc$$

(c) $x^2 - 3x^2 + 5 - 14 + 8x^2 = (1 - 3 + 8)x^2 + (5 - 14)$

$$= 6x^2 - 9$$

(d) $9(x + y)$ and $-2(x + y)$ are considered to be like terms. We can combine them as follows:

$$9(x + y) - 2(x + y) = (9 - 2)(x + y)$$

$$= 7(x + y)$$

Sometimes the distributive property must be used first before combining like terms in an algebraic expression.

Example 2

Simplify by using the distributive property, then combine like terms.

(a) $7(x - 2) - 2x + 14$ (b) $15z^2 - a - 3(z^2 - 5a)$

(c) $2[s - 3(t - 2s) + 4] - (s + t)$

Solution

(a) $7(x - 2) - 2x + 14 = 7x - 14 - 2x + 14$

$$= (7 - 2)x - 14 + 14$$

$$= 5x \qquad\qquad \text{Combine like terms.}$$

(b) $15z^2 - a - 3(z^2 - 5a) = 15z^2 - a - 3z^2 + 15a$

$$= (15 - 3)z^2 + (15 - 1)a$$

$$= 12z^2 + 14a$$

(c) We start by working on the inside parentheses.

$2[s - 3(t - 2s) + 4] - (s + t) = 2[s - 3t + 6s + 4] - s - t$

$$= 2[7s - 3t + 4] - s - t$$

$$= 14s - 6t + 8 - s - t$$

$$= 13s - 7t + 8 \qquad \text{Combine like terms.}$$

▲

CALCULATOR CROSSOVER

Use a calculator to simplify the following algebraic expressions.

1. $7.328x + 9.542x + 5.913$

2. $6.62(3.5a - 7.1)$

3. $58(32 - 64z^2) - 61(24 + 30z^2)$

4. $0.78(2.02 + 0.36uv) - 0.41(3uv + 0.9)$

Answers   **1.** $16.87x + 5.913$ **2.** $23.17a - 47.002$ **3.** $-5542z^2 + 392$
4. $-0.9492uv + 1.2066$

When a specific value of the variable is selected, then an algebraic expression can be evaluated to obtain a number. This number is called a **value of the expression.**

STRATEGY

To find the value of an expression

1. Replace each variable by its given value.
2. Simplify the resulting numerical expression.

Example 3   Find the value of $3x + 10(6 - y)$ when $x = 5$ and $y = -1$.

Solution   Replace x by 5 and y by -1, then simplify the resulting numerical expression.

$$3x + 10(6 - y) = 3(5) + 10[6 - (-1)]$$
$$= 15 + 10(6 + 1)$$
$$= 15 + 70$$
$$= 85 \qquad \blacktriangle$$

Example 4   Evaluate $\dfrac{r - 2s}{x + y}$ for the following values.

(a) $r = 1$, $s = 0$, $x = 6$, and $y = -3$.
(b) $r = 1$, $s = 3$, $x = -2$, and $y = 2$.

Solution   Replace the variables by the specified values and then simplify.

(a) $\dfrac{r - 2s}{x + y} = \dfrac{1 - 2(0)}{6 + (-3)}$

$= \dfrac{1}{3}$

(b) $\dfrac{r - 2s}{x + y} = \dfrac{1 - 2(3)}{-2 + 2}$

$= \dfrac{1 - 6}{-2 + 2}$

$= \dfrac{-5}{0}$

which is undefined. (Recall, division by zero is meaningless.) Therefore, the algebraic expression is undefined when x is -2 and y is 2. $\qquad \blacktriangle$

CALCULATOR   CROSSOVER

Consider the can of soup shown in the figure. The radius of the circular top (and bottom) is 1.25 inches and the height of the can is about 3.75 inches. The following problems will show how to find the amount, in terms of area, of metal used to manufacture the can.

1. The area of a circle of radius r is πr^2. Find the amount of metal needed to make the top of the can. Use 3.14 for π. Find the amount of metal needed to make the bottom of the can.

2. To find the area of the lateral surface (the area around the side), we can think of this surface as being formed from a rectangular piece of metal. The sheet of metal is curved to form a cylinder as shown in the following figure.

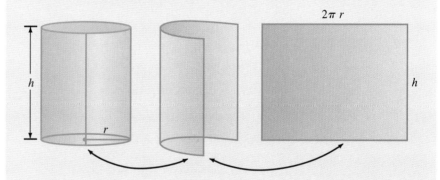

The width of the sheet is the height h of the resulting cylinder, and the length is the circumference of the circular top. The circumference of a circle of radius r is $2\pi r$. Find the area of the sheet of metal for this can.

3. What is the total amount of metal, in square inches, needed to manufacture the can?

4. If the metal costs \$0.01 per square inch, what is the cost of the metal to make the can?

Answers   **1.** 4.90625 in.², 4.90625 in.²   **2.** 29.4375 in.²   **3.** 39.25 in.²
4. \$0.3925

In order to use mathematics in applications, word phrases must be translated into algebraic expressions. We start by translating word phrases that involve the four basic operations of addition, subtraction, multiplication, and division. The following table lists some common word phrases and the equivalent algebraic expression. Let x represent the unknown number.

Word phrase	*Algebraic expression*
Addition	
The sum of a number and five	$x + 5$
Eight more than a number	$x + 8$
A number increased by 2.3	$x + 2.3$
Subtraction	
A number minus three	$x - 3$
Five and one half less than a number	$x - 5\frac{1}{2}$
A number decreased by 4.7	$x - 4.7$
A number subtracted from three	$3 - x$
Ten fewer than a number	$x - 10$
The difference of a number and nine	$x - 9$
The difference of nine and a number	$9 - x$
Multiplication	
Ten times a number	$10x$
Twice a number	$2x$
6.92 multiplied by a number	$6.92x$
Three fourths of a number	$\frac{3}{4}x$
Twenty-eight percent of a number	$0.28x$
Division	
A number divided by 8.1	$\frac{x}{8.1}$
The quotient of 12 and a number	$\frac{12}{x}$
The quotient of a number and 12	$\frac{x}{12}$
The reciprocal of a number	$\frac{1}{x}$

Example 5   Convert the following word phrases into algebraic expressions. Use x to represent a number.

(a) A number subtracted from 2.4.

(b) The quotient of twice a number and five.

(c) Three eighths of the sum of a number and seven.

(d) The sum of a number and two is divided by four times the number.

(e) Thirty-five percent of the sum of a number and 170.

Solution   (a) $2.4 - x$

(b) $\dfrac{2x}{5}$

(c) The sum of a number and 7 is $x + 7$. Three eighths of this sum is $\dfrac{3}{8}(x + 7)$.

(d) $\dfrac{x + 2}{4x}$

(e) The sum of a number and 170 is $x + 170$. Thirty-five percent of this sum is $0.35(x + 170)$.   ▲

Sometimes the word phrase to be converted to an algebraic expression does not contain words like sum or difference to indicate the operation. However, from the information given, we can determine the correct operation.

Example 6   Write as an algebraic expression.

(a) The number of cents in n quarters.

(b) The number of acres a construction company needs to develop x 3-acre lots.

(c) The number of yards in s feet.

Solution   (a) Since there are 25 cents in a quarter, then there are $25n$ cents in n quarters.

(b) Each lot needs 3 acres, therefore x lots requires $3x$ acres.

(c) Since 3 feet is 1 yard, we divide the number of feet by 3. Therefore, s feet is $\dfrac{s}{3}$ yards.   ▲

In some word problems, we must "introduce the variable" by letting x, or some other letter, stand for an unknown number. It is important to clearly state what the variable represents. In the next example, we illustrate this technique.

Example 7   Represent one number by a variable. Then express the other number in terms of the variable.

(a) The sum of two numbers is 12.

(b) Two thousand computer chips are separated into two groups—defective and nondefective.

Solution   **(a)** We introduce the variable x by letting

$$x = \text{one of the numbers}$$

Since the sum of the two numbers is 12, then

$$12 - x = \text{the other number}$$

Note that this is the correct expression for the other number, since $x + (12 - x)$ does equal 12.

(b) We introduce the variable n by letting

$$n = \text{the number of defective computer chips}$$

Since there is a total of 2,000 computer chips, then

$$2{,}000 - n = \text{the number of nondefective computer chips} \quad \blacktriangle$$

Some number problems in algebra deal with the set of integers. **Consecutive integers** are integers that immediately follow one after the other. For example, 5, 6, and 7 are three consecutive integers. We can classify an integer as being either even or odd:

$$\{\ldots -4, -2, 0, 2, 4, \ldots\} \text{ is } \textbf{the set of even integers}$$

and

$$\{\ldots -5, -3, -1, 1, 3, 5, \ldots\} \text{ is } \textbf{the set of odd integers}$$

Notice that the members of each set are listed consecutively.

Suppose we wanted to talk about two consecutive even (or odd) integers. We introduce the variable n by letting

$$n = \text{the first even(odd) integer}$$

Since the two integers are consecutive and even(odd), then

$$n + 2 = \text{the next consecutive even(odd) integer}$$

Example 8   Express the sum of two consecutive odd integers as an algebraic expression.

Solution   We introduce the variable n by letting

$$n = \text{the first odd integer}$$

Then the next consecutive odd integer is given by $n + 2$.
The sum of these two consecutive odd integers is

$$n + (n + 2) \qquad \text{or} \qquad 2n + 2 \qquad\qquad \blacktriangle$$

L E A R N I N G   A D V A N T A G E   *When doing homework or taking a test, read the directions carefully. Furthermore, neatness really does count. Always use a pencil and never a pen when doing mathematics. When you make a mistake, erase rather than cross out. If your work is messy, you may have trouble retracing steps or checking your answers.*

EXERCISE SET 1.5

For Exercises 1–22, simplify by combining like terms.

1. $2x + 9x$

2. $5s + 8s$

3. $-15(a + b) + 2(a + b)$

4. $8(x^2 + y) + 12(x^2 + y)$

5. $16x - 20x + 5 - 3x$

6. $23 - 4y + 7y - 16$

7. $8c^2 - 5 + c^2 - 14$

8. $19 + 4q^3 - 2 - 5q^3$

9. $\dfrac{5}{2}s^2 - \dfrac{7}{4} - \dfrac{9}{2}s^2 + \dfrac{11}{4}$

10. $\dfrac{3}{15} - \dfrac{2}{3}t^2 + \dfrac{7}{3}t^2 - \dfrac{2}{15}$

11. $8.9b - 0.6ay + 1.1b - 0.4ay$

12. $-0.7mn + 1.3x - 7.2x + 3.8mn$

13. $-25a + 33b - 7a + 14b + 52$

14. $16u - 12r - 13u + 15r - 1$

15. $-7qp^2 - 4 + 3q^2p - 8 + 2qp^2$

16. $-3rc^2 + r^2c - 6 - 2r^2c + 6rc^2$

17. $\dfrac{1}{2}z^2 + \dfrac{4}{3}z^3 - \dfrac{3}{4}z^2 - \dfrac{1}{6}z^3$

18. $\dfrac{3}{2}d^4 - \dfrac{11}{3}d^2 + \dfrac{1}{3}d^4 + \dfrac{11}{6}d^4$

19. $400v - 250w - 550v + 225w$

20. $600r^2 - 975b^2 + 750r^2 + 1{,}000b^2$

21. $rsw - sw - 8rsw + 4sw$

22. $-9z^2c + 5zc - 11z^2c - 9zc$

For Exercises 23–40, use the distributive property to simplify.

23. $4(2y - 4) - 7y$

24. $3(5x + 9) + 2x$

25. $14 - 4(2 - s^2) + 6$

26. $25 - 3(7 + z^2) - 35$

27. $y - 0.3(2 - 4y)$

28. $-a + 0.6(1 - 7a)$

29. $-\left(\dfrac{7}{2} - mn\right) + \dfrac{5}{2} - mn$

30. $-\left(2xy - \dfrac{5}{7}\right) + \dfrac{5}{14}$

31. $4(3r^3 - 5) - 5(1 - r^3)$

32. $5(-2 + 6h^4) - 4(3h^4 + 2)$

33. $6(x^3y^2 - 5) - 3(2 - 3x^3y^2)$

34. $-3(4 - s^2t^2) - 5(-1 + 7s^2t^2)$

35. $2(3 + r^2w) - 6(4 + 8r^2w)$

36. $4(x^2yz^4 - 10) - 9(-5 + 3x^2yz^4)$

37. $2[1 - (x^2t + 3)] - [-9x^2t - 2(8 - 2x^2t)]$

38. $4[1 - (4 - 2pq)] - [5 - 8(1 - pq)]$

39. $2x - [1 - (x - z^2)] + 3[-10z^2 - (x - 8z^2)]$

40. $5 + 2[-3 + 5(rt^3 - 1)] - [5rt^3 - (6 - 4rt^3)]$

For Exercises 41–52, find the value of the algebraic expression for the given values of the variables.

41. $4s + 2t$; $s = 3$ and $t = 8$

42. $-3x + 10z$; $x = 9$ and $z = \dfrac{1}{2}$

43. $6uv - 2v^2$; $u = \dfrac{1}{3}$ and $v = -1$

44. $x^2 + y^2$; $x = -2$ and $y = -3$

45. $(a + b)^2$; $a = \dfrac{1}{2}$ and $b = \dfrac{1}{2}$

46. $(x - y)^2$; $x = 3$ and $y = -4$

47. $a^2 + 2ab + b^2$; $a = \dfrac{1}{2}$ and $b = \dfrac{1}{2}$

48. $x^2 - 2xy + y^2$; $x = 3$ and $y = -4$

49. $\dfrac{r + 1}{r - 1}$; $r = -1$

50. $\dfrac{-2q + 1}{3q - 2}$; $q = 4$

51. $\dfrac{xz - 2}{x - 2z^2}$; $x = 6$ and $z = 3$

52. $\dfrac{y^2 - t}{yt - t}$; $y = -1$ and $t = 7$

For Exercises 53–70, convert the following word phrases into algebraic expressions. Use x to represent a number.

53. The sum of three and a number.

54. The difference of four and a number.

55. The difference of a number and 20.

56. A number divided by 100.

57. The quotient of a number and 34.

58. The reciprocal of a number.

59. Twelve minus a number.

60. One half of a number.

61. A number increased by 92.

62. Seven more than a number.

63. The sum of twice a number and 48.

64. Four times a number, the result increased by 11.

65. Six and one half percent of a number.

66. Ten and one-half percent of a number.

67. Ten times the difference of a number and three.

68. Five times the sum of four and a number.

69. Four fifths less than three times a number.

70. Fifty more than the quotient of six and a number.

For Exercises 71–76, write as an algebraic expression.

71. The amount of money received for working n hours at $30 per hour.

72. The number of cents in q quarters.

73. The cost of t red delicious apples at 45 cents per apple.

74. The cost of n compact discs at $12.95 each.

75. The number of feet in y inches.

76. The sales tax on a purchase that cost d dollars, if the sales tax rate is 7.5%.

For Exercises 77–88, represent one number by a variable. Then, express the other number in terms of the variable. Be sure to clearly state what your variable represents.

77. Debbie's age now and her age 12 years ago.

78. Andrew's age now and his age in $4\frac{1}{2}$ years.

79. The sum of two numbers is nine.

80. The sum of two numbers is $-\frac{3}{5}$.

81. The selling price of a lamp, in dollars, and 30% of the selling price decreased by $15.

82. The wholesale cost, in dollars, of a desk and the retail price, where the retail price is 150% of the wholesale cost, decreased by $20.

83. Forty-five items are separated in two groups—defective and nondefective.

84. A 12-foot board is cut into two pieces.

85. The student enrollment at Ridgemont High School is 1.75 times the enrollment at Salisbury High School.

86. John's bowling score and Richard's score, where Richard's score is two thirds of John's score.

87. The combined incomes of Mr. and Mrs. Kelly is $67,000 a year.

88. A parking meter contains 28 quarters and slugs.

For Exercises 89–94, write an algebraic expression from the given information. Be sure to clearly state what the variable represents.

89. The sum of two consecutive integers.

90. The sum of two consecutive odd integers.

91. The square of the sum of two consecutive odd integers.

92. The sum of the squares of two consecutive even integers.

93. The sum of the squares of three consecutive even integers.

94. The square of the sum of three consecutive odd integers.

ENRICHMENT EXERCISES

For Exercises 1–4, simplify.

1. $2 - \{3 - [x - (4y - 2x) + 1] - 1 + 4(1 - 8y)\}$

2. $x[5 - 6(4 - 3x)]$

3. $ab[a^2b - ab(b - a)]$

4. $8p^2q^3 + 4pq\{1 - 2[pq^2 + p^2q]\} + 8p^3q^2$

5. A pipe of length 12 feet is cut into three pieces. If the first piece is x feet long and the middle piece is $\dfrac{1}{2}$ foot less than twice the first piece, express the length of the third piece in terms of x.

Answers to Enrichment Exercises are on page A.3.

CHAPTER 1 **S**ummary and review

Examples

$$\frac{12}{18} = \frac{2 \cdot 6}{3 \cdot 6} = \frac{2}{3}$$

Fractions (1.1)

A fraction is in **lowest terms** if the numerator and denominator have no common factors other than one. To write a fraction in lowest terms, we make use of the fact

$$\frac{a \cdot c}{b \cdot c} = \frac{a}{b}, \qquad c \neq 0$$

$$\frac{12}{18} = \frac{\overset{2}{\cancel{12}}}{\underset{3}{\cancel{18}}} = \frac{2}{3}$$

A common practice for writing a fraction in lowest terms is to use the **cancellation method**.

Examples

$$\frac{2}{5} \cdot \frac{6}{7} = \frac{2 \cdot 6}{5 \cdot 7} = \frac{12}{35}$$

$$\frac{5}{4} \div \frac{2}{3} = \frac{5}{4} \cdot \frac{3}{2} = \frac{15}{8}$$

$$\frac{6}{5} - \frac{1}{5} = \frac{6-1}{5} = \frac{5}{5} = 1$$

$$\frac{4}{3} + \frac{5}{6} = \frac{4 \cdot 2}{3 \cdot 2} + \frac{5}{6}$$
$$= \frac{8}{6} + \frac{5}{6}$$
$$= \frac{13}{6}$$

Fractions (1.1)

To **multiply two fractions,** we multiply numerators and denominators.

$$\frac{a}{b} \cdot \frac{c}{d} = \frac{a \cdot c}{b \cdot d}$$

To **divide two fractions,** invert the second fraction and multiply.

$$\frac{a}{b} \div \frac{c}{d} = \frac{a}{b} \cdot \frac{d}{c}$$

To **add or subtract two fractions** with *like denominators,* add or subtract the numerators, keeping the common denominator.

$$\frac{a}{b} + \frac{c}{b} = \frac{a+c}{b}$$

$$\frac{a}{b} - \frac{c}{b} = \frac{a-c}{b}$$

If the two fractions have *unlike denominators,* first rewrite each fraction so that they have a common denominator. To do this, we use the property

$$\frac{a}{b} = \frac{a \cdot c}{b \cdot c}, \qquad c \neq 0$$

Sets (1.2)

A **set** is a collection of objects. It is common to use capital letters to denote sets. If A is a set, and a is a member or element of A, we write $a \in A$. If a is not a member of A, we write $a \notin A$.

We can either *list the elements* of a set or describe a set using *set-builder notation.*

Listing method:

$A = \{2, 4, 6\}$

Set-builder method:

$A = \{x \mid x$ is an even number between 1 and 7$\}$

The letter x used in the describing method represents a typical element of A. A letter used in this way is called a **variable.**

Two sets are **equal** if they have the same elements. A is a **subset** of B if each element of A is an element of B. We write $A \subseteq B$. The **empty set** $\varnothing$ is the set with no elements.

In addition to using variables in describing sets, variables are used in **variable expressions** such as $4x - 3$.

The real number system (1.2)

The set of **natural numbers, N,** is $N = \{1, 2, 3, \ldots\}$.

The set of **whole numbers, W,** is $W = \{0, 1, 2, 3, \ldots\}$.

The set of **integers, I,** is

$$I = \{\ldots, -3, -2, -1, 0, 1, 2, 3, \ldots\}$$

Examples

π and $\sqrt{2}$ are irrational numbers.

The members of the set $\left\{-2, -1.5, \dfrac{1}{2}\right\}$ are graphed on the number line below.

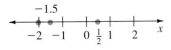

The opposite of 4 is -4.

$|4| = 4, \quad |0| = 0, \quad \left|-\dfrac{2}{5}\right| = \dfrac{2}{5}$

A **rational number** is any number of the form $\dfrac{a}{b}$, $b \neq 0$, where a and b are integers. A real number that is not rational is an **irrational number.**
The set of **real numbers** is comprised of rational and irrational numbers.
Real numbers can be graphed on a number line.
The **number line** displays an order of the real numbers.

$<$ means "is less than"

$>$ means "is greater than"

If $a < b$, then a lies to the left of b on the number line.
If $a > b$, then a lies to the right of b on the number line.

Operations on real numbers (1.3)

If a is a number, $-a$ is called the **opposite** of a.
For any real number a, we have

$$-(-a) = a$$

The **absolute value** of x is given by

$$|x| = \begin{cases} -x, & \text{if } x < 0 \\ x, & \text{if } x \geq 0 \end{cases}$$

The **four basic operations of real numbers** are addition, subtraction, multiplication, and division.

Addition

1. To add two real numbers having the same sign, add their absolute values. The sum has the same sign as the original numbers.
2. To add two real numbers of opposite signs, subtract the smaller absolute value from the larger absolute value. The sign of the answer is the sign of the original number having the larger absolute value.

Subtraction

$5 - 8 = 5 + (-8) = -3$

For any two real numbers a and b, $a - b = a + (-b)$.

Multiplication

$(-4)(-3) = 12,$
$5(-2) = -10,$
$0 \cdot 3.2 = 0$

To multiply two numbers a and b, first multiply their absolute values.

1. If the numbers have the same signs, the product is positive.
2. If the numbers have opposite signs, the product is negative.
3. If either number is zero, the product is zero.

$4^3 = 4 \cdot 4 \cdot 4 = 64,$
$x \cdot x \cdot x \cdot y \cdot y \cdot y = x^3 y^3$

Let a be any number and n a positive integer. The **nth power of a** is defined as $a^n = a \cdot a \cdot a \cdot \cdots \cdot a$, where a is used as a factor n times. The a is called the **base** and n is called the **exponent.**

Examples

$\dfrac{5}{2} = 2.5$, $\quad \dfrac{45}{-5} = -9$,

$\dfrac{3}{0}$ is undefined, $\quad \dfrac{0}{3} = 0$

Division

To divide two real numbers, divide their absolute values.

1. If the numbers have the same signs, the quotient is positive.
2. If the numbers have opposite signs, the quotient is negative.
3. Zero divided by any nonzero number is zero.

Division by zero is undefined.

Properties of real numbers (1.4)

Addition properties

Closure: For real numbers a and b, $a + b$ is a real number.
Associative: For real numbers a, b, and c,

$$a + (b + c) = (a + b) + c$$

Commutative: For real numbers a and b,

$$a + b = b + a$$

Additive identity: The number 0 is the additive identity. For any real number a,

$$a + 0 = 0 + a = a$$

Additive inverse: For any number a, $-a$ is the additive inverse of a,

$$a + (-a) = (-a) + a = 0$$

Multiplication properties

Closure: For any real numbers a and b, ab is a real number.
Associative: For any numbers a, b, and c,

$$a(bc) = (ab)c$$

Commutative: For any numbers a and b,

$$ab = ba$$

Multiplicative identity: The number 1 has the property that for any number a,

$$1 \cdot a = a \cdot 1 = a$$

Multiplicative inverse: For any nonzero number a, $\dfrac{1}{a}$ is the multiplicative inverse of a, and

$$a\left(\dfrac{1}{a}\right) = \left(\dfrac{1}{a}\right)a = 1$$

Examples

$3 + (6 + 2t) = (3 + 6) + 2t$
$\quad\quad\quad\quad = 9 + 2t$

$2^3 - 4 \cdot 2 + 12 \div 3 =$
$\quad\quad\quad 8 - 8 + 4 = 4$

$(6 - 8)(12 - 2) = (-2)(10)$
$\quad\quad\quad\quad\quad = -20$

Distributive properties

For any numbers a, b, and c,

$$a(b + c) = ab + ac$$
$$a(b - c) = ab - ac$$

We use the associative, commutative, and distributive properties to simplify expressions.

When simplifying an expression, we follow the **order of operations convention.**

1. Simplify all powers.
2. Perform all multiplications and divisions from left to right.
3. Perform all additions and subtractions from left to right.

When an expression contains grouping symbols, use the order of operations convention within each grouping starting with the innermost and working outward until you reach the final answer.

Algebraic expressions (1.5)

$3x - 7y + 2$ is an algebraic expression consisting of three terms:

$3x$, $-7y$, and 2

The coefficient of $3x$ is 3.

$-3x^2y$ and $5x^2y$ are like terms. $3xy^2$ and $-x^2y$ are unlike terms.

$5x - 8x = (5 - 8)x = -3x$

When $x = -1$,

$2x - 7 = 2(-1) - 7 = -9$

The algebraic expression for the difference of a number and 10 is $x - 10$.

An **algebraic expression** consists of numbers and one or more variables combined by the four basic operations of addition, subtraction, multiplication, and division. The products of numbers and variables that are separated by plus or minus signs are called **terms.**

The numerical factor in a term is called the **coefficient.**

Two terms that have the same variables raised to the same powers are called **like terms.**

We can **combine like terms** by using the distributive property.

When a specific value of the variable is selected, then an algebraic expression such as $2x - 7$ can be evaluated to obtain a number, called the **value of the expression.**

In order to use mathematics in applications, **word phrases must be translated into algebraic expressions.**

CHAPTER 1 REVIEW EXERCISE SET

Section 1.1

For Exercises 1–4, write each fraction in lowest terms.

1. $\dfrac{8}{12}$

2. $\dfrac{14}{18}$

3. $-\dfrac{5}{35}$

4. $-\dfrac{49}{14}$

For Exercises 5–10, combine as indicated and write the answer in lowest terms.

5. $\dfrac{4}{5} \cdot \dfrac{25}{8}$

6. $4\left(\dfrac{3}{22}\right)$

7. $\dfrac{5}{4} - \dfrac{1}{4}$

8. $8 \div \dfrac{4}{3}$

9. $\dfrac{3}{5} - \dfrac{2}{25}$

10. $\dfrac{1}{2} - \dfrac{1}{3} + \dfrac{5}{6}$

Section 1.2

11. Let A be the set of all odd integers between -2 and 4. Specify A by the listing method.

12. Replace the four numerical expressions by a single variable expression involving x, where x represents any member of the set $\{-1, 0, 1, 2\}$.

$$-2 - 3(-1), \qquad -2 - 3(0),$$
$$-2 - 3(1), \qquad -2 - 3(2)$$

For Exercises 13–17, let

$$S = \left\{-3\frac{1}{2}, \frac{4}{5}, -\sqrt{2}, 3\pi, 2, -\frac{9}{3}\right\}$$

Determine which members of S belong to the following sets.

13. The natural numbers.

14. The integers.

15. The irrational numbers.

16. The rational numbers.

17. The negative irrational numbers.

18. Draw a number line and graph the numbers of the set

$$\left\{-2\frac{2}{3}, -1, 0.25, 1, 5\right\}$$

For Exercises 19–22, write either $<$ or $>$ between the two numbers to make a true statement.

19. $-2\dfrac{1}{2}$ -2

20. 1 0

21. 0.55 0.551

22. -9.11 -9.10

Section 1.3

For Exercises 23–26, simplify.

23. $-(-2.7)$

24. $|-45|$

25. $-[-(-1)]$

26. $-|4 - 9|$

For Exercises 27–33, perform the indicated operations, when possible, and simplify.

27. $4.6 - 9.3$

28. $\left(3\dfrac{1}{2}\right)(0)$

29. $(53)(-0.1)$

30. $\dfrac{4}{0}$

31. $\dfrac{44}{3} \div \dfrac{11}{2}$

32. $\left(-\dfrac{2}{3}\right)\left(-\dfrac{9}{5}\right)$

33. $\dfrac{|3 - 8|}{|8 - 3|}$

34. Tom bought 10 pairs of socks at $3.65 per pair. How much did he pay for the socks?

35. The Topton Corporation makes electric ranges. The demand for their ranges is 1,350 per day, while their only plant has a 1,100 per day capacity. In 5 days by how many ranges has the demand exceeded supply?

Section 1.4

For Exercises 36–43, use the stated property to fill in the blank with the correct expression.

36. Associative property of addition: $24 + (3 + 2c) = $ _____

37. Associative property of multiplication: $2(3x) = $ _____

38. Commutative property of addition: $4 + (-2) = $ _____

39. Commutative property of multiplication: $x \cdot 3 = $ _____

40. Additive identity: $0 + 8t = $ _____

41. Additive inverse: $-3w + 3w = $ _____

42. Multiplicative identity: $1 \cdot 4s = $ _____

43. Multiplicative inverse: $5R\left(\dfrac{1}{5R}\right) = $ _____

For Exercises 44–48, use the appropriate associative property to rewrite each expression and then simplify.

44. $4 + (-3 + 2v)$

45. $(5q - 2) + (-5)$

46. $\left(\dfrac{5}{4}\right)\left[\left(\dfrac{4}{5}\right)b\right]$

47. $\left(-\dfrac{2}{3}\right)\left[\left(\dfrac{9}{4}\right)x\right]$

48. $\left(\dfrac{6}{14}\right)\left[\left(-\dfrac{7}{4}\right)y\right]$

For Exercises 49–52, use the distributive property and then simplify.

49. $4(2x - 1) + 3$

50. $2 - (2x - 1)5$

51. $4 + 7(x - z + 2)$

52. $\dfrac{2}{3}\left(6x + \dfrac{1}{4}\right) - \dfrac{5}{6}$

For Exercises 53–58, simplify each numerical expression.

53. $45 - 76 - 25$

54. $3^2 + 2^3 - 4^2$

55. $\dfrac{5}{3} - 4\left(\dfrac{1}{3}\right) + 1$

56. $36 \div (-4) + 20$

57. $(10 - 25)(55 - 45)$

58. $\dfrac{4^3 - 4^3 + 2}{2^4 - 15}$

Section 1.5

For Exercises 59–63, simplify by combining like terms.

59. $4z + 10z$

60. $4 - y + 3y - 18$

61. $\dfrac{3}{5} + \dfrac{7}{3}s^2 - \dfrac{4}{3}s^2 + \dfrac{2}{5}$

62. $\dfrac{2}{3}\left[\dfrac{9}{4}x^2\right] - \dfrac{3}{4}x^2 + 2$

63. $5r^2 - 6r^3 - 7r^3 + 2r^2$

For Exercises 64–67, use the distributive property to simplify.

64. $3(7t - 2) + 4t$

65. $-\left(\dfrac{2}{3} - x^3\right) - \left(-\dfrac{2}{3} + x^3\right)$

66. $2(-1 + 3t^2r^3) - 4(-2 - 5t^2r^3)$

67. $2[1 - (3 - z^4)] + 2$

68. Find the value of $3x - 2y - 4$ if $x = -1$ and $y = 3$.

For Exercises 69–75, convert each word phrase into an algebraic expression. Use x to represent a number.

69. The sum of a number and five.

70. The difference of -7 and a number.

71. One third of a number.

72. Three times a number, the result increased by $\dfrac{3}{2}$.

73. Thirty percent of a number.

74. Forty-three less than twice a number.

75. The reciprocal of the sum of two and a number.

For Exercises 76–79, represent one number by a variable. Then express the other number in terms of the variable. Be sure to clearly state what your variable represents.

76. Sam's age now and his age 10 years ago.

77. The selling price of a VCR, in dollars, and 35% of the selling price decreased by $150.

78. Eighty-four items are separated into two groups—defective and nondefective.

79. A 15-foot board is cut into two pieces.

For Exercises 80 and 81, write an algebraic expression from the given information. Be sure to clearly state what the variable represents.

80. The cube of the sum of two consecutive odd integers.

81. The sum of the cubes of two consecutive even integers.

CHAPTER 1 TEST

Section 1.1

1. Write the fraction $\dfrac{12}{21}$ in lowest terms.

For Problems 2 and 3, combine as indicated and write the answer in lowest terms.

2. $\dfrac{6}{5} \cdot \dfrac{35}{9}$

3. $\dfrac{6}{5} - \dfrac{1}{5}$

Section 1.2

For Problems 4–7, let

$$S = \left\{ -5, -\sqrt{2}, 0, \frac{1}{2}, 3 \right\}$$

Determine which members of S belong to the following sets.

4. The natural numbers

5. The integers

6. The rational numbers

7. The irrational numbers

8. Write either $<$ or $>$ between the two numbers to make a true statement.

$$-3\frac{1}{2} \qquad -4\frac{1}{2}$$

Section 1.3

For Problems 9–11, simplify

9. $-|-3|$

10. $7.4 - 3.2$

11. $\dfrac{|4 - 6|}{|13 - 5|}$

Section 1.4

For Problems 12 and 13, use the stated property to fill in the blank with the correct expression.

12. Associative property of addition: $3 + (2 + x) =$ _____

13. Associative property of multiplication: $4(3y) =$ _____

For Problems 14 and 15, use the appropriate associative property to rewrite each expression and then simplify.

14. $5 + (2 + w)$

15. $\dfrac{3}{2}\left(\dfrac{2}{3}x\right)$

16. Use the distributive property and then simplify. $3 - 2(1 - 4a)$

For Problems 17 and 18, simplify the numerical expression.

17. $4 - 2 \cdot 5$

18. $\dfrac{2^3 - 5 + 3}{3^2 - 11}$

Section 1.5

19. Simplify by combining like terms.

$$1 + 5x - 3x - 7$$

20. Use the distributive property to simplify.

$$4(1 - 2x) + 7x - 3$$

For Problems 21–23, convert each word phrase into an algebraic expression. Use x to represent a number.

21. The sum of a number and six.

22. The difference of twice a number and 2.

23. Forty percent of a number.

For Problems 24–26, represent one number by a variable. Then express the other number in terms of the variable. Be sure to clearly state what your variable represents.

24. Tom's age now and his age 12 years ago.

25. Twenty items are separated into two groups—defective and nondefective.

26. A 12-foot board is cut into two pieces.

Linear Equations and Inequalities

Overview

In this chapter, we will see how algebraic expressions are used to construct linear equations and inequalities. In the last chapter (Section 1.5), we simplified algebraic expressions. These techniques have continued use when we work with linear equations and inequalities. In this chapter we introduce an important application of algebra—solving word problems. As we progress through this book, we will learn techniques to solve equations, then put these equations to work in solving word problems.

2.1

Properties of Equality

OBJECTIVES

▶ *To solve equations using the addition-subtraction property of equality*

▶ *To solve equations using the multiplication-division property of equality*

In this section, we introduce properties of equality that will be used to solve an equation. Examples of equations are the following:

$$2x - 4 = 8, \qquad r^2 - t = 2rt, \qquad \text{and} \qquad z^3 - 4z^2 + z = 1$$

To **solve an equation** means to find all real numbers that make the equation a true statement. These numbers comprise the **solution set** of the equation. Any solution of an equation is said to *satisfy* the equation, since replacing the variable by a solution produces a true statement.

Given an equation such as $2x + 4 = -x + 13$, we can determine if any real number is a member of the solution set by replacing x by this real number in the equation to see if it produces a true statement. This *substitution method* is illustrated in Example 1.

Example 1

Check if any members of $\{-1, 0, 3\}$ belong to the solution set of the equation $2x + 4 = -x + 13$.

Solution

In the given equation, we replace x in turn by its values -1, 0, and 3 to determine which of these numbers, if any, make the equation a true statement.

$x = -1$	$x = 0$	$x = 3$
$2x + 4 = -x + 13$	$2x + 4 = -x + 13$	$2x + 4 = -x + 13$
$2(-1) + 4 \stackrel{?}{=} -(-1) + 13$	$2(0) + 4 \stackrel{?}{=} -0 + 13$	$2(3) + 4 \stackrel{?}{=} -3 + 13$
$-2 + 4 \stackrel{?}{=} 1 + 13$	$0 + 4 \stackrel{?}{=} 0 + 13$	$6 + 4 \stackrel{?}{=} 10$
$2 \neq 14$	$4 \neq 13$	$10 = 10$

Therefore, -1 and 0 are not solutions, since in each instance, the left side of the equation is not equal to the right side. However, 3 is a solution of the equation, since replacing x by 3 produces the true statement $10 = 10$. ▲

Our goal in this chapter is to find solution sets of certain equations. It would not be reasonable to use the substitution method to determine the solution set, since we would have to test all real numbers. Instead, we will develop efficient techniques for solving an equation. Then we will use the substitution method to check our answers.

Two equations are said to be **equivalent** if they both have the same solution set. For example,

$$2x + 1 = 4, \qquad 2x = 3, \qquad \text{and} \qquad x = \frac{3}{2}$$

are each equivalent equations, since the solution set for each of them is $\left\{\frac{3}{2}\right\}$.

When solving an equation, our goal is to rewrite this equation through a sequence of steps to obtain an equivalent form where the variable is isolated on one side.

STRATEGY

The basic goal when solving an equation

Through a sequence of steps, rewrite the original equation to ultimately obtain an equivalent equation where the variable is isolated on one side and has a coefficient of one.

To implement this goal, we use two properties of equality: the *addition-subtraction* and the *multiplication-division* properties. We introduce the first property with the following situation.

Suppose two basketball teams are tied at n points each at the end of regulation time. If each team scores 15 points in the first overtime period, then the score is still tied. Namely, adding 15 to both sides of the equation $n = n$ yields another equation $n + 15 = n + 15$. Therefore, adding the same quantity to both sides of an equation results in an equivalent equation. The same result holds when subtracting the same quantity from both sides of an equation.

P R O P E R T Y

The addition-subtraction property of equality

Let A, B, and C represent algebraic expressions. If

$$A = B$$

then

$$A + C = B + C$$

and

$$A - C = B - C$$

Therefore, given an equation $A = B$, we may add (or subtract) the same quantity to (or from) each side to obtain an equivalent equation.

In the next examples, we use the addition-subtraction property to reduce a given equation to an equivalent equation of the form where the variable is on one side and a constant is on the other side of the equal sign.

Example 2   Solve the equation

$$x + 49 = -25$$

then check your answer.

Solution   Remember that to solve an equation, the goal is to get the variable alone on one side of the equation. If we subtract a number from both sides, by the subtraction property of equality, we still maintain an equivalent equation. In particular, if we subtract 49 from both sides, we have

$$x + 49 - 49 = -25 - 49$$

Simplifying this equation,

$$x + 0 = -74$$

$$x = -74$$

We now check our answer in the original equation.

$$x + 49 = -25$$

$$-74 + 49 \stackrel{?}{=} -25$$

$$-25 = -25$$

Therefore, our answer satisfies the equation and so the solution of $x + 49 = -25$ is $x = -74$. Using set notation, the solution set of the original equation is $\{-74\}$. ▲

Example 3   Solve the equation

$$2t - 12 = 3t - 50$$

and check your answer.

Solution   Our plan is to isolate the variable to the right side of the equation. To achieve this, we first subtract $2t$ from both sides.

$$2t - 12 = 3t - 50$$

$$2t - 12 - 2t = 3t - 50 - 2t \qquad \text{Subtract } 2t \text{ from both sides.}$$

$$-12 = t - 50 \qquad \text{Simplify.}$$

$$-12 + 50 = t - 50 + 50 \qquad \text{Add 50 to both sides.}$$

$$38 = t \qquad \text{Simplify.}$$

Check $2t - 12 = 3t - 50$

$$2(38) - 12 \overset{?}{=} 3(38) - 50$$

$$76 - 12 \overset{?}{=} 114 - 50$$

$$64 = 64$$

The solution is $t = 38$. ▲

N O T E   *In Example 3, we applied the properties of equality twice. We can reduce these two steps into a single step:*

$$2t - 12 + 50 - 2t = 3t - 50 + 50 - 2t$$

Then we simplify both sides of this equation to obtain the answer, $38 = t$.

Example 4   Find the solution set of

$$-3\left(4z - \frac{1}{3}\right) + 5 = -15 - 6\left(\frac{2}{3} + 2z\right) - z$$

Solution   We first simplify both sides using the distributive property, then combine like terms.

$$-3\left(4z - \frac{1}{3}\right) + 5 = -15 - 6\left(\frac{2}{3} + 2z\right) - z$$

$$-12z + 1 + 5 = -15 - 4 - 12z - z \qquad \text{Distributive property.}$$

$$-12z + 6 = -19 - 13z \qquad \text{Combine like terms.}$$

Next, we apply the properties of equality in order to isolate z on the left side and then simplify both sides of the resulting equation.

$$-12z + 6 - 6 + 13z = -19 - 13z - 6 + 13z$$

$$z = -25$$

The solution set is $\{-25\}$.   ▲

C A L C U L A T O R   C R O S S O V E R

Solve each equation.

1. $0.234x - 4.329 + 7 = 5.524 - 0.766x$
2. $6.21 - 5.35 + 0.129w = 1.374w - 2.245w$
3. $27(34 - 29q) = 35(-12q + 6) - 4(27 + 91q)$
4. $2.1(13c - 51) + 7.9 = 0.2c + 3(8.7c + 5.3)$

Answers   **1.** 2.853   **2.** −0.86   **3.** −816   **4.** 115.1

The addition-subtraction property of equality enables us to solve certain equations. However, these properties would not apply to the equation $\dfrac{x}{3} = 8$ or to $5x = -1$. To solve equations like these, we need the *multiplication-division property of equality*.

P R O P E R T Y

The multiplication-division property of equality

Let A, B, and C represent algebraic expressions with $C \neq 0$. If

$$A = B$$

then

$$CA = CB$$

and

$$\frac{A}{C} = \frac{B}{C}$$

The multiplication-division property of equality states that we may multiply or divide both sides of an equation by the same nonzero quantity to obtain an equivalent equation.

Example 5   Solve the equation

$$\frac{1}{14}r = 2$$

and check your answer.

Solution   Remember that the goal in solving an equation is to isolate the variable on one side of the equation. On the left side of $\frac{1}{14}r = 2$, there is $\frac{1}{14}r$ instead of just r.

Realizing that $14\left(\frac{1}{14}\right) = 1$, we multiply both sides of the equation by 14.

$$\frac{1}{14}r = 2$$

$$14\left(\frac{1}{14}r\right) = (14)(2) \qquad \text{Multiply both sides by 14.}$$

$$1r = 28$$

$$r = 28$$

Check $$\frac{1}{14}r = 2$$

$$\frac{1}{14}(28) \stackrel{?}{=} 2$$

$$2 = 2$$

Our answer, $r = 28$, satisfies the equation.   ▲

N O T E   *When solving an equation like $3x = 5$, we can either multiply both sides by $\frac{1}{3}$ or divide both sides by 3. Either way results in the solution set $\left\{\frac{5}{3}\right\}$.*

Example 6   Find the solution set of

$$\frac{5}{4}t = \frac{35}{18}$$

Solution   To isolate t, we multiply both sides by $\frac{4}{5}$, the reciprocal of $\frac{5}{4}$.

$$\frac{5}{4}t = \frac{35}{18}$$

$$\frac{4}{5} \cdot \left(\frac{5}{4}t\right) = \frac{4}{5} \cdot \frac{35}{18}$$

$$1t = \frac{4}{5} \cdot \frac{35}{18}$$

$$t = \frac{\overset{2}{\cancel{4}}}{\underset{1}{\cancel{5}}} \cdot \frac{\overset{7}{\cancel{35}}}{\underset{9}{\cancel{18}}} \qquad \text{Reduce to lowest terms.}$$

$$t = \frac{14}{9}$$

The solution set is $\left\{\dfrac{14}{9}\right\}$. ▲

If an equation contains fractions, we can clear the fractions from the equation by first multiplying by the least common denominator (LCD) of the fractions.

Example 7   Solve the equation

$$t - \frac{8}{3} = \frac{1}{6}$$

Solution   The LCD of the fractions is 6, so our first step is to multiply both sides by 6.

$$t - \frac{8}{3} = \frac{1}{6}$$

$$6\left(t - \frac{8}{3}\right) = 6\left(\frac{1}{6}\right) \qquad \text{Multiply both sides by the LCD 6.}$$

$$6t - 6\left(\frac{8}{3}\right) = 1 \qquad \text{Distributive property.}$$

$$6t - 16 = 1$$

$$6t - 16 + 16 = 1 + 16 \qquad \text{Add 16 to both sides.}$$

$$6t = 17$$

$$t = \frac{17}{6} \qquad \text{Divide both sides by 6.}$$

Check $$t - \frac{8}{3} = \frac{1}{6}$$

$$\frac{17}{6} - \frac{8}{3} \overset{?}{=} \frac{1}{6}$$

$$\frac{17}{6} - \frac{16}{6} \overset{?}{=} \frac{1}{6}$$

$$\frac{1}{6} = \frac{1}{6}$$

The solution is $t = \dfrac{17}{6}$. ▲

N O T E   *Just as in Example 6 where we cleared fractions by multiplying both sides of the equation by the LCD, we can clear decimals from an equation that contains decimals by multiplying by an appropriate power of 10.*

C A L C U L A T O R   C R O S S O V E R

Solve the following equations.

1. $0.63b - 0.45b = 0.0504$
2. $0.193r + 0.056r = -3.543 + 20.475$
3. $-0.563a + 0.749a = -1.2709 + 1.2337$
4. $-0.81v + 0.77v = -0.453 + 0.521$

Answers   **1.** 0.28   **2.** 68   **3.** −0.2   **4.** −1.7

Applications of equations frequently involve translating a sentence into an equation. So far, we have converted word phrases into algebraic expressions. The only new thing is to look for key words or groups of words that describe equality. These key words, such as "is" or "is equal to," will tie together two word phrases that describe the two sides of an equation.

Example 8   Write an equation for the following sentence. Then solve the equation.

Twice a number, divided by three, is eight ninths.

Solution   Let x represent the number. The left side of the equation is "twice a number, divided by three" and the right side is "$\frac{8}{9}$." The two sides are connected by the word "is." Therefore, we symbolize the sentence as

$$\frac{2x}{3} = \frac{8}{9}$$

To solve this equation, first observe that $\frac{2x}{3}$ is the same as $\frac{2}{3}x$.

$$\frac{2x}{3} = \frac{8}{9}$$

$$\frac{2}{3}x = \frac{8}{9}$$

$$\frac{3}{2}\left(\frac{2}{3}x\right) = \frac{3}{2} \cdot \frac{8}{9} \qquad \text{Multiply both sides by } \frac{3}{2},$$

$$\text{the reciprocal of } \frac{2}{3}.$$

$$1x = \frac{\overset{1}{\cancel{3}}}{\underset{1}{\cancel{2}}} \cdot \frac{\overset{4}{\cancel{8}}}{\underset{3}{\cancel{9}}}$$

$$x = \frac{4}{3}$$

Therefore, the desired number is $\frac{4}{3}$. ▲

L E A R N I N G   A D V A N T A G E   *As you progress through the course, it is important to continue to review previous material. New concepts are based on mathematics studied in earlier parts of this textbook. Starting with this chapter, there are review problems at the end of each exercise set. These problems will help you review previously studied concepts that pertain to the next section.*

E X E R C I S E   S E T   2.1

1. Check if any members of $\left\{-\frac{1}{2}, 0, \frac{7}{2}\right\}$ belong to the solution set of $4x - 5 = 9$.

2. Check if any members of $\left\{-\frac{3}{2}, -1, 1, \frac{3}{2}\right\}$ belong to the solution set of $-2y + 14 = 11$.

3. Check if any member of $\{-1, 1, 4\}$ is a solution of $\frac{y-1}{3} + \frac{2y}{9} = -\frac{1}{3}$.

4. Check if any member of $\{-10, 0, 10\}$ is a solution of $t + 20 = -0.1t$.

For Exercises 5–16, solve the given equation. Check your answer.

5. $x - 6 = -3$

6. $w + 3 = 1$

7. $4.2 + z = 6.7$

8. $5.9 + u = -1.6$

9. $600 + h = -1,000$

10. $a - 39 = -91$

11. $\frac{5}{3} = \frac{7}{3} + n$

12. $\frac{3}{4} = -\frac{11}{4} + k$

13. $\frac{2}{3} + b = -\frac{7}{2}$

14. $s + \frac{4}{5} = \frac{3}{10}$

15. $s + 5\frac{2}{3} = 6\frac{1}{3}$

16. $4\frac{2}{7} + y = 3\frac{5}{7}$

For Exercises 17–22, find the solution set of each equation.

17. $4s - 3 = 7 + 3s$

18. $5x + 5 = 4x - 1$

19. $\frac{5}{8} + \frac{n}{6} = -\frac{5n}{6} - \frac{1}{8}$

20. $\frac{3c}{2} - \frac{4}{3} = \frac{7}{3} + \frac{c}{2}$

21. $3(5 - 2z) = 12 - 5z$

22. $13 + 15x = 4(3 + 4x)$

For Exercises 23–40, solve each equation. Check your answer.

23. $\dfrac{2}{9}t = -4$

24. $\dfrac{1}{10}r = 3$

25. $\dfrac{s}{12} = \dfrac{1}{4}$

26. $\dfrac{a}{6} = -\dfrac{3}{2}$

27. $4m = -20$

28. $-3n = 18$

29. $\dfrac{q}{10} = 4.9$

30. $\dfrac{x}{100} = 8.71$

31. $-12R = 28$

32. $-13w = -52$

33. $-x = -\dfrac{3}{5}$

34. $-s = \dfrac{4}{9}$

35. $\dfrac{3}{4}v = \dfrac{9}{16}$

36. $\dfrac{7}{3}u = -\dfrac{14}{9}$

37. $-3.8 = 1.9x$

38. $-5.2a = -10.4$

39. $\dfrac{4}{3} = \dfrac{w}{15}$

40. $-\dfrac{7}{11} = \dfrac{b}{44}$

For Exercises 41–48, solve each equation.

41. $3 - 17 = 12c - 7c$

42. $15 - 3w = 15 - 2$

43. $3z - 8z = 2 - 4$

44. $12x - 11x = \dfrac{4}{5} - \dfrac{3}{2}$

45. $-b + \dfrac{b}{2} = \dfrac{5}{7} - \dfrac{3}{14}$

46. $\dfrac{3x}{2} - x = \dfrac{1}{4} + 2$

47. $1.8s + 5.7s = 1.2 + 6.3$

48. $9.4 - 10.9 = 12.2q - 2.2q$

For Exercises 49–60, write an equation for each sentence. Solve the equation.

49. A number divided by five is equal to two fifths.

50. A number divided by three is equal to negative four twenty-firsts.

51. Three times a number, divided by seven, is negative four.

52. Twice a number, divided by four, is negative three.

53. The sum of twice a number and ten is eight less than the number.

54. The sum of six times a number and three is ten less than five times the number.

55. Five times a number, divided by two, is 1.5.

56. Ten times a number, divided by four, is 12.

57. Twenty plus five times a number is the sum of 15 and six times the number.

58. Twelve minus six times a number is the difference of 11 and five times the number.

59. The sum of $\dfrac{2}{3}$ times a number and $\dfrac{5}{2}$ is the difference of 10 and $\dfrac{1}{3}$ of the number.

60. The sum of three halves times a number and four is the sum of one half of the number and five.

REVIEW PROBLEMS

The following exercises review parts of Section 1.5. Doing these problems will help prepare you for the next section.

Simplify.

61. $5x - 3x + 2 - 9$

62. $\dfrac{2}{3}y + \dfrac{5}{6}y - \dfrac{3}{4} - 1$

63. $4(1 - 3z) + 2$

64. $5x - 6(1 - 2x)$

65. $10(0.2 - 0.1a) + 2(3a + 1) + 6a - 12$

66. Find the value of $2 - 4t$ when $t = -\dfrac{3}{2}$.

67. Find the value of $3(2x - 4) + 1$ when $x = -2$.

For Exercises 68–72, convert the following word phrases into algebraic expressions. Use x to represent a number.

68. Twice a number increased by two.

69. The difference of ten and a number.

70. The reciprocal of the sum of a number and three.

71. Forty-five percent of a number.

72. The quotient of a number and -32.

ENRICHMENT EXERCISES

For Exercises 1–3, solve each equation.

1. $3(4 - 2x) = 1 - (1 - x)$

2. $\dfrac{2}{7} - \dfrac{5y}{14} = \dfrac{3}{4} - \dfrac{y}{7}$

3. $\dfrac{a + 1}{a} = -2$

For Exercises 4–7, write an equation for the following sentence. Then solve the equation.

4. A number divided by the sum of three times the number and four is negative five.

5. Four times the sum of five and twice a number is three times the difference of three times the number and two.

6. Three times the reciprocal of a number is negative five sevenths.

7. The reciprocal of the sum of twice a number and three is one sixth.

Answers to Enrichment Exercises are on page A.5.

2.2

Solving Linear Equations

▶ *To apply techniques for solving linear equations*

In this section, we will use the techniques developed so far to solve a general *linear equation.*

D E F I N I T I O N

A **linear equation** is an equation that can be put into the standard form

$$ax + b = 0$$

where a and b are constants with $a \neq 0$.

Most linear equations that we will encounter will *not* be in this standard form. For example,

$$1 - \frac{3}{2} = 4x, \qquad 5y - y = 3 + 7y, \qquad -2.3 = \frac{z}{5}$$

are each linear equations that are not in standard form. Each one, however, *could* be put into the standard form.

Here are some guidelines to help us solve linear equations.

S T R A T E G Y

A strategy for solving linear equations

Step 1   **Clear the equation of any fractions or decimals,** if desired, by multiplying by the LCD of the fractions appearing or, in the case of decimals, by an appropriate power of 10.

Step 2   **Combine like terms on each side,** if necessary, to simplify the equation. You may have to use the distributive property first to separate terms.

Step 3   **Use the addition-subtraction property of equality,** if necessary, to move all terms containing a variable to one side and all constant terms to the other side.

Step 4 **Use the multiplication-division property of equality,** if necessary, to make the coefficient of the variable term equal to one. The equation is now in the form

$$x = \text{the solution} \quad \text{or} \quad \text{the solution} = x$$

Step 5 **(Optional) Check your answer,** by substituting it into the original equation.

Example 1   Solve the equation

$$3x - 5 = 11$$

and check your answer.

Solution   Steps 1 and 2   These steps do not apply here.

Step 3   Add 5 to both sides and simplify.

$$3x - 5 + 5 = 11 + 5$$
$$3x = 16$$

Step 4   Divide both sides by 3.

$$\frac{3x}{3} = \frac{16}{3}$$

$$x = \frac{16}{3}$$

Step 5   We check our answer by replacing x by $\frac{16}{3}$ in the original equation to determine if the resulting statement is true.

Check $x = \frac{16}{3}$: $3x - 5 = 11$

$$3 \cdot \frac{16}{3} - 5 \stackrel{?}{=} 11$$

$$16 - 5 \stackrel{?}{=} 11$$

$$11 = 11$$

The solution is $x = \frac{16}{3}$.   ▲

Example 2   Find the solution set of

$$0.2(z - 60) + 0.1z = 16 - 0.4z$$

Solution   Step 1   We clear the equation of decimals by multiplying both sides by 10.

$$0.2(z - 60) + 0.1z = 16 - 0.4z$$

$$10[0.2(z - 60) + 0.1z] = 10(16 - 0.4z)$$

$$10(0.2)(z - 60) + (10)(0.1z) = 10(16) - (10)(0.4z) \qquad \text{Distributive property.}$$

$$2(z - 60) + z = 160 - 4z$$

Step 2   Use the distributive property and then combine terms.

$$\overset{*}{2}(z - 60) + z = 160 - 4z$$

$$2z - 120 + z = 160 - 4z$$

$$3z - 120 = 160 - 4z$$

Step 3   Use the addition-subtraction property of equality, then simplify.

$$3z - 120 + 120 + 4z = 160 - 4z + 120 + 4z$$

$$7z = 280$$

Step 4   Use the multiplication-division property of equality, then simplify.

$$\left(\frac{1}{7}\right)(7z) = \left(\frac{1}{7}\right)(280)$$

$$z = 40$$

The solution set is {40}.   ▲

C A L C U L A T O R   C R O S S O V E R

Solve the following equations. Check your answers.

1. $0.525q + 0.4q - 0.045 = 1.25$

2. $4.08a - 9.903 = 5(6.4187 - 1.03a)$

3. $3(8790 + 875x) = 71(36x - 209) + 19{,}336$

4. $25(5t - 386.6) = 4(14.5t - 3940.5)$

Answers   **1.** 1.4   **2.** 4.55   **3.** −317   **4.** −91

When solving an equation that has fractions, you may want to first "clear" the equation of these fractions. This is done by multiplying both sides of the equation by the least common multiple of all the denominators. Recall that the least common multiple is the smallest positive integer that is divisible by each of the numbers. For example, the least common multiple of the three numbers

4, 6, and 9 is 36. The least common multiple of all the denominators is called the least common denominator or LCD. For example, the LCD of the three fractions $\dfrac{3}{4}$, $\dfrac{7}{6}$, and $\dfrac{1}{9}$ is 36.

Example 3   Solve the equation

$$\frac{2}{3}x - \frac{5}{4} = \frac{1}{6}$$

Solution   The LCD of $\dfrac{2}{3}$, $\dfrac{5}{4}$, and $\dfrac{1}{6}$ is 12. Therefore, multiplying both sides of the equation by 12 will clear the fractions.

$$12\left(\frac{2}{3}x - \frac{5}{4}\right) = 12\left(\frac{1}{6}\right)$$

$$12\left(\frac{2}{3}x\right) - 12\left(\frac{5}{4}\right) = 12\left(\frac{1}{6}\right) \qquad \text{Distributive property.}$$

$$8x - 15 = 2$$

$$8x = 17$$

$$x = \frac{17}{8} \qquad\qquad\qquad\qquad \blacktriangle$$

So far, all of our equations have had exactly one solution. An equation of this type is called a **conditional equation,** since the truth of the equation depends upon the value of the variable. In the next two examples we show two extremes for solution sets.

Example 4   Find the solution set of the equation

$$-5x + \frac{3}{2} = \frac{3}{2} - 5x$$

Solution   We add $5x$ to both sides and simplify.

$$-5x + \frac{3}{2} + 5x = \frac{3}{2} - 5x + 5x$$

$$\frac{3}{2} = \frac{3}{2}$$

The resulting equivalent equation $\dfrac{3}{2} = \dfrac{3}{2}$ is true for all real numbers. Therefore, the solution set of the original equation is the set of all real numbers. $\blacktriangle$

The original equation of Example 4 is called an **identity,** since it is equivalent to an equation that is true for all real numbers. When solving an equation that is an identity, the variable is eliminated, leaving an equivalent equation such as 2 = 2, which is a true statement. The original equation is therefore satisfied by any real number.

Example 5   Find the solution set of the equation

$$4 + 12a = 3 + 12a$$

Solution   We subtract $12a$ from both sides and simplify.

$$4 + 12a - 12a = 3 + 12a - 12a$$

$$4 = 3$$

The resulting equivalent equation, $4 = 3$, is false. Therefore, the solution set of the original equation is $\varnothing$, the empty set.   ▲

The original equation of Example 5 is called a **contradiction,** since it is equivalent to an equation that is false. An equation that is a contradiction has no solution. When attempting to solve, the variable is eliminated, leaving an equivalent equation that says that two unequal numbers are equal.

Given a linear equation, we will solve it by using our strategy for solving linear equations as stated above. If the solution is exactly one number, the original equation is conditional; otherwise it will turn out that the equation is either an identity or a contradiction.

Example 6   Write an equation for the following sentence. Then solve the equation.

The sum of three times a number and five is negative eight.

Solution   Let n represent the number. Then the sentence can be symbolized as

$$3n + 5 = -8$$

To solve this equation, first subtract 5 from both sides.

$$3n + 5 = -8$$

$$3n + 5 - 5 = -8 - 5$$

$$3n = -13$$

$$n = -\frac{13}{3} \qquad \text{Divide both sides by 3.}$$

The desired number is $-\dfrac{13}{3}$.   ▲

LEARNING ADVANTAGE *An algebra book cannot be read like a novel. You must **actively participate** by using pencil and paper. For example, if you are not sure how a step was obtained, write the information and do the calculations yourself. If you find that you are losing concentration, put aside your work and take a break.*

EXERCISE SET 2.2

For Exercises 1–44, use the strategy for solving linear equations to solve the given equation. Indicate those equations that are identities or contradictions.

1. $4x - 12 = 3$

2. $6y - 21 = -2$

3. $\dfrac{3}{4}t + \dfrac{5}{8} = \dfrac{3}{4}$

4. $\dfrac{4}{5}s - 1 = -\dfrac{3}{10}$

5. $3 - 8y = 7$

6. $-12a + 10 = -3$

7. $1.6w + 2.1 = -1.1$

8. $3.7 - 9.2z = 31.3$

9. $\dfrac{2}{3}u - \dfrac{4}{3} = 2$

10. $\dfrac{6}{5}c + \dfrac{3}{10} = -1$

11. $5 + 4z - 7z = 6 - 4$

12. $12 - 3x + 14 = 3x - 5x$

13. $3a - 14a = 17 - 11 - 12a$

14. $c - 5c + 6 - 8 = 3c - 1 - 15$

15. $-2 - 3x = 4$

16. $-5 - z = 2$

17. $-4t + 9t - 7 = 3 - t + 6$

18. $12s + s - 19 + 5 = 2s - s$

19. $\dfrac{1}{3}a - \dfrac{5}{6}a = \dfrac{2}{3} - \dfrac{1}{6}$

20. $\dfrac{3}{4} - \dfrac{1}{8}w = \dfrac{7}{8} + \dfrac{1}{4}w$

21. $1 - 2x = 3 - 2x$

22. $4 + 3z - 1 = 5 + 3z$

23. $3(5 + 2x) = 31 - 10$

24. $2(3z - 8) = 15 - 1$

25. $4(3 - 2T) - 1 = 7$

26. $7 - 3b = 5(b - 5)$

27. $10 - (5 + c) = 8$

28. $7 - (2 - C) = -17$

29. $a + 4 = -3(7 - a) - 7a$

30. $23 + 3y = -2y + 3(2y + 4)$

31. $-2(1 - 3x) = 2(3x - 1)$

32. $4(2 + 3y) - 1 = 2(3 + 6y) + 1$

33. $-(2x - 5) + 13 = 2(x + 2)$

34. $10 - (a - 3) = 3(5 - 3a)$

35. $\dfrac{2}{3}(2x - 6) = 2 - \dfrac{1}{3}$

36. $\dfrac{3}{5}(25 - 5t) = 12 - \dfrac{6}{5}$

37. $\dfrac{1}{3}(3p + 2) = 1 - \dfrac{1}{2}p$

38. $\dfrac{2}{9}(3 - 9r) - 1 = \dfrac{2}{3}(1 - r)$

39. $\dfrac{3R}{2} - \dfrac{1}{6} = \dfrac{2R}{3} + 1$

40. $\dfrac{5y}{4} - \dfrac{3}{2} = 1 - y$

41. $0.8 + w = 0.9w + 1 - 0.1$

42. $9.9 - 1.6 - 0.2v = 5.4 - 0.1v$

43. $0.6(z - 7) = 4(0.3z + 0.5)$

44. $3(5.8 + 0.7x) = 0.1(6 + 5x)$

45. What is the difference between a conditional linear equation and a linear equation that is an identity?

46. What is the difference between a linear equation that is a contradiction and one that is an identity?

For Exercises 47–54, write an equation from the information. Then solve the equation.

47. The sum of four times a number and five is three.

48. The sum of five times a number and seven is negative five.

49. Ten times a number plus 25 is -80.

50. Twice a number, increased by $\dfrac{4}{3}$, is six.

51. The sum of 2.6 and 1.3 times a number is zero.

52. The sum of five and 0.5 times a number is -20.

53. The sum of $\dfrac{3}{4}$ of a number and two is $-\dfrac{1}{4}$.

54. The sum of $\dfrac{2}{9}$ of a number and one is $-\dfrac{1}{3}$.

REVIEW PROBLEMS

The following exercises review parts of Section 1.5. Doing these problems will help prepare you for the next section.

For Exercises 55–58, write an algebraic expression from the given information.

55. The cost of x compact discs at $10.99 each.

56. The number of yards in d feet.

57. The sales tax on a purchase that cost z dollars, if the sales tax rate is 6%.

58. The amount of money received for working H hours at $14 per hour.

For Exercises 59–64, represent one number by a variable. Then express the other number in terms of the variable. Be sure to clearly state what your variable represents.

59. The sum of two numbers is 12.

60. The product of two numbers is three.

61. The selling price of a sofa, in dollars, and 25% of the selling price increased by $15.

62. A ten-foot pipe is cut into two pieces.

63. The sum of two consecutive even integers.

64. The sum of the squares of two consecutive odd integers.

ENRICHMENT EXERCISES

For Exercises 1–4, solve for x.

1. $ax + b = d$, where $a \neq 0$.

2. $ax + b = cx + d$, where $a \neq c$.

3. $a(x - 2) = d$, where $a \neq 0$.

4. $\dfrac{x + 1}{a} = cx$, where $ac \neq 1$.

5. Find a, if $x = -2$ is a solution of $ax + 2 = 7$.

6. Find a, if $x = -1$ is a solution of $ax - 3 = 11$.

7. Find c, if $3(x - 2) + c = 1 - 6\left(3 - \dfrac{x}{2}\right)$ is an identity.

Answers to Enrichment Exercises are on page A.5.

2.3

Solving Word Problems

O B J E C T I V E S

▶ *To convert word problems into mathematical problems*

▶ *To solve word problems*

Throughout the previous sections, we have touched upon translating word phrases and sentences into algebraic expressions and equations. In this section, we will convert a greater variety of sentences into equations. Then we will learn how to solve word problems.

There is no single method to solve the wide variety of word problems. However, the following strategy will help as a guide. The best way to achieve success at solving word problems comes from practice.

S T R A T E G Y

A strategy for solving word problems

1. **Read** the problem carefully. Take note of what is being asked and what information is given.
2. **Plan** a course of action. Represent the unknown number by a letter. If there is more than one unknown, represent one of them by a letter and express the others in terms of the letter.
3. **Create** an equation from the given information.
4. **Solve** this equation.
5. **Check** your solution in the original equation.
6. **Answer** the original question. Read the problem again to make sure you answered the question.

The rest of this section contains examples of solving word problems. As you read through them, keep in mind the six steps listed in the strategy.

Example 1

Ann is planning to buy a VCR that costs $650. This is $72 more than twice the amount she saved last month. How much did she save last month?

Solution

Let

$$x = \text{the amount of money that Ann saved last month}$$

From the information, we have

$$2x + 72 = 650$$

Solving this equation,

$$2x = 650 - 72$$

$$2x = 578$$

$$x = \frac{578}{2}$$

$$= 289$$

We check our answer. Does $72 more than twice 289 equal $650?

$$2(289) + 72 = 578 + 72 = 650$$

Therefore, our answer is correct. She had saved $289 last month.   ▲

Example 2   A pair of shoes is on sale at 25% off the original price. If the sale price is $45, what was the original price?

Solution   Let

$$p = \text{the original price of the shoes}$$

Since the pair of shoes is on sale at 25% off the original price,

the sale price = the original price − 25% of the original price

Therefore,

$$45 = p - 0.25p$$

To solve this equation, we start by multiplying both sides by 100.

$$(100)45 = 100(p - 0.25p)$$

$$4{,}500 = 100p - 25p$$

$$4{,}500 = 75p$$

$$\frac{4{,}500}{75} = p \qquad \text{Divide both sides by 75.}$$

$$60 = p$$

We check our answer with the original words. Is $60 minus 25% of 60 equal to 45?

$$60 - 0.25(60) = 60 - 15 = 45$$

Our answer satisfies the specified condition, and therefore, the original price of the pair of shoes was $60.

Alternate method   If the price of the shoes is marked down at 25% off the original price p, then the sale price is 75% of the original price. Therefore,

$$0.75p = 45$$

To solve for p, we divide both sides by 0.75.

$$p = \frac{45}{0.75}$$

$$= 60 \qquad \blacktriangle$$

Example 3   The larger of two numbers is 148 more than the smaller. If the larger is -3 times the smaller, find the two numbers.

Solution   Let

$$x = \text{the smaller number}$$

Then

$$x + 148 = \text{the larger number}$$

Since the larger number is -3 times the smaller number,

$$x + 148 = -3x$$

Next, we solve this equation.

$$x + 148 - 148 + 3x = -3x - 148 + 3x$$

$$4x = -148$$

$$x = -\frac{148}{4}$$

$$= -37$$

Therefore, the smaller number is -37, and the larger number is $-37 + 148 = 111$. To check our answers, is the larger number, 111, equal to -3 times the smaller number -37? The product $-3(-37)$ is 111, and our answers satisfy the conditions of the problem. $\qquad \blacktriangle$

Example 4   Jill must cut a 5-foot board into two pieces, so that the larger piece has a length that is $\dfrac{3}{2}$ the length of the smaller piece. Find the lengths of the two pieces.

Solution   Let

$$x = \text{the length of the smaller piece}$$

Since the total length of the board is 5 feet,

$$5 - x = \text{the length of the larger piece}$$

Now, the length of the larger piece is $\dfrac{3}{2}$ the length of the smaller piece. Therefore,

$$5 - x = \frac{3}{2}x$$

To solve this equation, we first multiply both sides by 2 to "clear" the equation of the fraction.

$$2(5 - x) = 2\left(\frac{3}{2}x\right)$$

$$10 - 2x = 3x$$

$$10 = 5x \qquad \text{Add } 2x \text{ to both sides.}$$

$$\frac{10}{5} = \frac{5x}{5} \qquad \text{Divide both sides by 5.}$$

$$2 = x$$

Therefore, the smaller piece is 2 feet long and the longer piece is $5 - 2$ or 3 feet long. We check our answers. Is the length of the larger piece, 3 feet, equal to $\dfrac{3}{2}$ the length of the smaller piece, 2 feet? The product $\dfrac{3}{2}(2)$ is 3 feet, and our answers satisfy the conditions of the problem. ▲

We finish the section with a coin problem and a mixture problem.

Example 5

(Coin problem) Reggie has $2.35 in nickels and dimes in his pocket. If he has seven more dimes than nickels, how many of each coin does he have in his pocket?

Solution

Let x represent the number of nickels. Then $x + 7$ represents the number of dimes. Since each nickel is worth 5 cents or 0.05 dollar, $0.05x$ dollars is the value of the x nickels. Since each dime is worth 10 cents or 0.1 dollar, $0.1(x + 7)$ dollars is the value of the $x + 7$ dimes. We summarize the above information in the following table.

Coin	Number of coins	Value per coin	Total value of coins
Nickel	x	$0.05	$0.05x
Dime	$x + 7$	$0.10	$0.10(x + 7)

Since the total value of the coins is \$2.35, we obtain the equation

$$\underset{x \text{ nickels}}{\text{the value of the}} + \underset{x + 7 \text{ dimes}}{\text{the value of the}} = \$2.35$$

$$0.05x \quad + \quad 0.10(x + 7) \quad = 2.35$$

To solve this equation, we first clear the equation of decimals by multiplying both sides by 100 to obtain

$$5x + 10(x + 7) = 235$$

$$5x + 10x + 70 = 235 \qquad \text{Distributive property.}$$

$$15x + 70 - 70 = 235 - 70 \qquad \text{Subtract 70 from both sides.}$$

$$15x = 165$$

$$x = \frac{165}{15}$$

$$= 11$$

Therefore, our answers are 11 nickels and $11 + 7$ or 18 dimes. We check our answers. Is the value of 11 nickels and 18 dimes equal to \$2.35?

$$0.05(11) + 0.10(18) = 0.55 + 1.80 = 2.35$$

and our answers satisfy the conditions of the problem. ▲

Example 6

(**Mixture problem**) A solution containing 8% salt is to be mixed with a solution containing 16% salt to obtain a 50-liter mixture with 10% salt. How many liters of each solution should be used?

Solution   Let

$$x = \text{the number of liters of the first (8\%) solution}$$

Since the total mixture is 50 liters, the amount of the second (16%) solution is given by $50 - x$. (See the figure.)

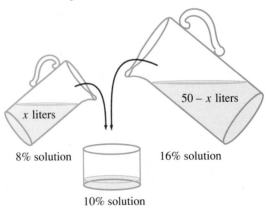

x liters

8% solution

50 − x liters

16% solution

10% solution

Now, the amount of salt in the x liters of the 8% solution is $0.08x$, and the amount of salt in the $50 - x$ liters of the 16% solution is $0.16(50 - x)$. Therefore,

$$\begin{array}{ccccc} \text{amount of salt} & + & \text{amount of salt} & = & \text{total amount} \\ \text{in the 8\% solution} & & \text{in the 16\% solution} & & \text{of salt} \\ 0.08x & + & 0.16(50 - x) & = & 0.10(50) \end{array}$$

To solve this equation, we first multiply both sides by 100 to clear the equation of decimal numbers.

$$0.08x + 0.16(50 - x) = 0.10(50)$$

$$100[0.08x + 0.16(50 - x)] = 100[0.10(50)]$$

$$8x + 16(50 - x) = 500$$

$$8x + 800 - 16x = 500 \qquad \text{Distributive property.}$$

$$-8x + 800 = 500 \qquad \text{Simplify the left side.}$$

$$-8x = -300 \qquad \text{Subtract 800 from both sides.}$$

$$x = \frac{-300}{-8} \qquad \text{Divide both sides by } -8.$$

$$x = \frac{300}{8}$$

$$= 37.5$$

Our answers are 37.5 liters of the 8% solution and $50 - 37.5 = 12.5$ liters of the 16% solution. We check our answers. Does 8% of 37.5 plus 16% of 12.5 equal 10% of 50?

$$0.08(37.5) + (0.16)(12.5) = 3 + 2 = 5$$

which is 10% of 50. Therefore, our answers satisfy the conditions of the problem. ▲

L E A R N I N G   A D V A N T A G E   *When solving a word problem, the information must be read carefully and slowly. Make sure to pay attention to each word. Do not attempt to digest the information too quickly; you may miss an important point.*

E X E R C I S E   S E T   2.3

For Exercises 1–4, solve the word problem.

1. Deb receives an 8% raise in salary. Her old salary was $34,200. What is her new salary?

2. Fran sold some stock for $6,720. This was $224 more than 112% of the amount she paid for it. What was the amount she paid for the stock?

3. The pro shop at Oakfield Tennis Club has a sale of 20% off the price of all graphite tennis rackets. If the sale price for a particular racket is $180, what was the original price of the racket?

4. Robert sold his baseball card collection for $85. This was $20 more than twice the original cost. What was the original cost of his collection?

For Exercises 5–18, represent one number by a variable. Then express the other number in terms of the variable. Be sure to clearly state what the variable represents.

5. Shawn's age now and his age nine years ago.

6. Kathy's age now and her age in $5\frac{1}{2}$ years.

7. The sum of two numbers is 40.

8. The sum of two numbers is 59.

9. Five hundred dollars is divided into two parts, one part is invested at 6% interest and the other part is invested at 7% interest.

10. Thirty-five students are separated into two groups.

11. The selling price, in dollars, of a desk and 30% of the selling price decreased by $20.

12. The wholesale cost, in dollars, of a computer, and the retail price, where the retail price is $60 more than 150% of the wholesale cost.

13. Jeff's bowling score and Tom's score, where Tom's score is $\frac{3}{4}$ of Jeff's score.

14. The student enrollment at Ridgemont High School is 2.5 times the enrollment at Central High School.

15. Eight hundred and seventy dollars is separated into two parts.

16. The combined incomes of Mr. and Mrs. Jones is $85,000 a year.

17. The number of cats and dogs at the animal shelter totaled 47.

18. A parking meter contains 32 coins consisting of dimes and quarters.

For Exercises 19–46, solve using the six-step strategy for solving word problems.

19. Guy is planning to buy a computer that costs $2,000. This is $800 more than four times the amount he saved last month. How much did he save last month?

20. Nadine sold a baseball card for $300. This is $27 less than three times the amount she paid for it. How much did Nadine pay for the baseball card?

21. The Garden Center is selling bags of humus at 20% off. If the sale price is $2.24, what was the original price?

22. The Carpet Outlet is selling room-size remnants at 25% off. If the sale price is $150, what was the original price?

23. John is paid $350 a week plus a commission of 7% on sales. Find the sales (in dollars) needed to give him a weekly total of $385.

24. Julie is paid $425 a week plus a commission of 8% on sales. Find the sales (in dollars) needed to give her a weekly total of $600.

25. The larger of two numbers is 27 more than the smaller. If the larger is four times the smaller, find the two numbers.

26. The larger of two numbers is ten more than the smaller. If the larger is six times the smaller, find the two numbers.

27. The larger of two numbers is 150 more than the smaller. If the larger is five times the smaller, find the two numbers.

28. The larger of two numbers is 3.2 more than the smaller. If the larger is $-\dfrac{1}{2}$ times the smaller, find the two numbers.

29. The sum of two numbers is -22. One number is four less than one half of the other number. Find the two numbers.

30. The sum of two numbers is -1.8. One number is 3.6 less than 0.2 times the other number. Find the two numbers.

31. The sum of two consecutive odd integers is -48. Find the two integers.

32. The sum of two consecutive even integers is -74. Find the two integers.

33. The sum of three consecutive even integers is 90. Find the three integers.

34. The sum of three consecutive odd integers is 129. Find the three integers.

35. Jack the plumber must saw a 21-foot pipe into two pieces. The length of the shorter piece is to be three fourths the length of the longer piece. Find the lengths of the two pieces.

36. Melinda is planning to cut a 7-foot board into two pieces. The length of the longer piece must be twice the length of the shorter piece. Find the lengths of the two pieces. Express your answer as mixed numbers.

37. Tracy is three years older than Samantha. Let x represent Samantha's age.
(a) Express Tracy's age in terms of x.
(b) Express Samantha's age four years ago in terms of x.
(c) Express Tracy's age four years ago in terms of x.
(d) If Samantha's age was two thirds of Tracy's age four years ago, find their (current) ages.

38. Gina is 30 years older than her son. Let x represent Gina's age.
(a) Express her son's age in terms of x.
(b) Express Gina's age six years ago in terms of x.
(c) Express her son's age six years ago in terms of x.
(d) If Gina's age was six times her son's age six years ago, find their (current) ages.

39. Joe is 30 years older than his daughter. Five years ago, Joe was three times as old as his daughter. How old is Joe? How old is his daughter?

40. Kristi is four years younger than her brother. In ten years, she will be $\dfrac{4}{5}$ times as old as her brother. How old is Kristi? How old is her brother?

41. Michael has $3.15 in nickels and dimes. If he has six more nickels than dimes, how many coins of each type does he have?

42. Mary has $340 in ten and twenty dollar bills. If she has a total of 26 bills, how many bills of each kind does she have?

43. One margarine/butter blend contains 15% butter. Another blend contains 65% butter. How many pounds of each blend should be used to make a 350-pound blend containing 45% butter?

44. How many liters of an insulin solution containing 8% zinc crystals should be mixed with another insulin solution that contains 15% zinc crystals to make a 700-liter solution that contains 11% zinc crystals?

45. The Fantasy Shop is planning to sell comic books by the box. Each box is to contain a mixture of 50-cent and 75-cent comics. The total value of the box of comics is to be $42, with twice as many 50-cent as 75-cent comics. How many comic books of each price should be placed in each box?

46. The Stationery Store is planning to sell greeting cards by the box. Each box is to contain a mixture of two kinds of cards. One kind costs $1.25 each and the other kind costs $1.75 each. Each box is to contain 50 cards with a value of $70. How many cards of each kind should be placed in each box?

REVIEW PROBLEMS

The following exercises review parts of Section 1.5. Doing these problems will help prepare you for the next section.

For Exercises 47–50, simplify.

47. $3(-2x^3 + y) + (4y + x^3)$

48. $2(x + 1) + 2(3x - 2)$

49. $\dfrac{(3h + 6)h}{2}$

50. $(3y + 1 + 7y + 5)\left(\dfrac{y}{2}\right)$

51. Evaluate $x^2 + 2ax + a^2$, when $x = -2$ and $a = 1$.

52. Evaluate $\dfrac{v - w}{t}$, when $v = 35$, $w = 55$, and $t = 4$.

53. Evaluate Prt, when $P = 1{,}000$, $r = 0.07$, and $t = 3$.

54. Evaluate $P(1 + rt)$, when $P = 500$, $r = 0.08$, and $t = 2$.

ENRICHMENT EXERCISES

1. A 31-foot rope is to be cut into three pieces. The length of the middle piece is $\frac{3}{2}$ times the length of the first piece. The length of the third piece, when reduced by 2 feet, is $\frac{3}{4}$ the length of the middle piece. Find the lengths of the three pieces.

2. Four people decide to equally share the cost of starting a small business. If they let a fifth person join this endeavor, the cost for each one would be reduced by $20,000. What is the total cost of starting this business?

Answers to Enrichment Exercises are on page A.6.

2.4

Formulas and Applications

OBJECTIVES

▶ *To use formulas in solving word problems*

▶ *To solve a formula for one variable in terms of the other variables*

Some problems require the application of a formula in their solution. A **formula** is an equation that contains two or more variables. A formula is a rule that expresses one quantity in terms of other quantity (or quantities).

For example, a car traveling at a rate of 65 miles per hour on an interstate highway goes a distance of 65 miles in one hour, $2 \cdot 65$ or 130 miles in two hours, $3 \cdot 65$ or 195 miles in three hours, and so on. In t hours, the car covers d miles, where $d = t \cdot 65$ or $65t$. Thus, $d = 65t$ is a formula that expresses the distance d the car travels in terms of the time t. The general formula for distance in terms of rate and time is the following,

$$d = rt$$

where d is the distance traveled in t hours at the rate of r mph.

Example 1

Tom and Joe leave school at the same time and bicycle in opposite directions. If Tom rides his bicycle at 4 mph and Joe rides his bicycle at 5 mph, how long will it take for them to be 3 miles apart?

Solution

When solving a motion problem, we first draw a diagram as shown in the figure below.

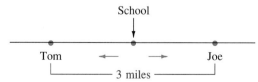

Let t be the time, in hours, it takes for Tom and Joe to be 3 miles apart. By the distance formula, the distance that Tom has traveled in t hours is

$$d = r \cdot t = 4t \text{ miles}$$

Similarly, the distance that Joe has traveled in t hours is

$$d = r \cdot t = 5t \text{ miles}$$

Therefore, after t hours, Tom has traveled $4t$ miles and Joe has traveled $5t$ miles. From the diagram, the total number of miles between the two at time t is $4t + 5t$. We set this expression equal to 3 miles and solve the resulting equation for t.

$$4t + 5t = 3$$
$$9t = 3$$
$$t = \frac{3}{9} \text{ or } \frac{1}{3} \text{ hour}$$

Therefore, in $\frac{1}{3}$ hour, or 20 minutes, Tom and Joe will be 3 miles apart. ▲

We next consider an application in business. If you deposit an amount P of money, called the **principal,** in a savings account, the bank will pay you **interest** on the principal. The interest I is determined by the **annual interest rate** r, expressed as a decimal, and the **time** t, in years, that the money remains on deposit. If the bank pays **simple interest,** it is given by the formula

$$I = Prt$$

Example 2   How much interest is made on $100 deposited in a savings account for 6 months that pays 7% interest?

Solution   To compute the interest at the end of 6 months, we change 7% to the decimal 0.07 and convert 6 months to $\frac{6}{12}$ or $\frac{1}{2}$ year. Next, we evaluate the formula $I = Prt$,

$$I = Prt$$
$$= (100)(0.07)\left(\frac{1}{2}\right)$$
$$= 7\left(\frac{1}{2}\right)$$
$$= \$3.50$$

Therefore, the $100 earned $3.50 in interest during the 6 months.   ▲

In the next example, we show how the interest formula can be used to solve an investment problem.

Example 3

Judy plans to invest $10,000, part at 12% and the other part at 8%. She wants the total interest for one year to be $950. How much should she invest at each interest rate?

Solution

Let x be the amount she invests at 12%. Since the total amount for investment is $10,000, then $10,000 - x$ is the amount that she invests at 8%.

We can now represent the interest earned by each part using the formula $I = Prt$. For the investment of x dollars at 12% for one year, $I = x(0.12)(1) = 0.12x$. For the investment of $10,000 - x$ dollars at 8% for one year, $I = (10,000 - x)(0.08)(1) = (0.08)(10,000 - x)$. Now, the sum of these two interests must be $950,

$$\begin{array}{ccc} \text{the interest} & \text{the interest} & \\ \text{at } 12\% & + & \text{at } 8\% & = \text{total interest} \\ 0.12x & + & (0.08)(10,000 - x) = & 950 \end{array}$$

Next, we solve this equation.

$$0.12x + 800 - 0.08x = 950 \qquad \text{Distributive property.}$$

$$0.04x = 150$$

$$x = 3,750 \qquad \text{Divide both sides by 0.04.}$$

Therefore, Judy should invest $x = \$3,750$ at 12% and $10,000 - x = 10,000 - 3,750 = \$6,250$ at 8%. ▲

There are many formulas associated with geometric figures. The inside front cover has a list of geometric figures and formulas. We may divide figures into two types: two dimensional and three dimensional. Examples of two dimensional figures are the following.

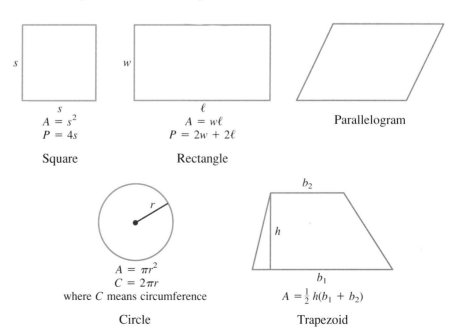

Square
$A = s^2$
$P = 4s$

Rectangle
$A = w\ell$
$P = 2w + 2\ell$

Parallelogram

Circle
$A = \pi r^2$
$C = 2\pi r$
where C means circumference

Trapezoid
$A = \frac{1}{2} h(b_1 + b_2)$

The **area** of a figure is the number of square units needed to cover the figure's enclosed region. The **perimeter** is the distance around the figure. We use the letter A to mean area and P to mean perimeter.

Some formulas use variables involving subscripts. A **subscript** is a number or other symbol placed below and to the right of the variable. It is used to explain what the (subscripted) variable means. For example, in the formula for area of a trapezoid, $A = \frac{1}{2}h(b_1 + b_2)$, there are two subscripted variables b_1 and b_2, which are pronounced "b sub 1" and "b sub 2." The term b_1 is the length of one of the parallel sides (which are called the bases), and b_2 is the length of the second parallel side or base.

Example 4   Find the area of the trapezoid shown below.

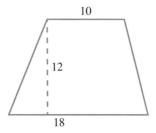

Solution   From the information given on the inside front cover of the book, the area A of a trapezoid is

$$A = \frac{1}{2}h(b_1 + b_2)$$

where $h = 12$ inches, $b_1 = 18$ inches, and $b_2 = 10$ inches.

$$A = \frac{1}{2}(12)(18 + 10)$$

$$= (6)(28) = 168 \text{ square inches} \qquad \blacktriangle$$

Example 5   A rectangular backyard, 60 feet by 80 feet, is to be enclosed by a fence costing $3 per yard. What is the cost of the fence?

Solution   First draw a picture as shown in the following figure.

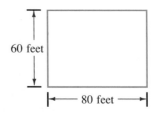

The perimeter of the backyard is

$$2(60) + 2(80) = 120 + 160 = 280 \text{ feet}$$

Now, 280 feet is $\dfrac{280}{3}$ or $93\dfrac{1}{3}$ yards. Therefore, the total cost of the fence is

$$\text{(cost per yard)(the number of yards)} = (3)\left(93\dfrac{1}{3}\right)$$

$$= 3\left(93 + \dfrac{1}{3}\right)$$

$$= 3 \cdot 93 + 3\left(\dfrac{1}{3}\right)$$

$$= 279 + 1$$

$$= \$280 \qquad \blacktriangle$$

N O T E   *In Example 5, notice that we converted the perimeter of the backyard from 280 feet to* $93\dfrac{1}{3}$ *yards, since the cost was given as $3 per* yard, *not $3 per foot.*

Three dimensional objects such as a brick have width, length, and height. The **volume** of an object, measured in cubic units, is the number of cubes needed to fill the object. Examples of three dimensional objects and the associated formula for the volume V are the following.

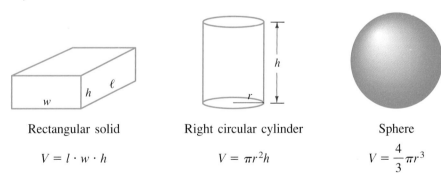

Rectangular solid	Right circular cylinder	Sphere
$V = l \cdot w \cdot h$	$V = \pi r^2 h$	$V = \dfrac{4}{3}\pi r^3$

Other three dimensional objects are shown on the inside front cover.

Example 6   Pronto Delivery Service plans to market a cylindrical container that has a radius of 4 inches. Find the height of the container, if the volume must be 400π (approximately 1,257) cubic inches.

Solution   The volume of a cylinder is given by $V = \pi r^2 h$. Replace V by 400π and r by 4, then solve the resulting equation for h.

$$V = \pi r^2 h$$

$$400\pi = \pi(4)^2 h$$

$$400\pi = \pi 16 h$$

$$\frac{400\pi}{16\pi} = h$$

$$25 = h$$

The height h is 25 inches.   ▲

Greek letters such as α (alpha), β (beta), and γ (gamma) are used to label angles. Consider the following triangle with angles α, β, and γ.

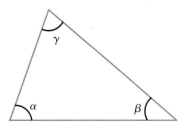

In a triangle, the sum of the measures of the three angles equals 180 degrees. We write

$$\alpha + \beta + \gamma = 180°$$

Two common triangles that are studied in trigonometry are the following:

Example 7   One angle of a triangle has a measure of 38 degrees. The measure of another (the second) angle is two degrees less than four times the measure of the third angle. Find the measures of the two unknown angles.

Solution   Let

$$\alpha = \text{the measure of the third angle}$$

Then,

$$4\alpha - 2 = \text{the measure of the second angle}$$

Since the sum of the measures of the three angles is 180 degrees,

$$38 + (4\alpha - 2) + \alpha = 180$$

We next solve this equation for α.

$$36 + 5\alpha = 180$$

$$5\alpha = 180 - 36$$

$$5\alpha = 144$$

Dividing both sides of the equation by 5, we have

$$\alpha = \frac{144}{5}$$

$$\alpha = 28.8$$

Therefore, the measure of the third angle is 28.8 degrees and the measure of the second angle is $4(28.8) - 2$ or 113.2 degrees. ▲

In the next example, we solve for one of the variables in a formula.

Example 8 The formula to convert temperatures from Celsius (C) to Fahrenheit (F) is given by $F = \dfrac{9}{5}C + 32$. Solve for C.

Solution To solve for C, we must isolate it on one side of the equation. We start by subtracting 32 from both sides, then multiplying by $\dfrac{5}{9}$.

$$F = \frac{9}{5}C + 32$$

$$F - 32 = \frac{9}{5}C + 32 - 32$$

$$F - 32 = \frac{9}{5}C$$

$$\frac{5}{9}(F - 32) = \frac{5}{9}\left(\frac{9}{5}C\right) \qquad \text{Multiply both sides by } \frac{5}{9}.$$

$$\frac{5}{9}(F - 32) = C$$

Therefore, the formula for C in terms of F is $C = \dfrac{5}{9}(F - 32)$. ▲

COMMON ERROR

The right side of the equation $C = \frac{5}{9}(F - 32)$ is $\frac{5}{9}(F - 32)$ and *not* $\frac{5}{9}F - 32$.

NOTE *The two formulas $F = \frac{9}{5}C + 32$ and $C = \frac{5}{9}(F - 32)$ both determine the same relationship between the variables C and F. The first one, however, is used to convert from Celsius to Fahrenheit and the second one is used to convert from Fahrenheit to Celsius.*

Example 9 Solve $y = \frac{2x - 1}{3x + 2}$ for x.

Solution This is an example of a rational equation that we will study in more detail at a later time. To solve for x, first multiply both sides by the denominator $3x + 2$.

$$y = \frac{2x - 1}{3x + 2}$$
$$y(3x + 2) = 2x - 1$$
$$3yx + 2y = 2x - 1 \quad \text{Distributive property.}$$
$$3yx - 2x = -1 - 2y$$
$$(3y - 2)x = -1 - 2y \quad \text{Distributive property.}$$
$$x = \frac{-1 - 2y}{3y - 2} \quad \text{Divide both sides by } 3y - 2. \quad ▲$$

CALCULATOR CROSSOVER

1. Find the area of a triangle if $b = 3.51$ feet and $h = 12.80$ feet.
2. A rectangle has a length that is the sum of 4.23 times the width and 5.545. If the perimeter is 18.412 inches, find the length and width.
3. Fran plans to invest $23,600, part at 11% and part at 9.5% for one year. If she wants to obtain a total interest of $2,410, how much should she invest at each interest rate?

Answers **1.** 22.464 square feet **2.** width = 0.7 inches, length = 8.506 inches **3.** $11,200 at 11% and $12,400 at 9.5%.

We finish this section with a business application.

Example 10 (**Cost-Revenue**) The Blackstone Company makes metal cabinets. Each week the company makes and sells x cabinets. The weekly **cost** C of making the x cabinets is

$$C = 2{,}000 + 40x$$

The company sells each cabinet for $80. The weekly revenue R from selling x cabinets is

$$R = 80x$$

The **break-even point** is that level of production x where the revenue equals the cost. Find the break-even point.

Solution We set $R = C$ and solve the resulting equation for x.

$$R = C$$
$$80x = 2{,}000 + 40x$$
$$40x = 2{,}000$$
$$x = 50$$

Therefore, the company must have a level of production of 50 cabinets per week to break even. ▲

L E A R N I N G   A D V A N T A G E   *Read the section or sections in the book that are assigned slowly, not once but twice—more often if you have trouble understanding it. Read it first in class if there is time available; read it again later in the day. Each time you read it, you will learn more or find questions to ask.*

E X E R C I S E   S E T   2 . 4

1. Two jeeps start toward each other at the same time from towns 75 miles apart. One jeep travels at 45 mph while the other one travels at 55 mph. How much time passes before they meet?

2. Two planes, flying in opposite directions, pass each other. One is flying at 200 mph and the other at 300 mph. How long will it take for them to be 60 miles apart?

3. A police car and a sports car are 15 miles apart and traveling towards each other. The sports car is traveling 10 mph faster than the police car. If they meet in 6 minutes $\left(\dfrac{1}{10}\text{ hour}\right)$, how fast is each car traveling?

4. Two boats, 22 miles apart, travel toward each other and meet 30 minutes later. If the speed of the first boat is 5 mph faster than the speed of the second boat, what is the speed of each boat?

5. How much interest is made on $10,000 invested at $8\frac{1}{2}\%$ for 2 years?

6. How much interest is made on $5,000 invested at 11% for 3 years?

7. How much interest is made on $2,500 invested at 12 percent for 9 months?

8. How much interest is made on $7,000 invested at 6% for 6 months?

9. Luther is planning to invest $15,000 for one year. He will use part of the money to buy a certificate of deposit paying 8% interest and will put the rest of the money in a savings account paying 6%. If he wants to obtain $1,000 in combined interest at the end of the year, how much money should he use for the certificate of deposit and how much should he put into the savings account?

10. Kathy is planning to invest $20,000 in two securities, one paying 8% and the other paying 12%. How much should she invest at each interest rate to realize an $1,800 yearly interest payment?

11. Find the area and perimeter of each figure shown below.

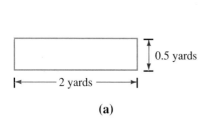

(a)

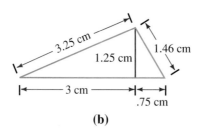

(b)

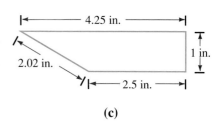

(c)

12. Find the area and perimeter of each figure shown below.

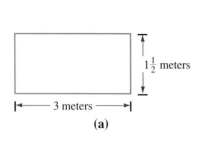

(a)

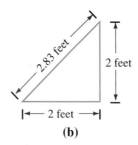

(b)

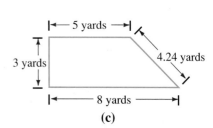

(c)

13. A rectangular garden, 55 feet by 40 feet, is to be enclosed by fencing costing $10 per foot. What is the total cost of the fence?

14. A field has dimensions 120 feet by 80 feet. What is the total cost of fencing to enclose it and divide it into three parts as shown if the fencing costs $12 per foot?

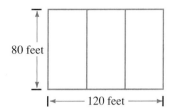

80 feet

|← —— 120 feet —— →|

15. One angle of a triangle has a measure of 48 degrees. The measure of another angle is three times the measure of the third angle. Find the measures of the two unknown angles.

16. One angle of a triangle has a measure of 32 degrees. The measure of another angle is four times the measure of the third angle increased by 5 degrees. Find the measures of the two unknown angles.

17. The measure of the first angle of a triangle is three times the measure of the second angle. The measure of the second angle is one-half the measure of the third angle. Find the measures of each angle.

18. The measure of the first angle of a triangle is two thirds the measure of the second angle, and the measure of the third angle is 37 degrees less than the measure of the first angle. Find the measure of each angle.

For Exercises 19–34, solve the formula for the indicated variable.

19. $C = 2\pi r$ for r.   Circumference of a circle.

20. $I = Prt$ for t.   Simple interest.

21. $E = \dfrac{I}{d^2}$ for I.   Illuminance of a light source.

22. $V = \pi r^2 h$ for h.   Volume of a right circular cylinder.

23. $A = P(rt + 1)$ for P.   Interest formula.

24. $E = \dfrac{\Delta L}{L(t_2 - t_1)}$ for ΔL.   Thermal expansion. (Δ is read "delta.")

25. $p_1 V_1 = p_2 V_2$ for V_2.   Boyle's law.

26. $\dfrac{V_2}{T_2} = \dfrac{V_3}{T_3}$ for T_3.   Charles' law.

27. $E = \dfrac{T_2 - T_1}{T_2}$ for T_1.   Efficiency of heat engines.

28. $V = V_0 + 0.61t$ for t.   Speed of sound.

29. $P = 2l + 2w$ for w.   Perimeter of a rectangle.

30. $h = vt - 16t^2$ for v.   Height of a projectile.

31. $\dfrac{1}{a} + \dfrac{1}{b} = \dfrac{1}{f}$ for b.   Object-image formula.

32. $\dfrac{1}{R} = \dfrac{1}{R_1} + \dfrac{1}{R_2}$ for R_1.   Resistance in an electrical circuit.

33. $a_n = a + (n - 1)d$ for d.   Arithmetic sequence.

34. $z = \dfrac{x - \mu}{\sigma}$ for x.   A conversion rule in statistics.

(μ is read "mu" and σ is read "sigma.")

For Exercises 35–42, solve the equation for x.

35. $y = 3x + 1$

36. $y = -5x + 3$

37. $4x - 2y - 1 = 0$

38. $-6x + 3y + 2 = 0$

39. $\dfrac{x - 1}{t + 3} = -\dfrac{2}{3}$

40. $\dfrac{x + 4}{t - 4} = 5$

41. $y = \dfrac{3x + 8}{2x - 1}$

42. $y = \dfrac{4x - 3}{x + 6}$

For Exercises 43–46, solve the equation for y.

43. $3x - 4y = 8$

44. $12x - 3y = 2$

45. $Ax + By + C = 0$

46. $\dfrac{x}{a} + \dfrac{y}{b} = 1$

47. If $2x - 3y = 6$, find y if $x = -2$.

48. If $-4x + 3y = 12$, find x if $y = 3$.

49. If $\dfrac{x}{5} - \dfrac{y}{3} = 1$, find x if $y = 6$.

50. If $\dfrac{y}{2} + \dfrac{x}{4} = -2$, find y if $x = -4$.

51. The Waltham Corporation manufactures piano rolls and has a weekly cost C of

$$C = 7{,}000 + 1.25x$$

where x is the number of piano rolls made in a week. The weekly revenue R of selling x piano rolls is given by

$$R = 4.75x$$

Find the break-even point for the company.

52. The Banner Wrench Company produces tool kits, where the weekly cost C is given by

$$C = 875 + 20x$$

where x is the number of tool kits made each week. The weekly revenue R of selling x tool kits is given by

$$R = 55x$$

Find the break-even point for the company.

53. A taxi cab company in Allentown charges $2.50 plus $2.20 per mile.
(a) Write a formula for the total fare F for a customer who rides a cab for t miles.
(b) What is the total cost for riding a cab for 25 miles?
(c) Emma has at most $24.50 to pay for cab fare. How many miles can she travel by cab?

54. Glen rents a car for $60 plus 20 cents per mile to tour Louisville.
(a) Write a formula for the total cost C of renting a car for t miles.
(b) What is the total cost for driving 45 miles?
(c) Glen wants to spend at most $100 for the car. How many miles can he travel?

55. What temperature is the same in Fahrenheit as in Celsius?

56. Suppose P dollars is deposited in an account that earns 6% interest. At the end of t years, the amount A of money in the account is given by

$$A = P(1 + 0.06t)$$

How long will it take the money to double?

REVIEW PROBLEMS

The following exercises review parts of Sections 1.2 and 1.5. Doing these problems will help prepare you for the next section.

Section 1.2

For Exercises 57–62, write either $<$ or $>$ between the two numbers to make a true statement.

57. $\dfrac{4}{3}$ $\dfrac{5}{3}$

58. -6 -7

59. $\dfrac{1}{2}$ -0.0001

60. $-3\dfrac{7}{8}$ $-3\dfrac{6}{7}$

61. 0.94 0.93

62. -1.115 -1.105

Section 1.5

Simplify.

63. $3(1 - 2x) + 5$

64. $6x - 12x + 3 - 10$

65. $3z - 4(6 - 2z)$

66. $5(3y + 1) - 3(2 - 7y) + 12$

E N R I C H M E N T   E X E R C I S E S

1. A triangle has sides whose lengths (in feet) are consecutive integers. A rectangle is formed so that its length is the sum of the shortest and longest sides of the triangle and its width is $\dfrac{1}{3}$ the remaining side of the triangle. If the perimeter of the rectangle is 28 feet, find the lengths of the sides of the triangle and of the rectangle.

2. A rectangular garden is 8 feet longer than it is wide. A 2-foot-wide sidewalk surrounds the garden. If the total area of the sidewalk is 176 square feet, what are the dimensions of the garden?

3. Circular tops for tin cans are to be cut from a sheet of tin that measures 72 inches by 126 inches. If each top is to have an area of 9π square inches, what is the maximum number of tops that can be cut from the sheet?

4. Solve for x: $y = \dfrac{ax + b}{cx + d}$.

Answers to Enrichment Exercises are on page A.6.

2.5

Solving Inequalities

OBJECTIVES

▶ *To solve linear inequalities using the addition property of inequality*

▶ *To solve linear inequalities using the multiplication property of inequality*

▶ *To solve general linear inequalities*

The inequality symbols introduced in Section 1.1 to compare numbers can also be used to compare algebraic expressions. For example, suppose that an unknown number x is greater than 2. We can symbolize that statement by writing $x > 2$.

Example 1   Let x represent the number of people waiting in line at a theater. Symbolize each statement with an inequality.

(a) There are more than 15 people waiting in line.
(b) There are no fewer than 10 people waiting in line.
(c) There are no more than 12 people waiting in line.

Solution   (a) If there are more than 15 people waiting in line, then the number x is greater than 15; that is, $x > 15$.

(b) No fewer than 10 people means 10 or more people are waiting in line. Therefore, the correct inequality is $x \geq 10$.

(c) No more than 12 people waiting in line means that 12 or less people are waiting in line. Therefore, the correct inequality is $x \leq 12$.   ▲

In this section, we will study *linear inequalities*.

DEFINITION

A **linear inequality** in one variable is any inequality that can be put in the form

$$ax + b < c, \quad \text{where } a \neq 0$$

The inequality symbol "$<$" can be replaced with any of the other three inequality symbols: $>$, $\leq$, or $\geq$.

Some examples of linear inequalities are

$$3x - 1 \geq 5 \quad \text{and} \quad 6 - \frac{5x - 2}{4} < 7x + 12$$

To **solve** a linear inequality, we find all the values for the variable that make the inequality true. When solving a linear *equation* we obtained a single number that made the equation a true statement. However, a linear *inequality* will have a solution set containing many numbers. Later in the section, we will develop a strategy to solve a linear inequality. Before doing that, however, let us first look at sets of numbers such as $\{x \mid x \geq -3\}$. We may graph this set by drawing a number line as shown in the figure below. Locate the number -3 and fill in a solid circle. Next, draw an arrow starting at -3 and extend it to the right. We drew a solid rather than hollow circle at -3 to indicate that x could be equal to -3.

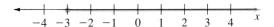

The set $\{x \mid x \geq -3\}$ drawn on the number line above is an example of an **interval.** Instead of using the set notation, we can use interval notation. For example, the interval $\{x \mid x \geq -3\}$ can be written as $[-3, +\infty)$. The symbol $+\infty$ is not a number, but indicates that the interval consists of all numbers greater than or equal to -3. If the set did not include -3, we would write $(-3, +\infty)$ to mean $\{x \mid x > -3\}$. That is, we use brackets when a number is included in the interval and parentheses when a number is not included in the interval.

Suppose a and b are numbers with $a < b$. The following table lists the types of intervals that we will encounter in this section.

Set notation	Interval notation	Graph
$\{x \mid x > a\}$	$(a, +\infty)$	
$\{x \mid x \geq a\}$	$[a, +\infty)$	
$\{x \mid x < b\}$	$(-\infty, b)$	
$\{x \mid x \leq b\}$	$(-\infty, b]$	
$\{x \mid a < x < b\}$	(a, b)	
$\{x \mid a \leq x < b\}$	$[a, b)$	
$\{x \mid a < x \leq b\}$	$[a, b]$	
$\{x \mid a \leq x \leq b\}$	$[a, b]$	

N O T E   *The two symbols $-\infty$ and $+\infty$ are never enclosed by a bracket, since that would imply that they are members of the interval.*

Example 2   Graph the following sets.

(a) $\{x \mid x < 6\}$ (b) $\{x \mid x \geq 1.5\}$
(c) $[1, 3)$ (d) $[-20, -5]$

Solution   (a) The inequality means all real numbers x less than 6, but x cannot be 6 it-self. On the number line, draw a hollow circle at 6 and extend an arrow to the *left*. The graph is shown below.

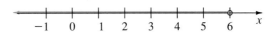

(b) The inequality means all real numbers x greater than or equal to 1.5. On the number line, fill in a solid circle at 1.5 and extend an arrow to the *right*. The graph is shown below.

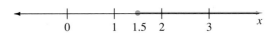

(c) This interval means all numbers greater than or equal to 1 and less than 3. On the number line, draw a solid circle at 1 and a hollow circle at 3. Then fill in the line segment between these two points. The graph is shown below.

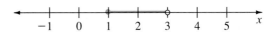

(d) This interval means all numbers between -20 and -5, *inclusive*. Since we include the endpoints, draw solid circles at both -20 and -5. Then fill in the line segment between these two points. The graph is shown below.

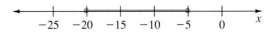

We will find that the interval notation is a convenient way to write the so-lution set of a linear inequality.

Our next topic in the section is solving linear inequalities. Solving an in-equality like $x - 6 < -8$ is similar to solving linear equations. We introduce the first technique with the following illustration.

From our study of the number line, we know that $a < b$ provided that a lies to the left of b

3 < 9 implies 3 + 2 < 9 + 2

For example, $3 < 9$ and 3 lies to the left of 9. Now if we add the number 2 to both sides of the inequality, we have $3 + 2$ and $9 + 2$. Now, $3 + 2$ or 5 lies to the left of $9 + 2$ or 11 as shown in the figure. Therefore, $3 < 9$ implies

$$3 + 2 < 9 + 2$$

This illustrates the *addition property of inequality.*

P R O P E R T Y

The addition property of inequality

Let A, B, and C represent algebraic expressions. If

$$A < B$$

then

$$A + C < B + C$$

Since subtraction is the addition of the opposite of a number, given an inequality, we may subtract the same quantity from each side and still maintain an inequality. The addition property remains true if $<$ is replaced by $\leq$, $>$, or $\geq$.

Example 3   Solve the inequality. Then graph the solution.

$$6 - 4x > -5x - 1$$

Solution   We want to isolate x on one side. Use the addition property to add $5x$ to both sides and to subtract 6 from both sides.

$$6 - 4x > -5x - 1$$

$$6 - 4x + 5x - 6 > -5x - 1 + 5x - 6 \qquad \text{Add } 5x \text{ and subtract 6.}$$

$$x > -7 \qquad \text{Simplify.}$$

Therefore, the solution set consists of all numbers greater than -7 and can be written as the interval $(-7, +\infty)$. The graph of this solution is shown in the figure below.

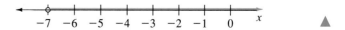

What happens when we multiply both sides of an inequality by a number? It depends upon whether the number is positive or negative. For example, starting with the inequality $2 < 4$, if we multiply both sides by 3, we retain the inequality,

$$2 \cdot 3 < 4 \cdot 3 \quad \text{or} \quad 6 < 12$$

If we multiply both sides of $2 < 4$ by -3, the inequality changes its sense,

$$2(-3) > 4(-3) \quad \text{or} \quad -6 > -12$$

Keeping these examples in mind, we state the multiplication property of inequality.

P R O P E R T Y

The multiplication property of inequality

Let A, B, and C represent algebraic expressions. Suppose $A < B$.

1. If $C > 0$, then $CA < CB$.
2. If $C < 0$, then $CA > CB$.

Therefore, we can multiply both sides of an inequality by a positive quantity and the inequality remains the same. On the other hand, if we multiply by a negative quantity, then the inequality is reversed. The multiplication property of inequality is also true if $<$ is replaced by $\leq$, $>$, or $\geq$.

N O T E   *Since division can be viewed as multiplication by a reciprocal, our new property applies also to division. Thus, if we divide both sides of an inequality by a positive quantity, the inequality remains the same; if we divide both sides of an inequality by a negative quantity, the inequality reverses its sense.*

We will use our new property to solve linear inequalities where the coefficient of the variable is not one.

Example 4   Solve $\frac{1}{3}x \geq -\frac{2}{9}$, then graph the solution set.

Solution   To isolate x on the left side, use the multiplication property of inequality with $C = 3$.

$$\frac{1}{3}x \geq -\frac{2}{9}$$

$$3\left(\frac{1}{3}x\right) \geq 3\left(-\frac{2}{9}\right)$$

$$x \geq -\frac{2}{3}$$

The solution set is the interval $\left[-\dfrac{2}{3}, +\infty \right)$ and its graph is shown below.

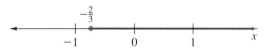

Example 5   Solve $-5x > 12$ and graph the solution set.

Solution   To make the coefficient of x equal to one, we multiply both sides by $-\dfrac{1}{5}$ (or divide both sides by -5).

$$-5x > 12$$

$$\left(-\frac{1}{5} \right)(-5x) < \left(-\frac{1}{5} \right)(12)$$

$$x < -\frac{12}{5}$$

So, the solution set is $\left(-\infty, -\dfrac{12}{5} \right)$ and is graphed below. As an aid to graphing, notice that we wrote the fraction $-\dfrac{12}{5}$ as the mixed number $-2\dfrac{2}{5}$.

We now give guidelines for solving a linear inequality. The steps are similar to those of solving a linear equation that were given in Section 2.2.

S T R A T E G Y

To solve a linear inequality

Step 1   **Remove fractions or decimals,** if necessary, by using the multiplication property of inequality.

Step 2   **Combine like terms on each side.** This may entail using the distributive property first to separate terms.

Step 3   **Use the addition property of inequality,** if necessary, to move all terms containing the variable to one side and all constant terms to the other side.   *(continued)*

S T R A T E G Y (*continued*)

Step 4 **Use the multiplication property of inequality,** if necessary, to make the coefficient on the variable term equal to one. The inequality now has one of the basic forms

$$x < b, \qquad x \le b, \qquad x > a, \qquad x \ge a$$

or a combination of these such as $a < x \le b$.

Example 6 Solve the inequality $3x - 2(x + 3) \ge 3(x - 4)$ and graph its solution set.

Solution First, we use the distributive property and then simplify.

$$3x - 2(x + 3) \ge 3(x - 4)$$

$$3x - 2x - 6 \ge 3x - 12$$

$$x - 6 \ge 3x - 12$$

$$x - 3x \ge -12 + 6 \qquad \text{Add 6 and subtract } 3x.$$

$$-2x \ge -6$$

$$\frac{-2x}{-2} \le \frac{-6}{-2} \qquad \begin{array}{l} \text{Divide by } -2 \text{ and reverse the sense} \\ \text{of the inequality.} \end{array}$$

$$x \le 3$$

Therefore, the solution set is the interval $(-\infty, 3]$ as shown in the figure below.

In the next example, we apply our techniques for solving inequalities to a slightly different kind of inequality.

Example 7 Solve $-5 < 3t + 7 < 10$ and graph its solution set.

Solution This inequality statement has three parts. Our goal is to isolate t in the middle part. To do this, we first subtract 7 from all three parts, realizing that the two inequality signs remain the same.

$$-5 - 7 < 3t + 7 - 7 < 10 - 7$$

$$-12 < 3t < 3$$

Next, we divide each part by 3. Since 3 is positive, again the inequalities remain the same.

$$\frac{-12}{3} < \frac{3t}{3} < \frac{3}{3}$$

$$-4 < t < 1$$

The solution set is the interval $(-4, 1)$ and is graphed below.

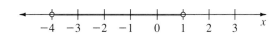

Example 8

Sheri received grades of 82%, 91%, and 84% on her first three algebra tests. What possible scores on the fourth test will give her a final average between 80% and 89%, inclusive?

Solution

Let x represent Sheri's grade on the fourth test. The average of the four tests is given by

$$\frac{82 + 91 + 84 + x}{4}$$

This average must fit between 80 and 89, *inclusive*. Therefore,

$$80 \leq \frac{82 + 91 + 84 + x}{4} \leq 89$$

Next, we find the solution set.

$$4(80) \leq 82 + 91 + 84 + x \leq 4(89) \qquad \text{Multiply by 4.}$$

$$320 \leq 257 + x \leq 356$$

$$320 - 257 \leq x \leq 356 - 257 \qquad \text{Subtract 257.}$$

$$63 \leq x \leq 99$$

Therefore, Sheri must score between 63% and 99%, inclusive, to achieve a final average that falls in the given range. ▲

The final topic of this section is solving a compound inequality. A **compound inequality** is two inequalities connected by the word "and" or "or."

Example 9

Solve the compound inequality:

$$x + 2 < 8 \qquad \text{and} \qquad x - 3 > -1$$

Solution

We first solve each inequality individually. The first inequality has the solution set $(-\infty, 6)$ and the second inequality has the solution set $(2, +\infty)$. Since the connecting word in the compound inequality is "and," the solution set consists of numbers common to $(-\infty, 6)$ and $(2, +\infty)$. As shown in the figure below, the solution set is $(2, 6)$.

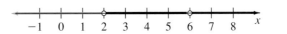

The set of elements common to two sets is called the *intersection* of the two sets. We use the symbol $\cap$ to indicate this intersection. In Example 9, the solution set is the intersection of $(-\infty, 6)$ and $(2, +\infty)$. That is,

$$(-\infty, 6) \cap (2, +\infty) = (2, 6)$$

In general, we have the following definition:

DEFINITION

If A and B are sets, the **intersection** of A and B is

$A \cap B = \{x \mid x$ is an element of A and x is an element of $B\}$

Besides the intersection of two sets, we can take the union of them.

DEFINITION

If A and B are sets, the **union** of A and B is

$A \cup B = \{x \mid$ either x is a member of A or x is a member of $B\}$

In the next example, we show four examples of unions and intersections. If you do not see the answer as stated, then graph the sets on a number line. Keep in mind that intersection, $\cap$, means the numbers common to both sets, and union, $\cup$, means all numbers either in the first set or in the second set (or both).

Example 10   (a) $(-\infty, 3) \cap [-1, +\infty) = [-1, 3)$ (b) $[2, 5) \cup (1, 3] = (1, 5)$
(c) $(-1, 3) \cap [0, 6) = [0, 3)$ (d) $(2, +\infty) \cap [5, +\infty) = [5, +\infty)$

In our final example, we solve a compound inequality involving the word "or." In this case, we write the solution set as the union of two intervals.

Example 11   Solve the compound inequality:

$$2x - 3 < 5 \qquad \text{or} \qquad x - 4 \geq 3$$

Solution   We solve each inequality individually. The solution set of $2x - 3 < 5$ is $(-\infty, 4)$ and the solution set of $x - 4 \geq 3$ is $[7, +\infty)$. Our final answer is the union of these two sets: $(-\infty, 4) \cup [7, +\infty)$ as shown in the figure below.

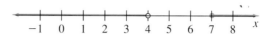

L E A R N I N G   A D V A N T A G E   *When reading this text-book, write down questions that need to be answered. You can get more help from your teacher or friends if you ask them specific questions. Read examples of a new procedure; try to do the examples without looking at the book.*

E X E R C I S E   S E T   2.5

For Exercises 1–6, let x represent the number of cars waiting in line at a tollbooth. Symbolize each statement with an inequality.

1. There are more than 10 cars waiting in line.

2. There are fewer than six cars waiting in line.

3. There are at least nine cars waiting in line.

4. There are no more than 12 cars waiting in line.

5. There are seven or fewer cars waiting in line.

6. There are 14 or more cars waiting in line.

For Exercises 7–20, graph the following sets.

7. $\{x \mid x < 7\}$

8. $\{x \mid x > -2\}$

9. $\left\{x \mid x \leq 3\frac{1}{2}\right\}$

10. $\{x \mid x < -0.5\}$

11. $[-2, +\infty)$

12. $(-\infty, 0]$

13. $\{x \mid 1 < x\}$

14. $\{x \mid -1 \geq x\}$

15. $(-1, 1)$

16. $(-2, 3)$

17. $[0, 2.5)$

18. $\left(-\frac{1}{3}, \frac{2}{3}\right]$

19. $\{x \mid -100 \leq x \leq -50\}$

20. $\{x \mid 20 < x < 50\}$

For Exercises 21–34, solve the inequality. Write your answer using interval notation. Then graph the solution.

21. $x - 3 \le 7$

22. $x - 5 < 1$

23. $x - \dfrac{3}{4} > \dfrac{3}{4}$

24. $x - 1.5 \ge 0.5$

25. $x + 30 < 50$

26. $x - 100 \ge 200$

27. $2 + 3x < 2x + 1$

28. $-5 + 6x \ge 5x - 2$

29. $-\dfrac{3}{4} + \dfrac{7}{8}x \ge -\dfrac{1}{2} - \dfrac{1}{8}x$

30. $-4.6 + 1.7x < 0.7x + 0.4$

31. $-5 \le x - 3 \le 0$

32. $-1 \le x - 2 \le 0$

33. $7 \le x + 7 < 13$

34. $6 < x + 11 \le 12$

For Exercises 35–64, solve the inequality.

35. $-\dfrac{3}{2} < x - \dfrac{3}{2} \le -\dfrac{1}{2}$

36. $\dfrac{2}{3} \le x - \dfrac{4}{3} < \dfrac{5}{3}$

37. $3x > -1$

38. $\dfrac{1}{2}x \le 2$

39. $-7n > 14$

40. $-5w < 15$

41. $\dfrac{1}{4}t < \dfrac{3}{2}$

42. $\dfrac{1}{3}y \ge -1$

43. $200z \le 4{,}000$

44. $150x > -4{,}500$

45. $2x - 5 + 3x \ge -5x + 15$

46. $16x - 21x + 7 < 3x + 14 + 2x + 8$

47. $4 \leq x + 7 < 5$

48. $-1 < y - 3 \leq 2$

49. $2 < \dfrac{2w - 1}{5} < 3$

50. $-3 < \dfrac{3m + 2}{4} < 4$

51. $20 < 2x - 10 \leq 40$

52. $-600 < 5r - 450 < -100$

53. $3(2 - x) < 5 - 2x$

54. $-4(s + 3) \geq 0$

55. $4(1 - y) \geq -2(1 + y)$

56. $3(z - 2) \leq -(z + 8)$

57. $4(t - 1) + 14t > 7(2t - 1)$

58. $2(4 - x) + x \leq 3(3 + x)$

59. $-9 \leq -3w < 5$

60. $22 < -4v < 28$

61. $-3 < 1 - 4y \leq -1$

62. $-6 \leq -1 - 3x < 2$

63. $2 - 3(2 + x) - x \geq 2(4 - 5x) - 2x$

64. $4x - (3x - 2) + 5 < x - 7(1 - 2x)$

For Exercises 65–78, write and solve an inequality from the given information.

65. The sum of a number and two is less than five.

66. The sum of a number and -3 is greater than -2.

67. The difference of a number and four is not more than one.

68. The difference of a number and two is not less than five.

69. The sum of three times a number and four is no more than the sum of twice a number and seven.

70. The sum of twice a number and six is not less than the sum of the number and two.

71. The difference of six and three times a number is less than negative two times the number.

72. The difference of 18 and four times a number is greater than six times the number.

73. The difference of four and a number does not exceed five.

74. Three more than four times a number is greater than negative one.

75. Twice a number reduced by $\dfrac{5}{2}$ is less than the sum of five times the number and $\dfrac{7}{2}$.

76. The sum of eight times a number and 0.5 is not less than the sum of 1.7 and twice the number.

77. The sum of three times a number and one is larger than two, but less than or equal to three.

78. The difference of four times a number and three is greater than or equal to −5, but does not exceed −1.

For Exercises 79–82, solve the word problem.

79. Wendy's goal is to have a bowling average of at least 180 for the tournament. Her scores so far are 166 and 189. What must she bowl in the last game to achieve this goal?

80. Kevin received grades of 66, 72, and 48 on his first three history tests. What possible scores on the fourth test will give him a final average of at least 70%?

81. A company that makes frozen chicken nuggets makes a profit of $0.45 per box sold. How many boxes must be sold to have a profit of at least $90,000?

82. A company that produces conveyer belts makes a profit of $250 per belt. How many belts must be sold to have a profit of at least $100,000?

For Exercises 83–90, find the union or intersection. Write your answer either as an interval, the empty set $\varnothing$, or the set of real numbers.

83. $(-\infty, 5) \cap (2, +\infty)$

84. $(-\infty, 8] \cap (0, +\infty)$

85. $[-1, 4) \cup (1, 5]$

86. $[-2, 7) \cup (-1, +\infty)$

87. $(-\infty, 4) \cup (2, +\infty)$

88. $(-\infty, 5] \cap (6, +\infty)$

89. $(-\infty, 5] \cap (-\infty, 2)$

90. $(-1, +\infty) \cap [1, +\infty)$

For Exercises 91–98, solve the compound inequality. Write your answer using interval notation.

91. $x + 2 < 12$ and $x - 3 \geq -1$

92. $x - 14 \leq -1$ and $x + 2 > 5$

93. $x - 3 < -2$ or $x + 4 > 9$

94. $2x + 1 \leq -1$ or $x - 6 > -2$

95. $3x + 1 \leq 7$ or $3x - 4 \geq 11$

96. $1 - 2x < 9$ or $x + 5 < -2$

97. $3 - 2(x - 1) > 10$ and $3(1 - 2x) + 7 > -1$

98. $7(2 + x) - 12 > 0$ and $8 + 2(1 - 3x) < 2$

REVIEW PROBLEMS

The following exercises review parts of Section 1.3. Doing these problems will help prepare you for the next section.

Simplify.

99. $|-41|$

100. $|95|$

101. $1 - |-1|$

102. $-|-5| + 2$

103. $\dfrac{|4 - 7|}{3}$

104. $\dfrac{|12 - 20|}{16}$

105. $\dfrac{|2-9|}{|4-6|}$

106. If $|x| = 4$, where on the number line would x be located?

107. If $|x| < 4$, where on the number line would x be located?

108. If $|x| > 4$, where on the number line would x be located?

E N R I C H M E N T   E X E R C I S E S

1. If $a < -cx < b$, show that $-\dfrac{b}{c} < x < -\dfrac{a}{c}$, where $c > 0$.

For Exercises 2–4, use the result of Exercise 1 to solve for x.

2. $-1 < -x < 2$

3. $-5 \leq -3x + 7 < -1$

4. $\dfrac{5}{3} < \dfrac{1}{6} - \dfrac{x}{3} \leq 2$

5. Show that $|x| < 2$ and $-2 < x < 2$ have the same solution set.

6. Determine if the following statement is true for all real numbers. If it is not, give a numerical example to show the statement is not true for all real numbers.

Statement: If $x > y$, then $x^2 > y^2$.

Answers to Enrichment Exercises are on page A.7.

2.6

Equations and Inequalities with Absolute Value

O B J E C T I V E S

▶ *To solve equations containing absolute value*

▶ *To solve inequalities containing absolute value*

In Section 1.3 we saw that the absolute value of a number x, $|x|$, is the distance from x to 0 on the number line. With this geometric meaning of absolute value, we can solve equations that contain absolute value.

Example 1   Solve for x: $|x| = 3$.

Solution   We want to find all numbers whose distance from 0 is 3. If x is 3 units from 0, then either $x = 3$ or $x = -3$.   ▲

In general, we have the following form to use.

> **R U L E**
>
> If $p > 0$, then
> $$|x| = p \text{ is equivalent to either } x = p \text{ or } x = -p.$$

Example 2   Solve for t: $|3t - 2| = 8$.

Solution   This equation is in the form $|x| = p$. Therefore, either $3t - 2 = 8$ or $3t - 2 = -8$. Next, we solve each of these two equations.

$$3t - 2 = 8 \quad \text{or} \quad 3t - 2 = -8$$
$$3t = 10 \quad \text{or} \quad 3t = -6 \qquad \text{Add 2.}$$
$$t = \frac{10}{3} \quad \text{or} \quad t = -2 \qquad \text{Divide by 3.}$$

We now check each solution by substituting back in the original equation.

Check $t = \dfrac{10}{3}$:   *Check* $t = -2$:

$$|3t - 2| = 8 \qquad\qquad |3t - 2| = 8$$
$$\left|3\left(\frac{10}{3}\right) - 2\right| \overset{?}{=} 8 \qquad |3(-2) - 2| \overset{?}{=} 8$$
$$|10 - 2| \overset{?}{=} 8 \qquad |-6 - 2| \overset{?}{=} 8$$
$$|8| \overset{?}{=} 8 \qquad\qquad |-8| \overset{?}{=} 8$$
$$8 = 8 \qquad\qquad 8 = 8$$

The solution set is $\left\{-2, \dfrac{10}{3}\right\}$.   ▲

Example 3   Solve the equation $|3s + 4| = -1$.

Solution   Since the absolute value can never be negative, there is no solution for this equation. Therefore, the solution set is $\varnothing$.   ▲

Example 4   Solve the equation $|4z + 3| - 2 = 7$.

Solution   We must first isolate the absolute value. Therefore, we add 2 to both sides of the equation to put it in the form $|x| = p$.

$$|4z + 3| - 2 = 7$$

$$|4z + 3| = 9 \qquad \text{Add 2 to both sides.}$$

$$\text{Either } 4z + 3 = 9 \qquad \text{or} \qquad 4z + 3 = -9$$

$$4z = 6 \qquad \text{or} \qquad 4z = -12 \qquad \text{Subtract 3.}$$

$$z = \frac{3}{2} \qquad \text{or} \qquad z = -3 \qquad \text{Divide by 4.}$$

The solution set is $\left\{-3, \dfrac{3}{2}\right\}$.   ▲

Suppose that the absolute value of two numbers is the same, $|a| = |b|$. What must be true of a and b? The two numbers are the same distance from 0. Therefore, they are either the same number or they are opposites. That is, either $a = b$ or $a = -b$.

Example 5   Solve $|3r + 2| = |5r - 1|$.

Solution   The quantities $3r + 2$ and $5r - 1$ have the same absolute value. Therefore, either

$$3r + 2 = 5r - 1 \qquad \text{or} \qquad 3r + 2 = -(5r - 1)$$

$$-2r = -3 \qquad\qquad\qquad 3r + 2 = -5r + 1$$

$$r = \frac{3}{2} \qquad\qquad\qquad\qquad 8r = -1$$

$$r = -\frac{1}{8}$$

The solution set is $\left\{-\dfrac{1}{8}, \dfrac{3}{2}\right\}$.   ▲

C O M M O N   E R R O R

When taking the opposite of a quantity such as $3x - 2$, be sure to distribute the -1 across *both* terms,

$$-(3x - 2) = -3x + 2$$

Example 6   Solve the equation $|2x - 3| = |2x + 5|$.

Solution   We get the following two equations.

$$2x - 3 = 2x + 5 \qquad \text{or} \qquad 2x - 3 = -(2x + 5)$$
$$-3 = 5 \qquad\qquad\qquad 2x - 3 = -2x - 5$$
$$4x = -2$$
$$x = -\frac{1}{2}$$

The first equation yields the false statement $-3 = 5$ and therefore has no solution. We next check $x = -\dfrac{1}{2}$.

Check $x = -\dfrac{1}{2}$:

$$|2x - 3| = |2x + 5|$$
$$\left|2\left(-\frac{1}{2}\right) - 3\right| \stackrel{?}{=} \left|2\left(-\frac{1}{2}\right) + 5\right|$$
$$|-1 - 3| \stackrel{?}{=} |-1 + 5|$$
$$|-4| \stackrel{?}{=} |4|$$
$$4 = 4$$

The solution is $x = -\dfrac{1}{2}$.   ▲

Inequalities involving absolute value

In Example 1, we saw that the solution set of $|x| = 3$ is $\{-3, 3\}$. Consider now the inequality $|x| < 3$. A number would belong to the solution set if its distance from 0 is less than 3. This includes all numbers between -3 and 3 as shown in the figure below.

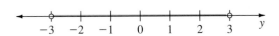

Therefore, the solution set of $|x| < 3$ is $\{x \mid -3 < x < 3\} = (-3, 3)$.

Example 7   Solve the inequality $|y| \le 4$.

Solution   This inequality will be satisfied by any number between -4 and 4, *inclusive*. Therefore, the solution set is $\{y \mid -4 \le y \le 4\} = [-4, 4]$. Its graph is shown below.

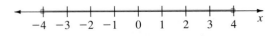

▲

The general form of the inequality in Example 7 is the following.

> ### R U L E
>
> Let $p > 0$. Then,
>
> $$|x| < p \qquad \text{is equivalent to} \qquad -p < x < p.$$

We can use this form to solve more complicated inequalities.

Example 8   Solve $|4t - 7| < 5$ and graph its solution set.

Solution   This inequality is of the form $|x| < p$ and is therefore equivalent to $-5 < 4t - 7 < 5$. Next, we solve this inequality by adding 7 to all sides, then dividing by 4.

$$-5 < 4t - 7 < 5$$

$$2 < 4t < 12$$

$$\frac{1}{2} < t < 3$$

The solution set is the interval $\left(\dfrac{1}{2}, 3\right)$, and its graph is shown below.

Example 9   Solve $|2c - 4| \le 3$ and graph its solution set.

Solution   This inequality means that the absolute value of $2c - 4$ is either less than 3 or equal to 3. Therefore, $2c - 4$ is between -3 and 3, inclusive. We write this as

$$-3 \le 2c - 4 \le 3$$

Solving this inequality,

$$1 \le 2c \le 7 \qquad \text{Add 4.}$$

$$\frac{1}{2} \le c \le \frac{7}{2} \qquad \text{Divide by 2.}$$

The solution set is the interval $\left[\dfrac{1}{2}, \dfrac{7}{2}\right]$ and its graph is shown below.

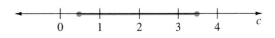

Next, we look at inequalities of the form $|x| > p$. Consider, for example, $|x| > 3$. A number would satisfy this inequality provided that its distance from 0 is greater than 3. Certainly, any number greater than 3 would work. Also numbers *less* than -3 have distances from 0 greater than 3. Therefore, $|x| > 3$ is satisfied by numbers either less than -3 or greater than 3. The solution set is graphed below.

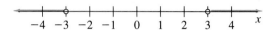

In general, we have the following result.

R U L E

Let $p > 0$. Then,

$$|x| > p \quad \text{is equivalent to} \quad x < -p \text{ or } x > p.$$

Example 10   Solve the inequality $|4z - 7| > 9$ and graph its solution set.

Solution   This inequality is of the form $|x| > p$. Therefore, $|4z - 4| > 9$ is equivalent to

$$4z - 7 < -9 \quad \text{or} \quad 4z - 7 > 9$$

We solve each inequality separately.

$$4z < -2 \quad \text{or} \quad 4z > 16 \qquad \text{Add 7.}$$

$$z < -\frac{1}{2} \quad \text{or} \quad z > 4 \qquad \text{Divide by 4.}$$

Therefore, the solution set is

$$\left\{ z \,\middle|\, z < -\frac{1}{2} \quad \text{or} \quad z > 4 \right\}$$

which can be written as

$$\left(-\infty, -\frac{1}{2}\right) \cup (4, +\infty)$$

This is graphed below.

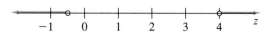

Example 11

Solve $|2r - 5| \geq 7$ and graph its solution set.

Solution

This inequality means that the absolute value of $2r - 5$ is greater than or equal to 7. Therefore,

$$2r - 5 \leq -7 \qquad \text{or} \qquad 2r - 5 \geq 7$$

Solving these inequalities,

$$2r \leq -2 \qquad \text{or} \qquad 2r \geq 12$$
$$r \leq -1 \qquad \text{or} \qquad r \geq 6$$

The solution set is

$$\{r \mid r \leq -1 \text{ or } r \geq 6\} = (-\infty, -1] \cup [6, +\infty)$$

This is graphed below.

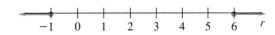

Keep in mind that the absolute value of any quantity is nonnegative. Therefore, some inequalities have special solution sets.

Example 12

Solve $|2t - 14| > -12$.

Solution

Since the absolute value of $2t - 14$ is automatically greater than or equal to zero, it is greater than -12 regardless of the value of t. Therefore, the solution set is R, the set of all real numbers. ▲

C A L C U L A T O R   C R O S S O V E R

Solve.

1. $|3.51x - 9.58| = 12.45$
2. $|35x + 532| - 721 = 129$
3. $|3.82x + 4.08| < 11.89$
4. $|5.91x - 3.87| \geq 15.94$

Answers   **1.** $\{-0.82, 6.28\}$ **2.** $\{-39.49, 9.09\}$
3. $\{x \mid -4.18 < x < 2.04\}$ **4.** $\{x \mid x \leq -2.04 \text{ or } x \geq 3.35\}$

L E A R N I N G   A D V A N T A G E   *Don't be afraid to talk to your instructor. Make an appointment with your teacher. Teachers are busy throughout the week, but if contacted early enough, they can arrange a meeting time. It doesn't take long to get most mathematics questions answered.*

E X E R C I S E S E T 2.6

For Exercises 1–28, solve each equation.

1. $|x| = 45$

2. $|y| = 6$

3. $|3z - 1| = 4$

4. $|2r + 1| = 2$

5. $|4 - 5s| = 2$

6. $|4 - 3v| = 6$

7. $|3a + 1| = -1$

8. $|9x + 3| = -2$

9. $|4x + 5| = 0$

10. $|2y - 3| = 0$

11. $|2x| - 5 = -2$

12. $|3s - 1| - 2 = 1$

13. $2 + |5r + 2| = 6$

14. $|4t + 1| - 3 = 2$

15. $-4 = 1 - |2x + 3|$

16. $3 = 1 + |4z - 2|$

17. $|2 + 3(y - 1)| = 5$

18. $|2(4k - 1) - 3| = 1$

19. $|2x + 1| = |3x - 2|$

20. $|3t + 4| = |1 - 2t|$

21. $|4 - 3w| = |3 - 5w|$

22. $|9 + 2x| = |6x + 4|$

23. $|x - 2| = |x - 3|$

24. $|2y - 1| = |2y - 5|$

25. $|4 - x| = |x - 4|$

26. $|7 - 3y| = |3y - 7|$

27. $|2 - 3(x - 1)| = |4 + 3(x + 1)|$

28. $|6 + 2(3z - 2)| = |5 - 3(1 - z)|$

For Exercises 29–42, solve each inequality, writing the answer using interval notation. Also graph the solution set.

29. $|y| < 2$

30. $|x| \le 5$

31. $|x - 4| \le 1$

32. $|z + 5| < 4$

33. $|2x + 3| \le 3$

34. $|4b + 7| < 5$

35. $|2x + 1| \ge 3$

36. $|3y - 2| > 1$

37. $|6x - 2| - 10 \le -6$

38. $|4a + 4| - 22 \le -10$

39. $|2 - x| > 1$

40. $|7 - 2z| > 5$

41. $|5v - 15| \le 0$

42. $|4t + 8| \le 0$

For Exercises 43–52, find the solution set. Write your answer using interval notation.

43. $|7x - 11| < 3$

44. $|12z + 4| \le 16$

45. $\left|2 - \dfrac{3}{2}x\right| \ge \dfrac{3}{4}$

46. $\left|1 - \dfrac{5}{6}t\right| > \dfrac{2}{3}$

47. $|x + 3| = x + 3$

48. $|r - 1| = r - 1$

49. $|z - 6| = 6 - z$

50. $|a + 1| = -a - 1$

51. $|3x - 10| > 0$

52. $|4z - 2| < 0$

For Exercises 53–60, write an inequality where x represents a number, then solve it.

53. The absolute value of a number is no more than five.

54. The absolute value of a number is greater than one.

55. Twice a number is added to four. The absolute value of this sum exceeds two.

56. The absolute value of the difference of a number and three is less than seven.

57. The opposite of six is added to the absolute value of the difference of a number and one. The result is more than -2.

58. Five is subtracted from the absolute value of the sum of twice a number and two. The result is larger than -1.

59. The absolute value of the difference of a number and three is the same as the absolute value of five times the number.

60. The absolute value of the sum of twice a number and one is the same as the absolute value of the sum of four times the number and three.

REVIEW PROBLEMS

The following exercises review parts of Section 1.4. Doing these problems will help prepare you for the next section.

Write with exponents.

61. $a \cdot a \cdot a$

62. $x \cdot x \cdot x \cdot x$

63. $3a \cdot a \cdot b \cdot b \cdot b \cdot c$

Write without exponents.

64. s^5

65. $5x^3y^2$

66. a^1

67. $\dfrac{2}{b^3}$

Evaluate.

68. 3^3

69. $\left(\dfrac{2}{3}\right)^2$

70. $(-4)^2$

71. -4^2

72. $(4.3028)^1$

73. $2^2 \cdot 2^3$

74. 2^5

75. $\dfrac{2^5}{2^2}$

76. 2^3

ENRICHMENT EXERCISES

1. Find values of a and b to show that in general

$$|a + b| \neq |a| + |b|$$

For Exercises 2–6, solve each inequality.

2. $3 < |x| < 5$

3. $1 \leq |x - 3| < 2$

4. $0 < |3 - 2x| \leq 7$

5. $-2 + 6|x| < 3 - 4|x|$

6. $3|2x + 5| + 4 \geq |2x + 5| + 8$

7. Solve for x in terms of the other variables. Assume that $\delta > 0$. $0 < |x - a| < \delta$. The symbol δ is pronounced "delta."

Answers to Enrichment Exercises are on page A.8.

Summary and review

Examples

Properties of equality (2.1)

To **solve** an equation means to find all values of the variable that make the equation a true statement. These values comprise the **solution set** of the equation. Two equations are **equivalent** if they both have the same solution set.

The Basic Goal When Solving an Equation

Through a sequence of steps, rewrite the original equation to ultimately obtain an equivalent equation where the variable is isolated to one side.

To achieve this goal we have two properties of equality.

The Addition-Subtraction Property of Equality

$x + 4 = 9$,
$x + 4 - 4 = 9 - 4$,
$x = 5$

Let A, B, and C represent algebraic expressions. If $A = B$, then

$$A + C = B + C \qquad \text{and} \qquad A - C = B - C$$

The Multiplication-Division Property of Equality

$3x = 5$,
$\dfrac{1}{3}(3x) = \dfrac{1}{3}(5)$
$x = \dfrac{5}{3}$

Let A, B, and C represent algebraic expressions with $C \neq 0$. If $A = B$, then

$$AC = BC \qquad \text{and} \qquad \frac{A}{C} = \frac{B}{C}$$

Solving linear equations (2.2)

A **linear equation** is an equation that can be put into the standard form $ax + b = 0$. Our goal is to solve linear equations.

A Strategy for Solving Linear Equations

$4(x - 2) + 1 = 7x$
$4x - 8 + 1 = 7x$
    Distributive property.
$4x - 7 = 7x$
$4x - 7x = 7$
    Add 7 and subtract 7x.
$-3x = 7$
$x = -\dfrac{7}{3}$

Step 1   Combine like terms on each side, if necessary, to simplify the equation.

Step 2   Use the addition-subtraction property of equality, if necessary, to move all terms containing a variable to one side and all constant terms to the other side.

Step 3   Use the multiplication-division property of equality, if necessary, to make the coefficient of the variable term equal to one.

Step 4   **(Optional)** Check your answer, by substituting it into the original equation.

Solving word problems (2.3)

There is no single method for solving word problems. However, the following strategy will help as a guide.

Examples

A Strategy for Solving Word Problems

1. **Read** the problem carefully. Take note of what is being asked and what information is given.
2. **Plan** a course of action. Represent the unknown number by a letter. If there is more than one unknown, represent one of them by a letter and express the others in terms of the letter.
3. **Create** an equation from the given information.
4. **Solve** this equation.
5. **Check** your solution in the original equation.
6. **Answer** the original question. Read the problem again to make sure you answered the question.

Formulas and applications (2.4)

Solve for w: $P = 2w + 2l$.

$P - 2l = 2w$

$w = \dfrac{P}{2} - l$

A **formula** is an equation that contains two or more variables. For example, $A = lw$ is the formula for the area A of a rectangle of length l and width w. On the inside front cover, there are some common formulas from geometry stated.

The perimeter of a rectangle is 44 cm and the width is 2 cm less than the length. Find the dimensions of the rectangle. Let x be the length, then $x - 2$ is the width. The formula for perimeter is $P = 2w + 2l$. Therefore,

$44 = 2x + 2(x - 2)$
$44 = 4x - 4$
$48 = 4x$
so, $x = 12$ and $x - 2 = 10$

The width is 10 cm and the length is 12 cm.

Solving inequalities (2.5)

A **linear inequality** in one variable is any inequality that can be put in the form $ax + b < c$, where the less than symbol can be replaced by $\leq$, $>$, or $\geq$. Examples of linear inequalities are $3 - 4x < -1$ and $4t + 6 \geq 0$. To **solve** a linear inequality means to find all numbers that, when replacing the variable, make the inequality a true statement.

The Addition Property of Inequality

Subtracting $3x$ from both sides of $4x < 3x - 1$ gives $x < -1$.

Let A, B, and C represent algebraic expressions. If $A < B$, then

$$A + C < B + C$$

The Multiplication Property of Inequality

Dividing both sides by -2, $-2x < 6$ becomes $x > -3$.

Let A, B, and C represent algebraic expressions. Suppose $A < B$.

1. If $C > 0$, then $CA < CB$.
2. If $C < 0$, then $CA > CB$.

Examples

Solve:
$4x - 3(1 - 2x) > 12(x - 1)$.

$4x - 3 + 6x > 12x - 12$
$10x - 3 > 12x - 12$
$-2x > -9$ Add 3 and subtract $12x$.

$x < \dfrac{9}{2}$ Divide by -2.

The solution set $\left(-\infty, \dfrac{9}{2}\right)$ is graphed below.

Solve: $|4 - 5t| = 6$.
Either
$4 - 5t = -6$ or $4 - 5t = 6$
$-5t = -10$ or $-5t = 2$

$t = 2$ or $t = -\dfrac{2}{5}$

Solve: $|2x - 3| < 3$.

$|2x - 3| < 3$
$-3 < 2x - 3 < 3$
$0 < 2x < 6$
$0 < x < 3$

The graph of the solution set
$\{x \mid 0 < x < 3\} = (0, 3)$ is
shown below.

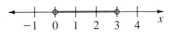

Solve: $|2y + 1| > 5$.

$2y + 1 < -5$ or $2y + 1 > 5$
$2y < -6$ or $2y > 4$
$y < -3$ or $y > 2$

The solution set is $(-\infty, -3) \cup (2, +\infty)$ and is shown below.

A Strategy for Solving Linear Inequalities

Step 1   Combine like terms on each side.

Step 2   Use the addition-subtraction property of inequality, if necessary, to move all terms containing the variable to one side and all constant terms to the other side.

Step 3   Use the multiplication property of inequality, if necessary, to make the coefficient on the variable term equal to one.

Equations and inequalities with absolute value (2.6)

If $p > 0$, then

1. $|x| = p$ is equivalent to either $x = p$ or $x = -p$.
2. $|x| < p$ is equivalent to $-p < x < p$.
3. $|x| > p$ is equivalent to $x < -p$ or $x > p$.

CHAPTER 2   REVIEW EXERCISE SET

Sections 2.1 and 2.2

For Exercises 1–8, solve each equation. Indicate those equations that are identities or contradictions.

1. $\dfrac{3}{2}x = -9$

2. $3r = -5 + 2r$

3. $4t - 3(1 - t) = 4$

4. $5(3z - 2) + 2 = 4(z - 5)$

5. $8x - 12 = 4(2x - 3)$

6. $5 - 6z - 2 = 2(1 - 3z)$

7. $\dfrac{3c}{2} - \dfrac{2c}{3} = 5$

8. $3(2 + 3a) - 2(1 - 5a) = 9a - 6$

For Exercises 9–14, write an equation for each sentence. Solve the equation.

9. A number divided by three is equal to $\dfrac{1}{5}$.

10. The quotient of three times a number and two is -6.

11. Three plus a number is the sum of 12 and four times the number.

12. The sum of three times a number and one is four less than six times the number.

13. The sum of three times a number and 10 is -2.

14. Twice a number, increased by $\dfrac{3}{2}$, is -1.

Sections 2.3 and 2.4

15. Lynne receives a 5% raise in salary. If her old salary was $25,000, what is her new salary?

16. The larger of two numbers is one more than the smaller. If the smaller is $-\dfrac{1}{3}$ times the larger, find the two numbers.

17. Negative one times the sum of three consecutive odd integers is 105. Find the three integers.

18. Doris is planning to cut an 8-foot board into two pieces. The length of one piece must be one and a half feet more than the other piece. Find the lengths of the two pieces.

19. Creighton plans to invest $5,000, part at 12% and the other part at 8%. He wants the total interest for one year to be $560. How much should he invest at each interest rate?

20. One angle of a triangle has a measure of 47 degrees. The measure of another (the second) angle is 9 degrees less than the measure of the third angle. Find the measures of the two unknown angles.

21. Solve $\dfrac{1}{a} + \dfrac{1}{b} = \dfrac{1}{f}$ for a.

22. Solve $I = Prt$ for P.

23. Solve $y = -4x + 7$ for x.

24. Solve $y = \dfrac{2x - 1}{4x + 3}$ for x.

Section 2.5

For Exercises 25–28, graph the followings sets.

25. $\{x \mid x > -1\}$

26. $(-\infty, 4]$

27. $(-2.5, 1.5]$

28. $(-\infty, 0) \cup (2, +\infty)$

For Exercises 29–32, solve the inequality, writing the answer using interval notation. Also graph the solution set.

29. $x + 4 < 7$

30. $2 - 4x < 2x - 10$

31. $5(y - 2) \geq 2(y - 8)$

32. $\dfrac{2}{3} < x + \dfrac{2}{3} \leq 1\dfrac{1}{3}$

For Exercises 33–35, write and solve an inequality from the given information.

33. The sum of a number and negative five is no more than -2.

34. The difference of three and a number does not exceed seven.

35. The sum of three times a number and eight is more than negative one but does not exceed four.

Section 2.6

For Exercises 36–44, solve the equation or inequality.

36. $|x| = 48$

37. $|5t - 3| = 12$

38. $|2 + 4z| + 4 = 1$

39. $|3w - 2| = |4w + 6|$

40. $|y| < 5$

41. $|3x + 9| \leq 3$

42. $|6y + 5| > 3$

43. $|5a - 4| \geq 7$

44. $\left| \dfrac{2z - 3}{5} + 9 \right| + 12 \leq 12$

CHAPTER 2 TEST

Sections 2.1 and 2.2

For Problems 1–4, solve each equation. Indicate those equations that are identities or contradictions.

1. $-3x = 12$

2. $4y - 2 = 3y + 5$

3. $3(2t - 1) - 2 = 3t + 7$

4. $5 + 8x = 2(1 + 4x) + 3$

5. Write an equation for the sentence, then solve the equation.

The sum of 4 and twice a number is 10.

Sections 2.3 and 2.4

6. The value of a painting increased by 25% from last year. If the value last year was $30,000, what is the current value of the painting?

7. The sum of four consecutive even integers is -4. Find the four integers.

8. One angle of a triangle has a measure of 75 degrees. The measure of another (the second) angle is twice the measure of the third angle. Find the measures of the two unknown angles.

9. Solve $I = Prt$ for r.

10. Solve $y = 3x - 12$ for x.

11. Solve $y = \dfrac{2}{1 - 5x}$ for x.

Section 2.5

For Problems 12–14, graph the given set on a number line.

12. $\{x \mid x < 4\}$

13. $[2, +\infty)$

14. $(-\infty, -1] \cup (1, +\infty)$

For Problems 15–18, solve the inequality, writing the answer using interval notation. Also graph the solution set.

15. $x - 3 > -2$

16. $3 - 2x \geq 3x + 8$

17. $-3 \leq 1 + 2x < 5$

18. $3(1 - 3y) < -2(5y - 2)$

For Problems 19 and 20, write and solve an inequality from the given information.

19. The difference of a number and 3 is larger than -4.

20. The sum of three times a number and -4 is more than 1 but does not exceed 3.

Section 2.6

For Problems 21–24, solve the equation or inequality.

21. $|x| = 6$

22. $|4y + 2| = 6$

23. $|x| \leq 9$

24. $|2x - 1| > 1$

Polynomials and Factoring

C H A P T E R

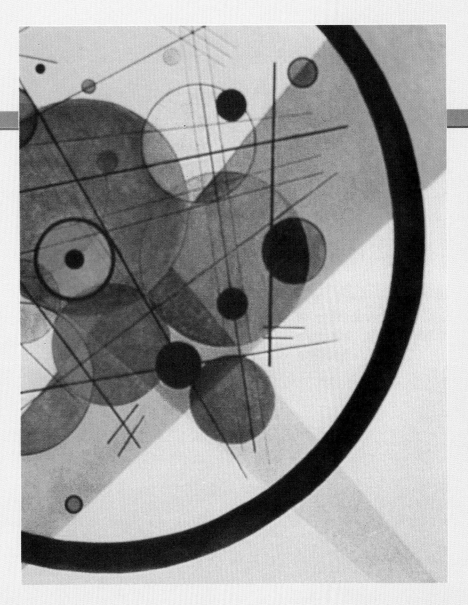

Overview

In this chapter we become more deeply involved with algebra and its applications. Our goal is to develop skills in working with algebraic expressions called polynomials. These skills in turn will enable us to solve certain quadratic equations and inequalities. These equations and inequalities themselves will be used as a tool to solve word problems.

3.1

Properties of Exponents; Scientific Notation

OBJECTIVES

▶ *To simplify expressions using multiplication and division properties of exponents*

▶ *To use scientific notation to simplify numerical expressions*

An expression such as $a^3 \cdot a^2$ can be simplified in the following way. Since a^3 means $a \cdot a \cdot a$ and a^2 means $a \cdot a$, then

$$a^3 \cdot a^2 = \underbrace{(a \cdot a \cdot a)(a \cdot a)}_{3 \text{ factors } 2 \text{ factors}} = \underbrace{(a \cdot a \cdot a \cdot a \cdot a)}_{3 + 2 \text{ factors}} \text{ or } a^5$$

Therefore,

$$a^3 \cdot a^2 = a^{3+2} = a^5$$

In general, if a is a number, m and n are positive integers, then

$$a^m \cdot a^n = \underbrace{(a \cdot a \cdot a \cdots a)}_{m \text{ factors}} \underbrace{(a \cdot a \cdot a \cdots a)}_{+ \quad n \text{ factors}}$$

$$= a^{m+n}$$

That is, to multiply powers with the same base, use the common base and add the exponents. This is stated as Exponent Rule 1.

R U L E

Exponent rule 1

If a is a number and m and n are positive integers, then

$$a^m \cdot a^n = a^{m+n}$$

Example 1   Simplify each of the following.

(a) $x^4 \cdot x^7$ **(b)** $r^6 r^3 r^{12}$ **(c)** $st^5 s^2 t^9$

Solution   **(a)** $x^4 \cdot x^7 = x^{4+7} = x^{11}$

(b) To simplify the product of three powers with the same base, add all three exponents.

$$r^6 r^3 r^{12} = r^{6+3+12} = r^{21}$$

(c) First, group together powers with the same base, then combine each group using $a^m a^n = a^{m+n}$.

$$st^5 s^2 t^9 = (ss^2)(t^5 t^9) = s^{1+2} t^{5+9} = s^3 t^{14} \qquad \blacktriangle$$

C O M M O N   E R R O R

Be careful not to mistake $a^6 \cdot a$ as a^6. Recall that a means a^1. Therefore,

$$a^6 \cdot a = a^{6+1} = a^7$$

If the rule $a^m a^n = a^{m+n}$ is to hold for $n = 0$, then

$$a^m a^0 = a^{m+0} = a^m$$

If $a \neq 0$, we may divide both sides of $a^m a^0 = a^m$ by a^m.

$$a^0 = \frac{a^m}{a^m} = 1$$

Therefore, for $a \neq 0$, we define

$$a^0 = 1$$

For example, $\left(\dfrac{1}{2}\right)^0 = 1$, $(3a)^0 = 1$, $(xy^3)^0 = 1$. We are assuming that a, x, and y are each nonzero.

The next rule is used to simplify the mth power of a number that is itself raised to a power. For example, the expression $(5^2)^3$ means $5^2 \cdot 5^2 \cdot 5^2$. But,

$$5^2 \cdot 5^2 \cdot 5^2 = 5^{2+2+2} = 5^6$$

Therefore,

$$(5^2)^3 = 5^{2 \cdot 3}$$

In general, to simplify $(a^m)^n$, use the same base a with the exponent that is the product of m and n.

R U L E

Exponent rule 2

$$(a^m)^n = a^{m \cdot n}$$

Example 2   Simplify each of the following.

(a) $(x^5)^7$ (b) $(a^2)^6$ (c) $(x^3)^5(y^7)^2$

Solution   (a) $(x^5)^7 = x^{5 \cdot 7}$ (b) $(a^2)^6 = a^{2 \cdot 6}$ (c) $(x^3)^5(y^7)^2 = x^{3 \cdot 5}y^{7 \cdot 2}$

$\qquad\qquad = x^{35}$ $\qquad = a^{12}$ $\qquad\qquad = x^{15}y^{14}$ ▲

Consider the expression $(ab)^3$. This may be written as $(ab)(ab)(ab)$. But, by regrouping the terms,

$$(ab)(ab)(ab) = aaabbb = a^3b^3$$

The following rule is suggested by this.

R U L E

Exponent rule 3

$$(ab)^n = a^nb^n$$

That is, $(ab)^n$ is the same as a^n times b^n.

Example 3   Simplify each of the following.

(a) $(xy)^5$ (b) $(7t^8)^2$

(c) $(abd)^5$ (d) $(-z^6)^3$

Solution   **(a)** $(xy)^5 = x^5 y^5$

    (b) $(7t^8)^2 = 7^2 (t^8)^2$ $(ab)^n = a^n b^n.$

              $= 49 t^{8 \cdot 2}$ $(a^m)^n = a^{mn}.$

              $= 49 t^{16}$

(c) To simplify $(abd)^5$, raise each variable to the fifth power. To justify this, rewrite $(abd)^5$ as $[(ab)d]^5$ and raise ab and d to the fifth power.

$$(abd)^5 = [(ab)d]^5 = (ab)^5 d^5 = a^5 b^5 d^5$$

(d) $(-z^6)^3 = [(-1)z^6]^3 = (-1)^3 (z^6)^3 = (-1)z^{6 \cdot 3} = -z^{18}.$ ▲

We want the rule $a^m a^n = a^{m+n}$ to hold for negative integers as well as positive integers. If we set $m = -n$, then

$$a^{-n} a^n = a^{-n+n} = a^0 = 1$$

Dividing both sides by a^n gives the reason for the following definition.

$$a^{-n} = \frac{1}{a^n}, \qquad \text{where } a \neq 0.$$

For example, $2^{-5} = \dfrac{1}{2^5} = \dfrac{1}{32}$ and $14^{-1} = \dfrac{1}{14^1} = \dfrac{1}{14}.$

C O M M O N   E R R O R

The term 2^{-3} is *not* -8. Using the definition of negative exponent, we have $2^{-3} = \dfrac{1}{2^3} = \dfrac{1}{8}$. A negative exponent does not make the quantity negative.

Example 4   Simplify and write each expression without negative exponents.

    (a) $(xy)^{-12}$ **(b)** $(st^2)^{-9}$ **(c)** $(2m^3 n)^{-3}$

Solution   **(a)** $(xy)^{-12} = \dfrac{1}{(xy)^{12}}$ **(b)** $(st^2)^{-9} = \dfrac{1}{(st^2)^9}$ **(c)** $(2m^3 n)^{-3} = \dfrac{1}{(2m^3 n)^3}$

                  $= \dfrac{1}{x^{12} y^{12}}$ $= \dfrac{1}{s^9 t^{18}}$ $= \dfrac{1}{2^3 (m^3)^3 n^3}$

                                                                               $= \dfrac{1}{8 m^9 n^3}$

                                                                                                ▲

We next develop a rule for the division of a^m by a^n. For example, consider the quotient $\dfrac{a^5}{a^2}$. This quotient can be written as a product.

$$\frac{a^5}{a^2} = a^5\left(\frac{1}{a^2}\right) = a^5 a^{-2} = a^{5-2}$$

So,

$$\frac{a^5}{a^2} = a^{5-2} \text{ or } a^3$$

This answer is correct, since

$$\frac{a^5}{a^2} = \frac{a \cdot a \cdot a \cdot a \cdot a}{a \cdot a}$$

$$= a \cdot a \cdot a$$

$$= a^3$$

Next, consider the quotient $\dfrac{a^3}{a^5}$. Notice that the exponent in the denominator (5) is larger than the exponent in the numerator (3). Nonetheless, we can write this quotient as a product.

$$\frac{a^3}{a^5} = a^3\left(\frac{1}{a^5}\right) = a^3 a^{-5} = a^{3-5}$$

So, $\dfrac{a^3}{a^5} = a^{3-5}$ or a^{-2}, which can be written as $\dfrac{1}{a^2}$. This answer is correct, since

$$\frac{a^3}{a^5} = \frac{a \cdot a \cdot a}{a \cdot a \cdot a \cdot a \cdot a}$$

$$= \frac{1}{a \cdot a}$$

$$= \frac{1}{a^2}$$

A division rule for powers is illustrated by these examples.

R U L E

Exponent rule 4

$$\frac{a^m}{a^n} = a^{m-n}$$

where $a \neq 0$, m and n are integers.

That is, a^m divided by a^n can be simplified by replacing the quotient by the common base a raised to the $m - n$ power.

For example,

$$\frac{5^{23}}{5^{21}} = 5^{23-21} = 5^2 = 25$$

Example 5   Simplify and write your answer without negative exponents.

(a) $\dfrac{x^{12}}{x^{42}}$ (b) $\dfrac{2b^3}{b^{-2}}$ (c) $\dfrac{4^{-2}m^5}{m^{-2}}$

Solution   (a) $\dfrac{x^{12}}{x^{42}} = x^{12-42}$ (b) $\dfrac{2b^3}{b^{-2}} = 2\left(\dfrac{b^3}{b^{-2}}\right)$ (c) $\dfrac{4^{-2}m^5}{m^{-2}} = 4^{-2}\left(\dfrac{m^5}{m^{-2}}\right)$

$= x^{-30}$ $= 2b^{3-(-2)}$ $= \left(\dfrac{1}{4^2}\right)m^{5-(-2)}$

$= \dfrac{1}{x^{30}}$ $= 2b^5$ $= \dfrac{1}{16}m^{5+2}$

$= \dfrac{1}{16}m^7$ ▲

COMMON ERROR

The quotient $\dfrac{a}{a^7}$ is *not* $\dfrac{1}{a^7}$. If we realize that a means a^1, then

$$\frac{a}{a^7} = \frac{a^1}{a^7} = a^{1-7} = a^{-6} = \frac{1}{a^6}$$

Consider the expression $\left(\dfrac{a}{b}\right)^3$. It means $\dfrac{a}{b} \cdot \dfrac{a}{b} \cdot \dfrac{a}{b}$. Therefore,

$$\left(\frac{a}{b}\right)^3 = \frac{a}{b} \cdot \frac{a}{b} \cdot \frac{a}{b} = \frac{a \cdot a \cdot a}{b \cdot b \cdot b} = \frac{a^3}{b^3}$$

In general, if n is a positive integer, the following rule is true.

RULE

Exponent rule 5

$$\left(\frac{a}{b}\right)^n = \frac{a^n}{b^n}, \qquad b \neq 0$$

Next, consider a negative exponent like $n = -2$,

$$\left(\frac{a}{b}\right)^{-2} = \frac{1}{\left(\frac{a}{b}\right)^2}$$

$$= \frac{1}{\frac{a^2}{b^2}}$$

$$= 1 \div \frac{a^2}{b^2}$$

$$= 1 \cdot \frac{b^2}{a^2}$$

$$= \frac{b^2}{a^2}$$

$$= \left(\frac{b}{a}\right)^2$$

Therefore,

$$\left(\frac{a}{b}\right)^{-2} = \left(\frac{b}{a}\right)^2$$

This property holds true for any integer n. We state this property as our next exponent rule.

R U L E

Exponent rule 6

$$\left(\frac{a}{b}\right)^{-n} = \left(\frac{b}{a}\right)^n = \frac{b^n}{a^n}, \qquad a \neq 0, b \neq 0$$

That is, to simplify $\left(\frac{a}{b}\right)^{-n}$, form the reciprocal $\frac{b}{a}$ of $\frac{a}{b}$ and raise it to the nth power.

Example 6   Simplify and write the answer without negative exponents.

(a) $(x^2)^3(y^{-4}z)^5$ **(b)** $\left(\frac{s^2}{t^3}\right)^{-6}$ **(c)** $\frac{(a^{-5})^{-2}}{a^7}$

Solution   **(a)** $(x^2)^3(y^{-4}z)^5 = x^{2 \cdot 3}y^{(-4)5}z^5$ $(a^m)^n = a^{m \cdot n}$ and $(ab)^n = a^n b^n$.

$\qquad\qquad\quad = x^6 y^{-20} z^5$

$\qquad\qquad\quad = \dfrac{x^6 z^5}{y^{20}}$ $a^{-n} = \dfrac{1}{a^n}$

(b) $\left(\dfrac{s^2}{t^3}\right)^{-6} = \left(\dfrac{t^3}{s^2}\right)^6$ Use $\left(\dfrac{a}{b}\right)^{-n} = \left(\dfrac{b}{a}\right)^n$.

$\qquad\qquad\quad = \dfrac{t^{3 \cdot 6}}{s^{2 \cdot 6}}$ Use $\left(\dfrac{a}{b}\right)^n = \dfrac{a^n}{b^n}$.

$\qquad\qquad\quad = \dfrac{t^{18}}{s^{12}}$

(c) $\dfrac{(a^{-5})^{-2}}{a^7} = \dfrac{a^{(-5)(-2)}}{a^7}$ Use $(a^m)^n = a^{mn}$.

$\qquad\qquad\quad = \dfrac{a^{10}}{a^7}$

$\qquad\qquad\quad = a^{10-7}$

$\qquad\qquad\quad = a^3$ ▲

CALCULATOR   CROSSOVER

Simplify. Then, evaluate for the given values of the variables.

1. $(2ab^2)^2$ for $a = 2.1$ and $b = 0.6$

2. $(s^2 t)^2(st^2)$ for $s = 2$ and $t = 1.5$

3. $(-3p^3 q)^{-3}(p^2 q)^6$ for $p = 3$ and $q = 0.3$

4. $\left(\dfrac{2xy^2}{x^2 y}\right)^{-2}$ for $x = 3.2$ and $y = 4$

Answers   **1.** $4a^2 b^4$, 2.286144   **2.** $s^5 t^4$, 162   **3.** $-\dfrac{1}{27}p^3 q^3$, -0.027

4. $\dfrac{x^2}{4y^2}$, 0.16

Scientific Notation

Numbers such as 12,000, 0.0003, and 529,210,000 are said to be written in **decimal notation.** A number is written in **scientific notation** if it is the product of a number between 1 and 10 and a power of 10. For example, in scientific notation, the numbers 12,000, 0.0003, and 529,210,000 are the following.

$$12{,}000 = 1.2 \times 10^4, \quad 0.0003 = 3 \times 10^{-4}, \quad \text{and} \quad 529{,}210{,}000 = 5.2921 \times 10^8$$

We use "$\times$" to symbolize multiplication, since "$\cdot$" could be mistaken for a decimal point.

To express a number in scientific notation use the following method.

STRATEGY

To express a number in scientific notation

1. Move the decimal point either to the left or to the right to make a number between 1 and 10.
2. Count the number of decimal places moved from where the original decimal point occurred, and raise ten to either this number or the negative of this number. We determine if this exponent is positive or negative in the following way:
   (a) If we moved the decimal point to the left, the exponent is positive.
   (b) If we moved the decimal point to the right, the exponent is negative.
3. Multiply the number in Step 1 by this power of 10.

Example 7

(a) Express 43,921 in scientific notation.
(b) Express 0.00086 in scientific notation.
(c) Express 4.9221×10^{-2} in decimal notation.

Solution

(a) First, locate the decimal point. It is understood to be to the right of the 1. This decimal point in 43,921 must be moved four places to the left to make a number between 1 and 10, and, therefore, the exponent of 10 is 4. Here are the details:

Place the decimal point here.

$$43{,}921 = 4.3921 \times 10^4$$

Move 4 places to the *left*.

The exponent indicates the number of places to the *left* that we moved the decimal point.

(b) The decimal point must be moved four places to the right to make a number between 1 and 10 and therefore the exponent of 10 is -4. Here is the pattern:

Place the decimal point here.

$$0.00086 = 8.6 \times 10^{-4}$$

Move 4 places to the *right*.

The exponent gives the number of places to the *right* that we moved the decimal point.

(c) To write 4.9221×10^{-2} in decimal notation, observe that the exponent is -2. Therefore, move the decimal point two places to the left:

$$4.9221 \times 10^{-2} = 0.049221$$ ▲

Some more examples of numbers in decimal notation and written again in scientific notation are given in the following table. Verify that the two numbers on each line are equal.

The number in decimal notation	The same number in scientific notation
913,000	9.13×10^5
80,510	8.051×10^4
1,700	1.7×10^3
400	4×10^2
16	1.6×10^1 or 1.6×10
5.7	5.7×10^0 or 5.7
0.3002	3.002×10^{-1}
0.07	7×10^{-2}
0.0018	1.8×10^{-3}
0.0009	9×10^{-4}
0.0000162	1.62×10^{-5}

N O T E   *Any number between 1 and 10 that is expressed as a decimal is already in scientific notation. For example, the number 5.7 is the same in both columns of the previous table.*

Example 8   Write the given number in scientific notation.

(a) One astronomical unit (au) is the average distance of the earth from the sun, approximately 93,000,000 miles.

(b) The probability of winning the Pennsylvania lottery is 0.00000009.

Solution   (a) 9.3×10^7 (b) 9×10^{-8}

Example 9   Evaluate $(2.6 \times 10^5)(4 \times 10^{-2})$. Express the answer in scientific notation and decimal notation.

Solution   Regroup the terms so that the last two factors are the powers of 10.

$$(2.6 \times 10^5)(4 \times 10^{-2}) = (2.6)(4) \times (10^5)(10^{-2})$$
$$= 10.4 \times 10^{5-2}$$
$$= 10.4 \times 10^3$$

Therefore, the answer is

$$1.04 \times 10^4 \qquad \text{In scientific notation.}$$

or

$$10,400 \qquad \text{In decimal notation.} \qquad \blacktriangle$$

Example 10   Use scientific notation to find the value of

$$\frac{(0.004)(90,000)}{0.0002}$$

Solution   We convert each of the three numbers into scientific notation, then use rules of exponents to simplify.

$$\frac{(0.004)(90,000)}{0.0002} = \frac{(4 \times 10^{-3})(9 \times 10^4)}{2 \times 10^{-4}}$$

$$= \frac{4 \cdot 9}{2} \times \frac{10^{-3} 10^4}{10^{-4}}$$

$$= 18 \times 10^{-3+4-(-4)} \qquad \text{Exponent rules.}$$

$$= 18 \times 10^5$$

$$= 1,800,000 \qquad \blacktriangle$$

CALCULATOR   CROSSOVER

Some calculators can be used to perform products and quotients of numbers in scientific notation. Texas Instruments calculators have a key labeled $\boxed{\text{EE}}$. For example, to enter the number 5.926×10^7, first enter the number 5.926, then press $\boxed{\text{EE}}$, then enter 7. The display will read $\boxed{5.926\ 07}$. If your calculator does not have an EE key, check the manual. The key may be labeled $\boxed{\text{SCI}}$ or $\boxed{\text{EXP}}$.

Perform the indicated operations.

1. $(5.9821 \times 10^4)(1.9826 \times 10^7)$

2. $\dfrac{9.82 \times 10^3}{7.92 \times 10^5}$

3. $\dfrac{493,720,000}{981,023,700}$

4. $\dfrac{5.819 \times 10^9}{3.881 \times 10^{-5}}$

5. A circular loop of wire has a radius of 2.3×10^{-3} meters. Find the area of the loop and write your answer in scientific notation.

Answers   **1.** 1.186×10^{12} **2.** 1.2399×10^{-2} **3.** 5.0327×10^{-1}
4. 1.4994×10^{14} **5.** 1.66×10^{-5} square meters

L E A R N I N G   A D V A N T A G E   *Be sure to always do the homework because you* **learn mathematics by doing it.** *The main purposes of homework are (1) to help reinforce what you learned in class, (2) to help identify those parts of the lesson that you do not understand, and (3) to increase the number of correctly solved problems to which you can refer when reviewing for a test.*

E X E R C I S E   S E T   3.1

For Exercises 1–28, simplify each expression. Assume that the variables are nonzero.

1. $x^2 \cdot x^7$

2. $b^3 \cdot b^4$

3. $y^{12}y^4zz^2$

4. $aa^3b^5b^3$

5. $a^3x^3a^4x^4$

6. $b^3z^2b^3z^9$

7. r^0

8. 3^0

9. $(x^2)^3$

10. $(a^5)^{10}$

11. $(b^4)^4$

12. $(z^3)^5$

13. $(5a)^2$

14. $(2b)^4$

15. $(x^2)^4x^3$

16. $a^7(a^3)^6$

17. $(c^4)^{10}(z^3)^9$

18. $(v^3)^2(r^3)^2$

19. $(-x)^5$

20. $(-t)^4$

21. $-(-a)^4$

22. $-(-r)^5$

23. $(3w^3)^3$

24. $(2v^3)^4$

25. $(abc)^6$

26. $(4x^2y)^2$

27. $\left(\dfrac{2}{3}a^3z^2\right)^2$

28. $\left(\dfrac{3}{4}bc^6\right)^2$

For Exercises 29–34, write the expression without negative exponents.

29. a^{-20}

30. x^{-15}

31. $(uv)^{-1}$

32. $(2x)^{-2}$

33. $(p^2q)^{-3}$

34. $(2ac^2)^{-1}$

For Exercises 35–66, simplify each expression, and write your answer without negative exponents.

35. $\dfrac{a^5}{2a^3}$

36. $\dfrac{4y^3}{y^2}$

37. $\dfrac{s}{s^2}$

38. $\dfrac{3p}{p^3}$

39. $\dfrac{18n^7}{6n^7}$

40. $\dfrac{21z^{10}}{3z^{10}}$

41. $\dfrac{m^3}{m^{-3}}$

42. $\dfrac{x^5}{x^{-6}}$

43. $\dfrac{x^3}{3^{-2}}$

44. $\dfrac{st^2}{12^{-1}}$

45. $\dfrac{(az)^4}{(az)^5}$

46. $\dfrac{(b^3c^2)^{10}}{(b^3c^2)^{11}}$

47. $\left(\dfrac{2}{7}\right)^{-2}$

48. $\left(\dfrac{3}{2}\right)^{-3}$

49. $\left(\dfrac{1}{4}\right)^{-1}$

50. $\left(\dfrac{1}{9}\right)^{-1}$

51. $\left(\dfrac{a}{b}\right)^{-5}$

52. $\left(\dfrac{c}{n}\right)^{-7}$

53. $\left(\dfrac{2x}{z}\right)^{-3}$

54. $\left(\dfrac{3y}{u}\right)^{-2}$

55. $(x^{-2})^{-3}$

56. $(a^{-3})^{-4}$

57. $(5^{-3})^{-1}$

58. $(4^{-1})^{-2}$

59. $(y^3)^2(z^{-2}x)^4$

60. $(rv^3)^5(r^2v^{-1})^2$

61. $(t^{-2}w)^{-3}(tw^4)^{-1}$

62. $(m^2n^{-2})^{-3}mn^2$

63. $\left(\dfrac{s}{t^2}\right)^{-2}\left(\dfrac{t}{s^2}\right)^{-2}$

64. $\left(\dfrac{u^2}{v}\right)^{-3}\left(\dfrac{v^{-4}}{u}\right)^{-1}$

65. $(ac^2)^{-5}\left(\dfrac{1}{ac^2}\right)^{-5}$

66. $(b^2m)^{-1}\left(\dfrac{1}{b^2m}\right)^{-1}$

For Exercises 67–74, express each number in scientific notation.

67. 0.000391   **68.** 0.000000283   **69.** 91,000,000   **70.** 765,000,000

71. 935,920   **72.** 29,105   **73.** 0.38×10^4 **74.** $9,154 \times 10^{-6}$

For Exercises 75–78, express each number in decimal notation.

75. 4.9×10^3 **76.** 9.23×10^5 **77.** 4×10^{-4} **78.** 8.003×10^{-6}

For Exercises 79–86, write each number in scientific notation.

79. A second is defined as the duration of 9,192,631,770 cycles of radiation associated with the cesium atom.

80. The moon revolves around the earth in a nearly circular orbit with a radius of 380,000 km.

81. The mass of the earth is approximately

$$620,000,000,000,000,000,000,000 \text{ kg}$$

82. A satellite is in a circular orbit about the earth with a radius of 6,829,000 meters.

83. The human thyroid contains about 0.0002822 ounce of iodine.

84. In four months a pair of houseflies could produce about 190,000,000,000,000,000,000 descendents—if all of them lived.

85. The mass of an oxygen molecule is 0.00000000000000000000531 mg.

86. The approximate frequency range of FM radio waves is between 88,000,000 and 108,000,000 Hz.

For Exercises 87–94, evaluate. Express your answer in scientific notation and in decimal notation.

87. $(2.1 \times 10^3)(4 \times 10^4)$ **88.** $(2 \times 10^2)(3.5 \times 10^6)$

89. $(6 \times 10^{-4})(4 \times 10^7)$ **90.** $(4 \times 10^{-2})(8 \times 10^{-2})$

91. $(4 \times 10^8)(5 \times 10^{-12})$ **92.** $(5 \times 10^{-5})(6 \times 10^4)$

93. $\dfrac{4.6 \times 10^4}{2 \times 10^6}$ **94.** $\dfrac{3.3 \times 10^7}{3 \times 10^6}$

For Exercises 95–102, evaluate using scientific notation. Write your answer in decimal notation.

95. $(0.00000002)(45,000)$ **96.** $(821,000)(0.00001)$

97. $(125,000) \div (0.00005)$ **98.** $(0.0000224) \div (4,000)$

99. $\dfrac{(0.0018)(1,400)}{(60,000)(0.07)}$ **100.** $\dfrac{(800,000)(0.00003)}{(0.00004)(3,000)}$

101. $\dfrac{(0.002)(11,000)}{0.022}$ **102.** $\dfrac{(40)(0.015)}{(0.03)(200)}$

103. One light-hour is the distance that light travels in one hour, which is about 6.7×10^8 miles. Find the number of miles in one light-year.

104. Sirius, one of the nearest stars, is about 51,156,000,000,000 miles from Earth. How far, in light-years, is Sirius from Earth? See Exercise 103.

105. A unit of time, called the nanosecond, is equal to 0.000000001 second. In a computer, an electronic signal travels from chip A to chip B at a rate of 300,000,000 meters per second. If the trip takes 1.5 nanoseconds, how far apart are the two chips? Recall the formula: $d = rt$.

106. In the human body, an impulse travels from one nerve ending to another at a rate of 25,000,000,000 centimeters per second. If the trip takes 3.6 nanoseconds, what is the distance between the two nerve endings? See Exercise 105.

REVIEW PROBLEMS

The following exercises review parts of Section 1.5. Doing these problems will help prepare you for the next section.

Simplify by combining like terms.

107. $2 - 5x + x^2 - 4x^2$

108. $3a^3 - 5a^2 + a^3 - 7a^2 + 1$

109. $R^2 - 5R^2 + R + 9R + 3$

110. $y^4x - 6y^4x^3 + 2y^4x + 4y^4x^3$

111. $4ax^2 - 9ax^2 - 8a^2x + 3a^2x$

Use the distributive property to separate terms, then simplify.

112. $3(2x - 4) + 4x - 8$

113. $12 - 2(X^3 - X^2 + 2X) - X^2 + 1$

114. $4(2r^3 + 1) - 5(3 - 7r^3)$

115. $3[2 - (3x - 5x^2)] - 4(7x^2 - 8x + 9)$

116. $\frac{2}{3}(6 - 9z^2) - \frac{1}{3}(3z^2 + 12)$

ENRICHMENT EXERCISES

Simplify each expression.

1. $(x^n)^2(x^2)^n$

2. $(x^{2r})^s(x^2)^{s-rs}$

3. $(x^{1-m})^n(x^{n+1})^m$

Simplify, and write the expression with positive exponents only.

4. $x^{n-m}, \quad 0 < n < m$

5. $x^{n-m}, \quad n < m < 0$

Solve for n.

6. $3^{6n} = 3^4(3^n)^2$

7. $2^n = 4^n 2^3$

Answers to Enrichment Exercises are on page A.10.

3.2

Sums and Differences of Polynomials

O B J E C T I V E S

▶ *To determine the degree of a polynomial*

▶ *To simplify polynomials by combining like terms*

▶ *To add and subtract polynomials*

The following are examples of *monomials.*

4 is a constant.

y is a variable.

$3y^5$ is a product of a constant and a variable.

$-2xy^3z^2$ is a product of a constant and several variables raised to positive integer powers.

> **D E F I N I T I O N**
>
> A **monomial** is a term that is either a constant, a variable, or a product of a constant and one or more variables raised to positive integer powers.

The **degree of a monomial** is the sum of the exponents of all of its variables. A nonzero constant has degree zero, and the constant 0 has no degree.

Example 1   (a) The degree of $4s^3$ is 3.

(b) The degree of $-3r^3t^2$ is $3 + 2$ or 5.

(c) The degree of $-4xyz^2$, which can be written as $-4x^1y^1z^2$, is $1 + 1 + 2$ or 4.

(d) The degree of $2\sqrt{3}$ is 0.   ▲

> **D E F I N I T I O N**
>
> A **polynomial** is a monomial or the sum of monomials.

For example, $4x^3 + 7x^2 + (-5x) + (-11)$, usually written as $4x^3 + 7x^2 - 5x - 11$, is a polynomial having four terms.

Polynomials of *two* or *three* terms have special names: *binomial* and *trinomial*. A *mono*mial can be thought of as a polynomial of *one* term.

Example 2 Classify each polynomial as a monomial, binomial, or trinomial.

(a) $3b^2 + 3b - 1$ has 3 terms, $3b^2$, $3b$, and -1; it is a trinomial.
(b) $4x^3z - 1$ has two terms; it is a binomial.
(c) $-3.1w^2v^4$ has one term; it is a monomial.
▲

In the monomial $-3a^4b$, -3 is called the **coefficient.** Two monomials that are exactly alike or that differ only in their (numerical) coefficients are called **like** or **similar.** Polynomials such as $2x^2 - 5x + 7x^2 - 12x - 6$ can be *simplified* or *reduced* by first grouping like terms and then combining them by using the distributive property.

Example 3 Simplify $2x^2 - 5x + 7x^2 - 12x - 6$.

Solution First group the like terms (monomials), then apply the distributive property to combine these similar terms. As indicated, some of these steps can be done mentally.

$$2x^2 - 5x + 7x^2 - 12x - 6 = (2x^2 + 7x^2) + (-5x - 12x) - 6$$

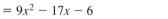

$$= (2 + 7)x^2 + (-5 - 12)x - 6$$

$$= 9x^2 - 17x - 6 \qquad ▲$$

The polynomial $3x^2 - 7x + 2$ is written in **descending order of exponents.** The highest exponent, 2, comes first, then comes the next highest exponent, 1, then the constant.

The **degree of a polynomial** is the greatest of the degrees of its terms after it has been simplified. For example, $3x^2y - 8x^2y^4 + 2x^5$ is of degree 6. Its terms are of degrees 3, 6, and 5, respectively, and 6 is the greatest of these degrees.

S T R A T E G Y

To find the degree of a polynomial:

1. Simplify the polynomial.
2. Find the degree of each of its terms.
3. The term with the highest degree is the degree of the polynomial.

N O T E   The degree of $3x^2 + 5x^3 - 5x^3 + 4$ is 2, *not* 3, since the polynomial simplifies to $3x^2 + 4$.

Example 4   Simplify $3a^2 - 4 - a^3 + 2a^2 - 8a^3 + 8$. Write the result in descending order of exponents. What is the degree of the polynomial?

Solution   To simplify, we group like terms and combine.

$$3a^2 - 4 - a^3 + 2a^2 - 8a^3 + 8 = (3a^2 + 2a^2) + (-a^3 - 8a^3) + (-4 + 8)$$
$$= 5a^2 - 9a^3 + 4$$

Next, we write this trinomial in descending order of exponents.

$$5a^2 - 9a^3 + 4 = -9a^3 + 5a^2 + 4$$

Its degree is 3.   ▲

Example 5   Simplify $4(1 - x + x^2) - 2(x^2 + 3 + 5x)$. Write the answer in descending powers.

Solution   First use the distributive property twice, then group like terms and combine.

$$4(1 - x + x^2) - 2(x^2 + 3 + 5x) = 4 - 4x + 4x^2 - 2x^2 - 6 - 10x$$
$$= (4x^2 - 2x^2) + (-4x - 10x) + (4 - 6)$$
$$= 2x^2 - 14x - 2 \quad ▲$$

We can evaluate a polynomial when a value of the variable is given. Consider, for example, the polynomial $3x^2 - 5x + 4$. It is common to use $P(x)$, read "P of x," to represent the polynomial. Therefore, we write

$$P(x) = 3x^2 - 5x + 4$$

The notation $P(x)$ tells us that the name of the polynomial is P and that x is the variable used to express the polynomial.

The $P(x)$ notation is especially useful when evaluating the polynomial P when a specific value of x is given. For example, $x = 2$, then $P(x) = 3x^2 - 5x + 4$ has the value of

$$P(2) = 3(2)^2 - 5(2) + 4 = 3 \cdot 4 - 10 + 4 = 12 - 6 = 6$$

Example 6   Let $T(x) = 3x^2 + 2x - 5$. Find

(a) $T(-1)$ **(b)** $T(2)$

Solution   **(a)** Replace x by -1.

$$T(x) = 3x^2 + 2x - 5$$
$$T(-1) = 3(-1)^2 + 2(-1) - 5$$
$$= 3 - 2 - 5$$
$$= -4$$

(b) Replace x by 2.

$$T(x) = 3x^2 + 2x - 5$$
$$T(2) = 3(2)^2 + 2(2) - 5$$
$$= 12 + 4 - 5$$
$$= 11 \qquad \blacktriangle$$

CALCULATOR   CROSSOVER

Many calculators have store $\boxed{\text{STO}}$ and recall $\boxed{\text{RCL}}$ keys. They are helpful when evaluating polynomials.

1. If $P(x) = -8.096 + 2.1x^2$, find $P(7.2)$.
2. Evaluate $45 - 2a^3$ for $a = 39$.
3. Find $f(3)$, if $f(s) = 7s^2 - 6s^4 + 12$.
4. If $Q(b) = (8.3 - 5b^2)^2$, find $Q(2.4)$.

Answers   **1.** 100.768   **2.** −118,593   **3.** −411   **4.** 420.25

Example 7 A baseball is thrown vertically upward so that the distance $D(t)$, in feet, from the ground after t seconds is

$$D(t) = 5 + 56t - 16t^2$$

How far above the ground is the ball after 2 seconds?

Solution To answer this question, we must find $D(2)$.

$$D(t) = 5 + 56t - 16t^2$$
$$D(2) = 5 + 56(2) - 16(2)^2$$
$$= 5 + 112 - 64$$
$$= 53$$

Therefore, the baseball will be 53 feet above the ground after 2 seconds. $\blacktriangle$

When adding or subtracting polynomials, we rely on the commutative, associative, and distributive properties to combine like terms. The operations of adding or subtracting polynomials can be done by two methods—the horizontal or the vertical method.

Example 8   Use the horizontal method to add $x^3 + 4x^2 + 5$ and $-7x^2 + 3x + 9$.

Solution   Write the sum of the two polynomials using parentheses. Then, group together like terms and simplify.

$$(x^3 + 4x^2 + 5) + (-7x^2 + 3x + 9) = x^3 + (4x^2 - 7x^2) + 3x + (5 + 9)$$
$$= x^3 - 3x^2 + 3x + 14 \quad \blacktriangle$$

The vertical method involves writing the two polynomials in columns.

Example 9   Use the vertical method to add $3x - 4y + 5$ and $9x + y - 12$.

Solution   Write the polynomials one below the other, so that like terms are in the same column. Then add the terms in each column.

$$\begin{array}{r} 3x - 4y + \ 5 \\ 9x + \ y - 12 \\ \hline 12x - 3y - \ 7 \end{array} \quad \blacktriangle$$

Recall that $-a = -1 \cdot a$. We can apply this property to simplifying a quantity such as $-(2x^2 - 5x + 3)$.

Example 10   Simplify $-(2x^2 - 5x + 3)$.

Solution   We write $-(2x^2 - 5x + 3)$ as $-1(2x^2 - 5x + 3)$, and then distribute the -1 through the polynomial.

$$-(2x^2 - 5x + 3) = -1(2x^2 - 5x + 3)$$
$$= (-1)2x^2 + (-1)(-5x) + (-1)3$$
$$= -2x^2 + 5x - 3 \quad \blacktriangle$$

N O T E   *From Example 10, we see that to simplify the negative of a polynomial, change the sign of* every *term of the polynomial.*

To subtract one polynomial from another, we again have the two methods available.

Example 11   Use the horizontal method to subtract $3x^3 - 4x^2 + 3x$ from $7x^3 - x^2 + 2x$.

Solution   For two quantities A and B, recall that $A - B$ means $A + (-B)$. Therefore, we can rewrite the subtraction problem as an addition problem.

$$7x^3 - x^2 + 2x - (3x^3 - 4x^2 + 3x) = 7x^3 - x^2 + 2x + [-(3x^3 - 4x^2 + 3x)]$$
$$= 7x^3 - x^2 + 2x - 3x^3 + 4x^2 - 3x$$
$$= (7x^3 - 3x^3) + (-x^2 + 4x^2) + (2x - 3x)$$
$$= 4x^3 + 3x^2 - x \quad \blacktriangle$$

To use the vertical method when subtracting, change the sign of each term of the polynomial that is being subtracted and then add.

Example 12   Use the vertical method to subtract $8c^2 - 3c + 1$ from $-3c^2 + 9c - 14$.

Solution   Write the polynomials one below the other as shown. Be sure that like terms are in the same column. Then subtract the bottom terms from the top terms by changing the sign and adding.

$$
\begin{array}{r}
-3c^2 + 9c - 14 \quad \longrightarrow \quad -3c^2 + 9c - 14 \\
(-) \quad \underline{(8c^2 - 3c + 1)} \quad \longrightarrow \quad \underline{-8c^2 + 3c - 1} \\
-11c^2 + 12c - 15
\end{array}
$$
▲

Example 13   Simplify and write the answer in descending powers of x.

$$(3x^2 - x + 5) - (4x^2 - 3x + 7) + (x^3 - x^2)$$

Solution $(3x^2 - x + 5) - (4x^2 - 3x + 7) + (x^3 - x^2)$

$$= 3x^2 - x + 5 - 4x^2 + 3x - 7 + x^3 - x^2$$

$$= x^3 + (3x^2 - 4x^2 - x^2) + (-x + 3x) + (5 - 7)$$

$$= x^3 - 2x^2 + 2x - 2$$
▲

L E A R N I N G   A D V A N T A G E   *When doing your homework, be honest with yourself. Having the correct answer written on your paper is not important—knowing the correct procedure for obtaining the correct answer* is *important.*

E X E R C I S E   S E T   3.2

For Exercises 1–8, find the degree of each polynomial.

1. $5z^3$ **2.** $-2r^8$ **3.** $4x^2y^3z$ **4.** rst^4

5. -3 **6.** 10^2 **7.** 0 **8.** $3mn^3q$

For Exercises 9–14, classify each polynomial as a monomial, binomial, or trinomial.

9. $8c^3 - 1$ **10.** 5.52 **11.** $x^2 + x + 4$

12. $r^4 - 3 - r$ **13.** wrt^3 **14.** u^2vw^2

For Exercises 15–24, simplify each polynomial. Write the result in descending order of exponents. What is the degree of the polynomial?

15. $7 - 3x^3 + x^3 - 2x^2 + 12$ **16.** $12a^3 - 17a + 5a^3 + 12a$

17. $3z - 2z^2 + 4z^2 + 9z$ **18.** $7t^2 - t - 5t^2 - 3t$

19. $4\left(1 - 7c^2 + \dfrac{1}{2}c\right) + 14\left(2c^2 + \dfrac{1}{7}c\right)$

20. $2(3a^3 - a^2 + 1) - (6a^3 + a^2 - 1)$

21. $w^2 - 8w + w^2 - 4w$

22. $3a - 4a^2 + 5a + a^2$

23. $4(2x - 3x^2) - (x^3 + x)$

24. $2(6t^2 - t^3 + 1) - [(5)(1 + 2t^3) + 2]$

25. If $P(x) = 3x^3 - x^4 + 1$, find $P(1)$ and $P(-1)$.

26. If $P(z) = -10 + z^3$, find $P(-2)$ and $P(0)$.

27. If $f(s) = \dfrac{1}{2}s^2 + \dfrac{3}{4}s - 6$, find $f(4)$.

28. If $g(t) = 1.6t^2 - 290 + 15t$, find $g(10)$.

29. The sum $S(n)$ of the first n positive integers is given by $S(n) = \dfrac{1}{2}n^2 + \dfrac{1}{2}n$. Find the sum of the first 60 positive integers.

30. The daily cost $C(x)$, in dollars, of making x storage buildings is given by

$$C(x) = 500x - 12x^2 + \frac{1}{4}x^3 + 600$$

Find the cost, if 20 storage buildings are made in a day.

For Exercises 31–34, use the horizontal method to add the polynomials. Write the answer in descending powers.

31. $(x^4 - 3x^2 + x - 1) + (2x^4 + 4x^2 - 2x + 5)$

32. $(2y^3 + 4y^2 - 7y + 8) + (4y^3 - 7y^2 + 3y - 8)$

33. $(5a^4 - 3a^2 + 7) + (a^4 - 3a^2 + a)$

34. $(3t^4 - 2t + 1) + (t^4 - t^2 + 5t - 14)$

For Exercises 35–38, use the vertical method to add the polynomials. Write the answer in descending powers.

35. $(v^5 + 6v^2 - 7v + 3) + (2v^4 - v^3 + 9v^2 - 12)$

36. $(2n^4 - 8n^2 - 5n + 14) + (4n^3 - 4n^2 + 9n - 21)$

37. $(4 - 5t + 3t^2 + 6t^3) + (t^4 + t^3 + 8t^2 + 1)$

38. $(10 + 7r - 2r^3 + 9r^4) + (-r^5 + 4r^2 - 6r + 7)$

For Exercises 39–42, simplify. Write the answer in descending powers.

39. $-(3u^2 - 3u + 5)$

40. $-(-z^2 + 15z - 7)$

41. $-(2 - 6m - 9m^4 + m^3)$

42. $-(10 + s^3 + 6s - 4s^6)$

For Exercises 43–46, use the horizontal method to subtract the first polynomial from the second. Write the answer in descending powers.

43. $4t^3 + 7t^2 + 16, \quad 5t^3 + 17t^2 - 23$

44. $7x^4 + 5x - 4, \quad 12x^4 - 6x + 9$

45. $-5x^5 - 19x^3 + 4x, \quad -2x^5 - 13x^3 + 15$

46. $-4a^3 + 6a^2 - 3a, \quad -a^3 + 11a - 30$

For Exercises 47–50, use the vertical method to subtract the first polynomial from the second. Write the answer in descending powers.

47. $-4y^3 - 5y^2 + 2y, \quad 5y^3 + 8y^2 - 4$

48. $2b^4 - 6b^3 + 8b, \quad b^5 - 7b^4 + 4b^3 + 9$

49. $5c^5 + 4c^3 - c,\quad 2c^6 + 8c^3 + 2c - 8$

50. $z^4 - 3z^2 + 5,\quad 6z^3 + 2z^2 + 1$

For Exercises 51–62, simplify. Write the answer in descending powers.

51. $(3a^2 - a + 5) - (6 - 2a + a^2) + (7a^2 - 4a + 12)$

52. $(u^5 - 7u^4 + u^2) - (u^3 - 2u^2) - (4u^5 - u^4)$

53. $(r^5 - r^3 + 7r^2) - (5 - 3r^2) - (r^4 + 2)$

54. $(-5x^4 + 7x^3) - (-3x^4 - 2x^2) - (x^4 + 9x^3 + x^2)$

55. $(s^3 - 7s^2 + 5) - (4s^3 - 8s) - (5s^3 + 7s^2 + 7)$

56. $(-v^4 - 4v^5) - (6 - 5v^5) - (14 + 8v^4 + v^2)$

57. $-[7 - (3 - a^4)]$

58. $-[15 - (14 + s^3)]$

59. $2n^3 - 4n^2 - [3n^2 - (n^3 - 4n^2)]$

60. $-3a^5 + a^4 - [-6a^5 - (a^4 - a^5)]$

61. $1 - x^5 - \{1 - [x^5 - (1 - x^5)]\}$

62. $v^4 - v^3 - [-(2v^2 + 9v^4) - 5]$

63. Subtract the sum $(3x^2 - x + 2) + (-5x^2 + x + 3)$ from $10x^2 - 5x + 6$.

64. Subtract the sum $(-2t^4 - t^3 + t) + (4t^2 - 8t)$ from $5t^4 + 9t^3 - 6t^2$.

For Exercises 65–68, simplify.

65. $(3a^2 - b^2 + 2ab) - (ab - a^2 - b^2)$

66. $ab^2 - (-ab + 2ab^2 + 2) + (5ab^2 - ab + 9)$

67. $rst - rs^2t + r^2st - (2r^2st - 3rst + 8rs^2t)$

68. $n^3m^4 + 5n^4m^3 - (-2nm + 5n^4m^3 - 12n^3m^4)$

69. What must be added to $3x - 4$ to make $16x + 5$?

70. What must be added to $4x^2 - 5x$ to make $7x^2 - 8x + 1$?

71. The perimeter of a triangle is $6x^3 + 12x^2 + 10x - 2$. If one side is $2x^3 + 7x^2$ and another is $4x^3 + 7x + 3$, what is the third side?

72. The sides of a triangle are $2d + 1$, $3d + 1$, and $5d$. Find the perimeter.

73. In a triangle, the measures of two angles are $2a + 30$ and $3a^2 - 20$ degrees. What is the measure of the third angle?

74. In a triangle, the measures of two angles are $4A - 55$ and $8A - 20$ degrees. What is the measure of the third angle.

REVIEW PROBLEMS

The following exercises review parts of Section 3.1. Doing these problems will help prepare you for the next section.

Simplify using properties of exponents. Write the answer without negative exponents.

75. x^2x^{-5}

76. $\left(\dfrac{2c}{3b}\right)^2$

77. $(5r^4t)^2$

78. $\left(\dfrac{b}{a^7}\right)^{-2}$

79. $\left(\dfrac{x^2}{z^3}\right)^{-4}$

80. $\left(\dfrac{2}{c}\right)^{-3}$

81. $\left(\dfrac{x^2y^{-1}}{x^3y^5}\right)^{-2}$

82. $\left(\dfrac{a^3b^{-1}c^{-2}}{a^{-4}b^2c^{-1}}\right)^{-5}$

ENRICHMENT EXERCISES

1. What must be subtracted from $5a^2 - 3a + 7$ to make $2a^3 - a^2 + 4a - 3$?

2. The sum of $-3 + 2z - 5z^2 + z^3$ and $4z^3 - 3z^2 - z$ is subtracted from the sum of $4 + 2z^3 - z$ and $2 - 5z^3 + 3z^2$. What is the difference?

3. Find values for a, b, c, and d so that the given equation is true.
$$3ax^4 - 5x^3 + x^2 - cx + 2 = 9x^4 - bx^3 + x^2 - 2d + 1$$

4. Find values for a, b, c, and d so that the given equation is true.
$$(4ax^3 - 3bx^2 - 10) - 3(x^3 + 4x^2 - cx - d) = x^2 - 6x + 8$$

5. Evaluate $3x^2y - 6xy^2 + 9$ for $x = 3a$ and $y = -2b$.

6. Evaluate $3x^3y^2 - 7x^2y^3 + 8xy^4 - 7y^5$ for $x = -2a^2$ and $y = a^2$.

7. A **multiple** of a real number x is of the form nx where n is an integer. Find four consecutive multiples of $\frac{1}{2}$ so that one fourth of the sum of the two greatest multiples is $\frac{3}{8}$ less than one half of the sum of the two smallest.

Answers to Enrichment Exercises are on page A.10.

3.3

Multiplying Polynomials

OBJECTIVES

▶ *To multiply and divide monomials*

▶ *To multiply a monomial and a polynomial*

▶ *To multiply two polynomials*

Suppose we wanted to multiply two monomials such as $7xy^3$ and $2x^2y$. The product $(7xy^3)(2x^2y)$ can be simplified by grouping the coefficients together as well as powers with the same base. Then multiply the coefficients and multiply the powers with the same base using the rule $a^m a^n = a^{m+n}$.

$$(7xy^3)(2x^2y) = (7 \cdot 2)xx^2y^3y$$
$$= 14x^{1+2}y^{3+1}$$
$$= 14x^3y^4$$

We summarize the steps for multiplying two monomials.

STRATEGY

Steps for multiplying two monomials
Step 1 Group together the coefficients.

Step 2 Group together the powers with the same base.

Step 3 Multiply the coefficients.

Step 4 Multiply the powers with the same base by using the rule

$$a^m a^n = a^{m+n}$$

Example 1   Multiply $-2a^3b^6c^2$ and $5a^2b^4$.

Solution   We first group the coefficients and powers with the same base, then multiply.

$$(-2a^3b^6c^2)(5a^2b^4) = (-2)(5)(a^3a^2)(b^6b^4)c^2$$
$$= -10a^{2+3}b^{6+4}c^2$$
$$= -10a^5b^{10}c^2 \qquad \blacktriangle$$

COMMON ERROR

When multiplying two monomials like $3a^2$ and $5a^7$, the answer is *not* $3 \cdot 5a^{2\cdot7}$. Make sure that you multiply the coefficients, but *add* the exponents.

$$(3a^2)(5a^7) = 3 \cdot 5a^{2+7}$$

When dividing a monomial by a monomial, first divide the coefficients. Then divide the factors with a common base using the rule $\dfrac{a^m}{a^n} = a^{m-n}$. For example, suppose we divide $6x^4y^{12}$ by $3x^3y^8$.

$$\frac{6x^4y^{12}}{3x^3y^8} = \left(\frac{6}{3}\right)\left(\frac{x^4}{x^3}\right)\left(\frac{y^{12}}{y^8}\right)$$
$$= 2x^{4-3}y^{12-8}$$
$$= 2xy^4$$

We summarize the steps for dividing two monomials.

Steps for dividing two monomials

Step 1 Divide the coefficients.

Step 2 Divide the powers with the same base using the rule $\dfrac{a^m}{a^n} = a^{m-n}$.

Example 2 Divide $4.5s^2t^5r^3$ by $-9st^2r^6$. Write the answer without negative exponents.

Solution We first divide the coefficients and then divide the factors with the same base.

$$\frac{4.5s^2t^5r^3}{-9st^2r^6} = \left(\frac{4.5}{-9}\right)\left(\frac{s^2}{s}\right)\left(\frac{t^5}{t^2}\right)\left(\frac{r^3}{r^6}\right)$$

$$= -0.5s^{2-1}t^{5-2}r^{3-6} \qquad \frac{a^m}{a^n} = a^{m-n}$$

$$= -0.5st^3r^{-3}$$

$$= -\frac{0.5st^3}{r^3} \qquad\qquad a^{-n} = \frac{1}{a^n} \qquad \blacktriangle$$

N O T E   *As shown in this example, the quotient of two monomials need not be a monomial.*

In the next example, we show how the various rules of exponents can be used to simplify products and quotients of monomials.

Example 3 Simplify each expression. Write the answer without negative exponents.

(a) $\dfrac{(2ab)(3a^2b^3)}{6a^2b^5}$
(b) $\dfrac{(2xy^3z^2)^4}{(3xyz)(4y^3)^2}$

Solution (a) $\dfrac{(2ab)(3a^2b^3)}{6a^2b^5} = \left(\dfrac{2\cdot3}{6}\right)\left(\dfrac{a\cdot a^2}{a^2}\right)\left(\dfrac{b\cdot b^3}{b^5}\right)$

$$= ab^{-1}$$

$$= \frac{a}{b}$$

(b) $\dfrac{(2xy^3z^2)^4}{(3xyz)(4y^3)^2} = \dfrac{2^4x^4y^{3\cdot4}z^{2\cdot4}}{(3xyz)4^2y^{3\cdot2}}$

$= \left(\dfrac{16}{3\cdot16}\right)(x^{4-1})(y^{12-7})(z^{8-1})$

$= \dfrac{1}{3}x^3y^5z^7$ ▲

We are now ready to multiply a monomial and a polynomial. This multiplication makes use of the distributive properties

$$A(B + C) = AB + AC \quad\text{and}\quad (B + C)A = BA + CA$$

Example 4

Do the indicated multiplication, then simplify.

(a) $a^2(3a^3 - 2a^2 + a)$

(b) $-2y^3(2y^5 + 7y^2 - 1)$

(c) $\left(\dfrac{1}{14} - 3st\right)(-7s^2t^3)$

Solution

(a) Distribute the a^2 through the three terms of the second polynomial, then simplify.

$$a^2(3a^3 - 2a^2 + a) = (a^2)(3a^3) + (a^2)(-2a^2) + (a^2)(a)$$
$$= 3a^2a^3 - 2a^2a^2 + a^2a$$
$$= 3a^{2+3} - 2a^{2+2} + a^{2+1}$$
$$= 3a^5 - 2a^4 + a^3$$

(b) $-2y^3(2y^5 + 7y^2 - 1) = (-2y^3)(2y^5) + (-2y^3)(7y^2) + (-2y^3)(-1)$
$$= -4y^{3+5} - 14y^{3+2} + 2y^3$$
$$= -4y^8 - 14y^5 + 2y^3$$

(c) $\left(\dfrac{1}{14} - 3st\right)(-7s^2t^3) = \left(\dfrac{1}{14}\right)(-7s^2t^3) + (-3st)(-7s^2t^3)$

$$= -\dfrac{1}{2}s^2t^3 + 21s^{1+2}t^{1+3}$$

$$= -\dfrac{1}{2}s^2t^3 + 21s^3t^4$$ ▲

To multiply two polynomials such as $x + 3$ and $x + 7$, the distributive property is used. To find the product $(x + 3)(x + 7)$, treat $x + 7$ as a single quantity and distribute it through $x + 3$.

$$(x + 3)(x + 7) = x(x + 7) + 3(x + 7)$$

Next, distribute x through $x + 7$ and 3 through $x + 7$.

$$x(x + 7) + 3(x + 7) = x \cdot x + x \cdot 7 + 3 \cdot x + 3 \cdot 7$$
$$= x^2 + 7x + 3x + 21$$
$$= x^2 + 10x + 21$$

Notice that the product is obtained in the following way.

STRATEGY

To multiply two polynomials

Multiply each term of the second polynomial by each term of the first polynomial.

Example 5 Multiply $2a + b$ and $3a - 5b$.

Solution $$(2a + b)(3a - 5b) = 2a(3a - 5b) + b(3a - 5b)$$
$$= 6a^2 - 10ab + 3ab - 5b^2$$
$$= 6a^2 - 7ab - 5b^2 \qquad \blacktriangle$$

Instead of using the horizontal method as in Example 5 for multiplying polynomials, this multiplication can also be done using the column form. For example, suppose we want to find the product $(x + 7)(x + 3)$. First, write one polynomial above the other.

$$x + 7$$
$$x + 3$$

Multiply each term of the top row by 3, then each term of the top row by x. Be sure to line up like terms, since we add them to obtain the answer.

$$
\begin{array}{r}
x + 7 \\
x + 3 \\
\hline
\end{array}
$$

$$3x + 21 \qquad \leftarrow \text{The top row multiplied by 3.}$$

Add like terms that are $\quad \dfrac{x^2 + 7x}{} \qquad \leftarrow \text{The top row multiplied by } x.$

lined up in columns. $\rightarrow$ $x^2 + 10x + 21$

Therefore, $(x + 3)(x + 7) = x^2 + 10x + 21$.

Example 6 Find $(s^3 - 4s)(s^4 - 2s^2 + 5s)$ by the column method.

Solution   We set up the problem for the column method.

$$s^4 - 2s^2 + 5s$$
$$s^3 - 4s$$

$-4s^5 \qquad\qquad + 8s^3 - 20s^2 \leftarrow$ The top row multiplied by $-4s$.

Add like terms that are $\quad\dfrac{s^7 - 2s^5 + 5s^4}{}\qquad\qquad\quad \leftarrow$ The top row multiplied by s^3.

lined up in columns. $\rightarrow \quad s^7 - 6s^5 + 5s^4 + 8s^3 - 20s^2$ ▲

While the column method is a quick way to multiply polynomials as opposed to the horizontal method of using the distributive property, it does have its drawbacks. In the next sections, for example, we will be factoring polynomials using techniques that are based on the horizontal method of multiplication.

Certain products are so common that methods for obtaining the answers have been developed. For example, the square of the sum of a and b, $(a + b)^2$, can be found by the usual method of "multiplying it out."

$$(a + b)^2 = (a + b)(a + b) = a^2 + ab + ba + b^2$$
$$= a^2 + 2ab + b^2$$

This gives us a formula that can be memorized to find the square of the binomial $a + b$.

R U L E

_____

The square of a sum formula

$$(a + b)^2 = a^2 + 2ab + b^2$$

N O T E   *Avoid the error $(a + b)^2 = a^2 + b^2$. Keep in mind that there is a "middle" term, 2ab.*

Example 7   (a) $(a + 3)^2 = a^2 + 2(a)(3) + 3^2 = a^2 + 6a + 9$

(b) $(2y + 5z)^2 = (2y)^2 + 2(2y)(5z) + (5z)^2$
$$= 4y^2 + 20yz + 25z^2$$ ▲

There is also a formula to find the square of the difference, $(a - b)^2$.

R U L E

_____

The square of the difference formula

$$(a - b)^2 = a^2 - 2ab + b^2$$

Example 8   (a) $(x - 4)^2 = x^2 - 2(x)(4) + 4^2$

$$= x^2 - 8x + 16$$

(b) $(9p - 3q)^2 = (9p)^2 - 2(9p)(3q) + (3q)^2$

$$= 81p^2 - 54pq + 9q^2 \quad \blacktriangle$$

Consider the product $(a + b)(a - b)$. If we multiply it out,

$$(a + b)(a - b) = a^2 - ab + ba - b^2 = a^2 - b^2$$

This gives us the following formula for the difference of two squares.

R U L E

The difference of two squares formula

$$(a + b)(a - b) = a^2 - b^2$$

Example 9   (a) $(a + 7)(a - 7) = a^2 - 7^2$ (b) $(2s + 3t)(2s - 3t) = (2s)^2 - (3t)^2$

$$= a^2 - 49 \qquad\qquad\qquad\qquad\qquad = 4s^2 - 9t^2$$

$$\blacktriangle$$

N O T E   *There is a "middle" term in the product $(a + b)(a - b)$. It is*
$0 \cdot ab$, which is zero.

The products of special trinomials can be found using the difference of two
squares formula.

Example 10   Find the product: $(x + y + z)(x + y - z)$.

Solution   If we group the first two terms together, we can apply the difference of two
squares formula.

$$(x + y + z)(x + y - z) = [(x + y) + z][(x + y) - z]$$

$$= (x + y)^2 - z^2$$

$$= x^2 + 2xy + y^2 - z^2 \quad \blacktriangle$$

The formulas for $(a + b)^2$ or $(a - b)^2$ can be used when cubing a binomial.
This technique is shown in the next example.

Example 11
$$(x + 6)^3 = (x + 6)(x + 6)^2$$
$$= (x + 6)(x^2 + 12x + 36)$$
$$= x(x^2 + 12x + 36) + 6(x^2 + 12x + 36)$$
$$= x^3 + 12x^2 + 36x + 6x^2 + 72x + 216$$
$$= x^3 + 18x^2 + 108x + 216 \qquad \blacktriangle$$

Example 12 Juicy Orchards has discovered that if there are n additional trees planted per acre, then the number of bushels of apples produced per acre, B, is given by

$$B = (50 + n)(30 - 6n)$$

Find the number of bushels of apples produced per acre if four additional trees are planted per acre.

Solution We replace n by 4 in the equation to find the answer.

$$B = (50 + n)(30 - 6n)$$
$$= (50 + 4)(30 - 6 \cdot 4)$$
$$= (54)(6) = 324$$

Therefore, if four additional trees were planted per acre, there would be 324 bushels of apples produced per acre. $\blacktriangle$

L E A R N I N G   A D V A N T A G E   *When doing your homework, keep it neat and organized. You are the one who will have to use your homework for review at the end of the chapter. If you are careful and complete, you will not have any regrets at the last minute.*

E X E R C I S E   S E T   3.3

1. Multiply $7x^2y^3$ by $2xy^2$.

2. Multiply $2ab^3$ by $5a^4b$.

3. Multiply $-(u^2v)^3$ by $(3uv^2)^2$.

4. Multiply $(rst)^3$ by $(-r^2st^2)^2$.

5. Divide $4x^3y^2$ by $2x^2y$.

6. Divide $-6ab^3$ by ab.

7. Divide $4rs^2$ by $2r^2$.

8. Divide $3u^3v$ by $6u^3v^2$.

For Exercises 9–14, simplify each expression.

9. $(2xy^2)(-9x^2y)$

10. $(-3ab^4)(5a^2b^3)$

11. $(6uv^2w)(-u^2vw^2)\left(\frac{1}{2}\right)(uv^2w^3)$

12. $(14x^2y^3z^3)\left(\frac{1}{7}\right)(xyz^4)(-x^3y^4z^2)$

13. $(2mn^2)^3(4m^2n)^2$

14. $(-5p^3q^4)^2(2pq)^3$

For Exercises 15–22, simplify each expression. Write your answer without negative exponents.

15. $\dfrac{3x^3y^2}{24x^6y}$

16. $\dfrac{-2ab^3}{6a^2b}$

17. $\dfrac{15ac^2z}{-3a^2c^2}$

18. $\dfrac{4mn^2p(-3)mnp^2}{2m^3np^4}$

19. $\dfrac{\frac{6}{5}(uvw^3)^2\left(\frac{25}{3}\right)(u^2v^3w^2)}{(uvw)^3}$

20. $\dfrac{4(ab^2c)^3(-6)(ab^3c^2)}{(6a^2bc)^2}$

21. $\dfrac{16(x^2y^3z^2)^3}{8(xy^3z^2)^2(x^2z)}$

22. $\dfrac{24(r^2s^3t^5)^3}{4(s^2t)^4(st^2)}$

For Exercises 23–28, do the indicated multiplication, then simplify.

23. $z^3(z^2 - 2z + 18)$

24. $w^2(w^5 + 4w^4 - w)$

25. $2b^4\left(1 - \dfrac{3}{2}b^5\right)$

26. $3v^5\left(\dfrac{2}{3} - \dfrac{1}{6}v^2\right)$

27. $-\dfrac{3}{7}t^3\left(\dfrac{14}{9}t^2 - 7t + \dfrac{2}{3}\right)$

28. $-\dfrac{4}{3}n^2\left(6 - \dfrac{1}{2}n - \dfrac{15}{8}n^3\right)$

For Exercises 29–42, perform the indicated multiplication.

29. $(x - 2)(x + 3)$

30. $(y - 5)(y - 4)$

31. $(t^2 - 7)(t^2 - 5)$

32. $(x^3 + 4)(x^3 + 2)$

33. $(2v + 3)(3v - 1)$

34. $(5y - 4)(6y - 2)$

35. $(2b^2 + 3)(b^2 - 4)$

36. $(3p^4 - 1)(p^4 - 9)$

37. $(5 - 2x^2)(7 - x^2)$

38. $(2 - 6r^2)(10 - 3r^2)$

39. $(2a - 3b)(4a + 5b)$

40. $(7n + 3m)(4n - 2m)$

41. $(x^2 + 2y^2)(x^2 - 2y^2)$

42. $(st - 3v^2)(st + 3v^2)$

For Exercises 43–50, use the column method to multiply.

43. $(2z + 1)(-4z + 3)$

44. $(5s + 9)(-s + 2)$

45. $(x + 1)(3x^2 + x - 5)$

46. $(t - 3)(t^3 - 2t^2 + 6t)$

47. $(2z + 5)(3z^4 - z^2 + z)$

48. $(3v - 7)(4v^4 + 3v^2 - 8)$

49. $(x^2 - x + 2)(3x^2 + 2x - 9)$

50. $(y^3 + y^2 + 7)(-2y^3 - 3y^2 + 2)$

For Exercises 51–60, use the square of a sum formula to find the following.

51. $(y + 4)^2$

52. $(a + 3)^2$

53. $(2x + 10)^2$

54. $(3b + 4)^2$

55. $(a^3 + c)^2$

56. $(x^2 + 2y)^2$

57. $(3u + 4v^2)^2$

58. $(5p + 2q^3)^2$

59. $\left(\dfrac{2}{5}n^2 + \dfrac{5}{3}m\right)^2$

60. $\left(\dfrac{7}{3}r + \dfrac{3}{7}t^2\right)^2$

For Exercises 61–70, use the square of the difference formula to find the following.

61. $(x - 5)^2$

62. $(d - 6)^2$

63. $(4z - a)^2$

64. $(b - 3c)^2$

65. $\left(\dfrac{3}{2}h - a^2\right)^2$

66. $\left(b^2 - \dfrac{5}{2}k\right)^2$

67. $(x^2 - 3u^2)^2$

68. $(5n^2 - m^2)^2$

69. $\left(\dfrac{2}{5}s^2 - \dfrac{15}{4}t^5\right)^2$

70. $\left(\dfrac{4}{3}u^3 - \dfrac{9}{2}v^2\right)^2$

For Exercises 71–86, use the difference of two squares formula to find the following.

71. $(c + 2)(c - 2)$

72. $(x + 4)(x - 4)$

73. $\left(2x - \dfrac{4}{5}\right)\left(2x + \dfrac{4}{5}\right)$

74. $\left(6a - \dfrac{1}{3}\right)\left(6a + \dfrac{1}{3}\right)$

75. $\left(z^2 + \dfrac{1}{2}t\right)\left(z^2 - \dfrac{1}{2}t\right)$

76. $(w + b^2)(w - b^2)$

77. $(f^2 - 3g)(f^2 + 3g)$

78. $(8h^2 - 3k^4)(8h^2 + 3k^4)$

79. $(r + t + 3)(r + t - 3)$

80. $(3x + z - 4)(3x + z + 4)$

81. $[5 - (s - t)][5 + (s - t)]$

82. $[a + (x - 2)][a - (x - 2)]$

83. $\left(4z + \dfrac{2y - r}{2}\right)\left(4z - \dfrac{2y - r}{2}\right)$

84. $\left(7p - \dfrac{3q + t}{3}\right)\left(7p + \dfrac{3q + t}{3}\right)$

85. $(4 - 2s + 3v)(4 + 2s - 3v)$

86. $(u - 6v + c)(u + 6v - c)$

For Exercises 87–94, do the indicated operations and simplify. Use special product rules where possible.

87. $(a - 2z)^2 - (a + 2z)^2$

88. $(y + r)^2 + (y - r)^2$

89. $(z + b)^3$

90. $(n - m)^3$

91. $(2ax - 3by^2)^2 + 12abxy^2$

92. $(4p^2 + 5q)^2 - 40p^2q$

93. $(2x + a)(2x - a)(4x^2 - a^2)$

94. $(w - 3t)(w + 3t)(w^2 - 9t^2)$

95. Express the area A of the triangle in the figure as a polynomial in x.

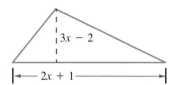

96. Express the area A of the rectangle in the figure as a polynomial in x.

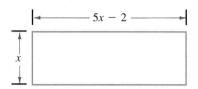

97. The manager at the Bates Motel knows that if the rent per room is increased by p dollars above \$30, then the total income I for the day is given by

$$I = (30 + p)(85 - 4p)$$

What is the total income for the day, if the manager rents rooms at \$40 each?

98. The Nut Hut sells mixed nuts at \$3.50 per can. If the store increases the price by c cents per can, they predict that the weekly revenue R, in dollars, from selling cans of mixed nuts is given by

$$R = (25 - 0.1c)(3.50 + 0.01c)$$

What is the weekly revenue R, if the store raises the price to \$4 per can?

99. Find two consecutive even integers so that the difference of their square is 196.

100. Find two consecutive odd integers so that the difference of their squares is 104.

101. Find two consecutive even integers so that the square of their sum is 20 more than four times the square of the larger integer.

102. Find two consecutive odd integers so that the square of their sum is 20 less than four times the square of the smaller integer.

REVIEW PROBLEMS

The following exercises review parts of Section 3.2. Doing these problems will help prepare you for the next section.

Simplify. Write the answer in descending powers.

103. $(a^3 - 2a + 8) + (1 - 3a + 2a^2)$

104. $(-x^5 - 2x^3 + 2) - (x^4 + 6x^3 - 12)$

105. $3(1 - 2c + c^2) - (c^3 - c^2 + 14c - 2)$

106. $-5(1 - x^2 - x^4) + 4x^3 + 12$

107. $\dfrac{2}{3}\left(\dfrac{3}{2}z^2 - 3z - 6\right) - (2z - 1)$

108. Evaluate $2x^3 - 4x^2 + 7$ for $x = -1$.

ENRICHMENT EXERCISES

1. Verify that the following two formulas are correct.
  (a) $(a + b)^3 = a^3 + 3a^2b + 3ab^2 + b^3$
  (b) $(a - b)^3 = a^3 - 3a^2b + 3ab^2 - b^3$

Use these formulas to cube each of the following.

2. $\left(c + \dfrac{1}{2}\right)^3$

3. $(x^2 - 1)^3$

4. $(2v + a^2)^3$

5. Using the formula $(a + b)^2 = a^2 + 2ab + b^2$, show that
$$(a + b)^4 = a^4 + 4a^3b + 6a^2b^2 + 4ab^3 + b^4$$

6. Without doing the multiplication, determine the number of terms in the resulting polynomial.
$$(x - a)^2(y + b)^2$$

7. Multiply: $(2x^{3n} - 1)(2x^{3n} + 1)$.

8. Multiply: $(x^r + y^s)(3x^r - 5y^s)$.

Answers to Enrichment Exercises are on page A.11.

3.4

Greatest Common Factor and Factoring by Grouping

OBJECTIVES

▶ To determine if a number is prime or composite

▶ To factor a number into primes

▶ To factor a monomial

▶ To factor out the greatest common monomial from a polynomial

▶ To factor by grouping

In the last section, we developed methods of multiplying polynomials. In this section, we will study ways to reverse that process, which is called *factoring*. A major use of factoring is to solve a quadratic equation (see Section 3.8).

We start our discussion with factoring real numbers. If a number such as 84 is written as a product of other numbers, then these numbers are called **factors** of 84. For example, since $84 = 2 \cdot 42$, 2 and 42 are each factors of 84. Also,

$$84 = 2 \cdot 6 \cdot 7, \qquad 84 = \left(\frac{1}{3}\right)(252), \qquad 84 = (-4)(-3)(7)$$

So there are many ways of **factoring** a number. We will be interested in integer factors that are *prime numbers*.

DEFINITION

A positive integer p greater than 1 is a **prime number** if the only positive integer factors are 1 and p. Otherwise, p is called a **composite number.**

N O T E   *The number 1 is considered neither prime nor composite.*

Examples of prime numbers:

$$2, 3, 5, 7, 11, 13, 17, 19, \ldots$$

Examples of composite numbers:

$$4, 6, 8, 9, 10, 12, 14, 15, \ldots$$

Both lists are infinite.

Example 1   Determine if the number is composite or prime.

(a) 27   (b) 1,074   (c) 23

Solution   (a) 27 is a composite number, since $27 = 3 \cdot 9$.

(b) 1,074 is a composite number, since it is even and any even number has 2 as a factor.

(c) 23 is a prime number, since no positive integer greater than 1 but less than 23 is a factor.   ▲

We say that a composite number is **factored into primes,** if it is written as a product of prime numbers. To factor into primes, we begin by using any two factors of the number.

Example 2   Factor 84 into primes.

Solution   We may start with any two factors of 84.

$$
\begin{array}{lll}
84 = 2 \cdot 42 & 84 = 4 \cdot 21 & 84 = 3 \cdot 28 \\
 = 2 \cdot 3 \cdot 14 & = 4 \cdot 3 \cdot 7 & = 3 \cdot 4 \cdot 7 \\
 = 2 \cdot 3 \cdot 2 \cdot 7 & = 2 \cdot 2 \cdot 3 \cdot 7 & = 3 \cdot 2 \cdot 2 \cdot 7
\end{array}
$$

Since the order in which the factors appear is not important, we get the same answer by all three ways. Namely,

$$84 = 2 \cdot 2 \cdot 3 \cdot 7 = 2^2 \cdot 3 \cdot 7$$ ▲

Recall that a number can also be considered a constant monomial. In addition to factoring constant monomials, we can also factor *variable* monomials.

For example, suppose we want to find the missing factor in the expression

$$x^5 = x^3 x^?$$

That is, x^5 is equal to x^3 times some power of x. Since $x^3 \cdot x^2 = x^5$, the missing factor is x^2.

Example 3   Find the missing factor.

(a) $x^7 = x^2 x^?$ (b) $42y^9 = 6y^5(?)$

Solution   (a) Since $2 + 5 = 7$, the missing factor is x^5.

(b) $42y^9 = 6(?)y^5(y^?)$. Therefore, $42 = 6(?)$ and $y^9 = y^5 y^?$. By inspection, $42 = 6(7)$ and $y^9 = y^5 y^4$. Therefore, the missing factor is $7y^4$.   ▲

Next, we look at the problem of factoring a monomial from a polynomial. We have used the distributive property to multiply a polynomial by a monomial. For example,

$$4(2x^3 + 5x^2 - 10) = 8x^3 + 20x^2 - 40$$

On the other hand, to factor $8x^3 + 20x^2 - 40$, we use the distributive property in reverse.

$$8x^3 + 20x^2 - 40 = 4(2x^3 + 5x^2 - 10)$$

The number 4 has been **factored out** from the polynomial. Four is called a **common factor,** since it is a factor common to each term of the polynomial $8x^3 + 20x^2 - 40$.

Example 4   Factor out the common factor 6 from the polynomial $12a^2 - 6a + 18$.

Solution   The polynomial has three terms, $12a^2$, $-6a$, and 18. We factor 6 out of each term,

$$12a^2 = 6(2a^2), \qquad -6a = 6(-a), \qquad 18 = 6(3)$$

Therefore,

$$12a^2 - 6a + 18 = 6(2a^2) + 6(-a) + 6(3)$$

$$= 6(2a^2 - a + 3) \qquad \text{Use the distributive property in reverse.}$$

▲

In the polynomial $8x^2 + 4x + 16$, 2 is a common factor as

$$8x^2 + 4x + 16 = 2(4x^2 + 2x + 8)$$

However, 4 is also a common factor, larger than 2.

$$8x^2 + 4x + 16 = 4(2x^2 + x + 4)$$

Since no integer bigger than 4 can be factored out, we call 4 the **greatest common factor** (GCF) of $8x^2 + 4x + 16$.

Sometimes, the greatest common factor may not be immediately recognized. If this happens, factor the monomial coefficients of each term into primes.

Example 5   Factor the greatest common factor from

$$18t^2 + 90t + 36$$

Solution   The polynomial $18t^2 + 90t + 36$ can be rewritten as

$$2 \cdot 3 \cdot 3t^2 + 2 \cdot 3 \cdot 3 \cdot 5t + 2 \cdot 2 \cdot 3 \cdot 3$$

Since 2 occurs at most one time and 3 occurs at most two times in each term, the GCF is $2 \cdot 3 \cdot 3$ or 18. Therefore, we factor out 18 from the polynomial.

$$18t^2 + 90t + 36 = 18t^2 + 18 \cdot 5t + 18 \cdot 2$$
$$= 18(t^2 + 5t + 2) \quad \blacktriangle$$

The greatest common factor of a polynomial can also be a variable monomial.

Example 6   Factor the greatest common factor from

$$4y^5 - 9y^3 + 7y^2$$

Solution   We can rewrite the polynomial as

$$4y \cdot y \cdot y \cdot y \cdot y - 9y \cdot y \cdot y + 7y \cdot y$$

Notice that at most the product of two y's, $y \cdot y$, appears in *each* of the three terms. Therefore, the greatest common factor is $y \cdot y = y^2$. So, we factor out y^2 from the polynomial.

$$4y^5 - 9y^3 + 7y^2 = y^2(4y^3) - (y^2)(9y) + (y^2)(7)$$
$$= y^2(4y^3 - 9y + 7) \quad \blacktriangle$$

Notice that the GCF, y^2, in Example 6 is the variable y raised to the smallest exponent that appears in the polynomial. This holds true in general. If a polynomial has a greatest common variable factor, then the following statement is true:

The greatest common variable factor is the variable raised to the smallest exponent appearing in the polynomial.

In the next example, we show how the GCF of a polynomial can be made of a numerical factor and a variable factor.

Example 7   Factor the GCF from $6x^4 + 24x^3 + 36x^2$.

Solution   We first look for the greatest common numerical factor of the coefficients.

$$6x^4 + 24x^3 + 36x^2 = 2 \cdot 3x^4 + 2 \cdot 2 \cdot 2 \cdot 3x^3 + 2 \cdot 2 \cdot 3 \cdot 3x^2$$

$$= (2 \cdot 3)(x^4 + 2 \cdot 2x^3 + 2 \cdot 3x^2)$$

$$= 6(x^4 + 4x^3 + 6x^2) \qquad \text{Factor out } 2 \cdot 3, \text{ the greatest}$$
$$\text{numerical factor.}$$

Next, we look for the greatest common variable factor. Since x^2 is the power of the variable with the lowest exponent, it is the greatest common variable factor. We, therefore, factor out x^2 to get the complete factorization.

$$6x^4 + 24x^3 + 36x^2 = 6(x^4 + 4x^3 + 6x^2)$$

$$= 6x^2(x^2 + 4x + 6) \qquad \blacktriangle$$

We summarize the steps taken in Example 7 to factor out the greatest common factor (GCF) from a polynomial.

S T R A T E G Y

Factoring the greatest common factor (GCF) from a polynomial

Step 1   Factor out the greatest common numerical factor, if other than 1. When necessary, first factor each coefficient into primes.

Step 2   Factor out the greatest common variable factor(s), if any. Each such factor is a variable raised to the smallest exponent appearing in the polynomial.

Example 8   Factor the GCF from $3a^3x^4 + 6a^3x + 12a^4x^2$.

Solution   Step 1   The greatest common numerical factor is 3.

Step 2   There are two variables, a and x, common to all terms of the polynomial. The smallest exponent for a is 3 and the smallest exponent for x is 1. Therefore, the GCF is $3a^3x$, and

$$3a^3x^4 + 6a^3x + 12a^4x^2 = 3a^3x(x^3 + 2 + 4ax) \qquad \blacktriangle$$

Example 9   Factor the greatest common factor from

$$3t^2(2a + b) - 2t(2a + b) + 12(2a + b)$$

Solution   The greatest common factor is $2a + b$. We factor it from each term to obtain

$$(2a + b)(3t^2 - 2t + 12) \qquad \blacktriangle$$

If a polynomial having four terms has no greatest common factor other than one, it may still be factored using the distributive property.

For example, consider the polynomial

$$ax + ay + bx + by$$

Observe that a is common to the first two terms and b is common to the last two terms. Therefore, we group the terms as shown and use the distributive property to factor a from the first two terms and b from the last two terms.

$$ax + ay + bx + by = (ax + ay) + (bx + by)$$
$$= a(x + y) + b(x + y)$$

In this form, $x + y$ is a common factor. We factor it from the two terms,

$$a(x + y) + b(x + y) = (x + y)(a + b)$$

This method is called **factoring by grouping.**

Example 10   Factor: $s^3r^2 - 4r^2 + s^3 - 4$.

Solution   The first two terms have r^2 as a common factor.

$$s^3r^2 - 4r^2 + s^3 - 4 = r^2(s^3 - 4) + (s^3 - 4)$$
$$= (s^3 - 4)(r^2 + 1) \qquad \text{Factor } s^3 - 4 \text{ from the}$$
$$\text{two terms.} \qquad \blacktriangle$$

Example 11   Factor: $12a - 15b^2 - 4ac + 5b^2c$.

Solution   The first two terms have a common factor of 3 and the last two terms have a common factor of c.

$$12a - 15b^2 - 4ac + 5b^2c = 3(4a - 5b^2) + c(-4a + 5b^2)$$

Notice that $4a - 5b^2$ and $-4a + 5b^2$ are not equal and we cannot factor further. However, if we factor $-c$ from the last two terms,

$$12a - 15b^2 - 4ac + 5b^2c = 3(4a - 5b^2) - c(4a - 5b^2)$$
$$= (4a - 5b^2)(3 - c) \qquad \blacktriangle$$

Example 12   Factor: $xa^2 - 5b^2 - xb^2 + 5a^2$.

Solution   The first two terms do not have a common factor, nor do the last two terms. However, if we group the first and third together as well as the second and fourth terms, we can factor by grouping.

$$xa^2 - 5b^2 - xb^2 + 5a^2 = (xa^2 - xb^2) + (5a^2 - 5b^2)$$
$$= x(a^2 - b^2) + 5(a^2 - b^2)$$
$$= (a^2 - b^2)(x + 5)$$

Although we will not do so now, in Section 3.6 we will find that the factor $a^2 - b^2$ can itself be factored further.   ▲

In our last example, we show how to write a formula for the area of a geometric figure.

Example 13   Write a formula for the area A of the shaded region in the diagram below.

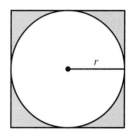

Solution   First, we see that the side of the square is $2r$. Now,

area of shaded region = area of the square − area of the circle

$$A = 2r \cdot 2r - \pi r^2$$
$$= 4r^2 - \pi r^2$$
$$= (4 - \pi)r^2$$ ▲

L E A R N I N G   A D V A N T A G E   *When doing your homework, write it all down; it is sometimes easier to do a mathematics problem in your head and then jot down a few of the steps and the answer. But the more you have on paper, the easier it is for another person to figure out what it is you are doing wrong and what it is you are doing right. So if others will be helping you later, they will need to be able to read your homework.*

E X E R C I S E   S E T   3.4

For Exercises 1–6, determine if the number is composite or prime.

1. 9

2. 16

3. 29

4. 31

5. 1,674

6. 2,985

For Exercises 7–28, factor all composite numbers. If the number is prime, write ''prime.''

7. 42

8. 90

9. 36

10. 180

11. 525

12. 315

13. 37

14. 41

15. 70

16. 105

17. 231

18. 154

19. 140   **20.** 270   **21.** 43   **22.** 47

23. 252   **24.** 84   **25.** 98   **26.** 75

27. 54   **28.** 48

For Exercises 29–54, find the missing factor.

29. $x^4 = x^2(x^?)$

30. $a^5 = a^3(a^?)$

31. $z^8 = z^4(z^?)$

32. $t^7 = t^3(t^?)$

33. $c^{10} = c^3(c^?)$

34. $u^{12} = u^9(u^?)$

35. $a^5 = a^4(a^?)$

36. $p^6 = p^5(p^?)$

37. $n^{12} = n(n^?)$

38. $s^{14} = s(s^?)$

39. $4y^6 = 2y^3(?)$

40. $8b^4 = 4b^2(?)$

41. $12s^5 = 3s^4(?)$

42. $15x^8 = 5x^4(?)$

43. $27q^{10} = 9q^8(?)$

44. $40r^{12} = 8r^4(?)$

45. $63C^8 = 21C^7(?)$

46. $22V^9 = 11V^8(?)$

47. $-18x^{20} = 9(?)$

48. $-44t^{16} = 11(?)$

49. $-28z^5r^3 = -7zr(?)$

50. $-62w^{13}t^5 = -2w^{12}t^2(?)$

51. $-54q_1^8 = -6q_1^7(?)$

52. $-96x_1^{24} = -16x_1^{15}(?)$

53. $-56v^{25}k^{22}m^{20} = -14v^{17}k^{18}m^{19}(?)$

54. $-72p^7q^4r^9 = -36p^3q^2r^5(?)$

55. Factor out the common factor 5 from $25x^2 - 5x + 20$.

56. Factor out the common factor 6 from $12y^2 + 24y - 36$.

57. Factor out the common factor x^2 from $3x^4 - 2x^3 + x^2$.

58. Factor out the common factor y^3 from $y^5 - y^4 + y^3$.

For Exercises 59–86, factor the greatest common factor, if other than one, from each polynomial.

59. $2x^2 + 8x + 4$

60. $3y^2 + 21y + 9$

61. $5t^2 - 25t + 10$

62. $4v^2 - 16v - 20$

63. $12b^2 + 28b + 72$

64. $36p^3 - 12p^2 + 30$

65. $45y^4 + 75y^2 - 30$

66. $2a^3 + 3a^2 - 11a$

67. $5z^6 - 3z^5 + 2z^3$

68. $14t^3 - 7t^2 + 49t$

69. $c^6 - c^3 - c^2$

70. $x^7 + x^5 - x^3$

71. $5s^6 - 45s^5$

72. $7m^4 + 489m^3$

73. $35 - 15F^2$

74. $27 - 24a_1^2$

75. $36c^6d^3 - 108c^4d^5 - 252c^3d^4$

76. $45y^7z^4 - 315y^5z^6 - 75y^5z^7$

77. $5b^2z^3t^2 - 10b^2z^4t^3 + 5b^3z^2t^2$

78. $4a^3u^2z - 2a^4u^3z + 8a^5u^4z$

79. $27m_1{}^3n_1{}^4 - 9m_1{}^4n_1{}^3 + 15m_1{}^5n_1{}^5$

80. $12p_1{}^2q_1{}^4 - 15p_1{}^3q_1{}^2 + 12p_1{}^2q_1{}^2$

81. $2s(z + t) + 3(z + t)$

82. $c^2(a^2 + b^2) - 5d(a^2 + b^2)$

83. $2r(h - t) - 3b(t - h)$

84. $4G(w - v^3) + 3H(v^3 - w)$

85. $20(a - x)^3y^2 - 15(a - x)^2y^3$

86. $32(b + z)^5v^3 - 40(b + z)^3v^5$

87. The sum S of the first n positive integers is given by

$$S = \frac{1}{2}n^2 + \frac{1}{2}n$$

Write this formula in factored form.

88. If P dollars is deposited in an account earning r percent in simple interest, then the amount A at the end of one year is $A = P + Prt$. Write this formula in factored form.

For Exercises 89–102, factor by grouping.

89. $2z^2 + 2y^2 + tz^2 + ty^2$

90. $7s^3 - 7t^2 + bs^3 - bt^2$

91. $3ms - 2ns + 3mz - 2nz$

92. $5yu + 4tu + 5yv + 4tv$

93. $14x - 7y - 2xz + yz$

94. $6a - 15b - 2ac^2 + 5bc^2$

95. $4N^2z - 2M^2z - 6N^2d + 3M^2d$

96. $18sr^3 - 12t^2r^3 - 15sw + 10t^2w$

97. $x^3yz + 2y^2z - x^3a - 2ya$

98. $np - 3npq + 3q - 1$

99. $ay - bz + az - by$

100. $qr - 3pv - 3pr + qv$

101. $tu^2 + 2sv^2 - v^2u^2 - 2st$

102. $m_1a - 6bm_2 + 3m_2a - 2bm_1$

103. If P dollars is deposited in an account earning r percent interest compounded twice a year, then the amount A in the account at the end of one year is

$$A = P\left(1 + \frac{r}{2}\right) + P\left(1 + \frac{r}{2}\right)r\left(\frac{1}{2}\right)$$

By factoring, show that $A = P\left(1 + \frac{r}{2}\right)^2$.

104. The sum S of the squares of the first n positive integers is given by

$$S = \frac{2n(n^2 + n) + (n^2 + n)}{2}$$

Rewrite this formula by factoring the numerator.

For Exercises 105–108, write a formula for the area A of the shaded region.

105.

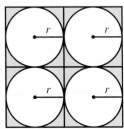

106.

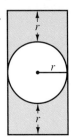

107.

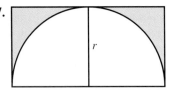

108.

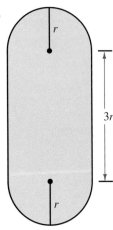

REVIEW PROBLEMS

The following exercises review parts of Section 3.3. Doing these problems will help prepare you for the next section.

Multiply.

109. $(a - 2)(a + 4)$

110. $(x + 12)(x - 3)$

111. $(2w + 1)(3w + 2)$

112. $(3k - 4)(2k - 5)$

113. $(z^2 + 1)(5z^2 - 7)$

114. $(3v^2 + 1)(4v^2 + 9)$

E N R I C H M E N T   E X E R C I S E S

Factor the greatest common factor from each polynomial.

1. $54t^5 + 36t^4 - 72t^3 + 126t^2$

2. $x^{3n+2} + x^{3n}$

3. $12y^{5n+7} - 84y^{5n+9}$

4. If P dollars is invested at an annual interest rate r compounded n times a year, then the amount A at the end of one year is

$$A = P\left(1 + \frac{r}{n}\right)^{n-1} + P\left(1 + \frac{r}{n}\right)^{n-1} r\left(\frac{1}{n}\right)$$

Factor the GCF from the right side and simplify.

Answers to Enrichment Exercises are on page A.11.

3.5

Factoring General Trinomials

OBJECTIVES

▶ *To factor a trinomial of the form* $ax^2 + bx + c$

▶ *To factor out the GCF and then factor the resulting trinomial*

In Section 3.3, we multiplied two binomials such as $x + 2$ and $x + 4$ to obtain a trinomial,

$$(x + 2)(x + 4) = x^2 + 6x + 8$$

In this section, we will start with a trinomial such as $x^2 + 6x + 8$ and factor it into the product of two binomials.

$$x^2 + 6x + 8 = (x + 2)(x + 4)$$

Notice that factoring this trinomial is the reverse of multiplying $x + 2$ and $x + 4$. Given a trinomial, we must find two binomials that are factors.

Going back to the example $x^2 + 6x + 8 = (x + 2)(x + 4)$, we see that the numbers 2 and 4 have two properties.

1. The product of 2 and 4 is 8, the constant term of $x^2 + 6x + 8$.
2. The sum of 2 and 4 is 6, the coefficient of x in $x^2 + 6x + 8$.

From this, we suggest a strategy for factoring a trinomial of the form $x^2 + bx + c$.

STRATEGY

Factoring a trinomial of the form $x^2 + bx + c$

1. Find the two integers whose product is c and whose sum is b.
2. Then,

$$x^2 + bx + c = (x + \text{first integer})(x + \text{second integer})$$

This strategy is illustrated in the next example.

Example 1   Factor $x^2 + 5x + 6$.

Solution   We are looking for two integers whose product is 6 and sum is 5. We list all pairs of positive integers with a product of 6, then check their sum.

	Two integers whose product is 6	Check the sum of the two integers
Trial 1:	1 and 6	$1 + 6 = 7$
Trial 2:	2 and 3	$2 + 3 = 5$

The numbers from the first trial do not work as their sum is 7, not 5. However the numbers 2 and 3 from the second trial do work. Their product is 6 and their sum is 5. Therefore,

$$x^2 + 5x + 6 = (x + 2)(x + 3)$$

We can check our answer by multiplying.

$$(x + 2)(x + 3) = x^2 + 3x + 2x + 6$$
$$= x^2 + 5x + 6$$

Example 2   Factor the following trinomials.

(a) $x^2 + 4x - 12$ **(b)** $a^2 - 9a + 20$ **(c)** $x^2 + 2x + 6$

Solution   **(a)** We want to find two integers whose product is -12 and sum is 4. We list all pairs of integers whose product is -12, then check their sum.

	Two integers whose product is -12	Check the sum of the two integers
Trial 1:	1 and -12	$1 + (-12) = -11$
Trial 2:	-1 and 12	$-1 + 12 = 11$
Trial 3:	2 and -6	$2 + (-6) = -4$
Trial 4:	-2 and 6	$-2 + 6 = 4$

Looking through the four trials, we see that -2 and 6 are the correct numbers. Therefore,

$$x^2 + 4x - 12 = [x + (-2)](x + 6)$$
$$= (x - 2)(x + 6)$$

(b) We are looking for two integers whose product is 20 and sum is -9. Since the product is positive and the sum is negative, the two integers must both be negative. We list all pairs of negative integers whose product is 20, then check their sum.

	Two integers whose product is 20	Check the sum of the two integers
Trial 1:	-1 and -20	$-1 + (-20) = -21$
Trial 2:	-2 and -10	$-2 + (-10) = -12$
Trial 3:	-4 and -5	$-4 + (-5) = -9$

Looking at the list, we see that -4 and -5 are the factors of 20 whose sum is -9. Therefore,

$$a^2 - 9a + 20 = (a - 4)(a - 5)$$

(c) We want two integers whose product is 6 and having a sum of 2. There are two possibilities.

Trial 1:	1 and 6	$1 + 6 = 7$
Trial 2:	2 and 3	$2 + 3 = 5$

Since neither trial gave the correct sum, we conclude that $x^2 + 2x + 6$ cannot be factored using integers. ▲

A polynomial such as $x^2 + 2x + 6$ of Example 2(c) that cannot be factored using integers is called a **prime polynomial.**

Example 3   Factor $x^2 + 2ax - 3a^2$.

Solution   Instead of looking for two integers, we are looking for two *monomials* whose product is $-3a^2$ and whose sum is $2a$. There are two possibilities.

Trial 1:	a and $-3a$	$a + (-3a) = -2a$
Trial 2:	$-a$ and $3a$	$-a + 3a = 2a$

The terms $-a$ and $3a$ are the correct monomials, and

$$x^2 + 2ax - 3a^2 = (x - a)(x + 3a)$$ ▲

In the next example, we show how to factor trinomials of the form $ax^2 + bx + c$, where $a \neq 1$.

Example 4   Factor $2x^2 + 7x + 3$.

Solution   We determine the possible factors of the squared term $2x^2$ and the possible factors of the constant term 3. Then, we list the possible factors of the trinomial and multiply to see which product gives the proper middle term of $7x$.

The factors of $2x^2$ are $2x$ and x. Since the middle term is positive, we use only the positive factors of the constant term 3, which are 1 and 3.

	Possible factors	**Product**
Trial 1:	$(2x + 3)(x + 1)$	$2x^2 + 5x + 3$
Trial 2:	$(2x + 1)(x + 3)$	$2x^2 + 7x + 3$

Since trial 2 gives the correct middle term of $7x$,

$$2x^2 + 7x + 3 = (2x + 1)(x + 3) \qquad \blacktriangle$$

Rather than using the trial and error method for factoring a trinomial, another method is available. In order to factor $ax^2 + bx + c$, we must find integers p, q, r, and s such that

$$ax^2 + bx + c = (px + r)(qx + s)$$

Multiplying the product on the right,

$$ax^2 + bx + c = pqx^2 + (ps + qr)x + rs$$

Comparing the coefficients, we must have

$$a = pq, \qquad b = ps + qr, \qquad \text{and} \qquad c = rs$$

Since $a = pq$ and $c = rs$, we have $ac = (pq)(rs) = (ps)(qr)$.

Therefore, $ax^2 + bx + c$ can be factored only if there are integers whose product is ac and sum is b. We summarize this method.

S T R A T E G Y

Alternate method to factor $ax^2 + bx + c$

Step 1   Multiply a and c.

Step 2   Find a pair of integers whose product is ac and sum is b.

Step 3   Split the middle term bx as the sum of two terms whose coefficients are the two integers found in Step 2.

Step 4   Factor by grouping.

We illustrate this technique in the next example.

Example 5   Factor $2x^2 + 7x + 3$.

Solution   We seek two integers whose product is $ac = 2 \cdot 3 = 6$ and whose sum is $b = 7$. Clearly, 1 and 6 satisfy these conditions. Thus the middle term $7x$ can be expressed as $1 \cdot x + 6x$, or simply $x + 6x$, giving

$$2x^2 + 7x + 3 = 2x^2 + x + 6x + 3$$

Next, factor by grouping.

$$2x^2 + x + 6x + 3 = x(2x + 1) + 3(2x + 1)$$
$$= (x + 3)(2x + 1) \qquad \blacktriangle$$

We now have two methods for factoring a trinomial of the form $ax^2 + bx + c$: the trial and error method shown in Example 4 or the more systematic method shown in Example 5. As you become proficient at factoring, the trial and error method may become the method of choice.

In the next example, we first factor out the GCF, then factor the resulting trinomial.

Example 6   Factor completely: $2z^3 + 8z^2 - 42z$.

Solution   For this polynomial, first factor out the GCF, which is $2z$.

$$2z^3 + 8z^2 - 42z = 2z(z^2 + 4z - 21)$$

Next, we want to factor $z^2 + 4z - 21$, if possible. Using the factoring strategy of this section, we want two integers whose product is -21 and sum is 4. The numbers -3 and 7 work. Therefore,

$$2z^3 + 8z^2 - 42 = 2z(z - 3)(z + 7) \qquad \blacktriangle$$

N O T E   *When factoring a polynomial, always check first to factor out the greatest common factor, if other than one.*

A trinomial of degree higher than 2 can sometimes be factored by making a substitution of a new variable in terms of the given variable.

Example 7   Factor $15x^4 + 7x^2 - 2$.

Solution   Setting $t = x^2$, then $x^4 = (x^2)^2 = t^2$ and the original trinomial becomes

$$15t^2 + 7t - 2$$

Using methods of this section,

$$15t^2 + 7t - 2 = (3t + 2)(5t - 1)$$

Replacing t by x^2,

$$15x^4 + 7x^2 - 2 = (3x^2 + 2)(5x^2 - 1) \qquad \blacktriangle$$

E X E R C I S E   S E T   3.5

For Exercises 1–28, factor. If a polynomial is prime, write "prime."

1. $x^2 + 3x + 2$

2. $y^2 + 6y + 8$

3. $t^2 + 10t + 21$

4. $a^2 + 9a + 14$

5. $c^2 + 5c + 3$

6. $w^2 + 7w + 4$

7. $r^2 + 12r + 35$

8. $x^2 + 10x + 16$

9. $x^2 + x - 6$

10. $y^2 + y - 2$

11. $t^2 + 7t - 8$

12. $u^2 + 2u - 15$

13. $m^2 + 7m + 7$

14. $z^2 + 5z + 4$

15. $a^2 - 7a + 12$

16. $s^2 - 10s + 30$

17. $n^2 - 11n + 24$

18. $y^2 - 4y + 3$

19. $x_1^2 + 7x_1 - 8$

20. $t_2^2 + 7t_2 + 12$

21. $x^2 + 5ax + 6a^2$

22. $z^2 + 7bz + 12b^2$

23. $a^2 + va - 12v^2$

24. $n^2 - 3mn - 4m^2$

25. $p^2 - 11pq + 24q^2$

26. $s^2 - 7ts + 6t^2$

27. $C^2 - 11RC + 30R^2$

28. $d_1^2 - 4d_1t + 3t^2$

For Exercises 29–40, factor as completely as possible.

29. $y^3 - 5y^2 + 6y$

30. $b^5 - 8b^4 + 12b^3$

31. $5a^4 + 5a^3 - 10a^2$

32. $3x^3 - 12x^2 + 9x$

33. $4z^7 + 12z^6 + 8z^5$

34. $6m^4 - 30m^3 + 36m^2$

35. $7D^6 - 14D^5 - 21D^4$

36. $4Y^9 - 20Y^8 + 24Y^7$

37. $6x_1^7 - 60x_1^6 + 96x_1^5$

38. $10v_1^3 - 120v_1^2 + 360v_1$

39. $9u^4 + 9u^3 + 18u$

40. $11k^5 - 11k^4 + 44k^3$

For Exercises 41–68, factor. If a polynomial is prime, write "prime."

41. $2x^2 + 5x + 3$

42. $3y^2 + 5y + 2$

43. $5a^2 + 13a + 6$

44. $2z^2 + 11z + 15$

45. $2c^2 - 11c - 21$

46. $5n^2 - 13n - 6$

47. $3k^2 - 8k + 4$

48. $2m^2 - 13m + 15$

49. $5p^2 + 3p - 14$

50. $3q^2 - q - 2$

51. $2T^2 - 13T + 6$

52. $3B^2 - 11B + 10$

53. $6a_1^2 + 7a_1 - 5$

54. $8x_3^2 - 14x_3 + 3$

55. $10x^2 - 27x + 5$

56. $12r_2^2 - 11r_2 + 2$

57. $10y_2^2 + 3y_2 - 7$

58. $8a^2 - 17a + 2$

59. $11m^2 - 21m + 10$

60. $13y^2 - 15y - 22$

61. $2y^2 + 5ay + 2a^2$

62. $3z^2 - 10cz + 3c^2$

63. $4n_1^2 + 19n_1m_1 - 5m_1^2$

64. $6p_1^2 + 5q_1p_1 - q_1^2$

65. $4x^2 - 5hx - 9h^2$

66. $6a^2 + 5ca - 4c^2$

67. $4s^2 + 16ts + 15t^2$

68. $8u^2 + 18vu - 5v^2$

For Exercises 69–76, factor completely.

69. $4a^5 - 6a^4 - 70a^3$

70. $9y^4 - 42y^3 + 24y^2$

71. $30w^7 + 55w^6 + 15w^5$

72. $24n^3 + 66n^2 + 36n$

73. $42C^{12} - 37C^{11} + 5C^{10}$

74. $5T^{13} + 47T^{12} - 30T^{11}$

75. $30u^3 - 56u^2 + 24u$

76. $12b^4 + 46b^3 + 40b^2$

77. Which pair of factors, $(3x + 2)$ and $(x + 4)$ or $(3x - 2)$ and $(x - 4)$, does *not* give $3x^2 - 14x + 8$ as the product?

78. What polynomial can be factored as $x^3(5x - 4)(2x + 3)$?

79. A commercial peach orchard estimates that if there are n additional trees planted per acre, then the number of bushels of peaches produced per acre, B, is given by

$$B = -(2n^2 - n - 55)$$

Factor the right side of this formula and then find the value of B when four additional trees are planted per acre.

80. A pellet was fired upward from the top of a tower 50 meters tall. After t seconds, its height is h meters from the ground, where

$$h = -5(t^2 - 3t - 10)$$

Factor the right side of this formula and then find h when $t = 3\frac{1}{2}$ seconds.

REVIEW PROBLEMS

The following exercises review parts of Section 3.3. Doing these problems will help prepare you for the next section.

Multiply.

81. $(x - 2)(x + 2)$

82. $(c + 5)(c - 5)$

83. $\left(z + \frac{1}{3}\right)\left(z - \frac{1}{3}\right)$

84. $\left(h + \frac{5}{2}\right)\left(h - \frac{5}{2}\right)$

85. $(s^2 + a)(s^2 - a)$

86. $(R^3 + k)(R^3 - k)$

87. $(x - 2)(x^2 + 2x + 4)$

88. $(a - 3)(a^2 + 3a + 9)$

89. $(x - y)(x^2 + xy + y^2)$

90. $(a + 2b)(a^2 - 2ab + 4b^2)$

ENRICHMENT EXERCISES

For Exercises 1–5, factor by first factoring out a fraction or decimal.

1. $\dfrac{1}{2}x^2 + x - \dfrac{15}{2}$

2. $z^2 + \dfrac{7}{2}z - 2$

3. $0.1a^2 - 0.7a + 1$

4. $w^2 + \dfrac{1}{6}w - \dfrac{1}{6}$

5. $4u^2 + \dfrac{8}{3}u - \dfrac{5}{9}$

6. If $3x^2 - (2a + 4)x + 12$ factors into $(3x - 4)(x - 3)$, what is the value of a?

7. Factor $6x^{2n} - 13x^n - 5$.

Factor completely by first factoring the GCF from the polynomial, then factoring the result.

8. $6a^2x^3y - 26a^2x^2y + 8a^2xy$ **9.** $ca^2 - cb^2$ **10.** $12ay^2 - 2aby - 2ab^2$

Answers to Enrichment Exercises are on page A.12.

3.6

Factoring Special Polynomials

OBJECTIVES

▶ To factor the difference of two squares, $a^2 - b^2$

▶ To recognize and factor perfect square trinomials

▶ To factor the sum of two cubes, $a^3 + b^3$

▶ To factor the difference of two cubes, $a^3 - b^3$

Recall from Section 3.3, the equation for the product of $a + b$ and $a - b$,

$$(a + b)(a - b) = a^2 - b^2$$

We can reverse this equation to obtain the factoring pattern for the difference of two squares.

RULE

Factoring the difference of two squares

$$a^2 - b^2 = (a + b)(a - b)$$

We can use the above formula to factor any binomial that is the difference of two squares. In the first example, we illustrate this technique.

Example 1   Factor each of the following.

(a) $x^2 - 4 = x^2 - 2^2$ **(b)** $4a^2 - 9 = (2a)^2 - 3^2$

$\qquad\qquad = (x + 2)(x - 2)$ $= (2a + 3)(2a - 3)$

(c) $k^2 - 16b^2 = k^2 - (4b)^2$ **(d)** $81C^{10} - 1 = (9C^5)^2 - 1^2$

$\qquad\qquad = (k + 4b)(k - 4b)$ $= (9C^5 + 1)(9C^5 - 1)$

(e) $m^2 - \dfrac{1}{4} = m^2 - \left(\dfrac{1}{2}\right)^2$

$\qquad\qquad = \left(m + \dfrac{1}{2}\right)\left(m - \dfrac{1}{2}\right)$ ▲

The difference of squares formula can also be applied to factor polynomials such as $(3x - 1)^2 - 16$.

Example 2   Factor $(3x - 1)^2 - 16$.

Solution   This polynomial has the form of $a^2 - b^2$, where $a = 3x - 1$. Therefore,

$$(3x - 1)^2 - 16 = (3x - 1 + 4)(3x - 1 - 4)$$

$$= (3x + 3)(3x - 5) \qquad\qquad \text{Simplify.}$$

$$= 3(x + 1)(3x - 5) \qquad\qquad \text{Factor out a 3.} \ \ ▲$$

In the next example, we use two techniques to factor the polynomial completely.

Example 3   Factor $2m^3 - 8m - m^2 + 4$.

Solution   We first factor by grouping, then use the difference of two squares formula.

$$2m^3 - 8m - m^2 + 4 = 2m(m^2 - 4) - (m^2 - 4)$$

$$= (2m - 1)(m^2 - 4)$$

$$= (2m - 1)(m + 2)(m - 2) \qquad\qquad ▲$$

N O T E   *The difference of two squares such as $x^2 - 9$ can be factored into $(x + 3)(x - 3)$. However, the sum of two squares such as $x^2 + 9$ cannot be factored. It is a prime polynomial.*

Two other multiplication formulas from Section 3.3 are the following:

$$(a + b)^2 = a^2 + 2ab + b^2 \qquad \text{and} \qquad (a - b)^2 = a^2 - 2ab + b^2$$

If we reverse these equations, we obtain the factoring forms.

R U L E

Factoring perfect squares

$$a^2 + 2ab + b^2 = (a + b)^2$$

$$a^2 - 2ab + b^2 = (a - b)^2$$

The left side of each equation is called a *perfect square trinomial*. Perfect square trinomials can be factored using our previous techniques; however, if we recognize a trinomial as being a perfect square, we may immediately factor it. Therefore, always keep these two formulas in mind when attempting to factor a trinomial.

Example 4   Factor $16t^2 - 72t + 81$.

Solution   Since the first and last terms are perfect squares, this trinomial may be a perfect square. Since the middle term is negative, we think of the formula

$$(\text{first})^2 - 2(\text{first})(\text{last}) + (\text{last})^2$$

Now,

$$16t^2 - 72t + 81 = (4t)^2 - 2(4t)(9) + (9)^2$$

Therefore, $16t^2 - 72t + 81$ is indeed a perfect square trinomial, and

$$16t^2 - 72t + 81 = (4t - 9)^2$$

We check our answer by squaring $4t - 9$.

$$(4t - 9)^2 = (4t)^2 - 2(4t)(9) + 9^2$$

$$= 16t^2 - 72t + 81$$

so our factoring is correct.   ▲

Example 5   Factor the following trinomials.

(a) $16 + 40p + 25p^2 = 4^2 + 2(4)(5p) + (5p)^2$

$$= (4 + 5p)^2$$

(b) $9w^2 - 42uw + 49u^2 = (3w)^2 - 2(3w)(7u) + (7u)^2$

$$= (3w - 7u)^2$$

(c) $t^4 + 8t^2 + 16 = (t^2)^2 + 8t^2 + 4^2$

$$= (t^2 + 4)^2$$

(d) $(3t + 7)^2 - 10(3t + 7) + 25 = [(3t + 7) - 5]^2$

$$= (3t + 2)^2 \qquad \text{Simplify.} \qquad \blacktriangle$$

There are two final formulas for factoring the sum and difference of two cubes. Consider the product $(a - b)(a^2 + ab + b^2)$. Performing the multiplication, we have

$$(a - b)(a^2 + ab + b^2) = a^3 + a^2b + ab^2 - a^2b - ab^2 - b^3$$

$$= a^3 - b^3$$

This verifies the formula for factoring the difference of two cubes.

R U L E

Factoring the difference of two cubes

$$a^3 - b^3 = (a - b)(a^2 + ab + b^2)$$

If we multiply $(a + b)(a^2 - ab + b^2)$, we obtain the formula for factoring the sum of two cubes.

R U L E

Factoring the sum of two cubes

$$a^3 + b^3 = (a + b)(a^2 - ab + b^2)$$

Example 6   Factor $27 - r^3$.

Solution   The first term is the cube of 3 and the second term is the cube of r. Therefore,

$$27 - r^3 = 3^3 - r^3 = (3 - r)(3^2 + 3r + r^2)$$

$$= (3 - r)(9 + 3r + r^2) \qquad \blacktriangle$$

Example 7   Factor.

(a) $8w^3 + 1 = (2w)^3 + 1^3$

$$= (2w + 1)[(2w)^2 - (2w)(1) + 1^2]$$

$$= (2w + 1)(4w^2 - 2w + 1)$$

(b) $\dfrac{1}{27} - m^3 = \left(\dfrac{1}{3}\right)^3 - m^3$

$$= \left(\dfrac{1}{3} - m\right)\left[\left(\dfrac{1}{3}\right)^2 + \dfrac{1}{3}m + m^2\right]$$

$$= \left(\dfrac{1}{3} - m\right)\left(\dfrac{1}{9} + \dfrac{1}{3}m + m^2\right)$$

(c) $125C^3 + 64D^3 = (5C)^3 + (4D)^3$

$$= (5C + 4D)[(5C)^2 - (5C)(4D) + (4D)^2]$$

$$= (5C + 4D)(25C^2 - 20CD + 16D^2)$$

(d) $s^3 - t^6 = s^3 - (t^2)^3$

$$= (s - t^2)(s^2 + st^2 + t^4)$$ ▲

We summarize the factoring forms that have been introduced in this section.

Special factoring forms

The difference of two squares

$$a^2 - b^2 = (a + b)(a - b)$$

Perfect squares

$$a^2 + 2ab + b^2 = (a + b)^2$$

$$a^2 - 2ab + b^2 = (a - b)^2$$

The sum of two cubes

$$a^3 + b^3 = (a + b)(a^2 - ab + b^2)$$

The difference of two cubes

$$a^3 - b^3 = (a - b)(a^2 + ab + b^2)$$

In the next example, we combine factoring methods.

Example 8   Factor each polynomial completely.

(a) $2v^7 - 16v^4$ **(b)** $3s^3 - 6s^2t + 3st^2$

Solution   **(a)** We factor out the greatest common factor, $2v^4$, from the polynomial, then use the difference of two cubes form.

$$2v^7 - 16v^4 = (2v^4)(v^3 - 8)$$

$$= (2v^4)(v - 2)(v^2 + 2v + 4)$$

(b) We factor out the greatest common factor, $3s$, from the polynomial, then use a perfect squares form.

$$3s^3 - 6s^2t + 3st^2 = (3s)(s^2 - 2st + t^2)$$

$$= (3s)(s - t)^2 \qquad \blacktriangle$$

CALCULATOR CROSSOVER

Factor each trinomial.

1. $x^2 - 320x - 1,625$
2. $a^2 - 43a + 252$
3. $2z^2 + 77z - 294$

Answers **1.** $(x - 325)(x + 5)$ **2.** $(a - 7)(a - 36)$ **3.** $(2z - 7)(z + 42)$

LEARNING ADVANTAGE   *Each day of a new chapter brings you one day closer to a chapter test. Be prepared for it by saving your assignments, correcting your mistakes, saving copies of quizzes, and asking questions in and out of class. The textbook, class notes, and corrected homework sheets should supply you with an abundance of solved problems to help prepare you for the test.*

EXERCISE SET 3.6

For Exercises 1–74, factor as completely as possible.

1. $x^2 - 25$

2. $a^2 - 9$

3. $t^2 - r^2$

4. $u^2 - w^2$

5. $n^2 - p^4$

6. $m^6 - q^2$

7. $s^2 - 9t^2$

8. $4u^2 - v^2$

9. $4y^2 - 1$

10. $16z^2 - 9$

11. $K^2 - 25T^2$

12. $Q^2 - 100R^2$

13. $v^2 - \dfrac{1}{9}$

14. $n^2 - \dfrac{1}{16}$

15. $9m_1{}^2 - \dfrac{1}{49}$

16. $49v_1{}^2 - \dfrac{1}{100}$

17. $x^2 + 1$

18. $y^2 + 9$

19. $c^2 - 81d^2$

20. $64w^2 - b^2$

21. $9x^2 + 16$

22. $49a^2 + 100$

23. $16s^2 - 9y^2$

24. $25r^2 - 49s^2$

25. $(q + 1)^2 - 25p^2$

26. $(n - 7)^2 - 16$

27. $\dfrac{1}{100}t^2 - \dfrac{1}{121}v^2$

28. $\dfrac{1}{49}w^2 - \dfrac{1}{144}u^2$

29. $\dfrac{4}{9}q^2 - v^4$

30. $y^2 - \dfrac{16}{81}z^2$

31. $(4x - 3)^2 - 9$

32. $(5y + 2)^2 - 4$

33. $k^2 - (4k + 3)^2$

34. $(2 - u)^2 - 9u^2$

35. $(-2 + h)^2 - (-2)^2$

36. $(x + h)^2 - x^2$

37. $4(3r + 4)^2 - 25(2r + 1)^2$

38. $16(5t + 1)^2 - 9(7 + 3t)^2$

39. $m^3 - 16m + m^2 - 16$

40. $z^3 - z + 2z^2 - 2$

41. $x^4 + 4x^2 - x^2 - 4$

42. $y^4 + 3y^2 - 9y^2 - 27$

43. $2 - 2w^2 - w + w^3$

44. $27 - 3a^2 - 18a + 2a^3$

45. $-2b^3 + 3b^2 + 8b - 12$

46. $36z^3 + 9z^2 - 16z - 4$

47. $y^2 + 6y + 9$

48. $a^2 + 10a + 25$

49. $z^2 + 12z + 36$

50. $c^2 + 4c + 4$

51. $h^2 + h + 1$

52. $n^2 + 21n + 49$

53. $x^2 - 10x + 25$

54. $m^2 - 6m + 9$

55. $n^2 - 2n + 4$

56. $t^2 - 10t + 36$

57. $v^2 - 14v + 49$

58. $s^2 - 2s + 1$

59. $x^2 + 2ax + a^2$

60. $y^2 + 2by + b^2$

61. $z^2 + 6vz + 9v^2$

62. $w^2 + 4tw + 4t^2$

63. $s^2 - 2cs + c^2$

64. $t^2 - 2mt + m^2$

65. $16k^2 - 8kz + z^2$

66. $4q^2 - 4qy + y^2$

67. $q^4 - 8q^2 + 25$ **68.** $p^6 - 6p^3 + 9$ **69.** $(2d - 3)^2 - 4(2d - 3) + 4$

70. $(1 - 4q)^2 - 8(1 - 4q) + 16$ **71.** $9 + 6(5 - 8p) + (5 - 8p)^2$ **72.** $81 - 18(3y + 2) + (3y + 2)^2$

73. $(2m_1{}^2 + 1)^2 - 6(2m_1{}^2 + 1) + 9$ **74.** $(3Y^2 - 14)^2 + 4(3Y^2 - 14) + 4$

For Exercises 75–86, factor each binomial completely.

75. $z^3 + 27$ **76.** $a^3 + 8$ **77.** $w^3 + \dfrac{1}{8}$

78. $s^3 + \dfrac{1}{27}$ **79.** $64t^3 + 1$ **80.** $8p^3 + 27$

81. $8b^3 - 1{,}000$ **82.** $27n_1{}^3 - 216n_2{}^3$ **83.** $27d^3 + 8v^3$

84. $64X^3 - 27Z^3$ **85.** $a^3b^3 - c^3d^3$ **86.** $x^3y^3 + \dfrac{1}{1{,}000}$

For Exercises 87–118, factor each polynomial completely. Remember to first check for a greatest common factor.

87. $7y^2 - 63$ **88.** $3a^2 - 48$

89. $-2w^2 + 50$ **90.** $-4N^2 + 36$

91. $x^5 - 81x^3$ **92.** $w^6 - 9w^4$

93. $10k^2 - 80k + 160$ **94.** $7h^2 - 14h + 7$

95. $6m^2 + 36m + 54$ **96.** $4c^2 + 16c + 16$

97. $4y^7 - 16y^5$ **98.** $5p^5 - 125p^3$

99. $z^4 - zy^3$ **100.** $2w^3 + 16t^3$

101. $x^4 - y^4$ **102.** $a^6 + b^9$

103. $2s^3r^4 - 12s^4r^3 + 18s^5r^2$ **104.** $0.01x^2 - 625$

105. $3z^4 + 12z^3 + 12z^2$ **106.** $4n^5 + 8n^4 + 4n^3$

107. $25X^8 - 50X^7 + 25X^6$ **108.** $3T^7 - 36T^6 + 108T^5$

109. $a^3 - 2a^2b + ab^2$ **110.** $st^2 - 2s^2t + s^3$

111. $2uv^2 + 4u^2v + 2u^3$ **112.** $3pq^2 + 6p^2q + 3p^3$

113. $hx^3 - 2h^2x^2 + h^3x$ **114.** $yz^3 - 2y^2z^2 + y^3z$

115. $y(3x + 2)^2 - 2y(3x + 2) + y$

116. $m^2(4z - 7)^2 + 8m^2(4z - 7) + 16m^2$

117. $9x^2 + 12xy + 4y^2 - 16$

118. $4x^2 - 4xy + y^2 - 9$

119. Find k so that $x^2 + (2k + 5)x + 16 = (x - 4)^2$.

120. Find k so that $x^2 - ky^2 = (x + 3y)(x - 3y)$.

121. Multiply $a^2 - ab + b^2$ by $a + b$ to verify the factoring form

$$a^3 + b^3 = (a + b)(a^2 - ab + b^2)$$

122. Write a formula in factored form for the area A of the shaded region.

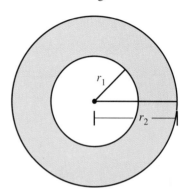

REVIEW PROBLEMS

The following exercises review parts of Section 3.3. Doing these problems will help prepare you for the next section.

Multiply.

123. $2x(4x^2 - 3x + 9)$

124. $(x + 7)(x - 7)$

125. $(2x - 5)(3x - 1)$

126. $(a + 3)^2$

127. $(2b - 5)^2$

128. $(r - 1)(r^2 + r + 1)$

129. $(t + 2)(t^2 - 2t + 4)$

130. $3t^2(2t - 5)(4t - 1)$

131. $(4x - 2a)(4x + 2a)$

132. $c^3(2w - b)(w + 4b)$

133. $4y^2(y - 3)(y^2 + 3y + 9)$

E N R I C H M E N T   E X E R C I S E S

Factor.

1. $4t^3 - t + 2t^2 - \dfrac{1}{2}$

2. $x^6 - 4x^2 - x^4 + 4$

3. $[x^2 - (a^2 + a)]^2 + 2[x^2 - (a^2 + a)]a + a^2$

4. $(a + b)^3 - (a - b)^3$

For Exercises 5 and 6, factor the polynomial into the product of two binomials. The letter n represents a positive integer.

5. $x^{2n} - y^{2n}$ **6.** $t^{4n} - 4t^{2n} + 3$

Answers to Enrichment Exercises are on page A.12.

3.7

General Factoring

O B J E C T I V E

▶   *To use the strategy for general factoring*

We have encountered various methods of factoring polynomials throughout the chapter. The exercise set at the end of this section has been designed for you to develop the ability to properly choose the correct or best method for factoring a polynomial. You will use various factoring methods introduced throughout the chapter. Therefore, we summarize these methods in the following strategy for general factoring.

S T R A T E G Y

A strategy for general factoring

1. **Does the polynomial have a greatest common factor?** Factor out the greatest common factor, if other than one.
2. **How many terms are in the polynomial?** If the polynomial has
    (a) **two terms,** check for the difference of two squares or the sum or difference of two cubes.
    (b) **three terms,** check for a perfect square trinomial. If the coefficient of the square term is not one, use the factoring methods of Section 3.5.
    (c) **four terms,** check for factoring by grouping.
3. **Has the polynomial been factored completely?** Check to see if the factors themselves can be factored further.

L E A R N I N G   A D V A N T A G E   *Giving quizzes during the teaching of a chapter is an instructor's way of forcing you to organize every-thing you have studied so far. Both you and the instructor can measure how much of it you have learned. Although quizzes are not something to look for-ward to with pleasure, they generally have a long-term positive effect because they encourage you to review your notes and homework each day.*

EXERCISE SET 3.7

Use the strategy for general factoring to factor each polynomial completely. If a polynomial is prime, write "prime." The even-numbered exercises are not necessarily matched to the preceding odd-numbered exercises.

1. $x^3 - x^2$

2. $a^5 + a^3$

3. $2y^2 - y - 6$

4. $3t^2 - 14t + 15$

5. $5r^2 + 125$

6. $4h^2 - 36$

7. $y^2 - 4$

8. $k^2 - 16$

9. $2s^7 + 2s^4$

10. $3t^7 + 6t^5 - 15t^4$

11. $v^3 - 8$

12. $s^3 + 27$

13. $t^2 + 4$

14. $h^2 + 121$

15. $4t^2 - 9t + 2$

16. $8h^2 - 25h + 3$

17. $4x_3^2 - 16$

18. $7u_1^2 + 7$

19. $D^2 + 2D + 1$

20. $y^2 - 7y - 8$

21. $18w^2y + 12wy^2 + 30wy$

22. $50n^3m^2 + 20n^2m^3 + 70n^2m^2$

23. $q_1^2 - 6q_1 + 9$

24. $R^2 - 18R + 81$

25. $ta + td + a + d$

26. $c^2 - 10c + 25$

27. $d^2 + 14d + 49$

28. $bx + 7x + by + 7y$

29. $au^2 - av^2 + ku^2 - kv^2$

30. $10a^2 + 37a + 7$

31. $x^3y^3 + 8z^3$

32. $2t^4 - 2t^2$

33. $s^2x + s^2y - t^2x - t^2y$

34. $36y^4 - 114y^3 + 48y^2$

35. $uv^3 - uv$

36. $xz^3 - 4xz$

37. $x^2z + x^2t - y^2z - y^2t$

38. $m^2q + m^2Z - p^2q - p^2Z$

39. $150b^3 + 175b^2 - 125b$

40. $56s^3 - 70s^2 + 21s$

41. $z + z^2 - 72$

42. $r^2 - r - 90$

43. $H^3 - 1$

44. $x^2 - 2ax + a^2$

45. $z^2 - 2hz + h^2$

46. $K^3 + 1$

47. $p^2 + 2pq + q^2$

48. $3x^2 + 3xy + y^2$

49. $2y^4 - 2y^2$

50. $t^2 - 22t + 40$

51. $81x^2 - 100r^2$

52. $8a^7 - 4ya^6 - 4y^2a^5$

53. $12 - 6r + 10s - 5sr$

54. $5t^4 - 5ty^3$

55. $b^2 + 4bh + 4h^2$

56. $4z^2 + 12kz + 9k^2$

57. $c^3n^3m - c^3nm^3 - d^3n^3m + d^3nm^3$

58. $30r^6 - 104r^5 + 90r^4$

59. $9d^2 + 24yd + 16y^2$

60. $x^3ya^3 + x^3yb^3 - xy^3a^3 - xy^3b^3$

61. $16b^5 - 54b^2$

62. $45q^5 - 51q^4 - 12q^3$

63. $r^2t^2 - 100r^2 - 9t^2 + 900$

64. $x^2y^2 - 16x^2 - 4y^2 + 64$

65. $f^6 + g^3$

66. $5 - 8(x + 2t) - 4(x + 2t)^2$

67. $15 - 2T - T^2$

68. $15(r - 2)^2 + 29(r - 2) + 12$

69. $3tw^2 - 29tw - 10t$

70. $2n^4 - 7n^2 - 4$

71. $t^4 - 2t^2 + 1$

72. $(a - 1)^3 - 1$

73. $p^3q^2 - 4p^2q^3 + 4pq^4$

74. $3x^2 - \dfrac{1}{3}$

REVIEW PROBLEMS

The following exercises review parts of Section 2.2. Doing these problems will help prepare you for the next section.

Solve each equation.

75. $a - \dfrac{2}{3} = 0$

76. $h + \dfrac{4}{5} = 0$

77. $2x - 3 = 0$

78. $8t + 24 = 0$

79. $5v + 75 = 0$

80. $3r - 27 = 0$

81. $3y + 4 = 0$

82. $4s - 9 = 0$

83. $\dfrac{2}{3}x + \dfrac{4}{9} = 0$

84. $\dfrac{1}{4}x - 2 = 0$

85. $\dfrac{5}{2}x + 35 = 0$

86. $(0.1)x + 8.3 = 0$

ENRICHMENT EXERCISES

1. Factor $a^2 + 2a^2b^2 + 4b^2 + 4ab - b^4 - a^4$.

2. Factor $4x^4 + 1$ by writing it as $4x^4 + 4x^2 + 1 - 4x^2$.

3. Factor $64t^4 + 1$.

4. Factor $\dfrac{81}{4}w^4 + 1$.

Answers to Enrichment Exercises are on page A.13.

3.8

Solving Quadratic Equations by Factoring

O B J E C T I V E S

▶ *To solve quadratic equations by factoring*

▶ *To solve cubic equations by factoring*

In Chapter 1, we studied the multiplication property of zero, for example, $5 \cdot 0 = 0$, $(-4)0 = 0$, $0 \cdot \left(-\dfrac{1}{2}\right) = 0$, and $0 \cdot 0 = 0$. Notice a common feature in all of these examples. When the product of two numbers is zero, at least one of the factors in the product is zero. The statement is also true for variable expressions.

P R O P E R T Y

The zero-product property

Suppose A and B are two expressions such that $A \cdot B = 0$, then either $A = 0$ or $B = 0$ (or both are zero).

That is, if the product of two factors is zero, at least one of them is zero.

We can use this property to solve an equation like

$$(x - 6)(x - 2) = 0$$

In the product on the left side, think of $x - 6$ as A and $x - 2$ as B. By the Zero-Product Property, $A \cdot B = 0$ means that either A is zero or B is zero (or both). That is, either

$$x - 6 = 0 \qquad \text{or} \qquad x - 2 = 0$$

Solving each of these equations, either $x = 6$ or $x = 2$. We check our two answers by seeing if they each make the original equation true.

Check $x = 6$:	*Check* $x = 2$:
$(x - 6)(x - 2) = 0$	$(x - 6)(x - 2) = 0$
$(6 - 6)(6 - 2) \overset{?}{=} 0$	$(2 - 6)(2 - 2) \overset{?}{=} 0$
$(0)(4) \overset{?}{=} 0$	$(-4)(0) \overset{?}{=} 0$
$0 = 0$	$0 = 0$

Therefore, both answers satisfy the original equation and the solution set is $\{2, 6\}$.

We will use the Zero-Product Property to solve certain *quadratic equations*.

D E F I N I T I O N

A **quadratic equation** is an equation that can be put in the form

$$ax^2 + bx + c = 0$$

where a, b, and c are constants and a is not zero.

The form $ax^2 + bx + c = 0$ is called the **standard form** of a quadratic equation.

Examples of quadratic equations are the following:

$$x^2 - 4x + 12 = 0, \qquad (t - 2)(2t + 1) = 0, \qquad \text{and} \qquad 3x = x^2$$

Notice that the first equation is already in the standard form with $a = 1$, $b = -4$, and $c = 12$. The other two equations are not in standard form, but *could* be put in standard form and, therefore, are quadratic equations.

If the polynomial on the left side of the standard form can be factored, then we can solve the quadratic equation using the Zero-Product Property.

Example 1 Solve the equation $6t^2 + 13t - 5 = 0$.

Solution We begin by factoring the left side.

$$6t^2 + 13t - 5 = 0$$

$$(3t - 1)(2t + 5) = 0$$

Next, use the Zero-Product Property to obtain two linear equations. Therefore, either

$$3t - 1 = 0 \qquad \text{or} \qquad 2t + 5 = 0$$

$$t = \frac{1}{3} \qquad \text{or} \qquad t = -\frac{5}{2}$$

We check our two solutions in the original equation.

Check $t = \frac{1}{3}$:

$$6t^2 + 13t - 5 = 0$$

$$6\left(\frac{1}{3}\right)^2 + 13\left(\frac{1}{3}\right) - 5 \overset{?}{=} 0$$

$$6\left(\frac{1}{9}\right) + \frac{13}{3} - 5 \overset{?}{=} 0$$

$$\frac{2}{3} + \frac{13}{3} - \frac{15}{3} \overset{?}{=} 0$$

$$\frac{2 + 13 - 15}{3} \overset{?}{=} 0$$

$$0 = 0$$

Check $t = -\frac{5}{2}$:

$$6t^2 + 13t - 5 = 0$$

$$6\left(-\frac{5}{2}\right)^2 + 13\left(-\frac{5}{2}\right) - 5 \overset{?}{=} 0$$

$$6\left(\frac{25}{4}\right) - \frac{65}{2} - 5 \overset{?}{=} 0$$

$$\frac{75}{2} - \frac{65}{2} - \frac{10}{2} \overset{?}{=} 0$$

$$\frac{75 - 65 - 10}{2} \overset{?}{=} 0$$

$$0 = 0$$

Both answers satisfy the original equation, so either $t = \frac{1}{3}$ or $t = -\frac{5}{2}$. ▲

Example 2   Solve the equation $(x + 6)(x + 5) = 2$.

Solution   This equation is not in standard form; notice that the right side is 2, not 0. Therefore, we first multiply the left side.

$$(x + 6)(x + 5) = 2$$
$$x^2 + 11x + 30 = 2$$
$$x^2 + 11x + 28 = 0 \qquad \text{Subtract 2 from both sides.}$$
$$(x + 4)(x + 7) = 0 \qquad \text{Factor the left side.}$$

Therefore, either $x + 4 = 0$ or $x + 7 = 0$. Thus, either $x = -4$ or $x = -7$. We check these answers in the original equation.

Check $x = -4$:	*Check* $x = -7$:
$(x + 6)(x + 5) = 2$	$(x + 6)(x + 5) = 2$
$(-4 + 6)(-4 + 5) \stackrel{?}{=} 2$	$(-7 + 6)(-7 + 5) \stackrel{?}{=} 2$
$(2)(1) \stackrel{?}{=} 2$	$(-1)(-2) \stackrel{?}{=} 2$
$2 = 2$	$2 = 2$

The solution set is $\{-4, -7\}$.   ▲

We summarize the steps used in the preceding example for solving a quadratic equation.

S T R A T E G Y

Steps for solving a quadratic equation

Step 1   Put in standard form $ax^2 + bx + c = 0$, if necessary.

Step 2   Factor the polynomial.

Step 3   Use the Zero-Product Property and set each factor equal to zero.

Step 4   Solve the resulting linear equations.

Step 5   Check each answer from Step 4 in the ORIGINAL equation.

The equation in the next example is not quadratic; however, it can be solved by the methods of this section.

Example 3

Solve for t: $2t^3 - 8t^2 - 10t = 0$.

Solution

We factor the left side of this equation.

$$2t(t^2 - 4t - 5) = 0 \qquad \text{GCF is } 2t.$$

$$2t(t + 1)(t - 5) = 0$$

Next, we use the Zero-Product Property and set each variable factor equal to zero. Either

$$2t = 0, \qquad t + 1 = 0, \qquad \text{or} \qquad t - 5 = 0$$

Therefore, either $t = 0$, $t = -1$, or $t = 5$.

It is up to you to check these answers in the original equation.   ▲

In the next example, we show how certain word problems can be solved using quadratic equations.

Example 4

The sum of two numbers is 15 and their product is 50. Find the numbers.

Solution

Let n represent one of the numbers. Since the sum of the two numbers is 15, then $15 - n$ is the other number. The product of the two numbers is 50. Therefore,

$$n(15 - n) = 50$$

We solve this equation using the five-step strategy.

$$15n - n^2 - 50 = 0$$

$$n^2 - 15n + 50 = 0 \qquad \text{Multiply both sides by } -1.$$

$$(n - 5)(n - 10) = 0 \qquad \text{Factor.}$$

Therefore, either $n = 5$ or $n = 10$.

When $n = 5$, $15 - n = 15 - 5 = 10$.
When $n = 10$, $15 - n = 15 - 10 = 5$.

Therefore, in either case, the two numbers are 5 and 10. We check these numbers to see if they satisfy the conditions of the problem. The sum of the two numbers is $5 + 10$ or 15 and the product is $5(10)$ or 50. The conditions are satisfied and the two correct numbers are 5 and 10.   ▲

C A L C U L A T O R   C R O S S O V E R

The sum of two numbers is 144 and their product is 5,040. Find the two numbers.

Answer   60 and 84.

L E A R N I N G   A D V A N T A G E   *If your instructor gives quizzes, take advantage of the situation by preparing for them; every minute you study and review now can save two minutes of study time at the end of the chapter. Also, you may have the impression that (1) you do not know much of the mathematics that the class has studied, or (2) you can do all the problems pretty well. The results of a quiz will tell you whether your impression is right or wrong.*

E X E R C I S E   S E T   3.8

For Exercises 1–70, solve the equation.

1. $x^2 + x - 6 = 0$

2. $t^2 - t - 30 = 0$

3. $2a^2 - 15a + 7 = 0$

4. $3n^2 + 14n - 24 = 0$

5. $2z^2 + 19z - 10 = 0$

6. $7c^2 - 39c + 20 = 0$

7. $x^2 - 16 = 0$

8. $z^2 - 25 = 0$

9. $a^2 = 36$

10. $h^2 = 64$

11. $u^2 - \dfrac{1}{4} = 0$

12. $r^2 - \dfrac{1}{9} = 0$

13. $s^2 = \dfrac{4}{25}$

14. $y^2 = \dfrac{16}{81}$

15. $x^2 - 6x + 9 = 0$

16. $w^2 - 10w + 25 = 0$

17. $b^2 + 4b + 4 = 0$

18. $u^2 + 20u + 100 = 0$

19. $b^2 - 2b = 0$

20. $y^2 + 10y = 0$

21. $6u^2 = 2u$

22. $10w^2 = -35w$

23. $2y^2 + 8y - 42 = 0$

24. $3v^2 + 6v - 9 = 0$

25. $3y^2 - 14y = 5$

26. $4z^2 - 15 = 17z$

27. $5a^2 - 17a = 12$

28. $b^2 = 16b - 48$

29. $5c = 2c^2$

30. $12m = 5m^2$

31. $18s = 6s^2$

32. $30t = 6t^2$

33. $2c(c - 2) = 5(c - 2)$

34. $3z(z + 5) = 10(z + 5)$

35. $(x + 1)^2 = x + 7$

36. $(r - 3)^2 = 4$

37. $(k + 1)(k - 2) = 28$

38. $(t - 2)(t + 5) = 18$

39. $(2s - 3)(s + 5) = -11$

40. $(3m - 10)(m + 7) = -76$

41. $(z - 2)(2z - 3)(z + 4) = 0$

42. $(2y + 5)(3y - 7)(y - 6) = 0$

43. $6u^3 - u^2 - u = 0$

44. $3t^3 + 16t^2 + 5t = 0$

45. $12m^3 + 15m^2 - 18m = 0$

46. $10q^3 - 34q^2 + 12q = 0$

47. $(a^2 - 4)(a^2 - 9) = 0$

48. $(2w^2 - 50)(w^2 - 1) = 0$

49. $q^3 - 16q + q^2 - 16 = 0$

50. $2r^3 - 2r + 3r^2 - 3 = 0$

51. $20t - 5t^3 - 4 + t^2 = 0$

52. $3C - 3C^3 + 1 - C^2 = 0$

53. $2(a + 1)^2 - (a + 1) - 6 = 0$

54. $4(2t - 1)^2 - 9(2t - 1) + 2 = 0$

55. $2s^4 - s^2 - 1 = 0$

56. $3x^4 - 11x^2 - 4 = 0$

57. $(3a - 1)^2 - 9 = 0$

58. $(4z + 3)^2 - 25 = 0$

59. $5(x + 2)(x - 2) = 8 - 31x$

60. $(6t - 1)(t + 1) = 3$

61. $(2z + 5)(z - 3) = -9$

62. $(2b + 1)(b - 3) = b(b - 3)$

63. $(2v + 1)(v - 2) + v(v + 2) = 0$

64. $(3r - 2)(r + 1) - (5r - 3)(r + 1) = 0$

65. $\dfrac{u^2}{2} - \dfrac{7u}{6} + \dfrac{1}{3} = 0$

66. $x(1 - x) = \dfrac{1}{4}$

67. $(3x - 2)^2 - 9x^2 = 0$

68. $(1 - x)^3 - 16(1 - x) = 0$

69. $2x^2(4x^2 - 9) - x(4x^2 - 9) - (4x^2 - 9) = 0$

70. $20a^2(3 - a) = 16a(3 - a) + 4a^3(3 - a)$

For Exercises 71–74, solve the word problem.

71. The sum of two numbers is $\dfrac{13}{6}$ and their product is one. Find the two numbers.

72. The sum of two numbers is 20 and their product is 96. Find the two numbers.

73. One number is 18 more than another number and their product is -32. Find the two numbers. (There are two answers; that is, two pairs of numbers that satisfy the conditions.)

74. One number is 20 more than another number and their product is -51. Find the two numbers. (There are two answers; that is, two pairs of numbers that satisfy the conditions.)

REVIEW PROBLEMS

The following exercises review parts of Section 1.5. Doing these problems will help prepare you for the next section.

Represent one number by a variable. Then, express the other number in terms of the variable. Be sure to clearly state what your variable represents.

75. Tom's age now and his age 15 years ago.

76. Two hundred items are separated into two groups—defective and nondefective.

77. A ten-foot board is cut into two pieces.

78. The wholesale cost of a bed and the retail price, where the retail price is 120% of the wholesale cost, decreased by $50.

79. The sum of two consecutive integers.

80. The sum of two consecutive even integers.

81. The sum of the squares of two consecutive odd integers.

82. The square of the sum of two consecutive odd integers.

E N R I C H M E N T   E X E R C I S E S

Solve.

1. $2(1 - 3x)^2(x^2 - 9) - 5(1 - 3x)(x^2 - 9) - 3(x^2 - 9) = 0$

2. $2\left(\dfrac{1}{x}\right)^2 - \dfrac{1}{x} - 1 = 0$

3. $x^5 + 9 - x^2 - 9x^3 = 0$

4. If one solution of $7x^2 + (2k - 5)x - 20 = 0$ is -4, find the value of k. What is the other solution of the quadratic equation?

Write each equation in the form $(x - h)^2 + (y - k)^2 = c$.

5. $x^2 - 2x + 1 + y^2 - 4y + 4 = 9$ **6.** $x^2 + y^2 - 6x - 8y = 0$ **7.** $3x^2 + 3y^2 - 12x + 6y - 6 = 0$

Answers to Enrichment Exercises are on page A.13.

3.9

Applications of Quadratic Equations

O B J E C T I V E

▶ *To solve word problems using quadratic equations*

In this section, we solve word problems that involve quadratic equations in their solutions. Keep in mind the strategy for solving word problems. For your convenience, we restate this strategy that was first given in Chapter 2.

S T R A T E G Y

A strategy for solving word problems

1. **Read** the problem carefully. Take note of what is being asked and what information is given.
2. **Plan** a course of action. Represent the unknown number by a letter. If there are two or more unknowns, represent one of them by a letter and express the others in terms of the letter.
3. **Create** an equation from the given information.
4. **Solve** this equation.
5. **Check** your solution using the original statement of the problem.

Example 1

The square of a number is 28 more than three times the number. Find the number.

Solution

Let $n =$ the number. From the information given, we write an equation.

The square of a number is 28 more than three times the number.

$$n^2 = 3n + 28$$

To solve this quadratic equation, we first put it in standard form.

$$n^2 - 3n - 28 = 0 \qquad \text{Subtract } 3n + 28 \text{ from both sides.}$$

$$(n + 4)(n - 7) = 0 \qquad \text{Factor the left side.}$$

By the Zero-Product Property, either

$$n = -4 \quad \text{or} \quad n = 7$$

Next, we check each number.

Check $n = -4$: The square of -4 is $(-4)^2 = 16$. Three times -4 is -12, and 16 is 28 more than -12,

$$-12 + 28 = 16$$
$$16 = 16$$

Check $n = 7$: The square of 7 is $7^2 = 49$. Three times 7 is 21, and 49 is 28 more than 21,

$$21 + 28 = 49$$
$$49 = 49$$

Therefore, both numbers, -4 and 7, satisfy the conditions of the problem. There are two correct answers. ▲

An important theorem from geometry is the Pythagorean Theorem, which gives a relationship among the sides of a right triangle. A right triangle is a triangle with a 90 degree angle.

T H E O R E M

Pythagorean theorem

Consider the right triangle with the longest side (hypotenuse) having length c and the other two sides (legs) having lengths a and b.

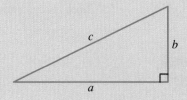

Then,

$$c^2 = a^2 + b^2$$

Example 2   The lengths of the sides of a right triangle are three consecutive even integers. Find the lengths of the three sides.

Solution Let x = the length of the shortest side. Since the lengths are consecutive even integers, then $x + 2$ and $x + 4$ are the lengths of the other two sides. Since $x + 4$ is the largest length, it must be the length of the hypotenuse. By the Pythagorean Theorem,

$$(x + 4)^2 = (x + 2)^2 + x^2$$

To solve this equation, we simplify and put it in standard form.

$$x^2 + 8x + 16 = x^2 + 4x + 4 + x^2$$

$$x^2 - 4x - 12 = 0$$

$$(x - 6)(x + 2) = 0 \qquad\qquad \text{Factor the left side.}$$

$$x = 6 \qquad \text{or} \qquad x = -2$$

We discard the solution $x = -2$, since x is a length and cannot be negative. The shortest side, therefore, is 6 inches, the other two sides are 8 and 10 inches. As a check, does $10^2 = 6^2 + 8^2$?

$$10^2 = 100 \qquad \text{and} \qquad 6^2 + 8^2 = 36 + 64 = 100$$

so the answers are correct. ▲

Example 3 A rectangle has a width that is one inch more than one-half the length. If the area is 12 square inches, find the length and width of the rectangle.

Solution Let x be the length of the rectangle as shown in the figure. Then, the width is $\frac{1}{2}x + 1$.

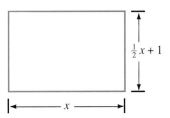

Now, length × width = area, and the area is 12 square inches. Therefore,

$$x\left(\frac{1}{2}x + 1\right) = 12$$

Next, we write this quadratic equation in standard form.

$$\frac{1}{2}x^2 + x - 12 = 0$$

We multiply both sides of the equation by 2 to clear the fraction,

$$2\left(\frac{1}{2}x^2 + x - 12\right) = 2 \cdot 0$$

$$x^2 + 2x - 24 = 0$$

Next, we solve this equation by factoring the left side to obtain

$$(x + 6)(x - 4) = 0$$

Therefore, either $x = -6$ or $x = 4$. Since x represents the length of a rectangle, x must be positive and therefore the number -6 is disqualified.

The length of the rectangle is $x = 4$ inches, and the width is $\frac{1}{2}x + 1 =$

$\frac{1}{2}(4) + 1 = 3$ inches.   ▲

N O T E   *Always check your answers in a word problem to see if they make sense. In Example 3, we obtained a negative value for x, but x represented a length.*

The next example is from business. If a company produces x items a day, then the daily cost C can sometimes be expressed as a polynomial of degree 2.

Example 4   A company that makes wooden mini blinds has a daily cost C (in dollars) of

$$C = 490 + 4x + 0.3x^2$$

where x is the number of mini blinds made each day. On one day the cost C was \$560. How many mini blinds were made that day?

Solution   Since the cost $C = \$560$, we have

$$490 + 4x + 0.3x^2 = 560$$

$$0.3x^2 + 4x + 490 - 560 = 0$$

$$0.3x^2 + 4x - 70 = 0$$

Next, we clear the equation of the decimal by multiplying both sides by 10.

$$10(0.3x^2 + 4x - 70) = 10 \cdot 0$$

$$3x^2 + 40x - 700 = 0$$

Next, we factor the left side.

$$(3x + 70)(x - 10) = 0$$

Either $x = -\dfrac{70}{3}$ or $x = 10$. We discard the negative answer. Therefore, the company made 10 mini blinds.   ▲

CALCULATOR   CROSSOVER

The sum S of the first n positive integers is given by

$$S = \frac{n(n + 1)}{2}$$

Find n, if $S = 1,830$.

Answer $n = 60$.

LEARNING   ADVANTAGE   *Each chapter ends with a Summary and Review that highlights the major points of the chapter. Also, there is a Review Exercise Set. Make these part of your homework at least two nights before the test; then you will still have time to see your instructor if any questions come up.*

EXERCISE SET 3.9

For Exercises 1–16, solve the problem, keeping in mind the strategy for solving word problems.

1. Find all numbers so that the square is equal to the difference of 48 and twice the number.

2. Find all numbers so that the square is equal to the difference of 60 and seven times the number.

3. The sum of two numbers is 20 and their product is 99. Find the two numbers.

4. The sum of two numbers is 19 and their product is 84. Find the two numbers.

5. The sum of two numbers is 22 and their product is 120. Find the two numbers.

6. The sum of two numbers is 60 and their product is 800. Find the two numbers.

7. Two consecutive positive integers have the property that the sum of the squares is 145. Find the two integers.

8. Two consecutive positive integers have the property that twice the square of the smaller integer plus three times the square of the other integer is 35. Find the two integers.

9. Two consecutive negative even integers have the property that the sum of the squares is 52. Find the two integers.

10. Two consecutive positive odd integers have the property that twice the square of the smaller integer plus four times the square of the other integer is 118. Find the two integers.

11. The difference of two negative numbers is eight and their product is 20. Find the two numbers.

12. The difference of two negative numbers is four and their product is 45. Find the two numbers.

13. The difference of two numbers is 13 and their product is -36. Find the two numbers. (There are two pairs of numbers satisfying the conditions.)

14. The difference of two numbers is 23 and their product is -90. Find the two numbers. (There are two pairs of numbers satisfying the conditions.)

15. Two positive numbers have the property that one of them is five less than twice the other. If their product is 12, find the two numbers.

16. Two positive numbers have the property that one of them is 13 less than three times the other. If their product is 10, find the two numbers.

For Exercises 17–40, solve each word problem. See the inside front cover for formulas involving geometric figures.

17. The lengths, measured in inches, of the three sides of a right triangle are three consecutive integers. Find these three lengths.

18. One leg of a right triangle is 1 inch shorter than the other leg. If the hypotenuse is 5 inches, find the length of the two legs.

19. The hypotenuse of a right triangle is 1 cm more than the longer of the two legs. The shorter leg is one less than half of the longer leg. Find the lengths of the three sides.

20. A rectangular box has a width that is 1 meter less than the length. If the height is 2 meters and the volume is 24 cubic meters, find the width and length.

21. A rectangle is shown in the figure. If the area is 33 square feet, find the length and the width.

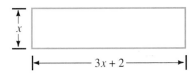

22. A rectangle is shown in the figure. If the area is 40 square feet, find the length and the width.

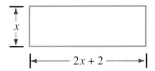

23. A rectangle has a width that is 2 feet more than two-thirds the length. If the area is 12 square feet, find the length and width of the rectangle.

24. A rectangle has a width that is 1 meter less than three-fourths the length. If the area is 8 square meters, find the length and width of the rectangle.

25. A triangle is shown in the figure below. If the area is 12 square cm, find the base and height.

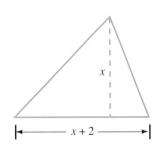

26. A triangle is shown in the figure below. If the area is 10 square inches, find the base and height.

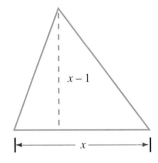

27. A triangle has a height that is two less than two-thirds the base. If the area is $1\frac{1}{3}$ square inches, find the base and the height of the triangle.

28. A triangle has a height that is four more than two fifths of the base. If the area is 15 square feet, find the base and the height of the triangle.

29. The triangular entrance of a tent has a base that is 3 feet less than twice the height. If the area is 27 square feet, find the base and the height of the entrance.

30. The rectangular top of a storage box has a length that is 5 feet less than three times the width. If the area of the top is 28 square feet, what are the length and width of the top?

31. Two ice skaters are at the same place on a lake, one skating south and the other skating east. When the person skating east had gone x miles, the person skating south had gone $\frac{3}{4}x$ miles, and the distance between them was $\frac{1}{4}$ mile more than the distance skated by the person going east. Find the distance between the two skaters at that time.

32. Two bicyclists start at the same point, one pedaling north and the other west. When the person going west has traveled x miles, the person going north has traveled $\frac{4}{3}x$ miles, and the distance between them is two thirds more than x miles. How far had the bicyclist going west traveled by this time?

33. Consider the trapezoid shown in the figure below. If the area is 9 square feet, find the lower base, upper base, and the height.

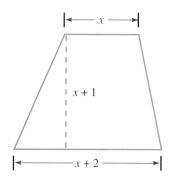

34. Consider the trapezoid shown in the figure below. If the area is 16 square feet, find the lower base, upper base, and the height.

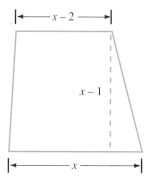

35. The Mellon Patch Company makes primitive dolls. The daily cost C of making x dozen dolls is given by

$$C = 20x^2 - 120x + 400$$

On one day, the daily cost was $1,200. How many dolls were made that day?

36. A crafts store makes sandals from natural palm leaves. It has a daily cost C from producing x dozen pairs of sandals given by

$$C = x^2 - 18x + 120$$

On one day the daily cost was $8,320. How many pairs of sandals were produced that day?

37. A projectile is fired vertically upward. The distance s that it is above the ground after t seconds is given by

$$s = 128t - 16t^2$$

How long does it take the projectile to reach 156 feet?

38. A projectile is fired vertically upward. The distance s that it is above the ground after t seconds is given by

$$s = 112t - 16t^2$$

How long does it take the projectile to reach 27 feet?

39. Juicy Orchards has discovered that if there are n additional trees planted per acre, then the number of bushels of apples produced per acre, B, is given by

$$B = (n + 50)(30 - 6n)$$

If the goal is to have a production of 324 bushels per acre, how many additional trees should be planted per acre?

40. The manager of the Bates Motel knows that if the rent per room is increased by p dollars above $30, then the total daily income I is given by

$$I = (p + 30)(85 - 4p)$$

If the desired daily income is $2,275, how much should the manager raise the rent?

REVIEW PROBLEMS

The following exercises review parts of Section 1.1. Doing these problems will help prepare you for the next chapter.

Simplify.

41. $\dfrac{2}{3} + \dfrac{5}{6}$

42. $\dfrac{8}{3} \cdot \dfrac{9}{2}$

43. $\dfrac{2}{3} - \dfrac{5}{6} + \dfrac{7}{12}$

44. $\dfrac{5}{2} + \dfrac{3}{4} - \dfrac{11}{8}$

45. $\left(\dfrac{3}{4}\right)\left(\dfrac{2}{5}\right)\left(\dfrac{8}{9}\right)$

46. $\dfrac{5}{2} \cdot \dfrac{12}{9} \cdot \dfrac{18}{20}$

47. $\dfrac{2}{3} \cdot \dfrac{6}{20} - \dfrac{3}{10} \cdot \dfrac{2}{3}$

48. $\dfrac{4}{3}\left(\dfrac{1}{2} - \dfrac{6}{5}\right)$

ENRICHMENT EXERCISES

1. Find three consecutive odd integers so that the cube of the largest reduced by the product of all three is 238.

2. A wire 20 inches long is cut into two pieces. Each piece is bent into a square. How should the wire be cut so that the sum of the areas of the two squares is $\dfrac{125}{8}$ square inches?

3. Find all numbers x such that the sum of twice x and one, squared, reduced by twice the sum of twice x and one, is three.

4. The kinetic energy is the energy an object possesses by virtue of its motion. An object of mass m kg moving at v meters per second has a kinetic energy of KE joules given by $KE = \dfrac{1}{2}mv^2$. An object having a mass of 8 kg has a kinetic energy of 3,600 joules. Find the velocity v of the object.

5. A baseball that weighs 0.16 kg is hit by the batter, giving the ball a kinetic energy of 128 joules. At what speed does the ball leave the bat?

6. Sandra wants to double the area of her rectangular flower garden by adding a strip of equal width to each of the four sides. If the original garden measures 20 feet by 30 feet, how wide a strip must be added?

Answers to Enrichment Exercises are on page A.14.

CHAPTER 3

Summary and review

Examples

$x^3 \cdot x^4 = x^7$, $2^0 = 1$, $(t^2)^5 = t^{10}$, $(xy)^6 = x^6 y^6$, $s^{-2} = \dfrac{1}{s^2}$,

$\dfrac{r^4}{r^3} = r$, $\left(\dfrac{z}{2}\right)^3 = \dfrac{z^3}{8}$, $\left(\dfrac{4}{3}\right)^{-2} =$

$\left(\dfrac{3}{4}\right)^2 = \dfrac{9}{16}$

Properties of exponents; Scientific notation (3.1)

You should be able to use properties of exponents to **simplify expressions.**

$\dfrac{(0.000045)(20,000)}{0.009}$

$= \dfrac{(4.5 \times 10^{-5})(2 \times 10^4)}{9 \times 10^{-3}}$

$= \dfrac{(4.5)(2)}{9} \times 10^{-5+4+3}$

$= 1 \times 10^2$

$= 100$

You should be able to simplify numerical expressions using **scientific notation.**

Sums and differences of polynomials (3.2)

$3x^2 y^5$ is a monomial of degree 7.
-2.8 is a monomial of degree zero.

You should be able to **determine the degree of a monomial.**

$5x - 3y$ is a binomial.
$r^2 - 7r + 2$ is a trinomial.

You should be able to **identify polynomials** that are monomials, binomials, or trinomials.

$x^3 + 3x^2 - 2 + 5x^2 + 7 - x^3 = 8x^2 + 5$.
This is a binomial of degree 2.

You should be able to **simplify polynomials** by combining like terms. Then, **determine the degree.**

Multiplying polynomials (3.3)

$(-2x^4 y^2)(3xy^5)$
$= (-2)(3)x^4 x y^2 y^5$
$= -6x^5 y^7$

You should be able to **multiply two monomials.**

Examples

$$\frac{6s^3t^5}{2st^2} = \left(\frac{6}{2}\right)\left(\frac{s^3}{s}\right)\left(\frac{t^5}{t^2}\right)$$
$$= 3s^2t^3$$

You should be able to **divide two monomials.**

$4x^2(5x^3 - 2x + 9)$
$$= 20x^5 - 8x^3 + 36x^2$$

You should be able to **multiply a monomial and a polynomial.**

$(3a - b)(2a + 5b)$
$$= 6a^2 + 15ab - 2ba - 5b^2$$
$$= 6a^2 + 13ab - 5b^2$$
$(2x + 3)^2 = 4x^2 + 12x + 9$
$(w^2 - 2z)^2 = w^4 - 4w^2z + 4z^2$
$(3p + q)(3p - q) = 9p^2 - q^2$

You should be able to **multiply two polynomials** using **special product formulas** when appropriate.

Greatest common factor (3.4)

The GCF of $6x^4 - 10x^3 + 12x^2$ is $2x^2$.
Factoring it out,

$6x^4 - 10x^3 + 12x^2$
$$= 2x^2(3x^2 - 5x + 6)$$

You should be able to **find the GCF** of a polynomial and **factor it out.**

Factoring by Grouping

$6x^2 - 4x + 3x - 2$
$$= 2x(3x - 2) + (3x - 2)$$
$$= (3x - 2)(2x + 1)$$

You should be able to **factor by grouping.**

Factoring trinomials (3.5)

$x^2 + 7x + 12 = (x + 3)(x + 4)$

$x^2 - x - 2 = (x + 1)(x - 2)$

$x^2 - 6x + 5 = (x - 1)(x - 5)$

$6x^2 + 7x - 3$
$$= (2x + 3)(3x - 1)$$

$x^2 + 4x + 1$ is a prime polynomial.

You should be able to **factor various types of polynomials.**

Factoring special polynomials (3.6)

$x^2 + 10x + 25 = (x + 5)^2$

$x^2 - 6x + 9 = (x - 3)^2$

$4x^2 - 9y^2$
$$= (2x + 3y)(2x - 3y)$$

$x^3 - 8 = (x - 2)(x^2 + 2x + 4)$

$x^3 + 1 = (x + 1)(x^2 - x + 1)$

You should be able to **use special forms** to factor polynomials.

Examples

$18x^4 - 39x^3 + 15x^2$
$= 3x^2(6x^2 - 13x + 5)$
$= 3x^2(2x - 1)(3x - 5)$

$x^6 - 8x^5 + 16x^4$
$= x^4(x^2 - 8x + 16)$
$= x^4(x - 4)^2$

$x^4 + 5x^3 - x^2 - 5x$
$= x(x^3 + 5x^2 - x - 5)$
$= x[x^2(x + 5) - (x + 5)]$
$= x(x + 5)(x^2 - 1)$
$= x(x + 5)(x + 1)(x - 1)$

Solve: $6x^2 + x - 2 = 0$.

$(2x - 1)(3x + 2) = 0$

Either $2x - 1 = 0$
$x = \dfrac{1}{2}$
or
$3x + 2 = 0$
$x = -\dfrac{2}{3}$

Check $x = \dfrac{1}{2}$:
$6x^2 + x - 2 = 0$
$6\left(\dfrac{1}{2}\right)^2 + \dfrac{1}{2} - 2 \overset{?}{=} 0$
$6\left(\dfrac{1}{4}\right) + \dfrac{1}{2} - 2 \overset{?}{=} 0$
$\dfrac{3}{2} + \dfrac{1}{2} - 2 \overset{?}{=} 0$
$\dfrac{4}{2} - 2 \overset{?}{=} 0$
$0 = 0$

Check $x = -\dfrac{2}{3}$:
$6x^2 + x - 2 = 0$
$6\left(\dfrac{-2}{3}\right)^2 - \dfrac{2}{3} - 2 \overset{?}{=} 0$
$6\left(\dfrac{4}{9}\right) - \dfrac{2}{3} - 2 \overset{?}{=} 0$
$\dfrac{8}{3} - \dfrac{2}{3} - 2 \overset{?}{=} 0$
$\dfrac{6}{3} - 2 \overset{?}{=} 0$
$0 = 0$

General factoring (3.7)

This is a **strategy for general factoring.**

1. **Does the polynomial have a greatest common factor?** Factor out the greatest common factor, if other than one.
2. **How many terms are in the polynomial?** If the polynomial has
   (a) **two terms,** check for the difference of two squares or the sum or difference of two cubes.
   (b) **three terms,** check for a perfect square trinomial. If the coefficient on the square term is not one, use the factoring methods of Sections 3.5 or 3.7.
   (c) **four terms,** check for factoring by grouping.
3. **Has the polynomial been factored completely?** Check to see if the factors themselves can be factored further.

Solving quadratic equations by factoring (3.8)

You should be able to **solve quadratic equations by factoring.**

Applications of quadratic equations (3.9)

When **solving a word problem,** keep in mind the following strategy.

A Strategy for Solving Word Problems

1. **Read** the problem carefully. Take note of what is being asked and what information is given.
2. **Plan** a course of action. Represent the unknown number by a letter. If there are two or more unknowns, represent one of them by a letter and express the others in terms of the letter.
3. **Create** an equation from the given information.
4. **Solve** this equation.
5. **Check** your solution using the original statement of the problem.

CHAPTER 3 REVIEW EXERCISE SET

Section 3.1

For Exercises 1–6, simplify and write your answers without negative exponents. Assume that all variables are positive.

1. $x^5 \cdot x^3$

2. $(2rs^3)^2 \cdot (-r^2s^4)^3$

3. $(a^{-2})^{-1}$

4. $\dfrac{3c^4d^2}{c^2d^3}$

5. $\left(\dfrac{w^2}{z^3}\right)^{-3}$

6. $\dfrac{(u^{-1}v^{-2})^{-1}}{(2uv^4)^3}$

For Exercises 7–10, express each number in scientific notation.

7. 5,900,000

8. 0.000492

9. 0.0000214

10. 23.16×10^4

For Exercises 11 and 12, evaluate. Express your answer in scientific notation and in decimal notation.

11. $(3.2 \times 10^4)(4 \times 10^{-3})$

12. $(0.00003)(62,000)$

Section 3.2

For Exercises 13–15, simplify each polynomial. Write the result in descending powers. What is the degree of the polynomial?

13. $5 - 3x^4 - 3x^5 + 8x^4 + 2x^3 - 8 + 3x^5$

14. $-(3 - 2y^2 - y) + 4y - 5y^2$

15. $4(3a^3 - 5a + 9) - 2(1 - 2a + 4a^3)$

16. The sum S of the first n positive integers is given by the formula $S = \dfrac{1}{2}n^2 + \dfrac{1}{2}n$. Find the sum of the first 40 positive integers.

17. Subtract $4z^3 - 5z^2 + 2z - 7$ from $2z^3 + 4z^2 - 7z + 1$.

18. In a triangle, the measures of two angles are $4A - 50$ and $5A + 20$ degrees. What is the measure of the third angle?

Section 3.3

For Exercises 19–21, simplify each expression. Write your answer without negative exponents.

19. $(4x^2y^3)(-3x^5y^2)$

20. $(3r^3t^2)^2(-r^2t^4)^3$

21. $\dfrac{6x^4y^3z^5}{24x^5y^2z}$

For Exercises 22–30, do the indicated multiplication, then simplify.

22. $r^3(9r^2 - 4r + 12)$

23. $-\dfrac{4}{3}x^2\left[1 - \dfrac{9}{8}x + \dfrac{x^2}{2}\right]$

24. $(z - 3)(z - 6)$

25. $(3c - d)(2c + 5d)$

26. $(s + 4v)(s - 4v)$

27. $(x^2 - 4)^2$

28. $(2z + 1)^2$　　　　　　**29.** $(b - 10)^3$　　　　　　**30.** $[3 - (w - r)][3 + (w - r)]$

Section 3.4

For Exercises 31–34, factor the greatest common factor, if other than one, from each polynomial.

31. $45x^2 - 15x + 20$　　　　　　　　　**32.** $2a^2 - 4a^3$

33. $27b^5z^2 - 15b^2z$　　　　　　　　　**34.** $28(x + a)^5 - 14(x + a)^3 + 77(x + a)^2$

For Exercises 35–38, factor by grouping.

35. $3x^3 - 4x + 6x^2 - 8$　　　　　　　　**36.** $2c^3 - 5c^2 - 6c + 15$

37. $3z^3 + 4yz - 6yz^2 - 8y^2$　　　　　　**38.** $2tv^2 + 3s^2w^2 + 2tw^2 + 3s^2v^2$

Section 3.5

For Exercises 39–42, factor completely. If a polynomial cannot be factored using integers, write "prime."

39. $2x^2 + 7x - 15$　　　**40.** $3a^2 - 15a + 12$　　　**41.** $5w^2 + 2w - 4$　　　**42.** $15u^2 + 7u - 2$

For Exercises 43–46, factor completely.

43. $6b^3 - b^2 - 5b$　　　**44.** $4u^3 + 4u^2v + uv^2$　　　**45.** $6x^5 + 32x^4 + 32x^3$　　　**46.** $6p^3 - 9p^2 + 3p$

Section 3.6

For Exercises 47–54, factor completely.

47. $D^2 - 9$　　　　　　　　　　**48.** $36w^2 - 49$

49. $m^2 - 16m + 64$　　　　　　**50.** $8s^3 - 1$

51. $36a^2 - 16b^2$　　　　　　　**52.** $4h^2 + 12hk + 9k^2$

53. $27b^4 + by^6$　　　　　　　　**54.** $2p^3 - 162p - 3p^2 + 243$

Section 3.7

For Exercises 55–64, factor completely.

55. $60b^2 - 15z^2$　　　　　　　**56.** $w^2 - 22w + 121$

57. $q^2 + 4v^2$　　　　　　　　　**58.** $48 + 18t + 3t^2$

59. $49s^4 - 42s^2r + 9r^2$　　　　**60.** $4ab - 2a + 2b - 1$

61. $T^2 + 4Ta - T - 4a$　　　　**62.** $3k^4 - 48$

63. $128nz^3 + 54nc^3$　　　　　　**64.** $ax^3 - ay^3 - 2bx^3 + 2by^3$

Section 3.8

For Exercises 65–72, solve each equation.

65. $2z(3z - 2)(5z + 8) = 0$

66. $x^3 - 9x = 0$

67. $c^2 + 12c + 36 = 0$

68. $b(b + 4) = 21$

69. $(x + 3)(x - 2) = 2(9 - 2x)$

70. $2n^3 - n^2 - 8n + 4 = 0$

71. $x^4 - 25x^2 + 144 = 0$

72. $y^4 - 8y + y^3 - 8 = 0$

Section 3.9

73. Find all numbers with the property that the sum of 15 and the square of the number is equal to eight times the number.

74. The sum of two negative numbers is -20 and their product is 96. Find the two numbers.

75. A rectangle has a length that is 2 inches more than three fourths of the width. If the area is 20 square inches, find the length and width of the rectangle.

76. The sum of one integer and the square of the next consecutive integer is 41. Find the two integers.

77. A flare is fired in the desert with a height h, in feet, after t seconds given by the equation

$$h = 144t - 16t^2$$

How many seconds will it take the flare to return to the ground?

78. White Birch Supplies makes picture frames. The daily cost C of making x dozen frames is given by

$$C = x^2 - 20x + 85$$

On one day, the daily cost was $885. How many picture frames were made that day?

CHAPTER 3 **TEST**

Section 3.1

For Problems 1 and 2, simplify and write your answers without negative exponents. Assume that all variables are positive.

1. $x^3 \cdot x^8$

2. $\dfrac{(a^{-2}b^2)^{-2}}{(2a^{-1}b)^3}$

3. Evaluate $(5.3 \times 10^{-5})(2 \times 10^3)$. Express your answer in scientific notation and in decimal notation.

Section 3.2

4. Simplify $3x^2 - 5x^3 + 2 + 6x^2 + x^4 + 6x^3$. Write the result in descending powers. What is the degree of the polynomial?

5. Subtract $-5x^3 + 2x^2 - 3x + 1$ from $x^4 + 5x^2 - 6x - 3$.

Section 3.3

For Problems 6–9, do the indicated multiplication, then simplify.

6. $3x^2(4x^3 - 2x + 8)$ **7.** $(y - 3)(y + 1)$ **8.** $(2a - 3)(3a - 1)$ **9.** $(4z - 3)^2$

For Problems 10 and 11, factor by grouping.

10. $4x^2 - 2 + 2x^3 - x$ **11.** $6rx^2 + 2sy + 3ry + 4sx^2$

Section 3.5

For Problems 12–15, factor completely. If a polynomial cannot be factored using integers, write "prime."

12. $3y^2 - 5y - 2$ **13.** $6x^2 + 9x + 3$

14. $3t^2 + 2t + 2$ **15.** $2x^4y + 3x^3y^2 + x^2y^3$

Section 3.6

For Problems 16–19, factor completely.

16. $R^2 - 49$ **17.** $p^2 + 10p + 25$ **18.** $8x^2 - 18y^2$ **19.** $8a^3 - t^3$

Section 3.7

For Problems 20–23, factor completely.

20. $2r^4 - 18r^2$ **21.** $6z^2 - 11z + 4$ **22.** $s^2 - 2s + 3$ **23.** $2xt - 3b + 6t - xb$

Section 3.8

For Problems 24–26, solve each equation.

24. $4y^2 + 4y + 1 = 0$ **25.** $(r - 2)(r - 6) = -3$ **26.** $2x(3x + 2) = 3(1 - x)$

Section 3.9

27. Find all numbers with the property that the sum of 6 and the square of the number is 5 times the number.

28. A model rocket is fired so that the height h, in feet, after t seconds is given by the equation $h = 96t - 16t^2$. How many seconds will it take the rocket to return to the ground?

Rational Expressions

CHAPTER

4

Overview

In this chapter we deal with rational expressions. Just as rational *numbers* play an important role in arithmetic, rational *expressions* play an analogous role in algebra. We will combine rational expressions using the four basic operations—addition, subtraction, multiplication, and division. Our work with factoring from the last chapter will play an important role in the algebra of rational expressions. In the last part of the chapter, we will solve equations and investigate applications involving rational expressions.

4.1

Rational Expressions

OBJECTIVES

▶ *To evaluate a rational expression*

▶ *To determine where a rational expression is undefined*

▶ *To simplify a rational expression*

Recall that rational numbers are fractions such as

$$\frac{3}{2}, \qquad \frac{-21}{7}, \qquad \frac{2}{1}, \qquad \text{and} \qquad \frac{0}{9}$$

In general, a rational number is a number of the form $\frac{a}{b}$, where a and b are integers with $b \neq 0$. Since the denominator of a fraction cannot be zero, an expression such as $\frac{5}{0}$ is undefined.

In this chapter, we deal with fractions where the numerator and denominator are polynomials. Such fractions are called *rational expressions*.

DEFINITION

A **rational expression** is a fraction of the form $\dfrac{P}{Q}$, where P and Q are polynomials with $Q \neq 0$.

Some examples of rational expressions are

$$\frac{7}{y}, \qquad \frac{6x^2y}{9xy^2}, \qquad \frac{3z + 2}{4}, \qquad \frac{n^2 + n}{n^2 - n + 6}$$

Notice that the polynomial $x^2 - 3x + 1$ can be written as $\dfrac{x^2 - 3x + 1}{1}$.

Any polynomial P could be written as $\dfrac{P}{1}$; therefore, all polynomials are also rational expressions.

A rational expression can be evaluated for any replacement value for the variable, except when it makes the denominator equal to zero. For example, the number 3 cannot be used as a replacement for the variable in $\dfrac{4x}{x - 3}$, since 3 would make the denominator equal to 0.

A rational expression $\dfrac{P}{Q}$ **is undefined for those values of the variable that make $Q = 0$.**

Example 1   For what values of t is $\dfrac{t - 5}{2t + 7}$ undefined?

Solution   We set the denominator, $2t + 7$, equal to zero and solve the resulting equation for t.

$$2t + 7 = 0$$
$$2t = -7$$
$$t = -\frac{7}{2}$$

Therefore, the rational expression is undefined for $t = -\dfrac{7}{2}$. ▲

Some rational expressions may be undefined for more than one value. For example, $\dfrac{2x - 9}{(2x - 1)(x + 5)}$ is undefined when x is $\dfrac{1}{2}$ or -5. On the other hand, the rational expression $\dfrac{4x-3}{x^2+1}$ is defined for all x, since x^2+1 is always positive and never zero regardless of what value for x is used.

We make the following convention.

> **Whenever we write a rational expression $\dfrac{P}{Q}$, it is assumed that no replacement value of the variable can be used that makes $Q = 0$.**

CALCULATOR CROSSOVER

1. Evaluate $\dfrac{4n+5}{5n^2-14n-24}$ for $n = 1.6$ and for $n = -2$.

2. What happens when you attempt to evaluate the rational expression in Question 1 for $n = -1.2$? Why?

Answers   **1.** -0.339; -0.125 **2.** An error message appears, because the denominator is 0 when $n = -1.2$.

In Chapter 1, we reduced a fraction to lowest terms by dividing out factors that were common to both the numerator and denominator. For example, $\dfrac{14}{21}$ can be reduced to $\dfrac{2}{3}$, since 7 is a common factor,

$$\frac{14}{21} = \frac{2 \cdot 7}{3 \cdot 7} = \frac{2}{3}$$

This method holds true for rational expressions as well.

RULE

The principle of rational expressions

If $\dfrac{P}{Q}$ is a rational expression and K is a nonzero polynomial, then

$$\frac{PK}{QK} = \frac{P}{Q}$$

That is, if a rational expression has a factor K, common to both numerator and denominator, then K can be divided out of the fraction,

$$\frac{PK}{QK} = \frac{P\overset{1}{\cancel{K}}}{Q\underset{1}{\cancel{K}}} = \frac{P}{Q}, \qquad \text{provided } K \neq 0.$$

We will use the notation of drawing lines through common factors to keep track in more complicated expressions as to which factors have been divided out.

We use the above principle to simplify rational expressions. A rational expression is **reduced to lowest terms** if the numerator and denominator have no common factors.

Example 2   Reduce $\dfrac{x^2 - 16}{x + 4}$ to lowest terms.

Solution   We factor the numerator to obtain

$$\frac{x^2 - 16}{x + 4} = \frac{(x + 4)(x - 4)}{x + 4}$$

$$= \frac{\overset{1}{\cancel{(x + 4)}}(x - 4)}{\underset{1}{\cancel{x + 4}}} \qquad \text{Divide the numerator and denominator by } x + 4.$$

$$= \frac{1 \cdot (x - 4)}{1}$$

$$= x - 4 \qquad\qquad\qquad\qquad\qquad \blacktriangle$$

N O T E   *In Example 2, we reduced $\dfrac{x^2 - 16}{x + 4}$ to lowest terms, $\dfrac{x^2 - 16}{x + 4} =$*
$x - 4$. Notice that the right side has a value of -8 when x is replaced by -4; however, the left side of this equation is undefined when $x = -4$, since the denominator is then zero. Therefore, $\dfrac{x^2 - 16}{x + 4} = x - 4$ is true for all x except $x = -4$.

More examples of reducing rational expressions to lowest terms follow.

Example 3   Reduce to lowest terms.

(a) $\dfrac{12x^3y^2}{15x^2y^4} = \dfrac{\overset{4 \ \ x \ \ 1}{\cancel{12}\cancel{x^3}\cancel{y^2}}}{\underset{5 \ \ 1 \ \ y^2}{\cancel{15}\cancel{x^2}\cancel{y^4}}}$

$$= \frac{4x}{5y^2}$$

(b) $\dfrac{3m^3 - 3m^2 - 6m}{6m^3 - 12m^2 - 18m} = \dfrac{3m(m^2 - m - 2)}{6m(m^2 - 2m - 3)}$

$$= \dfrac{3m(m - 2)(m + 1)}{6m(m + 1)(m - 3)}$$
$$= \dfrac{m - 2}{2(m - 3)}$$

(c) $\dfrac{t^2 - th - 2t + 2h}{2t^2 + th - 4t - 2h} = \dfrac{t(t - h) - 2(t - h)}{t(2t + h) - 2(2t + h)}$ Factor by grouping.

$$= \dfrac{(t - h)(t - 2)}{(2t + h)(t - 2)}$$
$$= \dfrac{t - h}{2t + h}$$ ▲

COMMON ERROR

Be sure not to use incorrect "shortcuts" when dividing out common factors.

Incorrect: $\dfrac{x + 4}{x^2 - 16} = \dfrac{\cancel{x} + \cancel{4}}{\cancel{x^2} - \cancel{16}} = \dfrac{2}{x - 4}$

Correct: $\dfrac{x + 4}{x^2 - 16} = \dfrac{\cancel{x + 4}}{(x + 4)(x - 4)}$

$$= \dfrac{1}{x - 4}$$

When factoring a polynomial such as $3x^2 - 5x - 2$, notice that the coefficient of x^2, 3, is positive. If a polynomial, such as $-2x^2 + 9x - 4$, has a negative coefficient on the term with the highest power, we use *the −1 technique* to put the polynomial in a better form for factoring purposes.

STRATEGY

The −1 technique

If the term with the highest power in a polynomial has a negative coefficient, factor out a negative one.

Example 4   Use the -1 technique, then factor.

(a) $-3z^2 + 8z - 5 = (-1)(3z^2 - 8z + 5)$
$$= (-1)(3z - 5)(z - 1)$$

(b) $9 - r^2 = (-1)(r^2 - 9) = (-1)(r + 3)(r - 3)$

(c) $12n^2 - 21n^3 - 6n^4 = -6n^4 - 21n^3 + 12n^2$
$$= (-1)(6n^4 + 21n^3 - 12n^2)$$
$$= (-1)(3n^2)(2n^2 + 7n - 4)$$
$$= (-1)(3n^2)(n + 4)(2n - 1) \qquad \blacktriangle$$

We use the -1 technique to reduce certain rational expressions to lowest terms.

Example 5   Reduce $\dfrac{3p^2 + 22p + 7}{49 - p^2}$ to lowest terms.

Solution   First, we put the denominator in standard form and then use the -1 technique.

$$\frac{3p^2 + 22p + 7}{49 - p^2} = \frac{3p^2 + 22p + 7}{-p^2 + 49}$$
$$= \frac{3p^2 + 22p + 7}{(-1)(p^2 - 49)}$$
$$= \frac{(p + 7)(3p + 1)}{(-1)(p + 7)(p - 7)}$$

Factor the numerator and denominator, then divide out the common factor $p + 7$.

$$= \frac{3p + 1}{(-1)(p - 7)} \qquad \blacktriangle$$

The answer in Example 5 can be rewritten as
$$\frac{3p + 1}{(-1)(p - 7)} = -\frac{3p + 1}{p - 7}$$

which is the most common form. In general, we have the following rule.

RULE

If $\dfrac{P}{Q}$ is a rational expression, then
$$\frac{P}{-Q} = -\frac{P}{Q} \quad \text{and} \quad \frac{-P}{Q} = -\frac{P}{Q}$$

N O T E   *Notice that both* $\dfrac{P}{-Q}$ *and* $\dfrac{-P}{Q}$ *are equal to the same expression,* $-\dfrac{P}{Q}$, *and therefore,*

$$\frac{P}{-Q} = \frac{-P}{Q}$$

Example 6   Write each rational expression in lowest terms.

(a) $\dfrac{5 - z}{z - 5}$ (b) $\dfrac{-36n^3 + 60n^2 + 24n}{36n - 36n^2 + 9n^3}$ (c) $\dfrac{h + 14}{14 - h}$

Solution   **(a)** Put the numerator into standard form, then use the -1 technique.

$$\frac{5 - z}{z - 5} = \frac{-z + 5}{z - 5}$$

$$= \frac{(-1)(z - 5)}{z - 5} \qquad \text{Divide out common factor } z - 5.$$

$$= -1$$

(b) We begin by using the -1 technique on the numerator and putting the denominator into standard form.

$$\frac{-36n^3 + 60n^2 + 24n}{36n - 36n^2 + 9n^3} = \frac{(-1)(36n^3 - 60n^2 - 24n)}{9n^3 - 36n^2 + 36n}$$

Next, factor both the numerator and denominator completely. The numerator has a greatest common factor of $12n$ and the denominator has a GCF of $9n$.

$$\frac{(-1)(36n^3 - 60n^2 - 24n)}{9n^3 - 36n^2 + 36n} = \frac{(-1)(12n)(3n^2 - 5n - 2)}{9n(n^2 - 4n + 4)}$$

$$= \frac{(-1)(12n)(3n + 1)(n - 2)}{9n(n - 2)^2}$$

$$= \frac{(-1)(4)(3n + 1)}{3(n - 2)} \qquad \begin{array}{l}\text{Divide out}\\ \text{common factors.}\end{array}$$

$$= -\frac{4(3n + 1)}{3(n - 2)} \qquad \frac{-P}{Q} = -\frac{P}{Q}.$$

(c) This expression is already in lowest terms. Notice that the -1 technique will not produce any common factors.   ▲

L E A R N I N G   A D V A N T A G E   *Your instructor is more willing to help you one week before a test than one day before a test. Ask for extra help early so that you and your instructor have more time to review methods or materials.*

E X E R C I S E   S E T   **4.1**

For Exercises 1–16, determine any values of the variable where the rational expression is undefined.

1. $\dfrac{7}{3x - 5}$

2. $\dfrac{12}{2t + 7}$

3. $\dfrac{m^3}{4m - 8}$

4. $\dfrac{q^2}{q + 1}$

5. $\dfrac{2p + 2}{3p - 6}$

6. $\dfrac{5r - 8}{4r - 9}$

7. $\dfrac{3s + 5}{s^2 + 9}$

8. $\dfrac{45 - 3n}{16 + n^2}$

9. $\dfrac{3}{h^2}$

10. $-\dfrac{4}{c^3}$

11. $\dfrac{k - 1}{2k^2 - k - 3}$

12. $\dfrac{2m - 3}{3m^2 - 5m + 2}$

13. $\dfrac{3t - 2}{t^3 - 8}$

14. $\dfrac{1 - 5b}{b^3 + 27}$

15. $\dfrac{1}{x^3 - 9x - x^2 + 9}$

16. $\dfrac{z - 1}{3z^3 - 12z + 2z^2 - 8}$

For Exercises 17–44, reduce the rational expression to lowest terms.

17. $\dfrac{10a^3b^2}{5a^2b^3}$

18. $\dfrac{7z^4y^3}{21z^3y^5}$

19. $\dfrac{-21(3a + 5)}{3(3a + 5)^2}$

20. $\dfrac{48(1 - 2x)^2}{4(1 - 2x)}$

21. $\dfrac{y^2 - 3y + 2}{y^2 - 5y + 6}$

22. $\dfrac{c^2 - 2c - 3}{c^2 - c - 6}$

23. $\dfrac{5x - 5}{4x - 4}$

24. $\dfrac{6z + 18}{7z + 21}$

25. $\dfrac{2w^2 - 5w + 3}{2w^2 + w - 6}$

26. $\dfrac{3s^2 + 11s + 6}{3s^2 - s - 2}$

27. $\dfrac{10z + 40}{5z + 20}$

28. $\dfrac{9a - 18}{3a - 6}$

29. $\dfrac{10v^2 - v - 2}{6v^2 + v - 2}$

30. $\dfrac{6b^2 - 17b + 5}{3b^2 + 11b - 4}$

31. $\dfrac{a - 2}{a^2 - 4a + 4}$

32. $\dfrac{z + 3}{z^2 + 6z + 9}$

33. $\dfrac{2x^2 + 20x + 50}{2x^2 + 3x - 35}$

34. $\dfrac{3y^2 - 24y + 48}{3y^2 - 13y + 4}$

35. $\dfrac{3v^3 + 9v^2 - 12v}{5v^3 + 40v^2 + 80v}$

36. $\dfrac{4n^3 - 16n^2 + 16n}{7n^3 + 7n^2 - 42n}$

37. $\dfrac{3d^2 + 6d + 12}{d^3 - 8}$

38. $\dfrac{s^3 + 64}{2s^3 - 8s^2 + 32s}$

39. $\dfrac{6x^3 - 2x - 3x^2 + 1}{6x^5 - 2x^3}$

40. $\dfrac{8a^4 + 4a}{6a^4 + 3a + 4a^3 + 2}$

41. $\dfrac{3x^2 - 8x + 5}{3x^3 - 3x + 4x^2 - 4}$

42. $\dfrac{2y^3 - 18y - 3y^2 + 27}{2y^2 + 3y - 9}$

43. $\dfrac{2a^6 - 16a^3 - 18a^4 + 144a}{4a^4 + 4a^3 - 24a^2}$

44. $\dfrac{3c^6 - 3c^3 - 3c^4 + 3c}{c^3 - 2c^2 + c}$

For Exercises 45–52, put the polynomial into standard form, if necessary. Then, use the -1 technique and factor.

45. $-x^2 + x + 2$

46. $-n^2 - 5n - 6$

47. $11y - 3 \quad 6y^2$

48. $1 - 8r^2 - 2r$

49. $-4h^3 - 10h^2 - 4h$

50. $-9k^3 + 21k^2 + 18k$

51. $25c^3 - 20c^5 - 95c^4$

52. $2z^2 - 12z^4 - 2z^3$

For Exercises 53–82, write each rational expression in lowest terms.

53. $\dfrac{x - 2}{-5x^2 + 9x + 2}$

54. $\dfrac{z + 5}{-2z^2 - 7z + 15}$

55. $\dfrac{3 - 2a}{6a^2 - 11a + 3}$

56. $\dfrac{7 - 5y}{5y^2 - 2y - 7}$

57. $\dfrac{-2s^2 + s + 3}{3s^2 + 2s - 1}$

58. $\dfrac{-2m^2 - 9m + 5}{2m^2 + 7m - 4}$

59. $\dfrac{6w^2 + w - 1}{-3w^2 - 14w + 5}$

60. $\dfrac{6c^2 + 13c + 2}{-5c^2 - 9c + 2}$

61. $\dfrac{h - 6}{6 - h}$

62. $\dfrac{3b - 30}{20 - 2b}$ **63.** $\dfrac{k - 7}{7 + k}$ **64.** $\dfrac{q + 12}{12 - q}$ **65.** $\dfrac{56 - 21s}{18s - 48}$

66. $\dfrac{5 - 2y}{6y - 15}$ **67.** $\dfrac{t - t^2}{2t^2 - 2t}$ **68.** $\dfrac{3n^2 - 12n}{4n - n^2}$ **69.** $\dfrac{3v - 4}{3v + 4}$

70. $\dfrac{2c - 1}{2c + 1}$ **71.** $\dfrac{3n - 7n^2 - 6n^3}{24n^3 - 14n^2 + 2n}$ **72.** $\dfrac{18x^3 + 39x^2 - 15x}{10x - x^2 - 2x^3}$ **73.** $\dfrac{-3k^3 - 6k^4 + 3k^2}{4k^2 - 4k^4 - 6k^3}$

74. $\dfrac{-16t^3 - 4t^2 - 12t^4}{3t^2 - 9t^4 + 6t^3}$ **75.** $\dfrac{a^3 + 9a}{a + 3}$ **76.** $\dfrac{2x^2 - x - 6}{2x^2 + x - 6}$ **77.** $\dfrac{9 - x^4}{x^2(x + 3) + 3(x + 3)}$

78. $\dfrac{2x^2(x + 2)^2 - 5x(x + 2)^2 + 2(x + 2)^2}{x^4 - 16}$ **79.** $\dfrac{5(b^2 - a^2)}{10a(a + b) - 10b(a + b)}$

80. $\dfrac{a^4 - 2a^2b^2 + b^4}{a^3 - b^3}$ **81.** $\dfrac{x^2a^2 + y^2b^2 - x^2b^2 - y^2a^2}{ax - by + bx - ay}$ **82.** $\dfrac{C^2Z^3 - CDZ + CDZ^2 - D^2}{C^2Z^2 - D^2}$

83. Suppose two resistors have resistances R_1 and R_2 ohms, respectively. If they are connected in parallel as shown in the figure, then the total resistance R is given by the equation

$$R = \dfrac{R_1 + R_2}{R_1R_2}$$

What is the total resistance R, if $R_1 = 12$ ohms and $R_2 = 8$ ohms?

84. Suppose three resistors have resistances R_1, R_2, and R_3, respectively. If they are connected in parallel as shown in the figure, then the total resistance R is given by

$$\dfrac{R_2R_3 + R_1R_3 + R_1R_2}{R_1R_2R_3}$$

What is the total resistance R, if $R_1 = 4$ ohms, $R_2 = 6$ ohms, and $R_3 = 10$ ohms?

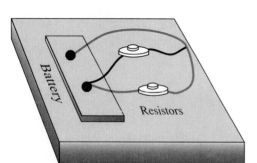

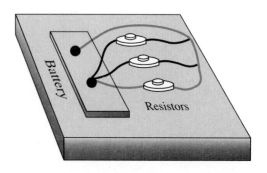

For Exercises 85–90, write each rational expression in lowest terms.

85. $\dfrac{x^{2n} + x^n - 6}{x^{2n} + 2x^n - 3}$

86. $\dfrac{16t^{4n} - 1}{2t^n + 1}$

87. $\dfrac{a^{2m} - 1}{1 - a^{2m}}$

88. $\dfrac{s^{4p} - 16}{32 - 2s^{4p}}$

89. $\dfrac{y^{2n} - z^{2n}}{y^{2n} + 2y^n z^n + z^{2n}}$

90. $\dfrac{a^{2m} - 3a^m b^m + 2b^{2m}}{a^{2m} - a^m b^m}$

REVIEW PROBLEMS

The following exercises review parts of Section 3.1. Doing these problems will help prepare you for the next section.

For Exercises 91–94, simplify and write your answers with positive exponents.

91. $\dfrac{12x^6}{2x^3}$

92. $\dfrac{9t^3 y^4}{-3t^3 y}$

93. $\dfrac{16x^3 z^2}{8x^5 z}$

94. $\dfrac{-21a^3 b^2}{-7a^4 b^3}$

For Exercises 95–97, perform the indicated operation. Write your answer in descending powers.

95. $(3x^3 - 4x^2 + 7x - 2) - (x^3 + 4x^2 + 4x - 2)$

96. $(z^4 - z^2 + 4z + 5) - (z^3 + 3z^2 - 2z - 6)$

97. $3x^5 + 7x^3 - 2x + 1) - (3x^5 + 6x^4 - 3x + 9)$

98. Subtract $y^3 - y^2 - 6y + 12$ from $y^3 + 4y^2 - y - 9$.

99. Subtract $9t^2 - 8t + 7$ from $9t^3 + 9t^2 + 14$.

100. Subtract $5x^4 - 3x^3 + 5x^2$ from $5x^4$.

ENRICHMENT EXERCISES

Simplify.

1. $\dfrac{x^5 - 4x + x^4 - 4}{4 - x^4 - 4x^2 + x^6}$

2. $\dfrac{0.6z^2 - 9.6}{z + 4}$

3. $\dfrac{3x^{-2} - 2x^{-1}}{3x - 2x^2}$

4. $\dfrac{6y - 2}{9y^{-1} - 3y^{-2}}$

5. $\dfrac{2a - 3 + a^{-1}}{a^2 - a}$

Answers to Enrichment Exercises are on page A.16.

4.2

Division of Polynomials

O B J E C T I V E S

▶   *To divide a polynomial by a monomial*

▶   *To divide a polynomial by a binomial*

▶   *To use synthetic division*

Recall from Chapter 1, the division $\dfrac{a}{b}$ can be expressed as multiplication of a

and $\dfrac{1}{b}$. For example, suppose we want to divide $12x + 15$ by 3.

$$\frac{12x + 15}{3} = (12x + 15)\left(\frac{1}{3}\right)$$

$$= 12x\left(\frac{1}{3}\right) + 15\left(\frac{1}{3}\right) \qquad \text{Distribute } \frac{1}{3}.$$

$$= \frac{12x}{3} + \frac{15}{3}$$

$$= 4x + 5$$

We illustrate this example with the following equation.

$$\frac{a + b}{c} = \frac{a}{c} + \frac{b}{c}, \qquad c \neq 0$$

We can use this equation to obtain the rule for dividing a polynomial by a monomial.

R U L E

To divide a polynomial by a monomial, divide *each term* of the polynomial by the monomial.

Example 1   Divide as indicated, then simplify.

(a) $\dfrac{3s^5 + 6s^4}{s^2} = \dfrac{3s^5}{s^2} + \dfrac{6s^4}{s^2}$

$$= 3s^{5-2} + 6s^{4-2}$$

$$= 3s^3 + 6s^2, \qquad s \neq 0$$

(b) $\dfrac{z^4 - z}{z^3} = \dfrac{z^4}{z^3} - \dfrac{z}{z^3}$

$\qquad\qquad = z - \dfrac{1}{z^2}$

(c) $\dfrac{12m^5 - 6m^4 + 3m^2}{3m^2} = \dfrac{12m^5}{3m^2} - \dfrac{6m^4}{3m^2} + \dfrac{3m^2}{3m^2}$

$\qquad\qquad\qquad\qquad = \dfrac{12}{3}m^{5-2} - \dfrac{6}{3}m^{4-2} + 1$

$\qquad\qquad\qquad\qquad = 4m^3 - 2m^2 + 1, \qquad m \neq 0$ ▲

Example 2   Divide: $\dfrac{2a^2 - ab - b^2}{2a + b}$.

Solution   This division can be performed by factoring the numerator and dividing out the common factor of $2a + b$ as we did in the last section.

$$\frac{2a^2 - ab - b^2}{2a + b} = \frac{(a - b)(2a + b)}{2a + b}$$

$$= a - b, \qquad 2a + b \neq 0 \quad ▲$$

In the division problem of Example 2, the denominator had to be a factor of the numerator. If this is not the case, then the method used to perform the division does not work. Another method is needed—a method very similar to long division of integers. In fact, let us review long division with the example of dividing 857 by 25.

INSTRUCTIONS

3
$25\overline{)857}$
$\underline{75}$
107

Round 1
Divide: 25 goes into 85, three times.
Multiply: $3 \cdot 25 = 75$.
Subtract: $85 - 75 = 10$.
Then bring down the 7.

The remainder from Round 1 is 107. Since 107 is larger than 25, we go through the Divide-Multiply-Subtract sequence again.

34
$25\overline{)857}$
$\underline{75}$
107
$\underline{100}$
7

Round 2
Divide: 25 goes into 107, four times.
Multiply: $4 \cdot 25 = 100$.
Subtract: $107 - 100 = 7$.

The remainder from Round 2 is 7. Since 7 is less than 25, we are finished. The quotient of $\dfrac{857}{25}$ is 34 with a remainder of 7. We can check our answer.

$$34 \cdot 25 + 7 = 850 + 7 = 857$$

Let us now apply this method to the division of polynomials.

Example 3   Divide $x^2 + x - 6$ by $x - 2$.

Solution   INSTRUCTIONS

$$
\begin{array}{r}
x \phantom{{}+x-6} \\
x-2\overline{)x^2 + x - 6} \\
\underline{x^2 - 2x}\phantom{{}-6} \\
3x - 6
\end{array}
$$

Round 1
Divide: x goes into x^2, x times.
Multiply: $x(x-2) = x^2 - 2x$.
Subtract: $x^2 + x - (x^2 - 2x) = 3x$.
Then, bring down -6.

The remainder from Round 1 is $3x - 6$. Note that the degree of $3x - 6$ is not less than the degree of $x - 2$. We therefore go through the Divide-Multiply-Subtract sequence again.

$$
\begin{array}{r}
x + 3 \\
x-2\overline{)x^2 + x - 6} \\
\underline{x^2 - 2x}\phantom{{}-6} \\
3x - 6 \\
\underline{3x - 6} \\
0
\end{array}
$$

Round 2
Divide: x goes into $3x$, three times.
Multiply: $3(x - 2) = 3x - 6$.
Subtract: $(3x - 6) - (3x - 6) = 0$.
The remainder is 0.

Therefore,

$$\frac{x^2 + x - 6}{x - 2} = x + 3$$

We check our answer by multiplication.

$$(x + 3)(x - 2) = x^2 + x - 6$$ ▲

Before we proceed to further examples, we introduce names for the polynomials involved in a division problem. In Example 2, we divided $x^2 + x - 6$ by $x - 2$. The polynomial $x - 2$ is called the **divisor** and $x^2 + x - 6$ is the **dividend.** The "answer" $x + 3$ is called the **quotient.** In this example, we had a **remainder** of 0. The relation among these polynomials is given by the equation

$$\frac{\textbf{dividend}}{\textbf{divisor}} = \textbf{quotient} + \frac{\textbf{remainder}}{\textbf{divisor}}$$

Multiplying both sides of this equation by the divisor, we have

dividend = (divisor)(quotient) + remainder

This last equation is used as a check of our division.

In performing a division involving two polynomials, keep in mind that both divisor and dividend must be in descending order of exponents and that if there are any missing terms, their "space" must be filled in using 0 as the coefficient.

Example 4   Divide $4x^3 - 3x + 4$ by $2x - 1$.

Solution   Since $4x^3 - 3x + 4$ is missing a term in x^2, we fill it in with $0x^2$, and then $4x^3 - 3x + 4 = 4x^3 + 0x^2 - 3x + 4$. Now, we proceed with the division process.

$$
\begin{array}{r}
2x^2 + x - 1 \\
2x-1\overline{)4x^3 + 0x^2 - 3x + 4} \\
\underline{4x^3 - 2x^2} \\
2x^2 - 3x \\
\underline{2x^2 - x} \\
-2x + 4 \\
\underline{-2x + 1} \\
3
\end{array}
$$

Therefore, the quotient is $2x^2 + x - 1$ with a remainder of 3. That is,

$$\frac{4x^3 - 3x + 4}{2x - 1} = 2x^2 + x - 1 + \frac{3}{2x - 1}$$

To check our answer, multiply the divisor $2x - 1$ and the quotient $2x^2 + x - 1$,

$$
\begin{array}{r}
2x^2 + x - 1 \\
2x - 1 \\
\hline
4x^3 + 2x^2 - 2x \\
-2x^2 - x + 1 \\
\hline
4x^3 \qquad -3x + 1
\end{array}
$$

Adding the remainder 3 yields the dividend $4x^3 - 3x + 4$.   ▲

In the next example, we will do Example 2 again. This time, however, we use long division rather than factoring the numerator and dividing out common factors.

Example 5   Divide: $\dfrac{2a^2 - ab - b^2}{2a + b}$.

Solution

$$
\begin{array}{r}
a - b \\
2a+b\overline{)2a^2 - ab - b^2} \\
\underline{2a^2 + ab} \\
-2ab - b^2 \\
\underline{-2ab - b^2} \\
0
\end{array}
$$

Therefore, the quotient is $a - b$ with a remainder of 0 and we have

$$\frac{2a^2 - ab - b^2}{2a + b} = a - b$$

As a check, multiply $2a + b$ and $a - b$,

$$(2a + b)(a - b) = 2a^2 - ab - b^2$$ ▲

The next example is a division problem where the divisor is a polynomial of degree 2.

Example 6   Divide $t^4 - t^3 - 2t^2 + t - 2$ by $t^2 + t$.

Solution

$$
\begin{array}{r}
t^2 - 2t \\
t^2 + t \overline{\smash{)}t^4 - t^3 - 2t^2 + t - 2} \\
\underline{t^4 + t^3} \\
-2t^3 - 2t^2 \\
\underline{-2t^3 - 2t^2} \\
0 + t - 2
\end{array}
$$

The division is completed when the degree of the remainder is less than the degree of the divisor. Therefore,

$$\frac{t^4 - t^3 - 2t^2 + t - 2}{t^2 + t} = t^2 - 2t + \frac{t - 2}{t^2 + t}$$ ▲

In the next example, we show how long division can help factor a polynomial when one of the factors is known.

Example 7   Factor $8x^3 + 2x^2 - 13x + 3$ completely, if one of the factors is $x - 1$.

Solution   Since $x - 1$ is one of the factors, it must divide the polynomial with a zero remainder. Therefore, the given polynomial is the product of $x - 1$ and some other polynomial. To find this other polynomial, we divide by $x - 1$.

$$
\begin{array}{r}
8x^2 + 10x - 3 \\
x - 1 \overline{\smash{)}8x^3 + 2x^2 - 13x + 3} \\
\underline{8x^3 - 8x^2} \\
10x^2 - 13x \\
\underline{10x^2 - 10x} \\
-3x + 3 \\
\underline{-3x + 3} \\
0
\end{array}
$$

Thus, the given polynomial is the product of $x - 1$ and $8x^2 + 10x - 3$. Now, it remains to try to factor $8x^2 + 10x - 3$. Using our factoring methods of the last chapter, we find that $8x^2 + 10x - 3$ factors into $(2x + 3)(4x - 1)$. Therefore,

$$8x^3 + 2x^2 - 13x + 3 = (x - 1)(8x^2 + 10x - 3)$$
$$= (x - 1)(2x + 3)(4x - 1) \quad \blacktriangle$$

If the divisor is of the form $x - c$, then the long division method can be simplified into a process known as **synthetic division.** Synthetic division is a convenient tool that we will use when studying polynomial functions.

Consider the following division problem.

$$
\begin{array}{r}
3x^2 - 2x + 4 \\
x - 2\overline{\smash{\big)}\,3x^3 - 8x^2 + 8x - 3} \\
\underline{3x^3 - 6x^2} \\
-2x^2 + 8x \\
\underline{-2x^2 + 4x} \\
4x - 3 \\
\underline{4x - 8} \\
5
\end{array}
$$

Notice that in the long division process, it is actually the coefficients that are important. The Divide-Multiply-Subtract sequence is performed on the coefficients. Therefore, we may rewrite the above division leaving out the variable. Keep in mind that all polynomials involved had been written in descending powers of x.

$$
\begin{array}{r}
3 - 2 + 4 \\
1 - 2\overline{\smash{\big)}\,3 - 8 + 8 - 3} \\
\underline{3 - 6} \\
-2 + 8 \\
\underline{-2 + 4} \\
4 - 3 \\
\underline{4 - 8} \\
5
\end{array}
$$

Now, some of the numbers in the display are repetitions of the numbers directly above them. These repetitions can be eliminated to obtain a more streamlined form.

$$
\begin{array}{r}
3 - 2 + 4 \\
-2\overline{\smash{\big)}\,3 - 8 + 8 - 3} \\
\underline{-6 + 4 - 8} \\
-2 + 4 + 5
\end{array}
$$

Notice that the coefficient on x in the divisor has also been eliminated. Now, the last line contains all the numbers of the first line except for 3, the first number of the first line. Therefore, we do not need the first line if we put the 3 at the start of the last line. Our format is changed to the following:

$$
\begin{array}{r}
-2\,\big|\,3 - 8 + 8 - 3 \\
\underline{-6 + 4 - 8} \\
3 - 2 + 4 + 5
\end{array}
$$

The last line now contains the coefficients of the quotient with the last number being the remainder. One final change—if we replace the constant -2 in the divisor by its opposite, 2, we can *add* the corresponding entries in the first two lines to obtain the entries in the last line. The final form for the synthetic division process looks like this:

$$\begin{array}{r} 2\,|\,3-8+8-3 \\ \underline{+6-4+8} \\ 3-2+4+5 \end{array}$$

In the next example, we will illustrate this process from beginning to end.

Example 8   Use synthetic division to divide $3x^3 + 7x^2 - 7x - 8$ by $x + 3$.

Solution   Step 1   Write the coefficients of the dividend as shown.

$$3 + 7 - 7 - 8$$

Step 2   The divisor is $x + 3$. Use -3 instead of 3 so we can add entries rather than subtract.

$$-3\,|\,3 + 7 - 7 - 8$$

Step 3   Draw the lines as shown and bring down the first coefficient, 3, of the dividend.

$$\begin{array}{r} -3\,|\,3+7-7-8 \\ \hline 3 \end{array}$$

Step 4   Multiply -3 times 3, the answer is -9, place -9 underneath the 7 and then add 7 and -9.

$$\begin{array}{r} -3\,|\,3+7-7-8 \\ \underline{-9} \\ 3-2 \end{array}$$

Step 5   Multiply -3 times -2, 6, and place it underneath the -7 and then add -7 and 6.

$$\begin{array}{r} -3\,|\,3+7-7-8 \\ \underline{-9+6} \\ 3-2-1 \end{array}$$

Step 6   Multiply -3 times -1, 3, and place it underneath the -8 and then add -8 and 3.

$$\begin{array}{r} -3\,|\,3+7-7-8 \\ \underline{-9+6+3} \\ 3-2-1-5 \end{array}$$

The last row gives the coefficients of the quotient with the last number being the remainder. *The degree of the quotient is one less than the degree of the dividend.* Therefore, the quotient is $3x^2 - 2x - 1$ and the remainder is -5. ▲

N O T E   *Notice that synthetic division can only be performed when the divisor is a binomial with the variable being of the first power and the coefficient of the variable being 1.*

Let us return to Example 8. In particular, when we divided $3x^3 + 7x^2 - 7x - 8$ by $x + 3$, we obtained a remainder of -5. Suppose we evaluate the polynomial $P(x) = 3x^3 + 7x^2 - 7x - 8$ when $x = -3$:

$$P(-3) = 3(-3)^3 + 7(-3)^2 - 7(-3) - 8$$

$$= 3(-27) + 7(9) + 21 - 8$$

$$= -81 + 63 + 21 - 8$$

$$= -5$$

which is the same number as the remainder. Therefore, dividing $P(x)$ by $x + 3$ resulted in a remainder that is $P(-3)$.

This observation is true in general.

T H E O R E M

The remainder theorem

If $P(x)$ is a polynomial, then for any number c, the remainder obtained from dividing $P(x)$ by $x - c$ is $P(c)$.

This theorem gives us a quick way to evaluate a polynomial $P(x)$ when x is a given number c. To find $P(c)$, use synthetic division to divide $P(x)$ by $x - c$. The remainder theorem tells us that the resulting remainder from this division is $P(c)$.

We illustrate this procedure with the next example.

Example 9   Use the remainder theorem to find $P(2)$, where

$$P(x) = x^4 + 3x^2 - 15x + 7$$

Solution   To find $P(2)$, we divide $x^4 + 3x^2 - 15x + 7$ by $x - 2$. By the remainder theorem, the remainder from this division will be the desired number.

Step 1   Write the coefficients of the dividend as shown. Remember to use zero as the coefficient of x^3.

$$1 + 0 + 3 - 15 + 7$$

Step 2   The divisor is $x - 2$. Use the opposite of -2, 2, so we can add entries rather than subtract.

$$2 \,|\, 1 + 0 + 3 - 15 + 7$$

Step 3 Draw straight lines as shown and bring down the first coefficient, 1, of the dividend.

$$2\,\rfloor\,1 + 0 + 3 - 15 + 7$$
$$\overline{}$$
$$1$$

Step 4 Multiply 2 times 1, or 2, and place it underneath the 0 and then add 0 and 2.

$$2\,\rfloor\,1 + 0 + 3 - 15 + 7$$
$$\,2$$
$$\overline{1 + 2}$$

Step 5 Multiply 2 times 2, or 4, and place it underneath the 3 and then add 3 and 4.

$$2\,\rfloor\,1 + 0 + 3 - 15 + 7$$
$$\,2 + 4$$
$$\overline{1 + 2 + 7}$$

Step 6 Multiply 2 times 7, or 14, and place it underneath the -15 and then add -15 and 14.

$$2\,\rfloor\,1 + 0 + 3 - 15 + 7$$
$$\,2 + 4 + 14$$
$$\overline{1 + 2 + 7 - 1}$$

Step 7 Multiply 2 times -1, or -2, and place it underneath the 7 and then add 7 and -2.

$$2\,\rfloor\,1 + 0 + 3 - 15 + 7$$
$$\,2 + 4 + 14 - 2$$
$$\overline{1 + 2 + 7 -\ 1 + 5}$$

The last row gives the coefficients of the quotient with the last number, 5, being the remainder. Therefore, by the remainder theorem, $P(2) = 5$. ▲

L E A R N I N G   A D V A N T A G E   *If your instructor provided review activities in class, take them seriously. The teacher may have just finished writing the test and may be creating review problems that are similar to those that will be on the test. However, do not assume that a topic will* not *be on your test just because it was not covered in the review lesson.*

E X E R C I S E   S E T   4 . 2

For Exercises 1–14, divide as indicated and simplify. Write the answer without negative exponents.

1. $\dfrac{7z^3 - 2z^2 - z}{z^2}$

2. $\dfrac{6n^4 + 2n^3 + n}{n^3}$

3. $\dfrac{12u^4 - 9u^3 + 3u^2}{3u^2}$

4. $\dfrac{36v^3 - 18v}{9v}$

5. $\dfrac{8m^3 - 4m^2 + 2m}{-2m^3}$

6. $\dfrac{15p^4 + 12p^3 - 3}{-3p^2}$

7. $\dfrac{\frac{3}{4}h^5 - \frac{5}{8}h}{\frac{3}{8}h^2}$

8. $\dfrac{\frac{2}{3}r^3 + \frac{7}{6}r^2}{\frac{1}{6}r^3}$

9. $\dfrac{3z^7 - 1.5z^3 + 3}{1.5z^2}$

10. $\dfrac{2.4k^3 - 3.6k^2}{1.2k}$

11. $\dfrac{s^2t^4 - 4st^2 + 15st}{5s^2t^3}$

12. $\dfrac{pq^3 - p^3q + 9p^2q^5}{3p^3q^4}$

13. $\dfrac{4r^2st^3 - 16r^3s^2t + 3r^4s^5t^3}{6r^4s^2t^3}$

14. $\dfrac{9x^3yz^4 + 18xy^3z^2 - 24x^2y^2z^3}{18x^2yz^2}$

For Exercises 15–24, divide by factoring the numerator and then dividing out common factors.

15. $\dfrac{2x^2 + xy - 3y^2}{x - y}$

16. $\dfrac{2a^2 + 5ab - 3b^2}{a + 3b}$

17. $\dfrac{a^3 + b^3}{a^2 - ab + b^2}$

18. $\dfrac{16s^2 - t^2}{4s + t}$

19. $\dfrac{c^2 + 4cd + 4d^2}{c + 2d}$

20. $\dfrac{u^2 - 6uv + 9v^2}{u - 3v}$

21. $\dfrac{2a^3 + 10a + a^2 + 5}{a^2 + 5}$

22. $\dfrac{3z^4 - 12z - 2z^3 + 8}{3z - 2}$

23. $\dfrac{x^3 + yx - 2yx^2 - 2y^2}{x^2 + y}$

24. $\dfrac{3st^2 + 12s - 2t^3 - 8t}{3s - 2t}$

For Exercises 25–38, use long division to divide. If the remainder is nonzero, write your answer in the form: quotient $+ \dfrac{\text{remainder}}{\text{divisor}}$. Check your answer by verifying the equation

$$\text{dividend} = \text{quotient} \cdot \text{divisor} + \text{remainder}$$

25. $\dfrac{x^2 + 3x - 10}{x + 5}$

26. $\dfrac{a^2 - a - 6}{a - 3}$

27. $\dfrac{6x^3 - 13x^2 + x + 2}{x - 2}$

28. $\dfrac{3z^3 - 8z^2 - 5z + 6}{z + 1}$

29. $\dfrac{z^2 - 3z - 40}{z - 8}$

30. $\dfrac{w^2 + 5w - 6}{w + 6}$

31. $\dfrac{x^2 - 8x - 15}{x - 2}$

32. $\dfrac{y^2 - 6y - 5}{y + 1}$

33. $\dfrac{2y^2 - 5y - 12}{2y + 3}$

34. $\dfrac{3x^2 - 16x + 5}{3x - 1}$

35. $\dfrac{3z^3 + 16z^2 + 12}{3z - 2}$

36. $\dfrac{2s^3 - 26s - 24}{s + 3}$

37. $\dfrac{3t^3 - 11t - 7}{t - 2}$

38. $\dfrac{6r^3 - 66r + 36}{r - 1}$

For Exercises 39–54, find the quotient and remainder.

39. $\dfrac{3x^4 + 48x}{2x + 4}$

40. $\dfrac{6z^4 + z^3 + z^2}{3z - 7}$

41. $\dfrac{2t^3 + 5t^2 - 2t - 5}{2t + 5}$

42. $\dfrac{3x^3 - 2x^2 + 12x - 8}{3x - 2}$

43. $\dfrac{4x^5 - 5x^3 - 6}{2x + 1}$

44. $\dfrac{9x^4 + 8x^2 - 11x - 12}{3x - 2}$

45. $\dfrac{9w^3 - 4w + 5}{3w - 1}$

46. $\dfrac{4c^3 + c - 3}{2c + 1}$

47. $\dfrac{x^4 - 2x^3 + 2x^2 + x - 1}{x^2 - x}$

48. $\dfrac{4z^5 + 8z^4 + z^3 - z^2 + 3z + 1}{z^2 + z}$

49. $\dfrac{2x^7 - 8x^5 + 6x^3 - x^4 + 4x^2 - 3}{2x^3 - 1}$

50. $\dfrac{9y^6 - 12y^5 - 15y^4 + 9y^3 - 4y^2 - 5y + 2}{3y^3 + 1}$

51. $\dfrac{2x^5 - x^4 + 3x^3 - x^2 + x}{2x^2 - x + 1}$

52. $\dfrac{x^8 - 1}{x^2 - x + 1}$

53. $\dfrac{3a^8 - 7a^6 + 7a^4 - 5a^2 + 2}{3a^4 - a^2 + 2}$

54. $\dfrac{2y^9 - 7y^7 + 8y^5 - 5y^3 + 2y}{2y^5 - y^3 + y}$

55. Factor $2x^3 - 11x^2 + 18x - 9$ completely, given that $x - 1$ is one of the factors.

56. Factor $8x^3 - 26x^2 + 3x + 9$ completely, given that $4x - 3$ is one of the factors.

57. Factor $12x^4 + 26x^3 - 16x^2 - 13x + 5$ completely, given that $2x^2 - 1$ is one of the factors.

58. Factor $x^4 + x^3 - x - 1$ completely, given that $x^2 + x + 1$ is one of the factors.

For Exercises 59–62, use synthetic division to find the quotient and remainder.

59. $\dfrac{2x^2 + 3x - 11}{x - 2}$

60. $\dfrac{3x^2 - 5x}{x - 1}$

61. $\dfrac{x^4 - 10x^2 - 7}{x + 3}$

62. $\dfrac{2x^3 - 35x - 7}{x - 4}$

For Exercises 63–70, find $P(c)$ by using the remainder theorem and synthetic division.

63. If $P(t) = t^3 - t^2 - 3t$, find $P(1)$.

64. If $P(z) = 2z^3 + 4z^2 + z + 6$, find $P(-2)$.

65. If $P(y) = y^5 + 5y^4 - y - 6$, find $P(-5)$.

66. If $P(x) = x^5 - x^4 + 3x^2 - 3x - 3$, find $P(1)$.

67. If $P(x) = 2x^4 - 3x^3 + 4x - 5$, find $P\left(\dfrac{1}{2}\right)$.

68. If $P(t) = 3t^4 + t^3 + 7t^2 + 5t$, find $P\left(\dfrac{2}{3}\right)$.

69. If $P(x) = x^3 - 2x^2 + x + 3$, find $P\left(-\dfrac{1}{3}\right)$.

70. If $P(x) = 2x^3 + x^2 - x + 1$, find $P\left(-\dfrac{1}{2}\right)$.

REVIEW PROBLEMS

The following exercises review parts of Section 1.3. Doing these problems will help prepare you for the next section.

Divide.

71. $\dfrac{2}{3} \div \dfrac{4}{9}$

72. $\dfrac{6}{7} \div \dfrac{2}{5}$

73. $\dfrac{5}{4} \div 10$

74. $\dfrac{2}{5} \div (-4)$

75. $4 \div \dfrac{1}{2}$

76. $2 \div \dfrac{1}{3}$

77. $(-30) \div \dfrac{4}{5}$

78. $(-21) \div \left(-\dfrac{3}{2}\right)$

79. $\left(-\dfrac{5}{2}\right) \div \left(-\dfrac{3}{2}\right)$

80. $\left(-\dfrac{5}{4}\right) \div \left(\dfrac{7}{-2}\right)$

E N R I C H M E N T   E X E R C I S E S

For Exercises 1–3, divide as indicated. Assume that n is a positive integer.

1. $\dfrac{a^{4n} - a^{3n} - a^n}{a^{2n}}$

2. $\dfrac{2x^{3n} + 3x^{2n} - 2x^n - 3}{x^n - 1}$

3. $(2x^{2n+1} - 2x^{2n}y - x^{n+1}y^n + x^ny^{n+1} - 3xy^{2n} + 3y^{2n+1}) \div (x - y)$

4. Determine c so that when $2x^2 - 9x + c$ is divided by $x - 5$, the quotient is $2x + 1$ and the remainder is 3.

5. Determine b so that when $3x^2 + bx - 1$ is divided by $x + 1$, the quotient is $3x - 2$ and the remainder is 1.

6. Let $P(x)$ be a polynomial. Suppose that when $P(x)$ is divided by $x - c$, the remainder is zero. Show that $x - c$ is a factor of $P(x)$. Conversely, show that if $x - c$ is a factor of $P(x)$, then the remainder is zero.

7. Use Exercise 6 and synthetic division to determine which of the following are factors of

$$2x^4 - 6x^3 + 5x^2 - 3x + 2$$

(a) $x - 1$ **(b)** $x + 1$ **(c)** $x - 2$

Answers to Enrichment Exercises are on page A.17.

4.3

Multiplication and Division of Rational Expressions

O B J E C T I V E S

▶ *To multiply rational expressions and write them in lowest terms*

▶ *To divide rational expressions and write them in lowest terms*

The way to multiply two rational expressions is very similar to the method of multiplying two rational numbers. Recall from arithmetic, the product of two fractions is a fraction of the form

$$\frac{\text{product of the numerators}}{\text{product of the denominators}}$$

For example,

$$\frac{3}{5} \cdot \frac{2}{7} = \frac{3 \cdot 2}{5 \cdot 7} = \frac{6}{35}$$

The product of two rational expressions has the same form.

RULE

Multiplication of two rational expressions

If $\dfrac{A}{B}$ and $\dfrac{C}{D}$ are two rational expressions, then

$$\frac{A}{B} \cdot \frac{C}{D} = \frac{A \cdot C}{B \cdot D}$$

To multiply fractions whose numerators and denominators are monomials, we multiply numerators and multiply denominators, then reduce the resulting fraction to lowest terms.

Example 1 (a) $\dfrac{7}{x} \cdot \dfrac{4}{y} = \dfrac{7 \cdot 4}{x \cdot y}$

$$= \dfrac{28}{xy}$$

(b) $\dfrac{12a}{15} \cdot \dfrac{10}{4a^3} = \dfrac{12 \cdot 10a}{15 \cdot 4a^3}$

$$= \dfrac{\overset{4}{\cancel{12}} \cdot \overset{2}{\cancel{10}}\overset{1}{\cancel{a}}}{\underset{5}{\cancel{15}} \cdot \underset{1}{\cancel{4}}\underset{a^2}{a^3}}$$

$$= \dfrac{2}{a^2}$$

(c) $\dfrac{-24s^3}{v^4} \cdot \dfrac{v^6}{-8s} = \dfrac{-24s^3v^6}{-8sv^4}$

$$= \dfrac{\overset{3}{\cancel{-24}}\overset{s^2v^2}{\cancel{s^3}\cancel{v^6}}}{\underset{1\ \ 1\ \ 1}{\cancel{-8}\cancel{s}\cancel{v^4}}}$$

$$= 3s^2v^2 \qquad\qquad \blacktriangle$$

Notice in Example 1 that we use slashes to keep track when dividing out common factors.

In the next example, we find the product of more complicated rational expressions. Be sure to factor the numerators and denominators completely, then divide out common factors.

Example 2 Find the indicated product. Write the answer in lowest terms.

(a) $\dfrac{2t+1}{3t-4} \cdot \dfrac{3t-4}{2t^2-9t-5}$ 　　　　 (b) $(3r+2) \cdot \dfrac{5r+5}{3r^2-7r-6}$

(c) $\dfrac{a^2-1}{2a^4+5a^3+2a^2} \cdot \dfrac{4a^3+4a^2+a}{1-a}$

Solution (a) First, we factor the denominator of the second fraction, then multiply.

$$\dfrac{2t+1}{3t-4} \cdot \dfrac{3t-4}{2t^2-9t-5} = \dfrac{2t+1}{3t-4} \cdot \dfrac{3t-4}{(2t+1)(t-5)}$$

$$= \dfrac{\cancel{(2t+1)}\cancel{(3t-4)}}{\cancel{(3t-4)}\cancel{(2t+1)}(t-5)} \qquad \text{Divide out common factors.}$$

$$= \dfrac{1}{t-5}$$

(b) First, we treat $3r+2$ as a fraction by writing it as $\dfrac{3r+2}{1}$.

$$(3r+2) \cdot \dfrac{5r+5}{3r^2-7r-6} = \dfrac{3r+2}{1} \cdot \dfrac{5r+5}{3r^2-7r-6}$$

$$= \dfrac{3r+2}{1} \cdot \dfrac{5(r+1)}{(3r+2)(r-3)} \qquad \begin{array}{l}\text{Factor } 5r+5 \text{ and} \\ 3r^2-7r-6.\end{array}$$

$$= \dfrac{(3r+2)(5)(r+1)}{(1)(3r+2)(r-3)}$$

$$= \dfrac{\cancel{(3r+2)}(5)(r+1)}{(1)\cancel{(3r+2)}(r-3)} \qquad \begin{array}{l}\text{Divide out the} \\ \text{common factor.}\end{array}$$

$$= \dfrac{5(r+1)}{r-3}$$

(c) To begin, factor both numerators and denominators completely.

$$\frac{a^2 - 1}{2a^4 + 5a^3 + 2a^2} \cdot \frac{4a^3 + 4a^2 + a}{1 - a} = \frac{(a + 1)(a - 1)}{a^2(2a^2 + 5a + 2)} \cdot \frac{a(4a^2 + 4a + 1)}{(-1)(a - 1)}$$

$$= \frac{(a + 1)(a - 1)}{a^2(2a + 1)(a + 2)} \cdot \frac{a(2a + 1)^2}{(-1)(a - 1)}$$

Divide out common factors.
$$= \frac{(a + 1)\cancel{(a - 1)}}{\underset{a}{\cancel{a^2}(2a + 1)}(a + 2)} \cdot \frac{\cancel{a}(2a + 1)^{\overset{2a+1}{\cancel{2}}}}{(-1)\cancel{(a - 1)}}$$

$$= \frac{(a + 1)(2a + 1)}{a(a + 2)(-1)}$$

$$= -\frac{(a + 1)(2a + 1)}{a(a + 2)}$$

Leaving the answer in factored form is acceptable.   ▲

N O T E   *In Part (c) of the previous example, notice that we factored the numerators and denominators, then divided out common factors* before *multiplying.*

Example 3   Multiply:

$$\frac{a^2 - 2ab + b^2}{4a^3b^4} \cdot \frac{2a^4b^2}{a^2 - b^2}$$

Solution   We factor, divide out common factors, then multiply.

$$\frac{a^2 - 2ab + b^2}{4a^3b^4} \cdot \frac{2a^4b^2}{a^2 - b^2} = \frac{(a - b)^2}{4a^3b^4} \cdot \frac{2a^4b^2}{(a + b)(a - b)}$$

$$= \frac{\overset{a-b}{\cancel{(a - b)^2}}}{\underset{2 \quad b^2}{\cancel{4a^3b^4}}} \cdot \frac{\overset{a}{\cancel{2a^4b^2}}}{(a + b)\cancel{(a - b)}}$$ Divide out common factors.

$$= \frac{a(a - b)}{2b^2(a + b)}$$ ▲

Recall the method for dividing fractions in arithmetic.

$$\frac{3}{2} \div \frac{9}{8} = \frac{3}{2} \cdot \frac{8}{9} = \frac{3 \cdot 8}{2 \cdot 9} = \frac{4}{3}$$

The important step is to change the division problem into a multiplication problem. To divide $\frac{3}{2}$ by $\frac{9}{8}$, we multiply $\frac{3}{2}$ by the *reciprocal* of $\frac{9}{8}$. The reciprocal of $\dfrac{9}{8}$ is $\dfrac{8}{9}$. Dividing two rational expressions is similar to dividing rational numbers.

R U L E

Division of rational expressions

If $\dfrac{A}{B}$ and $\dfrac{C}{D}$ are two rational expressions, where $B, C, D \neq 0$,

$$\frac{A}{B} \div \frac{C}{D} = \frac{A}{B} \cdot \frac{D}{C}$$

That is, the first expression divided by the second is the first times the reciprocal of the second. This is correct, since

$$\frac{A \cdot D}{B \cdot C} \cdot \frac{C}{D} = \frac{A \cdot \cancel{D} \cdot \cancel{C}}{B \cdot \cancel{C} \cdot \cancel{D}} = \frac{A}{B}$$

Example 4   Divide and simplify the answer.

(a) $\dfrac{3s^2}{r^3} \div \dfrac{9s^5}{r^2}$

(b) $\dfrac{y+3}{2y^2+y-1} \div \dfrac{y^2+7y+12}{(y+1)^2}$

(c) $\dfrac{2-a}{a^3-2a^2-3a} \div \dfrac{a^2-4}{a^3-3a^2}$

Solution   (a) $\dfrac{3s^2}{r^3} \div \dfrac{9s^5}{r^2} = \dfrac{3s^2}{r^3} \cdot \dfrac{r^2}{9s^5}$

$$= \frac{3s^2 r^2}{9r^3 s^5}$$

$$= \frac{\cancel{3}\cancel{s^2}\cancel{r^2}}{\cancel{9}\cancel{r^3}\cancel{s^5}}_{3\ r\ s^3}$$ Divide out common factors.

$$= \frac{1}{3rs^3}$$

(b) $\dfrac{y+3}{2y^2+y-1} \div \dfrac{y^2+7y+12}{(y+1)^2} = \dfrac{y+3}{2y^2+y-1} \cdot \dfrac{(y+1)^2}{y^2+7y+12}$

$$= \frac{y+3}{(2y-1)(y+1)} \cdot \frac{(y+1)^2}{(y+3)(y+4)}$$

Divide out common factors.

$$= \frac{\cancel{y+3}}{(2y-1)\cancel{(y+1)}} \cdot \frac{\cancel{(y+1)}^{y+1}}{\cancel{(y+3)}(y+4)}$$

$$= \frac{y+1}{(2y-1)(y+4)}$$

(c) $\dfrac{2-a}{a^3-2a^2-3a} \div \dfrac{a^2-4}{a^3-3a^2} = \dfrac{2-a}{a^3-2a^2-3a} \cdot \dfrac{a^3-3a^2}{a^2-4}$

$$= \frac{(-1)(a-2)}{a(a^2-2a-3)} \cdot \frac{a^2(a-3)}{a^2-4}$$

$$= \frac{(-1)(a-2)}{a(a-3)(a+1)} \cdot \frac{a^2(a-3)}{(a+2)(a-2)}$$

Divide out common factors.

$$= \frac{(-1)\cancel{(a-2)}}{\cancel{a}\cancel{(a-3)}(a+1)} \cdot \frac{\overset{a}{\cancel{a^2}}\cancel{(a-3)}}{(a+2)\cancel{(a-2)}}$$

$$= \frac{(-1)a}{(a+1)(a+2)}$$

$$= -\frac{a}{(a+1)(a+2)} \qquad \blacktriangle$$

Multiplication and division of rational expressions are further illustrated in the following examples. Notice that the main skill needed to do these problems is factoring.

Example 5   Perform the indicated operations:

$$\frac{(x^2-xy+y^2)(2x-y)}{2x^2-xy-y^2} \div \frac{x^3+y^3}{x^2-y^2}$$

Solution   First rewrite the division as multiplication by the reciprocal, then factor and simplify.

$$\frac{(x^2-xy+y^2)(2x-y)}{2x^2-xy-y^2} \div \frac{x^3+y^3}{x^2-y^2} = \frac{(x^2-xy+y^2)(2x-y)}{2x^2-xy-y^2} \cdot \frac{x^2-y^2}{x^3+y^3}$$

Factor the numerators and the denominators.

$$= \frac{(x^2-xy+y^2)(2x-y)}{(x-y)(2x+y)} \cdot \frac{(x+y)(x-y)}{(x+y)(x^2-xy+y^2)}$$

Divide out common factors.

$$= \frac{\cancel{(x^2-xy+y^2)}(2x-y)}{\cancel{(x-y)}(2x+y)} \cdot \frac{\cancel{(x+y)}\cancel{(x-y)}}{\cancel{(x+y)}\cancel{(x^2-xy+y^2)}}$$

$$= \frac{2x-y}{2x+y} \qquad \blacktriangle$$

Another formula that can be used on certain multiplication problems is the following.

R U L E

If A, B, and C are polynomials with $B \neq 0$, then

$$C\left(\frac{A}{B}\right) = \frac{CA}{B}$$

The proof of this equation is in Exercise 6 of the Enrichment Problems at the end of this section. The key to the proof is to first write C as the fraction $\frac{C}{1}$. The next example uses this result.

Example 6   Multiply: $(3zt - zy + 6t - 2y)\left(\dfrac{z+2}{3t-y}\right)$.

Solution   We write this as one fraction using the equation $C\left(\dfrac{A}{B}\right) = \dfrac{CA}{B}$, then factor the polynomial with four terms by grouping.

$$(3zt - zy + 6t - 2y)\left(\frac{z+2}{3t-y}\right) = \frac{(3zt - zy + 6t - 2y)(z+2)}{3t-y}$$

$$= \frac{[z(3t-y) + 2(3t-y)](z+2)}{3t-y}$$

$$= \frac{(3t-y)(z+2)(z+2)}{3t-y}$$

$$= \frac{\cancel{(3t-y)}(z+2)(z+2)}{\cancel{3t-y}} \qquad \text{Divide out the common factor.}$$

$$= (z+2)^2 \qquad\qquad\qquad \blacktriangle$$

L E A R N I N G   A D V A N T A G E   *Some instructors allow time for your requests during a review lesson. If so, bring up some topics you would like the instructor to review. Above all, remember that a review lesson can be a most costly lesson for you to miss. If you are absent for a review, be sure to contact someone in your class so that you find out exactly what work was covered.*

EXERCISE SET 4.3

For Exercises 1–10, find each product or quotient. Write the answer in lowest terms.

1. $\left(\dfrac{2}{5}\right)\left(\dfrac{3}{16}\right)$

2. $\dfrac{7}{3} \cdot \dfrac{21}{2}$

3. $\left(\dfrac{x}{y}\right)\left(\dfrac{3x^2}{y^3}\right)$

4. $\left(\dfrac{2v^2}{w}\right)\left(\dfrac{v^3}{w^5}\right)$

5. $\dfrac{5a^3}{b^2} \div \dfrac{25a}{b^4}$

6. $\dfrac{6p^2}{7q^3} \div \dfrac{36p^5}{49q^4}$

7. $\left(\dfrac{-3v}{u}\right)^2\left(\dfrac{u^2}{v^5}\right)$

8. $\left(\dfrac{-b}{2a}\right)\left(\dfrac{4a}{b}\right)^2$

9. $\dfrac{12n^5}{3n^8} \div \dfrac{-16n}{15n^3}$

10. $\dfrac{35p^5}{7q^3} \div \dfrac{5p^4}{-8p^7}$

For Exercises 11–48, multiply or divide as indicated. Write the answer in lowest terms.

11. $\dfrac{3}{z-2} \cdot \dfrac{z+1}{15}$

12. $\dfrac{2s+3}{6s} \cdot \dfrac{30s}{s-1}$

13. $\dfrac{3c-24}{4c+1} \cdot \dfrac{8c+2}{6c-48}$

14. $\dfrac{4w-48}{2w-1} \cdot \dfrac{6w-3}{2w-24}$

15. $\dfrac{x-3}{2x+15} \cdot \dfrac{4x+30}{x^2-3x}$

16. $\dfrac{a^2+5a}{3a+15} \cdot \dfrac{2a^2-7a}{10a-35}$

17. $\dfrac{3s-12}{4s-4} \div \dfrac{9s-36}{12s-36}$

18. $\dfrac{5b+10}{12b+12} \div \dfrac{18b+36}{6b-6}$

19. $\dfrac{2y^2+y-3}{3y-2} \cdot \dfrac{3y^2+y-2}{y^2-1}$

20. $\dfrac{3c^2+8c-3}{2c+1} \cdot \dfrac{2c^2-c-1}{3c^2-4c+1}$

21. $\dfrac{t^2-7t}{2t+3} \div \dfrac{t^2-6t-7}{2t^2+3t}$

22. $\dfrac{m^3+m^2}{m-4} \div \dfrac{m^2-1}{m^2-16}$

23. $\dfrac{3-z}{12} \cdot \dfrac{-4}{z-3}$

24. $\dfrac{v-10}{20} \cdot \dfrac{5}{10-v}$

25. $\dfrac{2k-1}{3k^{12}} \div \dfrac{1-2k}{4k^5}$

26. $\dfrac{7x-5}{4x-4} \div \dfrac{5-7x}{2x-2}$

27. $\dfrac{5p-12}{7-3p} \cdot \dfrac{3p-7}{12-5p}$

28. $\dfrac{8-5q}{10q-1} \cdot \dfrac{1-10q}{5q-8}$

29. $\dfrac{s^2-9}{s^2-1} \cdot \dfrac{s+1}{s-3}$

30. $\dfrac{b^2-25}{b-2} \cdot \dfrac{b^2-4}{b+5}$

31. $\dfrac{81-h^2}{9h-h^2} \div \dfrac{h^3+9h^2}{3h}$

32. $\dfrac{1 - x^2}{2x^4 + 2x^3} \div \dfrac{x^2 - 2x + 1}{6x^4}$

33. $(2v + 6) \cdot \dfrac{v^2 - 3v - 4}{v + 3}$

34. $(3t - 1) \cdot \dfrac{t^2 - 4t - 5}{3t^2 - t}$

35. $(2s^2 - s - 3) \div (2s - 3)$

36. $(4a^2 - 13a + 3) \div (4a - 1)$

37. $\dfrac{x^2 - 2ax + a^2}{x^2 - a^2} \cdot \dfrac{x + a}{2x^2 - ax - a^2}$

38. $\dfrac{u^2 - v^2}{3u^2 + 6uv + 3v^2} \cdot \dfrac{u + v}{u - v}$

39. $\dfrac{2n^2 + 3mn + m^2}{n^2 - m^2} \div \dfrac{2n^2 - mn - m^2}{n - m}$

40. $\dfrac{p^2 + 2sp + s^2}{2p^2 - 3sp + s^2} \div \dfrac{p + s}{p - s}$

41. $\dfrac{x^3 - x^2 + x - 1}{2x^3 - 2x + 3x^2 - 3} \div \dfrac{x^3 + 3x^2 + x + 3}{x^3 + 2x + x^2 + 2}$

42. $\dfrac{a^3 - 4a + a^2 - 4}{a^2 - a - 2} \div \dfrac{2a^2 - a - 6}{2a^3 + 8a + 3a^2 + 12}$

43. $\dfrac{z^3 - 8}{z^3 + 2z^2 + 4z} \cdot \dfrac{z^4 - 4z^2}{4z^2 - 9z + 2}$

44. $\dfrac{2c^3 - 2c + 4c^2 - 4}{2c^2 - c - 1} \cdot \dfrac{2c^3 + c^2}{8c^3 + 64}$

45. $\dfrac{zx + 3 - 3x - z}{2zx - 4z + x - 2} \cdot \dfrac{3zx + x - 6z - 2}{3zx + 2z - 9x - 6}$

46. $\dfrac{2pq - 6 + 4q - 3p}{2pq + 8q - 3p - 12} \cdot \dfrac{3pq - p + 12q - 4}{7pq + 2p + 14q + 4}$

47. $\dfrac{2z^2 - cz - c^2}{2c^2z^2 + c^3z} \div \dfrac{z^2 - c^2}{cz^3 + c^2z^2}$

48. $\dfrac{3m^2 - 5mn + 2n^2}{7m^2 + 14mn} \div \dfrac{2m^2 - mn - n^2}{14m^2n + 28mn^2}$

For Exercises 49–62, write each expression as a single fraction in lowest terms.

49. $\dfrac{84}{-90} \cdot \dfrac{50}{35} \div \dfrac{-16}{18}$

50. $\dfrac{-56}{60} \div \left(\dfrac{36}{20} \cdot \dfrac{24}{-15} \right)$

51. $\dfrac{14}{9} \div \dfrac{21}{15} \div \dfrac{35}{24}$

52. $\dfrac{18}{21} \div \dfrac{8}{33} \div \dfrac{44}{7}$

53. $\dfrac{c^3d^4}{3} \div \left(\dfrac{4d^3}{c} \cdot \dfrac{c^3}{18d} \right)$

54. $\dfrac{3r^2s}{s^2} \div \left(\dfrac{r}{s^4} \cdot \dfrac{6r^2}{s} \right)$

55. $\dfrac{x^2 - 4x + 4}{2x^2 + 5x + 2} \cdot \dfrac{x + 4}{x + 1} \div \dfrac{x^2 + 2x - 8}{2x^2 - 5x - 3}$

56. $\dfrac{m^2 - 6m + 9}{3m^2 - 4m + 1} \cdot \dfrac{6m^2 - 5m + 1}{m^2 - 2m - 3} \div \dfrac{2m^2 + m - 1}{m^2 + 3m + 2}$

57. $\dfrac{u^3 - v^3}{u^3 + v^3} \cdot \dfrac{3u^2 + 2uv - v^2}{2u^2 - uv - v^2} \cdot \dfrac{2u^2 - 3uv - 2v^2}{2u^2 + 2uv + 2v^2}$

58. $\dfrac{6by - 8b + 3y - 4}{2by - 2b + 2y - 2} \cdot \dfrac{b^2 + 4b + 3}{3y^2 - y - 4} \div \dfrac{b + 3}{y - 1}$

59. $\dfrac{x^2 - 6x + 9}{x^3 - 27} \div \dfrac{2x^2 - 5x - 3}{2x^2 + 6x + 18}$

60. $\dfrac{a^4 + 8a}{3a^2 - 6a + 12} \div \dfrac{a^4 - a^3 - 6a^2}{6a - 12}$

61. $\dfrac{ax + 5x - 4a - 20}{ax + 2a + 5x + 10} \cdot \dfrac{ax + 2 + x + 2a}{ax + x - 4a - 4}$

62. $\dfrac{cd + c - d - 1}{3cd - 2d + 3c - 2} \cdot \dfrac{3cd - 2d + 9c - 6}{2cd - 6 + 6c - 2d}$

For Exercises 63–78, perform the given operations. Write your answer in lowest terms.

63. $\dfrac{3}{-5} \div \left(\dfrac{21}{-15} \cdot \dfrac{2}{-7} \right)$

64. $\dfrac{-4}{33} \div \left(\dfrac{6}{-11} \cdot \dfrac{1}{3} \right)$

65. $\dfrac{2x + 3}{3x - 1} \cdot \dfrac{x}{x - 1} \div \dfrac{x^2}{2 - 6x}$

66. $\dfrac{4z - 7}{z^2 - z} \cdot \dfrac{z^3}{9} \div \dfrac{7 - 4z}{3z - 3}$

67. $\dfrac{2y^2 + 5y + 3}{y^4 - 2y^3} \div \dfrac{y + 1}{y^4} \cdot \dfrac{y - 2}{y^2 - y}$

68. $\dfrac{a}{a - 3} \div \dfrac{a - 2}{a + 2} \cdot \dfrac{a^2 - 5a + 6}{a^3 - 4a^2}$

69. $(x^2 + 2x) \cdot \dfrac{x + 3}{x^3 + 2x^2}$

70. $(y^3 - y^2) \cdot \dfrac{y - 2}{y^2 - y}$

71. $(a - 2)(2a + 1) \left(\dfrac{4a}{a^2 - 4} \right)$

72. $(z^2 - 9) \left(\dfrac{z - 1}{z^2 + 3z} \right)$

73. $(2t - 2) \cdot \dfrac{t^3 + t^2 + t}{t^3 - 1}$

74. $(x^3 + 64) \cdot \dfrac{6x^2}{2x^3 - 8x^2 + 32x}$

75. $\left(\dfrac{1}{(x + h)^2} \right)(xk - 4h + hk - 4x)$

76. $\left(\dfrac{8z + 24}{4y - 8s} \right)(yz - 6s + 3y - 2zs)$

77. $(2x^2 - 3xa - 2a^2) \dfrac{3x + 3a}{12ax + 6a^2}$

78. $(2v^2 - wv - 3w^2) \dfrac{3w - 2v}{v^3 + v^2w}$

REVIEW PROBLEMS

The following exercises review parts of Sections 3.4 and 3.7. Doing these problems will help prepare you for the next section.

Section 3.4

Factor all composite numbers. If the number is prime, write "prime."

79. 24

80. 39

81. 144

82. 78

83. 41

84. 65

Section 3.7

Factor completely.

85. $x^2 + 2x - 15$

86. $2y^2 - 7y - 4$

87. $x^3 - 100x$

88. $z^2 - 4z + 4$

89. $n^2 + 2nm + m^2$

90. $x^4 + 4x^2$

E N R I C H M E N T   E X E R C I S E S

Simplify.

1. $\dfrac{x^4 + 2x^2y^2 + y^4}{x^2 - y^2} \cdot \dfrac{2x^4 - x^2y^2 - y^4}{2x^4 + 3x^2y^2 + y^4}$

2. $\dfrac{a^3b - b^4}{a^4 - b^4} \div \dfrac{a^3 + a^2b + ab^2}{a^3 - b^3 - a^2b + b^2a}$

3. $\dfrac{\dfrac{R + T}{(R^2 + T^2)^2}}{\dfrac{(R + T)^2}{R^4 - T^4}}$

4. $\dfrac{x^2 - (y - z)^2}{-4a^3} \cdot \dfrac{ay - a(x - z)}{y^2 - (x - z)^2}$

5. $\dfrac{x^6a - x^6b - y^6a + y^6b}{a^2 + b^2} \div \dfrac{xa^2 - xb^2 - ya^2 + yb^2}{a^8 - b^8}$

6. Let A, B, and C be polynomials with $B \neq 0$. Prove that

$$C\left(\frac{A}{B}\right) = \frac{CA}{B}$$

Answers to Enrichment Exercises are on page A.17.

4.4

Addition and Subtraction of Rational Expressions

OBJECTIVES

▶ *To find the least common denominator of two rational expressions*

▶ *To add and subtract rational expressions*

▶ *To simplify the result of combining rational expressions*

This section deals with addition and subtraction of rational expressions. In the case of rational *numbers* whose denominators are the same, we add or subtract the numerators and keep the common denominator. For example,

$$\frac{4}{11} + \frac{6}{11} = \frac{4+6}{11} = \frac{10}{11}$$

and

$$\frac{7}{10} - \frac{3}{10} = \frac{7-3}{10} = \frac{4}{10} = \frac{2}{5}$$

We combine rational expressions with like denominators in the same way.

RULE

Addition and subtraction of rational expressions with like denominators

If $\dfrac{A}{B}$ and $\dfrac{C}{B}$ are rational expressions with common denominator B, then

$$\frac{A}{B} + \frac{C}{B} = \frac{A+C}{B} \qquad \text{and} \qquad \frac{A}{B} - \frac{C}{B} = \frac{A-C}{B}$$

A justification of the first equation follows. Recall that $\dfrac{A}{B}$ means A times $\dfrac{1}{B}$, the reciprocal of B. Therefore,

$$\frac{A}{B} + \frac{C}{B} = A\left(\frac{1}{B}\right) + C\left(\frac{1}{B}\right)$$

$$= (A+C)\left(\frac{1}{B}\right) \qquad \text{Distributive property.}$$

$$= \frac{A+C}{B}$$

Notice the importance of the common denominator B. Without this common denominator, we could not have used the distributive property. Without the distributive property, we could not have added the two fractions.

Example 1   Add or subtract as indicated.

(a) $\dfrac{5}{a} + \dfrac{7}{a} = \dfrac{5+7}{a}$

$\qquad\quad = \dfrac{12}{a}$

(b) $\dfrac{14}{t} - \dfrac{50}{t} = \dfrac{14-50}{t}$

$\qquad\quad = -\dfrac{36}{t}$ ▲

Example 2   Subtract: $\dfrac{p^2+7}{3p^2-2} - \dfrac{p^2-1}{3p^2-2}$.

Solution

$\dfrac{p^2+7}{3p^2-2} - \dfrac{p^2-1}{3p^2-2} = \dfrac{(p^2+7)-(p^2-1)}{3p^2-2}$ Subtract numerators.

$\qquad\qquad = \dfrac{p^2+7-p^2+1}{3p^2-2}$ Remove parentheses.

$\qquad\qquad = \dfrac{8}{3p^2-2}$ Combine like terms.

▲

Sometimes the result from adding or subtracting rational expressions can be simplified by reducing the answer to lowest terms. This point is illustrated in the next example.

Example 3   Subtract, then reduce the answer to lowest terms.

$$\frac{3t}{t^2-3t+2} - \frac{6}{t^2-3t+2}$$

Solution   We subtract numerators, keeping the same denominator.

$\dfrac{3t}{t^2-3t+2} - \dfrac{6}{t^2-3t+2} = \dfrac{3t-6}{t^2-3t+2}$

$\qquad\qquad = \dfrac{3(t-2)}{(t-1)(t-2)}$ Factor numerator and denominator.

$\qquad\qquad = \dfrac{3\cancel{(t-2)}}{(t-1)\cancel{(t-2)}}$ Divide out the common factor.

$\qquad\qquad = \dfrac{3}{t-1}$ ▲

Now, suppose that the denominators of the rational expressions are different. We must first find a common denominator before adding or subtracting them. Even though any common denominator will suffice, it is more efficient to use the *least* common denominator.

DEFINITION

The **least common denominator,** abbreviated LCD, for a collection of fractions is the simplest polynomial that is divisible by each of the denominators.

The least common denominator is frequently obvious. For example, a common denominator of $\frac{2}{3}$ and $\frac{1}{2}$ is 6, since 6 is divisible by both 3 and 2. Furthermore, 6 is the smallest (least) positive number divisible by both 3 and 2. If the least common denominator is not obvious, it can be found as follows.

STRATEGY

To find the least common denominator (LCD) of two or more rational expressions

Step 1   Factor each denominator completely.

Step 2   Form the product of each different factor to the highest power that it occurs in any denominator. The least common denominator is this product.

The principle of rational expressions allows us to reduce an expression to lowest terms. We can use the same principle to "build up" fractions. Namely, for nonzero K,

$$\frac{A}{B} = \frac{A \cdot K}{B \cdot K}$$

That is, we can write a fraction in a different form by multiplying numerator and denominator by the same nonzero quantity K. In particular, if we change the denominator from B to BK, then the numerator changes from A to AK.

Example 4   Add: $\dfrac{5}{12} + \dfrac{7}{30}$.

Solution   We first find the LCD of the two fractions. To do this, we use the two-step method just stated.

Step 1   Factor each denominator into primes:

$$12 = 2^2 \cdot 3 \qquad \text{and} \qquad 30 = 2 \cdot 3 \cdot 5$$

Step 2   Take each factor to the highest power that it appears in either denominator. Therefore,

$$\text{LCD} = 2^2 \cdot 3 \cdot 5 = 60$$

Next, we change $\dfrac{5}{12}$ to a fraction with denominator 60.

$$\frac{5}{12} = \frac{5 \cdot 5}{12 \cdot 5} = \frac{25}{60}$$

Then, change $\dfrac{7}{30}$ to a fraction with denominator 60.

$$\frac{7}{30} = \frac{7 \cdot 2}{30 \cdot 2} = \frac{14}{60}$$

Now,

$$\frac{5}{12} + \frac{7}{30} = \frac{25}{60} + \frac{14}{60}$$

$$= \frac{25 + 14}{60}$$

$$= \frac{39}{60}$$

$$= \frac{13}{20} \qquad \blacktriangle$$

When adding or subtracting fractions with unlike denominators, we rewrite each fraction with the LCD for a denominator. When rewriting a fraction, be sure not to change the value of it, but multiply the *numerator* by that same quantity K that you have multiplied the denominator to obtain the LCD.

Example 5   Subtract: $\dfrac{2}{a^2 - 1} - \dfrac{1}{a^2 - 3a + 2}$.

Solution   To find the LCD, we factor both denominators.

$$a^2 - 1 = (a + 1)(a - 1)$$

$$a^2 - 3a + 2 = (a - 1)(a - 2)$$

Therefore, LCD $= (a + 1)(a - 1)(a - 2)$.

Next, we change *each* fraction to an equivalent expression that has the LCD for a denominator.

$$\frac{2}{a^2 - 1} = \frac{2}{(a + 1)(a - 1)} \cdot \frac{a - 2}{a - 2}$$

$$= \frac{2(a - 2)}{(a + 1)(a - 1)(a - 2)}$$

$$\frac{1}{a^2 - 3a + 2} = \frac{1}{(a - 1)(a - 2)} \cdot \frac{a + 1}{a + 1}$$

$$= \frac{a + 1}{(a - 1)(a - 2)(a + 1)}$$

Now, we have common denominators and all we do to finish the problem is subtract our new numerators and simplify.

$$\frac{2}{a^2 - 1} - \frac{1}{a^2 - 3a + 2} = \frac{2(a - 2)}{(a + 1)(a - 1)(a - 2)} - \frac{a + 1}{(a + 1)(a - 1)(a - 2)}$$

$$= \frac{2(a - 2) - (a + 1)}{(a + 1)(a - 1)(a - 2)}$$

$$= \frac{2a - 4 - a - 1}{(a + 1)(a - 1)(a - 2)}$$

$$= \frac{a - 5}{(a + 1)(a - 1)(a - 2)} \qquad \blacktriangle$$

We summarize the steps taken to add or subtract rational expressions with unlike denominators.

STRATEGY

Addition and subtraction of rational expressions with unlike denominators

Step 1   Find the least common denominator.

Step 2   Rewrite each fraction using the least common denominator.

Step 3   Add or subtract the numerators, keeping the same denominator.

Step 4   Simplify the numerator, then write the fraction in lowest terms.

Example 6   Add or subtract as indicated, then simplify.

$$\frac{1}{t+1} - \frac{1}{2t} + \frac{5t+3}{2t^2(t+1)}$$

Solution   The least common denominator is $2t^2(t+1)$. Therefore,

$$\frac{1}{t+1} - \frac{1}{2t} + \frac{5t+3}{2t^2(t+1)} = \frac{1 \cdot 2t^2}{2t^2(t+1)} - \frac{1 \cdot t(t+1)}{2t^2(t+1)} + \frac{5t+3}{2t^2(t+1)}$$

$$= \frac{2t^2 - t(t+1) + (5t+3)}{2t^2(t+1)}$$

$$= \frac{2t^2 - t^2 - t + 5t + 3}{2t^2(t+1)}$$

Simplify the numerator. $$= \frac{t^2 + 4t + 3}{2t^2(t+1)}$$

Factor the numerator. $$= \frac{(t+3)(t+1)}{2t^2(t+1)}$$

Divide out the common factor $t+1$. $$= \frac{t+3}{2t^2} \qquad \blacktriangle$$

In the next example, we show how to add a polynomial and a rational expression.

Example 7   Add: $(2w - 3) + \dfrac{w}{w+5}$.

Solution   A polynomial such as $2w - 3$ can be rewritten as $\dfrac{2w-3}{1}$ and the LCD of $\dfrac{2w-3}{1}$ and $\dfrac{w}{w+5}$ is $w + 5$. Therefore,

$$(2w-3) + \frac{w}{w+5} = \frac{2w-3}{1} + \frac{w}{w+5}$$

$$= \frac{(2w-3)(w+5)}{w+5} + \frac{w}{w+5}$$

$$= \frac{(2w-3)(w+5) + w}{w+5}$$

$$= \frac{2w^2 + 7w - 15 + w}{w+5}$$

$$= \frac{2w^2 + 8w - 15}{w+5}$$

This last fraction is already in lowest terms. $\blacktriangle$

In the next example, we use the property, $\dfrac{A}{B} = \dfrac{-A}{-B}$, to simplify the sum of two rational expressions, where the denominators are a pair of opposites.

Example 8   Add: $\dfrac{2}{x-1} + \dfrac{3}{1-x}$.

Solution   Since the denominators are opposites, multiply the numerator and denominator of the second fraction by -1.

$$\dfrac{2}{x-1} + \dfrac{3}{1-x} = \dfrac{2}{x-1} + \dfrac{(-1)\cdot 3}{(-1)(1-x)}$$

$$= \dfrac{2}{x-1} + \dfrac{-3}{x-1} \qquad \text{Note that } (-1)(1-x) = x-1.$$

$$= \dfrac{2-3}{x-1}$$

$$= \dfrac{-1}{x-1}$$

$$= -\dfrac{1}{x-1} \qquad\qquad\qquad\qquad ▲$$

Example 9   Let x represent a number. Write an expression for the sum of twice x and the reciprocal of $x-3$. Simplify this expression.

Solution   The reciprocal of $x-3$ is $\dfrac{1}{x-3}$. Therefore, the sum of twice x and the reciprocal of $x-3$ is

$$2x + \dfrac{1}{x-3}$$

To simplify this expression, rewrite $2x$ as $\dfrac{2x}{1}$ and combine with $\dfrac{1}{x-3}$ as one fraction.

$$2x + \dfrac{1}{x-3} = \dfrac{2x}{1} + \dfrac{1}{x-3}$$

$$= \dfrac{2x(x-3)}{x-3} + \dfrac{1}{x-3}$$

$$= \dfrac{2x^2 - 6x + 1}{x-3} \qquad\qquad\qquad ▲$$

L E A R N I N G   A D V A N T A G E   *If you have saved all your assignments and corrected and written the details of every problem, you will be amazed at how quickly you can review and remember all the things that seemed so difficult a few days ago. You can cut your time spent reviewing homework in half—if you have spent enough time on your homework each day that you studied the chapter.*

E X E R C I S E   S E T   4.4

For Exercises 1–20, add or subtract as indicated. Write the answer in lowest terms.

1. $\dfrac{15}{4} + \dfrac{9}{4}$

2. $\dfrac{2}{5} + \dfrac{8}{5}$

3. $\dfrac{9}{2} - \dfrac{7}{2}$

4. $\dfrac{12}{7} - \dfrac{5}{7}$

5. $\dfrac{6}{z} + \dfrac{11}{z}$

6. $\dfrac{14}{a} - \dfrac{12}{a}$

7. $-\dfrac{40}{z^2} + \dfrac{20}{z^2}$

8. $-\dfrac{18}{r^3} + \dfrac{13}{r^3}$

9. $\dfrac{3}{b + 1} + \dfrac{7}{b + 1}$

10. $\dfrac{17}{a^2 + 7} - \dfrac{8}{a^2 + 7}$

11. $\dfrac{3q^2 + 1}{q^2} - \dfrac{4q^2 + 7}{q^2}$

12. $\dfrac{5 - w^2}{2w} - \dfrac{8 + w^2}{2w}$

13. $\dfrac{3x + 7}{x^2 + 10} + \dfrac{x^2 - 3x}{x^2 + 10}$

14. $\dfrac{c^2 + 5}{c^5 - 1} + \dfrac{c^2 - 13}{c^5 - 1}$

15. $\dfrac{2y}{y^2 - 2y - 3} + \dfrac{2}{y^2 - 2y - 3}$

16. $\dfrac{3r}{r^2 - 3r + 2} - \dfrac{6}{r^2 - 3r + 2}$

17. $\dfrac{2t}{2t^2 - t - 3} - \dfrac{3}{2t^2 - t - 3}$

18. $\dfrac{x}{x^2 - 100} + \dfrac{10}{x^2 - 100}$

19. $\dfrac{4p}{16p^2 - 1} + \dfrac{1}{16p^2 - 1}$

20. $\dfrac{2n}{4n^2 - 9} - \dfrac{3}{4n^2 - 9}$

For Exercises 21–66, add or subtract as indicated. Write the answer in lowest terms.

21. $\dfrac{3}{5} + \dfrac{7}{10}$

22. $\dfrac{2}{3} + \dfrac{1}{6}$

23. $\dfrac{11}{12} - \dfrac{5}{9}$

24. $\dfrac{13}{15} - \dfrac{1}{6}$

25. $\dfrac{4}{5y} + \dfrac{6}{15y}$

26. $\dfrac{17}{18h} - \dfrac{7}{9h}$

27. $\dfrac{1}{r^2 - 4} + \dfrac{r}{r + 2}$

28. $\dfrac{4v + 3}{v^2 - 9} - \dfrac{v + 1}{v - 3}$

29. $\dfrac{3}{m^2 - 3m + 2} + \dfrac{2}{m^2 - m - 2}$

30. $\dfrac{4}{p^2 + 5p + 6} + \dfrac{1}{p^2 + 2p - 3}$

31. $\dfrac{z}{2z^2 - 5z + 3} - \dfrac{2z}{2z^2 - z - 3}$

32. $\dfrac{3c}{2c^2 + 7c - 4} - \dfrac{4c}{c^2 + 2c - 8}$

33. $\dfrac{5x + 1}{3x^2 + 11x - 4} + \dfrac{7x - 2}{3x^2 + 14x - 5}$

34. $\dfrac{9y + 2}{4y^2 + 3y - 1} + \dfrac{2y + 5}{y^2 - 5y - 6}$

35. $\dfrac{r + 2}{r^2 + r} + \dfrac{r - 1}{r^2 + 2r}$

36. $\dfrac{2h}{h^3 + 2h^2} + \dfrac{1}{h^3 + 3h^2}$

37. $\dfrac{3t - 3}{t^2 - t} - \dfrac{t^2 - 3t}{t^3 - t^2}$

38. $\dfrac{z + 1}{z^2 + z} - \dfrac{z - z^2}{z^3 + z^2}$

39. $\dfrac{y^2 - 19}{y^2 + 4y - 5} - \dfrac{y - 4}{y - 1}$

40. $\dfrac{P^2 - 39}{P^2 + 3P - 10} - \dfrac{P - 7}{P - 2}$

41. $z + 5 + \dfrac{2}{z + 3}$

42. $2t - 1 - \dfrac{3t}{t + 4}$

43. $\dfrac{3y}{5} - \dfrac{7y}{10} + \dfrac{6y}{15}$

44. $\dfrac{4q}{3} + \dfrac{6q}{12} - \dfrac{2q}{9}$

45. $\dfrac{3}{c} - \dfrac{5}{2c} + \dfrac{7}{5c^2}$

46. $\dfrac{4}{3Y} + \dfrac{5}{9Y^2} - \dfrac{7}{18Y}$

47. $\dfrac{7}{3z^2} - \dfrac{4}{z^2} - \dfrac{10}{2z}$

48. $\dfrac{4}{b} + \dfrac{7}{5b^2} - \dfrac{10}{4b}$

49. $\dfrac{1}{y - 1} + \dfrac{1}{3y} - \dfrac{1}{y^2(y - 1)}$

50. $\dfrac{2}{m(m + 2)} + \dfrac{1}{m + 2} - \dfrac{1}{4m^2}$

51. $\dfrac{3}{2x - 1} + \dfrac{5}{1 - 2x}$

52. $\dfrac{4r}{r - 12} + \dfrac{3}{12 - r}$

53. $\dfrac{x}{x^2 - 4} + \dfrac{2}{4 - x^2}$

54. $\dfrac{t}{t^2 - 100} - \dfrac{10}{100 - t^2}$

55. $\dfrac{1}{9y^2 - 1} + \dfrac{3y}{1 - 9y^2}$

56. $\dfrac{2}{4 - 25q^2} - \dfrac{5q}{25q^2 - 4}$

57. $\dfrac{-1}{x^3 - 1} - \dfrac{x}{1 - x^3}$

58. $\dfrac{b}{8 - b^3} + \dfrac{2}{b^3 - 8}$

59. $\dfrac{z}{z + 3} - \dfrac{2}{3 - z} - \dfrac{3 - 5z}{z^2 - 9}$

60. $\dfrac{2(q^2 - 10)}{2q^2 - q - 1} + \dfrac{4}{q - 1} - \dfrac{13}{2q + 1}$

61. $\dfrac{x^2 + y^2}{x^3 - y^3} - \dfrac{xy}{y^3 - x^3}$

62. $\dfrac{2ab}{8a^3 - b^3} - \dfrac{4a^2 + b^2}{b^3 - 8a^3}$

63. $\dfrac{2(u + 2v)}{u^2 - 4uv + 4v^2} - \dfrac{u + 2v}{u^2 - 2uv}$

64. $\dfrac{y(y - 18b)}{2y^2 + 3by - 2b^2} + \dfrac{y + 3b}{2y - b}$

65. $2(a + b)^{-1} - a^{-1} - b^{-1}$

66. $(x + 2y)^{-1} - (x - 2y)^{-1}$

For Exercises 67–70, let x represent a number. Write an expression from the information given, then simplify.

67. The sum of a number and three times its reciprocal.

68. Five times a number reduced by the reciprocal of the number.

69. The reciprocal of the sum of a number and four reduced by twice the number.

70. The sum of the reciprocal of a number and the reciprocal of twice the number.

71. Write an expression for the sum of the reciprocals of two consecutive even integers, where n represents the smaller of these two numbers. Simplify the answer into one fraction.

72. Write an expression for three times the sum of the reciprocals of two consecutive odd integers, where n represents the smaller odd integer. Simplify the answer into one fraction.

73. The total resistance R from two resistors with resistances R_1 and R_2, respectively, connected in parallel is given by $\dfrac{1}{R} = \dfrac{1}{R_1} + \dfrac{1}{R_2}$. Find R, if $R_1 = 3$ ohms and $R_2 = 5$ ohms.

74. The total resistance R from three resistors with resistances R_1, R_2, and R_3, respectively, connected in parallel is given by $\dfrac{1}{R} = \dfrac{1}{R_1} + \dfrac{1}{R_2} + \dfrac{1}{R_3}$. Find R if $R_1 = 6$, $R_2 = 4$, and $R_3 = 2$ ohms.

75. (a) Show that the formula in Exercise 73 can be rewritten as

$$R = \frac{R_1 R_2}{R_1 + R_2}$$

(b) Use this new form to find R when $R_1 = 8$ and $R_2 = 10$ ohms.

76. (a) Show that the formula in Exercise 74 can be rewritten as

$$R = \frac{R_1 R_2 R_3}{R_2 R_3 + R_1 R_3 + R_1 R_2}$$

(b) Use this new formula to find R when $R_1 = 4$, $R_2 = 8$, and $R_3 = 6$ ohms.

REVIEW PROBLEMS

The following exercises review parts of Section 3.3. Doing these problems will help prepare you for the next section.

Multiply.

77. $4\left(\dfrac{1}{2} + 3\right)$

78. $14\left(1 - \dfrac{1}{7}\right)$

79. $x^3(2 - 3x)$

80. $2s^2(2s - 5)$

81. $4x\left(\dfrac{3}{2}x - \dfrac{1}{4}\right)$

82. $6z^2\left(\dfrac{2}{3} + \dfrac{5}{6}z^2\right)$

83. $12a^2b\left(\dfrac{5}{4}ab - \dfrac{1}{3}b^2\right)$

84. $20z\left(\dfrac{2}{5}a - \dfrac{7}{4}z^2\right)$

85. $18t^2\left(\dfrac{11}{6}st + \dfrac{2}{9}s^3t\right)$

ENRICHMENT EXERCISES

For Exercises 1–5, add or subtract as indicated. Simplify, if possible. Assume that n and m represent positive integers.

1. $\dfrac{2}{x^4 - y^4} + \dfrac{1}{x^3 y - xy^3}$

2. $\dfrac{x^n}{x^n - 2} - \dfrac{6 - x^n}{x^{2n} - 2x^n}$

3. $\dfrac{1}{10x^{2m}} - \dfrac{2}{15x^m y^n} + \dfrac{1}{5y^{2n}}$

4. $\dfrac{6a^n - 6}{a^{2n} - 3a^n} + \dfrac{a^n - 7}{a^n - 3}$

5. $\dfrac{2}{(z-x)(z-y)} - \dfrac{2}{(x-z)(y-x)} + \dfrac{2}{(y-x)(y-z)}$

6. Simplify: $\left(\dfrac{3}{2}x - \dfrac{3}{4}\right) \div \dfrac{36x^2 - 9}{8}$.

Answers to Enrichment Exercises are on page A.18.

4.5

Complex Fractions

O B J E C T I V E

▶   *To simplify complex fractions*

A rational expression that has fractions in its numerator or denominator is called a **complex fraction.** Examples of complex fractions are the following.

$$\frac{\dfrac{3}{2}}{\dfrac{5}{8}}, \qquad \frac{7 - \dfrac{1}{x}}{5 + 3x}, \qquad \text{and} \qquad \frac{8 + 9a}{6 - \dfrac{5}{a}}$$

A complex fraction can be simplified to an ordinary fraction. There are two methods available to simplify complex fractions.

S T R A T E G Y

Two methods for simplifying a complex fraction

Method 1: If necessary, combine the terms in the numerator, combine the terms in the denominator, then divide.

Method 2: Find the least common denominator (LCD) of all the fractions appearing in the numerator and denominator of the complex fraction; then, multiply both numerator and denominator by this LCD to clear fractions.

Example 1   Use Method 1 to simplify $\dfrac{\dfrac{6}{x^2}}{\dfrac{9}{x}}$.

Solution   The numerator and denominator are already expressed as single terms. Therefore, all we need do is divide.

$$\frac{\dfrac{6}{x^2}}{\dfrac{9}{x}} = \frac{6}{x^2} \div \frac{9}{x}$$

$$= \frac{6}{x^2} \cdot \frac{x}{9}$$

$$= \frac{2}{3x} \qquad \blacktriangle$$

Example 2   Use Method 2 to simplify $\dfrac{\dfrac{1}{2} + \dfrac{7}{3}}{\dfrac{5}{12} - \dfrac{1}{6}}$.

Solution   We first find the LCD of the fractions appearing in the complex fraction. The
LCD of $\dfrac{1}{2}, \dfrac{7}{3}, \dfrac{5}{12}$, and $\dfrac{1}{6}$ is 12.

Next, we multiply numerator and denominator by 12 and simplify.

$$\frac{\dfrac{1}{2} + \dfrac{7}{3}}{\dfrac{5}{12} - \dfrac{1}{6}} = \frac{12\left(\dfrac{1}{2} + \dfrac{7}{3}\right)}{12\left(\dfrac{5}{12} - \dfrac{1}{6}\right)}$$

$$= \frac{12\left(\dfrac{1}{2}\right) + 12\left(\dfrac{7}{3}\right)}{12\left(\dfrac{5}{12}\right) - 12\left(\dfrac{1}{6}\right)}$$

$$= \frac{6 + 28}{5 - 2}$$

$$= \frac{34}{3} \qquad \blacktriangle$$

N O T E   *We could have done Example 2 using Method 1. However, when
a numerator or denominator is the sum or difference of fractions, Method 2 is
more efficient.*

Example 3   Simplify $\dfrac{\dfrac{5}{n+1} - \dfrac{1}{n-1}}{\dfrac{1}{n+1} - \dfrac{2}{n-1}}$.

Solution   The LCD of the four fractions

$$\frac{5}{n+1}, \qquad \frac{1}{n-1}, \qquad \frac{1}{n+1}, \qquad \text{and} \qquad \frac{2}{n-1}$$

is $(n+1)(n-1)$. Therefore,

$$\frac{\dfrac{5}{n+1} - \dfrac{1}{n-1}}{\dfrac{1}{n+1} - \dfrac{2}{n-1}} = \frac{(n+1)(n-1)\left(\dfrac{5}{n+1} - \dfrac{1}{n-1}\right)}{(n+1)(n-1)\left(\dfrac{1}{n+1} - \dfrac{2}{n-1}\right)}$$

$$= \frac{(n+1)(n-1)\left(\dfrac{5}{n+1}\right) - (n+1)(n-1)\left(\dfrac{1}{n-1}\right)}{(n+1)(n-1)\left(\dfrac{1}{n+1}\right) - (n+1)(n-1)\left(\dfrac{2}{n-1}\right)}$$

$$= \frac{5(n-1) - (n+1)}{(n-1) - 2(n+1)}$$

$$= \frac{5n - 5 - n - 1}{n - 1 - 2n - 2}$$

$$= \frac{4n - 6}{-n - 3}$$

$$= -\frac{4n - 6}{n + 3} \qquad \blacktriangle$$

Example 4   Simplify $\dfrac{2 + \dfrac{3}{r}}{\dfrac{1}{3} - \dfrac{5}{2r}}$.

Solution   The LCD of $\dfrac{3}{r}, \dfrac{1}{3},$ and $\dfrac{5}{2r}$ is $6r$. Therefore, multiply numerator and denominator by $6r$.

$$\frac{2 + \dfrac{3}{r}}{\dfrac{1}{3} - \dfrac{5}{2r}} = \frac{6r\left(2 + \dfrac{3}{r}\right)}{6r\left(\dfrac{1}{3} - \dfrac{5}{2r}\right)}$$

$$= \frac{12r + 18}{2r - 15} \qquad \blacktriangle$$

Example 5   Simplify $4x - \dfrac{2 - \dfrac{1}{x}}{\dfrac{3}{x} + 1}$.

Solution   We first simplify the complex fraction.

$$4x - \frac{2 - \dfrac{1}{x}}{\dfrac{3}{x} + 1} = 4x - \frac{x\left(2 - \dfrac{1}{x}\right)}{x\left(\dfrac{3}{x} + 1\right)}$$

$$= 4x - \frac{2x - 1}{3 + x}$$

To combine $4x$ with the fraction after the minus sign, write $4x$ as $\dfrac{4x}{1}$ and then multiply numerator and denominator by $3 + x$.

$$4x - \frac{2x - 1}{3 + x} = \frac{4x}{1} - \frac{2x - 1}{3 + x}$$

$$= \frac{4x(3 + x)}{1 \cdot (3 + x)} - \frac{2x - 1}{3 + x}$$

$$= \frac{12x + 4x^2 - (2x - 1)}{3 + x}$$

$$= \frac{4x^2 + 10x + 1}{3 + x}$$

This last expression cannot be simplified any further.   ▲

A complex fraction may have a form that involves negative exponents. To simplify such expressions, first change the negative exponents to positive exponents by using $a^{-n} = \dfrac{1}{a^n}$.

Example 6   Simplify $\dfrac{p^{-1} - q^{-1}}{p^{-1} + q^{-1}}$.

Solution   Using the definition of negative exponent, $p^{-1} = \dfrac{1}{p^1} = \dfrac{1}{p}$ and $q^{-1} = \dfrac{1}{q}$. Therefore,

$$\frac{p^{-1} - q^{-1}}{p^{-1} + q^{-1}} = \frac{\dfrac{1}{p} - \dfrac{1}{q}}{\dfrac{1}{p} + \dfrac{1}{q}}$$

$$= \frac{pq\left(\dfrac{1}{p} - \dfrac{1}{q}\right)}{pq\left(\dfrac{1}{p} + \dfrac{1}{q}\right)}$$

$$= \frac{q - p}{q + p} \qquad \blacktriangle$$

L E A R N I N G   A D V A N T A G E   *Quizzes you took during the study of a chapter can tell you some of the things your instructor will expect you to know at the end of a chapter. Save the quizzes and use them to review for the chapter test.*

E X E R C I S E   S E T   4.5

For Exercises 1–58, simplify by using either Method 1 or Method 2.

1. $\dfrac{\dfrac{1}{2}}{\dfrac{3}{4}}$

2. $\dfrac{\dfrac{5}{8}}{\dfrac{1}{4}}$

3. $\dfrac{-\dfrac{2}{3}}{\dfrac{5}{6}}$

4. $\dfrac{-\dfrac{4}{5}}{\dfrac{7}{10}}$

5. $\dfrac{\dfrac{11}{12}}{-\dfrac{7}{15}}$

6. $\dfrac{\dfrac{5}{4}}{-\dfrac{25}{32}}$

7. $\dfrac{\dfrac{1}{a}}{\dfrac{2}{b}}$

8. $\dfrac{\dfrac{3}{x}}{\dfrac{1}{r}}$

9. $\dfrac{\dfrac{15}{s^2}}{\dfrac{12}{s}}$

10. $\dfrac{\dfrac{46}{w^3}}{\dfrac{36}{w^2}}$

11. $\dfrac{-\dfrac{2}{x^2}}{\dfrac{6}{x^3}}$

12. $\dfrac{\dfrac{5}{t^3}}{-\dfrac{15}{t^7}}$

13. $\dfrac{-\dfrac{35}{z^2}}{-\dfrac{20}{z^4}}$

14. $\dfrac{-\dfrac{1}{42c^5}}{-\dfrac{1}{18c^2}}$

15. $\dfrac{\dfrac{2}{3}+\dfrac{1}{4}}{\dfrac{1}{2}+\dfrac{5}{6}}$

16. $\dfrac{\dfrac{2}{5}+\dfrac{1}{10}}{\dfrac{4}{5}+\dfrac{1}{2}}$

17. $\dfrac{\dfrac{1}{5}-\dfrac{3}{2}}{\dfrac{7}{10}+\dfrac{2}{5}}$

18. $\dfrac{\dfrac{3}{7}+\dfrac{4}{21}}{\dfrac{5}{7}-\dfrac{8}{3}}$

19. $\dfrac{\dfrac{5}{a}+\dfrac{3}{2}}{\dfrac{1}{a}+\dfrac{1}{2}}$

20. $\dfrac{\dfrac{7}{t}+\dfrac{1}{3}}{\dfrac{2}{3}+\dfrac{5}{t}}$

21. $\dfrac{\dfrac{4}{7}-\dfrac{3}{c}}{\dfrac{1}{14c}+\dfrac{3}{14}}$

22. $\dfrac{\dfrac{5}{n}-\dfrac{7}{9}}{\dfrac{4}{3}+\dfrac{10}{n}}$

23. $\dfrac{\dfrac{2}{w^2}+12}{\dfrac{3}{w}-\dfrac{2}{w^2}}$

24. $\dfrac{\dfrac{4}{h^3}+\dfrac{16}{h^2}}{\dfrac{20}{h^2}-\dfrac{36}{h^3}}$

25. $\dfrac{\dfrac{1}{x}+\dfrac{3}{2x^2}}{\dfrac{2}{x}-\dfrac{1}{2}}$

26. $\dfrac{\dfrac{2}{5}-\dfrac{3}{y}}{\dfrac{7}{10y^2}+\dfrac{9}{10}}$

27. $\dfrac{5-\dfrac{3}{s^3}}{\dfrac{1}{s^2}+\dfrac{1}{s}}$

28. $\dfrac{\dfrac{1}{z}+\dfrac{2}{z^2}}{1+\dfrac{2}{z}}$

29. $\dfrac{12+\dfrac{2}{q^2}}{\dfrac{3}{q}-\dfrac{2}{q^2}}$

30. $\dfrac{\dfrac{2}{r}-\dfrac{3}{r^2}}{2-\dfrac{3}{r}}$

31. $\dfrac{\dfrac{2}{3p}-\dfrac{5}{3}}{\dfrac{2}{p^2}-\dfrac{5}{p}}$

32. $\dfrac{\dfrac{4}{5h}+\dfrac{9}{5}}{\dfrac{4}{h^2}+\dfrac{9}{h}}$

33. $\dfrac{\dfrac{7}{c^2}-\dfrac{3}{c}}{\dfrac{7}{c^3}-\dfrac{3}{c^2}}$

34. $\dfrac{\dfrac{6}{b^3}+\dfrac{11}{b^2}}{\dfrac{6}{b^4}+\dfrac{11}{b^3}}$

35. $\dfrac{\dfrac{10}{a^2-9}}{\dfrac{5}{a+3}}$

36. $\dfrac{\dfrac{3}{z+4}}{\dfrac{15}{z^2-16}}$

37. $\dfrac{\dfrac{3}{x-2}-\dfrac{4}{x+2}}{\dfrac{1}{x^2-4}}$

38. $\dfrac{\dfrac{1}{t+5}}{\dfrac{12}{t^2-25}+\dfrac{2}{t-5}}$

39. $\dfrac{\dfrac{2}{(n-2)(n+1)}-\dfrac{3}{n+1}}{\dfrac{1}{n-2}-\dfrac{7}{(n-2)(n+1)}}$

40. $\dfrac{\dfrac{1}{(2r-1)(r+3)} - \dfrac{5}{2r-1}}{\dfrac{3}{(2r-1)(r+3)} - \dfrac{1}{r+3}}$

41. $\dfrac{\dfrac{5}{x^2-x} + \dfrac{10}{x}}{-\dfrac{15}{x-1} + \dfrac{5}{x^2-x}}$

42. $\dfrac{\dfrac{12}{p^2} - \dfrac{18}{p+1}}{\dfrac{6}{p^3+p^2} - \dfrac{12}{p^2}}$

43. $3 + \dfrac{\dfrac{1}{x} + 2}{5 - \dfrac{2}{3x^2}}$

44. $2 - \dfrac{3 - \dfrac{1}{2t}}{1 - \dfrac{1}{4t}}$

45. $\dfrac{2 + \dfrac{3-x}{x}}{4 + \dfrac{2}{x}} - \dfrac{1}{x+1}$

46. $\dfrac{2 + \dfrac{1+z}{z}}{1 - \dfrac{2}{3z}} - \dfrac{z}{3z-2}$

47. $\dfrac{6 + \dfrac{5}{a} + \dfrac{1}{a^2}}{2 + \dfrac{9}{a} + \dfrac{4}{a^2}}$

48. $\dfrac{1 + \dfrac{1}{w} - \dfrac{6}{w^2}}{2 + \dfrac{5}{w} - \dfrac{3}{w^2}}$

49. $\dfrac{\dfrac{x-2}{x+2} + \dfrac{x+2}{x-2}}{\dfrac{1}{x-2} - \dfrac{1}{x+2}}$

50. $\dfrac{\dfrac{r+1}{r-1} - \dfrac{r-1}{r+1}}{\dfrac{1}{r+1} - \dfrac{1}{r-1}}$

51. $\dfrac{6z - \dfrac{17z-1}{z+2}}{2z-1}$

52. $\dfrac{3t + \dfrac{3-14t}{2t+1}}{3t-1}$

53. $\dfrac{x^{-1} + x^{-2}}{3x^{-1} - x^{-2}}$

54. $\dfrac{y^{-3} + 2y^{-1}}{y^{-3} - 2y^{-1}}$

55. $\dfrac{5n^{-3} - 2n^{-2}}{10n^{-2} - n^{-3}}$

56. $\dfrac{1 - z^{-2}}{1 + z^{-2}}$

57. $\dfrac{3 - 14x^{-1} + 8x^{-2}}{3 + x^{-1} - 2x^{-2}}$

58. $\dfrac{4 - y^{-1} - 3y^{-2}}{3 - 4y^{-1} + y^{-2}}$

REVIEW PROBLEMS

The following exercises review parts of Sections 2.2 and 3.7. Doing these problems will help prepare you for the next section.

Section 2.2

Solve the equations.

59. $3x - 8x = 35$

60. $3(2t - 3) = -4$

61. $z^2 - 3z + 2 = z^2 - 5z + 8$

62. $\dfrac{w}{5} + \dfrac{3w + 8}{15} = 0$

63. $\dfrac{x + 1}{3} - x = \dfrac{3}{7}$

64. $\dfrac{a}{4} + \dfrac{2a - 3}{2} = 0$

Section 3.7

Solve the equations.

65. $2x^2 - 3x - 2 = 0$

66. $3z^2 - z - 2 = 0$

67. $6y^3 - 5y^2 + y = 0$

68. $8c^3 - 14c^2 + 3c = 0$

69. $12z^2 - 16z + 5 = 0$

70. $7x^2 - 4x - 3 = 0$

ENRICHMENT EXERCISES

Simplify the expressions.

1. $\dfrac{1 - \dfrac{1}{1 - \dfrac{1}{5}}}{\dfrac{3}{4} - \dfrac{1}{2}}$

2. $\dfrac{\dfrac{1}{3} - \dfrac{4}{9}}{\dfrac{1}{1 - \dfrac{5}{2}} + \dfrac{1}{9}}$

3. $\dfrac{1 - \dfrac{1}{1 - \dfrac{1}{1 - x}}}{1 - x}$

4. $\dfrac{\dfrac{1}{x + h} - \dfrac{1}{x}}{h}$

5. $\dfrac{\dfrac{1}{(x + h)^2} - \dfrac{1}{x^2}}{h}$

6. $\dfrac{\dfrac{1}{x - 3} - \dfrac{1}{a - 3}}{x - a}$

7. $\dfrac{\dfrac{1}{x + \Delta x + 1} - \dfrac{1}{x + 1}}{\Delta x}$ (Treat Δx as a single variable.)

Answers to Enrichment Exercises are on page A.19.

4.6

Equations with Rational Expressions

O B J E C T I V E S

▶ *To solve equations containing rational expressions*

▶ *To use formulas containing rational expressions*

An equation like $\dfrac{x}{x-2} - \dfrac{3x}{x-5} = \dfrac{1}{10}$ contains rational expressions. The best way to solve such an equation is to "clear fractions" by multiplying both sides by the least common denominator of all fractions that appear. A solution of the resulting equation will usually be a solution of the original equation. However, always check your answer in the original equation.

Example 1 Solve the equation. Check your answer.

$$\frac{r}{2} - \frac{3r}{5} = \frac{3}{10}$$

Solution The LCD of the three fractions is 10. Therefore, we multiply both sides of the equation by 10 and then simplify.

$$\frac{r}{2} - \frac{3r}{5} = \frac{3}{10}$$

$$10\left(\frac{r}{2} - \frac{3r}{5}\right) = 10\left(\frac{3}{10}\right)$$

$$10\left(\frac{r}{2}\right) - 10\left(\frac{3r}{5}\right) = 3$$

$$5r - 6r = 3$$

$$-r = 3$$

$$r = -3$$

We check the answer $r = -3$ in the *original* equation.

$$\frac{r}{2} - \frac{3r}{5} = \frac{3}{10}$$

$$\frac{-3}{2} - \frac{3(-3)}{5} \overset{?}{=} \frac{3}{10}$$

$$-\frac{3}{2} + \frac{9}{5} \overset{?}{=} \frac{3}{10}$$

$$-\frac{15}{10} + \frac{18}{10} \overset{?}{=} \frac{3}{10}$$

$$\frac{3}{10} = \frac{3}{10}$$

The answer checks, so $r = -3$ is the solution.   ▲

The next examples have equations with the variable appearing in denominators. After solving, it is very important to check your answer(s) in the original equation.

Example 2   Solve the equation, then check your answer.

$$\frac{t}{t-3} - \frac{2}{t+3} = 1$$

Solution   Multiply both sides by the LCD, $(t-3)(t+3)$, and simplify the result.

$$(t-3)(t+3)\left(\frac{t}{t-3} - \frac{2}{t+3}\right) = (t-3)(t+3) \cdot 1$$

$$(t-3)(t+3)\left(\frac{t}{t-3}\right) - (t-3)(t+3)\left(\frac{2}{t+3}\right) = (t+3)(t-3)$$

$$(t+3)(t) - (t-3)(2) = (t+3)(t-3)$$

$$t^2 + 3t - 2t + 6 = t^2 - 9$$

$$t + 6 = -9$$

$$t = -15$$

On your own, check $t = -15$ in the original equation.   ▲

Example 3   Solve the equation, then check your answer.

$$\frac{z+2}{z-2} = \frac{1}{z} + \frac{8}{z(z-2)}$$

Solution   The LCD of the fractions appearing in the equation is $z(z-2)$. Therefore, we multiply both sides of the equation by $z(z-2)$ and then simplify.

$$z(z-2)\left(\frac{z+2}{z-2}\right) = z(z-2)\left(\frac{1}{z} + \frac{8}{z(z-2)}\right)$$

$$z(z-2)\left(\frac{z+2}{z-2}\right) = z(z-2)\left(\frac{1}{z}\right) + z(z-2)\left(\frac{8}{z(z-2)}\right)$$

$$z(z+2) = (z-2) + 8$$

$$z^2 + 2z = z + 6$$

$$z^2 + z - 6 = 0 \qquad\qquad \text{Subtract } z+6 \text{ from both sides.}$$

To solve this last equation, factor the left side.

$$(z + 3)(z - 2) = 0$$

Therefore, either $z = -3$ or $z = 2$.

Next, we check these two answers in the original equation.

Check $z = -3$: $\dfrac{z + 2}{z - 2} = \dfrac{1}{z} + \dfrac{8}{z(z - 2)}$

$$\dfrac{-3 + 2}{-3 - 2} \overset{?}{=} \dfrac{1}{-3} + \dfrac{8}{-3(-3 - 2)}$$

$$\dfrac{-1}{-5} \overset{?}{=} -\dfrac{1}{3} + \dfrac{8}{15}$$

$$\dfrac{1}{5} \overset{?}{=} \dfrac{-5 + 8}{15}$$

$$\dfrac{1}{5} = \dfrac{3}{15} \quad \text{or} \quad \dfrac{1}{5}$$

Therefore, -3 is a solution.

Check $z = 2$: $\dfrac{z + 2}{z - 2} = \dfrac{1}{z} + \dfrac{8}{z(z - 2)}$

$$\dfrac{2 + 2}{2 - 2} \overset{?}{=} \dfrac{1}{2} + \dfrac{8}{2(2 - 2)}$$

$$\dfrac{4}{0} \overset{?}{=} \dfrac{1}{2} + \dfrac{8}{0}$$

Since the denominator of $\dfrac{4}{0}$ is 0, and division by zero is undefined, the number 2 is not a solution of the original equation. We conclude that the only solution is $z = -3$. ▲

In Example 3, the number 2 was not a solution of the original equation, even though we obtained it when solving. Such a number is called an **extraneous** solution. Notice that 2 made the denominator of a fraction in the equation zero.

Any number that makes the denominator of a fraction in an equation equal to zero cannot be a solution.

In previous sections, we have solved a formula for a particular variable. In the next example, we show how to solve for a variable in a formula involving rational expressions.

Example 4   Solve for a in the formula $S = \dfrac{a - ar^n}{1 - r}$.

Solution   We want to isolate a on one side of the equation.

$$(1 - r)S = a - ar^n \qquad \text{Multiply both sides by } 1 - r.$$

$$(1 - r)S = a(1 - r^n) \qquad \text{Factor out } a.$$

$$\frac{(1 - r)S}{1 - r^n} = a \qquad \text{Divide both sides by } 1 - r^n.$$

Therefore, the desired formula is

$$a = \frac{(1 - r)S}{1 - r^n}$$

CALCULATOR   CROSSOVER

Suppose A dollars is borrowed at 15% annual interest rate. If the loan is to be repaid in n monthly installments of R dollars each month, the amount R is given by the formula

$$R = \frac{0.0125A}{1 - (1.0125)^{-n}}$$

Find the monthly payment R, if \$2,500 is borrowed for 8 months.

Answer   \$330.33

I E A R N I N G   A D V A N T A G E   *Some instructors will not cover the same topics on a chapter test that appeared on an earlier quiz. Look for important ideas you have studied that have not appeared on tests or quizzes before. You may find them on the chapter test.*

E X E R C I S E   S E T   4 . 6

For Exercises 1–44, solve the equation. Check your answers for extraneous solutions.

1. $\dfrac{y}{3} + \dfrac{2y}{5} = \dfrac{22}{15}$

2. $\dfrac{z}{6} + \dfrac{5z}{18} = \dfrac{8}{3}$

3. $\dfrac{2c}{7} - \dfrac{c}{14} = -\dfrac{6}{7}$

4. $-\dfrac{5p}{4} + \dfrac{p}{8} = -\dfrac{3}{16}$

5. $4x - \dfrac{2x}{3} = \dfrac{25}{9}$

6. $8a - \dfrac{7a}{4} = -\dfrac{15}{2}$

7. $\dfrac{x+1}{2} + \dfrac{x}{8} = 3$

8. $\dfrac{2n-1}{3} + \dfrac{5n}{12} = 4$

9. $\dfrac{2t-1}{2} - \dfrac{2t}{5} = 1$

10. $\dfrac{3z+2}{3} - \dfrac{7z}{9} = 2$

11. $\dfrac{s+2}{6} - \dfrac{5s+8}{4} = s$

12. $\dfrac{2h-7}{5} + \dfrac{3h+1}{10} = h$

13. $\dfrac{8q-1}{12} - \dfrac{4q+5}{4} = q$

14. $\dfrac{r+5}{3} - \dfrac{5r-25}{9} = 2r$

15. $\dfrac{2}{x} + \dfrac{3}{x+1} = 0$

16. $\dfrac{3}{t-4} + \dfrac{1}{t} = 0$

17. $\dfrac{5}{w+1} = \dfrac{15}{w-1}$

18. $\dfrac{4}{y+3} = \dfrac{6}{y-5}$

19. $\dfrac{1}{r-1} + \dfrac{1}{r+1} = \dfrac{1}{r^2-1}$

20. $\dfrac{7}{a^2-16} - \dfrac{5}{a-4} = \dfrac{2}{a+4}$

21. $\dfrac{3}{v} - \dfrac{8}{3v} + \dfrac{4}{5v} = \dfrac{1}{15}$

22. $\dfrac{12}{7c} + \dfrac{1}{3c} - \dfrac{2}{21c} = \dfrac{2}{21}$

23. $\dfrac{25}{z} + \dfrac{21}{z-5} = \dfrac{1}{z^2-5z}$

24. $\dfrac{4}{w+1} - \dfrac{5}{w} = \dfrac{20}{w^2+w}$

25. $\dfrac{6}{x+4} - \dfrac{7}{x-3} = 0$

26. $\dfrac{3}{b-1} + \dfrac{5}{b+8} = 0$

27. $\dfrac{q-3}{2q-1} = \dfrac{q-5}{2q-5}$

28. $\dfrac{4t-7}{t+1} = \dfrac{4t+1}{t-2}$

29. $\dfrac{n+8}{n+2} = \dfrac{2}{n} - \dfrac{12}{n^2+2n}$

30. $\dfrac{s+10}{s+6} = \dfrac{5}{s} - \dfrac{24}{s^2+6s}$

31. $\dfrac{x-2}{x-4} + \dfrac{8}{x} = \dfrac{8}{x^2-4x}$

32. $\dfrac{2}{y} + \dfrac{20}{y^2+5y} = \dfrac{y+1}{y+5}$

33. $\dfrac{z+2}{z-12} - \dfrac{14}{z^2-12z} = \dfrac{1}{z}$

34. $\dfrac{a-3}{a+10} + \dfrac{57}{a^2+10a} = \dfrac{5}{a}$

35. $\dfrac{4}{c^2-1} - \dfrac{1}{c+1} = 1$

36. $\dfrac{3}{v+2} + \dfrac{6}{v^2-4} = -1$

37. $\dfrac{2}{u+1} - 1 = \dfrac{3}{u^2+4u+3}$

38. $1 + \dfrac{3}{m-1} = \dfrac{4}{m^2+m-2}$

39. $\dfrac{1}{1-x} + \dfrac{1}{x^2-1} = \dfrac{2}{3}$

40. $\dfrac{1}{2-y} - \dfrac{8}{y^2-4} = 3$

41. $\dfrac{1}{2z^2 - z - 1} + \dfrac{2}{2z^2 + 3z + 1} = \dfrac{3}{z^2 - 1}$

42. $\dfrac{3}{3r^2 + 7r + 2} - \dfrac{1}{3r^2 - 5r - 2} = \dfrac{2}{r^2 - 4}$

43. $3x + \dfrac{25}{x - 10} = \dfrac{5(5 - x)}{10 - x}$

44. $2a + \dfrac{12}{6 - a} = \dfrac{3(2 - a)}{a - 6}$

45. Solve $x^{-1} + 3 = 2x^{-1}$ by multiplying both sides by x.

46. Solve: $4 - 2x^{-1} = 7$.

47. Solve $6y^{-2} + y^{-1} - 2 = 0$ by multiplying both sides by y^2.

48. Solve: $2t^{-2} - 5t^{-1} + 3 = 0$.

49. Solve: $4 - (x + 1)^{-2} = 0$.

50. Solve: $1 - (2z - 3)^{-2} = 0$.

51. One number is three times another number. The sum of their reciprocals is nine. Find the two numbers.

52. One number is twice another number. The sum of their reciprocals is 12. Find the two numbers.

53. One number is four less than another number. The larger number divided by the smaller number is three. Find the two numbers.

54. The sum of a number and its reciprocal is two. Find the number.

55. The difference of the reciprocal of a number and five is three fourths. Find the number.

56. A number is equal to the quotient of negative nine and the sum of six and the number. Find the number.

57. In optics, an object is p units from a lens with focal length f as shown in the diagram. If q is the image distance, then p, f, and q are related by the *lens equation*

$$\frac{1}{f} = \frac{1}{p} + \frac{1}{q}$$

(a) Solve the lens equation for q.

(b) An object is located 2 cm from the lens having a focal length of 1.2 cm. Use Part (a) to find the image distance q.

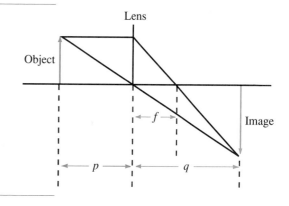

58. An electrical circuit has two resistors of resistance R_1 and R_2, respectively, in parallel with a total resistance R given by

$$\frac{1}{R} = \frac{1}{R_1} + \frac{1}{R_2}$$

(a) Solve this formula for R_1.

(b) Suppose an electrical circuit has two resistors in parallel with a total resistance of 2 ohms. If R_2 is 3 ohms, use Part (a) to find R_1.

For Exercises 59–70, solve the equation for the indicated variable.

59. $F = \dfrac{kqp}{r^2}$; for q.

60. $P = \dfrac{V^2}{R}$; for R.

61. $W = k\left(\dfrac{1}{a} - \dfrac{1}{b}\right)$; for b.

62. $C = kx\left(\dfrac{A}{d} - 1\right)$; for d.

63. $\dfrac{1}{R} = \dfrac{1}{R_1} + \dfrac{1}{R_2}$; for R.

64. $y = \dfrac{1}{1 - x}$; for x.

65. $z = \dfrac{t}{t + 1}$; for t.

66. $v = \dfrac{k}{p + a} - c$; for a

67. $P = \left(\dfrac{62.4n}{V}\right)T_k$; for V.

68. $z = 100\left(\dfrac{Cp^n}{1 + Cp^n}\right)$; for p^n

69. $\dfrac{y - xy'}{y^2} = 2x$; for y'. (y' is read "y prime.")

70. $\dfrac{y - 2xy'}{y^3} = 1$; for y'.

REVIEW PROBLEMS

The following exercises review parts of Section 1.5. Doing these problems will help prepare you for the next section.

Convert the following word phrases into algebraic expressions. Use x to represent the number.

71. The sum of a number and three fourths.

72. The difference of two thirds and a number.

73. Three fifths of the sum of a number and ten thirds.

74. The reciprocal of the sum of a number and four fifths.

75. The difference of the reciprocal of a number and three.

76. A number is increased by four, the result is then divided by three times the reciprocal of the number.

77. The reciprocal of the difference of seven and twice a number.

78. Five times a number is reduced by two. The result is divided by the sum of three and the reciprocal of the number.

ENRICHMENT EXERCISES

For Exercises 1–3, solve the equation.

1. $\dfrac{7}{n-2} + \dfrac{11}{n+3} = \dfrac{-24}{n-4}$

2. $\dfrac{\frac{3}{5}}{a+1} + \dfrac{\frac{4}{5}}{2a-3} + \dfrac{1}{a} = \dfrac{2a-1}{2a^2-a-3}$

3. $\dfrac{4(3v-7)}{v-1} + \dfrac{12v+19}{v+2} = \dfrac{3(8v-9)}{v}$

4. The total resistance R resulting from three resistors in parallel having resistances of R_1, R_2, and R_3 ohms, respectively, is given by

$$\frac{1}{R} = \frac{1}{R_1} + \frac{1}{R_2} + \frac{1}{R_3}.$$

Suppose that the first resistor has twice the resistance as the second, and the third has three times the resistance as the first. If the total resistance is 6 ohms, find the resistance of each of the three resistors.

A number x is the **harmonic mean** of a and b if $\dfrac{1}{x}$ is the average of $\dfrac{1}{a}$ and $\dfrac{1}{b}$. That is, $\dfrac{1}{x} = \dfrac{\frac{1}{a} + \frac{1}{b}}{2}$.

5. Find the harmonic mean of 4 and 6.

6. Find b, if the harmonic mean of 10 and b is 12.

7. The sum of two numbers is $\dfrac{3}{2}$ and their harmonic mean is $-\dfrac{7}{12}$. Find the two numbers.

Answers to Enrichment Exercises are on page A.19.

4.7

Applications of Rational Expressions

O B J E C T I V E

▶ *To use rational expressions in applications*

In this section, we deal with various applications involving rational expressions.

Example 1

Find two numbers whose sum is 25 and the sum of their reciprocals is $\frac{1}{6}$.

Solution

Let x be one of the numbers. Since the sum of the two numbers is 25, the second number is $25 - x$. Now, the sum of their reciprocals is $\frac{1}{6}$. Therefore,

$$\frac{1}{x} + \frac{1}{25 - x} = \frac{1}{6}$$

To solve this equation, multiply both sides by the least common denominator of the three fractions appearing, $6x(25 - x)$.

$$6x(25 - x)\left(\frac{1}{x} + \frac{1}{25 - x}\right) = 6x(25 - x)\left(\frac{1}{6}\right)$$

$$6(25 - x) + 6x = x(25 - x)$$

$$150 - 6x + 6x = 25x - x^2$$

$$x^2 - 25x + 150 = 0$$

$$(x - 10)(x - 15) = 0$$

Therefore, either $x = 10$ or $x = 15$. These two values of x actually give the same pair of numbers. For example, when $x = 10$, $25 - x = 15$ and when $x = 15$, $25 - x = 10$. We next check our pair of numbers 10 and 15 to see if they satisfy the requirements. Their sum, $10 + 15$, is 25. The sum of their reciprocals,

$$\frac{1}{10} + \frac{1}{15} = \frac{3}{30} + \frac{2}{30} = \frac{5}{30} = \frac{1}{6}$$

Therefore, 10 and 15 are the correct numbers.   ▲

There is a group of applications called work problems. Any job requires time for completion. We will assume that a person doing the job works at a constant rate.

Suppose Lisa can mow a lawn in 3 hours. Then in 1 hour, she can mow $\frac{1}{3}$ of the lawn. That is, her rate of work is $\frac{1}{3}$ of the lawn mowed per hour as shown in the diagram below.

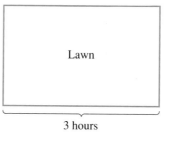

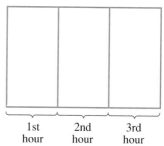

For example, after $\frac{1}{2}$ hour, she will have mowed $\left(\frac{1}{3}\right)\left(\frac{1}{2}\right) = \frac{1}{6}$ of the lawn.

That is, the fraction of the lawn mowed in $\frac{1}{2}$ hour is the product of the rate per hour and the time spent mowing.

The following work formula is illustrated by this problem.

R U L E

The fraction of work completed = (rate per hour) $\times$ $\left(\begin{array}{c}\text{number of}\\\text{hours worked}\end{array}\right)$.

Example 2 A pump can fill a tank in 18 hours. How full is the tank after

(a) 3 hours?   **(b)** 8 hours?   **(c)** t hours?

Solution   We restate the work formula for this problem.

The fraction of the tank filled = (rate of filling)$\left(\begin{array}{c}\text{the number of}\\\text{hours of filling}\end{array}\right)$.

Since the pump can fill the tank in 18 hours, then in 1 hour, $\frac{1}{18}$ of the tank is filled. That is, the rate per hour is $\frac{1}{18}$.

(a) The fraction of the tank filled in 3 hours is the product of $\frac{1}{18}$ and 3.

Therefore, the tank is $\left(\frac{1}{18}\right)(3)$ or $\frac{1}{6}$ full after 3 hours.

(b) After 8 hours, the tank is $\left(\frac{1}{18}\right)(8)$ or $\frac{4}{9}$ full.

(c) After t hours, the tank is $\left(\frac{1}{18}\right)(t)$ or $\frac{t}{18}$ full.   ▲

Suppose Lyn and Jay are both working on the same job. If Lyn completes $\frac{1}{3}$ of the job and Jay completes $\frac{2}{3}$ of the job, then the job is finished, since $\frac{1}{3} + \frac{2}{3} = 1$. That is, when the job is finished, the sum of the fractional parts done by Jay and Lyn must equal 1.

Example 3   It would take Nelson 3 hours to complete a drafting project. It would take Maria 2 hours to complete the same project. How long will it take to complete the project if they work together?

Solution   If Nelson and Maria work together, then in 1 hour $\frac{1}{3}$ of the project is completed by Nelson and $\frac{1}{2}$ of the project is completed by Maria.

Let t be the number of hours to complete the project working together. Then, the fraction of the project done by Nelson is

$$\left(\frac{1}{3}\right)(t) = \frac{t}{3}$$

and the fraction of the project done by Maria is

$$\left(\frac{1}{2}\right)(t) = \frac{t}{2}$$

Now, the sum of the fractions of the project done by Nelson and Maria is the total project. That is,

$$\begin{array}{ccc} \text{fraction of the} & & \text{fraction of the} \\ \text{project done} & + & \text{project done} & = 1 \quad \text{(the completed project)} \\ \text{by Nelson} & & \text{by Maria} \end{array}$$

This gives the equation

$$\frac{t}{3} + \frac{t}{2} = 1$$

To solve this equation, multiply both sides by the least common denominator 6, and then simplify.

$$6\left(\frac{t}{3} + \frac{t}{2}\right) = 6(1)$$

$$6\left(\frac{t}{3}\right) + 6\left(\frac{t}{2}\right) = 6$$

$$\overset{2}{6}\left(\frac{t}{\underset{1}{3}}\right) + \overset{3}{6}\left(\frac{t}{\underset{1}{2}}\right) = 6$$

$$2t + 3t = 6$$

$$5t = 6$$

$$t = \frac{6}{5}$$

Check this answer in the original equation.

Therefore, it takes $1\frac{1}{5}$ hours for Nelson and Maria to complete the project working together.   ▲

In the next example we deal with the formula $d = rt$, where d is the distance traveled at a given rate r and a given length of time t. This formula can be solved for t by dividing both sides by r,

$$\frac{d}{r} = \frac{rt}{r}$$

$$\frac{d}{r} = t$$

That is, the distance traveled divided by the rate is the amount of time for the trip. In the next example, we show how this formula is used when the time is the same for two different trips.

Example 4

It took Fred the same time to drive 450 miles as it did Jeff to drive 400 miles. Jeff's speed (rate) was 7 miles per hour slower than Fred's speed. How fast did each person drive?

Solution

Let x be Fred's rate. Then $x - 7$ is Jeff's rate. We make the following table using the information given in the problem to fill in the first two columns labeled distance and rate. The third column (time) is found by the formula $t = \frac{d}{r}$.

	Distance	Rate	Time
Fred	450	x	$\dfrac{450}{x}$
Jeff	400	$x - 7$	$\dfrac{400}{x - 7}$

Since

$$\text{Fred's time} = \text{Jeff's time},$$

$$\frac{450}{x} = \frac{400}{x - 7}$$

To solve this equation for x, first multiply both sides by $x(x - 7)$, then simplify the result.

$$x(x - 7)\left(\frac{450}{x}\right) = x(x - 7)\left(\frac{400}{x - 7}\right)$$

$$450(x - 7) = 400x$$

$$450x - 450(7) = 400x$$

$$450x - 400x = 450(7)$$

$$50x = 450(7)$$

$$x = \frac{450(7)}{50}$$

$$x = \frac{\overset{9}{\cancel{450}}(7)}{\underset{1}{\cancel{50}}}$$

$$x = 63$$

Therefore, Fred's rate is 63 miles per hour and Jeff's rate is $63 - 7$ or 56 miles per hour. ▲

L E A R N I N G   A D V A N T A G E   *When taking a test, remember that you will not earn points for blank paper. Always try to record as much information as possible in any problem that demands the use of more than one procedure or concept. Never hesitate to fill in any gaps in a multipart problem. However, to avoid causing confusion to the person grading your paper, it is very important that you explain what you are doing.*

E X E R C I S E   S E T   4.7

1. Find two numbers whose sum is 15 and the sum of their reciprocals is $\frac{5}{18}$.

2. Find two numbers whose sum is 14 and the sum of their reciprocals is $\frac{7}{24}$.

3. The sum of two numbers is $-\frac{1}{4}$ and the sum of their reciprocals is $\frac{2}{3}$. Find the two numbers.

4. The sum of two numbers is $-\frac{9}{20}$ and the sum of their reciprocals is $\frac{1}{2}$. Find the two numbers.

5. What number must be added to the numerator and denominator of $\frac{1}{4}$ to make a fraction equivalent to $\frac{2}{3}$?

6. The denominator of a fraction is 16 less than the numerator. Find this fraction, if it reduces to -3.

7. One number is 7 smaller than another number. The quotient of the larger number divided by the smaller number is $-2\frac{1}{2}$. Find the two numbers.

8. One number is 8 more than another number. The sum of the two numbers divided by the sum of the smaller number and 3 is -8. Find the two numbers.

9. Kelly can finish her homework assignment in $4\frac{1}{2}$ hours. How much of the assignment does she complete after

(a) $\frac{1}{2}$ hour?   **(b)** 2 hours?   **(c)** $1\frac{1}{2}$ hours?

10. Harlan can write a science fiction short story in $3\frac{1}{2}$ hours. How much is completed after

(a) 1 hour?   **(b)** $1\frac{1}{2}$ hours?   **(c)** $2\frac{1}{4}$ hours?

11. Sam can rake a lawn in 3 hours and Walter can rake the same lawn in 2.4 hours. How long would it take the two of them working together to rake the lawn?

12. It would take George 5 hours to paint the master bedroom and it would take his wife Phyllis 7 hours. How long would it take if they both worked together?

13. It takes Ted 4.8 hours to run the daily blood tests in the hospital's laboratory. Another technician, Pam, can do it in 2.4 hours. How long will it take to run the blood tests if they work together?

14. One pipe can fill a swimming pool in 6 hours. A second pipe can do it in $4\frac{1}{2}$ hours. How long would it take to fill the pool if both pipes are used?

15. Working together, Allison and John can landscape their property in 6 days. It would take Allison three times as long as it would take John to do it alone. Working alone, how long would it take each of them?

16. Working together, two fraternities at Faber College can build a football homecoming float in 8 days. It would take Lambda Lambda twice as long as it would take Delta House to do it alone. Working alone, how long would it take each club to build the float?

17. Mr. Campbell can prepare a company's federal income tax return in 15 hours and his associate can prepare it in 20 hours. If Mr. Campbell has been working on this tax return for 6 hours before he is joined by his associate, how long did it take the two of them working together to finish the tax return?

18. A farmer's new wheat combine can harvest his crop in 36 hours, and the old and new combines working together can complete the harvest in 26 hours. How long would it take the old combine, working alone, to harvest the crop?

19. At a Green Giant canning factory, green peas flow through a pipe into a holding tank at a rate that would fill the tank in $1\frac{1}{2}$ hours. Another pipe can empty a full tank in $1\frac{3}{4}$ hours. Starting with an empty tank, how long will it take to fill the tank?

    Hint: Let t be the time it takes to fill the empty tank. Then,

$$\begin{pmatrix} \text{fraction of the} \\ \text{tank filled in} \\ t \text{ hours} \end{pmatrix} - \begin{pmatrix} \text{fraction of the} \\ \text{tank emptied in} \\ t \text{ hours} \end{pmatrix} = 1 \begin{pmatrix} \text{the holding} \\ \text{tank is full} \end{pmatrix}$$

20. An oil storage tank can be filled by the intake pipe in $2\frac{1}{2}$ hours. An outtake pipe can empty the tank in 3 hours. Starting with an empty tank, with both pipes in operation, how long will it take to fill the tank?

21. At the Binney and Smith Company, hot wax being used to make Crayola crayons flows through a pipe into a heated tank at a rate that will fill the tank in $1\frac{1}{3}$ hours. Another pipe can empty a full tank in 2 hours. Starting with an empty tank, with both pipes being used, how long will it take to fill the tank?

22. At the Horbocker Brewery, a mixture of malt and hops flows through a pipe into a holding tank at a rate that will fill the tank in $2\frac{1}{2}$ hours. Another pipe can empty a full tank in $3\frac{3}{4}$ hours. Starting with an empty tank, with both pipes being used, how long will it take to fill the tank?

23. An oil storage tank has two intake pipes and one outtake pipe. Alone, each intake pipe could fill the tank in 3 and 4 hours, respectively. The outtake pipe can drain a full tank in 12 hours. Starting with an empty tank, with all three pipes working, how long will it take to fill the tank?

24. A holding tank has one intake pipe and two outtake pipes. The intake pipe could fill an empty tank in 2 hours. Alone, each outtake pipe can empty a full tank in 6 and 4 hours, respectively. Starting with an empty tank, with all three pipes working, how long will it take to fill the tank?

25. It took Fran the same time to drive 21 miles as it took Cary to drive 15 miles. Fran's speed was 10 miles per hour faster than Cary's speed. How fast did Fran drive?

26. Harry hiked 12 miles in the same time it took Mick to hike 14 miles. Harry's rate was 1 mile per hour less than Mick's. How fast did each person walk?

27. The Delaware River has a current of 4 miles per hour. A canoeist takes as long to go 7 miles downriver as to go 3 miles upriver. What is the rate of the canoeist in still water?

	Distance	Rate	Time
Downriver	7	$r + 4$	
Upriver	3	$r - 4$	

28. The Rock River in northern Illinois has a 3 km per hour current. Jennifer can paddle her kayak 2 km downriver in the same time that she can paddle $\frac{1}{2}$ km upriver. What is Jennifer's rate in still water?

	Distance	Rate	Time
Downriver	2	$r + 3$	
Upriver	$\frac{1}{2}$	$r - 3$	

29. A boat can travel at 18 km per hour in still water. It can travel 7 km upstream in the same time that it can travel 11 km downstream. What is the rate of the current?

30. Sue drove a three-wheel all-terrain vehicle 5 miles in the same time that it took Paul to drive a four-wheel all-terrain vehicle 8 miles. If Sue drove 4 miles per hour slower than Paul, how fast was each person traveling?

REVIEW PROBLEMS

The following exercises review parts of Section 1.4. Doing these problems will help prepare you for the next section.

Simplify each numerical expression.

31. 4^2

32. 2^4

33. 5^3

34. $\left(\dfrac{5}{2}\right)^2$

35. $(1.3)^2$

36. 1^{20}

37. $2^3 \cdot 2^2$

38. $\dfrac{5^2}{4^3}$

39. $(-3)^4$

40. -3^4

41. $(0.3)^3$

42. $(-2)^3(-2)^2$

43. $(-2)^5$

44. $(-1)^{34}$

E N R I C H M E N T   E X E R C I S E S

1. One scented candle will burn up completely in 3 hours. A second thicker candle of equal height will burn up completely in 6 hours. If the candles are lit at the same time, how long will it take the first candle to be $\frac{1}{4}$ the height of the second candle?

2. It takes one pump 3 hours to drain a tank. A second pump can drain the tank in 5 hours. The first pump works for 1 hour, then breaks down. The second pump is then used. How long will it take the second pump to finish draining the tank?

3. A holding tank is to be constructed with two intake pipes and one outtake pipe. Alone, each intake pipe can fill an empty tank in 4 and 8 hours, respectively. The outtake pipe must be large enough so that with all three pipes working, an empty tank is filled in 12 hours. How long would it take the outtake pipe, working alone, to completely drain a full tank?

4. A salesman must travel an interstate highway that is undergoing repairs the first 75% of the trip. He therefore anticipates averaging 45 mph on this part. If he wants to average 50 mph for the entire trip, what must be his average speed on the last 25% of the way?

Answers to Enrichment Exercises are on page A.20.

Summary and review

Rational expressions (4.1)

A rational expression is **reduced to lowest terms** if the numerator and denominator have no common factors. To reduce a rational expression to lowest terms, we use the principle of rational expressions.

The Principle of Rational Expressions

$$\frac{x^2 - 9}{x + 3} = \frac{\cancel{(x+3)}(x-3)}{\cancel{x+3}}$$
$$= x - 3$$

If $\dfrac{P}{Q}$ is a rational expression and K is a nonzero polynomial, then $\dfrac{PK}{QK} = \dfrac{P}{Q}$.

$$\frac{1 - x - 2x^2}{(2x - 1)^2}$$
$$= \frac{(-1)(2x^2 + x - 1)}{(2x - 1)^2}$$
$$= \frac{(-1)\cancel{(2x-1)}(x + 1)}{(2x - 1)^{\cancel{2}\,1}}$$
$$= -\frac{x + 1}{2x - 1}$$

If the term with the highest power in a polynomial has a negative coefficient, factor out a negative one.

Also, if $\dfrac{P}{Q}$ is a rational expression, then

$$\frac{P}{-Q} = -\frac{P}{Q} \quad \text{and} \quad \frac{-P}{Q} = -\frac{P}{Q}.$$

Division of polynomials (4.2)

Divide $2x^2 - 8x + 5$ by $x - 3$ using **(a)** long division and then using **(b)** synthetic division.

To **divide a polynomial by a monomial,** divide each term of the polynomial by the monomial,

$$\frac{a + b}{c} = \frac{a}{c} + \frac{b}{c}, \quad c \neq 0$$

(a)
$$\begin{array}{r} 2x - 2 \\ x - 3\overline{)2x^2 - 8x + 5} \\ \underline{2x^2 - 6x} \\ -2x + 5 \\ \underline{-2x + 6} \\ -1 \end{array}$$

(b) $3\,\lfloor\,2 - 8 + 5$
$$\underline{\quad 6 - 6}$$
$$\ 2 - 2 - 1$$

Therefore, the quotient is $2x - 2$ and the remainder is -1.

If the division cannot be done by dividing out common factors, we use a long division process. A streamlined version of this process, called **synthetic division,** can be used when the divisor is of the form $x - a$ or $x + a$.

Therefore, dividing $2x^2 - 8x + 5$ by $x - 3$ results in a quotient of $2x - 2$ and a remainder of -1. We can **check our division** by verifying the equation

$$\text{dividend} = (\text{divisor})(\text{quotient}) + \text{remainder}$$

where the dividend is $2x^2 - 8x + 5$ and the divisor is $x - 3$.

Multiplication and division of rational expressions (4.3)

$$\frac{x^2}{3} \cdot \frac{6}{x^5} = \frac{6x^2}{3x^5}$$
$$= \frac{2}{x^3}$$

To **multiply two rational expressions,** multiply numerators and multiply denominators:

$$\frac{A}{B} \cdot \frac{C}{D} = \frac{A \cdot C}{B \cdot D}$$

Examples

$$\frac{3x^2 - 4xy + y^2}{3x + 3y} \cdot \frac{21x + 21y}{(x - y)^3}$$

$$= \frac{\cancel{(x - y)}(3x - y)\overset{7}{\cancel{21}}\cancel{(x + y)}}{\underset{1}{\cancel{3}\cancel{(x + y)}(x - y)^{\cancel{3}\,2}}}$$

$$= \frac{7(3x - y)}{(x - y)^2}$$

$$\frac{3x^2 - 4x - 4}{2x^2 + 7x + 5} \div \frac{x - 2}{2x + 5}$$

$$= \frac{(3x + 2)\cancel{(x - 2)}}{\cancel{(2x + 5)}(x + 1)} \cdot \frac{\cancel{2x + 5}}{\cancel{x - 2}}$$

$$= \frac{3x + 2}{x + 1}$$

Find the LCD of $\dfrac{2}{x^3 - x}$ and $\dfrac{3}{x^3 - x^2}$.

$x^3 - x = x(x^2 - 1)$
$\qquad = x(x + 1)(x - 1)$
and
$x^3 - x^2 = x^2(x - 1)$

Therefore, the LCD is
$x^2(x + 1)(x - 1)$.

$$\frac{2x + 1}{x - 3} - \frac{2}{x^2 - 9}$$

$$= \frac{2x + 1}{x - 3} - \frac{2}{(x + 3)(x - 3)}$$

$$= \frac{(2x + 1)(x + 3)}{(x - 3)(x + 3)} - \frac{2}{(x + 3)(x - 3)}$$

$$= \frac{(2x + 1)(x + 3) - 2}{(x + 3)(x - 3)}$$

$$= \frac{2x^2 + 7x + 3 - 2}{(x + 3)(x - 3)}$$

$$= \frac{2x^2 + 7x + 1}{(x + 3)(x - 3)}$$

For **division of two rational expressions,** take the reciprocal of the second and multiply:

$$\frac{A}{B} \div \frac{C}{D} = \frac{A}{B} \cdot \frac{D}{C} = \frac{A \cdot D}{B \cdot C}$$

Addition and subtraction of rational expressions (4.4)

If rational expressions have **common denominators,** then

$$\frac{A}{B} + \frac{C}{B} = \frac{A + C}{B} \qquad \text{and} \qquad \frac{A}{B} - \frac{C}{B} = \frac{A - C}{B}$$

If rational expressions have **different denominators,** we must first find the least common denominator (LCD) before adding or subtracting. The **least common denominator** is the simplest polynomial that is divisible by each of the denominators.

To Find the Least Common Denominator (LCD)

Step 1   Factor each denominator completely.

Step 2   Form the product of each different factor to the highest power that it occurs in any denominator. The LCD is this product.

Addition and Subtraction of Rational Expressions with Unlike Denominators

Step 1   Find the least common denominator.

Step 2   Rewrite each fraction using the LCD.

Step 3   Add or subtract the numerators, keeping the same denominator.

Step 4   Simplify the numerator, then write the fraction in lowest terms.

Examples

Complex fractions (4.5)

A **complex fraction** is a rational expression that has fractions appearing somewhere in its numerator or denominator. Complex fractions can be simplified to ordinary fractions.

Two Methods for Simplifying a Complex Fraction

$$\frac{\frac{1}{a} - \frac{1}{b}}{\frac{1}{a} + \frac{1}{b}} = \frac{ab\left(\frac{1}{a} - \frac{1}{b}\right)}{ab\left(\frac{1}{a} + \frac{1}{b}\right)}$$

Method 1: If necessary, combine the terms in the numerator, combine the terms in the denominator, then divide.

$$= \frac{b - a}{b + a}$$

Method 2: Find the least common denominator of all the fractions appearing in the numerator and denominator of the complex fraction, then multiply both numerator and denominator by it to clear fractions.

Equations with rational expressions (4.6)

Solve: $1 + \dfrac{2}{x} = \dfrac{3}{x^2}$.

To **solve an equation with rational expressions,** first multiply by the LCD of all fractions appearing on either side of the equation. Then solve the resulting equation by the usual methods.

$$x^2\left(1 + \frac{2}{x}\right) = x^2\left(\frac{3}{x^2}\right)$$
$$x^2 + 2x = 3$$
$$x^2 + 2x - 3 = 0$$
$$(x - 1)(x + 3) = 0$$
$$x = 1 \quad \text{or} \quad x = -3$$

Check $x = 1$: *Check* $x = -3$:

$$1 + \frac{2}{x} = \frac{3}{x^2} \qquad 1 + \frac{2}{x} = \frac{3}{x^2}$$
$$1 + \frac{2}{1} \overset{?}{=} \frac{3}{1^2} \qquad 1 + \frac{2}{-3} \overset{?}{=} \frac{3}{(-3)^2}$$
$$1 + 2 \overset{?}{=} \frac{3}{1} \qquad 1 - \frac{2}{3} \overset{?}{=} \frac{3}{9}$$
$$3 = 3 \qquad \frac{1}{3} = \frac{1}{3}$$

The solution set is $\{1, -3\}$.

Applications of rational expressions (4.7)

Find two numbers whose sum is 4 and the sum of their reciprocals is $-\dfrac{1}{3}$. Let x be one of the numbers, then $4 - x$ is the other number.

The applications in this section involve solving equations containing rational expressions.

Since the sum of their reciprocals is $-\dfrac{1}{3}$,

$$\frac{1}{x} + \frac{1}{4 - x} = -\frac{1}{3}$$

Examples

Next, we solve this equation by first multiplying both sides by $3x(4 - x)$.

$$3x(4 - x)\left(\frac{1}{x} + \frac{1}{4 - x}\right)$$
$$= 3x(4 - x)\left(-\frac{1}{3}\right)$$
$$3(4 - x) + 3x = -x(4 - x)$$
$$12 - 3x + 3x = -4x + x^2$$
$$0 = x^2 - 4x - 12$$
$$0 = (x - 6)(x + 2)$$

Therefore, either $x = 6$ or $x = -2$. When $x = 6$, $4 - x = 4 - 6 = -2$; and when $x = -2$, $4 - x = 4 - (-2) = 6$. In either case, the two desired numbers are -2 and 6.

CHAPTER 4 REVIEW EXERCISE SET

Section 4.1

For Exercises 1–4, determine any values of the variable where the rational expression is undefined.

1. $\dfrac{3}{4x - 9}$

2. $\dfrac{(y + 1)^3}{y^3 - y}$

3. $\dfrac{4z}{z^2 + 16}$

4. $\dfrac{3x - 6}{x^3 - 8}$

For Exercises 5–10, reduce the rational expression to lowest terms.

5. $\dfrac{20x^3y^2}{5x^4y}$

6. $\dfrac{6(3 - 4z)^3}{8z - 6}$

7. $\dfrac{2a^2 - 5a + 2}{2a^3 - a^2}$

8. $\dfrac{r - 2}{2 - r}$

9. $\dfrac{12x^2 - 14x + 4}{1 - 4x^2}$

10. $\dfrac{z^2 - 2az + a^2}{z^3 - a^3}$

Section 4.2

For Exercises 11 and 12, divide as indicated and simplify. Write the answer without negative exponents.

11. $\dfrac{4z^4 - 6z^3 - 2z^2}{2z^3}$

12. $\dfrac{x^2y^4 - 12x^3y^5 + 18x^4y^4}{3x^3y^4}$

For Exercises 13 and 14, divide by factoring the numerator and then dividing out common factors.

13. $\dfrac{6x^2 + 5x + 1}{3x + 1}$

14. $\dfrac{x^2 - 4ax + 4a^2}{x - 2a}$

For Exercises 15 and 16, use long division to divide. Write your answer in the form: quotient $+ \dfrac{\text{remainder}}{\text{divisor}}$.

Check your answer by verifying the equation

$$\text{dividend} = (\text{quotient})(\text{divisor}) + \text{remainder}$$

15. $\dfrac{6x^2 - x - 12}{2x + 3}$

16. $\dfrac{24y^4 + 10y^3 b - 6y^2 b^2}{4y + 3b}$

For Exercises 17 and 18, use synthetic division to find the quotient and remainder.

17. $\dfrac{5x^2 - 6x + 11}{x - 3}$

18. $\dfrac{3x^4 - 8x^2 + 2x + 15}{x + 1}$

19. Factor $6x^4 - 9x^3 - 7x^2 - 12x - 20$ completely, given that $3x^2 + 4$ is one of the factors.

20. Factor $x^4 - x^3 - 8x^2 + 9x - 9$ completely, given that $x^2 - x + 1$ is one of the factors.

Section 4.3

For Exercises 21–26, multiply or divide. Write the answer in lowest terms.

21. $\dfrac{x^2 - 7x + 12}{2x + 5} \cdot \dfrac{4x^2 - 25}{x - 4}$

22. $\dfrac{6t - 3}{2t + 8} \cdot \dfrac{4t + 16}{1 - 2t}$

23. $\dfrac{p^5 - p^3}{3p^2 - 7p + 2} \cdot \dfrac{p - 2}{p^2 - p}$

24. $\dfrac{12x - 4}{x^2 + x - 2} \div \dfrac{9x^2 - 6x + 1}{x + 2}$

25. $\dfrac{z^5 - 6z^4 + 9z^3}{z^2 - 4z + 4} \div \dfrac{z^5 - 3z^4}{z - 2}$

26. $\dfrac{2x^2 - 3xy - 2y^2}{x^2 - y^2} \div \dfrac{2x^2 - xy - y^2}{x^2 + 2xy + y^2}$

For Exercises 27–30, perform the given operations. Write your answer in lowest terms.

27. $\dfrac{20}{18} \cdot \dfrac{21}{-50} \div \dfrac{98}{15}$

28. $\left(\dfrac{4s^2 t^3}{3st^4} \div \dfrac{-12}{6s^4 t} \right) \dfrac{-s^2 t}{2s^5 t^3}$

29. $\left[\dfrac{(x - 6)^2}{2x - 2} \cdot \dfrac{x^2 - 2x + 1}{4x + 7} \right] \div \dfrac{2x^2 - 13x + 6}{4x^3 + 7x^2}$

30. $\dfrac{2z^2 - 9z + 9}{z^3 - 3z^2 + 2z} \cdot \dfrac{3z^4 - 3z^3 - 6z^2}{z^2 - 2z - 3}$

Section 4.4

For Exercises 31–40, add or subtract as indicated. Write the answer in lowest terms.

31. $\dfrac{3}{x^2} + \dfrac{5}{x^2}$

32. $\dfrac{14 - z^3}{z^2 + 1} - \dfrac{10 - z^3}{z^2 + 1}$

33. $\dfrac{a}{a^2 - 4} - \dfrac{2}{a^2 - 4}$

34. $\dfrac{2}{3t^2 - 8t + 4} - \dfrac{3t}{3t^2 - 8t + 4}$

35. $\dfrac{3}{8} - \dfrac{5}{6}$

36. $\dfrac{2b}{4b^2 - 1} - \dfrac{2}{2b + 1}$

37. $\dfrac{9}{2x^2 - 5x + 2} + \dfrac{2x + 5}{2x - 1}$

38. $\dfrac{2}{z^2 - z} + \dfrac{2}{z^2 + z}$

39. $\dfrac{2r - 7}{2r^2 - 6r + 4} + \dfrac{2r + 5}{2r^2 - 2r - 4}$

40. $\dfrac{3}{y^2 + y - 2} - \dfrac{8}{3y^2 - 4y + 1}$

Section 4.5

For Exercises 41–48, simplify each complex fraction.

41. $\dfrac{\dfrac{3}{4}}{\dfrac{7}{8}}$

42. $\dfrac{\dfrac{-44x^3}{y^2}}{\dfrac{-24x^2}{y^4}}$

43. $\dfrac{\dfrac{3}{T} + \dfrac{4}{3}}{\dfrac{5}{3} - \dfrac{2}{T}}$

44. $\dfrac{14 - \dfrac{4}{y^2}}{16 + \dfrac{6}{y}}$

45. $\dfrac{2 - a}{\dfrac{a + 4}{a + 1} - a}$

46. $\dfrac{\dfrac{x^2 + 2x}{x + 1} - \dfrac{2x + 1}{x + 1}}{x - 2}$

47. $\dfrac{6 + x^{-1} - 12x^{-2}}{3 - 7x^{-1} + 4x^{-2}}$

48. $\dfrac{2k^{-1} - k^{-2}}{4 - k^{-2}}$

Section 4.6

For Exercises 49–54, solve each equation.

49. $\dfrac{6x}{5} - \dfrac{2x}{3} = \dfrac{7}{30}$

50. $\dfrac{4y + 3}{4} - \dfrac{5y}{3} = \dfrac{7y}{12}$

51. $\dfrac{4}{3r - 2} - \dfrac{9}{3r - 2} = 1$

52. $\dfrac{3}{z^2 - z} - \dfrac{1}{z} = \dfrac{2}{z - 1}$

53. $\dfrac{x + 1}{x - 3} = \dfrac{1}{x} + \dfrac{12}{x^2 - 3x}$

54. $\dfrac{4x^2 - 5x}{x^2 - 3x + 2} + \dfrac{x - 8}{x - 2} = 1$

For Exercises 55–60, solve the equation for the indicated variable.

55. $A = 2\pi r^2 + 2\pi rh$; for h.

56. $P = 2w + 2\ell$; for w.

57. $C = \dfrac{Tt}{T + t}$; for t.

58. $F = \dfrac{Gm_1m_2}{d^2}$; for d^2.

59. $y = \dfrac{m}{x_1 + h} - 7$; for x_1.

60. $\dfrac{y^2 - 2xyy'}{y^4} = 0$; for y'.

Section 4.7

61. Find two numbers whose sum is $\dfrac{3}{8}$ and the sum of their reciprocals is 12.

62. What number must be added to the numerator and denominator of $\dfrac{3}{2}$ to make a fraction equivalent to $\dfrac{8}{5}$?

63. Jenny can finish her algebra assignment in $3\dfrac{1}{2}$ hours. How much of the assignment does she complete after

(a) $\dfrac{1}{2}$ hour?

(b) 2 hours?

(c) $1\dfrac{1}{2}$ hours?

(d) t hours?

64. Creighton can complete his paper route in 3 hours. How much of his route is completed after

(a) $\dfrac{3}{4}$ hour?

(b) 2 hours?

(c) 1 hour and 20 minutes?

(d) t hours?

65. Two people must pick apples in an orchard. Working alone, one person could do the entire job in 4 hours, while the other person could do the entire job in 6 hours. How long will it take them if they work together?

66. It takes one pipe to fill a swimming pool 5 hours longer than another pipe. If the two pipes together can fill it in 6 hours, how long would it take each pipe to fill the pool alone?

67. The intake pipe can fill a tank in 2 hours when the outlet pipe is closed, but with the outlet pipe open it takes 5 hours. How long would it take the outlet pipe to empty a full tank?

68. It takes one crew 3 weeks to clean the windows of a skyscraper. It takes another crew 4 weeks to clean the windows of the same building. Working together, how long would it take the two crews to clean the windows of the skyscraper?

69. It took Sam the same time to drive 400 miles as it did John to drive 350 miles. Sam's speed was 8 miles per hour faster than John's speed. How fast did each person drive?

70. The Swan River has a current of 3 miles per hour. A canoeist takes as long to go 4 miles downriver as to go $1\dfrac{1}{2}$ miles upriver. What is the rate of the canoeist in still water?

C H A P T E R   4   **T E S T**

For problems 1–3, reduce the rational expression to lowest terms.

1. $\dfrac{54a^4x^3}{12a^5x^2}$

2. $\dfrac{4s - 12}{3 - s}$

3. $\dfrac{2x^2 - 10x + 12}{4x^2 - 24x + 36}$

4. Use long division to divide $6x^2 - 13x + 1$ by $3x - 2$. Write your answer in the form

$$\text{quotient} + \frac{\text{remainder}}{\text{divisor}}$$

Check your answer by verifying the equation

$$\text{dividend} = (\text{quotient})(\text{divisor}) + \text{remainder}$$

5. Use synthetic division to find the quotient and remainder: $\dfrac{2x^3 - 3x^2 - x + 1}{x - 2}$

Section 4.3

6. Multiply and write the answer in lowest terms: $\dfrac{2x^2 - 3x - 9}{9x^2 - 1} \cdot \dfrac{3x - 1}{2x + 3}$

7. Divide and write the answer in lowest terms: $\dfrac{t^2 - 2t - 3}{t + 3} \div \dfrac{t^2 - t - 6}{t^2 + 5t + 6}$

8. Perform the given operations. Write the answer in lowest terms:

$\left(\dfrac{3x^3y^4z^2}{5xy^5z^2} \cdot \dfrac{35xy}{9z^2}\right) \div \dfrac{x^5y^2}{6xz}$

Section 4.4

For problems 9–11, add or subtract as indicated. Write the answer in lowest terms.

9. $\dfrac{7}{c} - \dfrac{2}{c}$

10. $\dfrac{x}{x^2 - 9} + \dfrac{3}{x^2 - 9}$

11. $\dfrac{1}{y^2 + 2y} - \dfrac{1}{y^2 - 2y}$

Section 4.5

For problems 12–14, simplify each expression.

12. $\dfrac{\dfrac{15x^2}{y}}{\dfrac{-5x}{y^2}}$

13. $\dfrac{\dfrac{2}{t-2} + \dfrac{2t}{t-2}}{t+1}$

14. $2 + \dfrac{b}{1 + \dfrac{2}{b}}$

Section 4.6

For problems 15–19, solve each of the following equations.

15. $\dfrac{2}{h-2} = \dfrac{3}{h+3}$

16. $1 - \dfrac{5}{x} = \dfrac{6}{x^2}$

17. $3y + \dfrac{2}{y} = 7$

18. $\dfrac{1}{a-2} + \dfrac{2}{a^2-4} = \dfrac{3}{a+2}$

19. $1 + \dfrac{2}{(x-4)^2} = \dfrac{3}{x-4}$

For problems 20 and 21, solve for y.

20. $\dfrac{1}{x} - \dfrac{1}{y} = 3$

21. $\dfrac{y-2}{y+1} = 2x$

Section 4.7

22. Find two numbers whose sum is 1 and the sum of whose reciprocals is $\dfrac{16}{3}$.

23. The current of a river is 4 miles per hour. It takes a boat as long to go 3 miles downriver as to go 2 miles upriver. What is the rate of the boat in still water?

24. The intake pipe can fill a tank in 3 hours when the outlet pipe is closed, but with the outlet pipe open, it takes 7 hours. How long would it take the outlet pipe to empty a full tank?

Roots and Radicals

Overview

So far, we have raised numbers to integer powers. In this chapter, we define numbers raised to rational powers. We shall connect this concept with taking the nth root of a number. An expression involving taking a root is called a radical. Then we will learn the algebra of radical expressions and solve radical equations. The final topic is complex numbers, which will be used to solve the general quadratic equation in the next chapter.

5.1

Square Roots and Higher Roots

OBJECTIVES

▶ *To find the square roots of a number*

▶ *To find the nth roots of a number*

▶ *To use the notation for nth roots*

Given a number such as 7, we know how to find its square, 7^2,

$$7^2 = 7 \cdot 7 = 49$$

We say that the square of 7 is 49. We can also say that a *square root* of 49 is 7, since squaring 7 yields 49. Notice also that $(-7)^2 = (-7)(-7) = 49$. Therefore, -7 is also a square root of 49. The square roots of 49 are 7 and -7.

In general, a **square root** of a given number is any number whose square is the given number. Any positive number has two square roots, one positive and the other negative.

Example 1   Find the two square roots of each number.

(a) 25   (b) 100   (c) $\dfrac{4}{9}$ (d) 0.16

Solution   **(a)** Since $5^2 = 5 \cdot 5 = 25$ and $(-5)^2 = (-5)(-5) = 25$, the square roots of 25 are 5 and -5.

(b) Since $10^2 = 100$ and $(-10)^2 = 100$, the square roots of 100 are 10 and -10.

(c) Since $\left(\dfrac{2}{3}\right)^2 = \left(\dfrac{2}{3}\right)\left(\dfrac{2}{3}\right) = \dfrac{4}{9}$ and $\left(-\dfrac{2}{3}\right)^2 = \left(-\dfrac{2}{3}\right)\left(-\dfrac{2}{3}\right) = \dfrac{4}{9}$, the two square roots of $\dfrac{4}{9}$ are $\dfrac{2}{3}$ and $-\dfrac{2}{3}$.

(d) Since $(0.4)(0.4) = 0.16$ and $(-0.4)(-0.4) = 0.16$, the two square roots of 0.16 are 0.4 and -0.4. ▲

The positive square root of a number is symbolized by using the symbol $\sqrt{}$, which is called a **radical sign.** For example,

$$\sqrt{49} = 7, \qquad \sqrt{36} = 6, \qquad \sqrt{121} = 11$$

To express the negative square root, we use $-\sqrt{}$. For example, $-\sqrt{49} = -7$.

The expression within the radical sign is called the **radicand.** For example, the radicand is 4 in the expression $\sqrt{4}$. A **radical** is the total expression comprised of the radical sign together with the radicand. Using this symbolism, we have the following statements.

If a is a positive real number,

$\sqrt{a}$ means the positive square root of a.

$$\sqrt{a}\sqrt{a} = a$$

$-\sqrt{a}$ is the negative square root of a.

$$(-\sqrt{a})(-\sqrt{a}) = a$$

Furthermore, $\sqrt{a^2} = a$.

If a is zero, $\sqrt{0} = 0$.

Consider trying to find a square root of a negative number. For example, does -4 have square roots? If -4 had a square root, say b, then $b^2 = -4$. However, recall that the square of any number is nonnegative. That is, b^2 is nonnegative and therefore could not be equal to -4. We conclude that

A negative number has no real square roots.

Example 2   Evaluate each of the following.

(a) $\sqrt{64}$ **(b)** $-\sqrt{\dfrac{25}{4}}$ **(c)** $\sqrt{(0.79)^2}$

(d) $\sqrt{(-1)^2}$ **(e)** $\sqrt{6}\sqrt{6}$ **(f)** $\sqrt{-25}$

Solution   **(a)** $\sqrt{64}$ means the positive square root of 64. Therefore, $\sqrt{64} = 8$.

(b) $-\sqrt{\dfrac{25}{4}} = -\dfrac{5}{2}$ $\left(-\dfrac{5}{2}\right)^2 = \dfrac{25}{4}$.

(c) $\sqrt{(0.79)^2} = 0.79$ $\sqrt{a^2} = a$.

(d) $\sqrt{(-1)^2} = \sqrt{1} = 1$

(e) $\sqrt{6}\sqrt{6} = 6$ $\sqrt{a}\sqrt{a} = a$.

(f) Since -25 is negative, it has no real square roots. Therefore, $\sqrt{-25}$ is not a real number.   ▲

The numbers 144 and $\dfrac{4}{9}$ have rational square roots, since the square roots of 144 are 12 and -12 and the square roots of $\dfrac{4}{9}$ are $\dfrac{2}{3}$ and $-\dfrac{2}{3}$. The numbers 144 and $\dfrac{4}{9}$ are called *perfect squares*. In general, a number with rational square roots is called a **perfect square.** The following is a list of some positive *integers* that are perfect squares.

$$4, 9, 16, 25, 36, 49, 64, 81, 100, 121, 144, 169, \ldots$$

If a positive number a is not a perfect square then its square roots, $\sqrt{a}$ and $-\sqrt{a}$, are irrational numbers. Recall from Chapter 1, that the real numbers are split into two groups—rational and irrational numbers. A *rational number* can be expressed as a fraction where the numerator and denominator are integers, whereas an *irrational number* cannot be expressed as a quotient of integers. Furthermore, the decimal expansion of an irrational number neither ends nor has a repeating block of digits.

CALCULATOR CROSSOVER

A calculator with a $\boxed{\sqrt{x}}$ button can also be used to approximate numbers like $\sqrt{45}$. First enter 45, then press the $\boxed{\sqrt{x}}$ button. The (8 digit) display at the top is 6.7082039. If we round this decimal off to three decimal places, then $\sqrt{45}$ is approximately equal to 6.708. We write $\sqrt{45} \approx 6.708$.

Use the $\boxed{\sqrt{x}}$ button on your calculator to find approximate values on the following radicals. Round off to three decimal places. Then plot each radical on the same number line.

1. $\sqrt{33}$ **2.** $\sqrt{76.3}$ **3.** $\sqrt{12.91}$ **4.** $\sqrt{\dfrac{40}{7}}$ **5.** $-\sqrt{23.82}$

Answers   **1.** 5.745   **2.** 8.735   **3.** 3.593   **4.** 2.390   **5.** -4.881

Higher roots such as cube roots, fourth roots, and so on are defined in a manner similar to square roots.

D E F I N I T I O N

Let n be a positive integer. A number b is an ***n*th root** of a if $b^n = a$.

For example, 2 is a cube root of 8, since $2^3 = 8$. We use the notation $\sqrt[n]{}$, to represent nth roots. For example, 2 is the cube root of 8, which is written $2 = \sqrt[3]{8}$.

The positive square root of a, $\sqrt{a}$, has the property that $(\sqrt{a})^2 = a$.

The cube root of a, $\sqrt[3]{a}$, has the property that $(\sqrt[3]{a})^3 = a$.

The positive fourth root of a, $\sqrt[4]{a}$, has the property that $(\sqrt[4]{a})^4 = a$.

The fifth root of a, $\sqrt[5]{a}$, has the property that $(\sqrt[5]{a})^5 = a$.

$\cdot\ \cdot\ \cdot\ \cdot\ \cdot\ \cdot\ \cdot\ \cdot\ \cdot\ \cdot$

The nth root of a, $\sqrt[n]{a}$, has the property that $(\sqrt[n]{a})^n = a$.

The even roots—square, fourth, sixth, . . .—are not defined for negative numbers. Even roots of negative numbers are not real numbers.

The following are examples of common roots.

Square roots		Cube roots		Fourth roots	
$\sqrt{0} = 0$	$\sqrt{1} = 1$	$\sqrt[3]{0} = 0$	$\sqrt[3]{1} = 1$	$\sqrt[4]{0} = 0$	$\sqrt[4]{1} = 1$
$\sqrt{4} = 2$	$\sqrt{9} = 3$	$\sqrt[3]{-1} = -1$	$\sqrt[3]{8} = 2$	$\sqrt[4]{16} = 2$	$\sqrt[4]{81} = 3$
$\sqrt{16} = 4$	$\sqrt{25} = 5$	$\sqrt[3]{-8} = -2$	$\sqrt[3]{27} = 3$	$\sqrt[4]{256} = 4$	

In the symbol for the nth root, $\sqrt[n]{}$, n is called the **index.** Note that when $n = 2$, the index is usually not written, and we write $\sqrt{}$ for $\sqrt[2]{}$.

How many roots does a number have? It depends on the index as well as the number itself, as shown in the following table.

The number of real *n*th roots of a real number *a*

	a is positive	*a* is negative
n even	Two real *n*th roots	No real *n*th roots
	-3 and 3 are both square roots of 9.	-9 has no real square roots.
n odd	One real *n*th root	One real *n*th root
	2 is the only cube root of 8.	-2 is the only cube root of -8.

For example, 11.6 has two square roots, two fourth roots, two sixth roots, and so on. One of the roots is positive and the other is negative. In fact, the two roots are opposites of each other. The positive root is called the **principal nth root.** The number -5.1 has one cube root, one fifth root, one seventh root, and so on. The number -5.1 has no square root, no fourth root, no sixth root, and so on.

Example 3

(a) $\sqrt[3]{8} = 2$, because $2^3 = 8$.

(b) $\sqrt[3]{-8} = -2$, because $(-2)^3 = -8$.

(c) $\sqrt{-16}$ is not a real number, since there is no real number whose square is -16.

(d) $-\sqrt{16} = -4$, since $-\sqrt{16}$ is the negative square root of 16.

(e) $\sqrt[4]{-45}$ is not a real number, since there is no real number we can raise to the fourth power and obtain -45.

(f) $\sqrt[5]{-32} = -2$, because $(-2)^5 = -32$. ▲

Now, let us consider radicals of the form $\sqrt[n]{a^n}$, where n is a positive integer. If n is an even positive integer such as 2, $\sqrt{a^2}$ is *not* automatically a. For example, suppose a is -3. Then, $\sqrt{(-3)^2} = \sqrt{9} = 3$. If n is an odd positive integer such as 3, then it is true that $\sqrt[3]{a^3} = a$ for any real number a. In general, we have

RULE

If n is an even positive integer,

$$\sqrt[n]{a^n} = |a|$$

If n is an odd positive integer,

$$\sqrt[n]{a^n} = a$$

Example 4

(a) $\sqrt{10^2} = 10$ (b) $\sqrt[3]{6^3} = 6$

(c) $\sqrt[4]{(-5)^4} = |-5| = 5$ (d) $\sqrt[5]{(-2)^5} = -2$ ▲

NOTE   *Note that $\sqrt[4]{(-5)^4} \neq (\sqrt[4]{-5})^4$, since $\sqrt[4]{(-5)^4} = |-5| = 5$; whereas, $(\sqrt[4]{-5})^4$ is not a real number.*

Since we cannot take an even root of a negative number, we assume that all variables appearing under a radical sign represent nonnegative numbers.

Example 5 Simplify each expression.

(a) $\sqrt{49x^6y^2} = 7x^3y$, since $(7x^3y)^2 = 49x^6y^2$.

(b) $\sqrt[3]{-27a^6b^{15}} = -3a^2b^5$, since $(-3a^2b^5)^3 = -27a^6b^{15}$.

(c) $\sqrt[4]{16s^4t^8r^{16}} = 2st^2r^4$, since $(2st^2r^4)^4 = 16s^4t^8r^{16}$. ▲

In Chapter 3, we introduced the Pythagorean Theorem. We restate it here for convenience. The Pythagorean Theorem states that if a right triangle has sides with lengths a, b, and c, where a and b are the lengths of the two legs and c is the length of the hypotenuse, then

$$c^2 = a^2 + b^2$$

Example 6 For each right triangle, find the missing length.

(a)

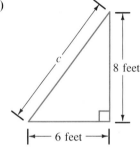

(b)

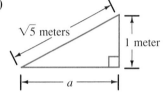

Solution (a) In this triangle, we are given lengths $a = 6$ feet and $b = 8$ feet. From this information, we find length c by using the Pythagorean Theorem.

$$c^2 = a^2 + b^2$$
$$= 6^2 + 8^2$$
$$= 36 + 64$$
$$= 100$$

Since c is the length of a side of a triangle, c is positive. Therefore, c is the positive square root of 100; that is, $c = \sqrt{100}$ or 10 feet.

(b) For this triangle, we are given lengths $b = 1$ meter and $c = \sqrt{5}$ meters. To find length a, use the Pythagorean Theorem.

$$c^2 = a^2 + b^2$$
$$(\sqrt{5})^2 = a^2 + 1^2$$
$$5 = a^2 + 1$$
$$5 - 1 = a^2$$
$$4 = a^2$$
$$2 = a \qquad a \text{ is the positive square root of 4.}$$

Therefore, a is 2 meters. ▲

In algebra, the radicand is frequently a variable or a variable expression. Some examples of nth roots with variable radicands are the following:

$$\sqrt{z}, \qquad \sqrt{x+7}, \qquad \sqrt[6]{y - \frac{3}{4}}, \qquad \sqrt[4]{2t - 1}$$

In the expression $\sqrt{x+7}$, suppose we replace x by -3. Then,

$$\sqrt{x+7} = \sqrt{-3+7} = \sqrt{4} = 2$$

If we replace x by -9, then

$$\sqrt{x+7} = \sqrt{-9+7} = \sqrt{-2}$$

which is not a real number since the radicand is negative. Therefore, we make the following rule.

R U L E

If n is even and we are taking the nth root of an expression involving a variable, it is understood that the variable is restricted so that the expression is nonnegative.

Example 7   Find the restriction on the variable so that the radicand is nonnegative.

(a) $\sqrt{x+7}$ **(b)** $\sqrt[6]{y - \frac{3}{4}}$ **(c)** $\sqrt[4]{2t - 1}$

Solution   **(a)** Set the radicand, $x + 7$, greater than or equal to zero and solve the resulting inequality for x.

$$x + 7 \geq 0$$

$$x \geq -7 \qquad \text{Subtract 7 from both sides.}$$

Therefore, the radical $x + 7$ is a real number whenever $x \geq -7$.

(b) Since $n = 6$ is even, set the radicand, $y - \frac{3}{4}$, greater than or equal to zero and solve the resulting inequality for y.

$$y - \frac{3}{4} \geq 0, \qquad \text{so } y \geq \frac{3}{4}$$

Therefore, the radical $y - \frac{3}{4}$ is a real number whenever $y \geq \frac{3}{4}$.

(c) We set $2t - 1 \geq 0$, since $n = 4$ is even, and solve for t.

$$2t - 1 \geq 0$$

$$2t \geq 1 \qquad \text{Add 1 to both sides.}$$

$$t \geq \frac{1}{2} \qquad \text{Divide both sides by 2.}$$

Therefore, $2t - 1$ is a real number whenever $t \geq \dfrac{1}{2}$.   ▲

We finish this section with an application from physics.

Example 8   A pendulum of length ℓ is shown in the rest position (see figure). The pendulum is displaced, then released.

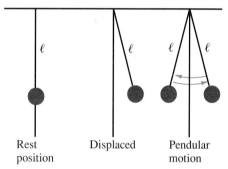

| Rest position | Displaced | Pendular motion |

The period T of a pendulum is the time required for one complete back and forth motion. If the length is measured in feet, the period T is given by the formula

$$T - 2\pi\sqrt{\frac{\ell}{32}}$$

Notice that the period depends only on the length ℓ and does not depend on the weight nor on the amount of displacement.

Find the period of a pendulum 8 feet in length. Use 3.14 for π.

Solution   To find the period T, replace ℓ by 8 and simplify.

$$T = 2(3.14)\sqrt{\frac{8}{32}}$$

$$= 6.28\sqrt{\frac{1}{4}}$$

$$= 6.28\left(\frac{1}{2}\right)$$

$$= 3.14 \text{ seconds} \qquad \text{▲}$$

LEARNING ADVANTAGE *Sometimes the solution to a problem is not always immediate. When working on a problem in which the answer is not obvious on the first effort, keep trying. Continue to work until you have the answer or until you are frustrated, then put it aside. In the time period during which you do not work on the problem, your subconscious continues to work on it, and sometimes the solution will appear to you when you least expect it.*

EXERCISE SET 5.1

For Exercises 1–8, find the two square roots of each number.

1. 9   **2.** 64   **3.** 121   **4.** 144

5. $\dfrac{25}{9}$ **6.** $\dfrac{81}{16}$ **7.** 0.81   **8.** 0.25

For Exercises 9–18, evaluate each expression.

9. $\sqrt{81}$ **10.** $\sqrt{16}$ **11.** $-\sqrt{\dfrac{4}{9}}$

12. $-\sqrt{\dfrac{100}{81}}$ **13.** $\sqrt{(0.18)^2}$ **14.** $\sqrt{\left(\dfrac{2}{3}\right)^2}$

15. $\sqrt{(-3)^2}$ **16.** $\sqrt{(-5)^2}$ **17.** $\sqrt{-36}$

18. $\sqrt{-4}$

For Exercises 19–46, find the value of each expression, if it exists.

19. $\sqrt[3]{27}$ **20.** $\sqrt{121}$ **21.** $\sqrt[4]{16}$

22. $\sqrt[3]{64}$ **23.** $\sqrt[7]{-1}$ **24.** $\sqrt[5]{1}$

25. $\sqrt{-16}$ **26.** $\sqrt[4]{-81}$ **27.** $\sqrt[7]{128}$

28. $\sqrt[3]{216}$ **29.** $-\sqrt[3]{8}$ **30.** $-\sqrt[4]{16}$

31. $\sqrt[3]{-\dfrac{1}{27}}$ **32.** $\sqrt[3]{-\dfrac{1}{8}}$ **33.** $\sqrt[6]{-1}$

34. $\sqrt{-81}$ **35.** $-\sqrt[4]{16}$ **36.** $-\sqrt[3]{-8}$

37. $\sqrt{9^2}$ **38.** $\sqrt[4]{51^4}$ **39.** $\sqrt[3]{48^3}$

40. $\sqrt[5]{652^5}$ **41.** $\sqrt[4]{(-1)^4}$ **42.** $\sqrt[6]{(-2)^6}$

43. $\sqrt[5]{(-8)^5}$ **44.** $\sqrt[3]{(-10)^3}$ **45.** $(\sqrt{-3})^2$

46. $(\sqrt[4]{-2})^4$

For Exercises 47–54, simplify each expression. Assume that all variables represent positive real numbers.

47. $\sqrt{16s^4t^2}$

48. $\sqrt{81w^6u^4}$

49. $\sqrt[3]{27x^6y^9}$

50. $\sqrt[3]{a^3c^{12}}$

51. $\sqrt[4]{16a^8b^4}$

52. $\sqrt[4]{81p^{16}q^{20}}$

53. $\sqrt[3]{-8a^9c^3z^{15}}$

54. $\sqrt[3]{-r^6s^3t^{12}}$

For Exercises 55–60, find the missing length of each right triangle.

55.

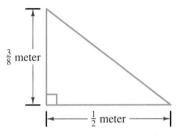

$\frac{3}{8}$ meter

$\frac{1}{2}$ meter

56.

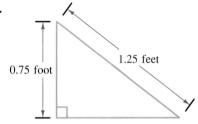

0.75 foot

1.25 feet

57.

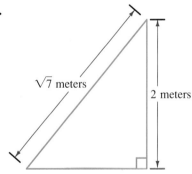

$\sqrt{7}$ meters

2 meters

58.

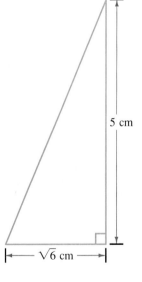

5 cm

$\sqrt{6}$ cm

59.

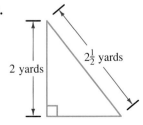

2 yards

$2\frac{1}{2}$ yards

60.

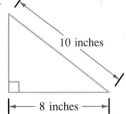

10 inches

8 inches

For Exercises 61–66, find the restriction on the variable so that the radicand is nonnegative.

61. $\sqrt{x-5}$

62. $\sqrt{t+12}$

63. $\sqrt{r+\dfrac{4}{5}}$

64. $\sqrt{2c-7}$

65. $\sqrt{3-9v}$

66. $\sqrt{4-16y}$

67. Find the period of a pendulum $4\dfrac{1}{2}$ feet long. (See Example 8.) Find an approximate decimal answer by replacing π by 3.14.

68. The formula for the period T of a pendulum when the length ℓ is measured in meters is given by

$$T = 2\pi\sqrt{\dfrac{\ell}{9.8}}$$

Find the period of a pendulum 2.45 meters in length.

69. The formula $v = \sqrt{64d}$ gives the velocity v an object will have in feet per second after falling d feet, if it starts from rest. What is the velocity of an object that has fallen 100 feet?

70. The radius r of a right circular cylinder with height h and volume V is given by $r = \sqrt{\dfrac{V}{\pi h}}$. A can in the shape of a right circular cylinder is to be constructed to have a volume of 32π cubic inches and a height of 8 inches. What should be the radius of such a can?

REVIEW PROBLEMS

The following exercises review parts of Section 3.1. Doing these problems will help prepare you for the next section.

Simplify. Write your answer without negative exponents.

71. $\dfrac{3^2}{3^4}$

72. x^4x^6

73. $\dfrac{a^5}{a^3}$

74. $\left(\dfrac{c}{2}\right)^3$

75. $(st)^5$

76. $(v^2u^3)^4$

77. $x^2(x^3 - x^4)$

78. $2c^3(3c^4 + 7c^2)$

79. $\left(\dfrac{s^2}{t^4v^2}\right)^4$

80. x^{-3}

81. $(a^{-1}b^2)^{-2}$

82. $\left(\dfrac{c^2}{d^3}\right)^{-1}$

83. $(x^{-3})^{-2}$

84. $(xz^{-2})^{-1}$

ENRICHMENT EXERCISES

Simplify the following expressions. Assume that m and n are positive integers greater than one and the base variables represent positive real numbers.

1. $\sqrt{x^{2n}}$ **2.** $\sqrt[3]{a^{3n}}$ **3.** $\sqrt[4]{z^{4n}}$

4. $\sqrt[m]{x^{mn}}$ **5.** $\sqrt[3]{-\sqrt[4]{1}}$ **6.** $\sqrt{\sqrt[3]{64}}$

7. Let m and n be positive integers greater than one. Prove that

$$\sqrt[n]{\sqrt[m]{a}} = \sqrt[nm]{a}$$

where all roots are real numbers.

For Exercises 8–11, use the result of Exercise 7 to simplify the term to an expression with a single radical.

8. $\sqrt{\sqrt[6]{7}}$ **9.** $\sqrt[4]{\sqrt[5]{12}}$ **10.** $\sqrt[3]{\sqrt[3]{(-2)}}$ **11.** $\sqrt[3]{\sqrt[4]{2}}$

Answers to Enrichment Exercises are on page A.22.

5.2

Rational Exponents

O B J E C T I V E S

▶ *To evaluate numbers of the form $a^{1/n}$*

▶ *To evaluate numbers of the form $a^{m/n}$*

▶ *To convert from radical notation to exponent notation and vice versa*

There is another notation to use for nth roots. This involves the use of exponents. In Chapter 1, we defined a^n where n is an integer. Let us now consider a power such as $8^{1/3}$. If the property for exponents $(a^m)^n = a^{mn}$ still holds, then

$$(8^{1/3})^3 = 8^{(1/3)3} = 8^1 = 8$$

This means that the number $8^{1/3}$ cubed is 8, and thus

$$8^{1/3} = \sqrt[3]{8}$$

Therefore, we define exponents of the form $1/n$, where n is a positive integer in terms of nth roots.

DEFINITION

If a is a real number and n is a positive integer, then,

$$a^{1/n} = \sqrt[n]{a}, \qquad \text{where } a \geq 0 \text{ when } n \text{ is even}$$

If a is positive and n is even, recall that there are two nth roots of a. The symbol $a^{1/n}$ represents the positive or principal nth root of a.

Using this definition, we can evaluate numbers raised to powers of the form $1/n$.

Example 1   Evaluate each expression.

(a) $64^{1/3} = \sqrt[3]{64} = 4$

(b) $100^{1/2} = \sqrt{100} = 10$

(c) $(-16)^{1/4} = \sqrt[4]{-16}$ is not a real number.

(d) $(-32)^{1/5} = \sqrt[5]{-32} = -2$

(e) $\left(\dfrac{125}{27}\right)^{1/3} = \sqrt[3]{\dfrac{125}{27}} = \dfrac{5}{3}$ ▲

We now define the more general power $a^{m/n}$, where m/n is any rational number.

D E F I N I T I O N

If m and n are positive integers with m/n written in lowest terms, then

$$a^{m/n} = (a^{1/n})^m,$$

provided that $a^{1/n}$ exists. If $a^{1/n}$ does not exist, then $a^{m/n}$ does not exist.

Therefore, given a rational exponent m/n, the denominator indicates the root to be taken and the numerator indicates the power:

power   root

$$32^{3/5} = (32^{1/5})^3 = (\sqrt[5]{32})^3 = 2^3 = 8$$

We define exponents of *negative* rational numbers as shown in the following equation:

D E F I N I T I O N

$$a^{-m/n} = \frac{1}{a^{m/n}}, \qquad a \neq 0$$

provided that $a^{m/n}$ exists.

Several illustrations of numbers raised to rational exponents are provided in the next example.

Example 2

(a) $16^{3/2} = (16^{1/2})^3 = (\sqrt{16})^3 = 4^3 = 64$

(b) $27^{2/3} = (27^{1/3})^2 = (\sqrt[3]{27})^2 = 3^2 = 9$

(c) $-9^{5/2} = -(9^{5/2}) = -(9^{1/2})^5 = -(\sqrt{9})^5 = -3^5 = -243$

(d) $(-64)^{4/3} = [(-64)^{1/3}]^4 = [\sqrt[3]{-64}]^4 = [-4]^4 = 256$

(e) $(-25)^{7/2}$ is not a real number since $(-25)^{1/2}$ is not a real number.

(f) $81^{-3/4} = \dfrac{1}{81^{3/4}} = \dfrac{1}{(81^{1/4})^3} = \dfrac{1}{(\sqrt[4]{81})^3} = \dfrac{1}{3^3}$

$= \dfrac{1}{27}$

(g) $(-8)^{-2/3} = \dfrac{1}{(-8)^{2/3}} = \dfrac{1}{[(-8)^{1/3}]^2}$

$= \dfrac{1}{(\sqrt[3]{-8})^2} = \dfrac{1}{(-2)^2} = \dfrac{1}{4}$

In Chapter 3, Section 1, we developed properties of exponents where the exponents were integers. Using the definition of rational exponent, it can be shown that all the properties still hold for rational exponents. We restate these properties without proving them.

P R O P E R T I E S

Properties of exponents

Suppose a and b are real numbers, where a and b are positive when even roots are taken, and r and s are rational numbers. Then,

1. $a^r a^s = a^{r+s}$
2. $(a^r)^s = a^{rs}$
3. $(ab)^r = a^r b^r$
4. $\left(\dfrac{a}{b}\right)^r = \dfrac{a^r}{b^r}, \quad b \neq 0$
5. $\dfrac{a^r}{a^s} = a^{r-s}, \quad a \neq 0$

We have defined $a^{m/n}$ to mean $(a^{1/n})^m$. However, using the properties of exponents, we have an alternate way of evaluating $a^{m/n}$.

$$a^{m/n} = a^{m(1/n)}$$

$$= (a^m)^{1/n} \qquad \text{Property 2.}$$

For example, we can evaluate $8^{2/3}$ in two different ways.

First way: $\qquad 8^{2/3} = (8^{1/3})^2 = 2^2 = 4$

Second way: $\qquad 8^{2/3} = (8^2)^{1/3} = 64^{1/3} = 4$

In most instances, it is easier to use the first way.

In the next example, we illustrate how the properties of exponents can be used to simplify expressions.

Example 3   Use the properties of exponents to simplify. Assume all variables represent positive real numbers. Write the answer with only positive exponents.

(a) $\quad 3^{1/4} \cdot 3^{1/2} = 3^{1/4+1/2}$

$\qquad\qquad\qquad = 3^{1/4+2/4}$

$\qquad\qquad\qquad = 3^{3/4}$

(b) $\quad \dfrac{7^{3/5}}{7^{9/5}} = 7^{3/5-9/5}$

$\qquad\qquad = 7^{-6/5}$

$\qquad\qquad = \dfrac{1}{7^{6/5}}$

(c) $\quad \dfrac{(a^{2/3}b^{1/2})^6}{a^2 b} = \dfrac{a^{(2/3)6}b^{(1/2)6}}{a^2 b}$

$\qquad\qquad = \dfrac{a^4 b^3}{a^2 b}$

$\qquad\qquad = a^{4-2}b^{3-1}$

$\qquad\qquad = a^2 b^2$

(d) $\quad \left(\dfrac{a^{-5/2}}{b^{3/4}}\right)^8 = \dfrac{(a^{-5/2})^8}{(b^{3/4})^8}$

$\qquad\qquad = \dfrac{a^{-20}}{b^6}$

$\qquad\qquad = \dfrac{1}{a^{20}b^6}$ $\qquad$ ▲

Example 4   Replace all radicals with rational exponents and simplify. Assume all variables represent positive real numbers.

(a) $\quad \sqrt[4]{x^3} = (x^3)^{1/4}$

$\qquad\qquad = x^{3/4}$

(b) $\quad \dfrac{\sqrt[3]{a^5}}{\sqrt{a^3}} = \dfrac{(a^5)^{1/3}}{(a^3)^{1/2}}$

$\qquad\qquad = \dfrac{a^{5/3}}{a^{3/2}}$

$$= a^{5/3 - 3/2}$$

$$= a^{10/6 - 9/6}$$

$$= a^{1/6}$$

(c) $\sqrt[3]{\sqrt{a}} = \sqrt[3]{a^{1/2}}$

$$= (a^{1/2})^{1/3}$$

$$= a^{1/6} \qquad \blacktriangle$$

In the next three examples, assume that the base variables represent positive real numbers.

Example 5   Simplify each expression using the properties of exponents.

(a) $\dfrac{(t^n)^{2/7}}{t^{n-2}} = \dfrac{t^{2n/7}}{t^{n-2}}$ Property 2.

$$= t^{2n/7 - (n-2)} \qquad \text{Property 5.}$$

$$= t^{-5n/7 + 2}$$

(b) $(a^{r^2} b^{3r})^{1/r} = (a^{r^2})^{1/r}(b^{3r})^{1/r}$ Property 3.

$$= a^r b^3 \qquad \text{Property 2.} \qquad \blacktriangle$$

We can use the distributive property to multiply expressions involving rational exponents, as illustrated in the next example.

Example 6   Multiply $s^{4/3}(s^{-2} - 5s^{9/2})$ and write the answer using only positive exponents.

Solution $s^{4/3}(s^{-2} - 5s^{9/2}) = s^{4/3}s^{-2} - 5s^{4/3}s^{9/2}$

$$= s^{4/3 - 2} - 5s^{4/3 + 9/2}$$

$$= s^{4/3 - 6/3} - 5s^{8/6 + 27/6}$$

$$= s^{-2/3} - 5s^{35/6}$$

$$= \dfrac{1}{s^{2/3}} - 5s^{35/6} \qquad \blacktriangle$$

Recall that the distributive property can also be used for factoring.

Example 7   Factor $x^{1/2}$ from $4x^{3/2} - 5x^{1/2}$.

Solution   We want to express $4x^{3/2} - 5x^{1/2}$ as the product of $x^{1/2}$ and a binomial.

$$4x^{3/2} - 5x^{1/2} = x^{1/2}(? - ?)$$

Since $4x^{3/2} = 4x \cdot x^{1/2}$, factoring $x^{1/2}$ out leaves $4x$ and factoring $x^{1/2}$ out from $5x^{1/2}$ leaves 5. Therefore,

$$4x^{3/2} - 5x^{1/2} = x^{1/2}(4x - 5)$$

CALCULATOR   CROSSOVER

A calculator can be used to find powers where the exponent is a rational number. For example, to find $1.4^{1/3}$, follow the sequence:

$$1.4 \boxed{y^x} 3 \boxed{1/x} =$$

or alternately:

$$1.4 \boxed{y^x} (1 \boxed{\div} 3) =$$

Either method gives

$$1.4^{1/3} = 1.1187 \qquad \text{(accurate to four decimal places)}$$

Here is another example. To find $52^{2/3}$, follow the sequence:

$$52 \boxed{y^x} (2 \boxed{\div} 3) =$$

The answer, accurate to two decimal places: $52^{2/3} = 13.93$

Find each of the following.

1. $842^{1/3}$ **2.** $5.92^{1/4}$ **3.** $190^{4/5}$ **4.** $8262^{-3/8}$

Answers   **1.** 9.4429   **2.** 1.5598   **3.** 66.5278   **4.** 0.0340

L E A R N I N G   A D V A N T A G E   *In order to study mathematics, it is important to also learn and use the notation and symbolism. Therefore, it is necessary to keep pencil and paper at hand to write any information from the book or notes with which you are having difficulty. By writing the mathematics, your mind will become more accustomed to the ''shorthand'' language of algebra.*

E X E R C I S E   S E T   5.2

For Exercises 1–26, evaluate each expression.

1. $100^{1/2}$

2. $64^{1/2}$

3. $(-1)^{1/5}$

4. $(-8)^{1/3}$

5. $(1,296)^{1/2}$

6. $(-2,401)^{1/4}$

7. $16^{3/2}$

8. $9^{5/2}$

9. $(-32)^{2/5}$

10. $(-8)^{4/3}$

11. $\left(\dfrac{25}{49}\right)^{3/2}$

12. $\left(\dfrac{121}{16}\right)^{3/2}$

13. $\left(\dfrac{1}{27}\right)^{2/3}$

14. $-25^{3/2}$

15. $(-25)^{3/2}$

16. $-81^{3/2}$

17. $25^{-3/2}$

18. $81^{-3/2}$

19. $\left(\dfrac{1}{32}\right)^{-1/5}$

20. $\left(\dfrac{1}{16}\right)^{-1/4}$

21. $\left(-\dfrac{64}{27}\right)^{-2/3}$

22. $\left(-\dfrac{1}{32}\right)^{-2/5}$

23. $25^{1/2} - 27^{1/3}$

24. $27^{-1/3} - 81^{-1/2}$

25. $-\left(-\dfrac{1}{64}\right)^{-2/3}$

26. $-\left(-\dfrac{1}{27}\right)^{-2/3}$

For Exercises 27–70, use properties of exponents to simplify. Assume that the variables represent positive real numbers. Write the answer using only positive exponents.

27. $4^{2/3} \cdot 4^{1/6}$

28. $5^{3/8} \cdot 5^{1/4}$

29. $t^{3/2} \cdot t^{-5/2}$

30. $z^{-3/4} \cdot z^{-7/4}$

31. $a^{1/12} \cdot a^{-1/12}$

32. $x^{-3/8} \cdot x^{3/8}$

33. $\dfrac{1}{a^{-1/2}}$

34. $\dfrac{2}{z^{-2/3}}$

35. $\dfrac{s^{4/5}}{s^{2/5}}$

36. $\dfrac{v^{5/4}}{v^{3/4}}$

37. $x^{1/3} \cdot x$

38. $q^2 \cdot q^{4/7}$

39. $b^{3/2} \cdot b^{5/2}$

40. $a^{4/5} \cdot a^{6/5}$

41. $d^{4/5} \cdot d^{1/5}$

42. $w^{1/4} \cdot w^{3/4}$

43. $p^{11/5} \cdot p^{5/6}$

44. $x^{2/3} \cdot x^{7/6}$

45. $\dfrac{y^{2/3}}{y^{3/4}}$

46. $\dfrac{s^{3/5}}{s^{1/10}}$

47. $(a^{3/2})^{4/9}$

48. $(z^{5/2})^{4/15}$

49. $\dfrac{z^2}{z^{9/4}}$

50. $\dfrac{c^{3/2}}{c^{9/4}}$

51. $(x^{3/7})^{-21/2}$

52. $(b^{-4/3})^{12/5}$

53. $(-8x^6y^2)^{1/3}$

54. $(9t^8r^4)^{1/2}$

55. $(9^8z^4)^{-3/16}$

56. $(4^{3/10}d^{-2/5})^{-5}$

57. $(3s^{1/2}t^{5/4})^2$

58. $(-2x^{2/3}y^2)^3$

59. $\left(\dfrac{w^{-2/5}}{v^{3/10}}\right)^{15}$

60. $\left(\dfrac{a^{-2/7}}{b^{1/7}}\right)^{14}$

61. $\left(\dfrac{m^{1/3}}{n^{-2/3}}\right)^6$

62. $\left(\dfrac{x^{3/2}}{z^{-4/3}}\right)^{12}$

63. $\left(\dfrac{x^{-2/3}}{a^{-4/5}}\right)^{-15}$

64. $\left(\dfrac{h^{-2/5}}{k^{-4/3}}\right)^{30}$

65. $\dfrac{(x^{-2}z^{3/4})^2}{x^{-5}z^2}$

66. $\dfrac{(a^{2/7}b^{-3/4})^{2/3}}{a^{3/7}b^{-3/2}}$

67. $\dfrac{(s^{2/3}t^{-4/3})^6}{(s^{-4/3}t^{-2/3})^2}$

68. $\dfrac{(c^{-4/3}d^{3/5})^{15}}{(c^{-2/5}d^{-1/10})^{10}}$

69. $\left(\dfrac{3^{-1/2}a^{-1/2}x^{3/4}}{3^{-1}a^{1/4}x^{-1/2}}\right)^{-4}$

70. $(r^{2/3}s^{-1/6})\left(\dfrac{r^{-4/9}}{s^{1/6}}\right)^3$

For Exercises 71–76, replace all radicals with rational exponents and simplify. Assume that all variables represent positive real numbers.

71. $\sqrt[3]{x^5}$

72. $\sqrt[4]{a^3}$

73. $\dfrac{\sqrt[4]{b^5}}{\sqrt[3]{b^4}}$

74. $\dfrac{\sqrt[3]{t^4}}{\sqrt[5]{t^3}}$

75. $\sqrt[3]{\sqrt{z^5}}$

76. $\sqrt[4]{\sqrt[3]{x^5}}$

For Exercises 77–82, simplify each expression using the properties of exponents. Assume that n is a natural number and that the base variables represent positive real numbers.

77. $\dfrac{(s^n)^{3/5}}{s^{-3-n}}$

78. $\dfrac{b^{n+2}}{(b^n)^{2/5}}$

79. $(x^{5n}t^{3n})^{1/n}$

80. $(u^{2n}v^n)^{3/n}$

81. $\dfrac{(a^nb^{2n})^{1/n}}{(a^{2n}b^{3n})^{2/n}}$

82. $\dfrac{(x^{2n}y^n)^{3/n}}{(x^{4n}y^{2n})^{2/n}}$

For Exercises 83–90, multiply. Write answers using only positive exponents. Assume that all variables represent positive real numbers.

83. $x^{1/2}(x^{1/4} - 3x^{1/3})$

84. $a^{2/3}(5a^{5/6} + 3a^{1/3})$

85. $4v^{3/2}(1 - 2v^{-6/5})$

86. $3q^{2/5}(q^{-2/5} - 2q^{-7/10})$

87. $-(2/3)x^{5/4}(6x^{-1/2} - 3x^{-1})$

88. $-4z^{3/5}(z^{-3/5} + \dfrac{1}{2}z^{-7/20})$

89. $2x^{3/2}(x^{-3/2} - 5x^{-3})$

90. $3x^{-4/3}(2x^4 - 7x^{1/6})$

91. Factor $x^{2/3}$ from $x^{4/3} + x^{2/3}$.

92. Factor $a^{1/5}$ from $a^{2/5} - 2a^{1/5}$.

93. Factor $3c^{2/5}$ from $6c^{2/5} - 9c^2$.

94. Factor $5t^{3/2}$ from $15t^{5/2} + 10t^{3/2}$.

95. Factor $q^{1/5}$ from $q^3 - q^{1/5}$.

96. Factor $r^{2/7}$ from $3r^{2/7} - r^2$.

97. Factor $z^{-5/3}$ from $z^{7/3} - z^{-5/3}$.

98. Factor $v^{-3/8}$ from $v^{-3/8} - v^{5/8}$.

99. Factor $y^{-2/3}$ from $6y^{-2/3} + 5y^{1/6}$.

100. Factor $z^{-4/7}$ from $2z^{3/14} - 3z^{-4/7}$.

REVIEW PROBLEMS

The following exercises review parts of Sections 3.3, 3.5, and 4.2. Doing these problems will help prepare you for the next section.

Section 3.3

Multiply.

101. $(2x + 5)(x - 2)$

102. $(a + 3)^2$

103. $(z - 3)(5z - 1)$

104. $(3c + 4)(2c + 3)$

Section 3.5

Factor.

105. $(x - 2)^2 - (x - 2)$

106. $x^2 + x - 12$

107. $6x^2 - 5x + 1$

108. $8a^2 - 26a + 15$

Section 4.2

Divide.

109. $\dfrac{12a^3b^2 - 8a^2b^3}{4a^2b^2}$

110. $\dfrac{15c^4z^4 + 5c^4z^3 - 10c^5z^4}{5c^3z^2}$

ENRICHMENT EXERCISES

Simplify. Assume that n is an integer greater than 1 and that the base variables represent positive real numbers.

1. $\sqrt[n]{x^{5n}}$

2. $\sqrt[4]{z^{12n-3}}$

3. $\sqrt[n]{a^{4n+3}}$

4. $z\sqrt[n]{\sqrt[3]{z^{4n}}} - \sqrt[n]{\sqrt{z^{14n/3}}}$

Answers to Enrichment Exercises are on page A.22.

5.3

The Algebra of Expressions with Rational Exponents

O B J E C T I V E S

▶ *To multiply and divide expressions with rational exponents*

▶ *To factor expressions that resemble polynomials but involve rational exponents*

▶ *To combine and simplify expressions with rational exponents*

In this section, we look at the algebra of expressions involving rational exponents. In particular, some of these expressions resemble polynomials, and, therefore, we can apply our knowledge of polynomial algebra and factoring techniques to these expressions. This material will be especially valuable to those of you who plan to take further courses in mathematics.

In the last section, we multiplied quantities such as $x^{2/3}(4x^{1/2} - 8x^3)$ using the distributive property. In this section, we continue this train of thought and look at other multiplications that can be done using other former techniques.

Throughout this section, assume that all variables represent positive integers.

Example 1   Multiply: $x^{3/4}y^{1/3}(3x^2y^{1/2} - 2x^3y^{-1/4} + 7x^{1/2}y^2)$.

Solution   We apply the distributive property, then use properties of exponents to simplify.

$$x^{3/4}y^{1/3}(3x^2y^{1/2} - 2x^3y^{-1/4} + 7x^{1/2}y^2)$$
$$= 3x^{3/4+2}y^{1/3+1/2} - 2x^{3/4+3}y^{1/3-1/4} + 7x^{3/4+1/2}y^{1/3+2}$$
$$= 3x^{11/4}y^{5/6} - 2x^{15/4}y^{1/12} + 7x^{5/4}y^{7/3}$$ ▲

Example 2   Multiply: $(x^{3/4} + 2)(x^{3/4} + 6)$.

Solution   This product has the form of multiplication of two binomials. Therefore, we multiply using our techniques for multiplying two binomials.

$$(x^{3/4} + 2)(x^{3/4} + 6) = x^{3/4}x^{3/4} + 6x^{3/4} + 2x^{3/4} + 12$$
$$= x^{6/4} + 8x^{3/4} + 12$$
$$= x^{3/2} + 8x^{3/4} + 12$$ Reduce the exponent 6/4 to lowest terms. ▲

Example 3   Multiply: $(\sqrt{c} - \sqrt{d})(2\sqrt{c} + \sqrt{d})$.

Solution   We multiply the two expressions as if they were two binomials.

$$(\sqrt{c} - \sqrt{d})(2\sqrt{c} + \sqrt{d}) = 2\sqrt{c}\sqrt{c} + \sqrt{c}\sqrt{d} - 2\sqrt{c}\sqrt{d} - \sqrt{d}\sqrt{d}$$
$$= 2c - \sqrt{c}\sqrt{d} - d$$ ▲

Example 4   Multiply: $(z^{5/2} + 3^{1/2})(z^{5/2} - 3^{1/2})$.

Solution   This product is of the standard form $(a + b)(a - b)$. Therefore, we may use the formula $(a + b)(a - b) = a^2 - b^2$ to multiply.

$$(z^{5/2} + 3^{1/2})(z^{5/2} - 3^{1/2}) = (z^{5/2})^2 - (3^{1/2})^2$$
$$= z^5 - 3 \qquad \blacktriangle$$

Example 5   Multiply: $(4\sqrt{r} + 3)^2$.

Solution   Here we can use the form $(a + b)^2 = a^2 + 2ab + b^2$.

$$(4\sqrt{r} + 3)^2 = (4\sqrt{r})^2 + 2(4\sqrt{r})(3) + 3^2$$
$$= 16r + 24\sqrt{r} + 9 \qquad \blacktriangle$$

Example 6   Multiply: $(t^{1/2} - 2)(2t^{3/2} - 3t + 5)$.

Solution   We can find this product by the column method for multiplying two polynomials.

$$
\begin{array}{r}
2t^{3/2} - 3t + 5 \\
\times\ t^{1/2} - 2 \\
\hline
2t^2 - 3t^{3/2} \qquad + 5t^{1/2} \\
-\ 4t^{3/2} + 6t \qquad\qquad - 10 \\
\hline
2t^2 - 7t^{3/2} + 6t + 5t^{1/2} \qquad - 10
\end{array}
$$

$\blacktriangle$

Example 7   Multiply: $(y^{1/3} + 4^{1/3})(y^{2/3} - y^{1/3}4^{1/3} + 4^{2/3})$.

Solution   We could find this product by the column method used in Example 6. However, recall the form

$$(a + b)(a^2 - ab + b^2) = a^3 + b^3$$

If we set $y^{1/3} = a$ and $4^{1/3} = b$, then $y^{2/3} = a^2$ and $4^{2/3} = b^2$ and our product fits this form. Therefore,

$$(y^{1/3} + 4^{1/3})(y^{2/3} - y^{1/3}4^{1/3} + 4^{2/3}) = (y^{1/3})^3 + (4^{1/3})^3$$
$$= y + 4 \qquad \blacktriangle$$

In the next examples, we deal with factoring expressions that are polynomial in form. For example, $a^{2/3} - 3a^{1/3} + 2$ can be written as $(a^{1/3})^2 - 3a^{1/3} + 2$ and is called a polynomial in $a^{1/3}$.

Example 8   Factor: $a^{2/3} - 3a^{1/3} + 2$.

Solution   As explained in the paragraph before this example, $a^{2/3} - 3a^{1/3} + 2$ is a polynomial in $a^{1/3}$. In fact, if we set $x = a^{1/3}$, then $x^2 = a^{2/3}$ and we get

$$x^2 - 3x + 2$$

Since this polynomial in x factors into $(x - 2)(x - 1)$, we can factor the original expression.

$$a^{2/3} - 3a^{1/3} + 2 = (a^{1/3} - 2)(a^{1/3} - 1) \qquad \blacktriangle$$

Some of these expressions can be factored if they fit a standard form. For convenience, we restate our factoring rules from Chapter 3.

R U L E S

Factoring forms

$a^2 - b^2 = (a + b)(a - b)$

$a^2 + 2ab + b^2 = (a + b)^2$

$a^2 - 2ab + b^2 = (a - b)^2$

$a^3 + b^3 = (a + b)(a^2 - ab + b^2)$

$a^3 - b^3 = (a - b)(a^2 + ab + b^2)$

Example 9   Factor: $2(x + 3)^{2/3}$ from $10(x + 3)^{5/3} - 6(x + 3)^{2/3}$.

Solution   We factor out $2(x + 3)^{2/3}$ as we would factor out the greatest common factor from a polynomial.

$$10(x + 3)^{5/3} - 6(x + 3)^{2/3} = 2(x + 3)^{2/3}[5(x + 3)^{3/3} - 3]$$

$$= 2(x + 3)^{2/3}[5x + 15 - 3]$$

$$= 2(x + 3)^{2/3}(5x + 12) \qquad \blacktriangle$$

In the next example, we show how division with expressions that contain rational exponents is similar to division of a polynomial by a monomial.

Example 10   Divide: $\dfrac{14a^{3/2}b^4 + 21a^2b^{5/3}}{7a^{1/2}b^{2/3}}$.

Solution   We divide each term of the numerator by the denominator, then use properties of exponents to simplify.

$$\frac{14a^{3/2}b^4 + 21a^2b^{5/3}}{7a^{1/2}b^{2/3}} = \frac{14a^{3/2}b^4}{7a^{1/2}b^{2/3}} + \frac{21a^2b^{5/3}}{7a^{1/2}b^{2/3}}$$

$$= 2a^{3/2-1/2}b^{4-2/3} + 3a^{2-1/2}b^{5/3-2/3}$$

$$= 2ab^{10/3} + 3a^{3/2}b \qquad \blacktriangle$$

In the final example of this section, we deal with combining two expressions into a single fraction. The technique is similar to combining rational expressions that we studied in Chapter 4.

Example 11   Combine into one fraction.

$$2 + \sqrt{1 - x^2} - \frac{x^2}{\sqrt{1 - x^2}}$$

Solution   To combine these expressions, we find the least common denominator, change to equivalent fractions, and then subtract numerators.

$$2 + \sqrt{1 - x^2} - \frac{x^2}{\sqrt{1 - x^2}} = \frac{2 + \sqrt{1 - x^2}}{1} - \frac{x^2}{\sqrt{1 - x^2}}$$

The least common denominator is $\sqrt{1 - x^2}$.

$$= \frac{(2 + \sqrt{1 - x^2})(\sqrt{1 - x^2})}{\sqrt{1 - x^2}} - \frac{x^2}{\sqrt{1 - x^2}}$$

$$= \frac{2\sqrt{1 - x^2} + 1 - x^2 - x^2}{\sqrt{1 - x^2}}$$

$$= \frac{2\sqrt{1 - x^2} + 1 - 2x^2}{\sqrt{1 - x^2}}$$ ▲

L E A R N I N G   A D V A N T A G E   *If you are not achieving your goals for this course, take some time to consider what you might do to improve your performance. Have you been doing the homework consistently? Are you taking advantage of the help available? Make an appointment to talk to your instructor. Possibly the two of you together can map out a strategy that will result in a successful completion of the course.*

E X E R C I S E   S E T   5.3

For Exercises 1–36, multiply. Assume that all variables represent positive numbers.

1. $x^{2/5}y^{2/3}(5x^2y^{1/3} - 4x^{1/10}y^{5/6} + 3x^{6/5}y^3)$

2. $u^{3/4}v^{4/3}(u^{3/8}v^{5/6} + 7u^3v^{1/2} + 8u^{5/4}v^{2/3})$

3. $2s^{4/7}t^{2/3}(s^{-3/7}t^{1/2} - 3s^{3/14}t^2 + 4s^{3/7}t^{1/6})$

4. $3r^{5/2}v^{3/8}(2r^{1/2}v^{1/2} - 4r^{-3/2}v^{-3/8} + 5r^4v^2)$

5. $(a^{1/2} - 1)(a^{3/2} + 5)$

6. $(z^{3/2} + 2)(z^{1/2} - 3)$

7. $(s^{3/4} + 6)(s^{1/4} - 1)$

8. $(h^{2/3} + 2)(h^{1/3} - 8)$

9. $(\sqrt{x} + \sqrt{y})(\sqrt{x} - 2\sqrt{y})$

10. $(3\sqrt{c} - \sqrt{d})(\sqrt{c} - 4\sqrt{d})$

11. $(2\sqrt{r} - 3\sqrt{s})(\sqrt{r} + 2\sqrt{s})$

12. $(\sqrt{t} - 2\sqrt{v})(3\sqrt{t} + 2\sqrt{v})$

13. $(w^{2/3} + 1)(w^{2/3} - 1)$

14. $(r^{3/2} - 2)(r^{3/2} + 2)$

15. $(2d^{1/2} - 1)^2$

16. $(3z^{5/2} + 2)^2$

17. $(q^{3/2} + a^{1/2})^2$

18. $(t^{5/2} - r^{1/2})^2$

19. $(x^{5/2} - y^{3/2})(x^{5/2} + y^{3/2})$

20. $(c^{7/2} - d^{1/2})(c^{7/2} + d^{1/2})$

21. $(x^{1/2} - 1)(3x^2 + 2x^{3/2} - 1)$

22. $(a^{3/2} + 2)(a^2 - a^{3/2} - a^{1/2})$

23. $(z^{1/3} + c^{1/3})(z^{2/3} - z^{1/3}c^{1/3} + c^{2/3})$

24. $(a^{1/3} - b^{1/3})(a^{2/3} + a^{1/3}b^{1/3} + b^{2/3})$

25. $(s^{1/3} - t^{1/3})(s^{2/3} + s^{1/3}t^{1/3} + t^{2/3})$

26. $(p^{1/3} + q^{1/3})(p^{2/3} - p^{1/3}q^{1/3} + q^{2/3})$

27. $(w^{1/3} - 2)(w^{2/3} + 2w^{1/3} + 4)$

28. $(y^{1/3} + 1)(y^{2/3} - y^{1/3} + 1)$

29. $(3 + d^{1/3})(9 - 3d^{1/3} + d^{2/3})$

30. $(2 - v^{1/3})(4 + 2v^{1/3} + v^{2/3})$

31. $(x^{1/2} - 1)(x^{1/2} + 1)(x^2 + 4)$

32. $(a^{2/3} - 2)(a^{1/3} + 3)(a^{1/3} - 1)$

33. $2s^{3/4}(s^{5/4} - 1)(s^2 + 1)$

34. $3z^{2/3}(z^{2/3} - 2)(z^{5/3} - 4)$

35. $a^{2/3}b^{1/2}(a^{1/2} + 3b^{2/3})(a^{1/6} - b^{1/3})$

36. $x^{5/2}y^{1/4}(2x^{1/2} + y^{1/4})(x^2 - y^{3/4})$

For Exercises 37–50, factor.

37. $x^{2/3} - 4$

38. $a^{2/5} - 9$

39. $z^{2/3} + 4z^{1/3} + 4$

40. $c^{2/5} - 6c^{1/5} + 9$

41. $y^{2/5} + 3y^{1/5} - 4$

42. $q^{2/3} - q^{1/3} - 6$

43. $3x^{2/7} + x^{1/7} - 4$

44. $5t^{2/5} + 14t^{1/5} - 3$

45. $9y^{2/3} - 4$

46. $16x^{2/9} - 1$

47. $6r^{2/3} - 7r^{1/3} + 1$

48. $6s^{2/5} - 5s^{1/5} + 1$

49. $2y^{2/7} - 11y^{1/7} + 9$

50. $16x^{2/3} + 8x^{1/3} + 1$

For Exercises 51–56, factor the indicated term from the expression.

51. Factor $7(x + 1)^{1/2}$ from
$21(x + 1)^{3/2} + 7(x + 1)^{1/2}$.

52. Factor $3(t - 6)^{2/3}$ from
$6(t - 6)^{5/3} - 15(t - 6)^{2/3}$.

53. Factor $4(q^2 - 3)^{2/5}$ from
$28(q^2 - 3)^{2/5} - 4(q^2 - 3)^{7/5}$.

54. Factor $5(m^3 - 7)^{3/7}$ from
$10(m^3 - 7)^{3/7} + 45(m^3 - 7)^{10/7}$.

55. Factor $(x^2 + 1)^{-1/2}$ from
$(x^2 + 1)^{1/2} + x^2(x^2 + 1)^{-1/2}$.

56. Factor $(4 - y)^{-1/3}$ from
$(4 - y)^{2/3} - (4 - y)^{-1/3}$.

For Exercises 57–62, divide.

57. $\dfrac{12a^{3/2} - 15a^{5/2}}{3a^{1/2}}$

58. $\dfrac{45x^{3/4} + 9x^{5/4}}{3x^{1/4}}$

59. $\dfrac{18x^{5/3} + 22x^3}{2x^{1/6}}$

60. $\dfrac{15z^{3/2} - 35z^2}{5z^{3/4}}$

61. $\dfrac{4w^{3/4}v^{3/2} - 6w^{1/2}v^{2/5}}{12w^{1/2}v^{1/3}}$

62. $\dfrac{2s^{9/5}t^{4/3} + 4s^{4/3}t^{3/4}}{8s^{1/4}t^{2/3}}$

For Exercises 63–72, combine into one fraction.

63. $\dfrac{2}{\sqrt{a}} + 3\sqrt{a}$

64. $2\sqrt{z} - \dfrac{3}{\sqrt{z}}$

65. $\dfrac{12}{x^{1/4}} - 11x^{3/4}$

66. $y^{1/3} - \dfrac{1}{y^{2/3}}$

67. $2(x^2 + 1)^{1/2} - \dfrac{2x^2}{(x^2 + 1)^{1/2}}$

68. $\dfrac{s}{(s + 3)^{1/2}} - (s + 3)^{1/2}$

69. $3 - \sqrt{1 - t^2} - \dfrac{t^2}{\sqrt{1 - t^2}}$

70. $4 + \sqrt{2 + y} + \dfrac{y}{\sqrt{2 + y}}$

71. $\dfrac{a^2}{(4 + a^2)^{3/4}} - (4 + a^2)^{1/4}$

72. $\dfrac{z^2}{(1 - z^2)^{2/3}} - (1 - z^2)^{1/3}$

REVIEW PROBLEMS

The following exercises review parts of Section 5.1. Doing these problems will help prepare you for the next section. Assume the variables represent nonnegative real numbers.

Simplify.

73. $\sqrt{9a^6b^2}$

74. $\sqrt{25x^4y^8}$

75. $\sqrt[3]{27x^3y^6z^9}$

76. $\sqrt[3]{8c^9d^3}$

77. $\sqrt[4]{16a^8b^{28}}$

78. $\sqrt[4]{x^{16}y^4z^{12}}$

ENRICHMENT EXERCISES

For Exercises 1 and 2, let m, n, r, and s represent natural numbers.

1. Multiply: $x^{2n/3}y^{3m/2}(x^{n/3}y^{m/4} + 3x^{n-1}y^{5m/8})$.

2. Divide: $\dfrac{a^{2r/5}b^{3s/7} - a^{3r/10}b^{5s/21}}{a^{r/2}b^{3s/14}}$.

3. Factor completely by first factoring $y^{4/3}$ from $2y^2 - 7y^{5/3} + 6y^{4/3}$.

4. Factor completely by first factoring $a^{1/6}b^{1/3}$ from $a^{5/6}b^{1/3} - 4a^{1/6}b$.

5. Factor completely by first factoring $x^{1/3}y^{4/3}$ from $x^{11/6}y^{4/3} - x^{1/3}y^{25/12}$.

6. Factor completely by first factoring $(z^2 - 3)^{1/4}$ from $(z^2 - 3)^{9/4} + 2(z^2 - 3)^{5/4}$.

7. Factor completely by first factoring $2(y + 4)^{1/5}$ from $4(y + 4)^{6/5} - 6(y + 4)^{11/5}$.

8. Let n be a natural number. Factor $3x^{2/n} - 10x^{1/n} + 3$.

Answers to Enrichment Exercises are on page A.23.

5.4

Simplifying Radical Expressions

▶ *To use the product rule for radicals to simplify radicals*

▶ *To use the quotient rule for radicals to simplify radicals*

▶ *To rationalize the denominator of a radical expression*

In this section, we consider properties that will help us simplify radical expressions. The first property is introduced by the following illustration. Notice that

$$\sqrt{4 \cdot 25} = \sqrt{100} = 10$$

and that

$$\sqrt{4} \cdot \sqrt{25} = 2 \cdot 5 = 10$$

Since both $\sqrt{4 \cdot 25}$ and $\sqrt{4} \cdot \sqrt{25}$ are equal to 10,

$$\sqrt{4 \cdot 25} = \sqrt{4} \cdot \sqrt{25}$$

This illustrates the *product rule for radicals.*

R U L E

Theorem 1.   The product rule for radicals

If a and b are numbers, and n is a positive integer, then

$$\sqrt[n]{a \cdot b} = \sqrt[n]{a} \cdot \sqrt[n]{b}$$

provided $\sqrt[n]{a}$ and $\sqrt[n]{b}$ are real numbers.

Proof $\quad \sqrt[n]{a \cdot b} = (ab)^{1/n}$ $\qquad$ Definition of a number raised to the $1/n$ power.

$\qquad = a^{1/n}b^{1/n}$ $\qquad$ Property of exponents.

$\qquad = \sqrt[n]{a} \cdot \sqrt[n]{b}$ $\qquad$ Definition of a number raised to the $1/n$ power.

The product rule for radicals states that the nth root of a times b is equal to the nth root of a times the nth root of b.

Example 1 Multiply.

(a) $\sqrt{5}\sqrt{3} = \sqrt{5 \cdot 3} = \sqrt{15}$

(b) $\sqrt{2}\sqrt{a} = \sqrt{2a}$

(c) $\sqrt[3]{3}\sqrt[3]{9} = \sqrt[3]{27} = 3$

(d) $\sqrt{\dfrac{5}{z}}\sqrt{\dfrac{11}{c}} = \sqrt{\dfrac{55}{zc}}$

(e) $\sqrt[3]{12}\sqrt{7}$ cannot be multiplied using the product rule for radicals since the two indexes are not the same.   ▲

N O T E   *The pattern of the product rule applies for the product of a and b but not for the* sum *of a and b.*

$$\sqrt[n]{a + b} \neq \sqrt[n]{a} + \sqrt[n]{b}$$

For example,

$$\sqrt{9 + 16} = \sqrt{25} = 5$$

whereas,

$$\sqrt{9} + \sqrt{16} = 3 + 4 = 7$$

Therefore,

$$\sqrt{9 + 16} \neq \sqrt{9} + \sqrt{16}$$

There is also a quotient rule for radicals whose proof is similar to that of the product rule.

R U L E

Theorem 2.   Quotient rule for radicals

If a and b are real numbers, with $b \neq 0$ and n a positive integer, then

$$\sqrt[n]{\dfrac{a}{b}} = \dfrac{\sqrt[n]{a}}{\sqrt[n]{b}}$$

provided $\sqrt[n]{a}$ and $\sqrt[n]{b}$ are real numbers.

That is, the nth root of the quotient is the quotient of the nth roots.

Example 2   Rewrite each expression by using the quotient rule for radicals.

(a) $\sqrt{\dfrac{9}{49}} = \dfrac{\sqrt{9}}{\sqrt{49}} = \dfrac{3}{7}$

(b) $\sqrt{\dfrac{11}{100}} = \dfrac{\sqrt{11}}{\sqrt{100}} = \dfrac{\sqrt{11}}{10}$

(c) $\sqrt[3]{\dfrac{27}{8}} = \dfrac{\sqrt[3]{27}}{\sqrt[3]{8}} = \dfrac{3}{2}$

(d) $\sqrt[4]{\dfrac{ab}{16}} = \dfrac{\sqrt[4]{ab}}{\sqrt[4]{16}} = \dfrac{\sqrt[4]{ab}}{2}$ ▲

Using the product and quotient rules for radicals, we can change the form and simplify radical expressions according to the following criteria.

S T R A T E G Y

To simplify a radical expression

A radical expression is *simplified* if:

1. The radicand (the quantity under the radical sign) contains no power greater than or equal to the index. This means that no perfect squares are factors under a square root sign; no perfect cubes are factors under a cube root sign; and so on.

Example: $\sqrt{x^5}$ does not satisfy this condition.

2. The exponents in the radicand and the index of the radical have no common factor.

Example: $\sqrt[4]{x^6}$ does not satisfy this condition.

3. The radicand has no fractions.

Example: $\sqrt[3]{\dfrac{2}{7}}$ does not satisfy this condition.

4. There are no radicals in denominators.

Example: $\dfrac{4}{\sqrt{3}}$ does not satisfy this condition.

Example 3   Simplify: $\sqrt{75}$.

Solution   The largest perfect square that is a factor of 75 is 25. We write 75 as $25 \cdot 3$, then apply Theorem 1.

$$\sqrt{75} = \sqrt{25 \cdot 3} = \sqrt{25} \cdot \sqrt{3} = 5\sqrt{3} \qquad \blacktriangle$$

Example 4   Simplify: $\sqrt{32x^4y^7}$, where $y \geq 0$.

Solution   The largest perfect square to be taken out of the radicand is $16x^4y^6$. Therefore,

$$\sqrt{32x^4y^7} = \sqrt{16x^4y^6 \cdot 2y} = \sqrt{16x^4y^6}\sqrt{2y} = 4x^2y^3\sqrt{2y} \qquad \blacktriangle$$

Example 5   Simplify: $\sqrt[3]{16r^7s^5}$.

Solution   We want to take out the largest perfect cube from the radicand. Namely, write $16r^7s^5$ as $8r^6s^3 \cdot 2rs^2$ and then use the product rule for radicals.

$$\sqrt[3]{16r^7s^5} = \sqrt[3]{8r^6s^3 \cdot 2rs^2}$$
$$= \sqrt[3]{8r^6s^3}\sqrt[3]{2rs^2}$$
$$= 2r^2s\sqrt[3]{2rs^2} \qquad \blacktriangle$$

Example 6   Simplify: $\sqrt[4]{48a^3b^8c^7}$, where $a \geq 0$ and $c \geq 0$.

Solution
$$\sqrt[4]{48a^3b^8c^7} = \sqrt[4]{16b^8c^4 \cdot 3a^3c^3}$$
$$= \sqrt[4]{16b^8c^4}\sqrt[4]{3a^3c^3}$$
$$= 2b^2c\sqrt[4]{3a^3c^3} \qquad \blacktriangle$$

Example 7   Simplify: $\sqrt[4]{x^6}$, where $x \geq 0$.

Solution   Notice that the index, 4, and the exponent in the radicand, 6, have a common factor of 2. This means that Condition 2 on simplifying radical expressions is not satisfied. Therefore, first write the radical in exponent form and simplify using properties of exponents.

$$\sqrt[4]{x^6} = (x^6)^{1/4}$$
$$= x^{6/4}$$
$$= x^{3/2} \qquad \text{Reduce 6/4 to lowest terms.}$$
$$= \sqrt{x^3} \qquad \text{Note that } x^{3/2} = x^{3(1/2)}.$$
$$= \sqrt{x^2 x}$$
$$= \sqrt{x^2}\sqrt{x}$$
$$= x\sqrt{x} \qquad \blacktriangle$$

N O T E   *Another way to simplify $\sqrt[4]{x^6}$ of Example 7 is to first write x^6 as x^4x^2 and use the product rule for radicals.*

$$\sqrt[4]{x^6} = \sqrt[4]{x^4x^2} = \sqrt[4]{x^4}\sqrt[4]{x^2} = x\sqrt[4]{x^2}$$

Notice, however, that $\sqrt[4]{x^2}$ is not simplified as the index and exponent have a common factor. Therefore, to finish the problem,

$$x\sqrt[4]{x^2} = x(x^2)^{1/4} = x \cdot x^{2/4} = x \cdot x^{1/2} = x\sqrt{x}$$

Notice that the method of Example 7 is shorter than this method.

Example 8   Simplify $\sqrt{\dfrac{3x}{8y}}$, where $x \geq 0$ and $y > 0$.

Solution   This radical does not satisfy Condition 3, as the radicand is a fraction. To simplify, multiply numerator and denominator by the "smallest" quantity that makes the denominator a perfect square. Therefore, we multiply numerator and denominator of the radicand by $2y$.

$$\sqrt{\frac{3x}{8y}} = \sqrt{\frac{3x \cdot 2y}{8y \cdot 2y}}$$

$$= \sqrt{\frac{6xy}{16y^2}}$$

$$= \frac{\sqrt{6xy}}{\sqrt{16y^2}}$$

$$= \frac{\sqrt{6xy}}{4y} \qquad \blacktriangle$$

N O T E   *In Example 8, we could have multiplied numerator and denominator by 8y instead of 2y. We would still attain the correct answer; however, it would have resulted in more work. The details follow:*

$$\sqrt{\frac{3x}{8y}} = \sqrt{\frac{3x \cdot 8y}{8y \cdot 8y}}$$

$$= \frac{\sqrt{24xy}}{8y}$$

$$= \frac{\sqrt{4 \cdot 6xy}}{8y}$$

$$= \frac{2\sqrt{6xy}}{8y}$$

$$= \frac{\sqrt{6xy}}{4y}$$

Example 9   Simplify: $\sqrt[3]{\dfrac{x}{2y}}$, where $y \neq 0$.

Solution   This radical does not satisfy Condition 3, as the radicand is a fraction. Therefore, we multiply numerator and denominator inside the radical by the "smallest" expression that will make the denominator a perfect cube. We use $4y^2$.

$$\sqrt[3]{\frac{x}{2y}} = \sqrt[3]{\frac{x \cdot 4y^2}{2y \cdot 4y^2}}.$$

$$= \sqrt[3]{\frac{4xy^2}{8y^3}}$$

$$= \frac{\sqrt[3]{4xy^2}}{\sqrt[3]{8y^3}}$$

$$= \frac{\sqrt[3]{4xy^2}}{2y} \qquad \blacktriangle$$

Example 10   Simplify: $\dfrac{3w}{\sqrt{6w}}$, where $w > 0$.

Solution   This radical does not satisfy Condition 4. Multiply numerator and denominator by the smallest expression that will make the denominator the square root of a perfect square.

$$\frac{3w}{\sqrt{6w}} = \frac{3w\sqrt{6w}}{\sqrt{6w}\sqrt{6w}} = \frac{3w\sqrt{6w}}{6w} = \frac{\sqrt{6w}}{2}$$ ▲

In Example 10, the process of eliminating the radical in the denominator is called *rationalizing the denominator*.

> # D E F I N I T I O N
>
> To **rationalize the denominator** of an expression is to rewrite the expression so that there are no radicals in the denominator.

In general the process is to multiply numerator and denominator by an expression so that the product in the denominator simplifies to something without radicals. For example, if the denominator is $\sqrt{8x}$, multiply by $\sqrt{2x}$:

$$\sqrt{8x}\sqrt{2x} = \sqrt{16x^2} = 4x$$

If the denominator is $\sqrt[3]{4x}$, multiply by $\sqrt[3]{2x^2}$:

$$\sqrt[3]{4x}\sqrt[3]{2x^2} = \sqrt[3]{8x^3} = 2x$$

Example 11   Rationalize the denominator.

(a) $\dfrac{3}{\sqrt{5}} = \dfrac{3}{\sqrt{5}} \cdot \dfrac{\sqrt{5}}{\sqrt{5}}$

$\qquad = \dfrac{3\sqrt{5}}{(\sqrt{5})^2}$

$\qquad = \dfrac{3\sqrt{5}}{5}$

(b) $\dfrac{4}{\sqrt{18z}} = \dfrac{4}{\sqrt{18z}} \cdot \dfrac{\sqrt{2z}}{\sqrt{2z}}$

$\qquad = \dfrac{4\sqrt{2z}}{\sqrt{36z^2}} = \dfrac{4\sqrt{2z}}{6z}$

$\qquad = \dfrac{2\sqrt{2z}}{3z}$

(c) $\dfrac{12ab^2}{\sqrt[3]{9a^2b}} = \dfrac{12ab^2}{\sqrt[3]{9a^2b}} \cdot \dfrac{\sqrt[3]{3ab^2}}{\sqrt[3]{3ab^2}}$

$\qquad = \dfrac{12ab^2\sqrt[3]{3ab^2}}{\sqrt[3]{27a^3b^3}}$

$\qquad = \dfrac{12ab^2\sqrt[3]{3ab^2}}{3ab}$

$\qquad = 4b\sqrt[3]{3ab^2}$ ▲

Many times a radical expression can be simplified by one of several ways. Therefore, it is important to think of a method before moving the pencil.

CALCULATOR   CROSSOVER

A spaceship attempting to leave our planet must have a minimum escape velocity. The escape velocity v is given by the following formula

$$v = \frac{13.34 \, M}{1011 \, r}$$

where M is the mass of the Earth and r is its radius. Find v, if Earth's mass is 5.98×10^{24} kg and its radius is 6.37×10^6 meters. The answer will be in meters per second.

Answer 1.2387×10^{16} m/s.

LEARNING   ADVANTAGE   *Be sure to work at a consistant pace throughout the course. Learning mathematics is a process that builds upon what you have previously learned. Therefore, the best way to achieve success in the course is to learn each lesson as it is presented and not get behind.*

EXERCISE   SET   5.4

For Exercises 1–10, use the product rule for radicals to multiply. Assume that the variables represent nonnegative real numbers.

1. $\sqrt{7}\sqrt{2}$

2. $\sqrt{3}\sqrt{5}$

3. $\sqrt[3]{16}\sqrt[3]{4}$

4. $\sqrt[4]{9}\sqrt[4]{9}$

5. $\sqrt{x}\sqrt{3y}$

6. $\sqrt{2r}\sqrt{t}$

7. $\sqrt[4]{3p}\sqrt[4]{2q}$

8. $\sqrt[3]{5s}\sqrt[3]{7t}$

9. $-\sqrt[3]{3y}\sqrt[3]{11z}$

10. $-\sqrt{0.1r}\sqrt{2w}$

For Exercises 11–18, rewrite each expression by using the quotient rule for radicals. Assume that all variables represent positive real numbers.

11. $\sqrt{\dfrac{3}{25}}$

12. $\sqrt{\dfrac{5}{49}}$

13. $\sqrt{\dfrac{2}{x^4}}$

14. $\sqrt{\dfrac{7}{b^6}}$

15. $\sqrt[3]{\dfrac{11}{64}}$

16. $\sqrt[4]{\dfrac{23}{16}}$

17. $\sqrt[3]{\dfrac{x}{8}}$

18. $\sqrt[3]{\dfrac{a}{27b^3}}$

For Exercises 19–74, simplify. Assume that the variables represent positive real numbers.

19. $\sqrt{48}$

20. $\sqrt{245}$

21. $\sqrt{12}$

22. $\sqrt{18}$

23. $\sqrt[3]{270}$

24. $\sqrt[4]{32}$

25. $\sqrt[3]{48c}$

26. $\sqrt[3]{80r}$

27. $-2\sqrt{8az}$

28. $-5\sqrt{45pq}$

29. $\sqrt{4^6}$

30. $\sqrt{5^4}$

31. $\sqrt{x^4}$

32. $\sqrt{a^6}$

33. $\sqrt[3]{16s^{10}}$

34. $\sqrt[3]{81t^{14}}$

35. $\sqrt{u^4v^2}$

36. $\sqrt{x^6z^{12}}$

37. $\sqrt[4]{32s^{17}}$

38. $\sqrt[4]{48c^9}$

39. $\sqrt[5]{64b^4c^6}$

40. $\sqrt{36y^8z^2}$

41. $\sqrt[3]{r^5}$

42. $\sqrt[3]{z^7}$

43. $\sqrt{b^3}$

44. $\sqrt{t^9}$

45. $\sqrt{49u^{11}}$

46. $\sqrt{81v^5}$

47. $\sqrt[4]{s^{13}t^5}$

48. $\sqrt[4]{b^7c^{11}}$

49. $\sqrt{100x^3y}$

50. $\sqrt{64r^5s}$

51. $-\sqrt{27a^7}$

52. $-\sqrt{40z^3}$

53. $-\sqrt{8x^3z^9}$

54. $-\sqrt{12u^5v^5}$

55. $\sqrt{x^{10}y^{11}}$

56. $\sqrt{a^3b^6}$

57. $-\sqrt{2z^5t^2}$

58. $-\sqrt{3p^4q^3}$

59. $\sqrt{49x^2y^7}$

60. $\sqrt{81c^3d^8}$

61. $\sqrt{\dfrac{9}{4}ab^2c^3}$

62. $\sqrt{\dfrac{16}{25}x^2y^4z^3}$

63. $\sqrt{\dfrac{2s}{5t}}$

64. $\sqrt{\dfrac{7z}{11w}}$

65. $\sqrt[3]{\dfrac{a}{4c}}$

66. $\sqrt[3]{\dfrac{u^2}{v}}$

67. $\sqrt{\dfrac{5p^3}{4q^5}}$

68. $\sqrt{\dfrac{16t^7}{3r^3}}$

69. $\sqrt[3]{\dfrac{4w}{9r^2}}$

70. $\sqrt[3]{\dfrac{5h}{4z^2}}$

71. $\sqrt[3]{\dfrac{2t^2}{x^7}}$

72. $\sqrt[3]{\dfrac{y^5}{z^{11}}}$

73. $\sqrt[4]{\dfrac{4s^5}{64st^3}}$

74. $\sqrt[4]{\dfrac{3h^4}{81h^3v^7}}$

For Exercises 75–90, simplify by rationalizing the denominator. Assume that the variables represent positive real numbers.

75. $\dfrac{2}{\sqrt{3}}$

76. $\dfrac{7}{\sqrt{5}}$

77. $\dfrac{x}{\sqrt{5x}}$

78. $\dfrac{a^2}{\sqrt{7a}}$

79. $\dfrac{c^3}{\sqrt{3c}}$

80. $-\dfrac{3z}{\sqrt{z}}$

81. $-\dfrac{5}{\sqrt{ab}}$

82. $\dfrac{\sqrt[3]{7}}{\sqrt[3]{32}}$

83. $\dfrac{\sqrt[3]{2}}{\sqrt[3]{9}}$

84. $\dfrac{\sqrt[4]{7}}{\sqrt[4]{24}}$

85. $\dfrac{\sqrt[4]{6}}{\sqrt[4]{54}}$

86. $\dfrac{2}{\sqrt[3]{x^2}}$

87. $\dfrac{7}{\sqrt[3]{y^5}}$

88. $\dfrac{2m}{\sqrt[4]{m^3}}$

89. $\dfrac{5p^2}{\sqrt[4]{p^7}}$

90. $\dfrac{12b^2}{\sqrt[5]{b^{14}}}$

For Exercises 91–98, simplify. Assume that all variables represent positive real numbers.

91. $\sqrt{\dfrac{48x^2}{7y^3}}$ **92.** $\sqrt{\dfrac{32a^4}{3b^3}}$ **93.** $\dfrac{\sqrt{20x^3y^4z}}{\sqrt{3xy^5z}}$

94. $\dfrac{\sqrt{24w^3s^5t^2}}{\sqrt{3w^5s^4t^3}}$ **95.** $\sqrt[3]{\dfrac{9a^5b^4}{4c^7}}$ **96.** $\sqrt[3]{\dfrac{32x^5y}{z^4}}$

97. $\dfrac{\sqrt[4]{8u^6v^7w^2}}{\sqrt[4]{2u^7v^4w^5}}$ **98.** $\dfrac{\sqrt[4]{3a^5b^2c^2}}{\sqrt[4]{9a^4b^5c^7}}$

For Exercises 99–104, let $D = \sqrt{b^2 - 4ac}$, where a, b, and c are the numbers from the trinomial $ax^2 + bx + c$. Evaluate D for the following trinomials.

99. $x^2 - 3x - 5$ **100.** $x^2 + 4x + 1$ **101.** $2x^2 - x - 1$

102. $-4x^2 + 3x + 6$ **103.** $-3x^2 - x + 2$ **104.** $x^2 - 10x + 4$

For Exercises 105–108, simplify and then approximate to the nearest tenth using 1.414 for $\sqrt{2}$ and 1.732 for $\sqrt{3}$.

105. $\sqrt{50}$ **106.** $\sqrt{27}$ **107.** $\sqrt{147}$ **108.** $\sqrt{300}$

109. The radius r of a spherical weather balloon with volume V is given by

$$r = \sqrt[3]{\dfrac{3V}{4\pi}}$$

What is the radius when the volume is $179\dfrac{2}{3}$ cubic yards? Use $\dfrac{22}{7}$ for π.

110. Find the area of a triangle whose base is $\sqrt{7}$ inches and whose altitude is $3\sqrt{28}$ inches.

REVIEW PROBLEMS

The following exercises review parts of Sections 3.2 and 4.4. Doing these problems will help prepare you for the next section.

Section 3.2

Combine.

111. $4x^2 + 7x^2 - 3x^2$ **112.** $5a^3 - 4a^3 + 9a^3 - a^3$

113. $\dfrac{1}{2}x^2y + \dfrac{7}{4}x^2y$ **114.** $3s^2 - 4s + 1 - (5s^2 - 2s - 4)$

Section 4.4

Combine.

115. $\dfrac{3}{x} - \dfrac{5}{x}$ **116.** $-\dfrac{6z}{x^2r} + \dfrac{5z}{x^2r} - \dfrac{12z}{x^2r}$

117. $\dfrac{4}{2x+1} - \dfrac{7}{2x+1}$

118. $\dfrac{3}{9-4x^2} - \dfrac{2x}{9-4x^2}$

119. $\dfrac{2}{y-3} + \dfrac{3}{y+2} - \dfrac{5y}{y^2-y-6}$

120. $\dfrac{4}{x-4} + \dfrac{x}{4-x}$

E N R I C H M E N T   E X E R C I S E S

For Exercises 1–3, for what values of x will the radical be a real number?

1. $\sqrt{x^2 - 2x - 3}$

2. $\sqrt{2x^2 + 5x - 18}$

3. $\sqrt{x^4 + 2x^2 + 1}$

4. Simplify: $\sqrt{r^2 - 2r + 1}, \quad r \geq 1.$

5. Simplify: $\sqrt{z^2 + 8z + 16}, \quad z \leq -4.$

6. Simplify: $\sqrt{y^4 + 6y^3 + 9y^2}, \quad y \geq 0$

For Exercises 7–9, find each square root.

7. $\sqrt{0.36}$

8. $\sqrt{0.09}$

9. $\sqrt{0.0144}$

10. If k, m, and n are positive integers, show that

$$\sqrt[km]{a^{kn}} = \sqrt[m]{a^n}, \quad \text{where } a \geq 0$$

Use the result of Exercise 10 to simplify the following radical expressions. Assume that all variables represent positive real numbers.

11. $\sqrt[4]{x^2}$

12. $\sqrt[6]{a^8}$

13. $\sqrt[10]{z^{25}}$

14. $\sqrt[8]{a^6 b^{10}}$

15. $\sqrt[9]{x^6 y^{12} z^{27}}$

Answers to Enrichment Exercises are on page A.23.

5.5

The Addition and Subtraction of Radical Expressions

O B J E C T I V E S

▶ *To simplify radical expressions by combining like radicals*

▶ *To combine the sum or difference of radical expressions as a single fraction*

Recall that the terms $4x^2$ and $-12x^2$ are like terms since their variable parts are the same. Similarly, $4\sqrt[3]{5}$ and $-12\sqrt[3]{5}$ are *like radicals* because their radicands and indexes are the same.

DEFINITION

Two radicals are said to be **like or similar radicals** if they have the same index and the same radicand.

For example, $2\sqrt{x}$, $-5\sqrt{x}$, and $\dfrac{3}{2}\sqrt{x}$ are like radicals, and $11\sqrt{7}$ and $6\sqrt{2}$ are not like radicals.

To combine like radicals we use the distributive property. For example, $5\sqrt{2} + 6\sqrt{2} = (5 + 6)\sqrt{2} = 11\sqrt{2}$.

Example 1   Simplify by combining like radicals.

(a) $3\sqrt{5} - 7\sqrt{5} - 8\sqrt{5}$ (b) $16\sqrt[3]{11} + 2\sqrt{7} - \sqrt[3]{11}$

(c) $4\sqrt{3} - 6\sqrt{27}$ (d) $2\sqrt{50} + 14\sqrt{8} - 20\sqrt{18}$

Solution   (a) All three radicals are similar. Using the distributive property,

$$3\sqrt{5} - 7\sqrt{5} - 8\sqrt{5} = (3 - 7 - 8)\sqrt{5} = -12\sqrt{5}$$

(b) The first and third radicals are similar. Therefore, we combine only those two.

$$16\sqrt[3]{11} + 2\sqrt{7} - \sqrt[3]{11} = 16\sqrt[3]{11} - \sqrt[3]{11} + 2\sqrt{7}$$
$$= (16 - 1)\sqrt[3]{11} + 2\sqrt{7}$$
$$= 15\sqrt[3]{11} + 2\sqrt{7}$$

Notice that we cannot combine any further, since $15\sqrt[3]{11}$ and $2\sqrt{7}$ are not like radicals.

(c) At first glance, it appears that we cannot combine the two terms. However, the square root in the second term is not simplified. Therefore, we first simplify, then combine the resulting similar radicals.

$$4\sqrt{3} - 6\sqrt{27} = 4\sqrt{3} - 6\sqrt{9 \cdot 3}$$
$$= 4\sqrt{3} - 6\sqrt{9}\sqrt{3}$$
$$= 4\sqrt{3} - 6 \cdot 3\sqrt{3}$$
$$= 4\sqrt{3} - 18\sqrt{3}$$
$$= -14\sqrt{3}$$

(d) First simplify the square roots, then combine any like radicals.

$$2\sqrt{50} + 14\sqrt{8} - 20\sqrt{18} = 2\sqrt{25 \cdot 2} + 14\sqrt{4 \cdot 2} - 20\sqrt{9 \cdot 2}$$
$$= 2\sqrt{25}\sqrt{2} + 14\sqrt{4}\sqrt{2} - 20\sqrt{9}\sqrt{2}$$

$$= 2 \cdot 5\sqrt{2} + 14 \cdot 2\sqrt{2} - 20 \cdot 3\sqrt{2}$$

$$= 10\sqrt{2} + 28\sqrt{2} - 60\sqrt{2}$$

$$= (10 + 28 - 60)\sqrt{2}$$

$$= -22\sqrt{2} \qquad \blacktriangle$$

N O T E   *In Example 1, we used the distributive property to combine like radicals. We can combine like radicals immediately as we would like terms as in $3x + 5x = 8x$. For example,*

$$3\sqrt{2} + 5\sqrt{2} = 8\sqrt{2}$$

In the next example, we show how to combine like radicals when the radicands involve variables.

Example 2   Simplify.

(a) $4\sqrt{x} - 3\sqrt{x} + 12\sqrt{x}$ **(b)** $5a^3\sqrt[3]{a} - 8a^3\sqrt[3]{a}$

(c) $\sqrt{z^5} + 3z^2\sqrt{z}$ **(d)** $r^2\sqrt{r^3} - 3r\sqrt{4r^5} + 2\sqrt{r^7}$

Solution   **(a)** $4\sqrt{x} - 3\sqrt{x} + 12\sqrt{x} = (4 - 3 + 12)\sqrt{x}$

$$= 13\sqrt{x}$$

(b) $5a^3\sqrt[3]{a} - 8a^3\sqrt[3]{a} = (5a^3 - 8a^3)\sqrt[3]{a}$

$$= -3a^3\sqrt[3]{a}$$

(c) First simplify $\sqrt{z^5}$, then combine.

$$\sqrt{z^5} + 3z^2\sqrt{z} = \sqrt{z^4 \cdot z} + 3z^2\sqrt{z}$$

$$= \sqrt{z^4}\sqrt{z} + 3z^2\sqrt{z}$$

$$= z^2\sqrt{z} + 3z^2\sqrt{z}$$

$$= (z^2 + 3z^2)\sqrt{z}$$

$$= 4z^2\sqrt{z}$$

(d) First, simplify all the radicals before combining them.

$$r^2\sqrt{r^3} - 3r\sqrt{4r^5} + 2\sqrt{r^7} = r^2\sqrt{r^2 \cdot r} - 3r\sqrt{4r^4 \cdot r} + 2\sqrt{r^6 \cdot r}$$

$$= r^2\sqrt{r^2}\sqrt{r} - 3r\sqrt{4}\sqrt{r^4}\sqrt{r} + 2\sqrt{r^6}\sqrt{r}$$

$$= r^2 r\sqrt{r} - 3r \cdot 2r^2\sqrt{r} + 2r^3\sqrt{r}$$

$$= r^3\sqrt{r} - 6r^3\sqrt{r} + 2r^3\sqrt{r}$$

$$= -3r^3\sqrt{r} \qquad \blacktriangle$$

In Example 2 we can see how to simplify radicals first before attempting to add or subtract them. We summarize this method in the following strategy.

> ### STRATEGY
>
> **Strategy for adding or subtracting radicals**
> 1. Put each radical in simplified form.
> 2. Apply the distributive property to combine similar radicals.

Keep in mind that all radicals must be first simplified before we can tell which ones are similar.

Example 3 Combine: $7\sqrt[3]{s^4t^5} - 4s\sqrt[3]{8st^5}$.

Solution We write each radical in simplified form and combine similar terms.

$$7\sqrt[3]{s^4t^5} - 4s\sqrt[3]{8st^5} = 7\sqrt[3]{s^3t^3}\sqrt[3]{st^2} - 4s\sqrt[3]{8t^3}\sqrt[3]{st^2}$$
$$= 7st\sqrt[3]{st^2} - 4s \cdot 2t\sqrt[3]{st^2}$$
$$= 7st\sqrt[3]{st^2} - 8st\sqrt[3]{st^2}$$
$$= -st\sqrt[3]{st^2}$$ ▲

In radical expressions involving fractions, we may be able to combine terms after first simplifying by rationalizing the denominator.

Example 4 Combine.

(a) $2\sqrt{\dfrac{5}{4}} - 10\sqrt{\dfrac{1}{5}} + 3\sqrt{20}$ **(b)** $\dfrac{2x}{\sqrt{x^3}} + \dfrac{3x^2}{\sqrt{x^5}}, \quad x > 0$

Solution **(a)** $2\sqrt{\dfrac{5}{4}} - 10\sqrt{\dfrac{1}{5}} + 3\sqrt{20} = 2\dfrac{\sqrt{5}}{\sqrt{4}} - 10\dfrac{\sqrt{1}}{\sqrt{5}} + 3\sqrt{4 \cdot 5}$

$$= 2\dfrac{\sqrt{5}}{2} - 10\dfrac{\sqrt{1}\sqrt{5}}{\sqrt{5}\sqrt{5}} + 3\sqrt{4}\sqrt{5}$$

$$= \sqrt{5} - 10\dfrac{\sqrt{5}}{5} + 3 \cdot 2\sqrt{5}$$

$$= \sqrt{5} - 2\sqrt{5} + 6\sqrt{5}$$

$$= 5\sqrt{5}$$

(b) $\dfrac{2x}{\sqrt{x^3}} + \dfrac{3x^2}{\sqrt{x^5}} = \dfrac{2x}{\sqrt{x^2 \cdot x}} + \dfrac{3x^2}{\sqrt{x^4 \cdot x}}$

$$= \dfrac{2x}{x\sqrt{x}} + \dfrac{3x^2}{x^2\sqrt{x}}$$

$$= \frac{2}{\sqrt{x}} + \frac{3}{\sqrt{x}}$$

$$= \frac{2 + 3}{\sqrt{x}}$$

$$= \frac{5}{\sqrt{x}}$$

$$= \frac{5\sqrt{x}}{\sqrt{x}\sqrt{x}}$$

$$= \frac{5\sqrt{x}}{x}$$

In algebra, it is sometimes necessary to combine two radical expressions as a single fraction. We illustrate this technique in the next example.

Example 5 Combine as a single fraction and simplify. Assume that the variables represent positive real numbers.

(a) $\dfrac{3}{\sqrt{5}} - \dfrac{2}{\sqrt{5}}$ **(b)** $\dfrac{4}{\sqrt[3]{s}} + \dfrac{7}{\sqrt[3]{s}}$ **(c)** $\dfrac{5}{\sqrt{x}} - 1$

Solution **(a)** $\dfrac{3}{\sqrt{5}} - \dfrac{2}{\sqrt{5}} = \dfrac{3 - 2}{\sqrt{5}}$

$$= \frac{1}{\sqrt{5}}$$

$$= \frac{1 \cdot \sqrt{5}}{\sqrt{5}\sqrt{5}}$$

$$= \frac{\sqrt{5}}{5}$$

(b) $\dfrac{4}{\sqrt[3]{s}} + \dfrac{7}{\sqrt[3]{s}} = \dfrac{4 + 7}{\sqrt[3]{s}}$

$$= \frac{11}{\sqrt[3]{s}}$$

$$= \frac{11\sqrt[3]{s^2}}{\sqrt[3]{s^3}} \qquad \text{Rationalize the denominator.}$$

$$= \frac{11\sqrt[3]{s^2}}{s}$$

(c) To combine these two terms as a single fraction, we first rewrite $\dfrac{5}{\sqrt{x}}$ as $\dfrac{5\sqrt{x}}{x}$.

$$\frac{5}{\sqrt{x}} - 1 = \frac{5\sqrt{x}}{x} - 1$$

$$= \frac{5\sqrt{x}}{x} - \frac{x}{x}$$

$$= \frac{5\sqrt{x} - x}{x} \qquad \blacktriangle$$

L E A R N I N G   A D V A N T A G E   *When doing homework, be sure to read the directions carefully. If it is an odd-numbered problem, compare your answer with the one in the back of the book. If your answer differs, then go back to make sure you have copied the problem correctly.*

E X E R C I S E   S E T   **5.5**

For Exercises 1–26, simplify by combining like radicals. Assume that the variables represent positive real numbers.

1. $2\sqrt{3} - 4\sqrt{3} - \sqrt{3}$

2. $6\sqrt{2} - 12\sqrt{2} + 19\sqrt{2}$

3. $3\sqrt{18} - 2\sqrt{2}$

4. $8\sqrt{6} - 2\sqrt{24}$

5. $3\sqrt{14} - \sqrt{56} + \sqrt{7}$

6. $\sqrt{12} - 7\sqrt{3} + 2\sqrt{6}$

7. $4\sqrt[3]{24} - 20\sqrt[3]{3}$

8. $2\sqrt[3]{32} + \sqrt[3]{108}$

9. $5\sqrt[4]{32} + \sqrt[4]{2} - \sqrt[4]{162}$

10. $\sqrt[5]{64} - 2\sqrt[5]{486} - 3\sqrt[5]{2}$

11. $4\sqrt{c} + 3\sqrt{c}$

12. $7\sqrt{xy} + 5\sqrt{xy}$

13. $3a\sqrt{a^3} - 4a^2\sqrt{a} + 2a$

14. $-3\sqrt{b^5} + b\sqrt{b^3} + \sqrt{b^2}$

15. $10\sqrt{z^7} + 3\sqrt{z^6} + 4z\sqrt{z^4}$

16. $\sqrt{r^{11}} - r^4\sqrt{r^3} + 2\sqrt{r^2}$

17. $10\sqrt{a^5} - 4a\sqrt{9a^3}$

18. $7\sqrt{16b^7} - 15b^2\sqrt{4b^3}$

19. $6\sqrt{\dfrac{12}{4}} + 21\sqrt{\dfrac{3}{9}}$

20. $6\sqrt{\dfrac{7}{4}} - 2\sqrt{\dfrac{28}{4}}$

21. $y\sqrt[3]{x^4y} - x\sqrt[3]{xy^4}$

22. $5a\sqrt[3]{a^5z^4} + 7za\sqrt[3]{a^5z}$

23. $2u^2\sqrt[3]{16v^8} - v^2\sqrt[3]{2u^6v^2}$

24. $\sqrt[4]{81z^{11}} - z\sqrt[4]{16z^7}$

25. $4xy\sqrt[3]{54xy^3} + y^2\sqrt[3]{16x^3y}$

26. $\sqrt[5]{243a^6b} + 2\sqrt[5]{32a^6b}$

For Exercises 27–54, combine as a single fraction and simplify. Assume that the variables represent positive real numbers.

27. $\dfrac{1}{\sqrt{3}} - \dfrac{5}{\sqrt{3}}$

28. $\dfrac{2}{\sqrt{5}} + \dfrac{7}{\sqrt{5}}$

29. $\dfrac{3\sqrt{2}}{\sqrt{7}} + \dfrac{4\sqrt{8}}{\sqrt{7}}$

30. $\dfrac{5\sqrt{3}}{\sqrt{11}} - \dfrac{2\sqrt{27}}{\sqrt{11}}$

31. $\sqrt{\dfrac{8}{3}} - \sqrt{\dfrac{18}{6}}$

32. $\dfrac{\sqrt{12}}{4} + \dfrac{\sqrt{27}}{6}$

33. $\dfrac{x\sqrt{x^3}}{12} - \dfrac{2\sqrt{x^5}}{9}$

34. $\dfrac{c\sqrt{c^5}}{20} + \dfrac{\sqrt{c^7}}{15}$

35. $\dfrac{\sqrt{3}}{\sqrt{2}} + 1$

36. $\dfrac{\sqrt{6}}{\sqrt{7}} - 2$

37. $\dfrac{\sqrt{8}}{\sqrt{3}} + \sqrt{6}$

38. $\dfrac{\sqrt{12}}{\sqrt{5}} - \sqrt{15}$

39. $7\sqrt{\dfrac{5}{16}} - 3\sqrt{20}$

40. $2\sqrt{\dfrac{3}{25}} - 9\sqrt{\dfrac{15}{5}}$

41. $\dfrac{1}{2}\sqrt{\dfrac{x^3}{x^5}} + \sqrt{\dfrac{x^5}{x^7}}$

42. $3\sqrt{\dfrac{a^4}{a^8}} + \dfrac{1}{4}\sqrt{\dfrac{a^5}{a^9}}$

43. $\dfrac{2\sqrt{3}}{\sqrt{8}} + \dfrac{2\sqrt{2}}{\sqrt{12}}$

44. $\dfrac{3\sqrt{12}}{\sqrt{18}} - \dfrac{10\sqrt{3}}{\sqrt{50}}$

45. $\dfrac{x^3\sqrt{y}}{3\sqrt{x^2y}} - \dfrac{x\sqrt{y^3}}{3\sqrt{xy^2}}$

46. $\dfrac{c^5\sqrt{d}}{\sqrt{cd^3}} - \dfrac{c\sqrt{d^5}}{\sqrt{c^3d}}$

47. $\dfrac{2}{\sqrt[3]{4}} + \dfrac{5}{\sqrt[3]{4}}$

48. $\dfrac{3}{\sqrt[4]{8}} - \dfrac{2}{\sqrt[4]{8}}$

49. $\sqrt[4]{\dfrac{1}{27}} - \sqrt[4]{\dfrac{16}{27}}$

50. $7\sqrt[3]{\dfrac{8}{9}} - 6\sqrt[3]{\dfrac{1}{9}}$

51. $\dfrac{2}{\sqrt[3]{a}} - \dfrac{5}{\sqrt[3]{a}}$

52. $\dfrac{1}{\sqrt[3]{b^2}} + \dfrac{4}{\sqrt[3]{b^2}}$

53. $\dfrac{12}{\sqrt[4]{x^3}} - \dfrac{8}{\sqrt[4]{x^3}}$

54. $\dfrac{5}{\sqrt[5]{z^2}} + \dfrac{4}{\sqrt[5]{z^2}}$

For Exercises 55–62, evaluate $\dfrac{-b - \sqrt{b^2 - 4ac}}{2a}$ for each trinomial $ax^2 + bx + c$.

55. $x^2 - 3x + 2$

56. $x^2 - x - 6$

57. $x^2 - 8$

58. $x^2 - 12$

59. $4x^2 - 4x - 17$

60. $x^2 - 4x - 8$

61. $4x^2 - 12x + 9$

62. $9x^2 - 6x + 1$

REVIEW PROBLEMS

The following exercises review parts of Section 3.3. Doing these problems will help prepare you for the next section.

Multiply.

63. $2ab(3az - 4bz)$

64. $x^3y^2(3xy - 2x^2y - 5)$

65. $(2x + 3)(x - 1)$

66. $(5b + d)(7b - 3d)$

67. $(2a - 2b)^2$

68. $(7h - 5k)^2$

69. $(5p - 4q)(p - q)$

70. $(3c^2 + 2z^2)^2$

71. $(4v - y^2)(2v + 5y^2)$

72. $2a^2z(6z^2 + a^2)(2z^2 - a^2)$

E N R I C H M E N T   E X E R C I S E S

For Exercises 1–4, simplify.

1. $\dfrac{\dfrac{4\sqrt{2}}{\sqrt{7}} + \dfrac{3}{\sqrt{7}}}{\dfrac{3}{\sqrt{28}} + \dfrac{2\sqrt{2}}{\sqrt{7}}}$

2. $\dfrac{\dfrac{1}{\sqrt{x}} - \dfrac{3}{\sqrt{x}}}{\dfrac{2}{\sqrt{x}} + \dfrac{7}{\sqrt{x}}}, \; x > 0.$

3. $\dfrac{\dfrac{2\sqrt{a}}{\sqrt{b}} - \dfrac{2\sqrt{b}}{\sqrt{a}}}{\dfrac{\sqrt{b}}{\sqrt{a}} - \dfrac{\sqrt{a}}{\sqrt{b}}}, \;$ a and b are positive with $a \neq b$.

4. $\dfrac{\dfrac{1}{2\sqrt{c}} - \dfrac{5}{3\sqrt{c}}}{\dfrac{2}{3\sqrt{c}} + \dfrac{1}{2\sqrt{c}}}, \; c > 0.$

For Exercises 5–8, rewrite each expression without radicals in the denominator, then simplify. (*Hint:* Make use of the fact that $(a + b)(a - b) = a^2 - b^2$.)

5. $\dfrac{1}{\sqrt{7} + 1}$

6. $\dfrac{2}{\sqrt{3} - \sqrt{2}}$

7. $\dfrac{2x}{\sqrt{x} - 1}, \; x > 0$ and $x \neq 1$

8. $\dfrac{a - b}{\sqrt{a} + \sqrt{b}}, \;$ a and b positive with $a \neq b$

For Exercises 9 and 10, simplify. Assume that n is an integer greater than 2, x and y represent positive real numbers.

9. $\sqrt[n]{x^{n+1}y^{n+2}} - 3xy\sqrt[n]{xy^2}$

10. $\sqrt[n]{x^{3n+2}y^{2n+1}} + x^2y\sqrt[n]{2^n x^{n+2}y^{n+1}}$

Answers to Enrichment Exercises are on page A.24.

5.6

Products and Quotients of Radical Expressions

OBJECTIVES

▶ *To multiply expressions that contain radicals*

▶ *To rationalize denominators containing sums and differences of radicals*

▶ *To factor expressions involving radicals*

Recall from Chapter 4, we used the distributive property to multiply algebraic expressions,

$$a(b + c) = ab + ac \qquad \text{and} \qquad (b + c)a = ba + ca$$

We may use the distributive property to multiply expressions that contain radicals. Keep in mind that radicands that contain variables are assumed to be nonnegative.

Example 1   Multiply and then simplify.

(a) $\sqrt{3}(14 + \sqrt{2})$ **(b)** $\sqrt{5}(\sqrt{15} - \sqrt{2})$

(c) $(\sqrt{6} - \sqrt{3})\sqrt{3}$ **(d)** $\sqrt{a}(\sqrt{b} + 5\sqrt{a})$

(e) $\sqrt[3]{r}(\sqrt[3]{r^5} - 3\sqrt[3]{r^2})$

Solution   **(a)** $\sqrt{3}(14 + \sqrt{2}) = \sqrt{3}(14) + \sqrt{3}\sqrt{2}$

$$= 14\sqrt{3} + \sqrt{3 \cdot 2}$$

$$= 14\sqrt{3} + \sqrt{6}$$

(b) $\sqrt{5}(\sqrt{15} - \sqrt{2}) = \sqrt{5}\sqrt{15} - \sqrt{5}\sqrt{2}$

$$= \sqrt{5}\sqrt{5 \cdot 3} - \sqrt{5 \cdot 2}$$

$$= \sqrt{5}\sqrt{5}\sqrt{3} - \sqrt{10}$$

$$= 5\sqrt{3} - \sqrt{10} \qquad \sqrt{a}\sqrt{a} = a.$$

(c) $(\sqrt{6} - \sqrt{3})\sqrt{3} = \sqrt{6}\sqrt{3} - \sqrt{3}\sqrt{3}$

$$= \sqrt{3 \cdot 2}\sqrt{3} - 3$$

$$= \sqrt{3}\sqrt{2}\sqrt{3} - 3$$

$$= 3\sqrt{2} - 3$$

(d) $\sqrt{a}(\sqrt{b} + 5\sqrt{a}) = \sqrt{a}\sqrt{b} + \sqrt{a}(5\sqrt{a})$

$$= \sqrt{ab} + 5a$$

(e) $\sqrt[3]{r}(\sqrt[3]{r^5} - 3\sqrt[3]{r^2}) = \sqrt[3]{r}\sqrt[3]{r^5} - 3\sqrt[3]{r}\sqrt[3]{r^2}$

$$= \sqrt[3]{r^6} - 3\sqrt[3]{r^3}$$

$$= r^2 - 3r$$

▲

Example 2   Multiply and then simplify.

(a) $(5\sqrt{2} - \sqrt{3})(\sqrt{2} + 4\sqrt{3})$ (b) $(3\sqrt{r} + \sqrt{t})^2$

(c) $(\sqrt{5} + \sqrt{3})(\sqrt{5} - \sqrt{3})$ (d) $(\sqrt{x} + 4)(\sqrt{x} - 4)$

Solution   (a) We expand the product using the distributive property.

$$\begin{aligned}
(5\sqrt{2} - \sqrt{3})(\sqrt{2} + 4\sqrt{3}) &= 5\sqrt{2}(\sqrt{2} + 4\sqrt{3}) - \sqrt{3}(\sqrt{2} + 4\sqrt{3}) \\
&= 5\sqrt{2}\sqrt{2} + 5\sqrt{2}(4\sqrt{3}) - \sqrt{3}\sqrt{2} - \sqrt{3}(4\sqrt{3}) \\
&= 5 \cdot 2 + 20\sqrt{2}\sqrt{3} - \sqrt{3}\sqrt{2} - 4\sqrt{3}\sqrt{3} \\
&= 10 + 20\sqrt{6} - \sqrt{6} - 4 \cdot 3 \\
&= 10 + 19\sqrt{6} - 12 \\
&= -2 + 19\sqrt{6}
\end{aligned}$$

(b) Recall the formula for squaring a binomial $(a + b)^2 = a^2 + 2ab + b^2$. Setting $a = 3\sqrt{r}$ and $b = \sqrt{t}$,

$$\begin{aligned}
(3\sqrt{r} + \sqrt{t})^2 &= (3\sqrt{r})^2 + 2 \cdot 3\sqrt{r}\sqrt{t} + (\sqrt{t})^2 \\
&= 3^2(\sqrt{r})^2 + 6\sqrt{r}\sqrt{t} + (\sqrt{t})^2 \qquad (\sqrt{a})^2 = (\sqrt{a})(\sqrt{a}) = a. \\
&= 9r + 6\sqrt{rt} + t
\end{aligned}$$

(c) In this part, we make use of the special product formula $(a + b)(a - b) = a^2 - b^2$, where a is $\sqrt{5}$ and b is $\sqrt{3}$.

$$\begin{aligned}
(\sqrt{5} + \sqrt{3})(\sqrt{5} - \sqrt{3}) &= (\sqrt{5})^2 - (\sqrt{3})^2 \\
&= 5 - 3 \\
&= 2
\end{aligned}$$

(d) Using the formula $(a + b)(a - b) = a^2 - b^2$, we have

$$\begin{aligned}
(\sqrt{x} + 4)(\sqrt{x} - 4) &= (\sqrt{x})^2 - 4^2 \\
&= x - 16 \qquad \blacktriangle
\end{aligned}$$

A way to rationalize a denominator that is the sum or difference of terms involving square roots is suggested in Parts (c) and (d) of Example 2. The first step is to form the *conjugate* of the denominator.

D E F I N I T I O N

The two binomial expressions, $a + b$ and $a - b$, are called **conjugates** of each other.

Some examples of binomial expressions and their conjugates are in the following table.

Expression	Its conjugate
$2 + \sqrt{5}$	$2 - \sqrt{5}$
$3 + \sqrt{10}$	$3 - \sqrt{10}$
$\sqrt{x} + 1$	$\sqrt{x} - 1$
$2s + 4\sqrt{t}$	$2s - 4\sqrt{t}$
$\sqrt{5} - \sqrt{12}$	$\sqrt{5} + \sqrt{12}$
$\sqrt{x} - 4$	$\sqrt{x} + 4$
$\sqrt{y} - \sqrt{2z}$	$\sqrt{y} + \sqrt{2z}$
$\sqrt{3u} - \sqrt{6v}$	$\sqrt{3u} + \sqrt{6v}$

Since the product of conjugates, $(a + b)(a - b)$, is the difference of two squares, $a^2 - b^2$, we have a means of rationalizing a denominator that involves the sum or difference of square roots.

Example 3   Multiply the given expression by its conjugate and simplify the result.

(a) $\sqrt{3} + 2$ (b) $\sqrt{6} - \sqrt{2}$

(c) $\sqrt{x} + \sqrt{y}$ (d) $\sqrt{2a} - 5\sqrt{b}$

Solution   (a) The conjugate of $\sqrt{3} + 2$ is $\sqrt{3} - 2$. Next, we multiply $\sqrt{3} + 2$ by its conjugate $\sqrt{3} - 2$ by using the special product formula $(a + b)(a - b) = a^2 - b^2$.

$$(\sqrt{3} + 2)(\sqrt{3} - 2) = (\sqrt{3})^2 - 2^2 = 3 - 4 = -1$$

(b) The conjugate of $\sqrt{6} - \sqrt{2}$ is $\sqrt{6} + \sqrt{2}$, and

$$(\sqrt{6} - \sqrt{2})(\sqrt{6} + \sqrt{2}) = (\sqrt{6})^2 - (\sqrt{2})^2 = 6 - 2 = 4$$

(c) The conjugate of $\sqrt{x} + \sqrt{y}$ is $\sqrt{x} - \sqrt{y}$, and

$$(\sqrt{x} + \sqrt{y})(\sqrt{x} - \sqrt{y}) = (\sqrt{x})^2 - (\sqrt{y})^2 = x - y$$

(d) The conjugate of $\sqrt{2a} - 5\sqrt{b}$ is $\sqrt{2a} + 5\sqrt{b}$, and

$$(\sqrt{2a} - 5\sqrt{b})(\sqrt{2a} + 5\sqrt{b}) = (\sqrt{2a})^2 - (5\sqrt{b})^2 = 2a - 25b$$

Now, consider a quotient such as $\dfrac{5}{\sqrt{3} + 1}$. In order to rationalize the denominator, we make use of the fact that

$$\frac{a}{b} = \frac{a \cdot c}{b \cdot c}$$

where b and c are nonzero.

That is, we may multiply the numerator and denominator by the same non-zero quantity c and the quotient remains unchanged in value. In particular, we *let c be the conjugate of the denominator.* Therefore, setting $c = \sqrt{3} - 1$,

$$\frac{5}{\sqrt{3} + 1} = \frac{5(\sqrt{3} - 1)}{(\sqrt{3} + 1)(\sqrt{3} - 1)}$$

$$= \frac{5(\sqrt{3} - 1)}{(\sqrt{3})^2 - 1^2}$$

$$= \frac{5(\sqrt{3} - 1)}{3 - 1}$$

$$= \frac{5(\sqrt{3} - 1)}{2} \quad \text{or} \quad \frac{5\sqrt{3} - 5}{2}$$

Either $\dfrac{5(\sqrt{3} - 1)}{2}$ or $\dfrac{5\sqrt{3} - 5}{2}$ is an acceptable answer.

Example 4 Rationalize the denominator and simplify.

(a) $\dfrac{3}{\sqrt{5} + 4}$

(b) $\dfrac{\sqrt{5}}{3\sqrt{3} - 2\sqrt{5}}$

(c) $\dfrac{\sqrt{x}}{\sqrt{x} - \sqrt{y}}$

(d) $\dfrac{2\sqrt{r} - 1}{2\sqrt{r} + 1}$

Solution (a) The conjugate of $\sqrt{5} + 4$ is $\sqrt{5} - 4$. Therefore, we multiply numerator and denominator by $\sqrt{5} - 4$ and simplify.

$$\frac{3}{\sqrt{5} + 4} = \frac{3(\sqrt{5} - 4)}{(\sqrt{5} + 4)(\sqrt{5} - 4)}$$

$$= \frac{3(\sqrt{5} - 4)}{(\sqrt{5})^2 - 4^2}$$

$$= \frac{3(\sqrt{5} - 4)}{5 - 16}$$

$$= \frac{3(\sqrt{5} - 4)}{-11}$$

$$= -\frac{3(\sqrt{5} - 4)}{11}$$

(b) The conjugate of $3\sqrt{3} - 2\sqrt{5}$ is $3\sqrt{3} + 2\sqrt{5}$. Therefore,

$$\frac{\sqrt{5}}{3\sqrt{3} - 2\sqrt{5}} = \frac{\sqrt{5}(3\sqrt{3} + 2\sqrt{5})}{(3\sqrt{3} - 2\sqrt{5})(3\sqrt{3} + 2\sqrt{5})}$$

$$= \frac{3\sqrt{5}\sqrt{3} + 2\sqrt{5}\sqrt{5}}{(3\sqrt{3})^2 - (2\sqrt{5})^2}$$

$$= \frac{3\sqrt{15} + 2 \cdot 5}{9 \cdot 3 - 4 \cdot 5} \qquad (ab)^2 = a^2 b^2.$$

$$= \frac{3\sqrt{15} + 10}{27 - 20}$$

$$= \frac{3\sqrt{15} + 10}{7}$$

(c) The conjugate of $\sqrt{x} - \sqrt{y}$ is $\sqrt{x} + \sqrt{y}$. Therefore,

$$\frac{\sqrt{x}}{\sqrt{x} - \sqrt{y}} = \frac{\sqrt{x}(\sqrt{x} + \sqrt{y})}{(\sqrt{x} - \sqrt{y})(\sqrt{x} + \sqrt{y})}$$

$$= \frac{\sqrt{x}\sqrt{x} + \sqrt{x}\sqrt{y}}{(\sqrt{x})^2 - (\sqrt{y})^2}$$

$$= \frac{x + \sqrt{xy}}{x - y}$$

(d) The conjugate of $2\sqrt{r} + 1$ is $2\sqrt{r} - 1$. Therefore,

$$\frac{2\sqrt{r} - 1}{2\sqrt{r} + 1} = \frac{(2\sqrt{r} - 1)(2\sqrt{r} - 1)}{(2\sqrt{r} + 1)(2\sqrt{r} - 1)}$$

$$= \frac{(2\sqrt{r} - 1)^2}{(2\sqrt{r})^2 - 1^2}$$

$$= \frac{4r - 4\sqrt{r} + 1}{4r - 1} \qquad \blacktriangle$$

We can rationalize a denominator that is the difference or sum of cube roots by using either

$$(a - b)(a^2 + ab + b^2) = a^3 - b^3$$

or

$$(a + b)(a^2 - ab + b^2) = a^3 + b^3$$

Example 5   Rationalize the denominator and simplify:

$$\frac{7}{\sqrt[3]{x} - \sqrt[3]{c}}$$

Solution   Letting $\sqrt[3]{x} = a$ and $\sqrt[3]{c} = b$ in the formula

$$(a - b)(a^2 + ab + b^2) = a^3 - b^3$$

we multiply the numerator and denominator of $\dfrac{7}{\sqrt[3]{x} - \sqrt[3]{c}}$ by $a^2 + ab + b^2$.

$$\frac{7}{\sqrt[3]{x} - \sqrt[3]{c}} = \frac{7[(\sqrt[3]{x})^2 + \sqrt[3]{x}\sqrt[3]{c} + (\sqrt[3]{c})^2]}{(\sqrt[3]{x} - \sqrt[3]{c})[(\sqrt[3]{x})^2 + \sqrt[3]{x}\sqrt[3]{c} + (\sqrt[3]{c})^2]}$$

$$= \frac{7(\sqrt[3]{x^2} + \sqrt[3]{xc} + \sqrt[3]{c^2})}{(\sqrt[3]{x})^3 - (\sqrt[3]{c})^3}$$

$$= \frac{7(\sqrt[3]{x^2} + \sqrt[3]{xc} + \sqrt[3]{c^2})}{x - c} \qquad \blacktriangle$$

Recall that we also used the distributive property to factor algebraic expressions.

$$ab + ac = a(b + c) \qquad \text{and} \qquad ba + ca = (b + c)a$$

Example 6   Factor completely.

(a) $4\sqrt{3} - 12\sqrt{5}$ **(b)** $5\sqrt{x} + 20x^2\sqrt{x}$
(c) $21z\sqrt{z^7} - 15z^2\sqrt{z^3}$

Solution   **(a)** $4\sqrt{3} - 12\sqrt{5} = 4(\sqrt{3} - 3\sqrt{5})$ **(b)** $5\sqrt{x} + 20x^2\sqrt{x} = 5\sqrt{x}(1 + 4x^2)$

(c) We first simplify the two square roots in the expression, then factor.

$$21z\sqrt{z^7} - 15z^2\sqrt{z^3} = 21z\sqrt{z^6 \cdot z} - 15z^2\sqrt{z^2 \cdot z}$$

$$= 21z \cdot z^3\sqrt{z} - 15z^2 z\sqrt{z}$$

$$= 21z^4\sqrt{z} - 15z^3\sqrt{z}$$

$$= 3z^3\sqrt{z}(7z - 5) \qquad \blacktriangle$$

Example 7   Simplify.

(a) $\dfrac{2\sqrt{3} - 10\sqrt{2}}{8}$ **(b)** $\dfrac{3\sqrt{5} + 6\sqrt{10}}{9\sqrt{2}}$ **(c)** $\dfrac{6\sqrt{x} - 18\sqrt{x^3}}{3\sqrt{x}}$, $x > 0$

Solution   **(a)** We first factor the numerator, then divide both numerator and denominator by any common factors.

$$\frac{2\sqrt{3} - 10\sqrt{2}}{8} = \frac{2(\sqrt{3} - 5\sqrt{2})}{8} = \frac{\sqrt{3} - 5\sqrt{2}}{4}$$

(b) $\dfrac{3\sqrt{5} + 6\sqrt{10}}{9\sqrt{2}} = \dfrac{3(\sqrt{5} + 2\sqrt{10})}{9\sqrt{2}}$

$$= \frac{\sqrt{5} + 2\sqrt{10}}{3\sqrt{2}}$$

$$= \frac{(\sqrt{5} + 2\sqrt{10})(\sqrt{2})}{3\sqrt{2}(\sqrt{2})}$$

$$= \frac{\sqrt{5}\sqrt{2} + 2\sqrt{10}\sqrt{2}}{3 \cdot 2}$$

$$= \frac{\sqrt{10} + 2\sqrt{2} \cdot 5\sqrt{2}}{6}$$

$$= \frac{\sqrt{10} + 2\sqrt{2}\sqrt{5}\sqrt{2}}{6}$$

$$= \frac{\sqrt{10} + 2 \cdot 2\sqrt{5}}{6}$$

$$= \frac{\sqrt{10} + 4\sqrt{5}}{6}$$

(c) $\dfrac{6\sqrt{x} - 18\sqrt{x^3}}{3\sqrt{x}} = \dfrac{6\sqrt{x} - 18\sqrt{x^2 \cdot x}}{3\sqrt{x}}$

$$= \frac{6\sqrt{x} - 18x\sqrt{x}}{3\sqrt{x}}$$

$$= \frac{6\sqrt{x}(1 - 3x)}{3\sqrt{x}}$$

$$= \frac{\overset{2}{\cancel{6}\cancel{\sqrt{x}}}(1 - 3x)}{\cancel{3}\cancel{\sqrt{x}}}$$

$$= 2(1 - 3x) \qquad \blacktriangle$$

L E A R N I N G   A D V A N T A G E   *Be a contributor to the class. Ask questions and take a genuine interest in class activity. By becoming involved, you will learn more from the time spent in class. If you are a passive spectator not only will class time move slowly but also the time will have been inefficiently used.*

E X E R C I S E   S E T   5 . 6

For Exercises 1–46, multiply and then simplify. Assume that the variables represent positive real numbers.

1. $\sqrt{2}(\sqrt{5} + \sqrt{11})$ **2.** $\sqrt{7}(\sqrt{2} + \sqrt{3})$

3. $(\sqrt{3} - 2\sqrt{5})\sqrt{13}$ **4.** $(2\sqrt{2} - 4\sqrt{7})\sqrt{3}$

5. $\sqrt{2}(\sqrt{2} - 3)$ **6.** $\sqrt{5}(1 - 6\sqrt{5})$

7. $(\sqrt{15} - 1)\sqrt{3}$ **8.** $(\sqrt{12} + 3)\sqrt{3}$

9. $\sqrt{12}(\sqrt{3} + \sqrt{6})$

10. $\sqrt{18}(\sqrt{9} - \sqrt{2})$

11. $\sqrt{z}(\sqrt{y} - 3\sqrt{z})$

12. $\sqrt{b}(\sqrt{b} - 2\sqrt{c})$

13. $\sqrt{x}(\sqrt{x} - 2\sqrt{x^3})$

14. $2\sqrt{b}(\sqrt{b^5} + 3\sqrt{b})$

15. $\sqrt[3]{z}(\sqrt[3]{z^2} - \sqrt[3]{z^5})$

16. $\sqrt[3]{2r}(\sqrt[3]{4r^8} + 1)$

17. $\sqrt[4]{8a^3b^2}(\sqrt[4]{2ab^6} + \sqrt[4]{a^2b^5})$

18. $\sqrt[4]{9x^2y^3}(2\sqrt[4]{x^6y^5} - \sqrt[4]{9x^2y})$

19. $7\sqrt{2u}(3\sqrt{8u} - 2\sqrt{2u})$

20. $3\sqrt{10z}(\sqrt{2z} + 4\sqrt{5z})$

21. $(3\sqrt{5} - 2)(2\sqrt{5} + 1)$

22. $(4\sqrt{2} + 3)(\sqrt{2} + 1)$

23. $(\sqrt{r} + 1)(2\sqrt{r} + 3)$

24. $(\sqrt{a} - 2)(3\sqrt{a} + 1)$

25. $(\sqrt{uv} + 4)(\sqrt{uv} + 2)$

26. $(\sqrt{xy} - 1)(2\sqrt{xy} + 7)$

27. $(\sqrt[3]{x^2} - 1)(3\sqrt[3]{x^2} - 4)$

28. $(2\sqrt[3]{4b^2} + 3)(\sqrt[3]{4b^2} - 1)$

29. $(\sqrt{3} - \sqrt{2})^2$

30. $(\sqrt{5} + \sqrt{6})^2$

31. $(\sqrt{x} + 2\sqrt{z})^2$

32. $(3\sqrt{a} - 4\sqrt{b})^2$

33. $(\sqrt{3} + \sqrt{5})(\sqrt{3} - \sqrt{5})$

34. $(\sqrt{7} + \sqrt{2})(\sqrt{7} - \sqrt{2})$

35. $(\sqrt[3]{4} + 1)(\sqrt[3]{4} - 1)$

36. $(3 - \sqrt[3]{9})(3 + \sqrt[3]{9})$

37. $(\sqrt{x} - 1)(\sqrt{x} + 1)$

38. $(\sqrt{r} - 5)(\sqrt{r} + 5)$

39. $(\sqrt[3]{9y} + 1)(\sqrt[3]{9y} - 1)$

40. $(\sqrt[3]{4a} - 3)(\sqrt[3]{4a} + 3)$

41. $(2\sqrt{7} - 3\sqrt{6})(2\sqrt{7} + 3\sqrt{6})$

42. $(5\sqrt{5} - 4\sqrt{2})(5\sqrt{5} + 4\sqrt{2})$

43. $\sqrt{6}\left(\dfrac{\sqrt{3}}{\sqrt{2}} - \dfrac{\sqrt{6}}{6}\right)$

44. $\sqrt{15}\left(\dfrac{\sqrt{5}}{\sqrt{3}} + \dfrac{\sqrt{3}}{\sqrt{5}}\right)$

45. $\sqrt{ab}\left(\dfrac{\sqrt{b}}{\sqrt{a}} + \dfrac{\sqrt{ab}}{4a}\right)$

46. $\sqrt{rs}\left(\dfrac{\sqrt{r}}{\sqrt{s}} - \dfrac{\sqrt{rs}}{3s}\right)$

For Exercises 47–54, multiply the given expression by its conjugate and simplify the result. Assume that the variables represent nonnegative real numbers.

47. $\sqrt{3} - 4$

48. $\sqrt{7} - 9$

49. $2\sqrt{11} + \sqrt{5}$

50. $\sqrt{6} + 4\sqrt{7}$

51. $5\sqrt{x} + 2\sqrt{y}$

52. $3\sqrt{a} + 3\sqrt{z}$

53. $7\sqrt{xy} - 4\sqrt{z}$

54. $6\sqrt{ab} - 3\sqrt{bc}$

For Exercises 55–78, rationalize the denominator and simplify. Assume that the variables represent positive real numbers.

55. $\dfrac{2}{\sqrt{3} - \sqrt{4}}$

56. $\dfrac{1}{\sqrt{7} - 2}$

57. $\dfrac{r}{\sqrt{r} + 1}$

58. $\dfrac{2z}{\sqrt{z}+3}$

59. $\dfrac{\sqrt{st}}{\sqrt{s}+\sqrt{t}}$

60. $\dfrac{\sqrt{cd}}{\sqrt{c}+2\sqrt{d}}$

61. $\dfrac{\sqrt{6}-2}{1-\sqrt{6}}$

62. $\dfrac{5-\sqrt{3}}{\sqrt{3}-4}$

63. $\dfrac{3\sqrt{2}-1}{\sqrt{2}+3}$

64. $\dfrac{4\sqrt{3}+3}{\sqrt{3}+1}$

65. $\dfrac{3+6\sqrt{3}}{\sqrt{3}+\sqrt{6}}$

66. $\dfrac{4-3\sqrt{5}}{\sqrt{5}+\sqrt{2}}$

67. $\dfrac{\sqrt{s}+2}{1+\sqrt{s}}$

68. $\dfrac{3+\sqrt{t}}{2\sqrt{t}+3}$

69. $\dfrac{2-\sqrt{xy}}{\sqrt{x}+\sqrt{y}}$

70. $\dfrac{\sqrt{b}-\sqrt{a}}{2\sqrt{a}+3\sqrt{b}}$

71. $\dfrac{\sqrt{r}-\sqrt{k}}{\sqrt{r}+4\sqrt{k}}$

72. $\dfrac{\sqrt{u}+2\sqrt{v}}{2\sqrt{u}+\sqrt{v}}$

73. $\dfrac{1}{\sqrt[3]{a}-\sqrt[3]{4}}$

74. $-\dfrac{2}{\sqrt[3]{z}+\sqrt[3]{6}}$

75. $\dfrac{y+5}{\sqrt[3]{y}+\sqrt[3]{5}}$

76. $\dfrac{x-4}{\sqrt[3]{x}-\sqrt[3]{4}}$

77. $\dfrac{2s^2-3s+1}{\sqrt[3]{2s}-1}$

78. $\dfrac{3z^2-4z-4}{\sqrt[3]{3z}+\sqrt[3]{2}}$

For Exercises 79–88, factor completely. Assume that the variables represent nonnegative real numbers.

79. $6-3\sqrt{5}$

80. $40\sqrt{3}+4$

81. $3x^2+15x\sqrt{x}$

82. $21t-7t\sqrt{t}$

83. $a^3\sqrt{a}-3\sqrt{a^5}$

84. $2\sqrt{s^3}-10\sqrt{s}$

85. $y\sqrt{y^3}-y^2\sqrt{y^3}$

86. $2v\sqrt{v^5}+v^2\sqrt{v^5}$

87. $r\sqrt[3]{r^2}-5r\sqrt[3]{r^8}$

88. $2t^2\sqrt[4]{t^5}+7t^2\sqrt[4]{t^9}$

For Exercises 89–100, simplify. Assume that the variables represent positive real numbers.

89. $\dfrac{3\sqrt{5}-15\sqrt{7}}{27}$

90. $\dfrac{14\sqrt{3}+28\sqrt{2}}{42}$

91. $\dfrac{18\sqrt{10}+4\sqrt{15}}{6\sqrt{5}}$

92. $\dfrac{21\sqrt{6}-42\sqrt{14}}{9\sqrt{2}}$

93. $\dfrac{28\sqrt{z^3}-7\sqrt{z}}{14\sqrt{z}}$

94. $\dfrac{8\sqrt{c}-12\sqrt{c^5}}{16\sqrt{c}}$

95. $\dfrac{44x\sqrt{x} + 77\sqrt{x^7}}{11\sqrt{x^3}}$

96. $\dfrac{64\sqrt{r^5} - 16\sqrt{r^3}}{4r^3\sqrt{r}}$

97. $\dfrac{7\sqrt[3]{x^4} + 14x\sqrt[3]{x}}{21\sqrt[3]{x^7}}$

98. $\dfrac{12\sqrt[3]{a^5} - 15a\sqrt[3]{a^2}}{18\sqrt[3]{a^8}}$

99. $\dfrac{2\sqrt[4]{a^5} - 6\sqrt[4]{a^9}}{8\sqrt[4]{a^{13}}}$

100. $\dfrac{3\sqrt[5]{v^{12}} - 9\sqrt[5]{v^7}}{12\sqrt[5]{v^{17}}}$

101. Are the conjugate of $2 + \sqrt{3}$ and the opposite of $2 + \sqrt{3}$ the same? Explain.

102. Are the conjugate of $1 - \sqrt{5}$ and the opposite of $1 - \sqrt{5}$ the same? Explain.

REVIEW PROBLEMS

The following exercises review parts of Section 3.8. Doing these problems will help prepare you for the next section.

Solve the following equations.

103. $x(x - 3) = 0$

104. $(2t - 1)(t + 4) = 0$

105. $3y^2 - 4y + 1 = 0$

106. $2x^2 - 7x + 6 = 0$

107. $a^2 - 2a + 1 = 0$

108. $9b^2 + 6b + 1 = 0$

109. $6z^3 - 13z^2 + 6z = 0$

110. $4t^4 - 4t^3 - 35t^2 = 0$

111. $x^4 - 1 = 0$

112. $y^4 - 16 = 0$

E N R I C H M E N T   E X E R C I S E S

For Exercises 1–3, simplify.

1. $\dfrac{\dfrac{10}{\sqrt{2}} - \dfrac{5}{\sqrt{3}}}{\dfrac{2}{\sqrt{3}} - \dfrac{4}{\sqrt{2}}}$

2. $\dfrac{\dfrac{3}{\sqrt{5}} + \dfrac{2}{\sqrt{10}}}{\dfrac{1}{\sqrt{10}} + \dfrac{1}{\sqrt{5}}}$

3. $\dfrac{\dfrac{1}{x} - \dfrac{3}{2x^2}}{-\dfrac{2}{x\sqrt{x}} + \dfrac{3}{x^2\sqrt{x}}}$

In calculus, a useful form for certain radical expressions is obtained by rationalizing the *numerator*. For Exercises 4–7, simplify by rationalizing the numerator.

4. $\dfrac{\sqrt{x + h} - \sqrt{x}}{h}, \quad h \neq 0$

5. $\dfrac{\sqrt{x} - \sqrt{a}}{x - a}, \quad x \neq a$

6. $\dfrac{\sqrt[3]{x+h} - \sqrt[3]{x}}{h}$, $h \neq 0$

7. $\dfrac{\sqrt[3]{(x+h)^2} - \sqrt[3]{x^2}}{h}$, $h \neq 0$

8. Let $x_1 = \dfrac{-b + \sqrt{b^2 - 4ac}}{2a}$ and $x_2 = \dfrac{-b - \sqrt{b^2 - 4ac}}{2a}$. Show that

(a) $x_1 + x_2 = -\dfrac{b}{a}$

and that

(b) $x_1 x_2 = \dfrac{c}{a}$

Answers to Enrichment Exercises are on page A.24.

5.7

Solving Radical Equations

OBJECTIVES

▶ *To solve equations containing radicals*

▶ *To solve word problems by using radical equations*

An equation that contains one or more radicals with variables in the radicands is called a **radical equation.** Examples of radical equations are

$$\sqrt{x} = 7, \qquad \sqrt{y^2 + 1} = y, \qquad \text{and} \qquad \sqrt[3]{r-1} = 2\sqrt[3]{r+9} + 3$$

To solve a radical equation requires changing the equation to one that does not contain radicals. For radical equations containing square roots, this is done using the *squaring property.*

RULE

The squaring property

Suppose A and B are two expressions. If

$$A = B$$

then,

$$A^2 = B^2$$

All solutions of the original equation will be solutions of the squared equation. However, upon squaring, we may introduce *extraneous solutions*. Extraneous solutions satisfy the squared equation but do not satisfy the original equation.

For example, consider the equation $x = 3$. The number 3, of course, is the only solution to this equation. Applying the squaring property to this equation, $x^2 = 9$. However, $x^2 = 9$ has *two* solutions, 3 and -3. But -3 is not a solution of the original equation and is therefore extraneous. Therefore, it is very important to check the numbers in the original equation for extraneous roots.

Example 1   Solve the equation $\sqrt{x} = 6$.

Solution   We use the squaring property with $A = \sqrt{x}$ and $B = 6$,

$$\sqrt{x} = 6$$

implies that

$$(\sqrt{x})^2 = 6^2$$

Since $(\sqrt{x})^2$ can be replaced by x (recall $(\sqrt{a})^2 = a$), we have $x = 6^2$ or 36. As we have done in all equation solving, we check our solution in the original equation.

$$\sqrt{x} = 6$$
$$\sqrt{36} \stackrel{?}{=} 6$$
$$6 = 6$$

Therefore, $x = 36$ is the solution.   ▲

To solve an equation like $\sqrt{2x + 5} + 10 = 4$, first isolate the radical to one side. Then, square both sides.

Example 2   Solve the equation $\sqrt{2x + 5} + 10 = 4$.

Solution   First, subtract 10 from both sides before using the squaring property.

$$\sqrt{2x + 5} + 10 = 4$$
$$\sqrt{2x + 5} + 10 - 10 = 4 - 10$$
$$\sqrt{2x + 5} = -6$$
$$(\sqrt{2x + 5})^2 = (-6)^2$$
$$2x + 5 = 36$$
$$2x = 36 - 5$$
$$x = \frac{31}{2}$$

Next, we check $\dfrac{31}{2}$ in the original equation.

$$\sqrt{2x + 5} + 10 = 4$$

$$\sqrt{2\left(\dfrac{31}{2}\right) + 5} + 10 \overset{?}{=} 4$$

$$\sqrt{31 + 5} + 10 \overset{?}{=} 4$$

$$\sqrt{36} + 10 \overset{?}{=} 4$$

$$6 + 10 \overset{?}{=} 4$$

$$16 \neq 4$$

Since $16 \neq 4$, $x = \dfrac{31}{2}$ is not a solution of the original equation. Thus, $\sqrt{2x + 5} + 10 = 4$ has no real solution. ▲

Example 3   Solve the equation $\sqrt{r} = -8$.

Solution   Before using our pencil, let us think what this equation means. We are to find a number whose *nonnegative* square root is -8. Since $\sqrt{r}$ is nonnegative no matter what r is, certainly $\sqrt{r}$ could never be equal to -8. Therefore, $\sqrt{r} = -8$ has no real solution. ▲

Example 4   Solve the equation $\sqrt{10y} = 2\sqrt{6}$.

Solution

$$\sqrt{10y} = 2\sqrt{6}$$

$$(\sqrt{10y})^2 = (2\sqrt{6})^2$$

$$10y = 4 \cdot 6 \qquad (ab)^2 = a^2 \cdot b^2.$$

$$y = \dfrac{4 \cdot 6}{10}$$

$$y = \dfrac{12}{5}$$

Check

$$\sqrt{10y} = 2\sqrt{6}$$

$$\sqrt{10\left(\dfrac{12}{5}\right)} \overset{?}{=} 2\sqrt{6}$$

$$\sqrt{2 \cdot 12} \overset{?}{=} 2\sqrt{6}$$

$$\sqrt{2 \cdot 2 \cdot 6} \overset{?}{=} 2\sqrt{6}$$

$$\sqrt{2^2 \cdot 6} \overset{?}{=} 2\sqrt{6}$$

$$2\sqrt{6} = 2\sqrt{6}$$

Therefore, $\dfrac{12}{5}$ is the solution. ▲

Example 5

Solution

Solve the equation $\sqrt{c^2 + 2c + 6} = c$.

$$\sqrt{c^2 + 2c + 6} = c$$

$$(\sqrt{c^2 + 2c + 6})^2 = c^2$$

$$c^2 + 2c + 6 = c^2$$

$$2c + 6 = 0 \qquad \text{Subtract } c^2 \text{ from both sides.}$$

$$c = -3$$

Check $c = -3$: $\sqrt{c^2 + 2c + 6} = c$

$$\sqrt{(-3)^2 + 2(-3) + 6} \overset{?}{=} -3$$

$$\sqrt{9 - 6 + 6} \overset{?}{=} -3$$

$$\sqrt{9} \overset{?}{=} -3$$

$$3 \neq -3$$

Therefore, -3 is not a solution and our original equation has no real solution.

▲

In the next example, squaring both sides of the radical equation results in a quadratic equation. The solutions of the quadratic equation must be checked in the original equation.

Example 6

Solution

Solve the equation $3t - 1 = \sqrt{4t}$.

$$(3t - 1)^2 = (\sqrt{4t})^2$$

$$9t^2 - 6t + 1 = 4t \qquad (a - b)^2 = a^2 - 2ab + b^2.$$

$$9t^2 - 10t + 1 = 0 \qquad \text{Subtract } 4t \text{ from both sides.}$$

$$(9t - 1)(t - 1) = 0 \qquad \text{Factor.}$$

Therefore, either $9t - 1 = 0$ or $t - 1 = 0$. (Recall: $A \cdot B = 0$ implies that either $A = 0$ or $B = 0$.)

Thus, either $t = \dfrac{1}{9}$ or $t = 1$.

Check $t = \dfrac{1}{9}$:

$$3t - 1 = \sqrt{4t}$$

$$3\left(\frac{1}{9}\right) - 1 \overset{?}{=} \sqrt{4\left(\frac{1}{9}\right)}$$

$$\frac{1}{3} - 1 \overset{?}{=} \sqrt{\frac{4}{9}}$$

$$-\frac{2}{3} \neq \frac{2}{3}$$

Check $t = 1$:

$$3t - 1 = \sqrt{4t}$$

$$3(1) - 1 \overset{?}{=} \sqrt{4(1)}$$

$$2 = 2$$

Therefore, $\dfrac{1}{9}$ is not a solution and 1 is a solution.   ▲

Example 7

Solve the equation $r - 2 = \sqrt{r - 2}$.

Solution   We square both sides and solve the resulting equation.

$$(r - 2)^2 = (\sqrt{r - 2})^2$$

$$r^2 - 4r + 4 = r - 2$$

$$r^2 - 5r + 6 = 0$$

$$(r - 2)(r - 3) = 0 \qquad\qquad \text{Factor.}$$

Therefore, either $r = 2$ or $r = 3$.

Check $r = 2$:	*Check* $r = 3$:
$r - 2 = \sqrt{r - 2}$	$r - 2 = \sqrt{r - 2}$
$2 - 2 \overset{?}{=} \sqrt{2 - 2}$	$3 - 2 \overset{?}{=} \sqrt{3 - 2}$
$0 = \sqrt{0}$	$1 = \sqrt{1}$

Therefore, both 2 and 3 are solutions of the original equation.   ▲

In the next example, we have an equation where all the radicals cannot be eliminated by applying the squaring property. To obtain a radical-free equation, we must isolate the remaining radical and square again.

Example 8

Solve the equation $\sqrt{1 - 3z} = \sqrt{3z} + 1$.

Solution   We begin by squaring both sides. When we do this, we eliminate the radical on the left side, but still have a radical on the right side.

$$\sqrt{1 - 3z} = \sqrt{3z} + 1$$

$$(\sqrt{1 - 3z})^2 = (\sqrt{3z} + 1)^2$$

$$1 - 3z = 3z + 2\sqrt{3z} + 1$$

Next, we isolate the remaining radical and square again.

$$-6z = 2\sqrt{3z}$$

$$-3z = \sqrt{3z} \qquad\qquad \text{Divide both sides by 2.}$$

$$(-3z)^2 = (\sqrt{3z})^2$$

$$9z^2 = 3z$$

$$9z^2 - 3z = 0$$

$$3z(3z - 1) = 0$$

Therefore, either $z = 0$ or $z = \dfrac{1}{3}$.

Next, we check our two answers in the original equation.

Check $z = 0$:

$$\sqrt{1 - 3z} = \sqrt{3z} + 1$$

$$\sqrt{1 - 3(0)} \overset{?}{=} \sqrt{3(0)} + 1$$

$$1 = 1$$

Check $z = \dfrac{1}{3}$:

$$\sqrt{1 - 3z} = \sqrt{3z} + 1$$

$$\sqrt{1 - 3\left(\frac{1}{3}\right)} \overset{?}{=} \sqrt{3\left(\frac{1}{3}\right)} + 1$$

$$\sqrt{1 - 1} \overset{?}{=} 1 + 1$$

$$0 \neq 2$$

Therefore, $z = 0$ is the only solution. ▲

Certain equations involving radicals where the index is greater than 2 can be solved by raising both sides to a power equal to the index. We need to check for extraneous solutions only when the power is even. However, it is always a good idea to check solutions in the original equation nonetheless. In the next example, we show how to solve a radical equation involving a cube root.

Example 9   Solve the equation $\sqrt[3]{5x - 14} = -4$.

Solution   Cubing both sides, we have

$$(\sqrt[3]{5x - 14})^3 = (-4)^3$$

$$5x - 14 = -64$$

$$5x = -50$$

$$x = -10$$

On your own, check that -10 is a solution of the original equation. ▲

We now give a sequence of steps for solving a radical equation. Refer to this strategy as you solve the equations in the exercise set.

S T R A T E G Y

A strategy for solving a radical equation

Step 1   If necessary, isolate a radical to one side of the equation.

Step 2   Raise both sides of the equation to the power equal to the index. If the resulting equation is free of radicals, go to Step 3. If the resulting equation is not free of radicals, go to Step 1.

Step 3   Solve the resulting equation.

Step 4   Check all solutions from Step 3 in the original equation.

LEARNING ADVANTAGE *Be alert in class and actively follow what the instructor is saying. Be ready to ask questions as well as answer questions asked by the instructor. If the class size is large, it is not always possible for the instructor to know everyone by name. It is then up to you to make yourself known by presenting a positive image.*

EXERCISE SET 5.7

For Exercises 1–66, solve, using the strategy for solving a radical equation.

1. $\sqrt{x} = 4$

2. $\sqrt{y} = 5$

3. $\sqrt{a} = -1$

4. $\sqrt{z} = -\dfrac{2}{3}$

5. $\sqrt{z-1} = 3$

6. $\sqrt{c+2} = 1$

7. $\sqrt{3t-1} = \dfrac{1}{2}$

8. $\sqrt{2w+3} = \dfrac{4}{3}$

9. $\sqrt{s-1} = 1$

10. $\sqrt{x+1} = 2$

11. $4 + 3\sqrt{s} = 13$

12. $5 + 2\sqrt{y} = 9$

13. $7 = \sqrt{p+12}$

14. $5 = \sqrt{r-2}$

15. $\sqrt{2u+1} = \dfrac{1}{3}$

16. $\sqrt{4b+2} = \dfrac{2}{3}$

17. $\sqrt{10c} = 2\sqrt{3}$

18. $\sqrt{6r} = 3\sqrt{5}$

19. $\sqrt{-2x} = 4x + 1$

20. $\sqrt{3t} = 2 - 3t$

21. $\sqrt{x+3} = x + 1$

22. $2a + 1 = \sqrt{6a+13}$

23. $2\sqrt{3y} + 1 = 7$

24. $3\sqrt{2s} + 1 = 5$

25. $3 + \sqrt{5x+1} = 5$

26. $7 + \sqrt{3z-4} = 13$

27. $x - 1 = \sqrt{x^2+3}$

28. $z + 4 = \sqrt{z^2+10}$

29. $2 - t = \sqrt{t^2+2}$

30. $4 - v = \sqrt{v^2+6}$

31. $\sqrt{a^2-2a} + 10 = a$

32. $\sqrt{s^2-3s} + 12 = s$

33. $\sqrt{x^2+4x} - 12 = -x$

34. $\sqrt{p^2+5p} - 10 = -p$

35. $\sqrt{b^2+3b+6} = 1 - b$

36. $\sqrt{w^2+5w+2} = 2 - w$

37. $\sqrt{4c^2+c-1} + 1 = 2c$

38. $\sqrt{9k^2-2k+1} - 3k = -1$

39. $3x + 2 = \sqrt{21x+2}$

40. $a + 2 = \sqrt{8a+1}$

41. $\sqrt{4w+1} - 1 = 2w$

42. $3 = 4v + \sqrt{3-2v}$

43. $\sqrt{3-b} + b = 3$

44. $7z = \sqrt{7z+2} - 2$

45. $\sqrt{7s-10} = 1 + \sqrt{5s-9}$

46. $2 - \sqrt{2r+3} = \sqrt{10r+19}$

47. $\sqrt{4-a}-a=2$

48. $x-3=\sqrt{x-2}-1$

49. $\sqrt[3]{2w+3}=-1$

50. $\sqrt[3]{3c-4}=-2$

51. $\sqrt[4]{x-2}=1$

52. $\sqrt[4]{t-5}=2$

53. $(2s+8)^{1/3}+3=0$

54. $(3a-1)^{1/3}-2=0$

55. $(5r-3)^{1/4}+5=0$

56. $(6t+1)^{1/4}+3=0$

57. $(x-2)^{-1/3}=-2$

58. $(y+4)^{-1/3}=-\dfrac{1}{2}$

59. $\sqrt[3]{b^2-7b+2}=-2$

60. $\sqrt[5]{x^2-8x+6}=-1$

61. $\sqrt[4]{3v+6}=\sqrt[4]{7v-2}$

62. $\sqrt[6]{9c+5}=\sqrt[6]{3c+2}$

63. $\sqrt[4]{x-2}+\sqrt[4]{x+3}=0$

64. $\sqrt[4]{3y+1}+\sqrt[4]{2y-2}=0$

65. $\sqrt[3]{4x^2-5x+1}=\sqrt[3]{x^2+5x-2}$

66. $\sqrt[3]{1-x-4x^2}=\sqrt[3]{x(4x+1)}$

67. The square root of the sum of a number and ten is three. Find the number.

68. The sum of the square root of a number and one is seven. Find the number.

69. The square root of the sum of the square of a number and three equals one more than the number. Find the number.

70. The product of two and the square root of six equals the square root of the difference of twice a number and four. Find the number.

71. The cube root of the sum of twice a number and five is negative 2. Find the number.

72. The cube root of the difference of a number and four is negative one. Find the number.

REVIEW PROBLEMS

The following exercises review parts of Sections 5.3 and 5.6. Doing these problems will help prepare you for the next section.

Section 5.3

Multiply.

73. $\sqrt{3}(\sqrt{2}-\sqrt{5})$

74. $\sqrt{5}(2\sqrt{15}-3\sqrt{10})$

75. $(\sqrt{x}-2)(3\sqrt{x}-1)$

76. $(\sqrt{a}-3)^2$

77. $(\sqrt{z}-\sqrt{2})(\sqrt{z}+\sqrt{2})$

78. $(4\sqrt{y}+\sqrt{3})^2$

Section 5.6

Rationalize the denominator.

79. $\dfrac{3\sqrt{x}-1}{\sqrt{2x}}$

80. $\dfrac{\sqrt{x}+3}{\sqrt{x}+2}$

81. $\dfrac{\sqrt{3}-\sqrt{5}}{2\sqrt{3}+\sqrt{5}}$

82. $\dfrac{4\sqrt{x}-1}{2\sqrt{x}-1}$

ENRICHMENT EXERCISES

For Exercises 1–6, solve the given equation.

1. $x^2 + 1 = 2\sqrt{3x^2 - 2}$

2. $a^2 - \sqrt{85(a^2 - 1)} + 1 = 20$

3. $\sqrt{16z^2 - 15} = 1 + 2\sqrt{3z^2 - 3}$

4. $\sqrt{x + 1} = \sqrt[4]{3x + 1}$

5. $\sqrt[4]{2y - 1} = \sqrt{y - 2}$

6. $\sqrt[6]{3 + x(1 - x)} = \sqrt[3]{2x + 3}$

7. Without solving the equation, show that

$$\sqrt{x - 5} = \sqrt{4 - x}$$

has no real solution. (*Hint:* Look at the restrictions on the variable.)

Answers to Enrichment Exercises are on page A.25.

5.8

Complex Numbers

OBJECTIVES

▶ *To write the square roots of negative numbers in terms of i*

▶ *To learn the algebra of complex numbers*

The equation $x^2 = -9$ has no solution in the set of real numbers, since any real number squared is nonnegative. In this section, we will introduce a new set of numbers, called *complex numbers,* so that every quadratic equation will have solutions. We start with the following definition.

DEFINITION

The number i is defined as a square root of -1.

$$i = \sqrt{-1}$$

Since i is a square root of -1,

$$i^2 = -1$$

Notice that i is not a real number, since its square is negative. Any nonzero real number multiple of i is called an **imaginary number.** Examples of imaginary numbers are the following:

$$-i, \quad 12i, \quad -3i, \quad i\sqrt{2}, \quad 5i\sqrt{15}$$

We can rewrite the square root of a negative number as an imaginary number. Let $\sqrt{-a}$ be the square root of a negative number. Then $\sqrt{-a}$ must have the property that its square is $-a$, that is,

$$(\sqrt{-a})^2 = -a$$

Now, consider the imaginary number $i\sqrt{a}$. Keeping in mind that $i^2 = -1$,

$$(i\sqrt{a})^2 = (i\sqrt{a})(i\sqrt{a}) = i^2(\sqrt{a})^2 + (-1)a = -a$$

Therefore, both $\sqrt{-a}$ and $i\sqrt{a}$ are square roots of $-a$, and

$$\sqrt{-a} = i\sqrt{a}$$

Example 1 Write each as an imaginary number.

(a) $\sqrt{-4} = i\sqrt{4} = 2i$ **(b)** $-\sqrt{-121} = -i\sqrt{121} = -i \cdot 11 = -11i$

(c) $-\sqrt{-15} = -i\sqrt{15}$ **(d)** $\sqrt{-24} = i\sqrt{24} = 2i\sqrt{6}$ ▲

N O T E   *It is common practice to write i before the radical in imaginary numbers such as $\sqrt{2}i$ to prevent a misunderstanding that i is part of the radicand. Therefore, we write $\sqrt{2}i$ as $i\sqrt{2}$. Of course, either form is correct.*

We now define the set of complex numbers.

D E F I N I T I O N

A **complex number** is any number that can be put in the form

$$a + bi$$

where a and b are real numbers. The number a is called the **real part** of $a + bi$ and the number b is called the **imaginary part.**

The quantity $a + bi$ is called the **standard form** for a complex number.

Example 2 Express each complex number in the standard form $a + bi$.

(a) $3 - \sqrt{-121} = 3 - i\sqrt{121} = 3 - 11i$

(b) $\sqrt{-20} = i\sqrt{20} = (2\sqrt{5})i = 0 + (2\sqrt{5})i$

(c) $6 = 6 + 0i$

(d) $0 = 0 + 0i$ ▲

Since any real number a is of the form $a + 0i$, the set of real numbers is a subset of the set of complex numbers. Recall in Chapter 1, we studied the set of real numbers along with its subsets of integers, rational numbers, and irrational numbers. Now, with the complex numbers, we have the following relationships existing in the family tree of numbers.

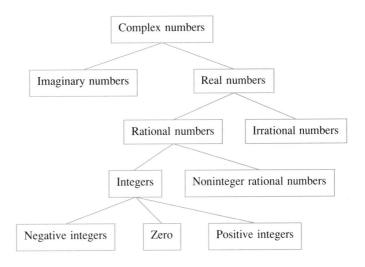

Two complex numbers are **equal** provided that their real parts and their imaginary parts are equal. That is,

$$a + bi = c + di$$

provided $a = c$ and $b = d$.

Example 3

Determine x and y if $4x - 2i = 2 + (3y - 1)i$.

Solution

Since the two complex numbers are equal, their real parts and their imaginary parts are equal.

$$4x = 2 \qquad \text{and} \qquad -2 = 3y - 1$$

$$x = \frac{1}{2} \qquad \text{and} \qquad y = -\frac{1}{3} \qquad \qquad \blacktriangle$$

An expression containing a sum or difference of two complex numbers can be simplified by combining the real and imaginary parts according to the following formulas.

D E F I N I T I O N

The **sum** and **difference** of two complex numbers $a + bi$ and $c + di$ are given by

$$(a + bi) + (c + di) = (a + c) + (b + d)i$$

and

$$(a + bi) - (c + di) = (a - c) + (b - d)i$$

Example 4   Add or subtract as indicated.

(a) $(4 - 2i) + (2 + 5i) = (4 + 2) + (-2 + 5)i = 6 + 3i$

(b) $(9 - 4i) - (5 - 2i) = (9 - 5) + (-4 + 2)i = 4 - 2i$

(c) $(-10 + \sqrt{-3}) - (-7 + \sqrt{-12}) = (-10 + i\sqrt{3}) - (-7 + 2i\sqrt{3})$

$$= -3 - i\sqrt{3} \qquad ▲$$

We now give the definition for multiplying two complex numbers.

D E F I N I T I O N

The **product** of the two complex numbers $a + bi$ and $c + di$ is given by

$$(a + bi)(c + di) = ac - bd + (ad + bc)i$$

We need not memorize this definition for multiplying two complex numbers. Instead, this definition allows us to apply the distributive property to complex numbers.

Example 5   Multiply: $(3 - 2i)(5 + 6i)$.

Solution   We multiply each term in the second complex number by each term in the first and then simplify the result.

$$(3 - 2i)(5 + 6i) = 3 \cdot 5 + 3 \cdot 6i - 5 \cdot 2i - (6i)(2i)$$

$$= 15 + 18i - 10i - 12i^2$$

$$= 15 + 8i - 12(-1) \qquad i^2 = -1.$$

$$= 27 + 8i$$

Therefore, the product of $3 - 2i$ and $5 + 6i$ is the complex number $27 + 8i$.

▲

Example 6   Multiply: $3i\left(2 - \dfrac{4}{3}i\right)$.

Solution   Using the distributive property, we have

$$3i\left(2 - \frac{4}{3}i\right) = 3i \cdot 2 - 3i\left(\frac{4}{3}i\right)$$

$$= 6i - 4i^2$$

$$= 4 + 6i \qquad \blacktriangle$$

In the next two examples, we show how to use formulas such as $(a + b)^2 = a^2 + 2ab + b^2$ to multiply complex numbers.

Example 7   Find: $(3 - 5i)^2$.

Solution   Using the formula $(a - b)^2 = a^2 - 2ab + b^2$,

$$(3 - 5i)^2 = 3^2 - 30i + (5i)^2$$

$$= 9 - 30i - 25$$

$$= -16 - 30i \qquad \blacktriangle$$

Example 8   Multiply: $(4 + 3i)(4 - 3i)$.

Solution   Recall the formula for binomials:

$$(a + b)(a - b) = a^2 - b^2$$

Letting $a = 4$ and $b = 3i$,

$$(4 + 3i)(4 - 3i) = 4^2 - (3i)^2$$

$$= 16 + 9$$

$$= 25 \qquad \blacktriangle$$

Notice in Example 8 that the product of the two complex numbers $4 + 3i$ and $4 - 3i$ is the real number 25. The two numbers $4 + 3i$ and $4 - 3i$ are called *complex conjugates*.

D E F I N I T I O N

The complex numbers $a + bi$ and $a - bi$ are called **(complex) conjugates**.

The following pairs of numbers are examples of complex conjugates.

Number	Complex conjugate
$1 - 4i$	$1 + 4i$
$8 + 3i$	$8 - 3i$
$2 - \dfrac{2}{5}i$	$2 + \dfrac{2}{5}i$
14	14
$-i$	i

A valuable property of complex conjugates is the fact that their product is a real number.

THEOREM

The complex conjugates $a + bi$ and $a - bi$ have the property that

$$(a + bi)(a - bi) = a^2 + b^2$$

Proof   It is easy to verify this formula.

$$(a + bi)(a - bi) = a^2 - (bi)^2 = a^2 + b^2$$

For example, $(1 + 8i)(1 - 8i) = 1^2 + 8^2 = 65$.

We use this theorem to convert division problems involving complex numbers into the standard form.

Example 9   Write $\dfrac{4 - 3i}{1 + 5i}$ in the standard form $a + bi$.

Solution   We want to find a complex number in the standard form $a + bi$ that is equal to the quotient $\dfrac{4 - 3i}{1 + 5i}$. To find this number, we multiply numerator and denominator by the complex conjugate of the denominator and then simplify.

$$\frac{4 - 3i}{1 + 5i} = \frac{(4 - 3i)(1 - 5i)}{(1 + 5i)(1 - 5i)}$$

$$= \frac{4 - 15 - 23i}{1^2 + 5^2}$$

$$= \frac{-11 - 23i}{26}$$

$$= -\frac{11}{26} - \frac{23}{26}i$$

Therefore, the division of $4 - 3i$ by $1 + 5i$ results in the complex number $-\frac{11}{26} - \frac{23}{26}i$. ▲

The general definitions for subtraction and division are the following.

$$A - B = C \qquad \text{provided that} \qquad A = B + C$$

$$\frac{A}{B} = C \qquad \text{provided that} \qquad A = B \cdot C$$

We did not use these definitions to find the number C in either subtraction or division problems. It can be shown, however, that the methods we used in the examples are consistent with the definitions. In Example 9, for instance, we established that $(4 - 3i) \div (1 + 5i) = -\frac{11}{26} - \frac{23}{26}i$. As a check, it could be verified that

$$(1 + 5i)\left(-\frac{11}{26} - \frac{23}{26}i\right) = 4 - 3i$$

The product property of square roots, $\sqrt{a}\sqrt{b} = \sqrt{ab}$ does not apply when both a and b are negative. Instead, convert first to imaginary numbers before multiplying.

$$\sqrt{-a}\sqrt{-b} = (i\sqrt{a})(i\sqrt{b}) = i^2\sqrt{ab} = -\sqrt{ab}$$

where $-a$ and $-b$ are negative real numbers.

Example 10

$$(\sqrt{-2})(\sqrt{-3}) = (i\sqrt{2})(i\sqrt{3})$$
$$= i^2(\sqrt{2 \cdot 3})$$
$$= -\sqrt{6}$$ ▲

Using the fact that $i^2 = -1$, we can find powers of i, i^n, where n is an integer. For example,

$$i^3 = i \cdot i^2 = i(-1) = -i$$
$$i^4 = i^2 \cdot i^2 = (-1)(-1) = 1$$
$$i^5 = i \cdot i^4 = i \cdot 1 = i$$
$$i^6 = i \cdot i^5 = i \cdot i = i^2 = -1$$
$$i^7 = i \cdot i^6 = i(-1) = -i$$
$$i^8 = i^4 \cdot i^4 = 1 \cdot 1 = 1$$

Notice that any power of i is either i, -1, $-i$, or 1.

Example 11   Simplify.

(a) $i^{18} = (i^2)^9 = (-1)^9 = -1$

(b) $i^{43} = i^{40}i^3 = (i^4)^{10}i^3 = (1)^{10}(-i) = -i$

(c) $i^{-4} = \dfrac{1}{i^4} = \dfrac{1}{1} = 1$ ▲

Example 12   Write i^{-3} in standard form.

Solution $i^{-3} = \dfrac{1}{i^3} = \dfrac{1}{-i}$. To write this quotient in standard form, we multiply numerator and denominator by i, the conjugate of $-i$.

$$\frac{1}{-i} = \frac{1 \cdot i}{(-i)i} = \frac{i}{-i^2} = \frac{i}{-(-1)} = \frac{i}{1} = i$$ ▲

N O T E   *A shorter way to simplify the quotient* $\dfrac{1}{i^3}$ *in Example 12 is to multiply numerator and denominator by i. The details follow:*

$$\frac{1}{i^3} = \frac{1 \cdot i}{i^3 \cdot i} = \frac{i}{i^4} = \frac{i}{1} = i$$

L E A R N I N G   A D V A N T A G E   *Before each class, spend some time reading the section of the book that will be covered in that class period. That way, you will be better prepared to understand the material that will be explained. Furthermore, you can be prepared to ask in class anything you did not understand from previewing the lesson.*

E X E R C I S E   S E T   5.8

For Exercises 1–10, write each as an imaginary number.

1. $\sqrt{-9}$ **2.** $\sqrt{-64}$ **3.** $-\sqrt{-100}$ **4.** $-\sqrt{-144}$

5. $-\sqrt{-17}$ **6.** $-\sqrt{-21}$ **7.** $\sqrt{-28}$ **8.** $\sqrt{-32}$

9. $-\sqrt{-50}$ **10.** $-\sqrt{-27}$

For Exercises 11–16, put each expression in standard form.

11. $4 - \sqrt{-100}$ **12.** $-5 + \sqrt{-9}$ **13.** $2 + \sqrt{-8}$

14. $-1 + \sqrt{-24}$ **15.** $1 - \sqrt{20}$ **16.** $\sqrt{-72}$

For Exercises 17–20, determine x and y.

17. $x - 2yi = \dfrac{1}{4} + 5i$

18. $3 + yi = 6x - 4i$

19. $3\sqrt{2} + 4yi = 2x - i\sqrt{20}$

20. $7\sqrt{2} - 3yi = 14x - yi$

For Exercises 21–52, perform the indicated operations and write the answer in standard form.

21. $(4 - 2i) + (6 + 3i)$

22. $(7 + 3i) + (-1 - 5i)$

23. $(5 - 2i) - (15 - 2i)$

24. $(7 - 4i) - (8 - 11i)$

25. $13 - (3 - 6i)$

26. $20i - (8 - 18i)$

27. $(4 - 2i) - 6i$

28. $(5 - 3i) - 7i$

29. $(1 + 2i)(2 - 5i)$

30. $(6 - 5i)(1 - 4i)$

31. $3i(7 - 3i)$

32. $4i(1 - 5i)$

33. $(3 - 2i)^2$

34. $(1 + 6i)^2$

35. $\left(\dfrac{1}{2} + \dfrac{1}{4}i\right)\left(\dfrac{1}{2} - \dfrac{1}{4}i\right)$

36. $(5 - 2i)(5 + 2i)$

37. $\dfrac{2 - 3i}{1 - 2i}$

38. $\dfrac{9 - 2i}{2 + 2i}$

39. $\dfrac{2}{1 + i}$

40. $\dfrac{1}{1 - i}$

41. $(4 - 3i) \div (2 - 3i)$

42. $(1 - 4i) \div (3 - i)$

43. $\dfrac{1 + 3i}{i}$

44. $\dfrac{7 - 8i}{i}$

45. $\dfrac{2 - 4i}{-6i}$

46. $\dfrac{9 - 3i}{-12i}$

47. $\dfrac{(10 - 2i)(1 + i)}{i}$

48. $\dfrac{(1 + 3i)(2 - i)}{5i}$

49. $2(1 - 3i)^2 + 3(5 - 8i) + (3 - 5i)$

50. $3(2 + i)^2 - 4(5 - 2i) + (2 - i)$

51. $(1 - i)^3$

52. $(1 + i)^3$

For Exercises 53–60, write square roots of negative numbers in terms of i, then perform the indicated operations.

53. $(-12 + \sqrt{-4}) + (20 - \sqrt{-49})$

54. $(-14 + \sqrt{-9}) - (4 - \sqrt{-121})$

55. $\dfrac{\sqrt{20}}{\sqrt{40}}$

56. $\dfrac{\sqrt{24}}{\sqrt{18}}$

57. $\dfrac{-4 - \sqrt{-28}}{-8}$

58. $\dfrac{9 + \sqrt{-27}}{-3}$

59. $\dfrac{1}{-3 - \sqrt{-8}}$

60. $\dfrac{3}{1 - \sqrt{-12}}$

For Exercises 61–68, simplify each power of i.

61. i^{16} **62.** i^{14} **63.** i^{25} **64.** i^{39}

65. i^{-24} **66.** i^{-13} **67.** i^{-33} **68.** i^{-15}

69. Evaluate $x^2 - 2x + 1$ for $x = 1 + i$.

70. Evaluate $x^2 + 3x - 1$ for $x = 2 - i$.

71. Evaluate $1 - 3x + 2x^2$ for $x = 1 - 3i$.

72. Evaluate $4 + 2x - 3x^2$ for $x = 3 + 2i$.

For Exercises 73–78, evaluate $\dfrac{-b \pm \sqrt{b^2 - 4ac}}{2a}$ for the given trinomial $ax^2 + bx + c$.

73. $x^2 + x + 1$ **74.** $2x^2 - x + 2$ **75.** $2x^2 - 2x + 1$

76. $3x^2 + 2x + 1$ **77.** $x^2 + 25$ **78.** $4x^2 + 9$

REVIEW PROBLEMS

The following exercises review parts of Sections 2.2 and 3.8. Doing these problems will help prepare you for the next section.

Section 2.2

Solve each equation.

79. $2x - 5 = 0$

80. $4x + 3 = 0$

Section 3.8

Solve each equation.

81. $(5t - 1)(3t + 7) = 0$

82. $(4z + 6)(8z - 3) = 0$

83. $x^2 - 49 = 0$

84. $z^2 = 121$

ENRICHMENT EXERCISES

1. The number -2 is a cube root of -8 since $(-2)^3 = -8$. Show that $1 + i\sqrt{3}$ is also a cube root of -8.

2. Show that $1 - i\sqrt{3}$ is another cube root of -8.

3. Let $z = a + bi$ and $w = c + di$. One notation for the complex conjugate is to place a "bar" above the letter. Therefore, $\bar{z} = a - bi$ and $\bar{w} = c - di$. Prove the following statements.

(a) $\overline{z + w} = \bar{z} + \bar{w}$ 　　　　　　(b) $\overline{zw} = \bar{z}\,\bar{w}$

4. Show that the reciprocal $\dfrac{1}{a + bi}$ of a nonzero complex number $a + bi$ is given by

$$\frac{1}{a + bi} = \frac{a}{a^2 + b^2} - \frac{b}{a^2 + b^2}i$$

5. Use the result of Problem 4 to find the reciprocals of the following numbers. Check your answers by showing that the product of the number and its reciprocal is one.

(a) $1 + i$ 　　　　　(b) i 　　　　　(c) $2 - i\sqrt{3}$

Answers to Enrichment Exercises are on page A.26.

C H A P T E R　5　　**S**ummary **and review**

Examples

$\sqrt{64} = 8$ and $-\sqrt{64} = -8$. The square roots of 64 are 8 and -8.

Square roots (5.1)

A square root of a given number is any number whose square is the given number. Any positive real number, a, has two square roots. The positive square root is given by $\sqrt{a}$, and the negative square root is given by $-\sqrt{a}$.

Both square roots of a, when squared, give a. That is,

$$(\sqrt{a})^2 = a$$

and

$$(-\sqrt{a})^2 = a \qquad \text{for } a > 0$$

The number zero has one square root, $\sqrt{0} = 0$. A negative number has no real square roots.

Examples

$\sqrt[4]{81} = 3, \sqrt[3]{8} = 2,$ and
$\sqrt[3]{-8} = -2$

Higher roots (5.1)

Let n be a positive integer. A number b is an nth root of a if $b^n = a$. The symbol $\sqrt[n]{a}$ means the nth root of a. The number n is the **index**, a is the **radicand**, and $\sqrt[n]{}$ is the **radical sign.**

In general,

$$(\sqrt[n]{a})^n = a, \qquad \text{where } a \geq 0 \text{ when } n \text{ is even}$$

Rational exponents (5.2)

$100^{1/2} = 10, (-8)^{1/3} = -2$

Rational exponents are another way to indicate roots. If a is a real number and n is a positive integer, then

$$a^{1/n} = \sqrt[n]{a}, \qquad \text{where } a \geq 0 \text{ when } n \text{ is even}$$

$16^{3/2} = (16^{1/2})^3 = 4^3 = 64$
$(-27)^{2/3} = [(-27)^{1/3}]^2$
$\qquad = [-3]^2 = 9$
$8^{-4/3} = \dfrac{1}{8^{4/3}} = \dfrac{1}{2^4} = \dfrac{1}{16}$

If m and n are positive integers, then

$$a^{m/n} = (a^{1/n})^m, \qquad \text{where } a \geq 0, \text{ when } n \text{ is even}$$

$$a^{-m/n} = \frac{1}{a^{m/n}}, \qquad \text{where } a \neq 0 \text{ and provided } a^{m/n} \text{ exists}$$

The algebra of rational exponents (5.2, 5.3)

$x^{2/3}(3x^{1/6}) = 3x^{2/3+1/6} = 3x^{5/6}$

All the properties of exponents still hold when the exponents are rational numbers.

$$a^r a^s = a^{r+s}, \qquad (a^r)^s = a^{rs}, \qquad (ab)^r = a^r b^r,$$

$$\left(\frac{a}{b}\right)^r = \frac{a^r}{b^r}, \, b \neq 0 \qquad \frac{a^r}{a^s} = a^{r-s}, \, a \neq 0$$

Using these properties, we can simplify expressions involving rational exponents.

Simplifying radical expressions (5.4)

$\sqrt[3]{24} = \sqrt[3]{8 \cdot 3} = \sqrt[3]{8}\sqrt[3]{3} = 2\sqrt[3]{3}$

The following rules are used to simplify radical expressions.

$$\sqrt[n]{a \cdot b} = \sqrt[n]{a} \cdot \sqrt[n]{b}, \text{ provided } \sqrt[n]{a} \text{ and } \sqrt[n]{b} \text{ are real numbers}$$

$$\sqrt[n]{\frac{a}{b}} = \frac{\sqrt[n]{a}}{\sqrt[n]{b}}, \text{ provided } \sqrt[n]{a} \text{ and } \sqrt[n]{b} \text{ are real numbers with } b \neq 0.$$

$\sqrt{\dfrac{7}{9}} = \dfrac{\sqrt{7}}{\sqrt{9}} = \dfrac{\sqrt{7}}{3}$

A radical expression is simplified if

1. The radicand contains no power greater than or equal to the index.
2. The exponents in the radicand and the index of the radical have no common factor.
3. The radicand has no fractions.
4. There are no radicals in denominators.

Examples

$$\frac{2}{\sqrt{3}} = \frac{2\sqrt{3}}{\sqrt{3}\sqrt{3}} = \frac{2\sqrt{3}}{3}$$

$$\frac{4}{5 - \sqrt{2}} = \frac{4(5 + \sqrt{2})}{(5 - \sqrt{2})(5 + \sqrt{2})}$$

$$= \frac{4(5 + \sqrt{2})}{5^2 - (\sqrt{2})^2}$$

$$= \frac{4(5 + \sqrt{2})}{23}$$

Rationalize the denominator (5.4, 5.6)

If an expression contains a radical in the denominator, we rationalize the denominator. There are two cases.

Case 1. There is one term in the denominator.

Case 2. There is the sum or difference of terms in the denominator.

$$4\sqrt[3]{5} - 7\sqrt[3]{5} = -3\sqrt[3]{5}$$

$$5\sqrt{12} + 2\sqrt{3}$$

$$= 5 \cdot 2\sqrt{3} + 2\sqrt{3}$$

$$= 10\sqrt{3} + 2\sqrt{3}$$

$$= 12\sqrt{3}$$

Addition and subtraction of radical expressions (5.5)

We add and subtract radical expressions by combining like radicals.

$$\sqrt{5}(4 - 2\sqrt{10})$$

$$= 4\sqrt{5} - 2\sqrt{5}\sqrt{10}$$

$$= 4\sqrt{5} - 10\sqrt{2}$$

Products and quotients of radical expressions (5.6)

We can multiply radical expressions using the distributive property and then simplifying.

$-2\sqrt{50}$

$5^{2 \cdot 2}$

$-10\sqrt{2}$

$$\sqrt{3x - 2} = 5$$

$$(\sqrt{3x - 2})^2 = 5^2$$

$$3x - 2 = 25$$

$$x = 9$$

Check $x = 9$: $\sqrt{3x - 2} = 5$

$$\sqrt{3(9) - 2} \stackrel{?}{=} 5$$

$$\sqrt{27 - 2} \stackrel{?}{=} 5$$

$$\sqrt{25} = 5$$

Therefore, $x = 9$ is the solution of $\sqrt{3x - 2} = 5$.

Solving radical equations (5.7)

To solve radical equations, we use the squaring property: We may square both sides of an equation provided we check the resulting solutions in the original equation.

$$12 - \sqrt{-20} = 12 - i\sqrt{20}$$

$$= 12 - 2i\sqrt{5}$$

Complex numbers (5.8)

A **complex number** is any number that can be put in the form $a + bi$, where a and b are real numbers and $i = \sqrt{-1}$. The number a is the **real part** and b is the **imaginary part**.

If $a + bi$ and $c + di$ are two complex numbers, we have the following properties.

Examples

$(4 - 2i) + (5 - i) = 9 - 3i$

$(5 + 7i) - (4 - 3i) = 1 + 10i$

$(2 - 5i)(3 + 4i)$
$\quad = 6 + 8i - 15i - 20i^2$
$\quad = 26 - 7i$

$\dfrac{1 - 4i}{2 + 3i} = \dfrac{(1 - 4i)(2 - 3i)}{(2 + 3i)(2 - 3i)}$

$\quad = \dfrac{-10 - 11i}{13}$

$\quad = -\dfrac{10}{13} - \dfrac{11}{13}i$

1. Equality.

$$a + bi = c + di, \quad \text{provided } a = c \text{ and } b = d$$

2. Addition and subtraction.

$$(a + bi) + (c + di) = (a + c) + (b + d)i$$
$$(a + bi) - (c + di) = (a - c) + (b - d)i$$

3. Multiplication.

$$(a + bi)(c + di) = (ac - bd) + (ad + bc)i$$

Rather than use the definition, it is easier to use the distributive property to multiply.

4. Division. To write a division problem in the standard form $a + bi$, rationalize the denominator using the complex conjugate.

C H A P T E R 5   R E V I E W   E X E R C I S E   S E T

Assume that all variables are positive.

Section 5.1

For Exercises 1–6, simplify each expression.

1. $\sqrt{64}$

2. $-\sqrt{\dfrac{9}{16}}$

3. $\sqrt[3]{-27}$

4. $-\sqrt[4]{16}$

5. $\sqrt{9x^8 y^4}$

6. $\sqrt[3]{-8a^6 b^9 c^{12}}$

Sections 5.2 and 5.3

For Exercises 7–12, simplify.

7. $3^{2/3} \cdot 3^{5/6}$

8. $z^{-4/5} \cdot z^{3/10}$

9. $(x^{3/5})^{-15/9}$

10. $(s^{2/3} t^{1/6})^{12}$

11. $\dfrac{w^{2/5}}{w^{1/15}}$

12. $\dfrac{(a^{-2/7} b^{3/2})^{14}}{a^{-4} b}$

For Exercises 13–17, multiply.

13. $x^{2/3} y^{5/2}(2x^3 y^{3/2} - 5x^{1/2} y + 7x^{1/3} y^{1/4})$

14. $(\sqrt{s} - 2\sqrt{w})(\sqrt{s} + \sqrt{w})$

15. $(c^{1/2} + b^{3/2})(c^{1/2} - b^{3/2})$

16. $(a^{5/2} - 2)^2$

17. $(z^{1/3} - h^{1/3})[z^{2/3} + (zh)^{1/3} + h^{2/3}]$

18. Factor $3t^{5/3}$ from $9t^{8/3} - 6t^{5/3}$.

19. Factor $3(a + 1)^{1/6}$ from $18(a + 1)^{1/3} - 3(a + 1)^{1/6}$.

20. Combine into one fraction: $\dfrac{t}{(t - 1)^{1/2}} - (t - 1)^{1/2}$.

Section 5.4

For Exercises 21–24, simplify.

21. $\sqrt{24}$ **22.** $\sqrt[3]{-16}$ **23.** $\sqrt[4]{z^8 t^{12}}$ **24.** $-\sqrt{3s^8 r^4}$

Sections 5.4 and 5.6

For Exercises 25–34, rationalize the denominator.

25. $\dfrac{2}{\sqrt{6}}$ **26.** $\dfrac{3}{\sqrt{x}}$ **27.** $\dfrac{4s}{\sqrt{2s}}$

28. $\dfrac{3k}{\sqrt[3]{k^2}}$ **29.** $\dfrac{4v}{3\sqrt[4]{v}}$ **30.** $\dfrac{3}{2 - \sqrt{5}}$

31. $\dfrac{4}{1 + \sqrt{3}}$ **32.** $\dfrac{\sqrt{x} - 1}{\sqrt{x} + 1}$ **33.** $\dfrac{\sqrt{st}}{2\sqrt{s} - \sqrt{t}}$

34. $\dfrac{c - 3}{c^{1/3} - 3^{1/3}}$

Section 5.5

For Exercises 35–40, simplify by combining like radicals.

35. $4\sqrt{3} + 5\sqrt{3} - 2\sqrt{3}$ **36.** $3\sqrt[4]{32} + 5\sqrt[4]{2}$ **37.** $2\sqrt{a^5} - 5a\sqrt{a^3}$

38. $5\sqrt[3]{c^2} - 19\sqrt[3]{c^2}$ **39.** $\sqrt[3]{u^4 v^5} - 4uv\sqrt[3]{uv^2}$ **40.** $2x\sqrt{x^3} - x^2\sqrt{x} + 3\sqrt{x^5}$

41. Add: $\dfrac{w^2}{\sqrt{w^3}} + \dfrac{3w}{\sqrt{w}}$

Section 5.6

For Exercises 42–46, multiply and then simplify.

42. $\sqrt{3}(\sqrt{3} - 2\sqrt{6})$ **43.** $\sqrt{s}(2\sqrt{s} - 3\sqrt{s^3})$ **44.** $\sqrt[3]{x}(\sqrt[3]{x^5} - 2\sqrt[3]{x^7})$

45. $(\sqrt{a} - 2\sqrt{b})^2$ **46.** $(\sqrt[3]{9y^2} + 2)(\sqrt[3]{9y^2} - 2)$

Section 5.7

For Exercises 47–55, solve each equation.

47. $\sqrt{x} = 8$ **48.** $\sqrt{t} = 2\sqrt{3}$ **49.** $\sqrt{z + 3} = 2$

50. $\sqrt{4z^2 - 2z + 9} = 2z$ **51.** $\sqrt{3 - 2x} = 3\sqrt{x}$ **52.** $3y - 1 = \sqrt{3y + 1}$

53. $\sqrt{2x + 9} = \sqrt{2x} + 3$ **54.** $\sqrt{1 - z} = \sqrt{-1 - 2z}$ **55.** $\sqrt[3]{3s} + 1 = -2$

Section 5.8

For Exercises 56–59, express each in standard form.

56. $\sqrt{-100}$ **57.** $2 - \sqrt{-9}$ **58.** $4 + \sqrt{-40}$ **59.** $1 - \sqrt{20}$

60. Determine x and y: $3x - (5y + 3)i = -5 + 22i$.

For Exercises 61–70, perform the indicated operations and write the answer in standard form.

61. $(4 - 2i) + (6 - 4i)$ **62.** $(6 + 9i) - (-2 - 4i)$

63. $2i(1 - i)$ **64.** $(3 - 2i)(1 - 5i)$

65. $(3 - 5i) \div (1 - i)$ **66.** $\dfrac{4 + 9i}{2i}$

67. $\dfrac{6 + 2i}{1 + 3i}$ **68.** $(6 - 2i)^2$

69. i^{52} **70.** i^{-19}

71. Evaluate: $x^2 - x + 3$ for $x = 1 + i$.   **72.** Evaluate: $(2x - 4)^3$ for $x = 2 - 3i$.

C H A P T E R   5   **T E S T**

Assume that all variables are positive.

Section 5.1

For problems 1–3, simplify each expression.

1. $-\sqrt{100}$ **2.** $\sqrt[3]{27}$ **3.** $\sqrt{16x^4y^2}$

Sections 5.2 and 5.3

4. Simplify $(a^{-4/3}a^{1/6})^6$

For problems 5 and 6, multiply.

5. $(\sqrt{x} + 3\sqrt{y})(\sqrt{x} - 3\sqrt{y})$ **6.** $(\sqrt{b} - 3)^2$

7. Combine into one fraction: $\dfrac{2}{(x + 2)^{1/2}} - (x + 2)^{1/2}$

Section 5.4

For problems 8–10, simplify.

8. $\sqrt{54}$ **9.** $\sqrt[3]{-54}$ **10.** $\sqrt[4]{32a^8b^4}$

Sections 5.4 and 5.6

For problems 11–14, rationalize the denominator.

11. $\dfrac{3}{\sqrt{5}}$ **12.** $\dfrac{2}{\sqrt{a}}$ **13.** $\dfrac{1}{1-\sqrt{3}}$ **14.** $\dfrac{\sqrt{c}}{\sqrt{c}-2}$

For problems 15 and 16, simplify by combining like radicals.

15. $3\sqrt{2}-2\sqrt{2}+6\sqrt{2}$ **16.** $5a\sqrt{a^3}-a^2\sqrt{a}+3\sqrt{a^5}$

17. Multiply and then simplify: $\sqrt[3]{t^2}(4\sqrt[3]{t}-9\sqrt[3]{t^5})$

Section 5.7

For problems 18–21, solve the equation. Be sure to check your answers.

18. $\sqrt{x}=4$ **19.** $\sqrt{1+y}=2\sqrt{y}$ **20.** $1-2x=\sqrt{x}$ **21.** $\sqrt[3]{2t}+3=1$

Section 5.8

22. Express in standard form: $-3-\sqrt{-24}$

For problems 23–26, perform the indicated operations and write the answer in standard form.

23. $(3+5i)-(1-2i)$ **24.** $-3i(1-2i)$ **25.** i^{17} **26.** $\dfrac{14+5i}{2-3i}$

Quadratic Equations

Overview

One of the major topics of this book is studying techniques for solving equations. We have solved linear equations, certain quadratic equations, rational equations, and radical equations. We return to the problem of solving quadratic equations. In Chapter 3, Section 8, we solved quadratic equations *that were factorable.* In this chapter, we develop methods of solving the general quadratic equation. Then, we investigate further applications of quadratic equations.

6.1

The Square Root Method to Solve Quadratic Equations

OBJECTIVES

▶ *To solve quadratic equations of the form $x^2 = a$*

▶ *To solve quadratic equations of the form $(ax + b)^2 = c$*

In Chapter 3, we solved quadratic equations such as

$$3x^2 + 7x + 2 = 0$$

by factoring. Many quadratic equations, however, cannot be factored but may still have real solutions. In this chapter, we will develop techniques of solving a general quadratic equation.

A quadratic equation like $x^2 = 49$ can be solved two ways.

Method 1. Solve it by the factoring technique.

$$x^2 - 49 = 0$$

$$(x - 7)(x + 7) = 0$$

Either

$$x = 7 \qquad \text{or} \qquad x = -7$$

Method 2. By using square root properties from the last chapter, $x^2 = 49$ means that either $x = \sqrt{49}$ or $x = -\sqrt{49}$. That is, either

$$x = 7 \qquad \text{or} \qquad x = -7$$

Notice that Method 2 is the faster way to solve $x^2 = 49$. The general solution of an equation of the form $x^2 = a$ is summarized as follows.

S T R A T E G Y

The square root method of solving $x^2 = a$

Let a be a nonnegative constant. If $x^2 = a$, then either

$$x = \sqrt{a} \qquad \text{or} \qquad x = -\sqrt{a}$$

The phrase "either $x = \sqrt{a}$ or $x = -\sqrt{a}$" can be written as $x = \pm\sqrt{a}$.

In the next several examples, we will use the square root method to solve equations of the form $x^2 = a$.

Example 1   Solve $z^2 = 32$.

Solution   Using the square root solution method,

$$z = \pm\sqrt{32} = \pm\sqrt{16 \cdot 2} = \pm 4\sqrt{2} \qquad \blacktriangle$$

The square root method can be used on equations such as $(t + 7)^2 = 4$, if we think of $t + 7$ as a single quantity.

Example 2   Solve $(t + 7)^2 = 4$.

Solution   Thinking of $t + 7$ as a single quantity, by the square root method,

$$t + 7 = \pm\sqrt{4} = \pm 2$$

If $t + 7 = 2$, then $t = 2 - 7 = -5$.
If $t + 7 = -2$, then $t = -2 - 7 = -9$.

We check our two answers in the original equation.

Check $t = -5$:   *Check* $t = -9$:

$(t + 7)^2 = 4$ $\qquad$ $(t + 7)^2 = 4$

$(-5 + 7)^2 \overset{?}{=} 4$ $\qquad$ $(-9 + 7)^2 \overset{?}{=} 4$

$2^2 \overset{?}{=} 4$ $\qquad\qquad$ $(-2)^2 \overset{?}{=} 4$

$4 = 4$ $\qquad\qquad\quad$ $4 = 4$

The two solutions are -5 and -9. $\qquad\qquad\qquad\qquad\qquad$ $\blacktriangle$

Example 3   Solve $(z - 1)^2 = 75$.

Solution   Using the square root method,

$$z - 1 = \pm\sqrt{75} = \pm\sqrt{25 \cdot 3} = \pm 5\sqrt{3}$$

Therefore, $z = 1 \pm 5\sqrt{3}$.

Check $z = 1 + 5\sqrt{3}$:   *Check* $z = 1 - 5\sqrt{3}$:

$$(z - 1)^2 = 75 \qquad\qquad (z - 1)^2 = 75$$

$$(1 + 5\sqrt{3} - 1)^2 \overset{?}{=} 75 \qquad (1 - 5\sqrt{3} - 1)^2 \overset{?}{=} 75$$

$$(5\sqrt{3})^2 \overset{?}{=} 75 \qquad\qquad (-5\sqrt{3})^2 \overset{?}{=} 75$$

$$5^2(\sqrt{3})^2 \overset{?}{=} 75 \qquad\qquad (-5)^2(\sqrt{3})^2 \overset{?}{=} 75$$

$$25 \cdot 3 \overset{?}{=} 75 \qquad\qquad 25 \cdot 3 \overset{?}{=} 75$$

$$75 = 75 \qquad\qquad 75 = 75$$

The two solutions are $1 \pm 5\sqrt{3}$.   ▲

   In the next example, we will rationalize the denominator of our answer for the final simplified form.

Example 4   Solve $3y^2 + 4 = 6$.

Solution   Before using the square root method, we first subtract 4 from both sides, then divide by 3.

$$3y^2 + 4 = 6$$

$$3y^2 + 4 - 4 = 6 - 4$$

$$3y^2 = 2$$

$$y^2 = \frac{2}{3}$$

$$y = \pm\sqrt{\frac{2}{3}} \qquad \text{Apply the square root method.}$$

$$= \pm\frac{\sqrt{2}}{\sqrt{3}}$$

$$= \pm\frac{\sqrt{2}\sqrt{3}}{\sqrt{3}\sqrt{3}} \qquad \text{Rationalize the denominator.}$$

$$= \pm\frac{\sqrt{6}}{3}$$

It is up to you to check these two answers.

The solutions are $\dfrac{\sqrt{6}}{3}$ and $-\dfrac{\sqrt{6}}{3}$.   ▲

Example 5   Solve $\dfrac{(2a+1)^2}{3} = \dfrac{12}{25}$.

Solution   First multiply both sides of the equation by 3, then use the square root method.

$$\frac{(2a+1)^2}{3} = \frac{12}{25}$$

$$(2a+1)^2 = 3\left(\frac{12}{25}\right)$$

$$(2a+1)^2 = \frac{36}{25}$$

$$2a+1 = \pm\sqrt{\frac{36}{25}}$$

$$2a+1 = \pm\frac{6}{5}$$

$$2a = -1 \pm \frac{6}{5} \qquad \text{Subtract 1 from both sides.}$$

$$a = \frac{1}{2}\left(-1 \pm \frac{6}{5}\right) \qquad \text{Multiply both sides by } \frac{1}{2}.$$

Therefore, either

$$a = \frac{1}{2}\left(-1 + \frac{6}{5}\right) = \frac{1}{2}\left(\frac{1}{5}\right) = \frac{1}{10}$$

or

$$a = \frac{1}{2}\left(-1 - \frac{6}{5}\right) = \frac{1}{2}\left(-\frac{11}{5}\right) = -\frac{11}{10}$$

It is left to you to check these answers.   ▲

DEFINITION

An **isosceles** triangle is a triangle that has two sides of equal length.

Example 6   Consider the isosceles right triangle as shown in the figure. Denote the common length of the two equal sides by s and the length of the hypotenuse by c.

(a) Express s in terms of c.

(b) If the hypotenuse is 8 inches long, find s.

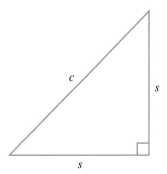

Solution

(a) By the Pythagorean property, $a^2 + b^2 = c^2$. For this particular right triangle, a and b are both equal to s. Therefore,

$$s^2 + s^2 = c^2$$

$$2s^2 = c^2$$

$$s^2 = \frac{c^2}{2}$$

For this equation, we apply the square root method. We only consider the positive square root, since s represents a length and therefore must be positive.

$$s = \sqrt{\frac{c^2}{2}} = \frac{\sqrt{c^2}}{\sqrt{2}} = \frac{c}{\sqrt{2}}$$

Rationalizing the denominator,

$$s = \frac{\sqrt{2}}{2}c$$

(b) We replace c by 8 in the formula developed in Part (a).

$$s = \frac{\sqrt{2}}{2}c = \frac{\sqrt{2}}{2}(8) = 4\sqrt{2} \text{ inches} \qquad \blacktriangle$$

In trigonometry, a special right triangle that is studied contains angles of 30 degrees and 60 degrees as shown in the figure.

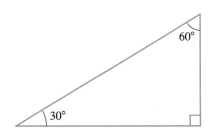

For this type of triangle, we have the following relationship.

R U L E

In a right triangle containing angles 30 degrees and 60 degrees, the length of the hypotenuse is twice the length of the side opposite the 30-degree angle.

We use this relationship in the next example.

Example 7   If the length of the side opposite the 60-degree angle in the triangle shown in the figure is 4 inches, find the lengths of the other two sides.

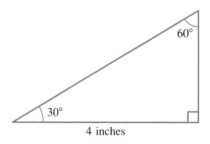

60°

30°

4 inches

Solution   Since the right triangle has angles of 30 degrees and 60 degrees, the length of the hypotenuse is twice the length of the side opposite the 30-degree angle. Therefore, if we let x be the length of the side opposite the 30-degree angle, $2x$ is the length of the hypotenuse. By the Pythagorean Theorem, $c^2 = a^2 + b^2$.
   Replacing c by $2x$, a by x, and b by 4, we have

$$(2x)^2 = x^2 + 4^2$$

$$4x^2 = x^2 + 16$$

$$3x^2 = 16$$

$$x^2 = \frac{16}{3}$$

$$x = \frac{4}{\sqrt{3}} = \frac{4\sqrt{3}}{3} \qquad x \text{ is positive. (Why?)}$$

The length of the side opposite the 30-degree angle is $\dfrac{4\sqrt{3}}{3}$ inches and the

length of the hypotenuse is $2\left(\dfrac{4\sqrt{3}}{3}\right)$ or $\dfrac{8\sqrt{3}}{3}$ inches.   ▲

CALCULATOR   CROSSOVER

If P dollars is deposited in a savings account at an annual interest rate of r (expressed as a decimal) and compounded twice a year, then the amount A in the account at the end of one year is given by the formula

$$A = P\left(1 + \frac{r}{2}\right)^2$$

1. Solve this formula for r.
2. If $P = \$500$ and $A = \$530.45$, find r.

Answers   1. $r = 2\left(\sqrt{\dfrac{A}{P}} - 1\right)$ 2. 6%

LEARNING   ADVANTAGE   **Is your answer reasonable?** *When solving a word problem, estimating and intelligent guessing are not improper and have their place. For example, intuitive guessing about the size or type of numbers that are reasonable answers is something to keep in mind as you solve the problem.*

EXERCISE   SET   6.1

For Exercises 1–25, solve each equation. Remember to check your answers.

1. $z^2 = 9$

2. $b^2 = 100$

3. $w^2 = \dfrac{4}{9}$

4. $r^2 = \dfrac{16}{81}$

5. $q^2 = 0.01$

6. $h^2 = 0.04$

7. $d^2 + 3 = 21$

8. $x^2 - 5 = 23$

9. $125 = t^2$

10. $50 = u^2$

11. $(s - 6)^2 = 16$

12. $(t + 15)^2 = 9$

13. $\left(z + \dfrac{5}{2}\right)^2 = \dfrac{25}{4}$

14. $\left(a - \dfrac{2}{3}\right)^2 = \dfrac{4}{9}$

15. $(y - 1)^2 = \dfrac{32}{9}$

16. $(a + 2)^2 = \dfrac{45}{16}$

17. $\dfrac{5}{y^2} - 2 = 28$

18. $\dfrac{8}{w^2} + 11 = 67$

19. $(2z - 3)^2 = 81$

20. $(3y + 1)^2 = 100$

21. $\dfrac{36}{(5r + 16)^2} = 3$

22. $4 = \dfrac{72}{(2c - 8)^2}$

23. $\left(\dfrac{2y + 3}{2}\right)^2 = 8$

24. $\dfrac{(6x + 12)^2}{3} = 12$

25. $(3 + 2y)^2 = 32$

26. $(8 - 3z)^2 = 128$

27. A positive number squared is 45. Find the number.

28. A negative number squared is 88. Find the number.

29. Twice a number is reduced by three. The result, squared, is equal to 18. Find the number. (There are two answers.)

30. Five is added to three times a number. The result, squared, is equal to 20. Find the number. (There are two answers.)

31. The area A of a circle of radius r is $A = \pi r^2$.
   (a) Solve this equation for r.
   (b) Find r, if the area is 20π square inches. Approximate your answer to the nearest tenth by using either a calculator or the table in the appendix.

32. The volume V of a right circular cylinder of radius r and height h is $V = \pi r^2 h$.
   (a) Solve this equation for r.
   (b) What is the radius of a right circular cylinder of volume 450π cubic feet and height 10 feet? Approximate your answer to the nearest tenth by using either a calculator or the table in the appendix.

33. The volume V of a right circular cone of radius r and height h is $V = \dfrac{1}{3}\pi r^2 h$. See the inside front cover for the figure.
   (a) Solve this equation for r.
   (b) What is the radius of a right circular cone of volume 252π cubic meters and height 7 meters? Approximate your answer to the nearest thousandth.

34. The volume V of a rectangular solid is given by $V = \ell w h$, where ℓ is the length, w the width, and h the height. See the inside front cover for the figure. Suppose a rectangular solid has a square base; that is, $\ell = w$.
   (a) Replace ℓ by w in the formula for the volume, then solve this formula for w.
   (b) Suppose a rectangular solid with a square base has a height of 5 inches and a volume of 120 cubic inches. What is the width of the square base? Approximate your answer to the nearest thousandth.

35. The formula for the volume V of a pyramid is $V = \dfrac{1}{3}Bh$, where B is the area of the base and h is the height. See the inside front cover for the figure. Suppose a pyramid has a rectangular base. This base has a length ℓ that is twice the width w.
   (a) Express the volume in terms of w and h, then solve for w.

   (b) Suppose the volume of this pyramid is 350 cubic yards and the height is 7 yards. Find the width and length of the base. Approximate your answers to the nearest hundredth.

36. The formula for the volume V of a pyramid is $V = \dfrac{1}{3}Bh$, where B is the area of the base and h is the height. See the inside front cover for the figure. Suppose a pyramid has a square base that measures s meters on each side.

    (a) Express the volume in terms of s and h, then solve for s.

(b) Suppose a pyramid with a square base has a volume of 12 cubic centimeters and a height of 5 centimeters. Find the width of the square base. Approximate your answer to the nearest tenth.

37. In baseball, the distance between bases is 90 feet. If the catcher throws the ball from home plate to second base, how far must the ball travel? Approximate your answer to the nearest foot by using either a calculator or the table in the appendix.

38. The braking distance d of a car depends upon its speed v. For one particular model, the braking distance is given by $d = \dfrac{3}{50}v^2$.

    (a) Solve this equation for v.

    (b) If the braking distance is 150 feet, how fast was the car traveling?

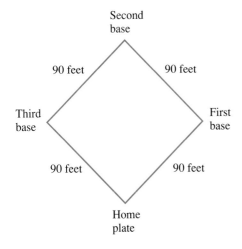

For Exercises 39–44, from the given right triangle, express s in terms of c.

39.

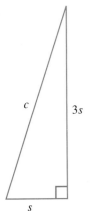

40.

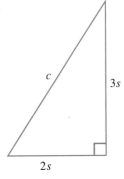

41.

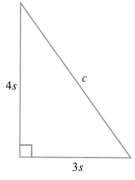

42.

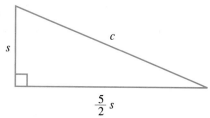

43.

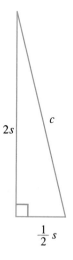

44.

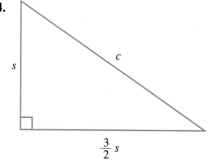

For Exercises 45–50, the length of one side is given. Find the lengths of the other two sides.

45.

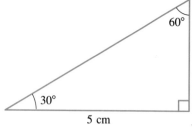

46.

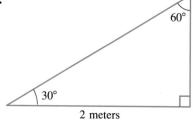

47.

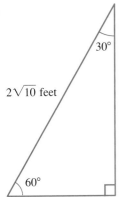

48.

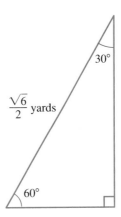

49.

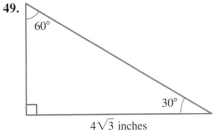

50.

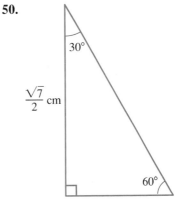

REVIEW PROBLEMS

The following exercises review parts of Section 3.3. Doing these problems will help prepare you for the next section.

Multiply.

51. $(x + 3)^2$

52. $(b + 7)^2$

53. $(y - 2)^2$

54. $(x - 1)^2$

55. $\left(x - \dfrac{2}{3}\right)^2$

56. $\left(z + \dfrac{5}{2}\right)^2$

57. $(3a + 1)^2$

58. $(2x_1 + 3)^2$

59. $(c^2 + 4)^2$

60. $(x^2 - 3)^2$

61. $(x^2 + y)^2$

62. $(x^2 - z^2)^2$

E N R I C H M E N T   E X E R C I S E S

For Exercises 1–4, solve using the square root property.

1. $(x + 1)^2 = (2x - 3)^2$

2. $(4a - 5)^2 = (7a + 3)^2$

3. $(3w - 5)^2 = (5 - 3w)^2$

4. $(3t + 7)^2 = (8 + 3t)^2$

For Exercises 5 and 6, solve for x using the square root property.

5. $\left(x + \dfrac{b}{4}\right)^2 = \dfrac{b^2 - 8c}{16}$

6. $\left(x + \dfrac{b}{2a}\right)^2 = \dfrac{b^2 - 4ac}{4a^2}$

Answers to Enrichment Exercises are on page A.28.

6.2

Completing the Square to Solve Quadratic Equations

O B J E C T I V E

▶ *To complete the square to solve a quadratic equation*

In Section 6.1, we could use the square root method to solve a quadratic equation in which the left side was a perfect square. For example, in the equation $(x + 3)^2 = 17$, the left side, $(x + 3)^2$, is the square of the binomial $x + 3$. We have from Chapter 3 two formulas involving squares of binomials:

$$(x + b)^2 = x^2 + 2bx + b^2$$

$$(x - b)^2 = x^2 - 2bx + b^2$$

The two trinomials from the right sides of the above equations,

$$x^2 + 2bx + b^2$$

and

$$x^2 - 2bx + b^2$$

are *examples of perfect square trinomials,* in which the coefficient of x^2 is one. Some examples of perfect square trinomials follow:

$$x^2 + 4x + 4, \qquad r^2 - 8r + 16, \qquad \text{and} \qquad z^2 + \frac{8}{3}z + \frac{16}{9}$$

Can you see the relationship between the coefficient of the middle term, $2b$ or $-2b$, and the constant term b^2?

R U L E

For the perfect square trinomials $x^2 + 2bx + b^2$ and $x^2 - 2bx + b^2$, the constant term is

$$\left(\frac{\text{coefficient of the middle term}}{2} \right)^2$$

That is, to produce the constant term, take $\dfrac{1}{2}$ of the coefficient of the middle term, then square.

N O T E   *This rule applies only to perfect square trinomials in which the coefficient of the squared variable term is one.*

Now, suppose we are given a binomial like $x^2 + 14x$. What constant should we add to this expression so that we obtain a perfect square trinomial? That is,

$$x^2 + 14x + ?$$

is a perfect square. The middle term of the proposed perfect square will be $14x$. Therefore, divide the coefficient of the middle term, 14, by 2, then square. The correct constant is $\left(\dfrac{14}{2} \right)^2 = (7)^2 = 49$. Thus, 49 must be added to $x^2 + 14x$ to make the perfect square trinomial $x^2 + 14x + 49$.

Example 1   What constant must be added to $x^2 - 6x$ to make a perfect square trinomial?

Solution   The coefficient of the x term is -6. Divide -6 by 2, then square. The correct constant is

$$\left(-\frac{6}{2} \right)^2 = (-3)^2 = 9 \qquad\qquad \blacktriangle$$

We can use this technique to solve quadratic equations.

Example 2   Solve $r^2 + 8r = -7$.

Solution   We first find the correct constant to make the binomial on the left side a perfect square trinomial. The coefficient of r is 8 and $\left(\dfrac{8}{2} \right)^2 = (4)^2 = 16$. Therefore, we add 16 to *both* sides.

$$r^2 + 8r + 16 = -7 + 16$$

$$(r + 4)^2 = 9 \qquad r^2 + 8r + 16 \text{ is a perfect square.}$$

The equation is now in the form where the square root method of the last section applies.

$$r + 4 = \pm 3 \qquad \text{The square root method.}$$

$$r = -4 \pm 3 \qquad \text{Subtract 4 from both sides.}$$

Therefore, either $r = -4 + 3 = -1$ or $r = -4 - 3 = -7$.

Check $r = -1$:   *Check* $r = -7$:

$$r^2 + 8r = -7 \qquad\qquad r^2 + 8r = -7$$

$$(-1)^2 + 8(-1) \overset{?}{=} -7 \qquad (-7)^2 + 8(-7) \overset{?}{=} -7$$

$$1 - 8 \overset{?}{=} -7 \qquad\qquad 49 - 56 \overset{?}{=} -7$$

$$-7 = -7 \qquad\qquad -7 = -7$$

The solutions are -1 and -7.   ▲

The method used in Example 1 is called **completing the square.** Completing the square will work on any quadratic equation.

Example 3   Solve $y^2 - 2y = 2$ by completing the square.

Solution   To complete the square, the correct constant to add is $\left(\dfrac{-2}{2}\right)^2 = (-1)^2 = 1$.

Therefore, we add 1 to both sides.

$$y^2 - 2y + 1 = 2 + 1$$

$$(y - 1)^2 = 3$$

$$y - 1 = \pm\sqrt{3}$$

$$y = 1 \pm \sqrt{3}$$

The two answers are $y = 1 + \sqrt{3}$ and $y = 1 - \sqrt{3}$.

On your own, check these two answers in the original equation.   ▲

If the coefficient of the squared variable term is not one, we first divide both sides of the equation by this number before completing the square. In the next three examples, we show this technique.

Example 4   Solve $4z^2 + 8z = 21$ by completing the square.

Solution   To make the coefficient of z^2 equal to one, divide both sides by 4.

$$4z^2 + 8z = 21$$

$$z^2 + 2z = \frac{21}{4}$$

$$z^2 + 2z + 1 = \frac{21}{4} + 1 \qquad \text{Complete the square.}$$

$$(z + 1)^2 = \frac{21}{4} + \frac{4}{4}$$

$$(z + 1)^2 = \frac{25}{4}$$

$$z + 1 = \pm\sqrt{\frac{25}{4}}$$

$$z + 1 = \pm\frac{5}{2}$$

$$z = -1 \pm \frac{5}{2}$$

Therefore, either $z = -1 + \dfrac{5}{2} = \dfrac{3}{2}$ or $z = -1 - \dfrac{5}{2} = -\dfrac{7}{2}$.   ▲

Example 5   Solve $3p^2 - 4p = 4$ by completing the square.

Solution   First divide both sides by 3 before completing the square.

$$3p^2 - 4p = 4$$

$$p^2 - \frac{4}{3}p = \frac{4}{3}$$

Now, to complete the square, take $\dfrac{1}{2}$ of the coefficient of p, then square:

$$\left[\frac{1}{2}\left(-\frac{4}{3}\right)\right]^2 = \left(-\frac{2}{3}\right)^2 = \frac{4}{9}$$

We therefore add $\dfrac{4}{9}$ to both sides of the equation.

$$p^2 - \frac{4}{3}p + \frac{4}{9} = \frac{4}{3} + \frac{4}{9}$$

$$\left(p - \frac{2}{3}\right)^2 = \frac{16}{9}$$

$$p - \frac{2}{3} = \pm\frac{4}{3}$$

$$p = \frac{2}{3} \pm \frac{4}{3}$$

Therefore, either $p = \dfrac{2}{3} + \dfrac{4}{3} = \dfrac{6}{3} = 2$ or $p = \dfrac{2}{3} - \dfrac{4}{3} = -\dfrac{2}{3}$. Check these two answers in the original equation.   ▲

Example 6   Solve $4v^2 - 24v = -40$ by completing the square.

Solution

$$4v^2 - 24v = -40$$

$$v^2 - 6v = -10 \qquad \text{Divide both sides by 4.}$$

$$v^2 - 6v + 9 = -10 + 9$$

$$(v - 3)^2 = -1$$

Notice that this equation cannot be solved, since the square root of -1 is not a real number. Therefore, the original equation has no solution in the real number system. It can be solved using complex numbers.   ▲

Sometimes the equation must first be rewritten so that the variable terms are on one side and the constant is on the other side.

Example 7   Solve $c^2 + 10 = -22 - 12c$ by completing the square.

Solution   We first rearrange terms before completing the square.

$$c^2 + 12c = -22 - 10$$

$$c^2 + 12c = -32$$

$$c^2 + 12c + 36 = -32 + 36 \qquad \text{Complete the square.}$$

$$(c + 6)^2 = 4$$

$$c + 6 = \pm 2$$

$$c = -6 \pm 2$$

Therefore, either $c = -6 + 2 = -4$ or $c = -6 - 2 = -8$.   ▲

We summarize the method of completing the square to solve quadratic equations.

S T R A T E G Y

A strategy for solving quadratic equations by completing the square

Step 1   If necessary, rewrite the equation so that the variable terms are on one side and the constant is on the other side.

Step 2   If the coefficient, a, of the squared variable term is not one, divide both sides of the equation by a.

Step 3   Complete the square on the side having the variables. Be sure to add the number that completes the square to *both* sides of the equation.

Step 4   Use the square root method to solve the equation in Step 3.

Step 5   Check your answers in the original equation.

CALCULATOR   CROSSOVER

Solve and round the answers to one decimal place.

1. $(3.2x - 4.7)^2 = 7.8$ **2.** $(10.1z + 6.9)^2 = 12.3$

Answers   **1.** 2.3 and 0.6   **2.** −0.3 and −1.0

LEARNING   ADVANTAGE   **Is your answer reasonable?** *When working on a word problem, make sure that the answer makes sense. If your answer is that* $4\frac{1}{2}$ *people bought tickets for the concert, or that the toolbox measured negative 5 feet on each side, stop, and review your calculations.*

EXERCISE   SET   6.2

1. What constant must be added to $x^2 - 8x$ to make a perfect square trinomial?

2. What constant must be added to $y^2 - 22y$ to make a perfect square trinomial?

3. What constant must be added to $r^2 + \frac{5}{3}r$ to make a perfect square trinomial?

4. What constant must be added to $z^2 - \frac{2}{7}z$ to make a perfect square trinomial?

For Exercises 5–34, solve the equation by completing the square.

5. $s^2 + 8s = 33$

6. $y^2 + 10y = -21$

7. $t^2 - 12t = -11$

8. $r^2 - 2r = 8$

9. $q^2 + 16q = -9$

10. $p^2 - 4p = 11$

11. $s^2 - 14s = 72$

12. $a^2 + 3a = -2$

13. $z^2 + 5z = -\frac{11}{2}$

14. $y^2 - 7y = -11$

15. $c^2 - 3c = -\dfrac{77}{36}$

16. $d^2 + d = -\dfrac{3}{16}$

17. $x^2 + \dfrac{4}{5}x = -\dfrac{3}{25}$

18. $y^2 + \dfrac{8}{3}y = -\dfrac{14}{9}$

19. $3w^2 + 2w = \dfrac{17}{3}$

20. $5m^2 + 4m = \dfrac{121}{5}$

21. $7v^2 - 6v = \dfrac{47}{7}$

22. $9b^2 - 3b = 12$

23. $5r^2 + 3r + \dfrac{2}{5} = 0$

24. $4k^2 - 7k + 3 = 0$

25. $z^2 - 100 = 6z + 12$

26. $a^2 - 64 = -12a - 36$

27. $2b^2 - 6b = -b + 7$

28. $5n^2 = \dfrac{27}{5} + 6n$

29. $32 = 9x^2 + 18x + 35$

30. $16z = 28z^2 + \dfrac{9}{7}$

31. $(x + 2)^2 = 2(x + 5)$

32. $y(y - 4) = 5$

33. $(2s - 4)(2s + 4) - (5s + 1)(s + 1) = -9$

34. $36(t - 1)^2 = 36t - 45$

REVIEW PROBLEMS

The following exercises review parts of Section 5.5. Doing these problems will help prepare you for the next section.

For Exercises 35–38, simplify.

35. $\dfrac{4 + 8}{10}$

36. $\dfrac{9 - 27}{6}$

37. $\dfrac{12 - 48}{6}$

38. $\dfrac{14 + 98}{21}$

Evaluate $\dfrac{-b \pm \sqrt{b^2 - 4ac}}{2a}$ for the given trinomial $ax^2 + bx + c$.

39. $x^2 - 6x + 9$

40. $4x^2 - 12x + 9$

41. $x^2 - 2x - 3$

42. $x^2 + 2x - 4$

43. $4x^2 + 4x - 1$

44. $x^2 + 4x - 11$

ENRICHMENT EXERCISES

For Exercises 1–4, solve by completing the square.

1. $x^4 - 8x^2 = -7$

2. $r^4 + 6r^2 = 7$

3. $\sqrt{y} + 2 = \dfrac{1}{4}y$

4. $\sqrt{a} + 3 + \dfrac{1}{\sqrt{3} + 3} = 3$

For Exercises 5–7, solve for x.

5. $x^2 + bx + 1 = 0$ **6.** $ax^2 + bx + c = 0$

7. $(\sqrt{x} - 1)^2 - 6(\sqrt{x} - 1) = 7$ (*Hint:* Set $z = \sqrt{x} - 1$ and then complete the square.)

8. Write the given equation in the form $(x - h)^2 + (y - k)^2 = r^2$, where h, k, and r are constants.

$$x^2 + 6x + y^2 - 4y + 9 = 0$$

(*Hint:* Complete the square twice, once on the x terms and again on the y terms.)

9. Write the given equation in the form $\dfrac{(x - h)^2}{a^2} + \dfrac{(y - k)^2}{b^2} = 1$, where a, b, h, and k are constants.

$$x^2 - 2x + 4y^2 + 16y = -1$$

Answers to Enrichment Exercises are on page A.28.

6.3

The Quadratic Formula

O B J E C T I V E

▶ *To solve a quadratic equation using the quadratic formula*

In the last section, we solved quadratic equations by completing the square. If we apply this method on the general equation, we will obtain a formula that will automatically generate any solutions. The **standard form for a quadratic equation** is

$$ax^2 + bx + c = 0, \qquad a \neq 0$$

It is important that $a \neq 0$, since if a is zero, we have $0x^2 + bx + c = 0$, or simply $bx + c = 0$. This is a linear equation, and linear equations were studied previously.

In order to solve a quadratic equation using the quadratic formula, it is important to write the quadratic equation in standard form and then to identify the values of a, b, and c.

Example 1 If necessary, write the quadratic equation in standard form so that a is positive. Identify the values of a, b, and c.

(**a**) $3x^2 - 5x + 2 = 0$ (**b**) $t^2 + t - 21 = 0$

(**c**) $3 + 7r = 2r^2$ (**d**) $(2w - 1)(w + 2) = \dfrac{3}{2}$

Solution   **(a)** The equation is already in standard form with $a = 3$, $b = -5$, and $c = 2$.

(b) Since $t^2 + t - 21 = 0$ is the same as $1 \cdot t^2 + 1 \cdot t - 21 = 0$, $a = 1$, $b = 1$, and $c = -21$.

(c) This equation is not in standard form, since one side is not equal to zero. Therefore, we subtract $2r^2$ from both sides of the equation.

$$3 + 7r - 2r^2 = 2r^2 - 2r^2$$

$$3 + 7r - 2r^2 = 0$$

$$-2r^2 + 7r + 3 = 0 \qquad \text{Rearrange terms.}$$

Since the coefficient on r^2 is negative, our final step is to multiply both sides of the equation by -1.

$$2r^2 - 7r - 3 = 0$$

The equation is now in standard form where $a = 2$, $b = -7$, and $c = -3$.

(d) We first do the indicated multiplication on the left side.

$$(2w - 1)(w + 2) = \frac{3}{2}$$

$$2w^2 + 3w - 2 = \frac{3}{2}$$

$$2w^2 + 3w - \frac{7}{2} = 0$$

This equation is in standard form; however, to avoid fractions, we have the option to multiply both sides by 2.

$$2\left(2w^2 + 3w - \frac{7}{2}\right) = 2 \cdot 0$$

$$4w^2 + 6w - 7 = 0$$

Therefore, $a = 4$, $b = 6$, and $c = -7$.   ▲

The quadratic formula is derived by completing the square on the general quadratic equation. We start with the standard form.

$$ax^2 + bx + c = 0, \quad a \neq 0$$

If a is not one, divide both sides by a.

$$x^2 + \frac{b}{a}x + \frac{c}{a} = \frac{0}{a}$$

Next, subtract $\dfrac{c}{a}$ from both sides of the equation.

$$x^2 + \frac{b}{a}x = -\frac{c}{a}$$

Now, to complete the square, divide $\dfrac{b}{a}$ by 2, then square.

$$\dfrac{\frac{b}{a}}{2} = \dfrac{b}{2a}, \qquad \text{so} \qquad \left(\dfrac{\frac{b}{a}}{2}\right)^2 = \left(\dfrac{b}{2a}\right)^2 = \dfrac{b^2}{4a^2}$$

Therefore, we add $\dfrac{b^2}{4a^2}$ to both sides of the equation.

$$x^2 + \dfrac{b}{a}x + \dfrac{b^2}{4a^2} = -\dfrac{c}{a} + \dfrac{b^2}{4a^2}$$

The left side is now a perfect square trinomial.

$$\left(x + \dfrac{b}{2a}\right)^2 = -\dfrac{c}{a} + \dfrac{b^2}{4a^2}$$

Next, we combine the two terms of the right side as one fraction.

$$\left(x + \dfrac{b}{2a}\right)^2 = \dfrac{-4ac}{4a^2} + \dfrac{b^2}{4a^2} \qquad \text{The common denominator is } 4a^2.$$

$$\left(x + \dfrac{b}{2a}\right)^2 = \dfrac{-4ac + b^2}{4a^2}$$

$$\left(x + \dfrac{b}{2a}\right)^2 = \dfrac{b^2 - 4ac}{4a^2} \qquad -4ac + b^2 = b^2 - 4ac.$$

Taking square roots,

$$x + \dfrac{b}{2a} = \pm\sqrt{\dfrac{b^2 - 4ac}{4a^2}}$$

$$x + \dfrac{b}{2a} = \pm\dfrac{\sqrt{b^2 - 4ac}}{\sqrt{4a^2}}$$

$$x + \dfrac{b}{2a} = \pm\dfrac{\sqrt{b^2 - 4ac}}{2a}$$

$$x = -\dfrac{b}{2a} \pm \dfrac{\sqrt{b^2 - 4ac}}{2a} \qquad \text{Subtract } \dfrac{b}{2a} \text{ from both sides.}$$

$$x = \dfrac{-b \pm \sqrt{b^2 - 4ac}}{2a}$$

This proves the quadratic formula.

> ## T H E O R E M
>
> ### Quadratic formula
>
> If $ax^2 + bx + c = 0$, where $a \neq 0$, then
>
> $$x = \frac{-b \pm \sqrt{b^2 - 4ac}}{2a}$$

N O T E   *Even though the quadratic formula works whether a is positive or negative, for convenience, we will always write the quadratic equation so that a is positive.*

In the next examples, we show how to find solutions of quadratic equations by using the quadratic formula.

Example 2   Solve $2x^2 - x - 3 = 0$ by the quadratic formula.

Solution   This equation is in standard form with $a = 2$, $b = -1$, and $c = -3$. Therefore, by the quadratic formula,

$$x = \frac{-b \pm \sqrt{b^2 - 4ac}}{2a}$$

$$= \frac{1 \pm \sqrt{(-1)^2 - 4(2)(-3)}}{2(2)}$$

$$= \frac{1 \pm \sqrt{1 + 24}}{4}$$

$$= \frac{1 \pm \sqrt{25}}{4}$$

$$= \frac{1 \pm 5}{4}$$

Therefore, either $x = \dfrac{1 + 5}{4} = \dfrac{6}{4} = \dfrac{3}{2}$ or $x = \dfrac{1 - 5}{4} = \dfrac{-4}{4} = -1$.   ▲

N O T E   *The solutions of the equation in Example 2 are rational numbers. This means that the original equation, $2x^2 - x - 3 = 0$, could have been solved by factoring. In fact, $2x^2 - x - 3$ factors into $(2x - 3)(x + 1)$. Therefore, $(2x - 3)(x + 1) = 0$, and so either $x = \dfrac{3}{2}$ or $x = -1$.*

Example 3   Solve $\dfrac{y^2}{4} + \dfrac{y}{2} - 4 = 0$ by the quadratic formula.

Solution   This equation is in standard form; however, it involves fractions. Therefore, before using the quadratic formula, first multiply both sides of the equation by 4 to clear fractions.

$$4\left(\frac{y^2}{4} + \frac{y}{2} - 4\right) = 4(0)$$

$$y^2 + 2y - 16 = 0$$

In this form, $a = 1$, $b = 2$, and $c = -16$. Using the quadratic formula,

$$y = \frac{-b \pm \sqrt{b^2 - 4ac}}{2a}$$

$$= \frac{-2 \pm \sqrt{2^2 - 4(1)(-16)}}{2(1)}$$

$$= \frac{-2 \pm \sqrt{4 + 64}}{2}$$

$$= \frac{-2 \pm \sqrt{68}}{2}$$

$$= \frac{-2 \pm \sqrt{4 \cdot 17}}{2}$$

$$= \frac{-2 \pm 2\sqrt{17}}{2}$$

The two solutions are $-1 + \sqrt{17}$ and $-1 - \sqrt{17}$.   ▲

To check our solutions from Example 3, we could substitute them into the original equation. However, there is a quicker way to check our solutions. Let us start with the general quadratic equation $ax^2 + bx + c = 0$. Suppose it has two solutions x_1 and x_2, where

$$x_1 = \frac{-b + \sqrt{b^2 - 4ac}}{2a} \qquad \text{and} \qquad x_2 = \frac{-b - \sqrt{b^2 - 4ac}}{2a}$$

Now, if we add the two solutions,

$$x_1 + x_2 = \frac{-b + \sqrt{b^2 - 4ac}}{2a} + \frac{-b - \sqrt{b^2 - 4ac}}{2a}$$

$$= \frac{-2b}{2a} = -\frac{b}{a}$$

If we take the product of the two solutions,

$$x_1 x_2 = \frac{-b + \sqrt{b^2 - 4ac}}{2a} \cdot \frac{-b - \sqrt{b^2 - 4ac}}{2a}$$

$$= \frac{b^2 - (b^2 - 4ac)}{4a^2}$$

$$= \frac{b^2 - b^2 + 4ac}{4a^2}$$

$$= \frac{4ac}{4a^2}$$

$$= \frac{c}{a}$$

Therefore, we have the following check.

THEOREM

If x_1 and x_2 are the solutions of $ax^2 + bx + c = 0$, then

1. $x_1 + x_2 = -\dfrac{b}{a}$

and

2. $x_1 x_2 = \dfrac{c}{a}$

Example 4   Solve $x^2 - 4x - 14 = 0$ and check the solutions.

Solution   Using the quadratic formula with $a = 1$, $b = -4$, and $c = -14$,

$$x = \frac{4 \pm \sqrt{16 - 4(1)(-14)}}{2}$$

$$= \frac{4 \pm \sqrt{72}}{2}$$

$$= \frac{4 \pm \sqrt{36 \cdot 2}}{2}$$

$$= \frac{4 \pm 6\sqrt{2}}{2}$$

$$= 2 \pm 3\sqrt{2}$$

Next, we check our solutions $x_1 = 2 + 3\sqrt{2}$ and $x_2 = 2 - 3\sqrt{2}$.

Sum: $x_1 + x_2 = (2 + 3\sqrt{2}) + (2 - 3\sqrt{2}) = 4$

and

$$-\frac{b}{a} = -\frac{-4}{1} = 4$$

Product: $x_1 x_2 = (2 + 3\sqrt{2})(2 - 3\sqrt{2}) = 2^2 - (3\sqrt{2})^2$

$$= 4 - 9 \cdot 2 = -14$$

and

$$\frac{c}{a} = \frac{-14}{1} = -14$$

▲

Example 5   Solve $2x^2 - 3x + 4 = 0$.

Solution   Using the quadratic formula with $a = 2$, $b = -3$, and $c = 4$,

$$x = \frac{3 \pm \sqrt{(-3)^2 - 4(2)(4)}}{2(2)}$$

$$= \frac{3 \pm \sqrt{9 - 32}}{4}$$

$$= \frac{3 \pm \sqrt{-23}}{4}$$

Notice that in this expression, we are to find the square root of a negative number. Since $\sqrt{-23}$ does not exist (as a real number), we conclude that the quadratic equation has no real solution.   ▲

C A L C U L A T O R   C R O S S O V E R

Solve, then round off your answers to one decimal place.

1. $2.4x^2 + 6.3x - 4.5 = 0$ **2.** $7.1a^2 - 9.8a + 1.2 = 0$
3. $x^2 - 4x - 35 = 0$

Answers   **1.** 0.6 and -3.2 **2.** 1.2 and 0.1   **3.** -4.2 and 8.2

L E A R N I N G   A D V A N T A G E   **Is your answer reasonable?** *If your answer to a word problem is unreasonable because the answer is negative when it should be positive, check each step in the solution where negative or positive numbers were combined through addition, subtraction, multiplication, or division.*

E X E R C I S E   S E T   6.3

For Exercises 1–10, if necessary, write the quadratic equation in the standard form $ax^2 + bx + c = 0$ so that a is positive. Identify the values of a, b, and c. Do not solve the equation.

1. $3x^2 + 4x - 5 = 0$

2. $6x^2 - 17x + 9 = 0$

3. $-y^2 + y - 1 = 0$

4. $-3z^2 + 4z + 1 = 0$

5. $3(v - 2)^2 = 7$

6. $(2r - 1)(r + 3) = 4$

7. $0 = 2y^2 - y(5y - 1)$

8. $7 = t(3 - 15t)$

9. $(x - 2)^2 = (7x + 3)x$

10. $(x - 3)(3x - 1) = x + 5$

For Exercises 11–42, use the quadratic equation to solve. Check your solutions.

11. $x^2 + 6x - 27 = 0$

12. $t^2 + 12t + 35 = 0$

13. $s^2 - 3s + 2 = 0$

14. $u^2 - 2u - 8 = 0$

15. $2v^2 + 7v - 4 = 0$

16. $3z^2 + 2z - 5 = 0$

17. $6c^2 + 13c + 6 = 0$

18. $4b^2 - 13b + 3 = 0$

19. $y^2 + y + 1 = 0$

20. $d^2 - 2d + 2 = 0$

21. $3m^2 + m - 1 = 0$

22. $4k^2 - k - 1 = 0$

23. $s^2 + 2s - 1 = 0$

24. $x^2 + 4x - 2 = 0$

25. $v^2 - 4v - 2 = 0$

26. $y^2 - 6y - 1 = 0$

27. $x^2 + 6x + 9 = 0$

28. $r^2 - 8r + 16 = 0$

29. $a^2 - 2a = -3$

30. $b^2 + 30 = -b$

31. $4z^2 - 7 = 4z$

32. $4d^2 + 12d = 5$

33. $0 = 20y^2 - 4y - 1$

34. $1 = 8y^2 - 2y$

35. $c^2 - 3c = 0$

36. $4z^2 + 7z = 0$

37. $s^2 - 12 = 0$

38. $a^2 - 28 = 0$

39. $\dfrac{x^2}{3} - \dfrac{x}{3} - \dfrac{1}{3} = 0$

40. $\dfrac{5y^2}{12} - y - \dfrac{2}{3} = 0$

41. $-x^2 + \dfrac{4}{3}x - \dfrac{1}{9} = 0$

42. $-y^2 + \dfrac{2}{3}y + \dfrac{2}{9} = 0$

For Exercises 43–48, solve using the quadratic formula. Then approximate the solutions to the nearest tenth using either a calculator or the table in the Appendix.

43. $2x^2 - 2x - 1 = 0$

44. $3z^2 + 3z - 2 = 0$

45. $4y^2 + 10y + 5 = 0$

46. $2p^2 - 8p + 7 = 0$

47. $8z^2 + 12z + 3 = 0$

48. $6s^2 + 11s + 2 = 0$

49. Solve $x^3 - 2x^2 - 2x = 0$.

50. Solve $x^3 - 4x^2 + 2x = 0$.

51. What is wrong with the following solution?

$$x^2 - 3x - 5 = 0$$

$$x = \frac{-3 \pm \sqrt{9 + 20}}{2}$$

$$= \frac{-3 \pm \sqrt{29}}{2}$$

What is the correct solution?

52. What is wrong with the following solution?

$$x^2 + 2x - 4 = 0$$

$$x = -2 \pm \frac{\sqrt{4 + 16}}{2}$$

$$= -2 \pm \frac{\sqrt{20}}{2}$$

$$= -2 \pm \sqrt{5}$$

What is the correct solution?

53. Solve $x(2x + 3) = 2$ in three ways: (1) the factoring method, (2) completing the square, and (3) the quadratic formula. Which method did you prefer? Why?

54. Solve $3(x^2 - 1) = 8x$ in three ways: (1) the factoring method, (2) completing the square, and (3) the quadratic formula. Which method did you prefer? Why?

The following exercises review parts of Section 3.9. Doing these problems will help prepare you for the next section.

55. The square of a positive number is $\dfrac{3}{4}$ more than the number. Find the number.

56. Find all numbers with the property that the square is 12 more than the number.

57. The length of a rectangle is one inch more than twice the width. Find the dimensions of the rectangle if the area is 21 square inches.

58. The height of a triangle is 2 cm less than the base. Find the height and base of the triangle, if the area is 24 square centimeters.

E N R I C H M E N T   E X E R C I S E S

For Exercises 1–5, solve using the quadratic formula.

1. $x^4 - 10x^2 + 9 = 0$

2. $(x - \sqrt{3})^2 + 6(x - \sqrt{3}) - 3 = 0$ (*Hint:* Let $z = x - \sqrt{3}$.)

3. $\dfrac{2}{y^2} + \dfrac{4}{y} - 3 = 0$

4. $x^2 + 2\sqrt{3}x - 9 = 0$

5. $r^2 - 4r\sqrt{2} - 2 = 0$

6. Replace a by $-a$, b by $-b$, and c by $-c$ in the quadratic formula to show that $-ax^2 - bx - c = 0$ has the same solutions as $ax^2 + bx + c = 0$.

Answers to Enrichment Exercises are on page A.29.

Which Method to Use?   We now have four methods available to solve a quadratic equation: factoring, the square root method, completing the square, and the quadratic formula. The completing the square method always works, but generally is not used. However, it did enable us to verify the quadratic formula and it does have uses in other parts of mathematics.

The following is a strategy for solving quadratic equations.

S T R A T E G Y

A strategy for solving the quadratic equation
$$ax^2 + bx + c = 0$$

1. If $b = 0$, use the square root method.
2. If $b \neq 0$, think first of the factoring method. If it appears that the trinomial does not factor, or if the factors are not apparent, use the quadratic formula.

To sharpen your ability to solve quadratic equations by whatever method seems appropriate, the following exercise set has been created. Keep in mind that if everything else fails, the quadratic formula will always work.

MISCELLANEOUS QUADRATIC EQUATIONS

Use the strategy for solving quadratic equations to find the solution set of each given equation.

1. $x^2 - 4x - 1 = 0$

2. $x^2 + 5x - 3 = 0$

3. $x^2 - x + 2 = 0$

4. $x^2 - 5x + 6 = 0$

5. $6r^2 + 7r - 3 = 0$

6. $3y^2 + 5y + 2 = 0$

7. $4x^2 - 4x - 5 = 0$

8. $8s^2 + s - 1 = 0$

9. $z^2 - 16 = 0$

10. $v^2 + 9 = 0$

11. $4(c^2 - 5c) + 25 = 0$

12. $\dfrac{1}{4}a^2 + \dfrac{1}{8}a - \dfrac{1}{2} = 0$

13. $\dfrac{1}{4}m^2 = m$

14. $\dfrac{2}{3}w = \dfrac{1}{3}w^2$

15. $t(t + 1) = 20$

16. $m(2m + 3) = 5$

17. $a^2 + 11a + 3 = 0$

18. $3n^2 - 4n + 5 = 0$

19. $0 = 10b^2 + 11b + 3$

20. $0 = -12x^2 + 11x - 2$

21. $y\left(3y + \dfrac{17}{2}\right) = 7$

22. $3(5k^2 + 2) = 19k$

23. $3c^2 = \sqrt{27}c$

24. $4z = \sqrt{48}z^2$

25. $d(d + 2) = 4$

26. $2(v^2 - 2) = -v$

27. $3c^2 + 8c - 1 = 0$

28. $9t^2 - 3t - 1 = 0$

29. $(2s + 1)(s - 5) = -2(s + 5)$

30. $(8u + 1)(u - 2) = 5u + 1$

31. $2s^2 - \sqrt{10}s = 0$

32. $6y^2 + \sqrt{3}y = 0$

33. $r + 2 = \dfrac{2}{r}$

34. $2n + 4 = \dfrac{5}{n}$

35. $\dfrac{2}{k - 1} + \dfrac{3}{k + 1} = 1$

36. $\dfrac{1}{2x + 1} + \dfrac{1}{2x - 1} = 1$

37. $\dfrac{2}{x} - \dfrac{3}{x-4} = 1$

38. $\dfrac{3}{t} + \dfrac{4}{t+1} = 2$

39. $\dfrac{\sqrt{8}}{z} + \dfrac{\sqrt{18}}{z} = \sqrt{2}z$

40. $-\dfrac{\sqrt{12}}{m} + \dfrac{\sqrt{27}}{m} = \sqrt{3}m$

6.4

Applications of Quadratic Equations

OBJECTIVE

▶ *To solve word problems using quadratic equations*

In this section, we present some applications of quadratic equations. Keep in mind the strategy for solving word problems. For your convenience, we restate this strategy which was first given in Chapter 2.

STRATEGY

Solving word problems

1. **Read** the problem carefully. Take note of what is being asked and what information is given.
2. **Plan** a course of action. Draw a figure, if appropriate. Represent the unknown number by a letter. If there are two or more unknowns, represent one of them by a letter and express the others in terms of the letter.
3. **Create** an equation from the given information.
4. **Solve** this equation.
5. **Check** your solution using the original statement of the problem.

Example 1   A projectile is launched from ground level. The height, h (in meters), above the ground t seconds later is given by

$$h = 50t - 4.9t^2$$

(a) How far above the ground will the projectile be after 2 seconds? 4 seconds?

(b) How many seconds will it take the projectile to hit the ground from the time it was launched?

Solution   **(a)** To find the height h after 2 seconds, replace t by 2 in the equation $h = 50t - 4.9t^2$.

$$h = 50t - 4.9t^2$$

$$= 50(2) - 4.9(2)^2$$

$$= 100 - 4.9(4)$$

$$= 100 - 19.6 = 80.4 \text{ meters}$$

To find the height after 4 seconds, replace t by 4.

$$h = 50t - 4.9t^2$$

$$= 50(4) - 4.9(4)^2$$

$$= 200 - 4.9(16)$$

$$= 200 - 78.4 = 121.6 \text{ meters}$$

(b) Remember that h is the height of the projectile after t seconds. When the projectile hits the ground, then the height is zero. Therefore, we want to find t when $h = 0$.

$$h = 50t - 4.9t^2$$

$$0 = 50t - 4.9t^2 \qquad \text{Setting } h = 0.$$

$$0 = t(50 - 4.9t) \qquad \text{Factor the right side.}$$

$$\text{either } t = 0 \text{ or } 50 - 4.9t = 0$$

$$-4.9t = -50$$

$$t = \frac{-50}{-4.9} \approx 10.2$$

The value of 0 seconds gives the time when the projectile was launched. The second value of 10.2 seconds gives the (approximate) time that the projectile hits the ground after being launched.   ▲

Example 2   A car that was involved in an accident left skid marks measuring 120 feet. The braking distance d for this particular make of car can be approximated by

$$d = \frac{1}{10}v^2 + v$$

where v is the speed of the car. What was the speed of the car at the time of the accident?

Solution   Set $d = 120$ and solve the resulting quadratic equation for v.

$$d = 120$$

$$\frac{1}{10}v^2 + v = 120$$

$$\frac{1}{10}v^2 + v - 120 = 0$$

Using the quadratic formula, set $a = \dfrac{1}{10}$, $b = 1$, and $c = -120$.

$$v = \frac{-1 \pm \sqrt{1^2 - 4\left(\dfrac{1}{10}\right)(-120)}}{2\left(\dfrac{1}{10}\right)}$$

$$= \frac{-1 \pm \sqrt{1 + 48}}{\dfrac{1}{5}}$$

$$= (-1 \pm \sqrt{49})(5)$$

$$= (-1 \pm 7)(5)$$

Therefore, either $v = (-1 - 7)(5) = -40$ or $v = (-1 + 7)(5) = 30$. Since v represents the speed of the car, we discard $v = -40$. The car was traveling 30 mph at the time of the accident. ▲

Example 3 The base of a triangle is 6 centimeters longer than the height as shown in the figure. If the area is 90 square centimeters, find the base and the height.

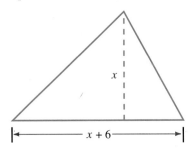

Solution From the figure, the height is x centimeters and the base is $x + 6$ centimeters. The area of a triangle is

$$\text{area} = \left(\frac{1}{2}\right)(\text{base})(\text{height})$$

$$= \left(\frac{1}{2}\right)(x + 6)(x)$$

Since the area is 90 square centimeters,

$$\left(\frac{1}{2}\right)(x + 6)(x) = 90$$

We solve this equation.

$$x^2 + 6x = 180$$

$$x^2 + 6x - 180 = 0$$

The trinomial does not factor, so we use the quadratic formula.

$$x = \frac{-6 \pm \sqrt{6^2 - 4(1)(-180)}}{2(1)}$$

$$= \frac{-6 \pm \sqrt{36 + 720}}{2}$$

$$= \frac{-6 \pm \sqrt{756}}{2}$$

$$= \frac{-6 \pm \sqrt{36 \cdot 21}}{2}$$

$$= \frac{-6 \pm 6\sqrt{21}}{2}$$

$$= -3 \pm 3\sqrt{21}$$

The two solutions of the equation are $x = -3 + 3\sqrt{21}$ and $x = -3 - 3\sqrt{21}$. However, since x represents a length, x is positive. We therefore discard the negative solution $-3 - 3\sqrt{21}$. The *height* of the triangle is

$$x = -3 + 3\sqrt{21} \text{ centimeters}$$

and the *base* is

$$x + 6 = -3 + 3\sqrt{21} + 6 = 3 + 3\sqrt{21} \text{ centimeters}$$

Using a calculator or the table in the Appendix, we have, accurate to three decimal places,

$$\text{height} \approx 10.748 \text{ centimeters}$$

$$\text{base} \approx 16.748 \text{ centimeters}$$ ▲

Example 4   A rectangular piece of glass, 10 inches wide and 16 inches long, is to be cut so that the area is reduced to 90 square inches (see the figure). What are the dimensions of the resulting piece of glass?

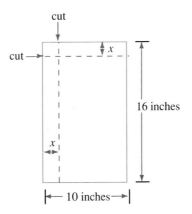

Solution

Let x be the number of inches cut from the length and width of the glass as shown in the figure. Then the dimensions of the reduced piece of glass is $10 - x$ inches wide and $16 - x$ inches long. The area of this piece of glass is to be 90 square inches. Therefore,

$$(\text{width}) \times (\text{length}) = 90$$

$$(10 - x)(16 - x) = 90$$

Next, we rewrite this equation in standard form and use the quadratic formula to solve it.

$$160 - 26x + x^2 = 90$$

$$160 - 26x + x^2 - 90 = 0$$

$$x^2 - 26x + 70 = 0$$

To solve this equation, we use the quadratic formula with $a = 1$, $b = -26$, and $c = 70$,

$$x = \frac{26 \pm \sqrt{(-26)^2 - 4(1)(70)}}{2(1)}$$

$$= \frac{26 \pm \sqrt{676 - 280}}{2}$$

$$= \frac{26 \pm \sqrt{396}}{2}$$

$$= \frac{26 \pm \sqrt{4 \cdot 9 \cdot 11}}{2}$$

$$= \frac{26 \pm 2 \cdot 3\sqrt{11}}{2}$$

$$= 13 \pm 3\sqrt{11}$$

Therefore, either $x = 13 + 3\sqrt{11}$ or $x = 13 - 3\sqrt{11}$. Now, since we are subtracting x from 10, $10 - (13 + 3\sqrt{11}) = -3 - 3\sqrt{11}$ is negative. Therefore, $13 + 3\sqrt{11}$ is not a sensible solution and is discarded. We use the other solution $x = 13 - 3\sqrt{11}$. Using a calculator or the table in the Appendix, $x \approx 13 - 3(3.317) = 13 - 9.951 = 3.049$ inches. Rounding off to the nearest tenth, $x \approx 3.0$ inches.

Therefore, the dimensions of the reduced piece of glass are approximately $10 - 3 = 7$ inches by $16 - 3 = 13$ inches. Notice that the resulting area is $7 \cdot 13 = 91$ square inches, not the requested 90 square inches. This is because of the rounding off error. ▲

Example 5

Beaver Falls Can Company has an order to make open cardboard containers. Each container must be 3 inches high and have a volume of 150 cubic inches. The company plans to make a container from a square piece of cardboard by cutting squares measuring 3 inches on a side from each corner of the cardboard and folding up the sides (see the figure). What size cardboard is needed to make the container?

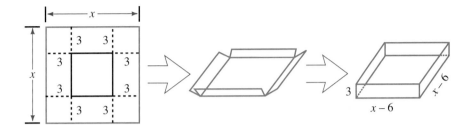

Solution   Let x be the common length of each side of the square piece of cardboard. The base of the box is a square, $x - 6$ inches on each side.

The volume of the box is

$$\text{volume} = \text{length} \times \text{width} \times \text{height}$$

$$= (x - 6)(x - 6)(3)$$

$$= 3(x - 6)^2$$

Since the volume must be 150 cubic inches, we have

$$3(x - 6)^2 = 150$$

We rewrite this equation by first dividing both sides by 3.

$$(x - 6)^2 = 50$$

$$x^2 - 12x + 36 = 50 \qquad (a - b)^2 = a^2 - 2ab + b^2.$$

$$x^2 - 12x - 14 = 0 \qquad \text{Subtract 50 from both sides.}$$

Since the trinomial does not factor, we use the quadratic formula.

$$x = \frac{12 \pm \sqrt{12^2 - 4(1)(-14)}}{2(1)}$$

$$= \frac{12 \pm \sqrt{144 + 56}}{2}$$

$$= \frac{12 \pm \sqrt{200}}{2}$$

$$= \frac{12 \pm \sqrt{100 \cdot 2}}{2}$$

$$= \frac{12 \pm 10\sqrt{2}}{2}$$

$$= 6 \pm 5\sqrt{2}$$

Therefore, the two solutions of the quadratic equation are $6 + 5\sqrt{2}$ and $6 - 5\sqrt{2}$. Now, the second solution is discarded because $x - 6$ is a length and replacing x by $6 - 5\sqrt{2}$ gives $x - 6 = 6 - 5\sqrt{2} - 6 = -5\sqrt{2}$. The length certainly cannot be a negative number. Therefore, the desired size of the cardboard is

$$6 + 5\sqrt{2} \text{ inches on each side}$$

Using a calculator or the table in the Appendix, the length is approximately equal to 13.1 inches. ▲

CALCULATOR   CROSSOVER

The supply and demand equations for Colombia coffee beans in the Winston-Salem area are the following.

$$\text{supply:} \quad p = \frac{25}{16}x^2 - 5x + 1$$

$$\text{demand:} \quad p = -x + 10$$

where p is price per pound in dollars and x is the quantity in thousands of pounds of coffee. Find the equilibrium quantity and the equilibrium price.

Answer   4,000 pounds at $6 per pound.

L E A R N I N G   A D V A N T A G E   **Is your answer reasonable?** *Your answer to a word problem may be unreasonable because you have substituted incorrectly in a formula. For example, the interest formula is $I = Prt$, where I represents the interest in dollars, P represents the amount deposited or borrowed, r represents the interest rate, and t represents the time. If $I = 5$, $P = 100$, and $t = 1$, an incorrect substitution could yield $100 = 5r$; then $r = 20$, which is equivalent to 2,000%. The correct substitution is $5 = 100r$.*

E X E R C I S E   S E T   6.4

1. A ball is thrown up into the air. The height, h (in feet), above the ground t seconds later is given by

$$h = 4 + 30t - 16t^2$$

(a) How far above the ground will the ball be after $\frac{1}{2}$ second? 1 second?

(b) How many seconds will it take the ball to hit the ground from the time it was thrown?

2. A projectile is launched from ground level. The height, h (in meters), above the ground t seconds later is given by

$$h = 48t - 4.9t^2$$

(a) How far above the ground will the projectile be after 2 seconds? 3 seconds?

(b) How many seconds will it take the projectile to hit the ground from the time it was launched?

3. The number of diagonals N of a polygon of n sides is given by $N = \dfrac{n^2 - 3n}{2}$. Squares and rectangles are examples of four-sided polygons. Figures below show four- and five-sided polygons with the diagonals drawn in blue.

If a polygon has 35 diagonals, how many sides does it have? (*Hint:* Set $N = 35$ and solve for n.)

4. The sum S of the first n positive integers is given by $S = \dfrac{n(n + 1)}{2}$. Find n if the sum S is 1,275.

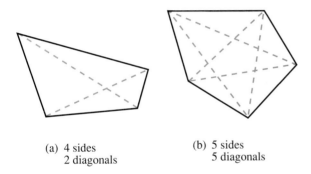

(a) 4 sides
    2 diagonals

(b) 5 sides
    5 diagonals

If an object is dropped from a tower, the distance d dropped in terms of the velocity v is $d = \dfrac{v^2}{64}$, where d is in feet and v is in feet per second. Refer to this formula when doing Exercises 5–8.

5. For a TV commercial, a car manufacturer wants to film a car dropping from the top of a building and hitting the ground. What must be the height of the building if the impact velocity is to be 80 feet per second (about 55 mph)?

6. Refer to Exercise 5. Suppose the manufacturer uses a building that is 200 feet tall. What will be the impact velocity of the car?

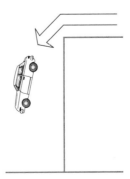

7. On a late night television program, a watermelon is dropped from the roof of a building. If the height of the building is 500 feet, what is the velocity of the watermelon when it hits the ground? Approximate your answer to the nearest tenth.

8. An object is dropped from an airplane flying at 1,280 feet. What is the velocity of the object when it hits the ground? Approximate your answer to the nearest tenth.

9. One number is five times another and their product is 40. Find the two numbers. (There are two answers.)

10. One number is twice another number. If their product is six, find the two numbers. (There are two answers)

11. One side of a rectangle is 3 centimeters longer than the other side. If the area is 36 square centimeters, find the dimensions of the rectangle. Approximate your answers to the nearest tenth.

12. The base of a triangle is 4 inches more than its height. If the area is 7 square inches, find the length of the base and height. Approximate your answers to the nearest tenth.

13. A rectangular piece of glass, 8 inches by 10 inches, is to be cut as shown in the figure so that the area is reduced to 47 square inches. What are the dimensions of the resulting piece of glass? Approximate the answers to the nearest tenth.

14. A rectangular piece of plywood, 8 inches by 11 inches, is to be cut as shown in the figure so that the area is reduced to 54 square inches. What are the dimensions of the resulting piece of plywood?

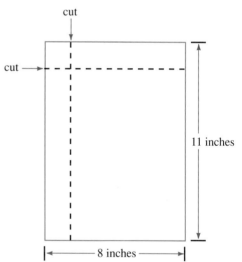

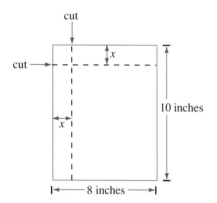

15. An open box is to be made from a square sheet of tin by cutting squares measuring 2 inches on a side from each corner and folding up the sides. If the required volume is 108 cubic inches, what size sheet of tin is needed to make the box? Approximate your answer to the nearest tenth.

16. An open box is to be made from a square sheet of cardboard by cutting squares measuring 3 inches on a side from each corner and folding up the sides. If the required volume is 144 cubic inches, what size piece of cardboard is needed to make the box? Approximate your answer to the nearest tenth.

17. New Hope Service makes and sells x dozen self-improvement audio tapes each day. The daily cost C is given by $C = 2x^2 + 54$ and the daily revenue R is given by $R = 24x$. Because of the time needed to make each tape, the company can produce no more than 4 dozen tapes each day. Find the break-even point for this company.

18. Valleycraft Company makes and sells x knitting machines each week. The daily cost C is given by $C = 9x^2 + 1,800$ and the daily revenue R is given by $R = 270x$. The company cannot make more than 15 machines each day. Find the break-even point for the company.

19. The sum of a number and its reciprocal is $\dfrac{10}{3}$. Find the number. (There are two answers.)

20. The sum of a number and its reciprocal is $\dfrac{17}{4}$. Find the number. (There are two answers.)

21. The difference of a number and its reciprocal is $-\dfrac{5}{6}$. Find the number. (There are two answers.)

22. The difference of a number and its reciprocal is $\dfrac{3}{2}$. Find the number. (There are two answers.)

23. Karna and Ed work for a construction company. On a surveying job, if each works alone, Karna can finish the job in 1 hour less than Ed. Working together they can finish the job in $1\dfrac{1}{5}$ hours. How long would it take Karna to do the job alone?

24. In a rope climbing contest, Kristie climbs a 15-meter rope 1 second faster than Stacey. If Kristie climbs $\dfrac{1}{2}$ meter per second faster than Stacey, what is Kristie's speed in climbing the rope?

REVIEW PROBLEMS

The following exercises review parts of Sections 3.5 and 4.4. Doing these problems will help prepare you for the next section.

Section 3.5

Factor.

25. $x^2 - 4x + 3$

26. $x^2 - 4x - 5$

27. $2y^2 - y - 6$

28. $3a^2 + 8a - 3$

29. $8x^2 - 11x + 3$

30. $6x^2 + 17x - 3$

31. $8t^2 + 2t - 6$

32. $18r^2 + 3r - 6$

Section 4.4

Combine into one fraction.

33. $\dfrac{1}{x-1} - \dfrac{2}{x+2}$

34. $\dfrac{3}{x+1} + \dfrac{4}{x-3}$

35. $\dfrac{4t}{2t-1} - \dfrac{2t}{t+3}$

36. $\dfrac{3y}{y-4} - \dfrac{3y}{y+2}$

37. $\dfrac{z}{z^2-1} + \dfrac{2z+1}{z+1}$

38. $\dfrac{4a}{a^2-4} - \dfrac{a+2}{a-2}$

39. $x + 1 + \dfrac{1}{x - 1}$ **40.** $2y - 3 - \dfrac{6}{y - 2}$

E N R I C H M E N T   E X E R C I S E S

1. The reaction time of a person can be measured by dropping a ruler as shown in the figure and asking the person to catch it as quickly as possible.

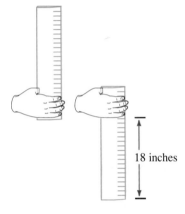

18 inches

For a particular person, the ruler dropped 18 inches. What is this person's reaction time? (*Hint:* Use the formula $d = 16t^2$.)

2. To move in a circular orbit, a satellite must have a horizontal velocity v, in meters per second, given by

$$v^2 = \frac{9.8R^2}{d + R}$$

where R meters is the radius of the earth, and d meters is the distance of the satellite above the earth. Find the horizontal velocity needed to keep a satellite located 600,000 meters (about 375 miles) above the earth in a circular orbit. Use 6,360,000 meters for the radius of the earth.

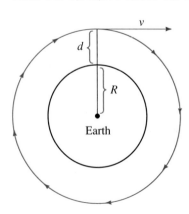

3. A wire 10 inches long is cut into two pieces. One piece is bent into a triangle with three equal sides and the other piece is bent into a square.

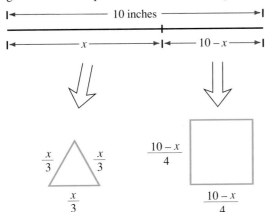

Let A be the sum of the areas of the triangle and square.

(a) Show that $A = \dfrac{1}{4}\left[\dfrac{(10-x)^2}{4} + \dfrac{\sqrt{3}}{9}x^2\right]$.

(b) Suppose the wire is cut so that one piece measures 6 inches and the other is 4 inches. Find A if the longer piece is used to form the triangle and the other piece is used to form the square.

Answers to Enrichment Exercises are on page A.30.

6.5

Quadratic and Rational Inequalities

O B J E C T I V E S

▶ *To solve quadratic inequalities*

▶ *To solve rational inequalities*

In this section, we develop a technique to solve quadratic inequalities in one variable. These inequalities will have one of the **standard forms:**

$$ax^2 + bx + c < 0 \qquad ax^2 + bx + c \le 0$$
$$ax^2 + bx + c > 0 \qquad ax^2 + bx + c \ge 0$$

where a, b, and c are constants, with $a \ne 0$.

Consider the quadratic inequality $x^2 + x - 2 > 0$. We will use a graphing method to determine its solution set. As we move x along the number line, the trinomial $x^2 + x - 2$ is sometimes positive, sometimes negative, and sometimes zero. To solve this inequality, we must find the values of x for which $x^2 + x - 2$ is positive. We make use of the following fact: *The intervals on which $x^2 + x - 2$ is positive are separated from intervals where it is negative by values of x for which it is zero.* Therefore, to find these intervals, we set $x^2 + x - 2$ equal to zero and solve the resulting equation.

$$x^2 + x - 2 = 0$$

$$(x + 2)(x - 1) = 0$$

Therefore, either $x = -2$ or $x = 1$. The two solutions, -2 and 1, divide the number line into three intervals as shown in the figure.

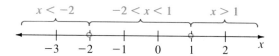

The trinomial $x^2 + x - 2$ cannot change sign over each of these intervals. If it is positive at one number in the interval, it is positive over the entire interval, or if it is negative at one number in the interval, it is negative over the entire interval. Therefore, to determine the sign of the trinomial on a given interval, we select a test number from that interval and use it to evaluate the trinomial. The resulting algebraic sign is the sign of the trinomial over the entire interval.

To find whether $x^2 + x - 2$ is positive or negative over the first interval, $x < -2$, we select any convenient test number in this interval, say, $x = -3$, and substitute it into $x^2 + x - 2$: $(-3)^2 + (-3) - 2 = 9 - 5 = 4 > 0$.

Since $x^2 + x - 2$ is positive at one number in this interval and cannot change its algebraic sign over this interval, we conclude that $x^2 + x - 2 > 0$ for all values of x where $x < -2$.

Next, we find the sign of $x^2 + x - 2$ on $-2 < x < 1$. We select 0 as our test number and evaluate $x^2 + x - 2$: $0^2 + 0 - 2 = -2 < 0$. Therefore, $x^2 + x - 2 < 0$ for all values of x in the interval $-2 < x < 1$. Finally, using another test number, say $x = 2$, in the interval $x > 1$, we find that $2^2 + 2 - 2 = 4 > 0$. Therefore, $x^2 + x - 2 > 0$ on $x > 1$.

We summarize this information in the following figure.

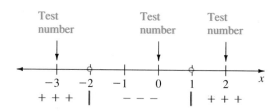

We conclude that $x^2 + x - 2 > 0$ provided that either $x < -2$ or $x > 1$. The solution set, $(-\infty, -2) \cup (1, +\infty)$, is shown on the following number line.

We summarize the steps taken to find the solution set of $x^2 + x - 2 > 0$ in the following strategy.

STRATEGY

Solving a quadratic inequality

Step 1   Put the inequality into standard form with zero on the right side and a quadratic on the left side.

Step 2   Set the quadratic equal to zero and solve the resulting equation.

Step 3   Arrange the solutions found in Step 2 in increasing order on a number line. These numbers will divide the number line into open intervals.

Step 4   The algebraic sign of the quadratic cannot change over any of the intervals found in Step 3.
Determine this sign for each interval by selecting a test number in the interval and substituting it for the unknown in the quadratic. The algebraic sign of the resulting value is the sign of the quadratic over the entire interval.

Step 5   Display the information obtained in Step 4 on the number line. The solution set of the inequality can be read from this figure.

If the quadratic equation of Step 2 has no real solutions, then just one interval is determined—the entire real line.

Example 1   Solve $2x^2 + x < 1$ and graph the solution set.

Solution   Step 1   The inequality is equivalent to $2x^2 + x - 1 < 0$.

Step 2   We set $2x^2 + x - 1$ equal to zero and solve the resulting equation.

$$2x^2 + x - 1 = 0$$

$$(x + 1)(2x - 1) = 0$$

Therefore, either $x = -1$ or $x = \dfrac{1}{2}$.

Step 3   We draw a number line and locate the solutions -1 and $\frac{1}{2}$. They divide

the number line into three intervals: $x < -1$, $-1 < x < \frac{1}{2}$, and $x > \frac{1}{2}$.

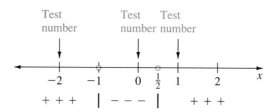

Step 4   We select convenient test numbers, one from each of the three intervals.

Interval	Test number	The sign of $2x^2 + x - 1$
$x < -1$	-2	$2(-2)^2 + (-2) - 1 = 8 - 3 = 5 > 0$
$-1 < x < \frac{1}{2}$	0	$2(0)^2 + 0 - 1 = -1 < 0$
$x > \frac{1}{2}$	1	$2(1)^2 + 1 - 1 = 2 > 0$

Step 5   We draw a figure showing the information from Step 4.

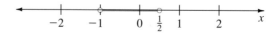

The solution set, $\left\{ x \,\middle|\, -1 < x < \frac{1}{2} \right\} = \left(-1, \frac{1}{2} \right)$, is graphed on the following number line.

Example 2   Solve $x^2 - 4x + 3 \geq 0$ and graph the solution set.

Solution   Set $x^2 - 4x + 3$ equal to zero and solve.

$$x^2 - 4x + 3 = 0$$

$$(x - 3)(x - 1) = 0$$

Therefore, either $x = 3$ or $x = 1$. Plotting these numbers on a line, we obtain the three intervals $x < 1$, $1 < x < 3$, and $x > 3$. Next, we select test numbers to evaluate $x^2 - 4x + 3$.

Interval	Test number	Evaluate $x^2 - 4x + 3$
$x < 1$	0	$0^2 - 4(0) + 3 = 3 > 0$
$1 < x < 3$	2	$2^2 - 4(2) + 3 = -1 < 0$
$x > 3$	4	$4^2 - 4(4) + 3 = 3 > 0$

We show this information on the following graph.

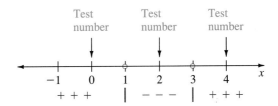

Therefore, either $x < 1$ or $x > 3$. Now, we are not done, since the inequality "$\geq$" means that either $x^2 - 4x + 3 > 0$ or $x^2 - 4x + 3 = 0$. Recall that the two numbers 1 and 3 make $x^2 - 4x + 3$ equal to zero, so they are included in the solution set.

The solution set, $\{x \mid \text{either } x \leq 1 \text{ or } x \geq 3\} = (-\infty, 1] \cup [3, +\infty)$, is graphed on the following number line.

Example 3   Solve $x^2 - 3x > x - 4$.

Solution   We first write this inequality in standard form, $x^2 - 4x + 4 > 0$, then solve the equation $x^2 - 4x + 4 = 0$.

$$x^2 - 4x + 4 = 0$$

$$(x - 2)^2 = 0$$

There is only one solution, $x = 2$. Notice that the inequality $x^2 - 4x + 4 > 0$ is of the special form $(x - 2)^2 > 0$, and this inequality is true for all x except $x = 2$. Therefore, the solution set is $\{x \mid x \neq 2\}$.   ▲

An inequality such as $\dfrac{2x - 3}{x + 2} < 0$ involving rational expressions is called a **rational inequality.** This inequality is in **standard form,** since the rational expression is on the left side and zero is on the right side.

We can find the solution set for rational inequalities by using a method similar to that for solving quadratic inequalities. A fraction can change its algebraic sign only if its numerator or denominator changes sign. Therefore, the following statement is true.

RULE

The intervals where a rational expression changes algebraic sign are separated by values of the variable for which either the numerator or denominator is zero.

Keep in mind that rational expression is undefined where the denominator is zero. In the next example, we show how to find the solution set of a rational inequality.

Example 4   Solve $\dfrac{x-3}{x+1} > 0$ and graph its solution set.

Solution   The numerator is zero when $x = 3$ and the denominator is zero when $x = -1$. We plot these two numbers in the following figure, dividing the number line into three intervals: $x < -1$, $-1 < x < 3$, and $x > 3$.

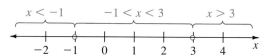

Next, we select the test numbers -2, 0, and 4 and evaluate $\dfrac{x-3}{x+1}$ at each number.

When $x = -2$, $\dfrac{x-3}{x+1} = \dfrac{-2-3}{-2+1} = \dfrac{-5}{-1} = 5 > 0$.

When $x = 0$, $\dfrac{x-3}{x+1} = \dfrac{0-3}{0+1} = -3 < 0$.

When $x = 4$, $\dfrac{x-3}{x+1} = \dfrac{4-3}{4+1} = \dfrac{1}{5} > 0$.

This information is shown in the following figure.

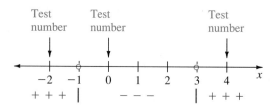

We conclude that the solution set is

$$\{x \mid \text{either } x < -1 \text{ or } x > 3\} = (-\infty, -1) \cup (3, +\infty)$$

and is graphed on the following number line.

Sometimes a rational inequality must first be put into standard form before solving it.

Example 5   Solve $\dfrac{1}{x - 1} \le \dfrac{2}{x + 1}$.

Solution   To write this inequality into standard form, first subtract $\dfrac{2}{x + 1}$ from both sides, then combine resulting two fractions on the right side.

$$\frac{1}{x - 1} \le \frac{2}{x + 1}$$

$$\frac{1}{x - 1} - \frac{2}{x + 1} \le 0$$

$$\frac{x + 1 - 2(x - 1)}{(x - 1)(x + 1)} \le 0$$

$$\frac{-x + 3}{(x - 1)(x + 1)} \le 0$$

Now the inequality is in standard form and we proceed as in the last example. The numerator is zero when $x = 3$ and the denominator is zero when $x = 1$ and when $x = -1$. We arrange these numbers on a number line and choose test numbers -2, 0, 2, and 4.

Interval	Test number	Algebraic sign of $\dfrac{-x + 3}{(x - 1)(x + 1)}$
$x < -1$	-2	Positive
$-1 < x < 1$	0	Negative
$1 < x < 3$	2	Positive
$x > 3$	4	Negative

This information is shown in the following figure.

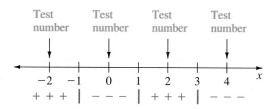

We conclude that the solution consists of all x such that either $-1 < x < 1$ or $x \geq 3$ and is graphed in the following figure.

COMMON ERROR

When putting an inequality such as $\dfrac{1}{x-1} \leq \dfrac{2}{x+1}$ into standard form, don't multiply both sides by the product of the two denominators. Remember that an inequality changes its sense if we multiply by a negative number, and a variable expression such as $x - 1$ is either positive or negative, depending on the value of x.

Example 6   Given: $y = \dfrac{3}{x^2 - 4}$. Find the values of x so that y is positive and the values of x so that y is negative.

Solution   Since $y = \dfrac{3}{x^2 - 4}$, y will be positive where $\dfrac{3}{x^2 - 4}$ is positive and y will be negative where $\dfrac{3}{x^2 - 4}$ is negative. Therefore, we find the intervals of x for which $\dfrac{3}{x^2 - 4}$ is positive and the intervals of x for which $\dfrac{3}{x^2 - 4}$ is negative.

Now, $\dfrac{3}{x^2 - 4} = \dfrac{3}{(x+2)(x-2)}$. The numerator is never zero; the denominator is zero when x is -2 and when x is 2. The number line, together with the analysis of the algebraic sign of $\dfrac{3}{x^2 - 4}$, is shown in the following figure.

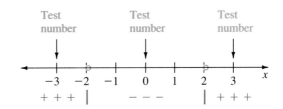

We conclude that y is positive when either $x < -2$ or $x > 2$ and y is negative when $-2 < x < 2$. ▲

L E A R N I N G   A D V A N T A G E   **Is the answer reasonable?** *A question on a test may relate to a practical situation in which the final answer will agree reasonably well with the real world. Suppose your answer is not reasonable. If you have time to review your calculations, consider the following possible error: If the answer is unreasonably large or unreasonably small, check the positions of any decimal points in your calculations.*

E X E R C I S E   S E T   6.5

For Exercises 1–26, solve the inequality and graph the solution set.

1. $x^2 + x - 2 < 0$

2. $x^2 - 2x - 3 > 0$

3. $x^2 - x - 12 \geq 0$

4. $x^2 - 3x + 2 \leq 0$

5. $2x^2 - 9x + 9 \geq 0$

6. $3x^2 - 4x + 1 \leq 0$

7. $6x^2 + x - 12 > 0$

8. $6x^2 - x - 1 > 0$

9. $x^2 - 6x + 9 \geq 0$

10. $x^2 + 8x + 16 \leq 0$

11. $t^2 - 4 > 0$

12. $y^2 - 9 < 0$

13. $(x - 3)(x + 3)(x - 1) > 0$

14. $x(x + 2)(x - 4) > 0$

15. $(x + 1)(3 - x) > (x - 2)(1 + x)$

16. $(x - 2)(1 - x) \leq (x - 2)(x - 5)$

17. $x^3 < x$

18. $x^5 \geq x^3$

19. $x^2(x - 3) < 0$

20. $(x - 1)^2(x + 1) > 0$

21. $-2x^2 + 4x + 6 \leq 0$

22. $-3x^2 + 9x + 12 \geq 0$

23. $x^3 - 6x^2 + 8x \geq 0$

24. $x^3 - x^2 - 6x \leq 0$

25. $(2x + 3)(x - 2) < x(x - 6)$

26. $(x - 4)(2x - 1) > (x - 3)(x - 4) - 5$

For Exercises 27–40, solve each rational inequality and graph the solution set.

27. $\dfrac{5}{x - 4} > 0$

28. $\dfrac{2}{3x - 6} < 0$

29. $\dfrac{2x + 8}{x + 3} \leq 4$

30. $\dfrac{x + 1}{x - 1} \leq 3$

31. $\dfrac{x + 2}{x - 1} \leq 2$

32. $\dfrac{1}{x + 3} \leq \dfrac{4}{x - 3}$

33. $\dfrac{4}{x - 2} + \dfrac{2}{x + 3} < 0$

34. $\dfrac{x}{x + 1} - \dfrac{x + 1}{x + 3} < 0$

35. $\dfrac{(x + 1)(x - 3)}{(x + 2)(x - 2)} < 0$

36. $\dfrac{(x - 1)(x + 4)}{(x - 3)(x + 3)} > 0$

37. $\dfrac{(x + 3)(1 - x)}{x - 2} \leq 0$

38. $\dfrac{(3 - 6x)(x - 2)}{2x + 1} \geq 0$

39. $\dfrac{(x - 2)(x - 4)(x + 1)}{x^2(x - 6)} > 0$

40. $\dfrac{(2x - 5)(x - 1)(x + 1)^2}{x(x - 4)} < 0$

For Exercises 41–50, find the values of x so that y is positive and the values of x so that y is negative.

41. $y = \dfrac{4}{x^2 - 25}$

42. $y = \dfrac{3}{x^2 - 16}$

43. $y = 3x^2 - 5x - 8$

44. $y = x^2 - 4x - 21$

45. $y = \dfrac{x - 6}{x + 2}$

46. $y = \dfrac{2x - 9}{3x + 2}$

47. $y = x(x - 4)(x + 10)$

48. $y = (x - 1)(x + 2)(x - 5)$

49. $y = (1 - x)^2(x - 2)$

50. $y = x^2(5x - 3)$

51. A ball is thrown straight upward where its height h feet above the ground after t seconds is given by $h = 80t - 16t^2$. Find the time interval when the ball is at least 96 feet above the ground.

52. Find all numbers so that the square plus the square of three more than the number is at least 45.

REVIEW PROBLEMS

The following exercises review parts of Section 5.8. Doing these problems will help prepare you for the next section.

Write each complex number in terms of i.

53. $\sqrt{-49}$

54. $\sqrt{-15}$

55. $\sqrt{-56}$

56. $-4 - \sqrt{-1}$

57. $-6 + \sqrt{-9}$

58. $\dfrac{7}{8} - \sqrt{-\dfrac{32}{4}}$

59. $\dfrac{1}{6} + \sqrt{-\dfrac{27}{3}}$

60. $\sqrt{3^2 - 4(3)(3)}$

61. $\sqrt{4^2 - 4(4)(9)}$

62. $\dfrac{-1 \pm \sqrt{1^2 - 4(2)(2)}}{2(2)}$

63. $\dfrac{2 \pm \sqrt{(-1)^2 - 4(4)(3)}}{2(4)}$

E N R I C H M E N T   E X E R C I S E S

For Exercises 1–4, solve each inequality and graph the solution set.

1. $x^4 - 5x^2 + 4 > 0$

2. $x^4 - 12x^2 + 27 > 0$

3. $x^3 - 4x - x^2 + 4 \geq 0$

4. $\dfrac{(x-3)^3 x}{(x-2)^2 x(x+1)} \leq 0$

5. Solve $x^3 - 13x + 12 > 0$, given that $x - 1$ is a factor of the left side.

Answers to Enrichment Exercises are on page A.30.

6.6

More Quadratic Equations

O B J E C T I V E S

▶ *To solve quadratic equations using complex numbers*

▶ *To solve equations that are quadratic in form*

Using the set of complex numbers, we can now solve quadratic equations that have no real solution. We will still use the strategy for solving quadratic equations on page 386.

Example 1 Solve $(3t + 2)^2 = -8$.

Solution The equation is of the form to use the square root method.

$$(3t + 2)^2 = -8$$

$$3t + 2 = \pm\sqrt{-8}$$

$$3t + 2 = \pm\sqrt{8}\sqrt{-1}$$

$$3t + 2 = \pm 2\sqrt{2}i$$

$$3t = -2 \pm 2\sqrt{2}i$$

$$t = -\frac{2}{3} \pm \frac{2\sqrt{2}}{3}i$$

The two numbers are $-\dfrac{2}{3} + \dfrac{2\sqrt{2}}{3}i$ and $-\dfrac{2}{3} - \dfrac{2\sqrt{2}}{3}i$. Check these numbers in the original equation. ▲

Example 2 Solve $x^2 - 3x + 5 = 0$.

Solution Since the trinomial cannot be factored, we use the quadratic formula with $a = 1$, $b = -3$, and $c = 5$.

$$x = \frac{-b \pm \sqrt{b^2 - 4ac}}{2a}$$

$$= \frac{3 \pm \sqrt{(-3)^2 - 4(1)(5)}}{2(1)}$$

$$= \frac{3 \pm \sqrt{9 - 20}}{2}$$

$$= \frac{3 \pm \sqrt{-11}}{2}$$

$$= \frac{3 \pm \sqrt{11}i}{2} \qquad\qquad \sqrt{-11} = \sqrt{11(-1)} = \sqrt{11}i.$$

$$= \frac{3}{2} \pm \frac{\sqrt{11}}{2}i$$

The two complex solutions are $\dfrac{3}{2} + \dfrac{i\sqrt{11}}{2}$ and $\dfrac{3}{2} - \dfrac{i\sqrt{11}}{2}$. Check these two numbers in the original equation. ▲

Example 3   Solve $r^2 + 6r + 11 = 0$.

Solution   Since $r^2 + 6r + 11$ cannot be factored, we use the quadratic formula with $a = 1$, $b = 6$, and $c = 11$.

$$r = \frac{-6 \pm \sqrt{6^2 - 4(1)(11)}}{2(1)}$$

$$= \frac{-6 \pm \sqrt{36 - 44}}{2}$$

$$= \frac{-6 \pm \sqrt{-8}}{2}$$

$$= \frac{-6 \pm \sqrt{8}i}{2}$$

$$= \frac{-6 \pm 2\sqrt{2}i}{2}$$

$$= -\frac{6}{2} \pm \frac{2\sqrt{2}i}{2}$$

$$= -3 \pm \sqrt{2}i$$

The two complex solutions are $-3 + i\sqrt{2}$ and $-3 - i\sqrt{2}$. ▲

In the quadratic formula $x = \dfrac{-b \pm \sqrt{b^2 - 4ac}}{2a}$, the expression $b^2 - 4ac$ is called the **discriminant.** The discriminant gives information about the nature and number of the roots (solutions) of the quadratic equation $ax^2 + bx + c = 0$. In particular, if $b^2 - 4ac > 0$, then $\sqrt{b^2 - 4ac} > 0$ and by the quadratic formula, there are two real roots. Similarly, if $b^2 - 4ac = 0$, there is exactly one real root. Finally, if $b^2 - 4ac < 0$, then $\sqrt{b^2 - 4ac}$ is a complex number and the quadratic formula yields two complex roots. We summarize these results:

R U L E

For the quadratic equation $ax^2 + bx + c = 0$, the discriminant gives the following information:

1. If $b^2 - 4ac > 0$, there are two real roots.
2. If $b^2 - 4ac = 0$, there is one real root.
3. If $b^2 - 4ac < 0$, there are two complex roots.

Example 4   Use the discriminant to determine if the roots are real or complex. Also, determine the number of roots.

(a) $2x^2 - x - 6 = 0$ **(b)** $4x^2 - 4x + 1 = 0$
(c) $t^2 - 2t + 3 = 0$

Solution

(a) Since $b^2 - 4ac = (-1)^2 - 4(2)(-6) = 1 + 48 = 49$ is positive, there are two real roots. Verify that they are 2 and $-\dfrac{3}{2}$.

(b) Since $b^2 - 4ac = (-4)^2 - 4(4)(1) = 0$, there is one real root. Verify that it is $\dfrac{1}{2}$.

(c) Since $b^2 - 4ac = (-2)^2 - 4(1)(3) = -8$ is negative, there are two complex roots. Verify that they are $1 + i\sqrt{2}$ and $1 - i\sqrt{2}$.   ▲

N O T E   *The discriminant of a quadratic equation does not give you the roots themselves, but instead gives information* about *the roots.*

Some equations that are not quadratic are nonetheless **quadratic in form.** By introducing a new variable, we can transform them into quadratic equations that can be solved by our usual techniques.

Example 5   Solve $(x + 4)^2 - 5(x + 4) + 6 = 0$.

Solution   We introduce a new variable—for instance y—by setting $y = x + 4$.

$$(x + 4)^2 - 5(x + 4) + 6 = 0$$
$$y^2 - 5y + 6 = 0$$

This resulting equation is quadratic in y and we solve it by factoring.

$$(y - 2)(y - 3) = 0$$

So, either $y = 2$ or $y = 3$. Next, we replace y by $x + 4$. Either

$$y = 2 \qquad \text{or} \qquad y = 3$$
$$x + 4 = 2 \qquad\qquad x + 4 = 3$$

So, either $x = -2$ or $x = -1$.   ▲

N O T E   *The equation $(x + 4)^2 - 5(x + 4) + 6 = 0$ of Example 5 could have been done by first simplifying to obtain a quadratic equation in x.*

$$(x + 4)^2 - 5(x + 4) + 6 = 0$$
$$x^2 + 8x + 16 - 5x - 20 + 6 = 0$$
$$x^2 + 3x + 2 = 0$$
$$(x + 2)(x + 1) = 0$$

So, either $x = -2$ or $x = -1$.

In the next example, we show how to solve another equation that is quadratic in form. For this one, however, there is only one method available to us.

Example 6   Solve $x^4 - 6x^2 - 16 = 0$.

Solution   If we rewrite this equation as $(x^2)^2 - 6x^2 - 16 = 0$, we see that it is quadratic in x^2. Setting $y = x^2$, we have

$$y^2 - 6y - 16 = 0$$

Solving this equation for y,

$$(y - 8)(y + 2) = 0$$

and therefore, either $y = 8$ or $y = -2$. Next, we replace y by x^2 and solve the two equations for x.

$$x^2 = 8 \qquad \text{or} \qquad x^2 = -2$$
$$x = \pm\sqrt{8} \qquad \text{or} \qquad x = \pm\sqrt{-2}$$
$$x = \pm 2\sqrt{2} \qquad \text{or} \qquad x = \pm i\sqrt{2}$$ ▲

Example 7   Solve $2t^{2/3} - t^{1/3} - 1 = 0$.

Solution   This equation is quadratic in $t^{1/3}$. To see this, rewrite it as $2(t^{1/3})^2 - t^{1/3} - 1 = 0$. Setting $x = t^{1/3}$,

$$2x^2 - x - 1 = 0$$

$$(x - 1)(2x + 1) = 0$$

Therefore, either

$$x = 1 \quad \text{or} \quad x = -\frac{1}{2}$$

Replacing x by $t^{1/3}$, either

$$t^{1/3} = 1 \quad \text{or} \quad t^{1/3} = -\frac{1}{2}$$

$$t = 1^3 = 1 \quad \text{or} \quad t = \left(-\frac{1}{2}\right)^3 = -\frac{1}{8}$$ ▲

Example 8   Solve $6x - 7\sqrt{x} + 2 = 0$.

Solution   If we rewrite x as $(\sqrt{x})^2$, then $6(\sqrt{x})^2 - 7\sqrt{x} + 2 = 0$, which is quadratic in $\sqrt{x}$. Set $y = \sqrt{x}$ and solve the resulting quadratic equation.

$$6y^2 - 7y + 2 = 0$$

$$(2y - 1)(3y - 2) = 0$$

So either

$$y = \frac{1}{2} \quad \text{or} \quad y = \frac{2}{3}$$

Replacing y by $\sqrt{x}$ and solving for x, either

$$\sqrt{x} = \frac{1}{2} \quad \text{or} \quad \sqrt{x} = \frac{2}{3}$$

$$x = \frac{1}{4} \quad \text{or} \quad x = \frac{4}{9} \qquad \text{Square both sides of each equation.}$$

We check our answers in the original equation.

Check $x = \frac{1}{4}$:   *Check* $x = \frac{4}{9}$:

$$6x - 7\sqrt{x} + 2 = 0 \qquad 6x - 7\sqrt{x} + 2 = 0$$

$$6\left(\frac{1}{4}\right) - 7\sqrt{\frac{1}{4}} + 2 \overset{?}{=} 0 \qquad 6\left(\frac{4}{9}\right) - 7\sqrt{\frac{4}{9}} + 2 \overset{?}{=} 0$$

$$\frac{3}{2} - 7\left(\frac{1}{2}\right) + 2 \overset{?}{=} 0 \qquad\qquad \frac{8}{3} - 7\left(\frac{2}{3}\right) + 2 \overset{?}{=} 0$$

$$0 = 0 \qquad\qquad\qquad\qquad 0 = 0$$

The two solutions are $\dfrac{1}{4}$ and $\dfrac{4}{9}$.   ▲

N O T E   *The equation $6x - 7\sqrt{x} + 2 = 0$ of Example 8 could have been solved by isolating the radical to one side of the equation, then squaring. This technique was explained in Section 5.7.*

In the next example, we show how to solve a formula for a variable when the formula is quadratic in that variable.

Example 9   A ball is thrown straight upward where the height h above the ground after t seconds is given by $h = 40t - 16t^2$. Solve this formula for t.

Solution   We first rewrite the formula as

$$16t^2 - 40t + h = 0$$

In this form, we see that it is quadratic in t. By using the quadratic formula with $a = 16$, $b = -40$, and $c = h$, we can solve for t.

$$t = \frac{40 \pm \sqrt{(-40)^2 - 4(16)(h)}}{2(16)}$$

$$= \frac{40 \pm \sqrt{1600 - 64h}}{32}$$

$$= \frac{40 \pm \sqrt{64(25 - h)}}{32}$$

$$= \frac{40 \pm 8\sqrt{25 - h}}{32}$$

$$= \frac{8(5 \pm \sqrt{25 - h})}{8 \cdot 4}$$

$$= \frac{5 \pm \sqrt{25 - h}}{4} \qquad\qquad ▲$$

L E A R N I N G   A D V A N T A G E   **Is your answer reason-able?** *Suppose on a test, you realize that your answer is unreasonable, but you do not have enough time to review your calculations. Write the following statement for the teacher.*

I know my answer is unreasonable (here you explain why it is unreasonable). However, I do not have time to locate my error.

E X E R C I S E   S E T   6 . 6

For Exercises 1–28, use the strategy for solving the quadratic equation on page 412 to solve each equation.

1. $x^2 + 49 = 0$

2. $y^2 + 121 = 0$

3. $(r + 1)^2 = -4$

4. $(t - 2)^2 = -9$

5. $(3s - 5)^2 - 128 = 0$

6. $(7q + 3)^2 - 8 = 0$

7. $v^2 - 3v - 1 = 0$

8. $2u^2 - 4u + 1 = 0$

9. $2p^2 - 5p + 3 = 0$

10. $2z^2 - z - 6 = 0$

11. $a^2 + 3a + 4 = 0$

12. $c^2 - 5c + 6 = 0$

13. $x^2 - 4x + 5 = 0$

14. $w^2 + 3w + 6 = 0$

15. $z^2 - 4z + 1 = 0$

16. $t^2 - 2t + 3 = 0$

17. $n^2 - 21 = 2n$

18. $p^2 - 4p = -11$

19. $(u - 1)^2 = 7u$

20. $(q - 2)(q + 3) = 6$

21. $(4y - 1)(y + 3) = 13y - 4$

22. $(4r + 1)(3r - 1) = 11r$

23. $16x(2x - 1) = -11$

24. $(3z + 7)(z + 1) = 13z$

25. $x^2 - 2\sqrt{3}x + 7 = 0$

26. $x^2 - 2\sqrt{2}x + 3 = 0$

27. $x^2 - 6\sqrt{2}x + 21 = 0$

28. $x^2 - 4\sqrt{3}x + 13 = 0$

For Exercises 29–38, use the discriminant to determine if the roots are real or complex. Also, determine the number of roots. Do not solve the equation.

29. $4x^2 + 4x - 3 = 0$

30. $2y^2 - 3y - 2 = 0$

31. $y^2 + 6y + 9 = 0$

32. $z^2 - 10z + 25 = 0$

33. $x^2 + 1 = 0$

34. $v^2 + 3 = 0$

35. $x^2 - 1 = 0$

36. $v^2 - 3 = 0$

37. $w^2 - 3w + 3 = 0$

38. $2r^2 - r + 1 = 0$

39. The difference of twice the square of a number and twice the number is negative three. Find the number. (There are two answers.)

40. The difference of a number and one, multiplied by the number, is negative two. Find the number. (There are two answers.)

For Exercises 41–64, solve each equation.

41. $9(x + 3)^2 - 15(x + 2) - 11 = 0$

42. $2(x - 2)^2 - (x - 2) - 6 = 0$

43. $(2t - 5)^2 - 4(2t - 5) + 5 = 0$

44. $(3x + 2)^2 + -6(3x + 2) + 11 = 0$

45. $6\left(\dfrac{x - 1}{2}\right)^2 - 7\left(\dfrac{x - 1}{2}\right) = 0$

46. $6\left(\dfrac{x + 2}{3}\right)^2 - 19\left(\dfrac{x + 2}{3}\right) = 0$

47. $x^4 - 9x^2 - 36 = 0$

48. $x^4 + 19x^2 + 18 = 0$

49. $t^4 + 13t^2 + 12 = 0$

50. $t^4 + 7t^2 - 8 = 0$

51. $4x^{2/3} - 4x^{1/3} - 3 = 0$

52. $4x^{2/3} - 7x^{1/3} + 3 = 0$

53. $5a^{2/3} - 7a^{1/3} + 2 = 0$

54. $3v^{2/3} + v^{1/3} - 10 = 0$

55. $15x - 8\sqrt{x} + 1 = 0$

56. $3x - 14\sqrt{x} + 8 = 0$

57. $4t - 8\sqrt{t} + 3 = 0$

58. $8z - 18\sqrt{z} + 9 = 0$

59. $2x^{2/5} - 3x^{1/5} + 1 = 0$

60. $t^{2/5} - t^{1/5} - 2 = 0$

61. $(2x + 1)^{2/3} - 4(2x + 1)^{1/3} + 3 = 0$

62. $(5x + 4)^{2/3} - 3(5x + 4)^{1/3} - 4 = 0$

63. $(x + 5) - 5\sqrt{x + 5} + 6 = 0$

64. $2(x - 3) - 3\sqrt{x - 3} + 1 = 0$

65. A ball is thrown straight upward where its height h feet above the ground after t seconds is given by $h = 80t - 16t^2$. Solve this formula for t.

66. An arrow is shot upward where its height h feet above the ground after t seconds is given by $h = 160t - 16t^2$. Solve this formula for t.

67. Solve $y = x^2 + 4x + 1$ for x.

68. Solve $y = 3x^2 - 2x - 1$ for x.

69. Solve $h = vt - 5t^2$ for t.

70. Solve $h = d + 32t - 16t^2$ for t.

REVIEW PROBLEMS

The following exercises review parts of Section 1.2. Doing these problems will help prepare you for the next section.

71. Replace the three numerical expressions by a single variable expression involving x, where x represents any member from the set $\left\{-\dfrac{1}{2}, 0, 1\right\}$.

$$\frac{3}{4} - 2\left(-\frac{1}{2}\right), \qquad \frac{3}{4}, \qquad \frac{3}{4} - 2(1)$$

72. Evaluate $3(2 - 4x)$ for $x = 2$.

73. Draw a number line and graph the numbers of the set $\left\{-3, -2, 0, \dfrac{1}{2}, \dfrac{4}{3}\right\}$.

ENRICHMENT EXERCISES

1. Verify that $4 = -(2i)^2$.

2. Use Exercise 1 to write $x^2 + 4$ as the difference of squares, then solve $x^2 + 4 = 0$ by factoring.

For Exercises 3–6, solve by factoring.

3. $x^2 + 25 = 0$

4. $r^2 + 100 = 0$

5. $t^2 + 8 = 0$

6. $z^2 + 27 = 0$

7. Solve $x^3 - 5x^2 + 9x - 5 = 0$, given that $x - 1$ is a factor of the left side.

8. Solve $x^3 - 4x^2 + 4x - 3 = 0$, given that $x - 3$ is a factor of the left side.

Answers to Enrichment Exercises are on page A.31.

C H A P T E R 6 Summary and review

Examples

The square root method to solve quadratic equations (6.1)

$(2x - 1)^2 = 16$
$2x - 1 = \pm 4$
$x = \dfrac{1 \pm 4}{2}$
$x = -\dfrac{3}{2}$ or $x = \dfrac{5}{2}$

We solve equations of the form $(ax + b)^2 = c$ by **taking square roots** of both sides.

Completing the square to solve quadratic equations (6.2)

$x^2 + 8x - 4 = 0$
$x^2 + 8x = 4$
$x^2 + 8x + 16 = 4 + 16$
$(x + 4)^2 = 20$
$x + 4 = \pm\sqrt{20}$
$x = -4 \pm 2\sqrt{5}$

In this section, we developed a strategy for solving quadratic equations by completing the square.

The quadratic formula (6.3)

Given: $2x^2 - 4x - 3 = 0$
Then,
$x = \dfrac{4 \pm \sqrt{(-4)^2 - 4(2)(-3)}}{2(2)}$
$= \dfrac{4 \pm \sqrt{16 + 24}}{4}$
$= \dfrac{4 \pm \sqrt{40}}{4}$
$= \dfrac{4 \pm 2\sqrt{10}}{4}$
$= \dfrac{2 \pm \sqrt{10}}{2}$

The quadratic formula: If $ax^2 + bx + c = 0$, where $a \neq 0$, then

$$x = \frac{-b \pm \sqrt{b^2 - 4ac}}{2a}$$

When solving a quadratic equation, your **answers can be checked** using the following property:

If x_1 and x_2 are the solutions of a quadratic equation, then

1. $x_1 + x_2 = -\dfrac{b}{a}$

2. $x_1 x_2 = \dfrac{c}{a}$

Examples

The two legs of a right triangle are the same length. If the hypotenuse is $3\sqrt{10}$ inches, find the length of the legs.

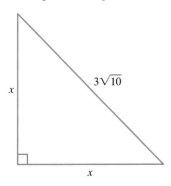

From the figure, and using the Pythagorean Theorem

$$x^2 + x^2 = (3\sqrt{10})^2$$
$$2x^2 = 9 \cdot 10$$
$$x^2 = 9 \cdot 5$$
$$x = 3\sqrt{5} \approx 6.7 \text{ inches}$$

Applications of quadratic equations (6.4)

There are many applications of quadratic equations. The strategy for solving word problems that was first given in Chapter 2 still applies.

Solve: $(x - 2)(x + 3) \le 0$

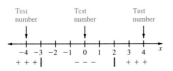

From the figure, the solution set is $-3 \le x \le 2$.

Quadratic and rational inequalities (6.5)

A quadratic inequality can be put into one of the **standard forms:**

$$ax^2 + bx + c < 0 \qquad ax^2 + bx + c \le 0$$
$$ax^2 + bx + c > 0 \qquad ax^2 + bx + c \ge 0$$

In this section, we developed a strategy for solving quadratic and rational inequalities.

Given: $x^2 - 2x + 5 = 0$
Then,

$$x = \frac{2 \pm \sqrt{(-2)^2 - 4(1)(5)}}{2(1)}$$
$$= \frac{2 \pm \sqrt{4 - 20}}{2}$$
$$= \frac{2 \pm \sqrt{-16}}{2}$$
$$= \frac{2 \pm 4i}{2}$$
$$= 1 \pm 2i$$

More quadratic equations (6.6)

The quadratic formula can be used to solve quadratic equations when the solutions are complex numbers. Complex solutions to the equation $ax^2 + bx + c = 0$ occur when the discriminant $b^2 - 4ac$ is negative.

The quadratic formula can be used to solve equations that are **quadratic in form.** By introducing a new variable, we can transform this type of equation into a quadratic equation.

CHAPTER 6 **REVIEW EXERCISE SET**

Section 6.1

For Exercises 1 and 2, use the square root method to solve.

1. $(3x - 7)^2 = 16$

2. $(4 - x)^2 - 8$

Section 6.2

For Exercises 3 and 4, complete the square to solve.

3. $x^2 + 2x = 36$

4. $x^2 - 3x = 3$

Section 6.3

For Exercises 5–8, use the quadratic formula to solve.

5. $x^2 - 2x - 2 = 0$

6. $r^2 - 4r + 1 = 0$

7. $3y^2 + 2y - 5 = 1$

8. $2s^2 - 3s = 9$

Section 6.4

9. A model rocket is launched from a pad 6 meters above ground level. The height, h (in meters), above the ground t seconds later is given by

$$h = 6 + 55t - 4.9t^2$$

  (a) What is the height of the rocket after 3 seconds? After 10 seconds?

  (b) How many seconds will it take the rocket to hit the ground from the time it was launched?

10. The braking distance d, in feet, for a particular car traveling v mph is approximated by

$$d = \frac{3}{10}v^2 + \frac{9}{10}v$$

The car required 516 feet to stop. Was the car traveling within the posted speed limit of 50 mph?

11. The base of a triangle is 5 inches longer than the height. If the area is 50 square inches, find the length of the base and the height. Approximate your answers to one decimal place using a calculator or the table in the Appendix.

12. A rectangular piece of plywood, 8 feet by 12 feet, is to be cut so that the area is reduced to 92 square feet as shown in the figure below. What are the dimensions of the resulting piece of plywood? Approximate your answers to the nearest tenth by using a calculator or the table in the Appendix.

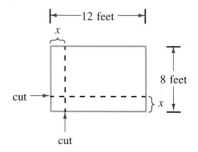

13. An open box is to be made from a square sheet of tin by cutting squares measuring 4 inches on a side from each corner and folding up the sides. If the required volume is 224 cubic inches, what size tin is needed to make the box? Approximate your answer to the nearest tenth.

Section 6.5

For Exercises 14–18, solve the inequality and graph the solution set.

14. $x^2 - 6x + 5 < 0$

15. $6x^2 + x - 1 > 0$

16. $\dfrac{x + 2}{x^2 + 5x} \geq 0$

17. $\dfrac{x - 4}{x^2 - x - 2} \leq 0$

18. $\dfrac{1}{x} \geq \dfrac{x - 1}{x^2 - 4}$

Section 6.6

For Exercises 19–22, solve each equation.

19. $y^2 - 6y + 25 = 0$

20. $4t^2 - 4t + 5 = 0$

21. $3x - 14\sqrt{x} + 15 = 0$

22. $x^4 + 11x^2 + 24 = 0$

23. Given $y = \dfrac{x - 2}{x + 1}$. Find the values of x so that y is positive and the values of x so that y is negative.

24. A pellet is shot straight up from the surface of the Earth so that its height h meters from the ground after t seconds is $h = 50t - 5t^2$. Solve this equation for t.

25. Solve $y = 3x^2 - 4x - 1$ for x.

Sections 6.1–6.3 and 6.6

For Exercises 26–45, solve the equation.

26. $9x^2 - 12x + 13 = 0$

27. $3x^2 + 5x - 2 = 0$

28. $t(t - 1) = t + 4$

29. $2y^2 - 6y + 5 = 0$

30. $(4x + 5)^2 = -12$

31. $x - 10\sqrt{x} + 9 = 0$

32. $x^4 - 4x^2 - 32 = 0$

33. $(1 - 5t)^2 = 16$

34. $y^2 - 4\sqrt{2}y + 6 = 0$

35. $t^4 - 13t^2 + 12 = 0$

36. $6r^2 - 19r + 10 = 0$

37. $(6x - 7)^2 = -48$

38. $3z^{2/3} - 5z^{1/3} - 2 = 0$

39. $8a^2 - 14a - 15 = 0$

40. $2x^2 - x + 1 = 0$

41. $(2y + 3)(y - 6) = y(2 - y) - 14$

42. $k^4 + 21k^2 - 72 = 0$

43. $2(3x - 7)^2 = -32$

44. $4x^{4/3} - 5x^{2/3} + 1 = 0$

45. $4t^{4/3} - 24t^{2/3} + 36 = 0$

C H A P T E R 6   **T E S T**

Section 6.1

For problems 1 and 2, use the square root method to solve the equation.

1. $(4x - 3)^2 = 9$

2. $(x + 2)^2 = 12$

Section 6.2

3. Complete the square to solve the equation:

$$x^2 + 4x = 14$$

Section 6.3

For problems 4–6, use the quadratic formula to solve the equation.

4. $2x^2 - 7x + 6 = 0$

5. $x^2 = 3(x + 1)$

6. $2x(x - 2) = -1$

Section 6.4

7. A ball is thrown up into the air. The height, h (in feet), of the ball above the ground t seconds later is

$$h = 6 + 46t - 16t^2$$

(a) What is the height of the ball after $\frac{1}{2}$ second? 2 seconds?

(b) How many seconds will it take the ball to hit the ground?

8. The base of a triangle is 3 inches longer than the height. If the area is 6 square inches, find the length of the base and the height. Approximate your answers to one decimal place.

Section 6.5

For problems 9–11, solve the inequality.

9. $x^2 + x - 6 > 0$

10. $3x^2 + 4x - 4 \leq 0$

11. $\dfrac{x + 3}{x^2 - 2x} \geq 0$

Section 6.6

For problems 12–14, solve the equation.

12. $x^2 - 2x + 2 = 0$ 　　　　　**13.** $x^2 + 2x + 4 = 0$ 　　　　　**14.** $8x^2 + 12x + 17 = 0$

Sections 6.1, 6.3, and 6.6

For problems 15–20, solve the equation.

15. $(2 + 3t)^2 = 16$ 　　　　　　　　　　**16.** $y^2 + 4y + 5 = 0$

17. $2x^2 - 4x - 3 = 0$ 　　　　　　　　　**18.** $2(3z^2 + 2) = 10z$

19. $x^2 + 6 = 4$ 　　　　　　　　　　　　**20.** $10x + (3x + 1)(x - 4) = 0$

Linear Equations in Two Variables and Their Graphs

Overview

Analytic geometry has its foundation based on associating points on a line with the set of real numbers. We continue this idea by associating points in a plane with pairs of real numbers. This in turn enables us to describe paths in a plane with equations. There are many applications that involve describing a path or curve. A communication satellite, for example, travels a path around the Earth. Once we have the connection between paths and mathematics, we can apply the powerful tool of mathematical analysis to help describe and understand the physical world.

7.1

Linear Equations in Two Variables and Their Graphs

OBJECTIVES

▶ *To plot points in a coordinate plane*

▶ *To find solutions of a linear equation*

▶ *To draw the graph of a linear equation*

▶ *To find the x- and y-intercepts of a linear equation*

In previous chapters, we have studied equations such as

$$4t - 5 = 2, \qquad x^2 - x + 2 = 0, \qquad \text{and} \qquad \frac{r}{r + 3} - \frac{1}{r^2 - 9} = 6$$

These equations have exactly *one* variable. In this section we will investigate equations having two variables. Examples of *linear* equations in *two* variables are

$$3y - 2x = 6, \qquad y = -\frac{1}{2}x + 1, \qquad \text{and} \qquad y - 2 = 3(x - 4)$$

D E F I N I T I O N

A **linear equation in two variables** is an equation that can be put in the form

$$ax + by = c$$

where a, b, and c are numbers and a and b are not both zero.

For example, $x^2 + y = 3$ is *not* linear, because the x variable is squared.

A linear equation in one variable such as $3x + 1 = 2$ has a solution that is one number. However, a solution to a linear equation in two variables requires two numbers. For example, the equation $7x + 3y = -17$ becomes a true statement when x is replaced by 1 and y is replaced by -8, since

$$7x + 3y = -17$$
$$7(1) + 3(-8) \overset{?}{=} -17$$
$$7 - 24 \overset{?}{=} -17$$
$$-17 = -17$$

We say that together $x = 1$ and $y = -8$ is a **solution** of the equation $7x + 3y = -17$, since the equation is satisfied when x and y are replaced by 1 and -8, respectively. We can write this solution as the ordered pair $(1, -8)$. In fact, $(1, -8)$ is not the only solution to the equation. We will see in the examples that linear equations in two variables will have many solutions.

N O T E *When an equation in two variables involves x and y, a solution, written as an ordered pair, has the form (x, y). That is, the x value of a solution is given first. If other letters are used in an equation, it will be made clear which variable is given first in the ordered pair form of a solution.*

Example 1

Check if the given ordered pair is a solution of the equation $3x + 2y = 12$.

(a) $(2, 3)$ **(b)** $(-4, 0)$

Solution

Remember that the first number in an ordered pair is the x value and the second number is the y value.

(a) Replace x by 2 and y by 3 in the equation.

$$3x + 2y = 12$$
$$3(2) + 2(3) \overset{?}{=} 12$$
$$6 + 6 = 12$$

Since the equation is satisfied, $(2, 3)$ is a solution.

(b) Replace x by -4 and y by 0.

$$3x + 2y = 12$$

$$3(-4) + 2(0) \overset{?}{=} 12$$

$$-12 + 0 \neq 12$$

Since the equation is not satisfied, $(-4, 0)$ is not a solution.   ▲

Ordered pairs of numbers can be visualized as points in a plane. Recall that we have used numbers to locate points on a number line. To locate a point in a plane, we need to draw a **coordinate plane.**

STRATEGY

To draw a coordinate plane

Step 1   Draw a horizontal number line.

Step 2   At the point with coordinate 0, draw a vertical number line that is perpendicular to the first number line. Each line has the same zero point, called the **origin.**

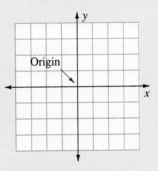

The horizontal number line is called the ***x*-axis.** The *positive* direction is to the *right* on the horizontal axis. The vertical number line is called the **y-axis.** The *positive* direction is *upward* on the vertical axis. The two axes form a coordinate system called the **Cartesian** or **rectangular coordinate system.** We use the coordinate system to locate points that lie in the plane.

For example, the origin is given by the ordered pair $(0, 0)$. The point P in the coordinate plane as shown in the figure is labeled with the ordered pair

(3, 4). The number 3 is called the **x-coordinate** and 4 is called the **y-coordinate** of P.

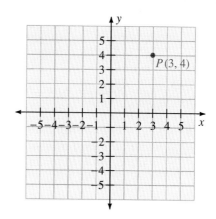

Example 2   Find the ordered pair for each point shown in the coordinate plane.

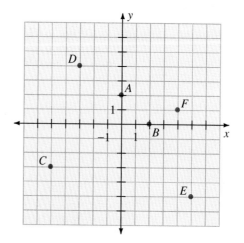

Solution   Be sure to always write the x-coordinate as the first entry of each ordered pair.

Point	Ordered Pair
A	$(0, 2)$
B	$(2, 0)$
C	$(-5, -3)$
D	$(-3, 4)$
E	$(5, -5)$
F	$(4, 1)$

When we write $P(x, y)$, we mean the point P with coordinates (x, y). Sometimes P is omitted and we simply write (x, y). The coordinate axes divide the plane into four parts or **quadrants.**

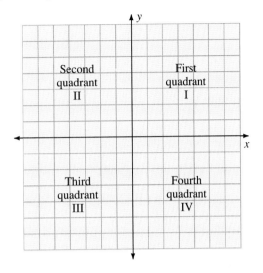

Any point in the coordinate plane either lies in one of the four quadrants or lies on a coordinate axis. As shown in the next figure, the sign of the coordinates determines in which quadrant the point lies.

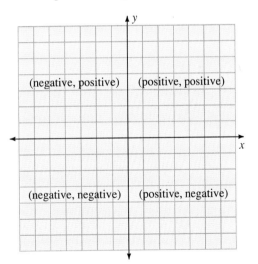

Example 3   Describe where in the coordinate plane the point (x, y) would be located.

(a) x is negative   (b) $x = 0$ (c) $y = 0$

Solution   **(a)** Since x is negative, the point (x, y) must lie to the left of the y-axis.

(b) $x = 0$ means that the point (x, y) is on the y-axis.

(c) $y = 0$ means that the point (x, y) is on the x-axis.   ▲

Given an ordered pair, we can **plot** or **graph** it in the coordinate plane.

Example 4   Plot each point in the same coordinate plane.

(a) $A(2, 4)$ **(b)** $B(-2, 4)$

(c) $C(-2, -4)$ **(d)** $D(2, -4)$

Solution   **(a)** To graph $(2, 4)$, start at the origin and move 2 units to the right, then 4 units up. We are now at the point A, which we mark with a dot and label $(2, 4)$.

(b) To plot $(-2, 4)$, start at the origin and move 2 units to the left, then 4 units up. We are now at the point B, which we mark with a dot and label $(-2, 4)$.

(c) and **(d)** These are done in a similar fashion.

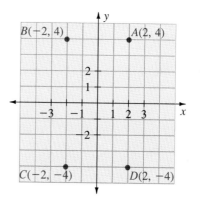

N O T E   *If the point has a zero x-coordinate, then there is no horizontal displacement. If the point has a zero y-coordinate, then there is no vertical displacement. Therefore, any point on the x-axis is of the form $(a, 0)$ and any point on the y-axis is of the form $(0, b)$.*

The solution set of a linear equation in two variables is a collection of ordered pairs of numbers. This solution set therefore can be visualized by graphing these ordered pairs in a coordinate plane.

D E F I N I T I O N

The **graph** of a linear equation $Ax + By = C$ is the set of points in the coordinate plane with coordinates (x, y) that are solutions of the equation.

The graph of the linear equation $Ax + By = C$ is a straight line.

A straight line is easy to draw because of the following statement from geometry.

Two points determine a straight line in the plane.

Therefore, to graph a linear equation we need find only two ordered pairs that are solutions. However, to serve as a check, we will also find a third pair.

S T R A T E G Y

To draw the graph of a linear equation

Step 1   Find three ordered pairs that are solutions to the equation.

Step 2   Plot these three pairs in the coordinate plane.

Step 3   Draw the line through these three points.

Example 5   Draw the graph of $4x - 3y = 12$.

Solution   Since the equation is linear, we know that the graph is a straight line. Therefore, we use the three-step method for drawing the graph of a linear equation.

For Step 1, we may choose any convenient points that are solutions of the equation. Setting $x = 0$,

$$4x - 3y = 12$$
$$4(0) - 3y = 12$$
$$y = -4$$

Therefore, $(0, -4)$ is a solution.

Next, setting $y = 0$,

$$4x - 3(0) = 12$$
$$4x = 12$$
$$x = 3$$

Therefore, $(3, 0)$ is another solution.

For a check point we may use any number. Let us choose $x = 6$. Then,

$$4x - 3y = 12$$
$$4(6) - 3y = 12$$
$$24 - 3y = 12$$
$$-3y = -12$$
$$y = 4$$

Therefore, our check point is $(6, 4)$.

For Steps 2 and 3, we plot these three points in the coordinate plane and draw the line through these points. See the following figure.

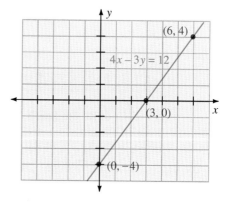

N O T E   *The three points in Example 5 were selected for convenience, because they were easier to plot than some other choices. For example, if we set $x = 1$ in $4x - 3y = 12$ and solved for y, then $y = -\dfrac{8}{3}$. The point $\left(1, -\dfrac{8}{3}\right)$ is a solution; however, it would be more difficult to plot with accuracy.*

Notice that the graph in the figure in Example 5 crosses the y-axis at $y = -4$. The number -4 is called the **y-intercept** of the line. Notice also that the graph crosses the x-axis at 3. The number 3 is called the **x-intercept** of the line. It is often convenient to graph linear equations by finding the intercepts.

S T R A T E G Y

To find the intercepts of a line

To find the **y-intercept**, set $x = 0$ in the equation and solve for y.
To find the **x-intercept**, set $y = 0$ in the equation and solve for x.

Example 6   Use the intercepts to draw the graph of $x + 3y = 6$.

Solution   To find the y-intercept, set $x = 0$ and solve for y.

$$0 + 3y = 6$$
$$y = 2$$

Therefore, $(0, 2)$ is a point on the line.
   To find the x-intercept, set $y = 0$ and solve for x.

$$x + 3(0) = 6$$
$$x = 6$$

Therefore, $(6, 0)$ is a point on the line.
   For a check point, set $y = 1$ in the equation and solve for x.

$$x + 3(1) = 6$$
$$x = 3$$

Our check point is $(3, 1)$.
   The graph of $x + 3y = 6$ is drawn in the following figure.

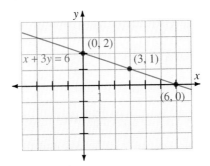

Example 7   Draw the graph of the equation $x = -3$.

Solution   The standard form for this equation is $x + 0y = -3$. Therefore, no matter what value of y is chosen, the x-coordinate is always -3. For example, $(-3, 0)$ and $(-3, 2)$ are both solutions. Plotting these two points, we obtain the vertical line as shown in the following figure.

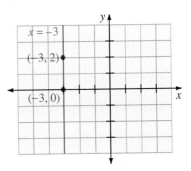

Example 8   Draw the graph of $y = 10$.

Solution   This equation in standard form is $0x + y = 10$. Therefore, no matter what value of x is chosen, $y = 10$. For example, $(0, 10)$ and $(5, 10)$ are both points on the graph. Plotting these two points, we obtain a horizontal line as shown in the following figure.

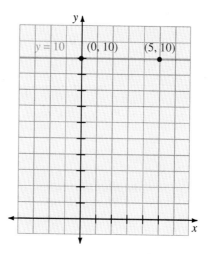

Examples 7 and 8 were illustrations of special linear equations.

R U L E

The graph of the equation $x = a$ is a vertical line passing through the x-axis at a.

The graph of the equation $y = b$ is a horizontal line passing through the y-axis at b.

In the following figure, we show these two special graphs.

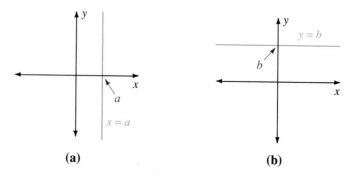

(a)   **(b)**

E X E R C I S E   S E T   7.1

For Exercises 1–6, state whether or not the equation is linear.

1. $\frac{x}{2} + \frac{2}{3}y = 12$

2. $y = x^3$

3. $\frac{2}{x} + y = -1$

4. $1.2x = 6.3y - 2.9$

5. $x - 400 = 0$

6. $y = 0$

7. Check if the given ordered pair is a solution to the equation $4x - 3y = -1$.

(a) $\left(-\frac{1}{4}, 0\right)$ (b) $\left(\frac{5}{3}, \frac{1}{3}\right)$

(c) $(2, 3)$ (d) $(-1, -2)$

8. Check if the given ordered pair is a solution to the equation $-7x + 5y = 10$.

(a) $(0, 2)$ (b) $\left(\frac{5}{7}, 3\right)$

(c) $(-4, -5)$ (d) $\left(\frac{2}{7}, 4\right)$

9. Find the ordered pair for each point shown in the coordinate plane.

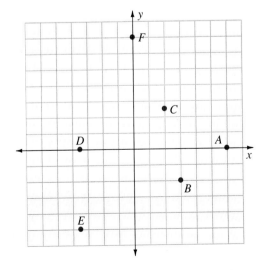

10. Find the ordered pair for each point shown in the coordinate plane.

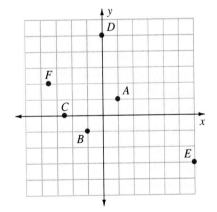

For Exercises 11–24, describe where in the coordinate plane the point (x, y) would be located.

11. x is positive and y is positive.

12. $y = 0$

13. x is not negative and $y = 0$.

14. $x = 0$ and $y = 0$

15. x is positive and y is negative.

16. x is negative and y is not positive.

17. The y-coordinate is zero.

18. The x-coordinate is zero.

19. The x-coordinate is -1.

20. The y-coordinate is 5.

21. $y = 2$ and x is positive.

22. $x = -10$ and y is negative.

23. x is twice y.

24. y is twice x.

25. Plot each point in the same coordinate plane.
  (a) $A(-3, 2)$ **(b)** $B(1, 0)$
  (c) $C(3, 5)$ **(d)** $D(-4, -1)$

26. Plot each point in the same coordinate plane.
  (a) $A(3, 3)$ **(b)** $B(-2, 0)$
  (c) $C(-1, -5)$ **(d)** $D(0, -1)$

27. Plot each point in the same coordinate plane.
  (a) $A(50, 100)$ **(b)** $B(50, -100)$
  (c) $C(-50, 100)$ **(d)** $D(-50, -100)$

28. Plot each point in the same coordinate plane.
  (a) $A\left(\frac{1}{4}, \frac{1}{2}\right)$ **(b)** $B\left(-\frac{1}{4}, \frac{1}{2}\right)$
  (c) $C\left(-\frac{1}{4}, -\frac{1}{2}\right)$ **(d)** $D\left(\frac{1}{4}, -\frac{1}{2}\right)$

For Exercises 29–34, find the ordered pair that is a solution to $8x - 3y = 12$ when

29. $x = 0$

30. $y = 0$

31. $x = 2$

32. $y = \dfrac{2}{3}$

33. $y = -10$

34. $x = 3$

For Exercises 35–42, draw the graph of the linear equation.

35. $3x - y = 9$

36. $2x + 3y = 12$

37. $y = -4x + 5$

38. $y = 2x - 6$

39. $x + y = 50$

40. $x - y = 100$

41. $x = 2y + 10$

42. $2x = -5y + 20$

For Exercises 43–54, use the x- and y-intercept to draw the graph of the linear equation.

43. $2x + 7y = 14$

44. $5x + 2y = 10$

45. $x - 2y = 1$

46. $-2x + y = 4$

47. $\dfrac{3}{4}x + \dfrac{1}{2}y = 1$

48. $\dfrac{2}{3}x - \dfrac{1}{2}y = 2$

49. $50x - 25y = 200$

50. $20x + 40y = 60$

51. $x + 2y = 50$

52. $3x + y = 75$

53. $-x + y = 30$ **54.** $-2x + 2y = 90$

For Exercises 55–76, draw the graph of the linear equation.

55. $x + y = 5$ **56.** $2x + y = 2$

57. $3x - 2y = 6$ **58.** $2x - 5y = 10$

59. $y = \dfrac{1}{2}x - 1$ **60.** $y = \dfrac{1}{3}x + 1$

61. $x = 4$ **62.** $y = 2$

63. $y = -\dfrac{3}{2}$

64. $x = -\dfrac{3}{4}$

65. $x = 0$

66. $y = 0$

67. $x = 40$

68. $y = -25$

69. $y - 5 = 0$

70. $x + 7 = 0$

71. $x = 3y + 1$

72. $x = -5y + 10$

73. $y = 1.5x$

74. $y = -0.5x$

75. $\dfrac{1}{4}x + \dfrac{2}{3}y = 1$

76. $\dfrac{2}{5}x + \dfrac{1}{3}y = 1$

For Exercises 77–80, write an equation from the given information. Then draw the graph of the equation.

77. The sum of the y-coordinate and twice the x-coordinate is negative two.

78. The difference of the x value and the y value is 12.

79. The y-coordinate is twice the difference of the x-coordinate and five.

80. The sum of the x-coordinate and y-coordinate, reduced by one, is zero.

81. The Tiller Corporation makes wicker products and finds that the cost y, in dollars, of making x dozen wicker baskets is given by

$$y = 0.5x + 25$$

Draw the graph of this cost equation. (*Note:* Since x is the number of units produced, $x \geq 0$.)

82. The cost y, in dollars, of making x electric coils is given by

$$y = 2.5x + 100$$

Draw the graph of this cost equation. (*Note:* Since x is the number of electric coils made, $x \geq 0$.)

83. The Oakmont Tennis Club buys a new ball machine for $1,000. The value of the machine decreases as time increases. The graph in the figure shows the relationship between time t (where $t = 0$ is when the ball machine was purchased) and the value V of the machine after t years.

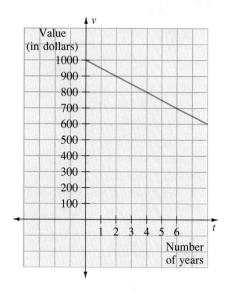

(a) What is the value of the machine when it is 3 years old?

(b) How old is the machine when the value is $800?

84. The value V, in dollars, of an antique car increases with time t, in years, as shown in the figure. For convenience, $t = 0$ means 1987.

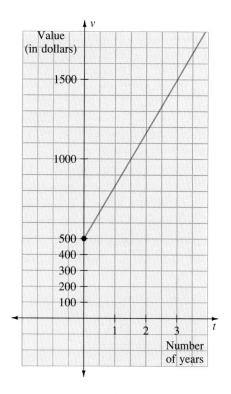

(a) What is the value of the car in 1987?

(b) In what year was the value of the car $1,500?

85. Tri-State Steel corporation has a cost y, in thousands of dollars, of producing x tons of steel. If 4 tons of steel are produced, the cost is $8,000. If 6 tons are produced, then the cost is $9,000.
   (a) Write the information given about production x, in tons, and the corresponding cost y, in thousands of dollars, as two ordered pairs (x, y).
   (b) Plot the two points from Part (a) in a coordinate plane.

86. In Oshkosh, Wisconsin, when the supply of coffee beans was 5,000 pounds, the cost p per pound was $4. When the supply increased to 7,000 pounds, the cost was $6 per pound.
   (a) Write the information given about supply x, in thousands of pounds, and price p per pound, in dollars, as two ordered pairs (x, p).
   (b) Plot the two points from Part (a) in a coordinate plane. Use the horizontal axis for x and the vertical axis for p.

   (c) Assuming that the relationship between production and cost is linear, draw a straight line through the two points.
   (d) From the graph, predict the cost of producing 10 tons of steel.

   (c) Assuming that the relationship between x and p is linear, draw a line through the two points.
   (d) Predict the price p, if the supply is 8,000 pounds.

87. In 1986, 0.6% of the viewing audience watched a music television channel. In 1989, 1.2% watched this channel, and in 1990, it increased to 1.4%.
   (a) Write this information as three ordered pairs (t, p), where t is the year and p is the corresponding percent of the viewing audience watching this music channel. Use $t = 0$ to mean 1986.
   (b) Plot these three points in a coordinate plane. Use the horizontal axis for t and the vertical axis for p. Is p linearly related to t for these three points?

88. In Pennsylvania, the fine for driving 10 miles over the speed limit is $65. The fine for driving 20 miles over the speed limit is $85, and the fine for driving 25 miles over the speed limit is $100.
   (a) Write this information as three ordered pairs (n, V), where n is the number of miles over the speed limit and V is the corresponding fine.
   (b) Plot these points in a coordinate system where n is the horizontal axis and V is the vertical axis. Is V linearly related to n for these three points?

89. What is the difference between (x, y) and $\{x, y\}$?

90. Is (x, y) ever the same as (y, x)?

REVIEW PROBLEMS

The following exercises review parts of Section 4.6. Doing these problems will help prepare you for the next section.

Solve each equation for the indicated variable.

91. $\dfrac{y - 6}{x - 3} = 5$; for y.

92. $\dfrac{y + 2}{x - 1} = -3$; for y.

93. $\dfrac{y - \dfrac{2}{3}}{x + 1} = \dfrac{4}{3}$; for y.

94. $\dfrac{y + 10}{x - \dfrac{5}{2}} = -2$; for y.

95. $\dfrac{x - 10}{t + \dfrac{3}{2}} = -12$; for x.

96. $\dfrac{x - 7}{t + \dfrac{4}{9}} = 18$; for x.

E N R I C H M E N T   E X E R C I S E S

1. Many computer graphics use a coordinate system as shown in the figure. The positive x-axis is to the right, but the positive y-axis faces downward. Using this coordinate system, graph the following points.

(a) $A(10, 0)$ (b) $B(0, 30)$
(c) $C(100, 80)$ (d) $D(50, 50)$

2. Find the value of m, if the graph of $y = mx + 5$ passes through $(3, 10)$.

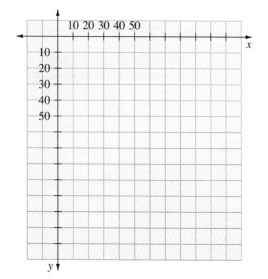

3. Find the value of b so that the graph of $3x - 2by = 7$ passes through $(-1, 14)$.

4. Determine c so that the point $(-2, c)$ lies on the graph of $2x + 3y = c$.

5. Determine d so that $(3, -1)$ lies on the graph of

$dx + (d - 5)y = 2$.

6. Find the coordinate of a point on the graph of $2x - 5y = -12$ if the y-coordinate is twice the x-coordinate.

7. Find the coordinate of a point on the graph of $x - 2y = 15$ if the y-coordinate is three more than the x-coordinate.

8. Let (x_1, y_1) and (x_2, y_2) be two points on the graph

of $y = mx + b$. Show that $\dfrac{y_2 - y_1}{x_2 - x_1} = m$.

Answers to Enrichment Exercises are on page A.36.

7.2

The Slope of a Line

O B J E C T I V E S

▶   *To find the slope of a line from the graph*

▶   *To find the slope of a line from the formula*

▶   *To understand the difference between zero slope and no slope*

▶   *To understand the geometric meaning of positive and negative slope*

Each nonvertical line has associated with it a number called the *slope* of the line. The slope is defined in such a way as to give us two important characteristics of the line. One characteristic is the steepness of the line and the other is whether the line rises or falls from left to right.

D E F I N I T I O N

Let (x_1, y_1) and (x_2, y_2) be any two points on a nonvertical line. The **slope** of the line, denoted by m, is

$$m = \frac{\text{change in } y}{\text{change in } x} = \frac{y_2 - y_1}{x_2 - x_1}$$

That is, the slope is found by taking the difference of the y-coordinates and dividing by the difference (in the same order) of the x-coordinates. See the figure below.

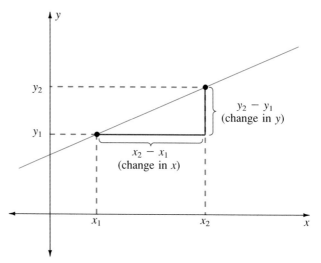

N O T E   *Recall that subscript notation such as x_1 is read "x sub-one."*
The variables x_1 and x_2 simply mean two choices for the x-coordinate.

Example 1   Find the slope of the given line.

(a)

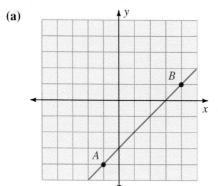

(b)

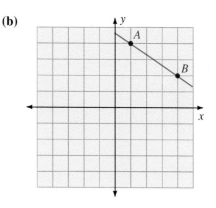

Solution   (a) Using the two points distinguished on the line, if we start at point A, move to the right 5 units, then up 5 units, we arrive at point B. The slope of the line is, therefore, $\dfrac{5}{5}$ or 1.

(b) Using the two points distinguished on the line, if we start at point A, move 3 units to the right, then *down* 2 units, we arrive at point B. The slope of the line is, therefore, $\dfrac{-2}{3}$ or $-\dfrac{2}{3}$.   ▲

It is illustrated in Example 1 that **positive slope** means an *upward slant* and **negative slope** means a *downward slant* of the line as the graph is scanned from left to right.

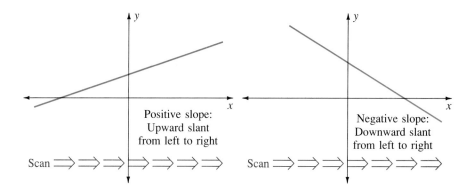

If the coordinates of two points on a line are known, then the slope of the line can be obtained by the slope formula given in the definition.

Example 2

Use the slope formula to find the slope of the line passing through $(-2, 1)$ and $(3, 4)$.

Solution

Let $(-2, 1)$ be (x_1, y_1) and $(3, 4)$ be (x_2, y_2) in the slope formula. Then, the slope m is given by

$$m = \frac{y_2 - y_1}{x_2 - x_1}$$

$$= \frac{4 - 1}{3 - (-2)}$$

$$= \frac{3}{5}$$

The line together with the two points is shown in the following figure.

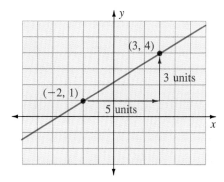

N O T E   *The order in which the two points are used in the slope formula is unimportant. If, in Example 2, we had interchanged the role of the two points by taking (3, 4) as (x_1, y_1) and (−2, 1) as (x_2, y_2), the slope formula would still yield the same number, $\dfrac{3}{5}$.*

Example 3   Find the slope of the line passing through the points $\left(\dfrac{1}{2}, -1\right)$ and $\left(-\dfrac{3}{4}, \dfrac{1}{3}\right)$.

Solution   Letting $(x_1, y_1) = \left(\dfrac{1}{2}, -1\right)$ and $(x_2, y_2) = \left(-\dfrac{3}{4}, \dfrac{1}{3}\right)$ in the slope formula,

$$m = \frac{y_2 - y_1}{x_2 - x_1}$$

$$= \frac{\dfrac{1}{3} - (-1)}{-\dfrac{3}{4} - \dfrac{1}{2}}$$

$$= \frac{12\left(\dfrac{1}{3} + 1\right)}{12\left(-\dfrac{3}{4} - \dfrac{1}{2}\right)} \qquad \text{Multiply numerator and denominator by 12.}$$

$$= \frac{4 + 12}{-9 - 6}$$

$$= \frac{16}{-15}$$

$$= -\frac{16}{15}$$ ▲

C A L C U L A T O R   C R O S S O V E R

Find the slope of the line through the two points.

1. (5.72, 6.69) and (3.10, 2.94)

2. (−1.06, 3.98) and (4.95, −10.21)

3. (4.14, −7.31) and (8.55, −9.15)

Answers   **1.** 1.43   **2.** −2.36   **3.** −0.417

In the next two examples, we look at slope as it relates to horizontal and vertical lines.

Example 4   Find the slope of the line passing through

(a) $(-3, -2)$ and $(2, -2)$ (b) $(4, 1)$ and $(4, 3)$

Solution   (a) The graph of this line is shown in Figure (a). Note that the line is horizontal; that is, it is parallel to the x-axis. Using the slope formula,

$$m = \frac{-2 - (-2)}{-3 - 2} = \frac{0}{-5} = 0$$

(b) The graph of this line is shown in Figure (b). Note that this line is vertical; that is, it is perpendicular to the x-axis. Using the slope formula,

$$m = \frac{3 - 1}{4 - 4} = \frac{2}{0}$$

Recall that division by zero is undefined, so the slope of a vertical line is undefined.

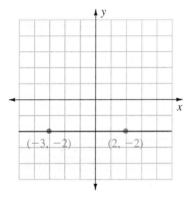

The slope is 0.

(a)

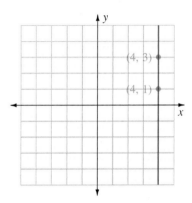

The slope is undefined.

(b)   ▲

The results of Example 4 are true in general.

R U L E

The slope of a horizontal line, $y = b$, is zero.
The slope of a vertical line, $x = a$, is undefined.

Two lines are **parallel** when they have no points in common. Since two nonvertical parallel lines slant in the same direction, they have the same slope.

In summary,

R U L E

Two nonvertical lines that are parallel have the same slope.

Example 5   By checking their slopes, determine if the line through the first pair of points is parallel to the line through the second pair of points.

(a) First pair of points: $(0, 5)$ and $(2, 4)$.

   Second pair of points: $\left(-1, \dfrac{3}{4}\right)$ and $\left(\dfrac{1}{2}, 0\right)$.

(b) First pair of points: $(0, 1)$ and $\left(-1, \dfrac{1}{3}\right)$.

   Second pair of points: $(3, 2)$ and $\left(\dfrac{1}{2}, 0\right)$.

Solution   (a) We use the slope formula to compute the slope of the two lines.

The slope of the line through $(0, 5)$ and $(2, 4)$ is

$$\frac{4 - 5}{2 - 0} = -\frac{1}{2}$$

The slope of the line through $\left(-1, \dfrac{3}{4}\right)$ and $\left(\dfrac{1}{2}, 0\right)$ is

$$\frac{0 - \dfrac{3}{4}}{\dfrac{1}{2} - (-1)} = \frac{-\dfrac{3}{4}}{\dfrac{3}{2}}$$

$$= -\frac{3}{4} \cdot \frac{2}{3}$$

$$= -\frac{1}{2}$$

Since the two slopes are the same, the two lines are parallel.

(b) The slope of the line through $(0, 1)$ and $\left(-1, \dfrac{1}{3}\right)$ is

$$\frac{\dfrac{1}{3} - 1}{-1 - 0} = \frac{-\dfrac{2}{3}}{-1} = \frac{2}{3}$$

The slope of the line through $(3, 2)$ and $\left(\dfrac{1}{2}, 0\right)$ is

$$\frac{0 - 2}{\dfrac{1}{2} - 3} = \frac{-2}{-\dfrac{5}{2}} = \frac{4}{5}$$

Since the slopes are not the same, $\dfrac{2}{3} \neq \dfrac{4}{5}$, the two lines are not parallel.

▲

L E A R N I N G   A D V A N T A G E   **Why study mathematics?**
Mathematics is an increasingly important tool for jobs of the future. Using mathematics in problem solving will be necessary for the typical college graduate. No longer can we say that the knowledge of mathematics is nice, but not necessary. To be a success in the world of tomorrow requires being able to use the powerful tool of mathematics.

E X E R C I S E   S E T   7.2

For Exercises 1–8, find the slope, if it exists, of each line.

1.

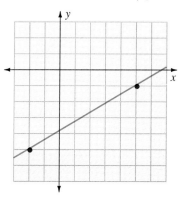

2.

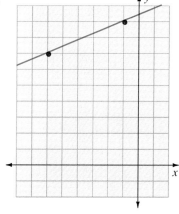

3.

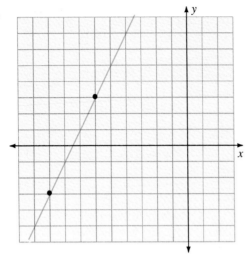

4.

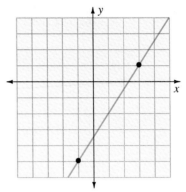

5.

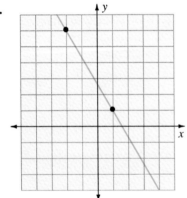

6.

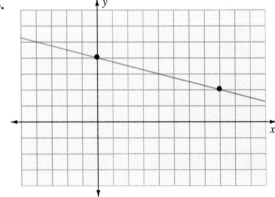

7.

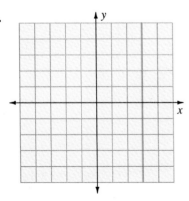

8.

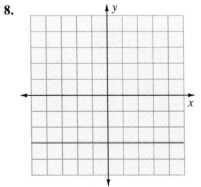

For Exercises 9–22, use the slope formula to find the slope, if it exists, of the line through the two points.

9. $(2, -5)$ and $(3, 4)$

10. $(-1, 0)$ and $(3, 5)$

11. $(0, 3)$ and $(2, 0)$

12. $(4, 1)$ and $(1, 0)$

13. $(6, -1)$ and $(-2, 4)$

14. $(6, -3)$ and $(-4, 7)$

15. $(2, -3)$ and $(6, 15)$

16. $(-4, -3)$ and $(2, 3)$

17. $(-2, 10)$ and $(1, 10)$

18. $(5, -3)$ and $(-4, -3)$

19. $(-2, 4)$ and $(-2, 6)$

20. $(4, -3)$ and $(4, -4)$

21. $\left(2, \frac{2}{3}\right)$ and $\left(\frac{1}{3}, -1\right)$

22. $\left(\frac{1}{2}, -\frac{1}{2}\right)$ and $\left(\frac{1}{4}, -2\right)$

For Exercises 23–32, determine if the line through the first pair of points is parallel to the line through the second pair of points.

23. First pair of points: $(1, 3)$ and $(-2, -3)$.
Second pair of points: $(0, -3)$ and $(-1, -5)$.

24. First pair of points: $(0, 10)$ and $(10, 0)$.
Second pair of points: $(-2, 3)$ and $(6, -5)$.

25. First pair of points: $(1, 3)$ and $(-5, 9)$.
Second pair of points: $(4, 1)$ and $(-2, -5)$.

26. First pair of points: $(2, -3)$ and $(-1, 3)$.
Second pair of points: $(0, 1)$ and $(4, -11)$.

27. First pair of points: $\left(\frac{1}{2}, -4\right)$ and $(-1, 2)$.
Second pair of points: $\left(\frac{1}{3}, 0\right)$ and $(3, -8)$.

28. First pair of points: $(2, 9)$ and $(-1, -6)$.
Second pair of points: $\left(\frac{1}{2}, \frac{3}{2}\right)$ and $\left(-\frac{1}{5}, -2\right)$.

29. First pair of points: $(-1, 4)$ and $(3, 2)$.
Second pair of points: $(-2, 5)$ and $(-4, 11)$.

30. First pair of points: $(1, 7)$ and $(-1, 4)$.
Second pair of points: $(0, -3)$ and $(2, 1)$.

31. First pair of points: $(-1, 5)$ and $(-1, -8)$.
Second pair of points: $(2, -5)$ and $(2, 7)$.

32. First pair of points: $(-20, -3)$ and $(10, -3)$.
Second pair of points: $(0, 45)$ and $(-33, 45)$.

Three points are **colinear** if they all lie on the same line. For Exercises 33–40, determine whether or not the three points are colinear.

33. $(1, 2)$, $(4, 3)$, and $(7, 4)$

34. $(3, 1)$, $(6, 3)$, and $(9, 5)$

35. $(0, 0)$, $(-2, -2)$, and $(-1, -1)$

36. $(0, 0)$, $(-3, 3)$, and $(1, 1)$

37. $(2, -2)$, $(2, 1)$, and $(2, -9)$

38. $(-7, 1)$, $(2, 1)$, and $(9, 1)$

39. $\left(0, \frac{3}{2}\right)$, $\left(\frac{1}{2}, \frac{5}{4}\right)$, and $\left(\frac{3}{4}, \frac{9}{8}\right)$

40. $\left(0, -\frac{5}{6}\right)$, $\left(\frac{3}{2}, -\frac{11}{6}\right)$, and $\left(-\frac{1}{2}, -\frac{1}{2}\right)$

41. In a coordinate plane, plot the point $(3, 4)$. Then plot two other points so that all three points lie on a line with a slope of $-\dfrac{1}{2}$.

42. In a coordinate plane, plot the point $(-1, -2)$. Then plot two other points so that all three points lie on a line with a slope of $\dfrac{3}{2}$.

43. In a coordinate plane, plot the point $(-1, 3)$. Then plot two other points so that all three points lie on a line with a slope of -2.

44. In a coordinate plane, plot the point $(3, -1)$. Then plot two other points so that all three points lie on a line with a slope of -1.

45. The limited edition of Stephen King's *The Eyes of the Dragon* was published in 1984 by Philtrium Press and sold for $120. By 1988, copies of this book were selling for $650. Assuming that the value of the book increases linearly with time, what is the predicted value of the book for 1992?

46. The National Transportation Safety Board reported 550 near collisions by airplanes in this country in 1982 and steadily increased to 790 in 1990. Assuming linearity, how many near collisions are predicted for 1992?

47. The area A of a circular oil spill is related to the radius r. An oil spill with a radius of 6 inches has an (approximate) area of 37.68 square inches. An oil spill with a radius of 8 inches has an area of 200.96 square inches. Is the area of this circular oil spill linearly related to its radius? (*Note:* When $r = 0$, then $A = 0$.)

48. The volume V of a hot air balloon is related to the diameter d. If the diameter is 10 feet, the volume is (approximately) 523 cubic feet. If the diameter is 18 feet, the volume is 3,052 cubic feet. Is the volume linearly related to the diameter? (*Note:* When $d = 0$, $V = 0$.)

REVIEW PROBLEMS

The following exercises review parts of Section 4.6. Doing these problems will help prepare you for the next section.

49. Solve $\dfrac{y - 3}{x - 2} = 4$ for y.

50. Solve $\dfrac{y + \dfrac{3}{4}}{x + \dfrac{5}{6}} = 6$ for y.

51. Solve $-\dfrac{2}{5} = \dfrac{x + \dfrac{3}{2}}{t - 4}$ for x.

52. Solve $\dfrac{y + 4}{x + \dfrac{3}{2}} = -\dfrac{10}{3}$ for y.

53. Solve $m = \dfrac{y - y_1}{x - x_1}$ for y.

ENRICHMENT EXERCISES

1. Find the value of k so that the line through $(-2k, -2)$ and $(k, 4)$ has a slope of 8.

2. Find the value of k so that the line through $\left(-\dfrac{1}{3}, 3k\right)$ and $(-k, 6k)$ has a slope of $\dfrac{3}{2}$.

3. Find the value(s) of k so that the line through $(-2k^2, 2)$ and $(-k, -7k^2)$ has a slope of -5.

4. Which line in the figure below has the larger slope?

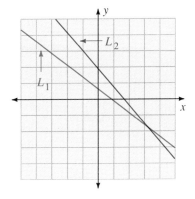

5. Which line in the figure below has the larger slope?

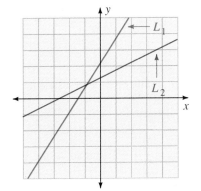

6. A roof truss is shown in the figure below.

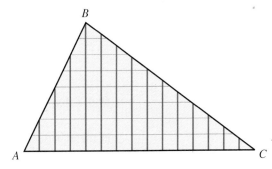

The slope of the line through A and B is 2 and the slope of the line through B and C is $-\dfrac{8}{11}$. The length from A to C is 45 feet. Find the sum of the lengths of the vertical boards (highlighted in blue) needed to build this roof truss.

Answers to Enrichment Exercises are on page A.37.

7.3

The Equation of a Line

O B J E C T I V E

▶ *To find the equation of a line given*
(a) the slope and y-intercept
(b) the slope and a point on the line
(c) two points on the line

There is a relationship between the slope of a line and the equation of the line. For example, consider the line given by the equation $y = \dfrac{3}{4}x + 2$. Two points on this line can be found by setting x equal to any number and then solving for y. Let us choose $x = 0$ and $x = 4$. When $x = 0$, $y = \dfrac{3}{4}(0) + 2 = 2$. When $x = 4$, $y = \dfrac{3}{4}(4) + 2 = 5$. Therefore, $(0, 2)$ and $(4, 5)$ are two points satisfying the equation $y = \dfrac{3}{4}x + 2$.

We now find the slope of the line using the slope formula with these two points.

$$m = \frac{y_2 - y_1}{x_2 - x_1}$$

$$= \frac{5 - 2}{4 - 0}$$

$$= \frac{3}{4}$$

Notice that the slope of the line, $\dfrac{3}{4}$, is the same number as the coefficient of x in the equation $y = \dfrac{3}{4}x + 2$.

Now, if we replace x by 0 in the equation $y = \dfrac{3}{4}x + 2$ and solve for y, we obtain the y-intercept.

$$y = \frac{3}{4}(0) + 2 = 2$$

Therefore, the y-intercept is 2. Notice that 2 is also the constant term in the equation $y = \dfrac{3}{4}x + 2$. The line is shown in the figure.

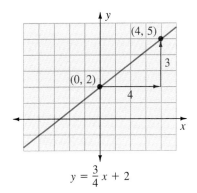

$$y = \frac{3}{4}x + 2$$

In summary, given the line with equation $y = \frac{3}{4}x + 2$, the coefficient of x, $\frac{3}{4}$, is the slope, and the constant term, 2, is the y-intercept. We can generalize this result.

R U L E

When the equation of a line is written as

$$y = mx + b$$

then the slope of the line is m and the y-intercept is b.

The equation $y = mx + b$ is called the **slope-intercept** form for the equation of a line.

N O T E   *To "read off" the slope and y-intercept from the equation of a line, the equation must be in the special form $y = mx + b$. For example, $2x - 3y = 1$ must first be solved for y, $y = \frac{2}{3}x - \frac{1}{3}$. From this form, we see that the slope is $\frac{2}{3}$ and the y-intercept is $-\frac{1}{3}$.*

Example 1   Find the slope and y-intercept of the line from the given equation.

(a) $y = -0.6x + 17$ (b) $5x - 3y = \frac{6}{5}$

Solution   (a) The equation is in the correct form to find the slope and y-intercept. The slope is the coefficient of x, which is -0.6, and the y-intercept is the constant term 17.

(b) We first solve the equation for y.

$$-3y = -5x + \frac{6}{5}$$

$$y = \frac{5}{3}x - \frac{2}{5} \qquad \text{Divide by } -3.$$

The coefficient of x is $\frac{5}{3}$. Therefore, the slope of the line is $\frac{5}{3}$. The constant term, $-\frac{2}{5}$, is the y-intercept.   ▲

Given the equation of a line, we can find the slope and y-intercept. Conversely, the slope and y-intercept determine a line.

Example 2   A line has a slope of $\frac{3}{4}$ and a y-intercept of -2.

(a) Draw the line.

(b) Write an equation for the line.

Solution   **(a)** Draw a coordinate plane as shown in the figure below. Since the y-intercept is -2, $(0, -2)$ is a point on the line. Plot this point. Next, the slope is $\frac{3}{4}$; therefore,

$$\frac{\text{change in } y}{\text{change in } x} = \frac{3}{4}$$

This means that for each horizontal change to the right of 4 units, there is a corresponding vertical change upward of 3 units.

Starting at the point $(0, -2)$, move to the right 4 units, then upward 3 units, ending at the point $(4, 1)$. The point $(4, 1)$ lies on the line. We graph the desired line by drawing the line determined by $(0, -2)$ and $(4, 1)$.

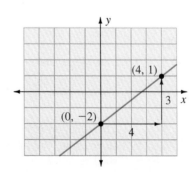

(b) Using the slope-intercept form, $y = mx + b$. We are given that $m = \dfrac{3}{4}$ and

$b = -2$. Therefore, the equation of the line is $y = \dfrac{3}{4}x + (-2)$ or simply

$y = \dfrac{3}{4}x - 2$.   ▲

N O T E   *In Example 2, there is another way to use the slope* $\dfrac{3}{4}$ *to find*

another point on the line. Starting at $(0, -2)$, *first move upward 3 units, then to the right 4 units. Notice that we still arrive at the point* $(4, 1)$ *as shown in the following figure.*

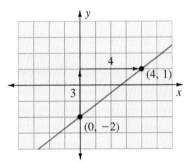

Example 3   A line has a slope of $-\dfrac{2}{3}$ and a y-intercept of 1.

(a) Draw the line.

(b) Write an equation for the line.

Solution   **(a)** Since the y-intercept is 1, $(0, 1)$ lies on the line. The slope $-\dfrac{2}{3}$ can be writ-

ten as $\dfrac{-2}{3}$. Therefore, starting at $(0, 1)$, move 3 units to the right, then 2

units *down*. We arrive at the point $(3, -1)$, which is another point on the line. Draw the line between $(0, 1)$ and $(3, -1)$ as shown in the following figure.

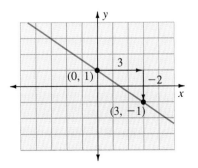

(b) To write the equation for this line, we use the slope-intercept form $y = mx + b$, where $m = -\dfrac{2}{3}$ and $b = 1$. Therefore, the desired equation is

$$y = -\frac{2}{3}x + 1$$ ▲

N O T E   *In Example 3, the slope* $-\dfrac{2}{3}$ *could be written as* $\dfrac{2}{-3}$. *In this form, move 3 units to the* left, *then upward 2 units.*

Given a point in the plane and a number m, there is a line having slope m passing through the point. In the next example, we show how to graph such a line.

Example 4   Draw the line through $(-3, 6)$ with slope $-\dfrac{5}{2}$.

Solution   We can write the slope $-\dfrac{5}{2}$ as $\dfrac{-5}{2}$. Starting at the point $(-3, 6)$, move 2 units to the right, then 5 units down. We arrive at the point $(-1, 1)$. Draw the line through $(-3, 6)$ and $(-1, 1)$ as shown in the figure.

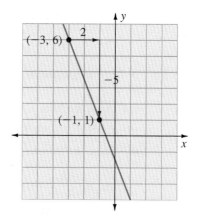

▲

If we are given a point (x_1, y_1) and the slope m, we may find the equation of the line. Let (x, y) be any other point on the line. Using the slope formula and substituting (x_2, y_2) with (x, y), we have

$$m = \frac{y - y_1}{x - x_1}$$

Multiplying both sides by the quantity $x - x_1$, we obtain

$$m(x - x_1) = y - y_1$$

that is,

$$y - y_1 = m(x - x_1)$$

This last equation is called the **point-slope** form of the equation of a line.

Example 5   Use the point-slope form to find the equation of the line through the point $\left(-1, \frac{1}{2}\right)$ with slope $-\frac{2}{5}$. Write the answer in the form $Ax + By = C$.

Solution   Letting $\left(-1, \frac{1}{2}\right)$ be (x_1, y_1) and $-\frac{2}{5}$ be m in the point-slope form, we have

$$y - y_1 = m(x - x_1)$$

$$y - \frac{1}{2} = -\frac{2}{5}[x - (-1)]$$

Next, simplify the equation.

$$10\left(y - \frac{1}{2}\right) = -(10)\left(\frac{2}{5}\right)(x + 1) \qquad \text{Multiply by 10.}$$

$$10y - 5 = -4(x + 1)$$

$$10y - 5 = -4x - 4$$

$$4x + 10y = 1 \qquad\qquad\qquad \text{Rearrange terms.} \quad \blacktriangle$$

Recall that two points determine a line. In the next example, we show how to find the equation of this line.

Example 6   Use the point-slope form to find the equation of the line through $(2, -1)$ and $(3, 7)$. Write the answer in the form $Ax + By = C$.

Solution   We must find the slope of the line in order to use the point-slope form. Since we are given two points, we can find the slope using the slope formula.

$$m = \frac{y_2 - y_1}{x_2 - x_1}$$

$$= \frac{7 - (-1)}{3 - 2}$$

$$= \frac{8}{1}$$

$$= 8$$

We now use the point-slope form using either given point and 8 as the slope. We arbitrarily choose $(2, -1)$.

$$y - (-1) = 8(x - 2)$$

$$y + 1 = 8x - 16$$

$$-8x + y = -17$$

Therefore, the line through the two given points has $-8x + y = -17$ as its equation.   ▲

In the next example, we show how the slope-intercept form may also be used to find the equation of a line when two points are given.

Example 7   Use the slope-intercept form to find the equation of the line through $(2, -5)$ and $(-3, -7)$.

Solution   Recall, the slope-intercept form is $y = mx + b$. We must find the two numbers m and b. Since m is the slope, we use the formula for slope

$$m = \frac{y_2 - y_1}{x_2 - x_1}$$

$$m = \frac{-7 - (-5)}{-3 - 2}$$

$$= \frac{-2}{-5}$$

$$= \frac{2}{5}$$

Our equation so far looks like this.

$$y = \frac{2}{5}x + b$$

Next, we find the number b. Since $(2, -5)$ lies on the line, it is a solution to the equation. Therefore, we replace x by 2 and y by -5.

$$-5 = \frac{2}{5}(2) + b$$

We solve this equation for b.

$$-5 = \frac{4}{5} + b$$

$$-5 - \frac{4}{5} = b \qquad \text{Subtract } \frac{4}{5} \text{ from both sides.}$$

$$-\frac{29}{5} = b$$

Therefore, we replace b by $-\dfrac{29}{5}$ in the equation $y = \dfrac{2}{5}x + b$ to obtain the desired equation. Namely,

$$y = \frac{2}{5}x - \frac{29}{5} \qquad\qquad ▲$$

N O T E   *In Example 7, the number b could have been found by using the point (−3, −7) instead of (2, −5). Either point will give the same value for b.*

C A L C U L A T O R   C R O S S O V E R

Find the equation of the line from the given information. Write your answer in the form $y = mx + b$.

1. The slope is −7.069 and the line passes through (−12.1, 10.3).
2. The slope is 10.92 and the line passes through (−7.95, −3.82).
3. The line passes through (1.35, −2.47) and (3.90, 1.06).
4. The line passes through (−4.56, −3.19) and (2.07, −4.48).

Answers   **1.** $y = -7.069x - 75.2349$ **2.** $y = 10.92x + 82.994$
3. $y = 1.384x - 4.338$ **4.** $y = -0.1946x - 4.0774$

Example 8   Find the equation of the line passing through

(a) (−2, 4) and (3, 4)   **(b)** (1, −2) and (1, 5)

Solution   **(a)** Using the slope formula,

$$m = \frac{4 - 4}{3 - (-2)} = \frac{0}{5} = 0$$

Since the slope is zero, the line is horizontal and we know that the equation of a horizontal line is of the form $y = b$. Since the common y value of the two given points is 4, we conclude that the desired equation is $y = 4$.

(b) Using the slope formula on these two points, we have

$$m = \frac{5 - (-2)}{1 - 1} = \frac{7}{0}$$

which is undefined. Since the slope is undefined, the line must be vertical and its equation is therefore of the form $x = a$. Since the common x value of the two given points is 1, the desired equation is $x = 1$.   ▲

Example 9   Find the equation of a line through (3, 6) and parallel to the graph of the line given by $y = 2x + 10$.

Solution   Since our line is parallel to the line given by $y = 2x + 10$, their slopes are the same. Since the slope of the given line is 2, the coefficient of x, we must find the equation of a line passing through $(3, 6)$ and having 2 as its slope. Using the point-slope form,

$$y - y_1 = m(x - x_1)$$
$$y - 6 = 2(x - 3)$$
$$y - 6 = 2x - 6$$
$$y = 2x$$
▲

L E A R N I N G   A D V A N T A G E   **Why study mathematics?**
Mathematics is used in areas such as pollution control. Mathematics was used to discover the relationship between the quantity of pollution that can be dissipated by an average sized tree. This in turn can be used to estimate the number of trees needed to counterbalance the exhaust from commuter cars during the rush hour in cities such as Seattle, Washington.

Many parts of environmental science utilize mathematics. The problems of pollution and toxic waste, for example, cannot be quantified and studied without a knowledge of mathematics.

E X E R C I S E   S E T   7.3

For Exercises 1–18, find the slope and y-intercept of the line from the given equation.

1. $y = -x - 5$

2. $y = -2x + 7$

3. $y = x + 2$

4. $y = -2x + 1$

5. $3x + 5y = 15$

6. $7x + 14y = 2$

7. $x - 2y = 18$

8. $4x - y = 17$

9. $y = x$

10. $y = -x + 1$

11. $y + x = 0$

12. $3y + x = 21$

13. $3x - 2y + 1 = 0$

14. $6x + 3y - 2 = 0$

15. $\frac{2}{5}y + x = 0$

16. $x + \frac{3}{4}y = 0$

17. $y = 0$

18. $y = -12$

19. Does the graph of $x = 7$ have a slope? A y-intercept? Why?

20. Does the graph of $x = -1$ have a slope? A y-intercept? Why?

For Exercises 21–24, the slope and *y*-intercept of a line are given. **(a)** Draw the line. **(b)** Write an equation for the line.

21. The slope is $\dfrac{2}{3}$ and the *y*-intercept is 1.

22. The slope is $\dfrac{1}{2}$ and the *y*-intercept is 3.

23. The slope is $-\dfrac{3}{2}$ and the *y*-intercept is 0.

24. The slope is $-\dfrac{4}{7}$ and the *y*-intercept is 10.

For Exercises 25–32, the slope *m* and *y*-intercept *b* for a line are given. **(a)** Draw the line. **(b)** Write an equation for the line.

25. $m = 1$ and $b = -2$

26. $m = 2$ and $b = 1$

27. $m = -\dfrac{4}{5}$ and $b = 2$

28. $m = -\dfrac{3}{7}$ and $b = 5$

29. $m = -\dfrac{2}{3}$ and $b = 0$ **30.** $m = -3$ and $b = 1$

31. $m = 0$ and $b = -3$ **32.** $m = 0$ and $b = 6$

For Exercises 33–42, draw the line through the given point P and having slope m.

33. $P(2, 1)$; $m = \dfrac{3}{5}$ **34.** $P(1, 2)$; $m = \dfrac{2}{7}$

35. $P(-2, -1)$; $m = 4$ **36.** $P(-4, -2)$; $m = 3$

37. $P(1, 1)$; $m = -\dfrac{2}{3}$

38. $P(2, 1)$; $m = -\dfrac{1}{2}$

39. $P(-4, -5)$; $m = \dfrac{5}{3}$

40. $P(7, -6)$; $m = -\dfrac{7}{4}$

41. $P(7, -4)$; the slope is undefined.

42. $P(-5, -5)$; the slope is zero.

For Exercises 43–48, use the point-slope form to find an equation of the line through the point P and having slope m. Write your answer in the form $Ax + By = C$.

43. $P(1, 5)$; $m = 3$

44. $P(2, -1)$; $m = 2$

45. $P(-2, -5)$; $m = \dfrac{2}{5}$

46. $P(-7, -1)$; $m = \dfrac{1}{3}$

47. $P\left(\dfrac{3}{16}, -4\right)$; $m = -32$

48. $P\left(-\dfrac{1}{4}, 3\right)$; $m = 8$

For Exercises 49–56, use the point-slope form to find an equation of the line through the two points P and Q. Write your answer in the form $Ax + By = C$.

49. $P(1, 5)$ and $Q(-1, 6)$

50. $P(-3, 4)$ and $Q(2, 1)$

51. $P(5, 7)$ and $Q(7, 7)$

52. $P(-3, -1)$ and $Q(4, -1)$

53. $P\left(\dfrac{3}{2}, \dfrac{1}{2}\right)$ and $Q(1, -1)$

54. $P(2, -1)$ and $Q\left(\dfrac{3}{4}, \dfrac{1}{3}\right)$

55. $P(-0.1, 0.3)$ and $Q(0.5, 1.5)$

56. $P(2.9, -3.7)$ and $Q(-0.1, -0.1)$

For Exercises 57–68, use the slope-intercept form to find an equation of the line through the two points P and Q. Keep your answer in the form $y = mx + b$.

57. $P(0, 0)$ and $Q(3, 10)$

58. $P(-1, -2)$ and $Q(0, 0)$

59. $P(1, 3)$ and $Q(-4, 1)$

60. $P(1, -2)$ and $Q(3, -1)$

61. $P(7, -1)$ and $Q(-4, -1)$

62. $P(10, 5)$ and $Q(-10, 4)$

63. $P\left(\dfrac{1}{2}, \dfrac{3}{2}\right)$ and $Q(1, 1)$

64. $P(3, -3)$ and $Q\left(\dfrac{4}{3}, -\dfrac{1}{3}\right)$

65. $P(7, -1)$ and $Q(-12, -1)$

66. $P(10, 4.5)$ and $Q(-10, 4.5)$

67. $P(0.6, 0.4)$ and $Q(-0.3, 0.7)$

68. $P(1.6, -0.7)$ and $Q(-0.6, 0.4)$

For Exercises 69–74, determine whether or not the two lines of the given equations are parallel.

69. $y = \dfrac{1}{2}x - 7$ and $3x - 6y = 5$

70. $4x - 7y = 6$ and $y = -\dfrac{5}{7}x + 12$

71. $\dfrac{2}{5}x + \dfrac{5}{2}y = -1$ and $y = \dfrac{4}{25}x + 14$

72. $\dfrac{1}{3}x + \dfrac{1}{4}y = 1$ and $\dfrac{3}{4}y - 5 = x$

73. $x = -23$ and $x = -46$

74. $y = 79$ and $y = -82$

For Exercises 75–80, find the equation of a line through the point P and parallel to the graph of the given equation.

75. $P(-8, -1)$; $3x - 8y = 10$

76. $P(6, -2)$; $4x + 3y = 0$

77. $P(-2, 0)$; $y = -\dfrac{1}{4}x$

78. $P(2, 0)$; $y = -4x$

79. $P(-50, -90)$; $x = 34$

80. $P(3, -7)$; $y = -80$

For Exercises 81–90, find the equation of the line described. When possible, write your answer in the form $y = mx + b$.

81. The y-intercept is -4 and the line passes through $(6, 16)$.

82. The y-intercept is 10 and the line passes through $(-12, 18)$.

83. The x-intercept is -2 and the line passes through $(2, 4)$.

84. The x-intercept is -1 and the line passes through $(1, -3)$.

85. The x-intercept is $-\dfrac{1}{2}$ and the y-intercept is 4.

86. The x-intercept is 0.1 and the y-intercept is -0.3.

87. The x-intercept is $\dfrac{3}{4}$ and the line is parallel to the graph of $4x - 5y = 0$.

88. The x-intercept is 2 and the line is parallel to the graph of $x - y = 0$.

89. The line passes through the points $(-1, 0)$ and $(-1, 3)$.

90. The line passes through $(3, 5)$ and is parallel to the y-axis.

REVIEW PROBLEMS

The following exercises review parts of Section 2.5. Doing these problems will help prepare you for the next section.

Write and solve an inequality from the given information.

91. The sum of a number and 4 is no greater than 3.

92. The sum of a number and -2 is less than 1.

93. The difference of 3 and a number is more than -4.

94. The difference of twice a number and $\dfrac{1}{2}$ is no less than $\dfrac{3}{2}$.

95. Two more than five times a number is greater than -3.

96. Twice a number reduced by 7 is less than the sum of four times the number and 6.

ENRICHMENT EXERCISES

1. Find k, if the line for $kx - 2y = 7$ is parallel to the line through $\left(\dfrac{1}{2}, -5\right)$ and $\left(1, -\dfrac{7}{4}\right)$.

2. Show that an equation of the line with x-intercept a and y-intercept b is

$$\frac{x}{a} + \frac{y}{b} = 1$$

3. Use the equation from Exercise 2 to find an equation of the line with x-intercept $-\dfrac{2}{7}$ and y-intercept $\dfrac{7}{2}$.

4. Show that the line through (x_1, y_1) and (x_2, y_2), with $x_1 \neq x_2$, has
$$\frac{y_1 x_2 - y_2 x_1}{x_2 - x_1}$$ as its y-intercept.

5. Use Exercise 4 to find the y-intercept of the line through $(5, -4)$ and $(7, 10)$.

6. Suppose a line has slope m and y-intercept b. Use the point-slope form of the equation of a line to show that $y = mx + b$.

Answers to Enrichment Exercises are on page A.39.

7.4

Graphing Linear Inequalities in Two Variables

O B J E C T I V E S

▶ *To graph linear inequalities in two variables*

▶ *To use linear inequalities in applications*

So far in this chapter, we have studied linear equations in two variables such as $-x + 2y = 6$. In this section, we investigate **linear inequalities in two variables** such as

$$-x + 2y \geq 6$$

This is read "negative x plus $2y$ is greater than or equal to 6." The inequality means two things: either (1) $-x + 2y = 6$ or (2) $-x + 2y > 6$.

The **graph** of a linear inequality is the set of points (x, y) in the plane that satisfy the inequality.

Now, the graph of $-x + 2y \geq 6$ includes the graph of the *equation* $-x + 2y = 6$. Furthermore, the graph of this equation is a straight line. Using the x- and y-intercepts, we graph the line as shown in the following figure.

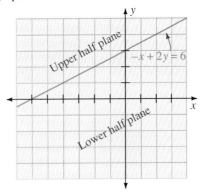

The line separates the plane into two regions—one above the line and the other below the line. These regions are called **half-planes.**

Exactly one of these half-planes belongs to the graph of the inequality. To determine the correct half-plane, we first solve $-x + 2y \geq 6$ for y.

$$2y \geq x + 6 \qquad \text{Add } x \text{ to both sides.}$$

$$y \geq \frac{1}{2}x + 3 \qquad \text{Divide by 2.}$$

Therefore, either $y = \frac{1}{2}x + 3$ or $y > \frac{1}{2}x + 3$. We already know that if the coordinates of a point (x, y) are related by the equation $y = \frac{1}{2}x + 3$, then the point lies on the line. Now, if $y > \frac{1}{2}x + 3$, then the y-coordinate is larger than $\frac{1}{2}x + 3$. This means that the point (x, y) lies in the half-plane above the line as shown in the figure below. For example, $(2, 4)$ lies on the line, but $(2, 5)$ and $(2, 6)$ lie above the line.

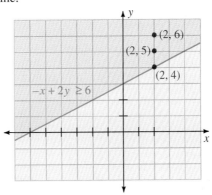

To indicate the graph of $-x + 2y \geq 6$, we shade the region above the line. Remember that the graph includes not only the upper half-plane but also the boundary line.

There is a faster way to determine which half-plane belongs to the graph of a linear inequality. Once the line has been drawn, select a convenient test point *not on the line* and see whether or not the coordinates of the point satisfy the original inequality. For example, using $(0, 0)$ as our test point for the inequality $-x + 2y \geq 6$, we replace x by 0 and y by 0 in the original inequality.

$$-x + 2y \geq 6$$

$$-0 + 2(0) \overset{?}{\geq} 6$$

Since 0 is not greater than or equal to 6, $(0, 0)$ is not a solution to the inequality $-x + 2y \geq 6$. Since $(0, 0)$ lies in the lower half-plane, no point from the lower half-plane satisfies the inequality. Our conclusion is that the graph includes the *upper* half-plane. This answer is the same as found before.

N O T E   *It is* not *automatic that the inequality symbol $\geq$ means that the graph will include the upper half-plane. A test point must be used to determine the correct half-plane.*

Example 1   Graph the inequality $x - y < 50$.

Solution   First, replace the less than symbol by an equal sign to obtain the equation $x - y = 50$. Using the intercepts, plot the two points $(0, -50)$ and $(50, 0)$ as shown in Figure (a). Draw the line using dashes, since the line is not in the graph of the inequality. (The inequality symbol is *strictly* less than.)

Next, choose a test point off the line. We use $(0, 0)$, since the origin is not on the line. Replacing x by 0 and y by 0 in the original inequality,

$$x - y < 50$$

$$0 - 0 < 50 \qquad \text{This is a true statement.}$$

Therefore, $(0, 0)$ is a solution of $x - y < 50$ and lies in the upper half-plane. We shade the region above the line as shown in Figure (b).

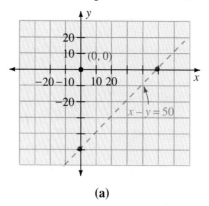

(a)

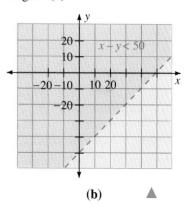

(b)

We summarize the strategy for graphing linear inequalities.

STRATEGY

To graph a linear inequality

Step 1   Replace the inequality symbol ($>$, $<$, $\geq$, or $\leq$) by an equal sign and draw the resulting straight line. If the inequality symbol is $\geq$ or $\leq$, use a solid line. If the symbol is $>$ or $<$, use dashes for the line. This line separates the plane into two regions called half-planes.

Step 2   Choose a convenient test point that is *not on the line.* Use $(0, 0)$ when possible, since the arithmetic would be easier.

Step 3   Determine whether or not the test point is a solution of the inequality.

    (a) If it is a solution, shade the half-plane in which the test point lies.

    (b) If it is not a solution, shade the other half-plane.

Example 2   Graph the inequality $3x + y \leq 9$.

Solution   Step 1   Replace "$\leq$" by "$=$" to obtain the equation $3x + y = 9$. We use the intercepts to graph the line. When $x = 0$, $y = 9$ and when $y = 0$, $x = 3$. We plot the two points $(0, 9)$ and $(3, 0)$ and draw the solid line as shown in Figure (a).

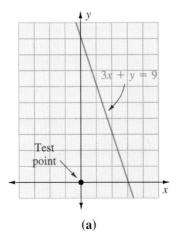

(a)

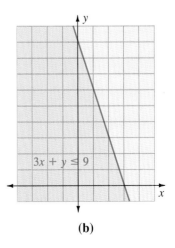

(b)

Step 2   We select $(0, 0)$ for our test point, as it is not on the line.

Step 3 Replace x by 0 and y by 0 in the inequality.

$$3x + y < 9$$

$$3(0) + 0 < 9$$

Since 0 is less than 9, the test point $(0, 0)$ is a solution of the inequality. $(0, 0)$ lies in the lower half-plane. Therefore, we shade the lower half-plane. The graph is shown in Figure (b). ▲

Example 3 Graph the inequality $x > y$.

Solution First, replace "$>$" by "$=$" to obtain the equation $x = y$. Letting $x = 0$, then $y = 0$, so the origin $(0, 0)$ is on the line. Notice that we cannot let $y = 0$ to find a second point, since $y = 0$ means that $x = 0$. Therefore, use another value for x, say $x = 5$. So, $(5, 5)$ is another point on the line. The line is drawn in Figure (a) using dashes. Next, choose a test point not on the line. In this case, $(0, 0)$ is on the line, so we must choose another point. We use $(1, 0)$. Replacing x by 1 and y by 0 in the original inequality.

$$x > y$$

$$1 > 0 \qquad \text{This is a true statement.}$$

Therefore, $(1, 0)$ is a solution and it lies below the line. Therefore, we shade the lower half-plane. The graph is shown in Figure (b).

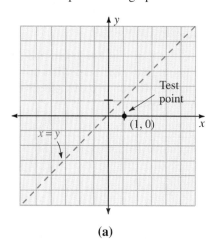

(a)

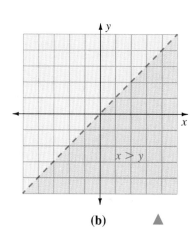

(b) ▲

We finish this section with an application of linear inequalities.

Example 4 Rawhide Leather Goods makes baseball gloves and hockey catch gloves. It costs $15 to make each baseball glove and $10 to make each hockey catch glove.

(a) Express the cost of making x baseball gloves.

(b) Express the cost of making y hockey catch gloves.

(c) The combined cost of making x baseball gloves and y hockey catch gloves cannot exceed $15,000. Express this restriction as a linear inequality.

(d) Graph the inequality of Part (c).

Solution

(a) If each baseball glove costs $15 to make, then the cost of making x gloves is

(cost per baseball glove)(the number of gloves)

or

$$15x$$

(b) If each hockey catch glove costs $10 to make, then the cost of making y hockey gloves is

(cost per hockey glove)(the number of hockey gloves)

or

$$10y$$

(c) The combined cost of making x baseball gloves and y hockey catch gloves is the sum of $15x$ and $10y$. This sum cannot exceed $15,000. Therefore,

$$15x + 10y \leq 15{,}000$$

(d) We use the three-step strategy for graphing linear inequalities. Replace the inequality symbol by the equal sign,

$$15x + 10y = 15{,}000$$

The intercepts give us the two points $(0, 1{,}500)$ and $(1{,}000, 0)$. The line determined by these two points is shown in Figure (a).

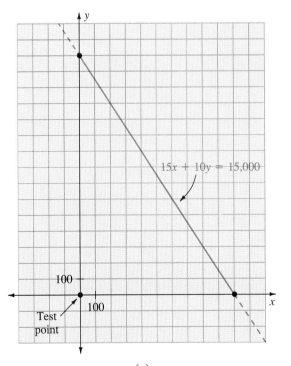

(a)

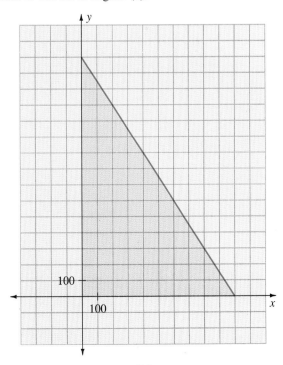

(b)

Next, use $(0, 0)$ as a test point. The inequality is satisfied by the origin, so the region to shade is below the line as shown in Figure (b).

Since x and y represent numbers of gloves, $x \geq 0$ and $y \geq 0$. Therefore, only that part of the lower half-plane in the first quadrant is shaded. ▲

L E A R N I N G   A D V A N T A G E   **Why study mathematics?**
To attain mathematical maturity means to be able to use mathematics as a language. Mathematical language is used to determine the truth of a statement as well as being able to predict the outcome of events. The language of mathematics is acquired through courses in elementary and intermediate algebra.

E X E R C I S E   S E T   7.4

For Exercises 1–22, graph by using the three-step strategy for graphing linear inequalities.

1. $x + 2y \leq 8$ **2.** $2x + y \geq 4$

3. $x - 3y > 9$ **4.** $4x - y < 8$

5. $3x - y \leq -3$ **6.** $5x + 2y \geq -10$

7. $x < 2y$

8. $y > 3x$

9. $y \le 0$

10. $x \le 0$

11. $x + y < 0$

12. $5x + 10y > 0$

13. $-10 < y$

14. $7 \le x$

15. $4x - 3y > 12$

16. $3x - 5y \le 15$

17. $6x - 4y \leq 10$

18. $4x + 3y \geq 14$

19. $y \leq -\dfrac{3}{2}x$

20. $y \geq \dfrac{3}{4}x$

21. $x \leq 1{,}000$

22. $y > -5{,}000$

For Exercises 23–32, write an inequality from the information, then graph it.

23. The sum of the x-coordinate and y-coordinate exceeds two.

24. The difference of the x-coordinate and y-coordinate does not exceed 50.

25. The difference of twice the x value and three times the y value is less than 18.

26. Five times the x value minus three times the y value is greater than -15.

27. The y-coordinate reduced by four times the x-coordinate is negative.

28. Ten times the x-coordinate minus 25 times the y-coordinate is positive.

29. Twice the x value is at most three times the y value.

30. Five times the y value is at least four times the x value.

31. The x-coordinate, reduced by one, is more than the y-coordinate divided by two.

32. The y-coordinate, subtracted from five, is less than the x-coordinate divided by ten.

33. Adrian is planning a meal of steak and salad. Each serving of steak contains 150 calories and each serving of salad contains 30 calories.
   (a) Express the number of calories in x servings of steak.
   (b) Express the number of calories in y servings of salad.
   (c) The combined number of calories from x servings of steak and y servings of salad must supply at least 450 calories. Express this restriction by a linear inequality.
   (d) Graph the linear inequality found in Part (c). Keep in mind that $x \geq 0$ and $y \geq 0$.

34. Keller Paper Company makes two types of paper: newsprint and paperboard. The company makes a profit of $2,000 from each ton of newsprint and $3,000 from each ton of paperboard it sells.
   (a) Express the profit from selling x tons of newsprint.
   (b) Express the profit from selling y tons of paperboard.
   (c) The combined profit from selling x tons of newsprint and y tons of paperboard must be at least $60,000. Express this restriction as a linear inequality.
   (d) Graph the linear inequality found in Part (c). Keep in mind that $x \geq 0$ and $y \geq 0$.

35. Washington Orchards makes apple cider and apple-cider vinegar. A masher machine is used 3 hours for each vat of apple cider and 2 hours for each vat of apple-cider vinegar.
   (a) What is the number of hours required of the masher machine to make x vats of apple cider?
   (b) What is the number of hours required of the masher machine to make y vats of apple-cider vinegar?
   (c) The combined time on the masher machine cannot exceed 12 hours a day. Express this restriction by a linear inequality.
   (d) Graph the inequality found in Part (c).

36. Financial Security Investments plans to invest in a money market fund and a municipal bond. The fund pays 10% annually and the bond pays 15% annually.
   (a) What is the annual interest if x million dollars is invested in the money market fund?
   (b) What is the annual interest if y million dollars is invested in the municipal bond?
   (c) For tax purposes, the combined interest from the two investments must not exceed $3 million. Express this restriction by a linear inequality.
   (d) Graph the inequality found in Part (c).

REVIEW PROBLEMS

The following exercises review parts of Section 1.5. Doing these problems will help prepare you for the next section.

Write an algebraic expression.

37. The amount of money received for working x hours at \$30 per hour.

38. The number of dollars in x quarters.

39. The cost of x tennis rackets, if each one costs \$250.

40. The cost of y pounds of coffee beans, if one pound costs \$6.99.

E N R I C H M E N T   E X E R C I S E S

1. In a coordinate plane, find the intersection of the graphs of $x \geq 0$ and $y \leq 3$.

2. In a coordinate plane, find the intersection of the graphs of $x \leq 5$ and $y \geq 2$.

3. Graph the inequalities $0 \leq x \leq 1$ and $0 \leq y \leq 4$.

4. Graph the inequalities $x + y \leq 10$ and $x \geq 2$.

5. Graph the inequalities $x \geq 0$, $y \geq 0$, and $2x - y \leq 6$.

Answers to Enrichment Exercises are on page A.4.

7.5

Applications of Linear Equations

OBJECTIVE

▶ *To use linear equations in applications*

Many applications involve a linear relationship between two variables. For example, the distance y traveled by a plane going 180 miles per hour is related to the time x by the linear equation $y = 180x$. Given a value for x, say 2 hours, then y is determined by the equation $y = 180(2)$ or 360 miles. We say that y is given in terms of x by the linear equation. That is, y is *linearly dependent* upon x. For that reason, x is called the **independent** variable and y is called the **dependent** variable.

Example 1

Depreciation   Meadowbrook Farms buys a wheat combine for $65,000. The value of the machine decreases with time. At the end of 12 years the machine has a salvage value of $20,000. Assume that the value V of the machine is linearly dependent upon the number of years after purchase.

(a) Write the linear equation that expresses V in terms of t, the number of years after purchase.

(b) Draw the graph of this linear equation.

(c) Use the graph to estimate the value of the combine after 6 years.

(d) Use the equation to find the exact value of the combine after 6 years.

Solution

(a) We are looking for an equation of the form $V = mt + b$, where V is the value, in thousands of dollars, of the combine t years after the purchase of the machine. From the information given, we can find two points of the form (t, V). The machine was purchased for $65,000. Therefore, when $t = 0$, $V = 65$. This gives the **data point** $(0, 65)$. Twelve years later, the machine is worth $20,000. Therefore, $(12, 20)$ is another data point.

The problem is now reduced to finding the equation of the line through the two points $(0, 65)$ and $(12, 20)$. We use the slope-intercept form $V = mt + b$. The slope m is found by using the two data points in the slope formula.

$$m = \frac{V_2 - V_1}{t_2 - t_1}$$

$$m = \frac{20 - 65}{12 - 0}$$

$$= \frac{-45}{12}$$

$$= -\frac{15}{4} \qquad \text{Divide numerator and denominator by } 3.$$

Replacing m by $-\dfrac{15}{4}$ in the equation $V = mt + b$,

$$V = -\frac{15}{4}t + b$$

Recall that the number b is the V-intercept. The data point $(0, 65)$ has a zero for the t-coordinate, so the V-coordinate 65 is the number b. Therefore, the linear equation that expresses V in terms of t is

$$V = -\frac{15}{4}t + 65$$

(b) The graph of this equation is shown in the figure below. Notice that the interval of values for the independent variable t is $0 \le t \le 12$. Therefore, we plot the two points $(0, 65)$ and $(12, 20)$ and draw the line segment between them.

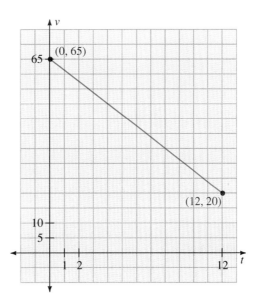

(c) To estimate the value of the combine after 6 years, first locate the number 6 on the t-axis as shown in the figure. Draw a vertical line up to the graph, then draw a horizontal line back to the V-axis. We estimate that the number that this line touches is 43. Therefore, the value of the combine after 6 years is approximately $43,000.

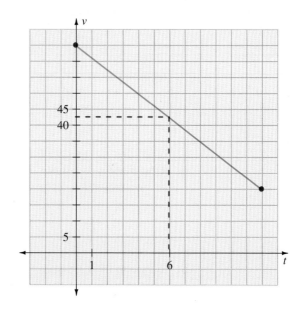

(**d**) The exact value of the combine after 6 years is found by replacing t by 6 in the equation.

$$V = -\frac{15}{4}t + 65$$

$$= -\frac{15}{4}(6) + 65$$

$$= 42.5$$

Therefore, the exact value of the combine after 6 years is $42,500.   ▲

Linear cost equations

The Powerstar Corporation makes x tennis rackets per day. The daily **total cost** y is the sum of the **variable cost** and the **fixed cost.** The variable cost is the product of the cost per tennis racket and the number x of tennis rackets made. The fixed cost (for example, the cost of machinery and insurance) does not depend upon the number of rackets produced. Now, the total cost is the sum of the variable cost and the fixed cost.

total cost = variable cost + fixed cost

In the next example, we illustrate how this equation can be used to express the total cost in terms of the number of items (tennis rackets) produced.

Example 2   The Powerstar Corporation makes x tennis rackets each day with a cost per racket of $25 and a fixed cost of $150 per day.

(a) Express the daily total cost C in terms of x.

(b) If 60 tennis rackets are made each day, what is the daily total cost?

(c) Draw the graph of the cost equation.

Solution   (a) The variable cost of making x rackets at $25 per racket is $25x$ and the fixed cost is $150. Using the formula for total cost,

total cost = variable cost + fixed cost

$$C \quad = \quad 25x \quad + \quad 150$$

Therefore, if the daily output is x tennis rackets, then the total cost C is given by

$$C = 25x + 150$$

(b) To find the total cost C when 60 rackets are made, we replace x by 60 in the cost equation.

$$C = 25x + 150$$
$$C = 25(60) + 150$$
$$= 1,500 + 150$$
$$= 1,650$$

The total cost for making 60 rackets per day is $1,650.

(c) We draw the graph by using the two points (0, 150) (60, 1,650) as shown in the figure below.

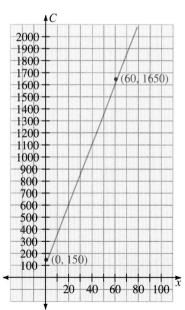

Linear demand equations

In the market place, the price p of an item is related to the amount sold x. The amount sold x is called the **demand** of the item. Suppose the price is related to the demand by the linear equation $p = mx + b$. We say that the price p is *linearly dependent* upon the demand x. The graph of this equation is a line. Since two points determine a line, we may find the values of m and b, if two data points are given.

Example 3

A store sells compact disc players. When the store had a demand of 200 compact disc players per week, the price per player was $500. When the demand rose to 300 players per week, the price was $450. Assume the price p per player is linearly dependent upon the demand x.

(a) Write the linear demand equation.
(b) What is the price of a compact disc player, if the weekly demand is 350 players?
(c) Draw the graph of the demand equation.

Solution

(a) We want to determine the linear equation $p = mx + b$. From the information given, we can find two data points (x, p). When $x = 200$, $p = 500$, so $(200, 500)$ is a data point. Also, when $x = 300$, $p = 450$, which means that $(300, 450)$ is another data point. The problem is now reduced to finding the equation of a line through the points $(200, 500)$ and $(300, 450)$. The number m is the slope of the line. Using the slope formula,

$$m = \frac{p_2 - p_1}{x_2 - x_1}$$

$$= \frac{450 - 500}{300 - 200}$$

$$= \frac{-50}{100}$$

$$= -\frac{1}{2}$$

Therefore, replacing m by $-\frac{1}{2}$, we have

$$p = -\frac{1}{2}x + b$$

Next, we use either data point to find b. Using $(200, 500)$, we replace x by 200 and p by 500.

$$500 = -\frac{1}{2}(200) + b$$

$$500 = -100 + b$$

$$600 = b$$

Therefore, m is $-\dfrac{1}{2}$, b is 600, and the equation is

$$p = -\frac{1}{2}x + 600$$

(b) Since the demand is 350 players, set $x = 350$ in the demand equation.

$$p = -\frac{1}{2}(350) + 600$$

$$= -175 + 600$$

$$= 425$$

The price per player is $425.

(c) To graph the demand equation, we use the two points $(0, 600)$ and $(350, 425)$. Remember that $x \geq 0$ and $p \geq 0$, so only the first quadrant portion of the line is drawn. (See the figure.)

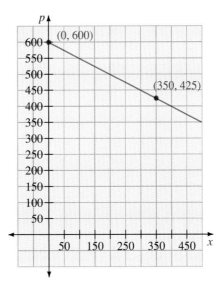

Linear supply equations

In addition to price-demand, there is another force occurring in the market place. It is the relationship between the price of an item and the amount of the item, or **supply,** that is available. Suppose the price p is related to the supply x by the linear equation $p = mx + b$. We say that the price p is *linearly dependent* upon the supply x. The slope m and the p-intercept b can be found if two data points are given.

Example 4

When the supply of dark roast coffee beans was 25,000 pounds in the Greater Cleveland area, the price was $6 per pound. When the supply rose to 35,000 pounds, the price per pound was $7. Assume the price p per pound is linearly dependent upon the supply x (in thousands of pounds).

(a) Write the linear supply equation.

(b) What is the price of coffee if the supply is 36,000 pounds?

(c) Draw the graph of the supply equation.

Solution

(a) We want to find m and b in the equation $p = mx + b$. From the information given, we obtain two data points in the form (x, p). When $x = 25$, then $p = 6$ and when $x = 35$, then $p = 7$. Therefore, the two data points are $(25, 6)$ and $(35, 7)$. We next use these two ordered pairs to determine the equation $p = mx + b$. Using the slope formula to find m,

$$m = \frac{p_2 - p_1}{x_2 - x_1} = \frac{7 - 6}{35 - 25} = \frac{1}{10} = 0.1$$

Replacing m by 0.1 in the equation $p = mx + b$,

$$p = 0.1x + b$$

Next, we find the value of b by using either data point. Using $(25, 6)$, replace x by 25 and p by 6, then solve for b.

$$6 = (0.1)(25) + b$$

$$6 = 2.5 + b$$

$$6 - 2.5 = b$$

$$3.5 = b$$

Therefore, b is 3.5, and the linear supply equation is

$$p = 0.1x + 3.5$$

(b) Replacing x by 36, the price of coffee is

$$p = 0.1(36) + 3.5$$

$$= \$7.10$$

(c) We graph the line as shown in the figure below using the two points $(0, 3.5)$ and $(25, 6)$.

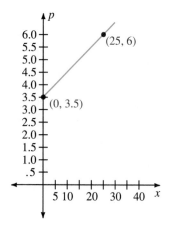

CALCULATOR CROSSOVER

Radon is a colorless, odorless radioactive gas that can seep into the basements of buildings located on the Reading Prong, an area that stretches from eastern Pennsylvania through New Jersey and into New York state. In this region, it is estimated that 250 out of 1,000 lung cancer deaths are caused by radon exposure when the amount of radon in the environment is 0.2 working levels. When the working level of radon is 0.5, it is estimated that 450 out of 1,000 lung cancer deaths are caused by radon. Assume that the number D of lung cancer deaths per 1,000 caused by radon exposure is linearly dependent upon the working level x of radon in the environment, $0.2 \le x \le 0.5$.

1. Write a linear equation that expresses this relationship.
2. If the working level of radon is 0.35, what is the estimated number of lung cancer deaths due to radon?

Answers   **1.** $D = 666.67x + 116.67$ **2.** 350

LEARNING ADVANTAGE   **Why study mathematics?**
Mathematics is not merely learning formulas, but looking for results and determining a repeated sequence of events that can be described by a mathematical formula. Too frequently, a person mistakes arithmetic for mathematics. Arithmetic and algebra are only the start toward using mathematics to solve problems.

EXERCISE SET 7.5

1. The Cola Company buys a new capping machine for $12,000. The machine has a useful life of 14 years at which time it has a scrap value of $4,000. Assume that the value V of the machine is linearly dependent upon time t.
   (a) Find the linear equation of V in terms of t.

   (b) Draw the graph of this linear equation.
   (c) Use the graph to estimate the value of the machine after 10 years.
   (d) Use the equation to find the exact value of the machine after 10 years.

2. The Bethlehem Foundry buys a new casting press for $50,000. This press has a useful life of 20 years with salvage value of $10,000. Assume that the value V of the press is linearly dependent upon time.
   (a) Write the linear equation that expresses V in terms of time t.
   (b) Draw the graph of this linear equation.
   (c) Use the graph to estimate the value of the press after 15 years.
   (d) Use the equation to find the exact value of the press after 15 years.

3. Frank's Supply Company makes trophies and has a daily fixed cost of $40 with a variable cost of $5.
   (a) If the company makes x trophies each day, express the daily total cost C in terms of x.

   (b) One day, 500 trophies were made. What was the corresponding total cost for that day?

5. Flex Products makes universal gyms. It has a daily fixed cost of $55. If the company makes 10 universal gyms in a day, the total cost is $5,305. Assume that the daily total cost C is linearly dependent upon the number of universal gyms made in a day.
   (a) If x universal gyms are made in a day, write the linear equation that expresses C in terms of x.
   (b) What is the corresponding total cost, if 8 universal gyms are made in a day?

7. The Thomas family rents a car for a 7-day vacation. If the rental agency charges $25 per day plus 20 cents per mile, express the cost of renting the car for the vacation if they drove a total of x miles.

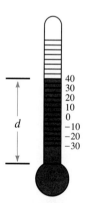

9. When the demand for flounder in Portland was 800 pounds, the price was $4.25 per pound. When the demand rose to 1,200 pounds, the price was $3.75 per pound. Assume that the price p per pound of flounder is linearly dependent upon the demand x of flounder.
   (a) Write a linear equation that expresses p in terms of x.
   (b) What is the price of flounder if the demand is 1,250 pounds?

4. Montana Leather Goods makes western saddles with a variable cost of $200 and a fixed cost of $100 a week.
   (a) If the company makes x saddles each week, express the weekly total cost C in terms of x.

   (b) One week, 7 saddles were made. What was the corresponding total cost for that week?

6. Caliber Instruments makes acoustic guitars and has a fixed cost of $45 a day. The total cost of making 8 guitars in a day is $2,573. Assume that the daily total cost C is linearly dependent upon the number of guitars made in a day.
   (a) If x guitars are made in a day, write the linear equation that expresses C in terms of x.

   (b) What is the corresponding total cost C if 16 guitars are made in a day?

8. The distance d that a column of mercury rises in a thermometer is linearly dependent upon the temperature F. In the thermometer shown below, when F is 20° below zero, the distance d is 12 cm. When the temperature is 10°, the distance is 15 cm.
   (a) Write the linear equation that expresses d in terms of F.
   (b) What is d when the temperature is 80°?

   (c) Draw the graph of the demand equation.

10. When the demand for red delicious apples in Bismarck was 20,000 pecks, the price was $5 per peck. When the demand rose to 24,000 pecks, the price was $4.80 per peck. Assume that the price p per peck is linearly dependent upon the demand x of red delicious apples.
  (a) Write the linear equation that expresses p in terms of x (in thousands of pecks).

  (b) What is the price of red delicious apples, if the demand is 30,000 pecks?
  (c) Draw the graph of the demand equation.

11. When the supply of California iceberg lettuce in a Midwestern region was 300,000 heads, the price was $1.00 per head. When the supply rose to 350,000 heads, the price was $1.20 per head. Assume that the price p per head is linearly dependent upon the supply x (in thousands of heads).
  (a) Write the linear equation that expresses p in terms of the supply x.
  (b) What is the price of lettuce, if the supply is 375,000 heads?
  (c) Draw the graph of the supply equation.

12. When the supply of vinyl for record companies in Memphis was 20,000 pounds, the price was $2 per pound. When the supply increased to 25,000 pounds, the price was $2.50 per pound. Assume that the price p per pound of vinyl is linearly dependent upon the supply of vinyl.
  (a) Write a linear equation that expresses p in terms of the supply x of vinyl.
  (b) What is the price of vinyl if the supply is 24,000 pounds?
  (c) Draw the graph of the supply equation.

REVIEW PROBLEMS

The following exercises review parts of Section 4.6. Doing these problems will help prepare you for the next section.

Solve for k.

13. $\dfrac{1}{3} = \dfrac{k}{5}$

14. $6 = \dfrac{k}{\sqrt{9}}$

15. $\dfrac{2}{5} = \dfrac{k}{\dfrac{1}{2}}$

16. $\dfrac{1}{8} = \dfrac{3k}{2^4}$

17. $5 = \dfrac{k}{2^3}$

18. $12 = \dfrac{6k}{\sqrt{16}}$

19. $6 = \dfrac{3\sqrt{8}k}{2^3}$

20. $15 = \dfrac{\sqrt{12}k}{4^2}$

21. $12 = \dfrac{2\sqrt{48}k}{\sqrt{12}}$

22. $4 = \dfrac{3\sqrt{18}k}{21}$

E N R I C H M E N T   E X E R C I S E S

1. The length of a steel rod is 10 centimeters at a room temperature of 25° centigrade. The rod expands with increased temperature and the length L of the rod is linearly dependent upon the temperature t. When t is 300°C, the length of the rod is 10.02 centimeters. Write a linear equation that expresses L in terms of t.

2. Blackstone Coal company receives shipments of coal every 3 weeks from its supplier. The amount y (in tons) of coal on hand at time t (in weeks) is shown in the following figure.

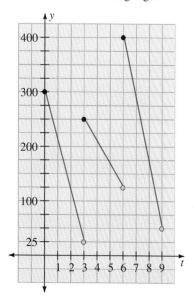

Write a linear equation for y in terms of t for
(a) $0 \le t < 3$ (b) $3 \le t < 6$ (c) $6 \le t < 9$

3. Sterling House makes ceramic figures. If it makes no more than 15 cases of the figures, the total cost is $200 plus $50 per case. If the company makes more than 15 cases, then the total cost is the sum of 200 and 50(15), the cost of the first 15 cases, plus $35 per case for each case over 15.
(a) Express the total cost C in terms of x, the number of cases of figures manufactured.

(b) Draw the graph of the expression obtained in Part (a).

Answers to Enrichment Exercises are on page A.42.

7.6

Variation

OBJECTIVE

▶ *To solve problems using the concept of variation*

An application that involves the relationship between two variables is called variation. Recall that in motion problems we use the formula $d = rt$. Suppose a car travels at a speed of 50 miles per hour for t hours. Then the distance d that it covers is $d = 50t$. We say that d *varies directly with t.*

DEFINITION

If there is a constant k such that

$$y = kx$$

holds for all values of x, then **y varies directly with x** (or as x). The number k is called the **constant of variation.**

If y varies directly as x, then $y = 0$ when $x = 0$. If $x \ne 0$, we can divide both sides by x to obtain

$$\frac{y}{x} = k$$

That is, y varies directly with x means that the quotient $\frac{y}{x}$ is constant for all nonzero values of x. Since the quotient $\frac{y}{x}$ is constant, an *increase* in one variable determines a proportional *increase* in the other variable.

Example 1

Suppose y is proportional to x, and $y = 8$ when $x = 24$. Find y when $x = 15$.

Solution

Since y varies directly with x, there is a constant k so that $y = kx$. When $x = 24$, $y = 8$, so $8 = k(24)$. Therefore, $k = \dfrac{8}{24}$ or $\dfrac{1}{3}$ and so $y = \dfrac{1}{3}x$. Now, to find y when $x = 15$, replace x by 15 in the equation $y = \dfrac{1}{3}x$. So, $y = \dfrac{1}{3}(15)$ or 5.

▲

N O T E   *The variation in Example 1 can also be seen as a proportion:* $\dfrac{y}{15} = \dfrac{8}{24}$. *Then, solve for y.*

Example 2

The distance between two towns is directly proportional to the map distance. On a roadmap, 9 miles is represented by $\dfrac{1}{4}$ inch. On this map, Milwaukee and Chicago are $2\dfrac{1}{2}$ inches apart. What is the actual distance between the two cities?

Solution

The actual distance y is directly related to the map distance x. So there is a constant k such that $y = kx$. When $x = \dfrac{1}{4}$, $y = 9$. Therefore, $9 = k\left(\dfrac{1}{4}\right)$. Multiplying by 4, we obtain $y = 36x$. On the map, Milwaukee and Chicago are $2\dfrac{1}{2}$ or $\dfrac{5}{2}$ inches apart. To find the real distance y, replace x by $\dfrac{5}{2}$ in $y = 36x$.

$$y = 36\left(\dfrac{5}{2}\right) = (18)(5) = 90$$

Therefore, the two cities are 90 miles apart.   ▲

There are other kinds of variations that relate two variables. Recall that the area A of a rectangle with length ℓ and width w is $A = \ell \cdot w$. Given a specified area, say 24 square inches, there are many different sized rectangles that contain 24 square inches. Some of them are pictured below.

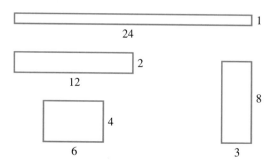

Each rectangle has the relationship that $\ell \cdot w = 24$. We say that the two variables are *inversely related*.

DEFINITION

If there is a constant k such that $y = \dfrac{k}{x}$, for all nonzero values of x, then **y varies inversely with x** (or as x). The number k is called the **constant of variation.**

If we multiply both sides of $y = \dfrac{k}{x}$ by x, we have

$$xy = k$$

Therefore, y varies inversely with x means that the product of x and y is constant. Since the product xy is constant, an *increase* in one variable determines a proportional *decrease* in the other variable.

Example 3   Suppose y varies inversely as x, and $y = \dfrac{1}{3}$ when $x = 4$. Find y when $x = \dfrac{1}{12}$.

Solution   Since y varies inversely with x, there is a constant k such that $y = \dfrac{k}{x}$. From the given information, y is $\dfrac{1}{3}$ when x is 4. Therefore, we replace y by $\dfrac{1}{3}$ and x by 4 in $y = \dfrac{k}{x}$ to determine k.

$$y = \frac{k}{x}$$

$$\frac{1}{3} = \frac{k}{4}$$

$$\frac{4}{3} = k$$

The constant of variation is $\dfrac{4}{3}$ and the equation is

$$y = \frac{\frac{4}{3}}{x}$$

To find y when x is $\dfrac{1}{12}$, replace x by $\dfrac{1}{12}$ and simplify.

$$y = \frac{\frac{4}{3}}{x}$$

$$= \frac{\frac{4}{3}}{\frac{1}{12}}$$

$$= \frac{4}{\overset{}{\underset{1}{3}}} \cdot \frac{\overset{4}{12}}{1}$$

$$= 16$$

Therefore, y is 16 when x is $\dfrac{1}{12}$.   ▲

In the next example, we illustrate an application of inverse variation.

Example 4   The number of vibrations per second a string makes at a given tension varies inversely as its length. If a 3-meter-long string vibrates 600 times a second, what length should the string be to vibrate 750 times a second?

Solution   Let y be the number of vibrations per second when the string is x meters long. Since y varies inversely as x, $y = \dfrac{k}{x}$. When $x = 3$, $y = 600$. Therefore,

$$600 = \frac{k}{3}$$

which means that the constant of variation k is 1,800. Replacing k by 1,800 in $y = \dfrac{k}{x}$, we have $y = \dfrac{1,800}{x}$. We want to find the length x when y is 750. So we replace y by 750 and solve for x.

$$y = \frac{1,800}{x}$$

$$750 = \frac{1,800}{x}$$

$$750x = 1,800$$

$$x = \frac{1,800}{750} \quad \text{or} \quad 2.4$$

The length of the string should be 2.4 meters.   ▲

There are cases where one variable varies either directly or inversely *as the square* of another variable. For example, the area A of a circle varies directly as the square of the radius r by the formula $A = \pi r^2$.

DEFINITION

(a) If there is a constant k such that

$$y = kx^2$$

holds for all values of x, then **y varies directly with the square of x.**

(b) If there is a constant k such that

$$y = \frac{k}{x^2}$$

holds for all values of x, then **y varies inversely with the square of x.**

In both cases, the number k is called the **constant of variation.**

Example 5   (a) Suppose y varies directly with the square of x and $y = 32$ when $x = 4$. Find y when $x = 6$.

(b) Suppose z varies inversely as the square of t and $z = \dfrac{2}{3}$ when $t = \dfrac{1}{2}$. Find z when $t = 9$.

Solution   (a) Since y varies directly as the square of x, there is a constant k such that $y = kx^2$. Replacing y by 32 and x by 4 in this equation,

$$y = kx^2$$
$$32 = k(4)^2$$
$$32 = 16k$$
$$\frac{32}{16} = k$$

Therefore, the constant of variation is 2 and the equation becomes

$$y = 2x^2$$

Now, we want to find y when x is 6. Therefore, we substitute 6 for x in the equation.

$$y = 2(6)^2$$
$$= 72$$

(b) Since z varies inversely with the square of t, there is a constant k such that $z = \dfrac{k}{t^2}$. Replacing z by $\dfrac{2}{3}$ and t by $\dfrac{1}{2}$,

$$\frac{2}{3} = \frac{k}{\left(\dfrac{1}{2}\right)^2}$$

$$\frac{2}{3} = \frac{k}{\dfrac{1}{4}}$$

$$\frac{2}{3} = 4k$$

$$\frac{1}{4}\left(\frac{2}{3}\right) = \frac{1}{4}(4k)$$

$$\frac{1}{6} = k$$

Therefore, $z = \dfrac{\dfrac{1}{6}}{t^2}$. Now, to find z when t is 9,

$$z = \frac{\dfrac{1}{6}}{9^2} = \frac{\dfrac{1}{6}}{81} = \frac{1}{486}$$

▲

Example 6

The brightness of an object varies inversely with the square of the distance from a light source. If the brightness of a sign is 30 luxes at a distance of 6 meters from a flashlight, find the brightness of the sign when the flashlight is 1 meter closer.

Solution

Let y denote the brightness of the sign x meters from the flashlight. Then there is a constant k such that

$$y = \frac{k}{x^2}$$

From the given information, $y = 30$ when $x = 6$. Therefore,

$$30 = \frac{k}{6^2}, \quad \text{so} \quad k = 30 \cdot 36$$

Thus,

$$y = \frac{30 \cdot 36}{x^2}$$

We are to find y when x is 5.

$$y = \frac{30 \cdot 36}{5^2} = \frac{\overset{6}{30} \cdot 36}{\underset{5}{5^2}} = \frac{6 \cdot 36}{5} = \frac{216}{5} = 43.2$$

The brightness of the sign at 5 meters is 43.2 luxes.   ▲

Other kinds of variation are given in the following table.

Variation statement	*Algebraic equation*
y varies directly with the square root of x.	$y = k\sqrt{x}$
y varies inversely as x^3.	$y = \dfrac{k}{x^3}$
y varies inversely with the fourth root of x.	$y = \dfrac{k}{\sqrt[4]{x}}$

Example 7 y varies inversely as the cube root of x, and $y = 2$ when $x = 8$. Find y when $x = 24$.

Solution   Since y varies inversely with the cube root of x, there is a constant k such that $y = \dfrac{k}{\sqrt[3]{x}}$. Replacing y by 2 and x by 8,

$$2 = \frac{k}{\sqrt[3]{8}}$$

$$2 = \frac{k}{2}$$

so, $k = 4$.

Now, replacing k by 4 in the original equation,

$$y = \frac{4}{\sqrt[3]{x}}$$

To find y when x is 24, replace x by 24,

$$y = \frac{4}{\sqrt[3]{24}} = \frac{4}{\sqrt[3]{8 \cdot 3}} = \frac{4}{2\sqrt[3]{3}} = \frac{2}{\sqrt[3]{3}}$$

$$= \frac{2\sqrt[3]{9}}{3}$$ Rationalize the denominator.

                                                                                              ▲

Many times one variable depends on the values of two or more other variables. If a variable z varies directly with two other variables x and y, we say that z varies *jointly* with x and y. In this case, there is a constant of variation k such that $z = kxy$. This is just one example of combined variation. In the following table we list several other examples.

Variation statement	Algebraic equation
z varies jointly with x^3 and $\sqrt[3]{y}$.	$z = kx^3\sqrt[3]{y}$
w varies jointly with r and t^2 and inversely with the square root of u.	$w = \dfrac{krt^2}{\sqrt{u}}$
T varies directly with a and inversely with b^4.	$T = \dfrac{ka}{b^4}$
X varies jointly with t_1 and t_2 and inversely with t^2.	$X = \dfrac{kt_1 t_2}{t^2}$

Example 8   Write the following variation statement as an algebraic equation:

The maximum load L that a horizontal beam supported at both ends can hold varies directly as the width w of the beam and the square of its depth d, and inversely as the length ℓ of the beam.

Solution   Letting k be the proportionality constant, then

$$L = \frac{kwd^2}{\ell}$$ ▲

Example 9   The term z varies jointly with x and the square root of y. When x is 3 and y is 8, $z = 4\sqrt{2}$. Find z when x is 21 and y is $\dfrac{1}{4}$.

Solution   The algebraic equation for the variation statement is

$$z = kx\sqrt{y}$$

Substituting $x = 3$, $y = 8$, and $z = 4\sqrt{2}$, we solve for k.

$$4\sqrt{2} = k(3)(\sqrt{8})$$
$$4\sqrt{2} = k(3)(2\sqrt{2})$$
$$\frac{4}{6} = k$$

So, k is $\dfrac{2}{3}$. Therefore, the equation becomes

$$z = \frac{2}{3}x\sqrt{y}$$

Setting $x = 21$ and $y = \dfrac{1}{4}$,

$$z = \frac{2}{3}(21)\sqrt{\frac{1}{4}} = \frac{2 \cdot \overset{7}{\cancel{21}}}{\cancel{3} \cdot \cancel{2}} = 7$$ ▲

L E A R N I N G   A D V A N T A G E   **Why study mathematics?**
*Mathematics helps me organize my concepts in a complete and logical manner.
If I am puzzled with a complex idea, I try to formulate its concepts mathematically. I know that any portion that I can't express through mathematics, I don't
yet comprehend.*

E X E R C I S E   S E T   7.6

For Exercises 1–18, solve the variation problem.

1. Suppose y varies directly with x, and $y = 15$ when $x = 5$. Find y when $x = 9$.

2. Suppose y varies directly with t, and $y = 27$ when $t = \dfrac{3}{2}$. Find y when $t = 10$.

3. If A varies inversely as v, and A is 30 when $v = \dfrac{1}{2}$, find A when $v = 5$.

4. Given that N varies inversely as m, and that $N = 1$ when $m = \dfrac{2}{5}$, find N when $m = 20$.

5. If z varies directly with the square of s, and if $z = 3$ when $s = 2$, find z when $s = \dfrac{1}{3}$.

6. Suppose B varies directly as the square of v, and when $v = \dfrac{1}{4}$, $B = \dfrac{3}{4}$, find B when $v = 2$.

7. If V varies inversely with the square of t, and $V = 189$ when $t = \dfrac{1}{3}$, find V when $t = 9$.

8. Given that W varies inversely with the square of m, and that W is $\dfrac{1}{7}$ when m is 7, find W when m is 21.

9. If y varies directly as the square root of z, and y is 12 when z is 4, find y when z is 9.

10. If r varies directly with the cube root of p, and r is 3 when p is 8, find r when p is $\dfrac{1}{8}$.

11. If z varies inversely with v^3, and $z = 4$ when $v = 2$, find z when $v = 3$.

12. Suppose N varies inversely with n^4, and N is 2 when n is 1. Find N when n is $\dfrac{1}{2}$.

13. If y varies inversely as the square root of x, and y is $4\sqrt{3}$ when x is 12, find y when x is 48.

14. If z varies inversely with the square root of y, and z is $2\sqrt{5}$ when y is 75, find z when y is 200.

15. If z varies jointly with x and y^2, and z is 4 when x is $\dfrac{1}{2}$ and y is 3, find z when x is -2 and y is $\dfrac{3}{2}$.

16. Suppose R varies jointly with s^2 and the square root of t, and R is 20 when s is -4 and t is 36. Find R when s is $\dfrac{1}{5}$ and t is 16.

17. Suppose w varies jointly with p and q and varies inversely with $\sqrt{r}$. If $w = 3$ when $p = 2$, $q = 2$, and $r = 8$, find w when $p = -3$, $q = -4$, and $r = 3$.

18. Suppose z varies jointly with y and t and varies inversely with $\sqrt[3]{x}$. If $z = \sqrt[3]{2}$ when $y = -1$, $t = 2$, and $x = 4$, find z when $y = 2$, $t = -3$, and $x = -27$.

19. Suppose y varies inversely as x. If x is tripled, how does y change?

20. Suppose y varies directly as x. If y is doubled, how does x change?

21. If A varies directly as the square of t, and t is halved, how does A change?

22. If V varies inversely as the square of p, and p is replaced by $4p$, how does V change?

23. For Susan, the amount A (in percent) of knowledge retained from a lecture varies inversely as the time t that has elapsed since the lecture. If 75% is retained after 2 days, how much is retained after 10 days?

24. The current C in an electrical circuit varies inversely as the resistance R. When the resistance is 20 ohms, the current is 32 amps. Find the current when the resistance is 12 ohms.

25. The cost of buying candy varies directly as the number of packages of candy bought. If 6 packages cost $10.62, find the cost of 20 packages of candy.

26. The number of pizzas that Al can make varies directly as time. If he can make 12 pizzas in 20 minutes, how many pizzas can he make in 1 hour and 15 minutes?

27. On a roadmap, $2\dfrac{1}{2}$ inches represents 255 miles. If Atlanta and Jacksonville are $3\dfrac{1}{16}$ inches apart, what is the actual distance between the two cities? Round your answer off to the nearest mile.

28. Los Angeles and San Diego are $\dfrac{15}{16}$ of an inch apart on a roadmap, where 5 inches represents 490 miles. What is the actual distance between the two cities? Round your answer off to the nearest mile.

29. How many bowls of Honey Sugar Flakes must be eaten to obtain the calcium equivalent to one bowl of Fortified Shredded Wheat, if two bowls of Honey Sugar Flakes supplies $1\frac{1}{2}\%$ and 3 bowls of Fortified Shredded Wheat supplies 36% of the U.S. Recommended Daily Allowance of calcium?

30. Three bottles of Zap cola has the same amount of caffeine as 8 bottles of another cola. How many bottles of the other cola must be drunk to obtain the same amount of caffeine from drinking 1 bottle of Zap cola?

31. The weight of an object on or above the Earth varies inversely as the square of the distance from the Earth's center. A satellite weighing 4,500 kg is to be launched into orbit. What will be the weight of the satellite when it is 300 km above the surface of the Earth? Use 6,500 km as the radius of the Earth.

32. The height of a cylinder of constant volume varies inversely as the square of the radius of the base. The height of a cylinder is 10 inches and the radius of the base is $\frac{2}{3}$ inch. Find the height of a cylinder with the same volume that has a base radius of 6 inches.

33. Sound travels in waves where the wave motion is measured by the frequency f, the number of waves per second. If d is the wavelength, then f varies inversely as d. For two notes, suppose the wavelength of the first note is $\frac{5}{4}$ that of the second note. How do their frequencies compare?

34. If two musical notes differ by an octave, then the frequency of the higher note is twice that of the lower note. What is the relationship between the wavelengths of the two notes? (See Exercise 33.)

ENRICHMENT EXERCISES

1. Suppose that y varies directly as x and that when x is x_1, then y is y_1 and when x is x_2, y is y_2. If x_1 and x_2 are neither zero, show that

$$\frac{y_1}{x_1} = \frac{y_2}{x_2}$$

2. Suppose y varies directly as x and when $x = \frac{3}{4}$, $y = 2$, use Exercise 1 to find y when $x = \frac{8}{5}$.

3. Suppose y varies inversely as x. If $y = y_1$ when $x = x_1$ and $y = y_2$ when $x = x_2$, show that $y_1 x_1 = y_2 x_2$.

4. If y varies inversely as x and $y = -2$ when $x = 5$, use Exercise 3 to find y when $x = \dfrac{20}{3}$.

5. An example of inverse variation is the law of the lever. A lever is a bar together with a fulcrum. If weights w_1 and w_2 are placed at distances d_1 and d_2 from the fulcrum, and the lever is balancing, then

$$w_1 d_1 = w_2 d_2$$

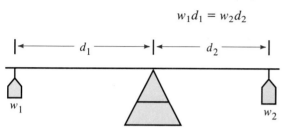

If a 120-pound weight is 6 feet from the fulcrum of a lever, how far from the fulcrum must a 100-pound weight be to balance the lever?

6. To obtain an estimate of the total population T of a species of wildlife in a given area, N animals are caught at random and tagged, then released. Later M animals are caught at random and it is observed that M_1 of them have tags. Assume the total population T varies directly as the number N, where the constant of variation is $\dfrac{M}{M_1}$.

To obtain an estimate of the deer population in the Horicon Marsh National Preserve, 72 deer are caught at random and tagged, then released. Later, 62 deer are caught at random and it is observed that 12 of them have tags. What is the estimated deer population of the preserve?

Answers to Enrichment Exercises are on page A.43.

C H A P T E R   7   # Summary and review

Linear equations in two variables and their graphs (7.1)

$(-1, 2)$ is a solution of $2x + 3y = 4$, since $2(-1) + 3(2) = -2 + 6 = 4$

A **linear equation in two variables** is an equation that can be put in the form $ax + by = c$, where a, b, and c are numbers and a and b are not both zero. A **solution** of a linear equation in two variables is any ordered pair (x, y) that satisfies the equation.

Ordered pairs can be visualized graphically by using a **coordinate plane.** To **graph** a linear equation means to plot its solution set in a coordinate plane. The graph of any linear equation is a straight line. Since two points determine a

Examples

Graph $2x - 3y = 6$ using the intercepts.
When $x = 0$, $y = -2$ and when $y = 0$, $x = 3$. Plot the two points $(0, -2)$ and $(3, 0)$, then draw the line as shown in the figure.

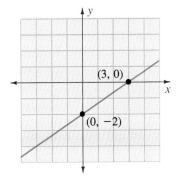

line, we find two ordered pairs that are solutions of the equation, then a third solution for a check. Plot these three points and draw the line through them. The **y-intercept** of a line is the y-coordinate of the point where the line crosses the y-axis. The **x-intercept** of a line is the x-coordinate of the point where the line crosses the x-axis.

To find the y-intercept, set $x = 0$ in the equation and solve for y. To find the x-intercept, set $y = 0$ in the equation and solve for x.

Two **special lines** are vertical and horizontal lines. The graph of the equation $x = a$ is a vertical line passing through the x-axis at a. The graph of the equation $y = b$ is a horizontal line passing through the y-axis at b.

The slope of a line (7.2)

Find the slope of the line passing through $(1, 2)$ and $(-3, 5)$. Using the formula for slope,

$$m = \frac{y_2 - y_1}{x_2 - x_1}$$

$$= \frac{5 - 2}{-3 - 1} = \frac{3}{-4} = -\frac{3}{4}$$

For a nonvertical line, its **slope m** is given by

$$m = \frac{\text{vertical change}}{\text{horizontal change}}$$

If (x_1, y_1) and (x_2, y_2) are any two points on the line, then

$$m = \frac{y_2 - y_1}{x_2 - x_1}$$

The slope of a line tells us two things:

1. the steepness of the line

2. whether the line is slanted upwards or downwards from left to right

A horizontal line has zero slope and a vertical line has no slope.

Two lines are **parallel** when they have no points in common. If two nonvertical lines are parallel, then they have the same slope. Conversely, if they have the same slope, then they are parallel.

The equation of a line (7.3)

Find the equation of the line whose slope is -2 and y-intercept is 3. Use the form $y = mx + b$, where $m = -2$ and $b = 3$. The desired equation is $y = -2x + 3$.

The **slope-intercept** form of a line is

$$y = mx + b$$

where m is the slope and b is the y-intercept.

Examples

Find the equation of the line with slope 4 and passing through $(-1, 3)$.

Using the point-slope form with $m = 4$ and $(x_1, y_1) = (-1, 3)$, $y - 3 = 4[x - (-1)]$, which simplifies to $-4x + y = 7$.

If (x_1, y_1) is a point on a line having slope m, then the **point-slope** form of the equation of this line is

$$y - y_1 = m(x - x_1)$$

Graphing linear inequalities in two variables (7.4)

The graph of $-x + 2y \geq 6$ is shown in the following figure.

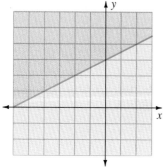

The **graph** of a linear inequality is the set of points (x, y) in the plane that satisfy the inequality.

Applications of linear equations (7.5)

Many applications involve linear equations. In particular, when y is given in terms of x, we say that y is *linearly dependent* upon x and call x the **independent variable** and y the **dependent variable.**

The applications studied in this section include depreciation, linear cost equations, and linear supply and demand equations.

Variation (7.6)

If y varies inversely as x, and $y = 2$ when $x = 3$, find y when $x = \dfrac{3}{5}$.

There is a constant k such that $y = \dfrac{k}{x}$. Replace y by 2 and x by 3.

$2 = \dfrac{k}{3}$,   so $k = 6$

Therefore,

$y = \dfrac{6}{x}$

When $x = \dfrac{3}{5}$,

$y = \dfrac{6}{\frac{3}{5}} = 6\left(\dfrac{5}{3}\right) = 10$

There are several types of variations. Some examples are given in the following table.

Variation statement	Equation
y varies directly with x.	$y = kx$
y varies inversely with x.	$y = \dfrac{k}{x}$
y varies directly with the square of x.	$y = kx^2$
y varies inversely with the cube root of x.	$y = \dfrac{k}{\sqrt[3]{x}}$
z varies jointly as x and y^3.	$z = kxy^3$
R varies jointly with s^2 and t and inversely with $\sqrt{h}$.	$R = \dfrac{ks^2t}{\sqrt{h}}$

CHAPTER 7 **REVIEW EXERCISE SET**

Section 7.1

For Exercises 1–4, describe where in the coordinate plane the point (x, y) would be located.

1. x is negative and y is positive.

2. The y-coordinate is negative.

3. The x-coordinate is -2.

4. y is three times x.

For Exercises 5–10, draw the graph of the linear equation.

5. $3x + 2y = 12$

6. $2x - y = 200$

7. $x = 3$

8. $y = -1$

9. $y = 2x - 3$

10. $y = -\dfrac{2}{3}x + 4$

Section 7.2

For Exercises 11–14, use the slope formula to find the slope, if it exists, of the line through the two points.

11. $(2, 3)$ and $(-4, 1)$

12. $(-1, 4)$ and $\left(\dfrac{3}{5}, 4\right)$

13. $(1, -3)$ and $(1, 5)$

14. $\left(-\dfrac{3}{2}, \dfrac{1}{4}\right)$ and $\left(-\dfrac{1}{6}, \dfrac{5}{12}\right)$

Section 7.3

For Exercises 15 and 16, find the slope and y-intercept of the line from the given equation.

15. $y = 4x - 5$

16. $\dfrac{3}{2}y - 2x = 1$

For Exercises 17–20, the slope m and y-intercept b of a line are given. **(a)** Draw the line, and **(b)** write an equation for the line.

17. $m = \dfrac{3}{2}$ and $b = -2$

18. $m = -2$ and $b = 3$

19. $m = 0$ and $b = \dfrac{3}{2}$

20. $m = -\dfrac{4}{3}$ and $b = 0$

For Exercises 21 and 22, use the point-slope form to find an equation of the line through the two points P and Q. Write your answer in the form $Ax + By = C$.

21. $P(2, 1)$ and $Q(-2, 4)$

22. $P(-4, -2)$ and $Q(0, 0)$

For Exercises 23–26, use the slope-intercept form to find an equation of the line through the two points P and Q. Write your answer in the form $y = mx + b$.

23. $P(5, 1)$ and $Q(-1, 6)$

24. $P(0, 2)$ and $Q(2, 0)$

25. $P\left(\dfrac{1}{2}, \dfrac{1}{2}\right)$ and $Q\left(-\dfrac{1}{2}, -\dfrac{1}{2}\right)$

26. $P(3, -4)$ and $Q(-1, 0)$

27. Find an equation of the line through $P(-2, 4)$ and parallel to the line $3x - 2y = 5$.

28. Find the equation of the line with x-intercept -3 and passing through the point $(-2, -1)$. Write your answer in the form $y = mx + b$.

Section 7.4

For Exercises 29–34, graph the linear inequality.

29. $2x + y > 10$

30. $3x - 2y \leq 6$

31. $x > 3y$

32. $x < 50$

33. $y \geq -2 - x$

34. $9x < 3y + 12$

Section 7.5

35. The sanitation department of Salisbury Township buys a water-carrying street sweeper for $28,000. The useful life of the sweeper is 10 years, at which time it has a salvage or trade-in value of $8,000. Assume that the value V, in thousands of dollars, of the machine is linearly dependent upon the number of years t after purchase.

   (a) Write a linear equation that expresses V in terms of t.

   (b) Draw the graph of this linear equation. (*Note:* Since the useful life is 10 years, draw the line for $0 \le t \le 10$.)

   (c) Use the graph to estimate the value of the sweeper after 6 years.

   (d) Use the equation to find the exact value of the sweeper after 6 years.

36. Custom Design Corporation makes x hundred radiator enclosures each day with a cost per 100 enclosures of $25 and a fixed cost of $275 per day.

   (a) Express the daily total cost C in terms of x.

   (b) If 500 radiator enclosures are made per day, what is the daily total cost?

   (c) Draw the graph of the cost equation.

37. A store sells pool tables. When the store had a demand of 550 pool tables per month, the price per table was $1,200. When the demand rose to 600 pool tables per month, the price per table was $800. Suppose that the price p per pool table linearly depends upon the demand x.

   (a) Find the linear demand equation.

   (b) What is the price of a pool table, if the monthly demand is 650 pool tables?

   (c) Draw the graph of the demand equation.

38. When the supply of codfish in San Francisco was 18,000 pounds, the price was $5 per pound. When the supply increased to 20,000 pounds, the price was $5.50 per pound. Assume that the price p per pound is linearly dependent upon the supply x in thousands of pounds.

   (a) Write the linear supply equation.

   (b) What is the price of codfish, if the supply is 22,000 pounds?

   (c) Draw the graph of the supply equation.

Section 7.6

39. Suppose y varies directly with x, and $y = 40$ when $x = 2$. Find y when $x = 5$.

40. If y varies inversely with z, and $y = 6$ when $z = -2$, find y when $z = 6$.

41. If r varies jointly with u^2 and v, and r is 12 when u is 2 and v is -1, find r when u is 3 and v is 4.

42. Suppose w varies jointly with x and y and inversely with $\sqrt{z}$. If $w = 2$ when $x = 1$, $y = 3$, and $z = 18$, find w when $x = \sqrt{2}$, $y = -1$, and $z = 8$.

C H A P T E R 7 **T E S T**

Section 7.1

For Problems 1–3, draw the graph of the linear equation.

1. $4x + 3y = 12$ **2.** $2x - 5y = 10$ **3.** $-x + 7y = 14$

Section 7.2

For Problems 4–6, use the slope formula to find the slope, if it exists, of the line through the two points.

4. $(2, 5)$ and $(1, 3)$ **5.** $(-3, 2)$ and $(-3, 4)$ **6.** $(4, -2)$ and $(-1, -2)$

Section 7.3

7. Find the slope and y-intercept of the line given by $3x - 2y = 12$.

For Problems 8–10, the slope m and y-intercept b of a line are given. **(a)** Draw the line, and **(b)** write an equation for the line.

8. $m = 2$ and $b = -1$.   **9.** $m = 0$ and $b = 3$

10. $m = -\dfrac{4}{3}$, $b = 0$

11. Use the point-slope form to find an equation of the line through the two points $P(-1, 4)$ and $Q(2, 1)$. Write your answer in the form $Ax + By = C$.

12. Use the slope-intercept form to find an equation of the line through $P(3, -3)$ and $Q(1, -4)$. Write your answer in the form $y = mx + b$.

Section 7.4

For Problems 13–15, graph the linear inequality.

13. $x + 4y < 12$ **14.** $2x - 3y \geq 18$ **15.** $y \leq 2x$

Section 7.5

16. The Davis Company makes wooden cabinets and has a fixed weekly cost of $5,000 and a variable cost of $300.
   (a) If the company makes x cabinets each week, express the weekly total cost C in terms of x.

   (b) In one particular week, 200 cabinets were made. What was the corresponding weekly total cost?

17. When the demand for swordfish steaks in the Hudson River Valley was 10,000 pounds, the price was $5 per pound. When the demand rose to 12,000 pounds, the price was $6 per pound. Assume that the price p (in dollars per pound) is linearly dependent upon the demand x (in thousands of pounds).
   (a) Write a linear equation that expresses p in terms of x.
   (b) What is the price of swordfish if the demand is 12,500 pounds?
   (c) Draw the graph of the demand equation.

Section 7.6

18. Suppose y varies directly with x and $y = 35$ when $x = 7$. Find y when $x = 3$.

19. If s varies inversely with t and $s = 4$ when $t = 6$, find s when $t = -2$.

20. If z varies jointly with x and y^2, and $z = 24$ when $x = 2$ and $y = 4$, find z when $x = 8$ and $y = \sqrt{3}$.

Systems of Equations

Overview

We continue our investigation of the connection between analytic geometry and applications. In the last chapter, we considered a single linear equation in two variables. Applications of a single equation are restricted to just one condition between two variables. With a system of two equations in two variables, we are able to study more complex applications. This trend is continued when we look at systems of three equations in three variables, which is the topic of Section 8.3.

8.1

Systems of Linear Equations in Two Variables

OBJECTIVES

▶ *To solve a system of equations by the graphing method*

▶ *To solve a system of equations by the elimination method*

▶ *To solve a system of equations by the substitution method*

In the last chapter, we saw that the graph of a linear equation in two variables is a straight line. Two linear equations taken together form a *system of linear equations*. For example,

$$-x + y = 1$$
$$-2x + y = -1$$

form a system of two linear equations in the two variables x and y. The solution set of a system is the set of all ordered pairs that satisfy both equations.

In our first example, we ask whether or not an ordered pair of numbers is a solution to a given system of equations.

Example 1   Determine if $(-1, 3)$ is a solution of the system of linear equations

$$-x + 4y = 13$$
$$5x - 2y = -11$$

Solution   To see if $(-1, 3)$ is a solution of the system, replace x by -1 and y by 3 in each equation.

$$
\begin{array}{ll}
-x + 4y = 13 & 5x - 2y = -11 \\
-(-1) + 4(3) \stackrel{?}{=} 13 & 5(-1) - 2(3) \stackrel{?}{=} -11 \\
1 + 12 \stackrel{?}{=} 13 & -5 - 6 \stackrel{?}{=} -11 \\
13 = 13 & -11 = -11
\end{array}
$$

The ordered pair $(-1, 3)$ satisfies both equations and, therefore, is a solution of the system.   ▲

By graphing the two equations in the same coordinate plane, we can estimate the solution. This is called solving a system by the **graphing method.**

Example 2   Solve the system by the graphing method:

$$-x + y = 1$$
$$-2x + y = -1$$

Solution   We graph the two lines in the same coordinate plane as shown in the figure.

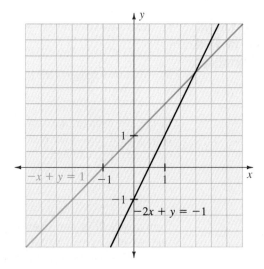

It appears that the point of intersection of the two lines is $(2, 3)$. If the coordinates of this point, $x = 2$ and $y = 3$, are the solution, they satisfy both equations, since the point lies on both lines.

Check $x = 2$ and $y = 3$:

$$-x + y = 1 \qquad -2x + y = -1$$

$$-2 + 3 \overset{?}{=} 1 \qquad -2(2) + 3 \overset{?}{=} -1$$

$$1 = 1 \qquad -1 = -1$$

Since both equations of the system are satisfied, we conclude that the solution of the system is $x = 2$ and $y = 3$. ▲

In general, the solution set of a system of two linear equations consists of points of intersection of two straight lines. There are three possibilities as shown in the following figure. Either the two lines (a) intersect in one point, (b) are parallel, or (c) coincide, that is, the graph of each equation in the system is the same line.

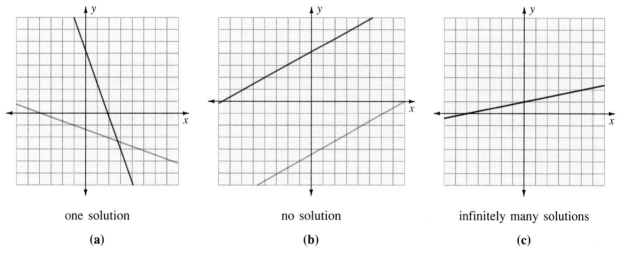

one solution	no solution	infinitely many solutions
(a)	**(b)**	**(c)**

We summarize these three possibilities in the following rule.

R U L E

The solution set of a system of linear equations

1. If the two lines intersect in exactly one point, the coordinates of that point are the *one solution* of the system.
2. If the two lines are parallel, then the lines do not intersect; and therefore, the system has *no solution*. The system is **inconsistent**.
3. If the two equations of the system define the same line, that is, the two lines coincide, then every point on this common line gives a solution to the system and therefore the system has *infinitely many solutions*. The system is **dependent**.

Given a system, we want to solve it. With the second example, we graphed the system and read the solution set from the graph. This graphing method has its limitations. For example, it would be difficult to estimate the coordinates of the point of intersection if the coordinates are not integers.

Fortunately, there are other algebraic methods that we will use that do not depend upon graphing. The first method is called the **elimination method.** This method is used in the next example.

Example 3   Solve the system:

$$x - y = 18$$
$$x + y = -2$$

Solution   The elimination method uses the idea of combining the two equations in such a way as to eliminate one of the variables. For this system, if we add corresponding sides of the two equations, notice that the y variable is eliminated. Here are the details:

$$
\begin{aligned}
x - y &= 18 \\
\underline{x + y} &= \underline{-2} \quad \text{Add.} \\
2x &= 16
\end{aligned}
$$

Solving this resulting equation,

$$2x = 16$$
$$x = 8$$

Next, we replace x by 8 in either of the original equations, then solve for y. We select the first equation.

$$x - y = 18$$
$$8 - y = 18$$
$$-y = 10$$
$$y = -10 \qquad \text{Multiply both sides by } -1.$$

Next, we check our solution, $x = 8$ and $y = -10$ in *both* equations of the original system.

$$\text{\textit{Check}} \quad x = 8 \text{ and } y = -10:$$

$$
\begin{array}{cc}
x - y = 18 & x + y = -2 \\
8 - (-10) \stackrel{2}{=} 18 & 8 + (-10) \stackrel{2}{=} -2 \\
18 = 18 & -2 = -2
\end{array}
$$

The solution to the system is $x = 8$ and $y = -10$, and can be written as the ordered pair $(8, -10)$. ▲

Notice that in Example 3, the two equations of the system were added together to obtain a single equation with only one variable. That is, we *eliminated* one of the variables by adding equations.

When using the elimination method, it is necessary to eliminate one of the variables when the equations are added together.

Example 4   Solve the system:

$$-3x + 2y = -7$$
$$6x + 5y = -4$$

Solution   Adding the two equations does not eliminate either variable, since the addition of the two equations gives the equation $3x + 7y = -11$. However, we can eliminate x if we use the following line of reasoning: Note that the coefficient of x in the first equation is -3 and the coefficient of x in the second equation is 6. Therefore, if we multiply both sides of the first equation by 2, this will result in making the coefficient of x equal to -6, the opposite of 6.

$$2(-3x + 2y) = 2(-7)$$
$$2(-3x) + 2(2y) = -14$$
$$-6x + 4y = -14$$

If we then add this new equation to the second equation, the x variable will be eliminated.

$$-3x + 2y = -7 \xrightarrow{\text{multiply both sides by 2}} -6x + 4y = -14$$
$$6x + 5y = -4 \xrightarrow{\text{leave as is}} \underline{6x + 5y = -4}$$
$$9y = -18$$

Solving $9y = -18$ for y, we have $y = -2$.

Next, replace y by -2 in either of the original equations. We select the first equation.

$$-3x + 2y = -7$$
$$-3x + 2(-2) = -7$$
$$-3x = -3$$
$$x = 1 \qquad \text{Divide both sides by } -3.$$

To check the solution $x = 1$ and $y = -2$, we must show that *both* equations of the original system are satisfied.

Check $x = 1$ and $y = -2$:

$$-3x + 2y = -7 \qquad\qquad 6x + 5y = -4$$
$$-3(1) + 2(-2) \overset{?}{=} -7 \qquad 6(1) + 5(-2) \overset{?}{=} -4$$
$$-3 - 4 \overset{?}{=} -7 \qquad\qquad 6 - 10 \overset{?}{=} -4$$
$$-7 = -7 \qquad\qquad\qquad -4 = -4$$

The solution is $x = 1$ and $y = -2$.   ▲

Sometimes it is necessary to multiply each equation by appropriate numbers so that the coefficients of one of the variables are opposites. This situation is illustrated in the next example.

Example 5   Solve the system:

$$3x + 5y = 7$$

$$2x - 12y = -26$$

Solution   We plan to eliminate x. To achieve this, multiply the first equation by 2 and the second equation by -3, since this will result in $6x$ in the first equation and $-6x$ in the second equation.

$$3x + 5y = 7 \quad \text{———multiply by 2———} \rightarrow \quad 6x + 10y = 14$$

$$2x - 12y = -26 \quad \text{——multiply by } -3 \rightarrow \quad \underline{-6x + 36y = 78}$$

$$46y = 92$$

$$y = 2$$

Next, we replace y by 2 in either equation of the original system. We choose the second equation.

$$2x - 12y = -26$$

$$2x - 12(2) = -26$$

$$2x = -2$$

$$x = -1$$

The solution is $(-1, 2)$. On your own, check this solution by substituting $x = -1$ and $y = 2$ into each equation of the original system.   ▲

N O T E   *After finding the value of one of the variables, we use one of the original equations to solve for the other variable. Since it makes no difference which equation we pick, choose the equation that is easier to solve.*

Recall that a system has no solution when the two lines are parallel. The elimination method can also be used on this kind of system.

Example 6   Solve the system:

$$2x - 3y = -1$$

$$-\frac{2}{3}x + y = -2$$

Solution   If the system has a solution, we can find it using the elimination method. We choose to eliminate x by multiplying the second equation by 3, then adding.

$$2x - 3y = -1 \xrightarrow{\text{leave as is}} 2x - 3y = -1$$

$$-\frac{2}{3}x + y = -2 \xrightarrow{\text{multiply by 3}} -2x + 3y = -6$$

$$0 = -7$$

The assumption that the system had a solution led to the contradiction that $0 = -7$. Therefore, the system has no solution.   ▲

N O T E   *In the previous example, the elimination method was used to conclude that the system had no solution. Keep in mind that geometrically, the two lines are parallel.*

The other "unusual" case is when a system has infinitely many solutions. Recall that graphically the two equations have the same line for their respective graphs.

Example 7   Solve the system:

$$\frac{1}{4}x + y = -1$$

$$-\frac{1}{2}x - 2y = 2$$

Solution   We choose to eliminate the variable y. If we multiply the first equation by 2, then add to the second equation, notice that the y variable will be eliminated.

$$\frac{1}{4}x + y = -1 \xrightarrow{\text{multiply by 2}} \frac{1}{2}x + 2y = -2$$

$$-\frac{1}{2}x - 2y = 2 \xrightarrow{\text{leave as is}} -\frac{1}{2}x - 2y = 2$$

$$0 = 0$$

The resulting true statement, $0 = 0$, means that the two equations describe the same line. In fact, multiplying the first equation of the system by -2 gives the second equation. Therefore, the two lines coincide and the system has infinitely many solutions.   ▲

Before using the elimination method on a system, each equation must be in the standard form $Ax + By = C$.

Example 8   Solve the system:

$$3x + y - 1 = 2(x - 1)$$

$$\frac{x}{2} = 2 + \frac{y}{3}$$

Solution   We rewrite each equation in the standard form. Starting with the first equation,

$$3x + y - 1 = 2(x - 1)$$

$$3x + y - 1 = 2x - 2 \qquad \text{Distribute the 2.}$$

$$3x + y - 1 - 2x + 1 = 2x - 2 - 2x + 1$$

$$x + y = -1 \qquad \text{Simplify.}$$

For the second equation, subtract $\dfrac{y}{3}$ from both sides.

$$\frac{x}{2} = 2 + \frac{y}{3}$$

$$\frac{x}{2} - \frac{y}{3} = 2$$

To avoid working with fractions, we also multiply both sides by the least common denominator, 6.

$$6\left(\frac{x}{2} - \frac{y}{3}\right) = 6(2)$$

$$6\left(\frac{x}{2}\right) - 6\left(\frac{y}{3}\right) = 12$$

$$3x - 2y = 12$$

Now, we use the elimination method on the system.

$$x + y = -1$$

$$3x - 2y = 12$$

We choose to eliminate x. Multiply the first equation by -3 and add the result to the second equation.

$$-3x - 3y = 3$$

$$\underline{3x - 2y = 12}$$

$$-5y = 15$$

$$y = -3$$

Replace y by -3 in the first equation,

$$x + y = -1$$

$$x + (-3) = -1$$

$$x = 2$$

The solution is $x = 2$ and $y = -3$. Check these numbers in the original system.

▲

C A L C U L A T O R   C R O S S O V E R

Solve each system by the elimination method.

1. $2.1x - 3.9y = 7.6$ **2.** $0.7x + 0.9y = 2.45$

$x + 4.5y = 8$ $-0.6x + 1.01y = 7.9$

Answers   **1.** $x = 4.9$, $y = 0.7$ **2.** $x = -3.7$, $y = 5.6$

We summarize the elimination method.

S T R A T E G Y

The elimination method

Step 1   If necessary, transform each equation to the standard form $Ax + By = C$. Also clear fractions in each equation, if desired, by multiplying by the LCD.

Step 2   Decide on one variable to eliminate.

Step 3   If necessary, multiply one or both equations by appropriate numbers so that the numerical coefficients of the variable to be eliminated are opposites.

Step 4   Add the two equations to obtain an equation with only one variable.

Step 5   Solve this equation for the variable.

Step 6   Take the answer from Step 5 and use either equation from the original system to find the value of the other variable.

Step 7   Check your solution in the original system.

We now consider another algebraic method to solve a system of equations. It is called the **substitution method** and is most convenient to use when a system contains an equation that expresses one variable in terms of the other. We illustrate this method in the next example.

Example 9   Solve the system:

$$y = -2x$$

$$4x + y = 7$$

Solution

In the first equation, y is $-2x$. We substitute $-2x$ for y in the second equation, then solve the resulting equation for x.

$$4x + y = 7$$

$$4x + (-2x) = 7$$

$$2x = 7$$

$$x = \frac{7}{2}$$

To find y, replace x by $\dfrac{7}{2}$ in the equation $y = -2x$.

$$y = -2x = -2\left(\frac{7}{2}\right) = -7$$

The solution is $x = \dfrac{7}{2}$ and $y = -7$.   ▲

Example 10

Solve the system:

$$x - \frac{2}{3}y = -1$$

$$-3x + 4y = 9$$

Solution

Since the first equation can be easily solved for x, we use the substitution method. Therefore, we start the method by solving the first equation for x.

$$x - \frac{2}{3}y = -1$$

$$x = \frac{2}{3}y - 1 \qquad \text{Add } \frac{2}{3}y \text{ to both sides.}$$

Substitute for x in the second equation:

$$-3x + 4y = 9$$

$$-3\left(\frac{2}{3}y - 1\right) + 4y = 9 \qquad \text{Substitute } \frac{2}{3}y - 1 \text{ for } x.$$

$$-2y + 3 + 4y = 9$$

$$3 + 2y = 9$$

$$2y = 6 \qquad \text{Subtract 3 from both sides.}$$

$$y = 3 \qquad \text{Divide both sides by 2.}$$

Next, replace y by 3 in the first equation of the original system.

$$x - \frac{2}{3}y = -1$$

$$x - \frac{2}{3}(3) = -1$$

$$x - 2 = -1$$

$$x = 1 \qquad \text{Add 2 to both sides.}$$

Therefore, the solution is $x = 1$ and $y = 3$.

We summarize the substitution method.

S T R A T E G Y

The substitution method

Step 1   Solve one of the equations for one of the variables if neither equation is already solved for one of its variables.

Step 2   Substitute for that variable in the other equation. We now have one equation in one unknown.

Step 3   Solve the resulting equation.

Step 4   Find the value of the other variable by substituting the number from Step 3 into the equation of Step 1.

Step 5   Check your answer in the original system.

C A L C U L A T O R   C R O S S O V E R

Solve each system by the substitution method. Round off your answers to two decimal places.

1. $0.21x - 3.5y = 7$ **2.** $0.34x - 3.9y = 5.44$

     $x + 0.21y = 0.56$ $y = 0.98x - 6.19$

Answers   **1.** $\{0.97, -1.94\}$ **2.** $\{5.37, -0.93\}$

When confronted with solving a system of equations, we have three methods available: graphing the equations and reading the solution from the graph, the elimination method, and the substitution method. Which method should we use? The graphing method is of limited use and works best when the solution is an ordered pair of integers. Therefore, one of the two algebraic methods is preferred. Of these two, the elimination method is favored. The substitution method is used only when one of the equations is already solved for one of the variables or if one of the equations can be easily solved for one variable in terms of the other variable.

In the exercise set, there is a section in which you decide which method to use. With some practice, you will become efficient at solving systems of equations.

L E A R N I N G   A D V A N T A G E   **Mathematics at Work in Society: Medical Careers.** *Whenever people think of medical careers they usually think of DOCTORS, NURSES, and DENTISTS. These important careers are the most visible careers in health care. But what are some other medical careers that are just as important? Here are two examples.*

A PHYSICIAN'S ASSOCIATE is trained to assist a doctor in the general practice of medicine. A physician's associate may examine and diagnose patients under the doctor's supervision. Training: 2 years of college plus 2 years of specialized study in an associate's program.

A PHARMACIST dispenses drugs and medicines prescribed by physicians and dentists. Training: 4 years of college plus 1 year of specialized training.

These careers require good technical training. You are achieving a good start by taking this course to develop your mathematical skills.

E X E R C I S E   S E T   8.1

For Exercises 1–4, decide whether or not the ordered pairs are solutions of the given system of equations.

1. $6x - 5y = 1$
$2x - y = 1$
(a) $(1, 1)$
(b) $(0, 2)$
(c) $(5, -1)$

2. $-3x + 2y = -8$
$4x + 7y = 1$
(a) $(2, -1)$
(b) $(-3, 0)$
(c) $(1, 2)$

3. $y = -2x$
$x = 0$
(a) $(1, 1)$
(b) $(0, 5)$
(c) $(0, 0)$

4. $2x - 4y = 6$
$3x + 2y = 7$
(a) $(2, 3)$
(b) $(-1, -2)$
(c) $(-1, -4)$

For Exercises 5–14, solve each system using the graphing method. If the system is inconsistent, write ''no solution.'' If the system is dependent, write ''infinitely many solutions.'' Use graph paper.

5. $y - 2x = 2$
$y + 2x = 6$

6. $x + y = 5$
$x - y = 7$

7. $2x + y = 4$
$x - y = -4$

8. $-x + y = -2$
$3x - 2y = 6$

9. $y = x + 1$
$x + y = 3$

10. $x - y = 7$
$2x + y = 2$

11. $y = 2x + 1$
$y = 2x + 3$

12. $3x - 2y = 6$
$-3x + 2y = 6$

13. $x + y = -2$
$-3x - 3y = 6$

14. $2x + 2y = 3$
$3x + 3y = 4.5$

For Exercises 15–38, solve the system by the elimination method.

15. $3x + y = 7$
$x - y = -11$

16. $2x - y = 18$
$3x + y = 2$

17. $2x + 3y = 1$
$-2x - 5y = 7$

18. $-7x + 6y = -11$
$7x - 4y = 13$

19. $2w + z = -1$
$5w + z = -13$

20. $-3s + 4t = 2$
$6s + 4t = -7$

21. $-5p - 4x = -\dfrac{10}{7}$

$5p + 4x = \dfrac{20}{7}$

22. $2x + v = \dfrac{1}{5}$

$-x - \dfrac{1}{2}v = -1$

23. $60r - 50y = 11$
$-35r + 5y = -4$

24. $25x - 31t = 25$
$50x + 29t = 50$

25. $-\dfrac{1}{2}m + \dfrac{1}{3}n = 0$

$\dfrac{1}{4}m - 7n = -41$

26. $2x - \dfrac{1}{6}y = -5$

$3x + \dfrac{1}{3}y = -4$

27. $8x - 18y = -4$
$-4x + 9y = -2$

28. $7x - 3y = 4$
$-14x + 6y = -7$

29. $3h - 4k = 2$
$9h + 8k = 1$

30. $2s + 3t = 1$
$4s + 15t = 8$

31. $\dfrac{x}{3} + \dfrac{y}{4} = \dfrac{2}{3}$

$5x - 14y = 10$

32. $\dfrac{x}{2} - \dfrac{y}{3} = 7$

$-x + y = -13$

33. $3(x + 1) - 5(y + 2) = 1$
$2(3x - 4) + 6y = y - 7$

34. $-2(3v + 4) + 3(1 - 2w) = 1$
$3(v - 1) - 2w = -5$

35. $\dfrac{2x + 3}{4} - \dfrac{5y - 2}{7} = \dfrac{9}{14}$
$14x - 20y = -11$

36. $\dfrac{3y - 1}{2} - \dfrac{2x - 3}{3} = \dfrac{1}{3}$
$9y - 4x = -1$

37. $\dfrac{3x}{5} - \dfrac{2y}{3} = -1$
$-\dfrac{7(x - 1)}{10} + \dfrac{y - 2}{12} = -\dfrac{37}{15}$

38. $\dfrac{4\left(1 - \dfrac{3}{2}x\right)}{3} + \dfrac{3\left(2 - \dfrac{1}{3}y\right)}{2} = \dfrac{5}{6}$
$3x + 2y = 6$

For Exercises 39–48, solve the system using the substitution method.

39. $y = 2x$
$x + y = 9$

40. $y = 5x$
$x + y = -12$

41. $2a - b = 10$
$a = 3b$

42. $3y - 2t = 8$
$y = -2t$

43. $\dfrac{1}{2}x + \dfrac{2}{3}y = 1$
$-x + y = 0$

44. $0.8u + 0.3v = 7$
$2u + v = 20$

45. $x - 3 = 0$
$3x - y = 5$

46. $4x + 13y = 6$
$x - 2 = 0$

47. $\dfrac{h}{2} - 3k = 8$
$2h = h + k + 6$

48. $y - \dfrac{x}{5} = -4$
$3y = 2y + x - 8$

For Exercises 49–62, solve the system by either the elimination method or the substitution method.

49. $x - 2y = 1$
$3x + 2y = 11$

50. $4r + 3t = 8$
$-4r + t = 8$

51. $x = \dfrac{1}{2} + 2t$
$-6x + 4t = -7$

52. $w - 2c = 0$
$3w + 5c = 1.1$

53. $4x + 5y = -2.2$
$x + 1.5y = 0$

54. $x - y = -0.1$
$5x - 4y = 0$

55. $\quad A - B = 0.3$
$\quad\quad 10A - 13B = 0$

56. $N = 3 - 6M$
$\quad -3M + 4N = 3$

57. $X - 6Y = 28$
$\quad 5Y = -12 - 2X$

58. $x + 2y = -9$
$\quad y = -7 - x$

59. $4x_1 + 3x_2 = 2$
$\quad 5x_1 - 7x_2 = -19$

60. $\quad 3x_1 - 4x_2 = 19$
$\quad\quad 7x_1 + 18x_2 = 17$

61. $\quad 0.1x - 0.2y = -0.5$
$\quad\quad -0.3x + 0.4y = 0.3$

62. $0.7u + 0.3v = -4.3$
$\quad 1.5u + 0.9v = -10.5$

For Exercises 63–68, let x be one number and y be the other number. Write a system of equations in the two variables x and y, then solve.

63. The sum of two numbers is 11 and the difference is 67. Find the two numbers.

64. The sum of four times the first number and five times the second number is negative seven. The difference of the first number and the second number, divided by four, is five fourths. Find the two numbers.

65. The sum of twice the first number and three times the second number is $\dfrac{17}{12}$. The difference of the first number and the second number, divided by 2, is $\dfrac{1}{24}$. Find the two numbers.

66. The sum of two numbers, divided by 2, is -1. Three times the first number minus four times the second number is $\dfrac{9}{2}$. Find the two numbers.

67. One number is two thirds of another number. The sum of the two numbers is -10. Find the two numbers.

68. One number is $\dfrac{1}{2}$ of another number. The sum of the two numbers is $-\dfrac{3}{2}$. Find the two numbers.

For Exercises 69–72, let x and y be the unknowns you are to find. Next, translate the given information into a system of equations and solve the system.

69. Two angles are complementary. The measure of one angle is 6 degrees less than the measure of the other angle. Find the measures of the two angles.

70. Two angles are complementary. The measure of one angle is five times the measure of the other angle. Find the measures of the two angles.

71. The length of a rectangle is 3 inches less than twice the width. If the perimeter is 24 inches, find the length and the width of the rectangle.

72. The length of a rectangle is 1 meter more than three times the width. If the perimeter is $7\dfrac{1}{3}$ meters, find the length and the width of the rectangle.

73. (a) Show that $(15, 16)$ is a solution of

$$-13x + 15y = 45$$
$$39x - 45y = -135$$

(b) Is $(15, 16)$ the only solution of the system? Why?

74. (a) Show that $(19, 8)$ is a solution of

$$3x - 19y = -95$$
$$-6x + 38y = 190$$

(b) Is $(19, 8)$ the only solution of the system? Why?

For Exercises 75 and 76, solve the systems using the elimination method. (*Hint:*   Let $\dfrac{1}{a} = x$ and $\dfrac{1}{b} = y$.)

75. $3\dfrac{1}{a} - 2\dfrac{1}{b} = -2$

$\quad 4\dfrac{1}{a} + 3\dfrac{1}{b} = 20$

76. $\dfrac{4}{a} - \dfrac{5}{b} = -7$

$\quad \dfrac{3}{a} + \dfrac{12}{b} = -30$

REVIEW PROBLEMS

The following exercises review parts of Section 1.5. Doing these problems will help prepare you for the next section.

Write as an algebraic expression.

77. The total number of dollars in x dimes and y quarters.

78. The total cost of x pounds of Colombian coffee beans and y pounds of Supremo coffee beans, if the Colombian beans cost \$5.49 per pound and the Supremo beans cost \$6.99 per pound.

79. The amount of money received for working x hours at \$12.50 per hour plus y hours of overtime at \$18 per hour.

80. The total cost of x tennis rackets at \$120 per racket and y cans of tennis balls at \$2.99 per can.

ENRICHMENT EXERCISES

For Exercises 1 and 2, determine the number of solutions by graphing the system from the given information.

1. $y = \dfrac{3}{2}x + b_1$

$\quad y = -\dfrac{3}{2}x + b_2$, where $b_1 < 0$ and $b_2 > 0$

2. $y = mx + b_1$
$\quad y = mx + b_2$, where $b_1 \neq b_2$

3. Consider the system

$$a_1 x + b_1 y = c_1$$

$$a_2 x + b_2 y = c_2, \text{ where } a_1 b_2 - b_1 a_2 \neq 0$$

(a) Eliminate y from the system to obtain

$$x = \frac{c_1 b_2 - b_1 c_2}{a_1 b_2 - b_1 a_2}$$

(b) Eliminate x from the system to obtain

$$y = \frac{a_1 c_2 - c_1 a_2}{a_1 b_2 - b_1 a_2}$$

(c) What can you say about the system if $a_1 b_2 - b_1 a_2 = 0$?

Use the formulas derived in Problem 3 to solve the following systems.

4. $3x + 7y = 12$
   $4x + 5y = 13$

5. $2x - 4y = -11$
   $7x + 6y = 8$

6. $\dfrac{1}{2}x - \dfrac{1}{3}y = 10$

   $\dfrac{1}{3}x - \dfrac{1}{2}y = 10$

7. $0.1x - 0.4y = 1$

   $0.3x + 0.2y = 1$

8. (a) Use the substitution method to solve for x and y in terms of A, B, and m.

$$Ax + By = 1$$

$$y = mx, \text{ where } A + Bm \neq 0$$

(b) What is the solution set when $A + Bm = 0$?

Answers to Enrichment Exercises are on page A.48.

8.2

Applications Using Systems of Equations

O B J E C T I V E S

▶　*To use two variables in application problems*

▶　*To generate a system of equations to solve an application problem*

Many applications that we have encountered in previous chapters can also be solved using two variables instead of just one variable. In applications where there are two items to find, two unknowns can be introduced to solve the problem. When solving word problems using two variables, use the following strategy.

S T R A T E G Y

Solving word problems using two variables

Step 1　Read the problem carefully and determine what two unknown values you are to find. Introduce two variables and write what each variable is to represent.

Step 2　Write two equations in the two variables from the given information. If necessary, first organize the information in a table.

Step 3　Solve the system algebraically by using either the elimination or the substitution method.

Step 4　Check your answers by using the *original words* of the problem.

Step 5　Be sure that you have answered the question asked in the problem.

Example 1　The sum of two numbers is -7. Six times the smaller number increased by three times the larger number is -45. Find the two numbers.

Solution　Step 1　Let x be the smaller number and y be the larger number.

Step 2　Since their sum is -7,

$$x + y = -7$$

Six times the smaller number x plus three times the larger number y is -45. Therefore,

$$6x + 3y = -45$$

Step 3   To find the two numbers, we solve the system.

$$x + y = -7$$

$$6x + 3y = -45$$

We choose to use the substitution method. From the first equation,

$$y = -7 - x$$

Replacing y by $-7 - x$ in the second equation,

$$6x + 3(-7 - x) = -45$$

$$6x - 21 - 3x = -45$$

$$3x = -24$$

$$x = -8$$

To find y, replace x by -8 in the equation $y = -7 - x$.

$$y = -7 - x$$

$$= -7 - (-8)$$

$$= 1$$

Step 4   Our answers are -8 and 1. We check these numbers using the original words of the problem. The sum of the two numbers, $(-8) + 1$, is -7. Six times the smaller number increased by three times the larger number is given by $6(-8) + 3(1) = -48 + 3 = -45$. Therefore, the two conditions are true for our two numbers.

Step 5   The two numbers are -8 and 1.   ▲

Mixture problems can also be solved by using two variables. This method is explained in the next example.

Example 2   A skin-bleaching (fade) cream is guaranteed to contain 2% hydroquinine. Two ingredients, A and B, are used to make the cream. Ingredient A contains 3% hydroquinine and ingredient B contains 1.5% hydroquinine. How much of each ingredient should be mixed to make 100 mg of the cream?

Solution   Let

$x =$ the amount (in milligrams) of ingredient A

$y =$ the amount (in milligrams) of ingredient B used to make 100 mg of the cream.

Since 100 mg of the cream are to be made,

$$x + y = 100$$

The other equation is obtained from the information about the percentage of hydroquinine. We summarize the data in the following table.

	Ingredient A	Ingredient B	Cream
Amount	x	y	100
Percent of hydroquinine	3	1.5	2
Amount of hydroquinine	$0.03x$	$0.015y$	$(0.02)(100)$

From the last row of the table,

$$\begin{matrix} \text{the amount} & & \text{the amount} & & \text{the amount} \\ \text{of hydroquinine} & + & \text{of hydroquinine} & = & \text{of hydroquinine} \\ \text{in ingredient A} & & \text{in ingredient B} & & \text{in the cream} \end{matrix}$$

$$0.03x \quad + \quad 0.015y \quad = \quad (0.02)(100)$$

Our next step is to solve the system:

$$x + \quad y = 100$$
$$0.03x + 0.015y = 2$$

Either the substitution or the elimination method is suitable for this system. It is up to you to choose the method and to solve this system. The answer is that 33.33 mg of ingredient A and 66.67 mg of ingredient B are to be used to make 100 mg of the cream. ▲

Example 3

A touring boat at the Wisconsin Dells can travel 18 miles downriver in 48 minutes and can travel 15 miles upriver in 1 hour and 12 minutes. Find the rate of the boat in still water and the rate of the current.

Solution   Let

$$r_b = \text{the rate, in mph, of the boat in still water}$$

and

$$r_c = \text{the rate, in mph, of the current}$$

Since we want time to be given in hours, we change 48 minutes to $\dfrac{48}{60} = \dfrac{4}{5}$ hour. Also, change 1 hour and 12 minutes, which is $60 + 12 = 72$ minutes, to $\dfrac{72}{60} = \dfrac{6}{5}$ hours.

Next, write the given information in a table.

	Rate	Time	Distance
Downriver	$r_b + r_c$	$\dfrac{4}{5}$	18
Upriver	$r_b - r_c$	$\dfrac{6}{5}$	15

Now, using the formula

$$(\text{rate})(\text{time}) = \text{distance}$$

we generate a system of two equations.

$$(r_b + r_c)\left(\frac{4}{5}\right) = 18$$

$$(r_b - r_c)\left(\frac{6}{5}\right) = 15$$

To simplify the equations of the system, we multiply the first equation by $\dfrac{5}{4}$ and the second equation by $\dfrac{5}{6}$,

$$r_b + r_c = 18\left(\frac{5}{4}\right)$$

$$r_b - r_c = 15\left(\frac{5}{6}\right)$$

Simplifying the right sides of each equation, we have

$$r_b + r_c = \frac{45}{2}$$

$$r_b - r_c = \frac{25}{2}$$

We solve this system by the elimination method to obtain $r_b = 17.5$ mph and $r_c = 5$ mph. It is left for you to check these numbers.

Therefore, the boat travels in still water at 17.5 mph and the rate of the current is 5 mph.   ▲

Systems of linear equations have applications in economics. A business that manufactures and sells a product has associated cost and revenue equations. These two equations form a system. The solution to the system gives the break-even point. The **break-even point** is that level of production x where cost equals revenue. The next example shows how to find and interpret the break-even point.

Example 4   The Powerstar Corporation makes and sells x tennis rackets each day with a total daily cost C, in dollars, given by

$$C = 25x + 150$$

The corporation sells each racket for $50.

(a) Express the daily revenue R in terms of x.
(b) Find the break-even point for the corporation.

Solution   **(a)** The daily revenue R is given by the selling price per racket times the number of rackets sold.

revenue = (selling price per racket)(the number of rackets sold)

that is,

$$R = 50x$$

(b) The break-even point is that level of production x where the revenue equals cost.

$$R = C$$
$$50x = 25x + 150$$

Solving this equation for x,

$$50x - 25x = 150$$
$$25x = 150$$
$$x = 6 \text{ tennis rackets} \qquad \blacktriangle$$

We can draw the cost and revenue equations of the previous example *in the same coordinate system.* In order to do this, we set both expressions for cost and revenue equal to y.

Cost: $y = 25x + 150$
Revenue: $y = 50x$

The cost and revenue graphs are shown in the following figure. Notice that the break-even point is the x-coordinate of the point of intersection of the cost equation and the revenue equation.

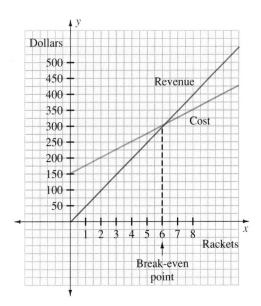

That is, the break-even point is the x value of the solution of the system

$$y = 25x + 150$$

$$y = 50x$$

In the last chapter, we studied supply and demand equations. When there is both a supply and a demand equation for the same item, we can solve this system of linear equations to obtain the **equilibrium point.** The equilibrium point occurs when the demand price is equal to the supply price.

Example 5

The supply and demand equations for dark roast coffee beans in the Greater Cleveland area are the following.

Supply: $p = 0.1x + 3.5$
Demand: $p = -0.08x + 8$

where p is the price per pound, in dollars, and x is the number of pounds, in thousands, of coffee.

(a) Find the equilibrium point.

(b) Draw the supply and demand equations on the same coordinate system.

Solution **(a)** To find the equilibrium point, we solve the system

$$p = 0.1x + 3.5$$

$$p = -0.08x + 8$$

Using the substitution method, replace p in the first equation by the expression for p given by the second equation. That is, *set the demand price equal to the supply price and solve for x.*

$$-0.08x + 8 = 0.1x + 3.5$$

$$-0.18x = -4.5$$

$$x = \frac{4.5}{0.18}$$

$$= 25$$

Therefore, the equilibrium quantity is 25,000 pounds. To find the equilibrium price, replace x by 25 in either the supply or the demand equation. Using the supply equation,

$$p = (0.1)(25) + 3.5 = 2.5 + 3.5 = 6$$

Therefore, the equilibrium price is $6 per pound.

(b) To draw the graphs, use the intersection point $(25, 6)$ as well as the p-intercepts of each line. The graphs are shown in the following figure.

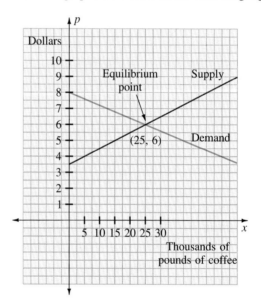

LEARNING ADVANTAGE **Mathematics at Work in Society: Optometry.** *An OPTOMETRIST examines eyes for vision problems and disease, prescribes lenses to correct vision problems, and when disease is discovered, refers patients to the appropriate physician. Training: 4 years of college plus 2 to 3 years of specialized study at an accredited college.*

Optometry requires good technical training. If you are interested in this career, continue to improve your mathematics background.

EXERCISE SET 8.2

For Exercises 1–22, solve the word problem using the five-step strategy given on page 558.

1. The sum of two numbers is -12. The difference of the larger number and twice the smaller number is 15. Find the two numbers.

2. One number reduced by twice a second number is two. The sum of three times the first number and seven times the second number is 45. Find the two numbers.

3. The sum of two numbers is $\frac{5}{4}$. The difference of twice one number and eight times the other number is -5. Find the two numbers.

4. The sum of two numbers is $\frac{7}{6}$. The difference of nine times the first number and twelve times the second number is -7. Find the two numbers.

5. A fraction reduces to $\frac{2}{5}$. One half of the numerator plus $\frac{1}{5}$ of the denominator is -4. Find the numerator and denominator.

6. A fraction reduces to 4. The sum of three times the numerator and six times the denominator is 3. Find the numerator and denominator.

7. The length x of a rectangle is 3 cm less than ten times its width y. The perimeter is 38 cm. Find the width and length of the rectangle.

8. The length of a rectangle is five times the width, reduced by 1 inch. If the perimeter is $7\frac{3}{5}$ inches, find the length and width of the rectangle.

9. Faber College plans to buy 18 additional computer terminals. There are two types of terminals—slow and fast. Each slow terminal costs $1,200 and each fast terminal costs $1,800. The college has $28,800 available for these terminals. How many terminals of each type can the college purchase?

10. The Nut Hut plans to make and sell a mixture of cashews and pecans in two types of tins—regular and super. Each regular tin contains 1 pound of cashews and 2 pounds of pecans. Each super tin contains 4 pounds of cashews and 3 pounds of pecans. The Hut has 74 pounds of cashews and 83 pounds of pecans available. Under these constraints, how many tins of each type can be made?

11. A bath soap is guaranteed to have 10% free fatty matter. Two ingredients, A and B, are used to make the soap. Ingredient A contains 8% and ingredient B contains 14.4% free fatty matter. How much of each ingredient should be mixed to make 500 mg of the bath soap?

12. A prescription calls for a cough syrup to have 2.5% codeine sulfate. The pharmacist only has cough syrups with 1% and 3% codeine sulfate. How many milliliters of each kind of cough syrup should be mixed to obtain 1,000 ml of cough syrup with 2.5% codeine sulfate?

13. A company that makes a salve for treatment of acne guarantees that the salve contains 0.5% hydrocortisone. Two ingredients, A and B, are used to make the salve. Ingredient A contains 0.8% and ingredient B contains 0.3% hydrocortisone. For each 50 ounces of the cream that is produced, how much of each ingredient should be used?

14. June has twice as many quarters as dimes. If she has a total of $3.60, how many coins of each type does she have?

15. A motorboat can travel 23 km downstream in 46 minutes and 25.5 km upstream in $1\frac{1}{2}$ hours. Find the rate of the motorboat in still water and the rate of the current in kilometers per hour. Start your solution by setting

r_b = the rate of the boat in still water
r_c = the rate of the current

Next, fill out the following table.

	Rate	Time	Distance
Downstream			
Upstream			

16. A boat can travel 12 miles downstream in 36 minutes and 14 miles upstream in 1 hour. Find the rate of the boat in still water and the rate of the current in miles per hour.

17. A company plane can fly 225 miles against the wind from Houston to San Antonio in 3 hours. The return trip (with the wind) takes 1 hour. Find the speed of the wind and the speed of the plane.

18. A sight-seeing boat on the Snake River can travel 10 miles downriver in 30 minutes and can travel 8 miles upriver in 2 hours. Find the rate of the boat in still water and the rate of the current.

19. The Lincoln Company manufactures cast-iron wood stoves. Each week the company makes and sells x stoves. There is a fixed cost of $180 per week plus a variable cost of $75 per stove. The company sells the stoves for $90 each. Find the break-even point for this company.

20. Dynasty Products makes and sells x miniature grandfather clocks each week with a total weekly cost C, in dollars, given by

$$C = 100x + 250$$

The clocks are sold for $125 each.
(a) Express the weekly revenue R in terms of x.

(b) Find the break-even point for the company.

21. The supply and demand equations for flounder in Bangor, Maine, are the following:

Supply: $p = \dfrac{11}{4000}x + 1.25$

Demand: $p = -\dfrac{1}{800}x + 5.25$

where p is the price per pound, in dollars, and x is the number of pounds of flounder.
(a) Find the equilibrium point.

(b) Draw the supply and demand equations on the same coordinate system.

22. The supply and demand equations for red delicious apples in Fargo, North Dakota, are the following:

Supply: $p = \dfrac{29}{220}x + 2$

Demand: $p = -\dfrac{1}{20}x + 6$

where p is the price per peck, in dollars, and x is the quantity, in thousands of pecks, of apples.
(a) Find the equilibrium point.

(b) Draw the supply and the demand equations on the same coordinate system.

REVIEW PROBLEMS

The following exercises review parts of Section 1.5. Doing these problems will help prepare you for the next section.

Simplify.

23. $-8x - 4y + z - 2(-4x + 2y - 5z)$

24. $5x + 3y - 6z - (5x - 6y - 10z)$

25. $9x + 12y - 7z - 6(3x + 2y - 9z)$

26. $4(5x + 3y - 9z) + (-20x - y + z)$

27. $7(-x - 3y - 4z) + (7x - 2y + 10z)$

28. $8(3x - 2y - 6z) + 2(-12x + 7y + 24z)$

ENRICHMENT EXERCISES

The Anthem Plus Company makes ceramic figures. The total cost C is given by $C = 50x + 200$, if $x \leq 15$, and $C = 35x + 950$, if $x > 15$, where x is the number of cases of figures made and sold each day. The company sells the figures for \$58.75 per case.

(a) Express the daily revenue R in terms of x.

(b) Find the break-even point for the company.

(c) On the same coordinate system, draw the cost and the revenue equations.

Answers to Enrichment Exercises are on page A.49.

8.3

Systems of Linear Equations in Three Variables

OBJECTIVE

▶ *To solve systems of linear equations in three variables*

In this section, we learn to solve a system containing three linear equations in three variables. Since there are three variables involved in these systems, a solution is an **ordered triple** (x, y, z) of real numbers. For example, a solution to the equation $3x - 2y + z = 10$ is $(1, -2, 3)$, since replacing x by 1, y by -2, and z by 3 in the equation results in a true statement. In fact, there are infinitely many ordered triples that are solutions of $3x - 2y + z = 10$.

Suppose we are given a system of three linear equations in three variables. A **solution** to this system is an ordered triple that satisfies *all three* equations.

The graph of an equation in two variables is a line in the coordinate plane. The graph of an equation of three variables is a plane in three-dimensional space. We will not attempt to graph planes; however, we can still visualize the solution set for a system of three equations in three variables.

Since the graph of each equation of the system is a plane, then the solution set is the intersection of the three planes. When this intersection is a single point P, as shown in the following figure, then the system has one ordered triple in its solution set.

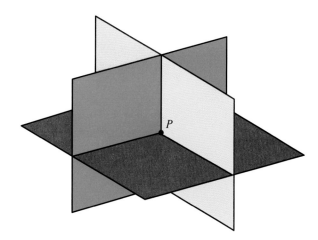

Later in this section, we will investigate other relative positions of three planes. At this time, however, we develop an algebraic method to find the solution of a system of three equations in three variables.

Example 1   Solve the system:

$$(1) \quad 2x - 3y - z = 2$$
$$(2) \quad 3x + 10y + 4z = 5$$
$$(3) \quad x - 2y + z = 6$$

Solution   We first want to eliminate one of the variables using any two of the equations. Notice that the variable z can be eliminated by adding Equations (1) and (3).

$$2x - 3y - z = 2$$
$$\underline{x - 2y + z = 6} \qquad \text{Add.}$$
$$(4) \quad 3x - 5y \quad\;\; = 8$$

Next, we take *any* other pair of equations from the system and again eliminate z. We choose Equations (1) and (2).

$$2x - 3y - z = 2 \;\; \longrightarrow \text{multiply by 4} \longrightarrow \;\; 8x - 12y - 4z = 8$$
$$3x + 10y + 4z = 5 \;\; \longrightarrow \text{leave as is} \longrightarrow \;\; \underline{3x + 10y + 4z = 5}$$
$$(5) \quad 11x - 2y \quad\;\; = 13$$

We now have derived a system of two equations in the two variables x and y:

$$(4) \quad 3x - 5y = 8$$
$$(5) \quad 11x - 2y = 13$$

We solve this system as in the last section.

$$3x - 5y = 8 \quad \text{---multiply by } -2 \longrightarrow \quad -6x + 10y = -16$$

$$11x - 2y = 13 \quad \text{---multiply by } 5 \longrightarrow \quad \underline{55x - 10y = 65}$$

$$49x \qquad = 49$$

Therefore, $x = 1$. To find y, replace x by 1 in either equation.

$$3x - 5y = 8$$

$$3(1) - 5y = 8$$

$$-5y = 5$$

So, $y = -1$. Next, we replace x by 1 and y by -1 in any one of the three equations of the original system and solve for z. Using the third equation,

$$x - 2y + z = 6$$

$$1 - 2(-1) + z = 6$$

$$1 + 2 + z = 6$$

So $z = 3$ and the solution of the system is $x = 1$, $y = -1$, and $z = 3$. Next, we check our solution in each of the three original equations.

Check $x = 1$, $y = -1$, and $z = 3$:

$$2x - 3y - z = 2 \qquad\qquad 3x + 10y + 4z = 5 \qquad\qquad x - 2y + z = 6$$

$$2(1) - 3(-1) - 3 \overset{?}{=} 2 \qquad 3(1) + 10(-1) + 4(3) \overset{?}{=} 5 \qquad 1 - 2(-1) + 3 \overset{?}{=} 6$$

$$2 + 3 - 3 \overset{?}{=} 2 \qquad\qquad 3 - 10 + 12 \overset{?}{=} 5 \qquad\qquad 1 + 2 + 3 \overset{?}{=} 6$$

$$2 = 2 \qquad\qquad\qquad\qquad 5 = 5 \qquad\qquad\qquad\qquad 6 = 6$$

The solution $x = 1$, $y = -1$, and $z = 3$ checks.   ▲

N O T E   *Notice that the pattern in solving the system of Example 1 is the following:*

3 equations		*2 equations*		*1 equation*
in 3 variables	*to*	*in 2 variables*	*to*	*in 1 variable*

Example 2   Solve the system:

$$-2x + 3z = -5$$

$$4x - 5y = 4$$

$$10y + 3z = 2$$

Solution   Notice that this system involves the three variables x, y, and z. We rewrite the system as follows.

$$-2x + 0y + 3z = -5$$

$$4x - 5y + 0z = 4$$

$$0x + 10y + 3z = 2$$

The third equation does not have the variable x. Therefore, to obtain another equation without x, we eliminate x from the first two equations.

$$-2x + 0y + 3z = -5 \quad\text{multiply by 2}\longrightarrow\quad -4x + 0y + 6z = -10$$

$$4x - 5y + 0z = 4 \quad\text{leave as is}\longrightarrow\quad \underline{4x - 5y + 0z = 4}$$

$$- 5y + 6z = -6$$

Next, we solve the system

$$-5y + 6z = -6$$

$$10y + 3z = 2$$

$$-5y + 6z = -6 \quad\text{multiply by 2}\longrightarrow\quad -10y + 12z = -12$$

$$10y + 3z = 2 \quad\text{leave as is}\longrightarrow\quad \underline{10y + 3z = 2}$$

$$15z = -10$$

Therefore, $z = -\dfrac{10}{15}$ or $-\dfrac{2}{3}$. To find the value of y, we replace z by $-\dfrac{2}{3}$ in $-5y + 6z = -6$ to obtain $y = \dfrac{2}{5}$. Finally, we use the first equation of the original system to obtain $x = \dfrac{3}{2}$. We can write the solution $x = \dfrac{3}{2}$, $y = \dfrac{2}{5}$, and $z = -\dfrac{2}{3}$ as the ordered triple $\left(\dfrac{3}{2}, \dfrac{2}{5}, -\dfrac{2}{3}\right)$.   ▲

In Section 8.1, we saw that a system of two equations in two variables will not have a solution if the two lines are parallel. In the same vein, a system of three equations in three variables will not have a solution if the three planes are parallel. There are other geometric configurations where the system will have an empty solution set. Two possibilities are shown in the following figures.

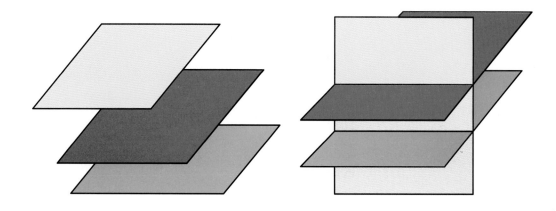

When attempting to solve a system, if we eliminate all the variables and obtain a false statement, then the solution set is empty. This case is shown in the next example.

Example 3   Solve the system:

$$3x - 5y + 4z = 11$$
$$-9x + 15y - 12z = 2$$
$$3x - 12y + 4z = 15$$

Solution   We eliminate the variable x from the first two equations.

$3x - 5y + 4z = 11$ ——multiply by 3——→ $9x - 15y + 12z = 33$

$-9x + 15y - 12z = 2$ ——leave as is——→ $\underline{-9x + 15y - 12z = 2}$

$$0 = 35$$

Since we have arrived at the false statement $0 = 35$, we conclude that the solution set is empty.   ▲

We summarize the method for solving systems that was developed in this section.

STRATEGY

Solving a system of equations in three variables

Step 1   Decide upon one of the three variables to eliminate. Pick any two equations and eliminate that variable.

Step 2   Choose any other pair of equations and eliminate the same variable.

Step 3   Treat the two equations in two variables obtained in Steps 1 and 2 as a system. Solve this system.

Step 4   Substitute the two numbers from Step 3 into any one of the original equations and solve for the third variable.

Step 5   Check your answer in all three equations of the original system.

LEARNING   ADVANTAGE   **Mathematics at Work in Society: Consumerism.** *CONSUMER is the name given to people who use goods and services to satisfy personal needs. You are a consumer and will continue to be a consumer for the rest of your life. Mathematics will aid you. For example,*

1. *Math prevents being overcharged or shortchanged;*
2. *By calculating the unit price, math determines the most economical buy;*
3. *Math helps you to read graphs and to interpret data in newspapers and magazines;*
4. *Math helps with do-it-yourself building and repair projects.*

E X E R C I S E   S E T   8.3

For Exercises 1–28, solve the system.

1. $x - 3y + z = -3$
$x + 2y + 5z = -8$
$-x + 3y + 2z = 3$

2. $3x + y - z = -5$
$2x - y + 2z = 4$
$6x + y + 3z = 8$

3. $4x - y + 3z = 4$
$2x - y - 2z = -1$
$5x + y + z = 0$

4. $x - 2y + 5z = -5$
$x + 3y + z = -2$
$-x + y - 2z = 0$

5. $x + 2y - 3z = -6$
$2x - y + 4z = 2$
$-x + 3y + 3z = 1$

6. $3x - y + 2z = 0$
$2x + 3y + 8z = 8$
$x + y + 6z = 0$

7. $2x + 6z = 6$
$2y + 3z - 2$
$x - y = -1$

8. $3x + y = 0$
$4x - 3z = -5$
$2y + 3z = -3$

9. $2x + 3y = 2$
$6y - 2z = 0$
$4x + 3z = 1$

10. $3x + 4z = 4$
$-3x + 2y = -3$
$6x - 8z = 0$

11. $3x + 2y + 4z = -3$
$9x - 2y + 8z = 2$
$-6x + 4y + 16z = 0$

12. $4x + 3y - 2z = 6$
$8x + 9y - 20z = 8$
$12x - 3y - 10z = 0$

13. $4x + 3y = -1$
$8x - 3y + 18z = 1$
$-4x + 3y - 18z = 0$

14. $5x - 3y + 4z = 1$
$-10x + 9y + 8z = 4$
$15x - 4z = 5$

15. $2x + 4y + 7z = 0$
$3x - 3y + 4z = 0$
$4x + 5y - 3z = 0$

16. $-3x + 5y + z = 0$
$2x + 7y + 7z = 0$
$5x + y + 9z = 0$

17. $x - 2y + 3z = -5$
$5x + 3y - 2z = 5$
$2x - 5y + 5z = -10$

18. $3x + y - 4z = -11$
$2x + 7y + 5z = 8$
$6x + 3y - 3z = -12$

19. $3x + 4z = 4$
$7y - 2z = -1$
$6x + 21y = 2$

20. $4x + 5y = 0$
$8x + 3z = -1$
$10y - 9z = -1$

21. $x - y + 4z = 9$
$5x + 3y - 6z = 3$
$3x + 7y + 4z = -1$

22. $5x - 2y + 4z = 23$
$2x + 7y - 6z = -14$
$3x + 3y + 8z = 11$

23. $\frac{1}{2}x + \frac{1}{3}y - z = -\frac{1}{3}$
$x - \frac{1}{3}y - \frac{1}{2}z = -\frac{1}{6}$
$x + \frac{1}{2}y - \frac{5}{12}z = \frac{1}{12}$

24. $\frac{1}{3}x - \frac{1}{6}y + \frac{1}{4}z = \frac{1}{3}$
$x + \frac{1}{2}y + z = 1$
$-\frac{1}{2}x + \frac{1}{12}y + \frac{1}{3}z = \frac{1}{4}$

25. $5x - 7y + 10z = 14$
$-20x + 28y - 40z = 55$
$-5x + 3y - 39z = 61$

26. $x - 5y + 3z = -1$
$-5x + 25y - 15z = 8$
$-5x - 11y + 8z = -14$

27. $2x - 4y + z = -3$
$3y - 2z = 3$
$5z = -10$

28. $2x - 6y + 5z = -4$
$y - 4z = 12$
$z = -4$

29. Find three numbers in increasing order so that their sum is 2, the difference of the smallest and largest numbers is -5, and the sum of twice the first number and five times the second number is 1.

30. Find three numbers in increasing order so that the difference of the first and second numbers is -1, the sum of the second number and three times the third number is 1, and the difference of twice the first number and the third number is -7.

31. Cedar Enterprises makes wooden storage bins and uses three finishing machines A, B, and C. Each day, machines A and C together can finish 27 bins, machines B and C together can finish 23 bins, and all three machines together can finish 35 bins. How many bins can each machine finish each day by itself?

32. Sam has $103 in $1, $5, and $10 bills. She has five more $1 bills than $5 bills. If she had eight more $10 bills, then she would have as many $10 bills as $1 bills. How many bills of each kind does she have?

REVIEW PROBLEMS

The following exercises review parts of Section 1.3. Doing these problems will help prepare you for the next section.

Simplify.

33. $(1)(3) - (-1)(2)$

34. $(2)(1) - (-3)(-4)$

35. $\left(-\dfrac{1}{2}\right)(2) - (1)(-10)$

36. $\left(\dfrac{2}{3}\right)\left(-\dfrac{3}{2}\right) - \left(\dfrac{1}{2}\right)(-6)$

37. $2[(-2)(-1) - (5)(3)] + [(4)(-1) - (5)(1)] + 3[(4)(3) - (-2)(1)]$

38. $2[(-1)(0) - (2)(5)] - (-3)[(4)(0) - (2)(3)] - [(4)(5) - (0)(3)]$

39. $-3[(1)(3) - (-2)(1)] - [(-3)(3) - (-2)(1)] - 2[(3)(-1) - (1)(1)]$

E N R I C H M E N T   E X E R C I S E S

Solve the following systems.

1. $3\left(\dfrac{1}{a}\right) - \dfrac{1}{b} + 4\left(\dfrac{1}{c}\right) = 2$

$\quad 6\left(\dfrac{1}{a}\right) + \dfrac{1}{b} - 8\left(\dfrac{1}{c}\right) = -1$

$\quad -3\left(\dfrac{1}{a}\right) + \dfrac{1}{b} - 12\left(\dfrac{1}{c}\right) = -4$

2. $4a^2 - 3b^2 + 8c^2 = -10$

$\quad a^2 + 5b^2 + 16c^2 = 5$

$\quad -a^2 + 3b^2 - 8c^2 = 4$

3. $x - 2y + 4z - t = 2$

$\quad 2x + y + 2z + t = 5$

$\quad -x + 3y - 4z + 2t = 0$

$\quad -x + 4y + 2z + 4t = 10$

4. Suppose a system of three equations in three variables has an empty solution set. Is it necessary that at least two of the planes of the system are parallel? If not, draw three planes, no two of them parallel and with no solution.

Answers to Enrichment Exercises are on page A.50.

8.4

Cramer's Rule

▶ *To evaluate determinants*

▶ *To solve systems of linear equations using Cramer's rule*

In this section, we develop another method to solve systems of linear equations. This method involves the use of *determinants*.

D E F I N I T I O N

Suppose a, b, c, and d are any four numbers. A **2 by 2 determinant**, denoted by

$$\begin{vmatrix} a & b \\ c & d \end{vmatrix}$$

has a value given by

$$\begin{vmatrix} a & b \\ c & d \end{vmatrix} = ad - bc$$

N O T E   *A **determinant** is a square array of numbers enclosed by two vertical lines. The **value** of a determinant is a number.*

Example 1   Evaluate the following 2 by 2 determinants.

(a) $\begin{vmatrix} 4 & 2 \\ 1 & -3 \end{vmatrix} = (4)(-3) - (2)(1) = -12 - 2 = -14$

(b) $\begin{vmatrix} \frac{1}{2} & 2 \\ -\frac{1}{2} & 4 \end{vmatrix} = \left(\frac{1}{2}\right)(4) - (2)\left(-\frac{1}{2}\right) = 2 + 1 = 3$ ▲

We now define a 3 by 3 determinant.

D E F I N I T I O N

A **3 by 3 determinant,** denoted by

$$\begin{vmatrix} a_1 & b_1 & c_1 \\ a_2 & b_2 & c_2 \\ a_3 & b_3 & c_3 \end{vmatrix}$$

has a value given by

$$\begin{vmatrix} a_1 & b_1 & c_1 \\ a_2 & b_2 & c_2 \\ a_3 & b_3 & c_3 \end{vmatrix} = a_1b_2c_3 + a_2b_3c_1 + a_3b_1c_2 - a_1b_3c_2 - a_2b_1c_3 - a_3b_2c_1$$

We need not find the value of a 3 by 3 determinant from the definition. There is a method called the **expansion of a determinant by minors** that is used to find the value of a determinant.

Before we explain this method, first observe that a 3 by 3 determinant has 3 rows and 3 columns.

$$\begin{array}{l} \text{Row 1} \longrightarrow \\ \text{Row 2} \longrightarrow \\ \text{Row 3} \longrightarrow \end{array} \begin{vmatrix} a_1 & b_1 & c_1 \\ a_2 & b_2 & c_2 \\ a_3 & b_3 & c_3 \end{vmatrix}$$

$$\text{Column 1} \quad \text{Column 2} \quad \text{Column 3}$$
$$\begin{vmatrix} a_1 & b_1 & c_1 \\ a_2 & b_2 & c_2 \\ a_3 & b_3 & c_3 \end{vmatrix}$$

Notice that each entry of the determinant is in exactly one row and one column.

A **minor** of an entry in a determinant is the 2 by 2 determinant that remains after mentally deleting the row and column in which the element appears. Some examples of minors are the following.

The minor of a_1 is $\begin{vmatrix} a_1 & b_1 & c_1 \\ a_2 & b_2 & c_2 \\ a_3 & b_3 & c_3 \end{vmatrix} = \begin{vmatrix} b_2 & c_2 \\ b_3 & c_3 \end{vmatrix}$

The minor of a_2 is $\begin{vmatrix} a_1 & b_1 & c_1 \\ a_2 & b_2 & c_2 \\ a_3 & b_3 & c_3 \end{vmatrix} = \begin{vmatrix} b_1 & c_1 \\ b_3 & c_3 \end{vmatrix}$

The minor of a_3 is $\begin{vmatrix} a_1 & b_1 & c_1 \\ a_2 & b_2 & c_2 \\ a_3 & b_3 & c_3 \end{vmatrix} = \begin{vmatrix} b_1 & c_1 \\ b_2 & c_2 \end{vmatrix}$

Example 2 Find the minors of the entries in the second row:

$$\begin{vmatrix} 2 & -4 & 1 \\ 3 & 5 & -2 \\ 4 & -1 & 6 \end{vmatrix}$$

Solution   The second row is 3   5 -2.

The minor of 3 is $\begin{vmatrix} 2 & -4 & 1 \\ 3 & 5 & -2 \\ 4 & -1 & 6 \end{vmatrix} = \begin{vmatrix} -4 & 1 \\ -1 & 6 \end{vmatrix}$

$$= (-4)(6) - (1)(-1)$$

$$= -23$$

The minor of 5 is $\begin{vmatrix} 2 & -4 & 1 \\ 3 & 5 & -2 \\ 4 & -1 & 6 \end{vmatrix} = \begin{vmatrix} 2 & 1 \\ 4 & 6 \end{vmatrix}$

$$= (2)(6) - (1)(4)$$

$$= 8$$

The minor of -2 is $\begin{vmatrix} 2 & -4 & 1 \\ 3 & 5 & -2 \\ 4 & -1 & 6 \end{vmatrix} = \begin{vmatrix} 2 & -4 \\ 4 & -1 \end{vmatrix}$

$$= (2)(-1) - (-4)(4)$$

$$= 14$$ ▲

There is one more concept before we evaluate a determinant using minors. The **sign array** of a 3 by 3 determinant is the following 3 by 3 array of signs.

$$\begin{vmatrix} + & - & + \\ - & + & - \\ + & - & + \end{vmatrix}$$

To form the sign array, start with a $+$ sign at the upper left, then alternate signs across the first row and alternate signs down each column.

We are now ready to evaluate a determinant by expansion of minors. The terms in the definition of determinant can be rearranged and then factored in the following way:

$$\begin{vmatrix} a_1 & b_1 & c_1 \\ a_2 & b_2 & c_2 \\ a_3 & b_3 & c_3 \end{vmatrix} = a_1 b_2 c_3 + a_2 b_3 c_1 + a_3 b_1 c_2$$
$$- a_1 b_3 c_2 - a_2 b_1 c_3 - a_3 b_2 c_1$$

$$= a_1 b_2 c_3 - a_1 b_3 c_2$$
$$+ a_2 b_3 c_1 - a_2 b_1 c_3$$
$$+ a_3 b_1 c_2 - a_3 b_2 c_1$$

$$= a_1(b_2 c_3 - b_3 c_2)$$
$$- a_2(b_1 c_3 - b_3 c_1)$$
$$+ a_3(b_1 c_2 - b_2 c_1)$$

The quantity in the first parentheses is the minor of a_1:

$$b_2c_3 - b_3c_2 = \begin{vmatrix} a_1 & b_1 & c_1 \\ a_2 & b_2 & c_2 \\ a_3 & b_3 & c_3 \end{vmatrix}$$

The quantity in the second parentheses is the minor of a_2:

$$b_1c_3 - b_3c_1 = \begin{vmatrix} a_1 & b_1 & c_1 \\ a_2 & b_2 & c_2 \\ a_3 & b_3 & c_3 \end{vmatrix}$$

The quantity in the third parentheses is the minor of a_3:

$$b_1c_2 - b_2c_1 = \begin{vmatrix} a_1 & b_1 & c_1 \\ a_2 & b_2 & c_2 \\ a_3 & b_3 & c_3 \end{vmatrix}$$

Therefore, we may write the 3 by 3 determinant as

$$\begin{vmatrix} a_1 & b_1 & c_1 \\ a_2 & b_2 & c_2 \\ a_3 & b_3 & c_3 \end{vmatrix} = +a_1 \begin{vmatrix} b_2 & c_2 \\ b_3 & c_3 \end{vmatrix} - a_2 \begin{vmatrix} b_1 & c_1 \\ b_3 & c_3 \end{vmatrix} + a_3 \begin{vmatrix} b_1 & c_1 \\ b_2 & c_2 \end{vmatrix} \tag{1}$$

the first column of the sign array

Equation (1) is called evaluating the determinant by *expanding by minors down the first column*. Notice that each of the three terms on the left side is given the sign $(+)$ or $(-)$ from the first column of the sign array.

A 3 by 3 determinant can be expanded down any column or across any row in a similar fashion. We list the steps in the following strategy.

S T R A T E G Y

Evaluating a determinant by expansion of minors

Step 1   Choose any row or column for expansion.

Step 2   Write the product of each element in the selected row or column with its minor.

Step 3   Find the selected row or column chosen in Step 1 in the 3 by 3 sign array. Connect the three products in Step 2 with the corresponding sign from the sign array.

Step 4   Evaluate each 2 by 2 determinant.

Step 5   Multiply and add or subtract as indicated.

Example 3   Evaluate by expanding minors across the first row.

$$\begin{vmatrix} 2 & -3 & 1 \\ 4 & 6 & -2 \\ -5 & 0 & 1 \end{vmatrix}$$

Solution   Step 1   We are given to expand by minors across the first row.

Step 2   The products of each element in the first row are the following.

$$(2)\begin{vmatrix} 6 & -2 \\ 0 & 1 \end{vmatrix} \qquad (-3)\begin{vmatrix} 4 & -2 \\ -5 & 1 \end{vmatrix} \qquad (1)\begin{vmatrix} 4 & 6 \\ -5 & 0 \end{vmatrix}$$

Step 3   The first row in the sign array is +   −   +. Therefore,

$$\begin{vmatrix} 2 & -3 & 1 \\ 4 & 6 & -2 \\ -5 & 0 & 1 \end{vmatrix} = +(2)\begin{vmatrix} 6 & -2 \\ 0 & 1 \end{vmatrix} - (-3)\begin{vmatrix} 4 & -2 \\ -5 & 1 \end{vmatrix} + (1)\begin{vmatrix} 4 & 6 \\ -5 & 0 \end{vmatrix}$$

$$= 2[(6)(1) - (-2)(0)] + 3[(4)(1) - (-2)(-5)] + (1)[(4)(0) - (6)(-5)]$$

$$= 2(6) + 3(-6) + 30 = 12 - 18 + 30 = 24 \qquad \blacktriangle$$

N O T E   *It makes no difference which row or column is used to expand a determinant. You will always get the same answer.*

Since the choice of the row or column by which to expand a determinant is up to us, a good choice would be a row or column that has at least one zero in it, if there is one. This will make the computation easier. This idea is illustrated in the next example.

Example 4   Evaluate:

$$\begin{vmatrix} -1 & 2 & 3 \\ 4 & 1 & 0 \\ 10 & -5 & 0 \end{vmatrix}$$

Solution   Since the third column has two zeros in it, we expand the determinant by that column.

$$\begin{vmatrix} -1 & 2 & 3 \\ 4 & 1 & 0 \\ 10 & -5 & 0 \end{vmatrix} = +(3)\begin{vmatrix} 4 & 1 \\ 10 & -5 \end{vmatrix} - (0)\begin{vmatrix} -1 & 2 \\ 10 & -5 \end{vmatrix} + (0)\begin{vmatrix} -1 & 2 \\ 4 & 1 \end{vmatrix}$$

$$= 3[(4)(-5) - (1)(10)] - 0 + 0$$

$$= 3[-20 - 10]$$

$$= 3(-30) = -90$$

Notice that since the last two minors were multiplied by zero, there was no need to evaluate them. $\qquad \blacktriangle$

We are now ready to solve systems of linear equations using Cramer's Rule. There are two rules, one for systems of two equations in two variables and the other for systems of three equations in three variables.

We start the discussion of Cramer's Rule with the following problem. Suppose we want to solve this system for x and y:

$$a_1 x + b_1 y = c_1$$

$$a_2 x + b_2 y = c_2$$

We use the elimination method. We first eliminate y by multiplying the first equation by b_2, the second equation by $-b_1$, then adding.

$$a_1 x + b_1 y = c_1 \quad\text{------multiply by } b_2 \longrightarrow \quad a_1 b_2 x + b_1 b_2 y = c_1 b_2$$

$$a_2 x + b_2 y = c_2 \quad\text{------multiply by } -b_1 \longrightarrow \quad \underline{-b_1 a_2 x - b_1 b_2 y = -b_1 c_2}$$

$$(a_1 b_2 - b_1 a_2)x = c_1 b_2 - b_1 c_2$$

Assuming that $a_1 b_2 - b_1 a_2 \neq 0$, we have

$$x = \frac{c_1 b_2 - b_1 c_2}{a_1 b_2 - b_1 a_2}$$

Similarly, we can solve for y by eliminating x. To do this, we multiply the first equation by $-a_2$ and the second equation by a_1, then add the results.

$$a_1 x + b_1 y = c_1 \quad\text{------multiply by } -a_2 \longrightarrow \quad -a_1 a_2 x - b_1 a_2 y = -c_1 a_2$$

$$a_2 x + b_2 y = c_2 \quad\text{------multiply by } a_1 \longrightarrow \quad \underline{a_1 a_2 x + a_1 b_2 y = a_1 c_2}$$

$$(a_1 b_2 - b_1 a_2)y = a_1 c_2 - c_1 a_2$$

Assuming that $a_1 b_2 - b_1 a_2 \neq 0$, we have

$$y = \frac{a_1 c_2 - c_1 a_2}{a_1 b_2 - b_1 a_2}$$

Notice that the solution to this system can be written using determinants:

$$x = \frac{c_1 b_2 - b_1 c_2}{a_1 b_2 - b_1 a_2} = \frac{\begin{vmatrix} c_1 & b_1 \\ c_2 & b_2 \end{vmatrix}}{\begin{vmatrix} a_1 & b_1 \\ a_2 & b_2 \end{vmatrix}}$$

$$y = \frac{a_1 c_2 - c_1 a_2}{a_1 b_2 - b_1 a_2} = \frac{\begin{vmatrix} a_1 & c_1 \\ a_2 & c_2 \end{vmatrix}}{\begin{vmatrix} a_1 & b_1 \\ a_2 & b_2 \end{vmatrix}}$$

This result is called Cramer's Rule for 2 by 2 systems and is stated in the following theorem.

THEOREM

Cramer's rule—for two equations in two variables

The solution to the system

$$a_1x + b_1y = c_1$$
$$a_2x + b_2y = c_2$$

is given by

$$x = \frac{D_x}{D} \quad \text{and} \quad y = \frac{D_y}{D}$$

where

$$D = \begin{vmatrix} a_1 & b_1 \\ a_2 & b_2 \end{vmatrix}, \qquad D_x = \begin{vmatrix} c_1 & b_1 \\ c_2 & b_2 \end{vmatrix}, \qquad D_y = \begin{vmatrix} a_1 & c_1 \\ a_2 & c_2 \end{vmatrix}$$

and $D \neq 0$.

Notice that the determinant D is taken from the coefficients of the variables x and y of the system. The determinants D_x and D_y are formed by replacing the coefficients of x or y by the constant terms from the system.

Example 5   Use Cramer's Rule to solve:

$$3x - 4y = 5$$
$$4x + 2y = 3$$

Solution   We start by forming and evaluating the determinants D, D_x, and D_y.

$$D = \begin{vmatrix} 3 & -4 \\ 4 & 2 \end{vmatrix} = (3)(2) - (-4)(4) = 6 + 16 = 22$$

$$D_x = \begin{vmatrix} 5 & -4 \\ 3 & 2 \end{vmatrix} = (5)(2) - (-4)(3) = 10 + 12 = 22$$

$$D_y = \begin{vmatrix} 3 & 5 \\ 4 & 3 \end{vmatrix} = (3)(3) - (5)(4) = 9 - 20 = -11$$

$$x = \frac{D_x}{D} = \frac{22}{22} = 1 \quad \text{and} \quad y = \frac{D_y}{D} = \frac{-11}{22} = -\frac{1}{2}$$

The solution is $x = 1$ and $y = -\dfrac{1}{2}$.   ▲

Example 6   Solve:

$$3x - 2y = -5$$
$$-6x + 4y = 10$$

Solution   To use Cramer's Rule, we form and evaluate the determinant of coefficients D.

$$D = \begin{vmatrix} 3 & -2 \\ -6 & 4 \end{vmatrix} = (3)(4) - (-2)(-6) = 12 - 12 = 0$$

Since D is zero, Cramer's Rule cannot be used. (Remember that we must divide by D, and division by zero is undefined.) This means that the system is either inconsistent (no solution) or dependent (infinitely many solutions). To determine which case occurs, we use the elimination method on the system.

$$3x - 2y = -5 \xrightarrow{\text{multiply by 2}} 6x - 4y = -10$$

$$-6x + 4y = 10 \xrightarrow{\text{leave as is}} \underline{-6x + 4y = 10}$$

$$0 = 0$$

Our conclusion is that the system has infinitely many solutions.   ▲

There is a version of Cramer's Rule for systems of three equations in three variables.

T H E O R E M

Cramer's rule—for three equations in three variables

The solution to the system

$$a_1 x + b_1 y + c_1 z = k_1$$

$$a_2 x + b_2 y + c_2 z = k_2$$

$$a_3 x + b_3 y + c_3 z = k_3$$

is given by

$$x = \frac{D_x}{D}, \qquad y = \frac{D_y}{D}, \qquad z = \frac{D_z}{D}$$

where

$$D = \begin{vmatrix} a_1 & b_1 & c_1 \\ a_2 & b_2 & c_2 \\ a_3 & b_3 & c_3 \end{vmatrix} \qquad D_x = \begin{vmatrix} k_1 & b_1 & c_1 \\ k_2 & b_2 & c_2 \\ k_3 & b_3 & c_3 \end{vmatrix}$$

$$D_y = \begin{vmatrix} a_1 & k_1 & c_1 \\ a_2 & k_2 & c_2 \\ a_3 & k_3 & c_3 \end{vmatrix} \qquad D_z = \begin{vmatrix} a_1 & b_1 & k_1 \\ a_2 & b_2 & k_2 \\ a_3 & b_3 & k_3 \end{vmatrix}$$

and $D \neq 0$.

Notice that the determinant D is formed from the coefficients of the variables. The other three determinants are formed from D as follows.

D_x is formed by replacing the first column in D by the constants of the system.

D_y is formed by replacing the second column in D by the constants of the system.

D_z is formed by replacing the third column in D by the constants of the system.

Example 7   Use Cramer's Rule to solve:

$$x + y + 2z = 2$$
$$4x + 2y - 2z = -1$$
$$3x - 3y = 3$$

Solution   We first form D from the coefficients of the system, then evaluate it by expansion down the third column. The "missing" z term in the third equation is actually $0 \cdot z$, and, therefore, the coefficient is zero.

$$D = \begin{vmatrix} 1 & 1 & 2 \\ 4 & 2 & -2 \\ 3 & -3 & 0 \end{vmatrix} = 2\begin{vmatrix} 4 & 2 \\ 3 & -3 \end{vmatrix} - (-2)\begin{vmatrix} 1 & 1 \\ 3 & -3 \end{vmatrix} + (0)\begin{vmatrix} 1 & 1 \\ 4 & 2 \end{vmatrix}$$

$$= 2[(4)(-3) - (2)(3)] + 2[(1)(-3) - (1)(3)] + 0$$
$$= 2(-18) + 2(-6) = -36 - 12 = -48$$

Next we form D_x, D_y, and D_z and evaluate them by expansion of minors.

$$D_x = \begin{vmatrix} 2 & 1 & 2 \\ -1 & 2 & -2 \\ 3 & -3 & 0 \end{vmatrix} = -24 \qquad D_y = \begin{vmatrix} 1 & 2 & 2 \\ 4 & -1 & -2 \\ 3 & 3 & 0 \end{vmatrix} = 24$$

$$D_z = \begin{vmatrix} 1 & 1 & 2 \\ 4 & 2 & -1 \\ 3 & -3 & 3 \end{vmatrix} = -48$$

Therefore,

$$x = \frac{D_x}{D} = \frac{-24}{-48} = \frac{1}{2}$$

$$y = \frac{D_y}{D} = \frac{24}{-48} = -\frac{1}{2}$$

$$z = \frac{D_z}{D} = \frac{-48}{-48} = 1$$

The solution is $x = \dfrac{1}{2}$, $y = -\dfrac{1}{2}$, $z = 1$.

N O T E   *When D = 0, Cramer's Rule for systems of three equations in three variables cannot be used. Either the solution set is empty or there are infinitely many solutions.*

CALCULATOR   CROSSOVER

Use Cramer's Rule to solve each system.

1. $4.1x - 3.9y = -46.74$

$2.6x + 8.5y = 60.6$

2. $0.2x - 1.5y + 7.8z = -4.4$

$-3.8x + 2.1y - 3.4z = 5.8$

$x - \quad y + \quad z = 0$

Answers   **1.** $x = -3.58,\ y = 8.22$ **2.** $x = -2.49,\ y = -3.70,\ z = -1.21$

L E A R N I N G   A D V A N T A G E   **Mathematics at Work in Society: Life Scientist.** *Life science is the study of living organisms and the relationship of plants and animals to their surroundings. Those who work in this area are called life scientists. They can be BIOLOGISTS, BOTANISTS, AGRONOMISTS, ZOOLOGISTS, or ECOLOGISTS.*

Much of their work involves measurement and estimation. To perform these tasks they need mathematics.

E X E R C I S E   S E T   8 . 4

For Exercises 1–8, find the value of each 2 by 2 determinant.

1. $\begin{vmatrix} 1 & 5 \\ 2 & 3 \end{vmatrix}$

2. $\begin{vmatrix} 4 & 7 \\ 2 & 0 \end{vmatrix}$

3. $\begin{vmatrix} -1 & 3 \\ -2 & 0 \end{vmatrix}$

4. $\begin{vmatrix} 4 & -6 \\ -1 & 1 \end{vmatrix}$

5. $\begin{vmatrix} \frac{2}{3} & 4 \\ 1 & 3 \end{vmatrix}$

6. $\begin{vmatrix} \frac{2}{9} & \frac{3}{2} \\ 2 & 9 \end{vmatrix}$

7. $\begin{vmatrix} -4 & -1 \\ -3 & -2 \end{vmatrix}$

8. $\begin{vmatrix} -3 & -5 \\ 1 & -1 \end{vmatrix}$

For Exercises 9–16, find the value of each 3 by 3 determinant.

9. $\begin{vmatrix} 1 & 0 & -1 \\ 2 & 3 & 5 \\ -1 & 0 & 3 \end{vmatrix}$

10. $\begin{vmatrix} 2 & -1 & 1 \\ 0 & 2 & 2 \\ 1 & -1 & 5 \end{vmatrix}$

11. $\begin{vmatrix} 2 & 1 & 4 \\ -1 & 0 & -1 \\ 3 & 0 & 0 \end{vmatrix}$

12. $\begin{vmatrix} -1 & 3 & -2 \\ 0 & 1 & 1 \\ -1 & 2 & -1 \end{vmatrix}$

13. $\begin{vmatrix} 3 & 1 & 4 \\ 1 & 0 & 0 \\ -3 & -2 & -1 \end{vmatrix}$

14. $\begin{vmatrix} 0 & 3 & -4 \\ 4 & -1 & 2 \\ 0 & 0 & 1 \end{vmatrix}$

15. $\begin{vmatrix} 1 & 0 & -2 \\ 0 & 1 & -3 \\ 9 & -1 & 0 \end{vmatrix}$ **16.** $\begin{vmatrix} 2 & -1 & 0 \\ -4 & 0 & 6 \\ 0 & -5 & 2 \end{vmatrix}$

For Exercises 17–36, solve using Cramer's Rule.

17. $3x + 2y = 3$
$\quad 4x - y = -7$

18. $2x + 3y = -1$
$\quad\quad x + 2y = -2$

19. $-x + 5y = 3$
$\quad 4x - 3y = 5$

20. $\quad 6x + 7y = 1$
$\quad -3x - 4y = -1$

21. $5x + y = -2$
$\quad\; x + 2y = 2$

22. $\quad x + y = 1$
$\quad 7x - 3y = -8$

23. $-2x + 3y = 9$
$\quad\; 3x + 6y = -\dfrac{13}{2}$

24. $2x + 3y = \dfrac{11}{6}$
$\quad 3x + 6y = 1$

25. $\dfrac{x}{3} - \dfrac{y}{2} = 6$
$\quad -2x + 3y = -1$

26. $-4x - 3y = 12$
$\quad \dfrac{x}{3} + \dfrac{y}{4} = -1$

27. $\quad x + y - z = 0$
$\quad 3x - 2y + z = 7$
$\quad -2x + y - 2z = 9$

28. $\quad x - y + z = -1$
$\quad 2x + y - z = 1$
$\quad\; x + y + z = 3$

29. $4x - 5y - 3z = 2$
$\quad -2x + 10y = -3$
$\quad\quad\; 3x + 2y = \dfrac{11}{10}$

30. $4x + y - 3z = -2$
$\quad 8x + 6z = 3$
$\quad -16x + 3z = 4$

31. $\quad 2x + y - 3z = 2$
$\quad 5x + 10y = -1$
$\quad 10x - 5y - 7z = 0$

32. $3x + 4y + 6z = -7$
$\quad 6x - 10y + 3z = -2$
$\quad 5x - 13y - 2z = 0$

33. $x - z = 2$
         $3y + 2z = 0$
         $3x + y = -2$

34. $y + 2z = 1$
         $-3x - y = 0$
         $x + z = -3$

35. $x - z = 5$
         $y + z = -1$
         $x - y = 6$

36. $2y + z = 0$
         $x + 2y = -1$
         $2x - 3z = 0$

For Exercises 37–42, solve each word problem by writing a system of equations and then using Cramer's Rule.

37. A triangle has a perimeter of $13\frac{1}{2}$ inches. The length of the largest side is $1\frac{1}{2}$ inches more than the sum of the lengths of the other two sides. The smallest side is 7 inches shorter than the largest side. Find the lengths of the three sides of the triangle.

38. The sum of the length, width, and height of a rectangular box is 27 meters. The width is one half the length and the height is 3 meters less than the sum of the width and length. Find the dimensions of the box.

39. At a Faber College football game, student tickets cost $4 each, children's tickets cost $2 each, and adult tickets cost $6 each. There were twice as many children's tickets sold as adult tickets and four times as many student tickets sold as children's tickets. If $1,050 were the total receipts for the game, how many tickets of each type were sold?

40. The measurement of the largest angle of a triangle is four times that of the smallest, and the other angle measures 45 degrees less than the largest. Find the measurements of the three angles.

41. Determine a, b, and c so that the solution set of $ax + by + cz = 1$ contains the following three ordered triples (x, y, z): $(1, 0, 1)$, $(2, -1, 0)$, and $(0, 3, -2)$.

42. Determine a, b, and c so that the solution set of $ax + by + cz = -2$ contains the following three ordered triples (x, y, z): $(0, 1, -1)$, $(2, 0, 3)$, and $(-1, -1, 0)$.

REVIEW PROBLEMS

The following exercises review parts of Sections 8.1 and 8.3. Doing these problems will help prepare you for the next section.

Section 8.1

Solve the system by the substitution method.

43. $2x - 3y = -2$
       $y = 3$

44. $-5x + 4y = -3$
       $y = -2$

Section 8.3

Solve the system by the substitution method.

45. $x - 2y + 4z = -12$
$\qquad y + 3z = 3$
$\qquad\qquad z = -1$

46. $x + 4y - 2z = -9$
$\qquad y + 2z = -3$
$\qquad\qquad z = -4$

47. $x - 6y = -6$
$\qquad y - 8z = -2$
$\qquad\qquad z = -3$

48. $x + 7y + 5z = 9$
$\qquad y - 4z = -4$
$\qquad\qquad z = 0$

E N R I C H M E N T   E X E R C I S E S

Solve using Cramer's Rule.

1. $\dfrac{4}{x} - \dfrac{3}{y} = -3$

$\dfrac{6}{x} + \dfrac{3}{2y} + \dfrac{1}{z} = -7$

$\dfrac{2}{x} - \dfrac{1}{z} = -3$

2. $3x^2 + 2y^2 + z^2 = -3$
$9x^2 + 16y^2 + 20z^2 = 5$
$2y^2 + z^2 = 1$

3. Evaluate

$$\begin{vmatrix} a & a & k_1 \\ b & b & k_2 \\ c & c & k_3 \end{vmatrix}$$

For Exercises 4 and 5, solve each word problem. Write a system of three equations in three variables (unknowns) and solve using Cramer's Rule. Be sure to indicate what each variable represents.

4. The Acme Music Associates makes tuning forks, pitch pipes, and mouth organs. Each item requires three machines in its manufacture. The number of hours needed on each machine to make a case of each item is shown in the following table.

	Hours needed		
	Tuning Forks	Pitch Pipes	Mouth Organs
Machine A	$\dfrac{2}{3}$	$\dfrac{1}{3}$	1
Machine B	$\dfrac{1}{3}$	$\dfrac{1}{9}$	1
Machine C	$\dfrac{1}{3}$	$\dfrac{2}{9}$	$\dfrac{2}{3}$

Each day, the amount of time available on each machine is 15 hours on machine A, 10 hours on machine B, and 9 hours on machine C. How many cases of each item can be made per day?

5. Amalgam Dental Supplies makes hand excavators, drill bits, and pinch valves. Each item requires machining, polishing, and finishing in its manufacture. The number of hours needed for each case of each item is shown in the following table.

Hours needed			
	Hand Excavators	Drill Bits	Pinch Valves
Machining	$\frac{1}{8}$	$\frac{1}{2}$	$\frac{1}{4}$
Polishing	$\frac{1}{4}$	$\frac{1}{2}$	$\frac{1}{2}$
Finishing	$\frac{1}{4}$	$\frac{1}{2}$	$\frac{1}{4}$

Each day, the amount of time available for machining, polishing, and finishing is, respectively, 10, 15, and 12 hours. How many cases of each item can be made per day?

Answers to Enrichment Exercises are on page A.50.

8.5

Matrix Methods for Solving Systems of Equations

OBJECTIVES

▶ To write the augmented matrix of a system

▶ To use matrix methods to solve a system with two equations

▶ To use matrix methods to solve a system of three equations

In this section, we develop another way to solve systems of equations. This involves using the concept of *matrix*. A **matrix** is a rectangular array of numbers. For example, $\begin{bmatrix} 4 & 0 & 2 \\ 2 & 1 & 5 \end{bmatrix}$ is a matrix. This matrix has two rows and three columns, and is called a 2 × 3, read "two by three," matrix. The matrix

$[-2 \quad 4 \quad 5 \quad 1]$ is a 1×4 matrix and the matrix $\begin{bmatrix} 9 & 3 & -1 \\ 2 & 1 & 9 \\ 3 & 0 & 6 \end{bmatrix}$ is a 3×3 matrix.

Any matrix that has the same number of rows as columns is called a **square** matrix.

Associated with a system of equations is a matrix called the **augmented matrix.** For example, the augmented matrix for the system

$$3x - 2y = 1$$
$$-2x + 3y = -4$$

is the matrix

$$\begin{bmatrix} 3 & -2 & | & 1 \\ -2 & 3 & | & -4 \end{bmatrix}$$

Notice that the vertical bar in the matrix separates the four numbers that are the coefficients from the two constants.

The augmented matrix is simply a compact way to write the two equations of the system. We will see that the following **row operations** performed on matrices will lead to the solution of the system.

R U L E

Three row operations that can be performed on a matrix

1. Interchange any two rows of the matrix.
2. Multiply any row by a nonzero number.
3. Add to any row a constant multiple of another row.

Since each row of an augmented matrix comes from an equation from the system, the three row operations reflect operations that can be performed on the equations of a system. For example, the first row operation of interchanging any two rows is equivalent to interchanging any two equations in the system.

Row operations can be used on the augmented matrix to change it into a matrix of a system whose solution can easily be obtained. Solving a system using **matrix methods** is illustrated in our first example.

Example 1   Solve the system

$$3x - 2y = 1$$
$$-2x + 3y = -4$$

using matrix methods.

Solution   We start with the augmented matrix, then use row operations to change it into a matrix of a system that is easier to solve than the original system. In particular, consider the augmented matrix of the original system:

$$\begin{matrix} R_1 \\ R_2 \end{matrix} \quad \left[\begin{array}{cc|c} 3 & -2 & 1 \\ -2 & 3 & -4 \end{array} \right]$$

Notice that this matrix has two rows, call them R_1 and R_2. Next, we focus our attention on the columns of this matrix that are to the left of the vertical line. Notice that in the first column, the first entry is 3. We want this entry to be 1. To achieve this, we use row operation 2, and divide the first row by 3 $\left(\text{or mul-} \right.$ tiply by $\left. \frac{1}{3} \right)$. Therefore, using this row operation, we can change the matrix as follows:

$$\left[\begin{array}{cc|c} 3 & -2 & 1 \\ -2 & 3 & -4 \end{array} \right] \overset{\frac{1}{3}R_1}{\Longrightarrow} \left[\begin{array}{cc|c} 1 & -\frac{2}{3} & \frac{1}{3} \\ -2 & 3 & -4 \end{array} \right]$$

The notation above the arrow tells us what we did to change from the first matrix to the second. That is, $\frac{1}{3}R_1$ means to write a new row 1 that is obtained by multiplying the old entries of row 1 by $\frac{1}{3}$.

Next, we want a zero below the 1 in the first column. In order to achieve this, we use row operation 3. In particular, if we multiply the first row by 2 and add the results to the corresponding entries of row R_2, we will have the desired zero. Here are the details:

$$\left[\begin{array}{cc|c} 1 & -\frac{2}{3} & \frac{1}{3} \\ -2 & 3 & -4 \end{array} \right] \overset{2R_1 + R_2}{\Longrightarrow} \left[\begin{array}{cc|c} 1 & -\frac{2}{3} & \frac{1}{3} \\ 0 & \frac{5}{3} & -\frac{10}{3} \end{array} \right]$$

Notice that the notation above the arrow, $2R_1 + R_2$, means "two times row 1 added to row 2." Next, look at column 2. We want a 1 in row 2, column 2. The existing entry is $\frac{5}{3}$. To make it a 1, use row operation 2 and multiply the second row by $\frac{3}{5}$, the reciprocal of $\frac{5}{3}$.

$$\left[\begin{array}{cc|c} 1 & -\frac{2}{3} & \frac{1}{3} \\ 0 & \frac{5}{3} & -\frac{10}{3} \end{array} \right] \overset{\frac{3}{5}R_2}{\Longrightarrow} \left[\begin{array}{cc|c} 1 & -\frac{2}{3} & \frac{1}{3} \\ 0 & 1 & -2 \end{array} \right]$$

This final augmented matrix gives the following system of equations:

$$x - \frac{2}{3}y = \frac{1}{3} \qquad \text{or} \qquad x - \frac{2}{3}y = \frac{1}{3}$$

$$0x + 1y = -2 \qquad\qquad\qquad y = -2$$

The second equation, $y = -2$, is the y value of the solution. To find the x value, substitute $y = -2$ into the first equation and solve for x.

$$x - \left(\frac{2}{3}\right)(-2) = \frac{1}{3}$$

$$x + \frac{4}{3} = \frac{1}{3}$$

$$x + \frac{4}{3} - \frac{4}{3} = \frac{1}{3} - \frac{4}{3}$$

$$x = \frac{1-4}{3}$$

$$= \frac{-3}{3}$$

$$= -1$$

We now check our answer, $x = -1$ and $y = -2$, in the original system.

Check $x = -1$ and $y = -2$:

$$3x - 2y = 1 \qquad\qquad -2x + 3y = -4$$
$$3(-1) - 2(-2) = 1 \qquad -2(-1) + 3(-2) = -4$$
$$-3 + 4 = 1 \qquad\qquad 2 - 6 = -4$$

The solution of the system is $x = -1$ and $y = -2$. ▲

The matrix method as shown in the previous example is another way to solve a system of equations. The three row operations reflect corresponding operations that are permissible to do on equations in a system. One advantage of this technique is that large systems can be solved by a computer using matrix methods.

The ways matrix methods can determine when the system is either inconsistent or dependent are shown in the next examples.

Example 2 Solve the system using matrix methods.

$$2x - 6y = \frac{1}{2}$$

$$-3x + 9y = \frac{1}{4}$$

Solution First, write the augmented matrix of the system.

$$\begin{bmatrix} 2 & -6 & | & \frac{1}{2} \\ -3 & 9 & | & \frac{1}{4} \end{bmatrix}$$

To simplify, we start by multiplying the first row by $\frac{1}{2}$.

$$\begin{bmatrix} 2 & -6 & \Big| & \tfrac{1}{2} \\ -3 & 9 & \Big| & \tfrac{1}{4} \end{bmatrix} \xrightarrow{\tfrac{1}{2}R_1} \begin{bmatrix} 1 & -3 & \Big| & \tfrac{1}{4} \\ -3 & 9 & \Big| & \tfrac{1}{4} \end{bmatrix} \xrightarrow{3R_1 + R_2} \begin{bmatrix} 1 & -3 & \Big| & \tfrac{1}{4} \\ 0 & 0 & \Big| & 1 \end{bmatrix}$$

The system of equations this last matrix represents is

$$x - 3y = \frac{1}{4}$$

$$0 = 1$$

This system has no solution and the original system is inconsistent.   ▲

Example 3   Use matrix methods to solve the system:

$$-3x + 4y = -12$$

$$\frac{1}{4}x - \frac{1}{3}y = 1$$

Solution   We write the augmented matrix and use row operations to simplify.

$$\begin{bmatrix} -3 & 4 & \Big| & -12 \\ \tfrac{1}{4} & -\tfrac{1}{3} & \Big| & 1 \end{bmatrix} \xrightarrow{-\tfrac{1}{3}R_1} \begin{bmatrix} 1 & -\tfrac{4}{3} & \Big| & 4 \\ \tfrac{1}{4} & -\tfrac{1}{3} & \Big| & 1 \end{bmatrix} \xrightarrow{-\tfrac{1}{4}R_1 + R_2} \begin{bmatrix} 1 & -\tfrac{4}{3} & \Big| & 4 \\ 0 & 0 & \Big| & 0 \end{bmatrix}$$

The system that corresponds to this last matrix is

$$x - \frac{4}{3}y = 4$$

$$0 = 0$$

This system is dependent and has infinitely many solutions. In fact, any point on the line given by $x - \dfrac{4}{3}y = 4$ is a solution of the system. Notice that if we multiply both sides of the equation $x - \dfrac{4}{3}y = 4$ by -3, we obtain the first equation of the original system. Therefore, we may write the solution set as $\{(x, y) \,|\, -3x + 4y = -12\}$.   ▲

Next, we consider using matrix methods to solve a system of three equations in three unknowns. The key is to transform the augmented matrix using row operations until we obtain a matrix that is in a special form. In the next example, we solve a system whose augmented matrix already is in that special form.

Example 4   Write the system of equations from the given augmented matrix, then solve the system.

$$\begin{bmatrix} 1 & -2 & 3 & \Big| & 2 \\ 0 & 1 & 5 & \Big| & 1 \\ 0 & 0 & 1 & \Big| & -1 \end{bmatrix}$$

Solution   From the augmented matrix, we can write the equations of the system. Remember that each of the three rows corresponds to an equation.

$$x - 2y + 3z = 2$$
$$y + 5z = 1$$
$$z = -1$$

Next, we solve this system. Notice that the third equation gives the z value of the solution. We replace z by -1 in the second equation and solve for y.

$$y + 5z = 1$$
$$y + 5(-1) = 1$$
$$y - 5 = 1$$
$$y = 6$$

To get the value for x, replace y by 6 and z by -1 in the first equation and solve for x.

$$x - 2y + 3z = 2$$
$$x - 2(6) + 3(-1) = 2$$
$$x - 12 - 3 = 2$$
$$x - 15 = 2$$
$$x = 17$$

The solution is $x = 17$, $y = 6$, and $z = -1$.   ▲

Notice the form of the augmented matrix of Example 4. It has 1's along the diagonal from left to right and 0's below each 1. Our plan is to use row operations on the augmented matrix of a general system to end up with an augmented matrix that has this same form.

The way to use the matrix method to achieve an augmented matrix with this simplified form is shown in the next example.

Example 5   Use matrix methods to solve the system:

$$2x + 7y = 5$$
$$x + 3y + 2z = -2$$
$$-x - y + 4z = -8$$

Solution   First, write the augmented matrix for the system.

$$\begin{bmatrix} 2 & 7 & 0 & | & 5 \\ 1 & 3 & 2 & | & -2 \\ -1 & -1 & 4 & | & -8 \end{bmatrix}$$

The best way to simplify this matrix is to start with the first column. We want a 1 in the first row, first column. We could achieve that 1 by multiplying the first row by $\frac{1}{2}$. However, that would introduce fractions into the matrix.

Notice that the first entry of row 2 is a 1. If we use row operation 1 and interchange row 1 and row 2, we will then have a 1 where we want it without creating any fractions. Therefore, as our first step, we interchange those two rows. The notation we use is $R_1 \leftrightarrow R_2$.

$$\left[\begin{array}{ccc|c} 2 & 7 & 0 & 5 \\ 1 & 3 & 2 & -2 \\ -1 & -1 & 4 & -8 \end{array}\right] \xrightarrow{R_1 \leftrightarrow R_2} \left[\begin{array}{ccc|c} 1 & 3 & 2 & -2 \\ 2 & 7 & 0 & 5 \\ -1 & -1 & 4 & -8 \end{array}\right]$$

Next, use row 1 to obtain 0's in the first column below this 1. To make the 2 in row 2, column 1 a zero, multiply the first row by -2 and add to the second row. To make the -1 in row 3, column 1 a 0, add row 1 to row 3. The details follow:

$$\left[\begin{array}{ccc|c} 1 & 3 & 2 & -2 \\ 2 & 7 & 0 & 5 \\ -1 & -1 & 4 & -8 \end{array}\right] \begin{array}{c} \xrightarrow{-2R_1 + R_2} \\ \xrightarrow{R_1 + R_3} \end{array} \left[\begin{array}{ccc|c} 1 & 3 & 2 & -2 \\ 0 & 1 & -4 & 9 \\ 0 & 2 & 6 & -10 \end{array}\right]$$

Next, we focus our attention on the second column. The diagonal entry in that column is in row 2. It is already a 1 and, therefore, the only thing remaining to do is to obtain a 0 below it. Since the number below the 1 is a 2, we multiply the second row by -2 and add the results to the third row.

$$\left[\begin{array}{ccc|c} 1 & 3 & 2 & -1 \\ 0 & 1 & -4 & 9 \\ 0 & 2 & 6 & -10 \end{array}\right] \xrightarrow{-2R_2 + R_3} \left[\begin{array}{ccc|c} 1 & 3 & 2 & -1 \\ 0 & 1 & -4 & 9 \\ 0 & 0 & 14 & -28 \end{array}\right]$$

The final step is to locate the diagonal entry in the third column. It is the number 14 and we want to make it a 1. To achieve this, multiply the third row by $\frac{1}{14}$.

$$\left[\begin{array}{ccc|c} 1 & 3 & 2 & -1 \\ 0 & 1 & -4 & 9 \\ 0 & 0 & 14 & -28 \end{array}\right] \xrightarrow{\frac{1}{14}R_3} \left[\begin{array}{ccc|c} 1 & 3 & 2 & -1 \\ 0 & 1 & -4 & 9 \\ 0 & 0 & 1 & -2 \end{array}\right]$$

This last matrix gives the following system of equations:

$$x + 3y + 2z = -1$$
$$y - 4z = 9$$
$$z = -2$$

Substitute $z = -2$ into the second equation and solve for y.

$$y - 4z = 9$$
$$y - 4(-2) = 9$$
$$y + 8 = 9$$
$$y = 1$$

Now, take the values of $z = -2$ and $y = 1$ and substitute them into the first equation and solve for x.

$$x + 3y + 2z = -1$$
$$x + 3(1) + 2(-2) = -1$$
$$x + 3 - 4 = -1$$
$$x - 1 = -1$$
$$x = 0$$

The solution of the original system is $x = 0$, $y = 1$, and $z = -2$. On your own, check this solution in the original system. ▲

L E A R N I N G   A D V A N T A G E   **Mathematics at Work in Society: Transportation Careers.** *The job of a TRAFFIC CONTROL ENGINEER involves monitoring traffic conditions; determining speed limits, the placement of stop signs, and traffic signals; as well as timing traffic signals to permit the best flow of cars and trucks. Mathematics is used to study how traffic signals should be timed with respect to each other to permit a car traveling at a constant speed to pass through each (green) signal at the same time of the cycle. Setting the signals like this permits a better flow of traffic and it encourages drivers to obey the posted speed limit.*

E X E R C I S E   S E T   8.5

For Exercises 1–12, complete the second matrix by performing the indicated row operation(s).

1. $\begin{bmatrix} 3 & -6 & | & 2 \\ 2 & 4 & | & -7 \end{bmatrix} \xrightarrow{\frac{1}{3}R_1} \begin{bmatrix} & & | & \\ 2 & 4 & | & -7 \end{bmatrix}$

2. $\begin{bmatrix} -2 & 1 & | & -4 \\ 5 & -3 & | & 1 \end{bmatrix} \xrightarrow{-\frac{1}{2}R_1} \begin{bmatrix} & & | & \\ 5 & -3 & | & 1 \end{bmatrix}$

3. $\begin{bmatrix} 1 & -2 & | & 3 \\ 3 & 7 & | & -1 \end{bmatrix} \xrightarrow{-3R_1 + R_2} \begin{bmatrix} 1 & -2 & | & 3 \\ & & | & \end{bmatrix}$

4. $\begin{bmatrix} 1 & 3 & | & -5 \\ 4 & -1 & | & -20 \end{bmatrix} \xrightarrow{-4R_1 + R_2} \begin{bmatrix} 1 & 3 & | & -5 \\ & & | & \end{bmatrix}$

5. $\begin{bmatrix} 1 & -4 & | & 2 \\ 0 & -3 & | & -9 \end{bmatrix} \xrightarrow{-\frac{1}{3}R_2} \begin{bmatrix} 1 & -4 & | & 2 \\ & & | & \end{bmatrix}$

6. $\begin{bmatrix} 1 & 3 & | & -1 \\ 0 & \frac{1}{4} & | & \frac{1}{2} \end{bmatrix} \xrightarrow{4R_2} \begin{bmatrix} 1 & 3 & | & -1 \\ & & | & \end{bmatrix}$

7. $\begin{bmatrix} \frac{1}{4} & -1 & \frac{1}{2} & | & 7 \\ 2 & 3 & -1 & | & 4 \\ 0 & -10 & -2 & | & -6 \end{bmatrix} \xrightarrow{4R_1} \begin{bmatrix} & & & | & \\ 2 & 3 & -1 & | & 4 \\ 0 & -10 & -2 & | & -6 \end{bmatrix}$

8. $\begin{bmatrix} 1 & -3 & 2 & | & -5 \\ 0 & -5 & 10 & | & -3 \\ 0 & 7 & 16 & | & 1 \end{bmatrix} \xrightarrow{-\frac{1}{5}R_2} \begin{bmatrix} 1 & -3 & 2 & | & -5 \\ & & & | & \\ 0 & 7 & 16 & | & 1 \end{bmatrix}$

9. $\begin{bmatrix} 1 & -4 & 3 & | & -1 \\ 0 & 5 & 4 & | & 2 \\ 2 & -1 & -1 & | & 5 \end{bmatrix} \xrightarrow{-2R_1 + R_3} \begin{bmatrix} 1 & -4 & 3 & | & -1 \\ 0 & 5 & 4 & | & 2 \\ & & & | & \end{bmatrix}$

10. $\begin{bmatrix} 1 & 4 & 2 & | & -3 \\ 0 & 1 & -2 & | & -5 \\ 0 & -\frac{1}{2} & 0 & | & \frac{11}{2} \end{bmatrix} \xrightarrow{\frac{1}{2}R_2 + R_3} \begin{bmatrix} 1 & 4 & 2 & | & -3 \\ 0 & 1 & -2 & | & -5 \\ & & & | & \end{bmatrix}$

11. $\begin{bmatrix} 1 & -5 & \frac{3}{2} & | & -2 \\ 1 & 4 & \frac{1}{2} & | & 3 \\ 3 & -2 & \frac{5}{4} & | & -4 \end{bmatrix} \begin{array}{c} \xrightarrow{-R_1 + R_2} \\ \xrightarrow{-3R_1 + R_3} \end{array} \begin{bmatrix} 1 & -5 & \frac{3}{2} & | & -2 \\ & & & | & \\ & & & | & \end{bmatrix}$

12. $\begin{bmatrix} 1 & -3 & 4 & | & -7 \\ 4 & \frac{1}{4} & 0 & | & \frac{1}{2} \\ -\frac{2}{3} & 1 & -2 & | & \frac{5}{2} \end{bmatrix} \begin{array}{c} \xrightarrow{-4R_1 + R_2} \\ \xrightarrow{\frac{2}{3}R_1 + R_3} \end{array} \begin{bmatrix} 1 & -3 & 4 & | & -7 \\ & & & | & \\ & & & | & \end{bmatrix}$

For Exercises 13–22, solve the system using matrix methods.

13. $x - 2y = 3$
$3x + 5y = -2$

14. $x + 3y = 1$
$-2x - 7y = -3$

15. $6x + y = 24$
$\frac{1}{3}x - 2y = -11$

16. $2x - 4y = 2$
$\frac{1}{2}x + y = -\frac{7}{2}$

17. $2x - 4y = -5$
$3x - y = 0$

18. $3x - 18y = -13$
$-\frac{1}{2}x - \frac{1}{4}y = 0$

19. $x - 2y = -4$
$-\frac{1}{2}x + y = 2$

20. $3x - 2y = -9$
$-x + \frac{2}{3}y = 3$

21. $-4x + 2y = -3$
$\quad\;\; 2x - \;\; y = 3$

22. $\quad 3x - 2y = \dfrac{1}{2}$
$\quad -6x + 4y = -2$

For Exercises 23–32, write the system from the given augmented matrix, then solve the system. (See Example 4.)

23. $\begin{bmatrix} 1 & -2 & 3 & | & 2 \\ 0 & 1 & 5 & | & 1 \\ 0 & 0 & 1 & | & -1 \end{bmatrix}$

24. $\begin{bmatrix} 1 & 1 & -2 & | & -1 \\ 0 & 1 & 3 & | & 5 \\ 0 & 0 & 1 & | & 2 \end{bmatrix}$

25. $\begin{bmatrix} 1 & 2 & -1 & | & 2 \\ 0 & 1 & -4 & | & 5 \\ 0 & 0 & 1 & | & -2 \end{bmatrix}$

26. $\begin{bmatrix} 1 & -1 & 1 & | & 0 \\ 0 & 1 & 5 & | & -7 \\ 0 & 0 & 1 & | & -3 \end{bmatrix}$

27. $\begin{bmatrix} 1 & 0 & -1 & | & -7 \\ 0 & 1 & 3 & | & -2 \\ 0 & 0 & 1 & | & 3 \end{bmatrix}$

28. $\begin{bmatrix} 1 & -2 & 0 & | & -1 \\ 0 & 1 & -1 & | & 2 \\ 0 & 0 & 1 & | & -1 \end{bmatrix}$

29. $\begin{bmatrix} 1 & \frac{1}{2} & -\frac{3}{4} & | & -\frac{1}{2} \\ 0 & 1 & -\frac{1}{2} & | & 5 \\ 0 & 0 & 1 & | & -4 \end{bmatrix}$

30. $\begin{bmatrix} 1 & \frac{2}{3} & -1 & | & \frac{4}{3} \\ 0 & 1 & \frac{1}{6} & | & -3 \\ 0 & 0 & 1 & | & 6 \end{bmatrix}$

31. $\begin{bmatrix} 1 & -2 & \frac{3}{2} & | & -\frac{1}{2} \\ 0 & 1 & -3 & | & 7 \\ 0 & 0 & 0 & | & 1 \end{bmatrix}$

32. $\begin{bmatrix} 1 & 4 & -2 & | & \frac{4}{3} \\ 0 & 1 & -1 & | & -3 \\ 0 & 0 & 0 & | & -1 \end{bmatrix}$

For Exercises 33–40, solve the system using matrix methods.

33. $x - 2y + 3z = 9$
$\quad -x + 3y + 4z = 4$
$\quad 2x - 4y - 3z = 0$

34. $x + 3y + 4z = -8$
$\quad 2x + 7y - 3z = -18$
$\quad -x - 6y + \;\; z = 14$

35. $\quad x + 2y - 4z = -7$
$\quad 4x + 10y + \;\; z = 4$
$\quad -2x + 4y + 6z = 2$

36. $\quad x - 2y + 4z = 3$
$\quad 2x + \;\; 4z = 2$
$\quad -5x - 9y + 18z = 42$

37. $2x - 6y + 3z = -5$
$\quad x - 2y + 5z = 1$
$\quad -x + 3y - 6z = -2$

38. $3x + 2y + \;\; z = -9$
$\quad x - 4y + 2z = 13$
$\quad 2x + \;\; y + 3z = 4$

39. $\quad 2x + 3y + 2z = 2$
$\quad -4x - 6y - 3z = -3$
$\quad x + y - \dfrac{1}{6}z = 0$

40. $3x + 2y + 2z = 0$
$\quad 6x - 4y - 3z = -3$
$\quad x - \dfrac{2}{3}z = -1$

REVIEW PROBLEMS

The following exercises review parts of Section 7.4. Doing these problems will help prepare you for the next section.

Graph the following linear inequalities.

41. $x + y < 0$

42. $2x - y \geq 0$

43. $3x - 2y \leq -6$

44. $5x + 4y \geq 20$

45. $x \geq -2y$

46. $2x < 3y$

E N R I C H M E N T   E X E R C I S E S

1. Use matrix methods to find the value(s) of a so that the system is inconsistent.

$$x + 3y = -2$$
$$-ax + y = 1$$

2. Use matrix methods to solve for x and y in terms of a.

$$x - ay = 2$$
$$ax + y = 3$$

For Exercises 3–6, use matrix methods to solve the system.

3. $x + y - z + t = 2$
$-x + y + z - t = -4$
$2x - 2y + z - 2t = 1$
$y - 2z + 3t = 4$

4. $x + y - z + t = -8$
$x + 2y + 3z - 4t = 7$
$-y + 2z + 5t = 3$
$-x + 4z + 10t = 4$

5. $x + 2y - 3z + s - t = 8$
$-x + y + 2z - s + t = -7$
$2y - 3z - 4s = -5$
$-4z + 4s + 2t = 8$
$2z + 3s + t = 2$

6. $2x + 3y + 2s + 3t = 0$
$x - 2y + 2z - 3s + 2t = -2$
$3x + y - 3z + s = -8$
$-x + y - z + s - t = 1$
$-3x - 2z + 6t = 10$

Answers to Enrichment Exercises are on page A.51.

8.6

Systems of Linear Inequalities

OBJECTIVES

▶ *To solve systems of linear inequalities by the graphing method*

▶ *To solve word problems using systems of linear inequalities*

In Chapter 7, we learned how to graph a linear inequality. To review this graphing technique, consider the inequality

$$-3x + 4y \geq -12$$

To graph this inequality, first replace the inequality by an equal sign and graph the resulting equation.

$$-3x + 4y = -12$$

When $x = 0$, $y = -3$ and when $y = 0$, $x = 4$. Plotting the two points $(0, -3)$ and $(4, 0)$, we draw the straight line as shown in the figure. Next, we use a test point that is not on the line. We choose $(0, 0)$. Replacing x and y by zero in the original inequality,

$$-3(0) + 4(0) \geq -12$$

we obtain the true statement $0 \geq -12$. Since $(0, 0)$ lies in the upper half-plane, we shade the half-plane above the line.

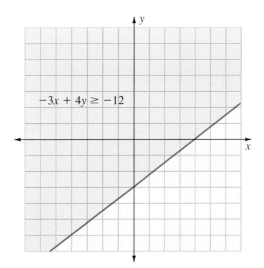

Graphically, the solution set of the inequality is the line together with the shaded region above the line. We can find the solution set of a system of linear inequalities by graphing each inequality one at a time.

Example 1   Graph the solution set of the system

$$x + 2y \geq 4$$
$$-x + y \leq 1$$

Solution   We first graph the inequality $x + 2y \geq 4$ using blue shading. (See the figure.) On the same coordinate system, we graph the inequality $-x + y \leq 1$ using red shading. The solution set is the red-blue shaded region common to both half-planes. Notice that this region includes parts of the two boundary lines.

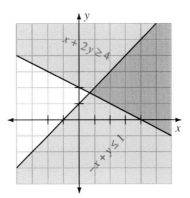

Example 2   Graph the solution set of the system

$$4x + y > 40$$

$$x + 3y > -15$$

Solution   We first graph $4x + y > 40$ using red shading. Next, we graph $x + 3y > -15$ using blue shading. The red-blue shaded region common to both half-planes is the solution of the system as shown in the following figure. Notice that the boundary lines are dashed since they are not part of the solution set.

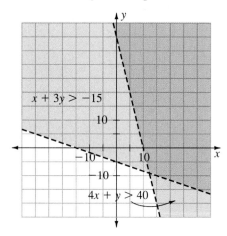

Example 3   Graph the solution set of the system

$$2x - 4y \le 16$$

$$y \ge -5$$

Solution   We first graph $2x - 4y \le 16$ using red shading. Next, we graph $y \ge -5$ using blue shading. The graph of the solution set is the red-blue shaded region shown in the figure.

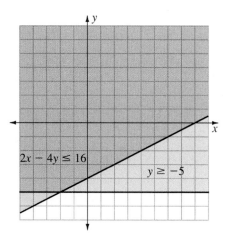

We now consider some applications involving a system of linear inequalities. In these applications, x and y represent quantities that are nonnegative. Therefore, the solution set will lie in the first quadrant.

Example 4   Michelle is planning a meal that is to consist of a casserole and salad. She wants the meal to supply at least 1,000 mg of calcium and 30 mg of protein. Each serving of the casserole supplies 100 mg of calcium and 5 mg of protein. Each serving of the salad supplies 200 mg of calcium and 3 mg of protein. Let

> x = The number of servings of casserole in the meal
> y = The number of servings of salad in the meal

(**a**) Express the restrictions on x and y as a system of linear inequalities.

(**b**) Graph the system of Part (a).

Solution   (**a**) First, we organize the information given in the problem. We construct the following table, in which the first column gives the number of milligrams of calcium and protein per serving of the casserole. The second column gives the number of milligrams of calcium and protein per serving of the salad.

	Casserole	*Salad*
Calcium	100	200
Protein	5	3

Now, if one serving of the casserole gives 100 mg of calcium, then x servings give $100x$ mg of calcium. Likewise, if one serving of the salad gives 200 mg of calcium, then y servings give $200y$ mg of calcium. Therefore, the total amount of calcium obtained from x servings of the casserole and y servings of the salad is given by $100x + 200y$. Since the meal is to consist of at least 1000 mg of calcium, we have

$$100x + 200y \geq 1000$$

Reading the information concerning the protein content, the amount of protein from x servings of the casserole and y servings of the salad is $5x + 3y$. Since the meal is to consist of at least 30 mg of protein, we have

$$5x + 3y \geq 30$$

There are two other inequalities in this problem. Since x and y represent numbers of servings, we have $x \geq 0$ and $y \geq 0$.

   In summary, the system of inequalities that expresses the restrictions on x and y is the following.

$$100x + 200y \geq 1,000$$
$$5x + 3y \geq 30$$
$$x \geq 0 \quad \text{and} \quad y \geq 0$$

(b) To graph the system of inequalities, remember that $x \geq 0$ and $y \geq 0$ mean that the region lies in the first quadrant. The graph is shown in the following figure.

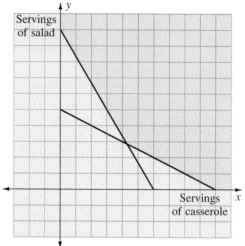

L E A R N I N G   A D V A N T A G E   **Mathematics at Work in Society: Physical Fitness Careers.** *Careers in physical fitness are careers that are devoted to helping people keep their bodies healthy and strong. Some of the many careers in physical fitness are PHYSICAL EDUCATION TEACHER, COACH, RECREATION DIRECTOR, AND PHYSICAL THERAPIST.*

PHYSICAL THERAPISTS, for example, help people to overcome physical disabilities received from disease or injury. They perform tests that measure motor development and muscle strength. A formula for measuring arm strength is the following:

$$\text{Arm strength} = (C + F)\left(\frac{W}{10} + H - 60\right)$$

where C = the number of chin-ups,
P = the number of push-ups,
W = your weight,
H = your height (in inches) and H $\geq$ 60.

If your height is less than 60, replace H − 60 by 0 in the formula.
You and a classmate should compute your respective arm strength. If the two of you arm wrestle, the winner should be the one with the larger arm strength. Try it out!

EXERCISE SET 8.6

For Exercises 1–22, graph the system of linear inequalities.

1. $5x + 4y < 20$
$2x + 7y > 14$

2. $x + y > 3$
$-6x + 5y < 30$

3. $2x + y > 6$
$4x + 5y < 20$

4. $-x + 2y \leq 4$
$6x + 5y \geq 30$

5. $3x + 2y \leq -6$
$-6x + 7y \leq 42$

6. $2x - 5y \leq -10$
$-7x - 5y \geq 35$

7. $y \geq 2x - 10$
$y \leq 3$

8. $2y \leq -3x + 12$
$y \geq -4$

9. $x \geq 0$ and $y \geq 0$

10. $x \geq 5$ and $y \geq 6$

11. $x \leq -2$
    $y \geq 0$

12. $y \geq -1$
    $x \geq 0$

13. $x + y \leq 10$
    $x \geq 0$
    $y \geq 0$

14. $2x + y \leq 8$
    $x \geq 0$
    $y \geq 0$

15. $4x - 3y \geq 12$
    $x \geq 0$ and $y \geq 0$

16. $-2x + 7y \leq 14$
    $x \geq 0$ and $y \geq 0$

17. $0 \le x \le 5$
$0 \le y \le 5$

18. $0 \le x \le 7$
$0 \le y \le 10$

19. $-2 < x < 4$
$-3 < y < -1$

20. $-6 < x < -2$
$-2 < y < 3$

21. $x \le 10$
$y \le 5$
$2y \ge -x + 10$

22. $x \le 12$
$y \le 9$
$4y \ge -3x + 36$

23. Laurie has up to $20,000 available to invest in a combination of two different types of stocks, one a low-risk and the other a high-risk stock. She plans to invest at most $8,000 in the high-risk stock. Let

$x =$ the amount of money invested in the low-risk stock

$y =$ the amount of money invested in the high-risk stock

(a) Express the restrictions on x and y as a system of linear inequalities.
(b) Graph the system of Part (a).

24. Kevin wants to invest up to $10,000 in a combination of treasury bills and municipal bonds. He wants to invest at least $3,000 in treasury bills. Let

x = the amount invested in treasury bills

y = the amount invested in municipal bonds

(a) Express the restrictions on x and y as a system of linear inequalities.
(b) Graph the system of Part (a).

25. Texas Farm Instruments makes two types of power saws, a chain saw and a circular saw. Two machines, A and B, are used in the production of each saw. The number of hours needed on each machine to produce one dozen of each type of saw is given in the following table.

	Chain saw	Circular saw
Machine A	3	4
Machine B	4	2

Each machine is available at most 12 hours each day. Let

x = the number (in dozens) of chain saws made each day

y = the number (in dozens) of circular saws made each day

(a) Express the restrictions on x and y as a system of linear inequalities.

(b) Graph the system of Part (a).

26. Iowa Feed Company guarantees at least 12 pounds of fat and 10 pounds of protein in each bag of its cattle feed. Two ingredients, A and B, are used to make the feed. Ingredient A is 60% fat and 40% protein, and ingredient B is 50% fat and 50% protein. Let

x = the number of pounds of ingredient A in each bag of cattle feed

y = the number of pounds of ingredient B in each bag of cattle feed

(a) Express the restrictions on x and y by a system of linear inequalities.

(b) Graph the system of Part (a).

REVIEW PROBLEMS

The following exercises review parts of Section 3.8. Doing these problems will help prepare you for Section 9.1.

Solve by factoring.

27. $x^2 + 2x - 3 = 0$

28. $x^2 - 6x + 8 = 0$

29. $2x^2 - 3x - 2 = 0$

30. $3x^2 + x - 2 = 0$

31. $-x^2 + 4x - 3 = 0$

32. $-x^2 + 4x + 21 = 0$

33. $-x^2 - 4x + 5 = 0$

34. $-x^2 - 4x + 12 = 0$

35. $-6x^2 + 5x + 6 = 0$

ENRICHMENT EXERCISES

Graph the following systems of inequalities.

1. $2x + 5y \le 25$
$3x + 2y \le 21$
$y \ge 1$
$x \ge 0, \ y \ge 0$

2. $y \le \dfrac{3}{7}x + 4$
$y \ge \dfrac{7}{4}x - \dfrac{21}{4}$
$2x + y \ge 2$
$x \ge 0, \ y \ge 0$

3. $12x + 5y \ge 60$
$4x + 5y \ge 40$
$2x + 5y \ge 30$
$x \ge 0, \ y \ge 0$

Answers to Enrichment Exercises are on page A.53.

CHAPTER 8 **Summary and review**

Examples

Solve the system:

$2x - y = 9$
$-2x - 2y = -3$

Systems of linear equations in two variables (8.1)

One algebraic method for solving a system of equations is the **elimination method.**

Certain systems are also easily solved by the **substitution method,** namely, those systems that either have one of the equations solved for one of the variables or have an equation that can easily be solved for one variable in terms of the other variable.

Examples

By adding the two equations, we can eliminate x.

$$\begin{array}{r} 2x - y = 9 \\ -2x - 2y = -3 \\ \hline -3y = 6 \\ y = -2 \end{array}$$

Replace y by -2 in the first equation of the system and solve for x.

$$\begin{aligned} 2x - (-2) &= 9 \\ 2x + 2 &= 9 \\ 2x &= 7 \\ x &= \frac{7}{2} \end{aligned}$$

The solution is $x = \dfrac{7}{2}$ and $y = -2$.

Applications using systems of equations (8.2)

When solving word problems using two variables, use the following strategy.

STRATEGY

Solving word problems using two variables

Step 1   Read the problem carefully and determine what two unknown values you are to find. Introduce two variables and write what each variable is to represent.

Step 2   Write two equations in the two variables from the given information. If necessary, first organize the information in a table.

Step 3   Solve the system algebraically using either the elimination or the substitution method.

Step 4   Check your answers by using the *original words* of the problem.

Step 5   Be sure that you have answered the question asked in the problem.

Systems of linear equations in three variables (8.3)

A solution of a system of three linear equations in three variables is an ordered triple (x, y, z) of real numbers that satisfy all three equations of the system.

Solve the system:

$$\begin{aligned} 3x + y - 2z &= 1 \\ 4x - y - 5z &= -1 \\ x - 3y - 7z &= 0 \end{aligned}$$

We eliminate y from the first two equations by adding them.

$$\begin{array}{r} 3x + y - 2z = 1 \\ 4x - y - 5z = -1 \\ \hline 7x - 7z = 0 \text{ or } x - z = 0 \end{array}$$

Next, multiply the first equation by 3 and add to the third equation.

$$\begin{array}{r} 9x + 3y - 6z = 3 \\ x - 3y - 7z = 0 \\ \hline 10x - 13z = 3 \end{array}$$

Now, we solve the system

$$\begin{aligned} x - z &= 0 \\ 10x - 13z &= 3 \end{aligned}$$

Examples

Replacing x by z in the second equation,

$$10z - 13z = 3$$
$$-3z = 3$$
$$z = -1$$

Therefore, $x = -1$.

To find y, use the first equation of the original system.

$$3x + y - 2z = 1$$
$$3(-1) + y - 2(-1) = 1$$
$$-3 + y + 2 = 1$$
$$y = 2$$

The solution is $x = -1$, $y = 2$, and $z = -1$.

Cramer's rule (8.4)

Cramer's Rule can be used to solve systems of equations. There are two versions of Cramer's Rule. One is for a system of two equations in two variables and the other for a system of three equations in three variables.

Matrix methods for solving systems of linear equations (8.5)

Matrix methods use row operations to simplify the augmented matrix of a system.

Graph the solution set of the system.

$$-3x + 6y \le 24$$
$$x \ge -4$$

We draw the two half-planes and shade the intersection as shown in the figure.

Systems of linear inequalities (8.6)

Systems of linear inequalities can be graphically visualized by drawing each linear inequality one at a time in the same coordinate plane, then shading the region that is the common intersection of the half-planes.

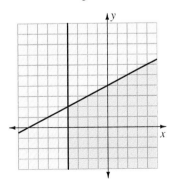

C H A P T E R   8   **R E V I E W   E X E R C I S E   S E T**

Section 8.1

For Exercises 1 and 2, solve each system by the graphing method. If the system is inconsistent, write "no solution." If the system is dependent, write "infinitely many solutions."

1. $2x + y = 4$
$-2x + y = 4$

2. $x + y = 3$
$y = -x + 1$

For Exercises 3–8, solve each system either by the elimination method or the substitution method.

3. $3x - y = 5$
$4x + y = 9$

4. $4x + 3y = -11$
$-3x - 4y = 10$

5. $3x + 2y = 5$
$2x + 3y = 20$

6. $9x - 2y = 0$
$3x + 4y = -7$

7. $y = -6x - 11$
$2x + 3y = 3$

8. $3x - y = 1$
$2x - y = \dfrac{1}{2}$

Section 8.2

9. The sum of two numbers is -16. The difference of twice the first number and the second number is 2. Find the two numbers.

10. The width of a rectangle is one fourth of the length. If the perimeter is 30 inches, find the length and width of the rectangle.

11. A boat can travel 12 miles downstream in $\dfrac{1}{2}$ hour. It takes $\dfrac{3}{4}$ hour to return to the starting point. Find the rate of the boat in still water and the rate of the current.

12. A chemist has two solutions, one containing 20% iodine and the other containing 60% iodine. How much of each solution should be mixed to obtain an 80-liter solution containing 35.5% iodine?

13. The Arkham Company makes and sells x brass beds each week with a total weekly cost C, in dollars, given by

$$C = 200x + 1{,}500$$

The beds are sold for $350 each.
(a) Find the weekly revenue R in terms of x.

(b) Find the break-even point for the company.

Section 8.3

Solve each system.

14. $3x + 5y - 2z = 2$
$x + y + 6z = -3$
$-7x + 3z = 2$

15. $2x + 4y - 7z = -5$
$4x + y - 14z = -3$
$x - 2y = \dfrac{1}{2}$

16. $2x + 2y + 5z = 11$
$2x + y + 4z = 8$
$7x - y + 3z = -5$

17. $5x + y - 8z = 12$
$2x - y + 2z = 1$
$-4x + y + z = -6$

Section 8.4

For Exercises 18–23, find the value of each determinant.

18. $\begin{vmatrix} -1 & 3 \\ 4 & 10 \end{vmatrix}$

19. $\begin{vmatrix} 4 & -\frac{1}{2} \\ 2 & -1 \end{vmatrix}$

20. $\begin{vmatrix} 4 & -\frac{1}{6} \\ -6 & 2 \end{vmatrix}$

21. $\begin{vmatrix} -2 & -1 & -2 \\ 4 & -4 & -3 \\ -3 & -2 & 4 \end{vmatrix}$

22. $\begin{vmatrix} 1 & -2 & 3 \\ 0 & 1 & -4 \\ 0 & 0 & 1 \end{vmatrix}$

23. $\begin{vmatrix} 5 & 1 & -2 \\ 6 & 0 & 4 \\ -7 & 0 & -3 \end{vmatrix}$

For Exercises 24–33, use Cramer's Rule to solve each system.

24. $4x - 7y = -2$
$-8x + 2y = 0$

25. $3x - 3y = -4$
$-6x + 4y = 3$

26. $5x + 3y = 4$
$-15x - 9y = -12$

27. $3x - 8y = -6$
$-\dfrac{3}{2}x + 4y = -4$

28. $5x - 2y + z = 10$
$x + 4y - 3z = 4$
$-3x + 6y - 2z = -8$

29. $-2x - 3y + z = \dfrac{1}{2}$
$-4x - y + 2z = -\dfrac{3}{2}$
$x + y - z = 0$

30. $x - 2y + z = -1$
$2x + y - z = 2$
$-x + 3y - 4z = 0$

31. $-x + y - 3z = 0$
$3x - 2y - 3z = -2$
$x - y - 2z = -1$

32. $x - y = 0$
$x + 2z = 1$
$-y + z = -1$

33. $y - 3z = -2$
$x + z = 0$
$-x - 3y = 1$

Section 8.5

For Exercises 34–39, use matrix methods to solve the system.

34. $x - 3y = -5$
$3x + 2y = -4$

35. $2x + 6y = 3$
$4x - 3y = 1$

36. $x - 2y + 3z = 7$
$2x + 3y - z = -7$
$4y + 3z = 2$

37. $x + 3y - 2z = -3$
$-x - y + 3z = -7$
$2x - 2y - z = 6$

38. $3x - 2y + 5z = 2$
$2x + 3y - 2z = -1$
$x - y + z = 3$

39. $x - y + 3z = 0$
$2x - 4y + 6z = 1$
$-2x + 2y + 3z = -3$

Section 8.6

For Exercises 40–45, graph the system of linear inequalities.

40. $x - 2y \leq -2$
$x + 2y \leq 6$

41. $x > -2$
$y < 3$

42. $x - y < 2$
$x + y > -2$

43. $y + x \geq 0$
$y - x \leq 0$

44. $5x - 3y \geq -5$
$0 \leq y \leq 5$ and $x \geq 0$

45. $3x + 4y \leq 12$
$x \geq 0$ and $y \geq 0$

CHAPTER 8   **T E S T**

Section 8.1

For problems 1–4, solve the system by either the elimination method or the substitution method.

1. $2x - 3y = 5$
$x + y = 0$

2. $4x - 3y = -3$
$-2x + 9y = 4$

3. $2x + 3y = -11$
  $3x - 4y = -8$

4. $5x - 3y = -2$
  $y = 10x - \dfrac{8}{3}$

Section 8.2

5. The sum of two numbers is -5. The difference of four times the first number and twice the second number is 1. Find the two numbers.

6. Jeff has two solutions of antifreeze, one containing 20% alcohol and the other containing 12% alcohol. How many quarts of each should he mix to obtain 16 quarts of antifreeze containing 15% alcohol?

Section 8.3

For problems 7–9, solve each system.

7. $x - 2y + 3z = 9$
  $-2x + y - 2z = -7$
  $x + y + z = 2$

8. $3x - 2y + z = -1$
  $x - y + 2z = -5$
  $x + 3y - z = 0$

9. $x - 3y + z = -1$
  $2x + 3y = 2$
  $2x + 4z = -1$

Section 8.4

For problems 10–12, find the value of each determinant.

10. $\begin{vmatrix} 2 & 3 \\ 4 & 1 \end{vmatrix}$

11. $\begin{vmatrix} 2 & -3 & 3 \\ -1 & 2 & 0 \\ 5 & -6 & 0 \end{vmatrix}$

12. $\begin{vmatrix} 3 & -1 & 4 \\ 4 & 2 & -3 \\ 1 & 0 & -1 \end{vmatrix}$

For problems 13–15, use Cramer's Rule to solve each system.

13. $3x - 2y = 5$
  $2x + 3y = -1$

14. $4x - 5y = 0$
  $-6x + 15y = \dfrac{3}{2}$

15. $2x - 3y - z = 1$
  $6y - 2z = -5$
  $x + y + z = -1$

Section 8.5

For problems 16–18, use matrix methods to solve each system.

16. $x - 2y = 1$
  $3x + 10y = -1$

17. $2x - 4y = 1$
  $-x + 2y = 2$

18. $x - 3y + 2z = -3$
  $-x + 4y - 3z = 6$
  $3x - 2y + 2z = 0$

Section 8.6

For problems 19 and 20, graph the system of linear inequalities.

19. $-2x + 5y \le 10$
  $3x + 5y \ge -15$

20. $4x + 7y \le 28$
  $x \ge 0$ and $y \ge 0$

Conic Sections

Overview

In this chapter, we study four curves called conic sections. They are called conic sections because geometrically, they each can be formed by the inter*section* of a plane and a double *cone* as shown in the following figures.

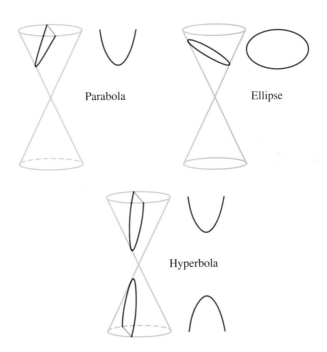

Parabola

Ellipse

Hyperbola

We will not be using these geometric origins to study conic sections; instead we will investigate conic sections by means of their equations and graphs in a coordinate plane. There are plentiful applications of conic sections and their related equations. In astronomy, the planets in our solar system, as well as Halley's comet, circle the sun in elliptical orbits. Some comets travel into our system in parabolic or hyperbolic orbits never to return again. Parabolas have reflective properties that are used to collect television signals from a communication satellite.

9.1

Parabolas

OBJECTIVES

▶ To graph parabolas of the form $y = x^2 + k$

▶ To graph parabolas of the form $y = ax^2$

▶ To graph parabolas of the form $y = (x - h)^2$

▶ To graph parabolas of the form $y = a(x - h)^2 + k$

▶ To write $y = ax^2 + bx + c$ in standard form by completing the square

So far, we have studied equations in two variables that were linear. In this section, we start the study of second-degree equations. One of the most basic of second-degree equations is $y = x^2$. The solution set of this equation will consist of ordered pairs (x, y) whose coordinates satisfy the equation. We generate several members of this solution set by letting x take on some convenient values, then finding the corresponding value of y from the equation.

Choose x	Determine y from $y = x^2$	(x, y)
−4	$y = (-4)^2 = 16$	(−4, 16)
−3	$y = (-3)^2 = 9$	(−3, 9)
−2	$y = (-2)^2 = 4$	(−2, 4)
−1	$y = (-1)^2 = 1$	(−1, 1)
0	$y = (0)^2 = 0$	(0, 0)
1	$y = (1)^2 = 1$	(1, 1)
2	$y = (2)^2 = 4$	(2, 4)
3	$y = (3)^2 = 9$	(3, 9)
4	$y = (4)^2 = 16$	(4, 16)

We plot these solutions in a coordinate plane and draw a smooth curve connecting them to obtain the graph of $y = x^2$ as shown in the following figure.

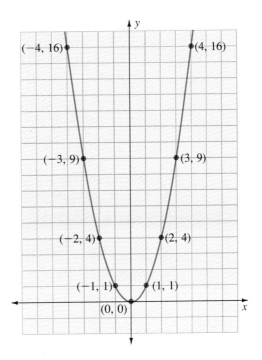

This graph is called a **parabola.** The point $(0, 0)$ that is the lowest point is called the **vertex** of this parabola. The vertical line through the vertex is called the **axis of symmetry.** Every point, except the vertex, corresponds to another point that is the reflection in the axis of symmetry. For example, the following pairs of points are reflections of each other: $(-1, 1)$ and $(1, 1)$; $(-2, 4)$ and $(2, 4)$; $(-3, 9)$ and $(3, 9)$; and $(-4, 16)$ and $(4, 16)$.

In general, we have the following statement:

All equations in two variables of the form

$$y = ax^2 + bx + c, \qquad a \neq 0$$

have graphs that are parabolas.

Notice that $y = x^2$ is of the form $y = ax^2 + bx + c$, where $a = 1$ and b and c are both zero.

To graph a parabola, we must be able to find the vertex and the axis of symmetry and determine if the parabola opens upward or downward.

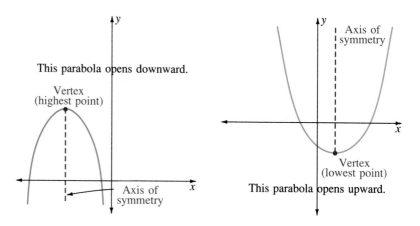

We can obtain this information by putting the general equation $y = ax^2 + bx + c$ into one of the four basic forms:

1. $y = x^2 + k$

2. $y = ax^2$

3. $y = (x - h)^2$

4. $y = a(x - h)^2 + k$

Furthermore, we can graph each of these four basic forms by comparing it to the parabola $y = x^2$. We start our graphing by considering equations of the form $y = x^2 + k$.

Example 1   Draw the graph of $y = x^2 + 1$. Compare it to the graph of $y = x^2$.

Solution   We choose convenient values for x, then find the corresponding values for y.

Choice of x	Determine $y = x^2 + 1$	(x, y)
−4	$y = (-4)^2 + 1 = 17$	$(-4, 17)$
−3	$y = (-3)^2 + 1 = 10$	$(-3, 10)$
−2	$y = (-2)^2 + 1 = 5$	$(-2, 5)$
−1	$y = (-1)^2 + 1 = 2$	$(-1, 2)$
0	$y = (0)^2 + 1 = 1$	$(0, 1)$
1	$y = (1)^2 + 1 = 2$	$(1, 2)$
2	$y = (2)^2 + 1 = 5$	$(2, 5)$
3	$y = (3)^2 + 1 = 10$	$(3, 10)$
4	$y = (4)^2 + 1 = 17$	$(4, 17)$

Next, we plot these ordered pairs in a coordinate plane and draw a smooth curve through them.

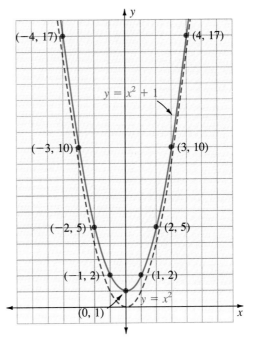

Notice that the parabola can be obtained by moving the parabola $y = x^2$ upward by 1 unit. This is because the y values for $y = x^2 + 1$ are one larger than the corresponding y values of $y = x^2$. ▲

Example 2   Draw the graph of $y = x^2 - 4$. Compare this graph to the graph of $y = x^2$.

Solution   We choose convenient values for x, then find the corresponding values of y.

Choice of x	Determine y from $y = x^2 - 4$	(x, y)
-3	$y = (-3)^2 - 4 = 5$	$(-3, 5)$
-2	$y = (-2)^2 - 4 = 0$	$(-2, 0)$
-1	$y = (-1)^2 - 4 = -3$	$(-1, -3)$
0	$y = (0)^2 - 4 = -4$	$(0, -4)$
1	$y = (1)^2 - 4 = -3$	$(1, -3)$
2	$y = (2)^2 - 4 = 0$	$(2, 0)$
3	$y = (3)^2 - 4 = 5$	$(3, 5)$

We plot the points from the preceding table and draw a smooth curve through them as shown in the following figure. Notice that the resulting parabola can be obtained by moving the parabola $y = x^2$ down 4 units. This is because the y values of $y = x^2 - 4$ are 4 less than the corresponding y values of $y = x^2$.

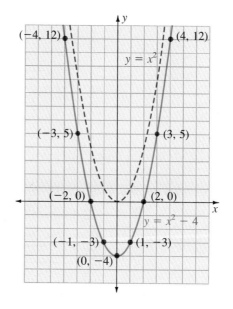

In general, we make the following observation:

The graph of $y = x^2 + k$ is the same as the graph of $y = x^2$, but it is moved up or down k units depending upon whether k is positive or negative.

Now, we look at equations of the form $y = ax^2$, where $a \neq 0$.

Example 3   Draw the graph of $y = 2x^2$. Compare it with the graph of $y = x^2$.

Solution   We again make a table of values.

Choice of x	Determine y from $y = 2x^2$	(x, y)
−2	$y = 2(-2)^2 = 8$	$(-2, 8)$
−1	$y = 2(-1)^2 = 2$	$(-1, 2)$
0	$y = 2(0)^2 = 0$	$(0, 0)$
1	$y = 2(1)^2 = 2$	$(1, 2)$
2	$y = 2(2)^2 = 8$	$(2, 8)$

Plotting the ordered pairs (x, y), we draw a smooth curve through them as shown in the following figure. Notice that this parabola has the same vertex, but it is narrower than the graph of $y = x^2$.

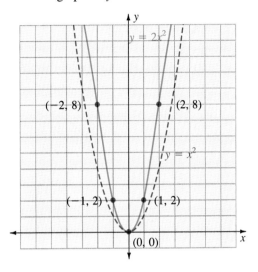

Example 4   Draw the graph of $y = -\frac{1}{2}x^2$. Compare it to the graph of $y = x^2$.

Solution   To graph this parabola, we again make a table.

Choice of x	Determine y from $y = -\frac{1}{2}x^2$	(x, y)
-3	$y = -\frac{1}{2}(-3)^2 = -\frac{9}{2}$	$\left(-3, -\frac{9}{2}\right)$
-2	$y = -\frac{1}{2}(-2)^2 = -2$	$(-2, -2)$
-1	$y = -\frac{1}{2}(-1)^2 = -\frac{1}{2}$	$\left(-1, -\frac{1}{2}\right)$
0	$y = -\frac{1}{2}(0)^2 = 0$	$(0, 0)$
1	$y = -\frac{1}{2}(1)^2 = -\frac{1}{2}$	$\left(1, -\frac{1}{2}\right)$
2	$y = -\frac{1}{2}(2)^2 = -2$	$(2, -2)$
3	$y = -\frac{1}{2}(3)^2 = -\frac{9}{2}$	$\left(3, -\frac{9}{2}\right)$

The graph is shown in the following figure. Notice that the parabola opens downward and has the same vertex, but it is wider than the graph of $y = x^2$.

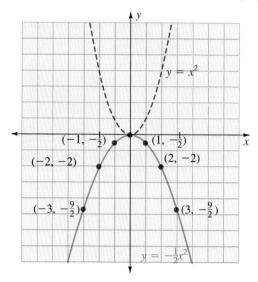

In general, we can make the following statement about $y = ax^2$:

The graph of $y = ax^2$ opens upward if $a > 0$ and opens downward if $a < 0$. Furthermore, it has the same vertex as $y = x^2$, but it is narrower if $|a| > 1$ and wider if $|a| < 1$.

The equation $y = (x - 2)^2$ can be put in the form $y = ax^2 + bx + c$, since squaring yields $y = x^2 - 4x + 4$. We now investigate the graphs of equations of the general form $y = (x - h)^2$.

Example 5   Draw the graph of $y = (x - 2)^2$. Compare it to the graph of $y = x^2$.

Solution   We choose convenient values for x and find the corresponding values for y.

Choice of x	Find y from $y = (x - 2)^2$	(x, y)
-1	$y = (-1 - 2)^2 = 9$	$(-1, 9)$
0	$y = (0 - 2)^2 = 4$	$(0, 4)$
1	$y = (1 - 2)^2 = 1$	$(1, 1)$
2	$y = (2 - 2)^2 = 0$	$(2, 0)$
3	$y = (3 - 2)^2 = 1$	$(3, 1)$
4	$y = (4 - 2)^2 = 4$	$(4, 4)$
5	$y = (5 - 2)^2 = 9$	$(5, 9)$

We plot the ordered pairs determined in the table and draw a smooth curve connecting them. Notice that this parabola is the same curve as $y = x^2$ but it is moved 2 units to the right. That is, the vertex of $y = (x - 2)^2$ is at the point $(2, 0)$ and the axis of symmetry is given by $x = 2$.

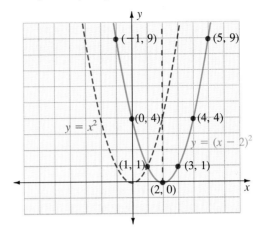

We can generalize the development shown in Example 5.

The graph of $y = (x - h)^2$ is a parabola with its vertex at $(h, 0)$. It is the parabola $y = x^2$ moved to the right $|h|$ units if h is positive or moved to the left $|h|$ units if h is negative.

Example 6

Draw the graph of $y = (x + 3)^2$.

Solution

We can rewrite the equation as $y = [x - (-3)]^2$. Therefore, the graph is a parabola with its vertex at $(-3, 0)$. We can think of it as the parabola $y = x^2$ moved to the left $|-3| = 3$ units. The desired graph is shown in the following figure.

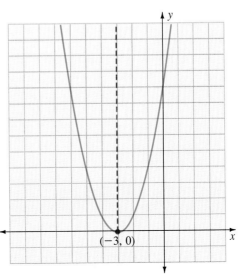

The three types of parabolas that we have graphed so far are shown in the following diagram.

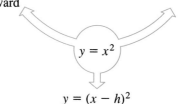

$$y = ax^2$$

changes the width:

$a > 0$, opens upward

$a < 0$, opens downward

$$y = x^2 + k$$

moves the parabola up or down $|k|$ units

$y = x^2$

$$y = (x - h)^2$$

moves the parabola right or left $|h|$ units

In the next example, we use the results obtained so far and put them together to graph equations of the form $y = a(x - h)^2 + k$.

Example 7   Draw the graph of $y = \dfrac{1}{2}(x - 3)^2 - 4$.

Solution

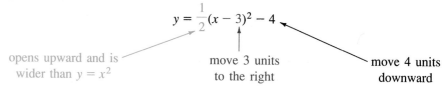

$$y = \frac{1}{2}(x - 3)^2 - 4$$

opens upward and is wider than $y = x^2$

move 3 units to the right

move 4 units downward

Observe that the vertex of the parabola is at $(3, -4)$, and, therefore, the axis of symmetry is $x = 3$. Since the number a is $\dfrac{1}{2}$, the parabola opens upward and is wider than $y = x^2$. We plot a few points, reflect them in the axis of symmetry to obtain more points, and then draw the graph. (See the figure.)

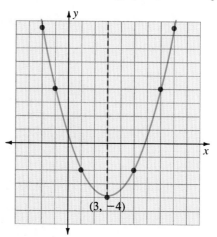

$(3, -4)$

Example 8   Draw the graph of $y = -2(x + 2)^2 + 5$.

Solution

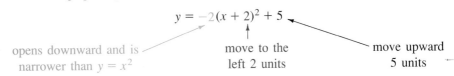

$$y = -2(x + 2)^2 + 5$$

opens downward and is
narrower than $y = x^2$

move to the
left 2 units

move upward
5 units

We plot the vertex $(-2, 5)$ and draw the axis of symmetry $x = -2$. Next, we plot a few points and reflect them in the axis of symmetry to obtain more points, then draw the parabola. (See the figure below.)

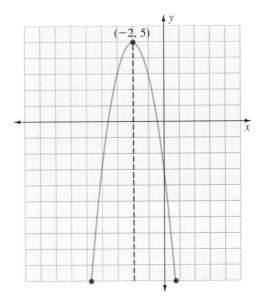

Notice that from the equation $y = a(x - h)^2 + k$, we can "read off" the coordinates of the vertex of the parabola. Namely, the vertex is located at (h, k). With this in mind, we have a strategy for graphing parabolas.

STRATEGY

Drawing the graph of $y = a(x - h)^2 + k$

Step 1   The vertex is located at (h, k) and the axis of symmetry is $x = h$. Plot the point (h, k) and draw a vertical dashed line through that point for the axis of symmetry.

Step 2   Plot two additional points by choosing values of x on each side of the axis of symmetry.

Step 3   Draw a smooth parabolic curve through the vertex and the two points found in Step 2. Keep in mind that the parabola opens upward if $a > 0$ and opens downward if $a < 0$.

We now consider drawing the graph of $y = ax^2 + bx + c$, where $a \neq 0$. We can rewrite this equation in the form $y = a(x - h)^2 + k$ by *completing the square*. The details are shown in the next example.

Example 9   Draw the graph of $y = x^2 + 2x - 2$.

Solution   We complete the square on the binomial $x^2 + 2x$. The square of one-half the coefficient of x is $\left(\dfrac{2}{2}\right)^2 = 1$.

$$y = x^2 + 2x - 2$$
$$y = x^2 + 2x + 1 - 2 - 1 \qquad \text{Add and subtract 1.}$$
$$y = (x + 1)^2 - 3 \qquad\qquad x^2 + 2x + 1 = (x + 1)^2.$$

The graph of $y = (x + 1)^2 - 3$ is the parabola shown in the following figure.

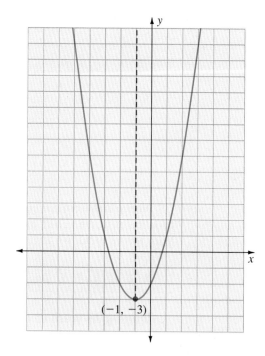

$(-1, -3)$

If the coefficient of x^2 is not 1, then it must be factored out before completing the square. This technique is shown in the next two examples.

Example 10   Draw the graph of $y = 2x^2 - 12x + 16$.

Solution   Since the coefficient of x^2 is 2, not 1, we first factor it from the first two terms on the right side before completing the square.

$$y = 2x^2 - 12x + 16$$

$$y = 2(x^2 - 6x) + 16$$

$$y = 2(x^2 - 6x + 9) + 16 - 18$$ Add 9 within the parentheses to complete the square; subtract 18 since 9 had been added inside the parentheses *times the factor 2.*

$$y = 2(x - 3)^2 - 2$$

The equation is now in standard form; its graph is shown in the following figure.

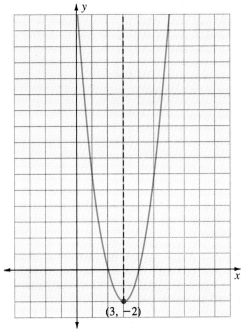

(3, −2)

Example 11

Draw the graph of $y = -3x^2 - 6x + 2$.

Solution

The coefficient of x^2 is -3, so we factor -3 from the first two terms before completing the square.

$$y = -3x^2 - 6x + 2$$

$$y = -3(x^2 + 2x) + 2$$

$$y = -3(x^2 + 2x + 1) + 2 + 3$$

Add 1 inside the parentheses to complete the square.

Add 3, since we added 1, *multiplied by a factor of* -3.

$$y = -3(x + 1)^2 + 5$$

The equation is now in the standard form. Its graph is shown in the figure below.

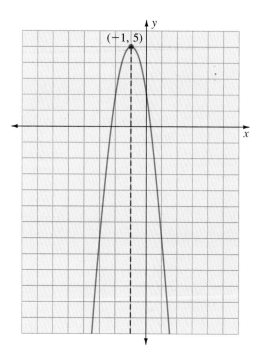

An equation of the form $x = a(y - k)^2 + h$ is a parabola with its vertex at (h, k) and opens to the right if $a > 0$ and opens to the left if $a < 0$. In either case, the axis of symmetry is the horizontal line given by $y = k$. See the following figures.

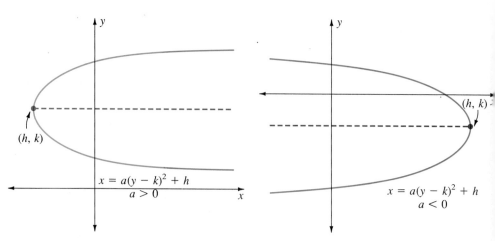

Example 12   Draw the graph of $x = -2(y + 2)^2 + 2$.

Solution   Since $a = -2$ is negative, the parabola opens to the left. The vertex is $(2, -2)$, with the axis of symmetry given by $y = -2$. Setting $y = -1$, $y = 0$, and $y = 1$ in the given equation, we obtain the three points $(0, -1)$, $(-6, 0)$, $(-16, 1)$. We plot these points, reflect them in the axis of symmetry, and draw the graph as shown in the following figure.

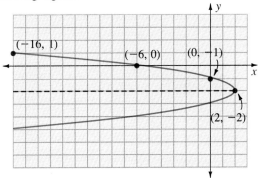

E X E R C I S E   S E T   9.1

For Exercises 1–50, use the graphing strategy on page 627 to draw the graphs of each equation.

1. $y = x^2 - 3$

2. $y = x^2 + 2$

3. $y = -x^2 + 2$

4. $y = -x^2 - 1$

5. $y = 3x^2$

6. $y = \dfrac{1}{2}x^2$

7. $y = -\dfrac{1}{4}x^2$

8. $y = -2x^2$

9. $y = (x - 4)^2$

10. $y = (x - 1)^2$

11. $y = -(x + 2)^2$

12. $y = -(x - 6)^2$

13. $y = -2(x + 1)^2$

14. $y = \dfrac{1}{4}(x - 3)^2$

15. $y = \dfrac{1}{4}x^2 + \dfrac{1}{4}$

16. $y = 2x^2 - 4$

17. $y = -2x^2 + 5$

18. $y = -\dfrac{1}{2}x^2 + \dfrac{1}{2}$

19. $y = (x - 2)^2 + 1$

20. $y = (x - 1)^2 - 1$

21. $y = (x + 3)^2 - 5$

22. $y = (x + 6)^2 + 1$

23. $y = -(x + 1)^2 + 3$

24. $y = -(x - 3)^2 + 2$

25. $y = -(x + 4)^2 + 6$

26. $y = -(x - 3)^2 - 1$

27. $y = x^2 - 4x$

28. $y = x^2 - 6x + 8$

29. $y = -x^2 - 2x - 2$

30. $y = -x^2 - 4x - 5$

31. $y = 2x^2 - 4x - 6$

32. $y = 3x^2 - 12x + 15$

33. $y = -3x^2 - 6x - 8$

34. $y = -2x^2 - 8x - 7$

35. $y = x^2 + 4x - 7$

36. $y = x^2 - 2x - 5$

37. $y = -4x^2 + 24x - 24$

38. $y = -3x^2 + 12x + 3$

39. $y = x^2 + 6x + 10$

40. $y = x^2 - 4x + 6$

41. $y = -3x^2 + 6x - 4$

42. $y = -2x^2 - 8x - 3$

43. $x = y^2$

44. $x = -y^2$

45. $x = 2y^2$

46. $x = -\dfrac{1}{4}y^2$

47. $x = (y - 1)^2$

48. $x = (y + 2)^2$

49. $x = 2(y + 3)^2 - 2$

50. $x = -\dfrac{1}{2}(y - 4)^2 + 5$

REVIEW PROBLEMS

The following exercises review parts of Section 6.3. Doing these problems will help prepare you for the next section.

Solve each equation.

51. $(x + 2)(x + 3) = 0$

52. $(2x - 3)(x - 4) = 0$

53. $(x - 1)^2 - 4 = 0$

54. $2(x + 2)^2 - 32 = 0$

55. $6x^2 - x - 2 = 0$

56. $3x^2 - 5x - 2 = 0$

57. $(x - 5)^2 - 8 = 0$

58. $x^2 - 2x - 12 = 0$

E N R I C H M E N T   E X E R C I S E S

For Exercises 1–4, draw the graph of the equation.

1. $x = 2y^2 - 4y - 1$

2. $x = -3y^2 - 6y + 1$

3. $y + x^2 + 3 = 4x$

4. $y = \dfrac{x^4 - x^2}{x^2 - 1}$

5. Determine the number a so that $(4, -2)$ lies on the graph of $y = ax^2$.

6. Determine the number k so that $(-\sqrt{5}, 5)$ lies on the graph of $y = x^2 + k$.

7. Find an equation of the parabola from the given information:
1. It opens upward;
2. it has the same width as $y = x^2$;
3. its axis of symmetry is $x = 3$; and
4. it contains the point $(2, 7)$.

8. Suppose that (x_1, y_0) and (x_2, y_0) with $x_1 \neq x_2$ are two points on the graph of $y = a(x - h)^2 + k$, $a \neq 0$. Show that

$$\frac{x_1 + x_2}{2} = h$$

Answers to Enrichment Exercises are on page A.57.

9.2

More Parabolas

OBJECTIVES

▶ *To find the intercepts of a parabola*

▶ *To find the highest or lowest point of a parabola*

▶ *To use parabolas in applications*

In this section, we consider other features that parabolas possess. In the last section, we developed the method of completing the square to rewrite the quadratic equation $y = ax^2 + bx + c$ as $y = a(x - h)^2 + k$. In this second form, we could "read off" the vertex, (h, k). Furthermore, if $a > 0$, the parabola opened upward, and if $a < 0$, the parabola opened downward. In this section, we shall show how to find the x-coordinate of the vertex directly from the original equation $y = ax^2 + bx + c$. Before we do this, let us look at the concept of intercepts as they apply to parabolas.

D E F I N I T I O N

The **y-intercept** of the parabola is the y-coordinate of the point where the parabola crosses the y-axis.

The **x-intercepts** of the parabola, if they exist, are the x-coordinate(s) of the point(s) where the parabola crosses the x-axis.

Now, from the general equation $y = ax^2 + bx + c$ for a parabola, the y-intercept is c, since the parabola crosses the y-axis where $x = 0$, and $y = (0)^2 + b(0) + c = c$.

The x-intercepts occur when $y = 0$, and, therefore, the x-intercepts are the real solutions of

$$ax^2 + bx + c = 0$$

Recall that a quadratic equation may have two real solutions, one real solution, or no real solution. This reflects the fact that a parabola may have two x-intercepts, one x-intercept, or no x-intercepts. (See the figures.)

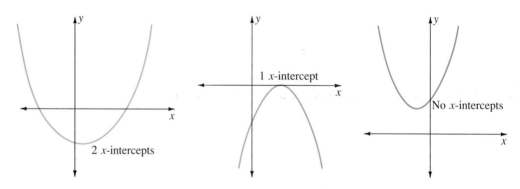

Example 1   Find the intercepts of the parabola given by

$$y = x^2 - 4x + 3$$

Solution   Setting $x = 0$, the y-intercept is

$$y = 0^2 - 4(0) + 3 = 3$$

To find the x-intercepts, we set $x^2 - 4x + 3$ equal to zero and solve the quadratic equation.

$$x^2 - 4x + 3 = 0$$

$$(x - 1)(x - 3) = 0$$

Therefore, the x-intercepts are $x = 1$ and $x = 3$.   ▲

Example 2 Find the x-intercepts, if they exist, of

$$y = 2x^2 - 12x + 2$$

Solution To find the x-intercepts, set

$$2x^2 - 12x + 2 = 0$$

$$x^2 - 6x + 1 = 0 \qquad \text{Divide by 2.}$$

Since the left side does not factor, we use the quadratic formula.

$$x = \frac{-b \pm \sqrt{b^2 - 4ac}}{2a}$$

$$= \frac{6 \pm \sqrt{(-6)^2 - 4(1)(1)}}{2(1)}$$

$$= \frac{6 \pm \sqrt{32}}{2}$$

$$= \frac{6 \pm \sqrt{16 \cdot 2}}{2}$$

$$= \frac{6 \pm 4\sqrt{2}}{2}$$

$$= 3 \pm 2\sqrt{2}$$

The x-intercepts are $3 + 2\sqrt{2}$ and $3 - 2\sqrt{2}$. ▲

Example 3 Find the x-intercepts, if they exist, of

$$y = -x^2 - 2x - 2$$

Solution The x-intercepts, if they exist, are the solutions of $-x^2 - 2x - 2 = 0$. Multiplying both sides by -1,

$$x^2 + 2x + 2 = 0$$

Since the left side is not factorable, we use the quadratic formula.

$$x = \frac{-2 \pm \sqrt{2^2 - 4(1)(2)}}{2(1)}$$

$$= \frac{-2 \pm \sqrt{-4}}{2}$$

Since the $\sqrt{-4}$ appears in the quadratic formula, we know that $x^2 + 2x + 2 = 0$ has no real solution. Therefore, the parabola has no x-intercepts, which means that it does not cross the x-axis. ▲

From the last section, recall that if the quadratic equation $y = ax^2 + bx + c$ is put into the form $y = a(x - h)^2 + k$, then the vertex is the point (h, k). In particular, the x-coordinate of the vertex is h.

THEOREM

Consider the graph of the equation $y = ax^2 + bx + c$:

1. The x-coordinate of the vertex of the graph is $x = -\dfrac{b}{2a}$.

2. If $a > 0$, the graph opens upward.
3. If $a < 0$, the graph opens downward.

Proof   We complete the square on the right side of the equation.

$$y = ax^2 + bx + c$$

$$y = a\left(x^2 + \frac{b}{a}x\right) + c \qquad \text{Factor } a \text{ from the first two terms.}$$

Now, to complete the square, take $\dfrac{1}{2}$ of $\dfrac{b}{a}$, then square:

$$\left(\frac{1}{2}\cdot\frac{b}{a}\right)^2 = \left(\frac{b}{2a}\right)^2 = \frac{b^2}{(2a)^2} = \frac{b^2}{4a^2}$$

Therefore, add $\dfrac{b^2}{4a^2}$ inside the parentheses to make a perfect square trinomial. To balance this, we subtract.

$$a\left(\frac{b^2}{4a^2}\right) = \frac{b^2}{4a}$$

$$y = a\left(x^2 + \frac{b}{a}x + \frac{b^2}{4a^2}\right) + c - \frac{b^2}{4a}$$

$$y = a\left(x + \frac{b}{2a}\right)^2 + \frac{4ac - b^2}{4a}$$

Comparing this last equation with $y = a(x - h)^2 + k$, we conclude that:

1. $h = -\dfrac{b}{2a}$. That is, the x-coordinate of the vertex is $-\dfrac{b}{2a}$.

2. If $a > 0$, the graph opens upward.
3. If $a < 0$, the graph opens downward.

We can use this theorem along with finding the intercepts to draw the graph of the equation $y = ax^2 + bx + c$.

Example 4   Find the coordinates of the vertex of the parabola given by $y = -2x^2 + 4x + 6$. Then, graph the parabola using the intercepts.

Solution   The x-coordinate of the vertex is

$$x = -\frac{b}{2a} = -\frac{4}{2(-2)} = 1$$

To find the y-coordinate, replace x by 1 in the original equation.

$$y = -2x^2 + 4x + 6$$

$$y = -2(1)^2 + 4(1) + 6 = 8$$

Therefore, the vertex is $(1, 8)$.

The y-intercept is 6 and the x-intercepts, if they exist, are solutions of $-2x^2 + 4x + 6 = 0$. To solve this quadratic equation, first divide both sides by -2.

$$x^2 - 2x - 3 = 0$$

$$(x - 3)(x + 1) = 0$$

The x-intercepts are 3 and -1.

Since the coefficient of x^2 is negative, the parabola opens downward. To draw the graph, we plot the vertex, along with $(0, 6)$, $(-1, 0)$, and $(3, 0)$. Using the axis of symmetry as our guide, we draw the parabola as shown in the following figure.

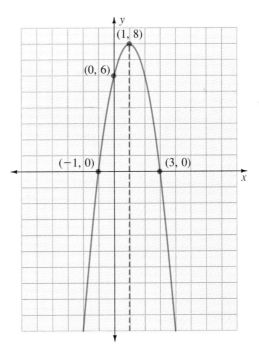

N O T E   *In the last example, the x-intercepts were used to graph the parabola. However, if the parabola has no x-intercepts, then use other points for graphing.*

Many problems in mathematics involve finding the largest or smallest value of a variable. If we are given the equation $y = ax^2 + bx + c$, then y will have either the smallest value or the largest value when $x = -\dfrac{b}{2a}$, as shown in the figure.

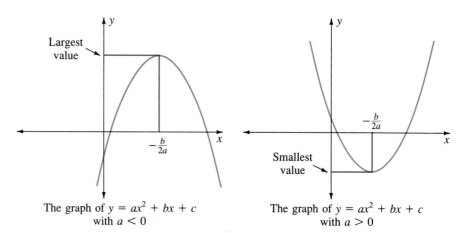

The graph of $y = ax^2 + bx + c$
with $a < 0$

The graph of $y = ax^2 + bx + c$
with $a > 0$

In either case, we say that y has an **extreme value** at $x = -\dfrac{b}{2a}$. We summarize these concepts.

R U L E

Given $y = ax^2 + bx + c$, where $a \neq 0$. Then y has an extreme value. This extreme value occurs when $x = -\dfrac{b}{2a}$. Furthermore,

1. If $a > 0$, then y has a smallest value at $x = -\dfrac{b}{2a}$.

2. If $a < 0$, then y has a largest value at $x = -\dfrac{b}{2a}$.

Example 5   Given: $y = -2x^2 + 16x - 15$.

(a) Show that y has a largest value.

(b) Find the x-coordinate where this largest value for y occurs.

(c) What is the largest value?

Solution   **(a)** Since $a = -2$ is negative, y has a largest value.

(b) This largest value occurs when

$$x = -\frac{b}{2a}$$

$$= -\frac{16}{2(-2)}$$

$$= 4$$

(c) To find the largest value of y, replace x by 4 in the equation.

$$y = -2x^2 + 16x - 15$$

$$= -2(4)^2 + 16(4) - 15$$

$$= -32 + 64 - 15$$

$$= 17 \qquad \blacktriangle$$

Example 6   The resistance R, in ohms, of an electrical resistor is given by $R = 3t^2 - 15t + 65$, where t is the temperature. At what temperature is the resistance the lowest? What is the resistance at this temperature?

Solution   This equation fits the standard form with $a > 0$. Therefore, we know that R will be the lowest when $t = -\frac{b}{2a}$. Reading the numbers from the equation, $a = 3$ and $b = -15$. Therefore, the lowest resistance occurs when $t = -\frac{-15}{2 \cdot 3} = \frac{5}{2}$ or 2.5° above zero. At that temperature, the resistor has a resistance of

$$R = 3\left(\frac{5}{2}\right)^2 - 15\left(\frac{5}{2}\right) + 65 = 3\left(\frac{25}{4}\right) - \frac{75}{2} + 65$$

$$= \frac{75}{4} - \frac{150}{4} + \frac{260}{4}$$

$$= \frac{185}{4} = 46.25 \text{ ohms} \qquad \blacktriangle$$

EXERCISE SET 9.2

For Exercises 1–14, find the intercepts of the parabola.

1. $y = x^2 + x - 6$

2. $y = x^2 + 3x - 10$

3. $y = 2x^2 + 7x - 4$

4. $y = 10x^2 - 11x - 6$

5. $y = -2x^2 + 6x$

6. $y = x^2 - 2x$

7. $y = 4x^2 + 4x + 1$

8. $y = 9x^2 - 12x + 4$

9. $y = x^2 - 2$

10. $y = -3x^2 + 9$

11. $y = x^2 - 4x + 1$

12. $y = x^2 + 2x - 1$

13. $y = x^2 + 2x + 2$

14. $y = 4x^2 - 12x + 10$

For Exercises 15–20, find the coordinates of the vertex of the parabola. Then draw the graph using the intercepts.

15. $y = x^2 + x - 12$

16. $y = x^2 + 3x - 10$

17. $y = -x^2 - x + 2$

18. $y = -x^2 - x + 6$

19. $y = 2x^2 + 11x + 12$

20. $y = 3x^2 - 2x - 8$

For Exercises 21–28, (a) Determine if y has a largest or smallest value. (b) Find the value of x at which this extreme value occurs. (c) Find the extreme value.

21. $y = 0.1x^2 - 2.4x + 4.6$

22. $y = -4x^2 + 20x - 15$

23. $y = 14 - 4x - 4x^2$

24. $y = 25 - 40x + 10x^2$

25. $y = (1 - 4x)(3 + 2x)$

26. $y = (5x - 3)(x - 1)$

27. $y = \dfrac{x^2}{4} - \dfrac{3}{2}x - 7$

28. $y = \dfrac{3x^2}{8} + \dfrac{3}{4}x + \dfrac{1}{8}$

29. The Mellon Patch Company makes primitive dolls. The daily cost C, in dollars, of making x dozen dolls is given by

$$C = 20x^2 - 120x + 400$$

How many dolls should be made each day to have the lowest cost? What is the lowest cost?

30. A crafts store makes sandals from natural palm leaves. It has a daily cost C, in dollars, from producing x dozen pairs of sandals given by

$$C = 10x^2 - 180x + 1{,}200$$

How many sandals should be made each day to obtain the lowest cost? What is the lowest cost?

31. A projectile is fired vertically upward. The height, h (in feet), that it is above the ground after t seconds is given by

$$h = 128t - 16t^2$$

When will the projectile reach its maximum height? What is the maximum height?

32. A projectile is shot vertically upward so that the height, h (in meters), above the ground after t seconds is given by

$$h = 80t - 5t^2$$

When will the projectile reach its maximum height? What is the maximum height?

33. Mr. Koehler plans to build a pig pen next to his barn using 72 feet of fencing for three sides and the barn wall on the fourth side (see the figure). If A is the area of the pen, then

$$A = x(72 - 2x)$$

Find the dimensions of the pen that will contain the largest area.

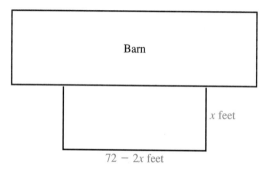

34. A rectangular field adjacent to a river is to be enclosed with 124 meters of fencing on three sides and using the river as the fourth side (see the figure). Let A be the area of the field. Then

$$A = x(124 - 2x)$$

Find the dimensions of the field that contains the largest area.

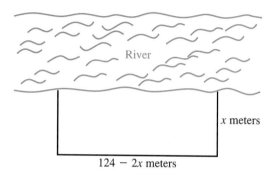

REVIEW PROBLEMS

The following exercises review parts of Section 5.4. Doing these problems will help prepare you for the next section.

Simplify.

35. $\sqrt{24 + 36}$

36. $\sqrt{80 + 20}$

37. $\sqrt{(4 - 2)^2 + (5 - 3)^2}$

38. $\sqrt{(1 - 5)^2 + (9 - 12)^2}$

39. $\sqrt{(15 - 8)^2 + (6 - 7)^2}$

40. $\sqrt{(2\sqrt{3} - 4\sqrt{3})^2 + (\sqrt{3} - 5\sqrt{3})^2}$

41. $\sqrt{\left(\dfrac{3}{2} - 5\right)^2 + \left(3 - \dfrac{1}{2}\right)^2}$

E N R I C H M E N T   E X E R C I S E S

1. Find a pair of numbers such that their sum is 137 and their product is maximum.

2. Find a pair of numbers such that their sum is 29 and their product is maximum.

3. Find the length and width of a rectangular field whose perimeter is 548 feet and whose area is the largest.

4. A projectile is fired vertically upward with an initial velocity of v. The height h that it is above the ground after t seconds is given by

$$h = vt - 16t^2$$

Show that if the projectile is given twice the initial velocity, it takes twice as long to reach its maximum height.

Answers to Enrichment Exercises are on page A.∎∎∎.

9.3

Circles and Ellipses

O B J E C T I V E S

▶ *To find the distance between two points in the plane*

▶ *To draw the graphs of circles*

▶ *To put equations of circles into the standard form $(x - h)^2 + (y - k)^2 = r^2$ by completing the square*

▶ *To draw the graphs of ellipses*

▶ *To put equations of ellipses into the standard form $\dfrac{(x - h)^2}{a^2} + \dfrac{(y - k)^2}{b^2} = 1$*

In this section, we look at finding the distance between two points in the plane. In Figure (a), the points $P(3, 2)$ and $Q(11, 8)$ are graphed. Let d be the distance between them. To find d, we draw a right triangle as shown in Figure (b). Notice that the coordinates of the point C are $(11, 2)$.

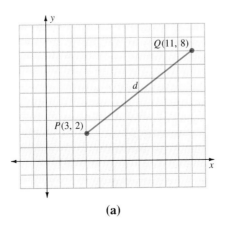

(a)

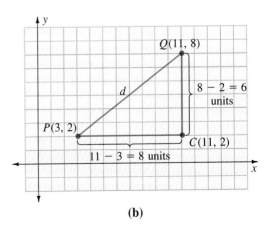

(b)

Now, the distance from P to C is $11 - 3 = 8$ units and the distance from C to Q is $8 - 2 = 6$ units. By the Pythagorean Theorem,

$$d^2 = 8^2 + 6^2 = 64 + 36 = 100$$

Therefore, $d = \sqrt{100}$ or 10 units. We generalize this result as the distance formula.

D E F I N I T I O N

The distance formula

Let $P(x_1, y_1)$ and $Q(x_2, y_2)$ be two points in the plane as shown in the figure, then the distance d between them is given by

$$d = \sqrt{(x_2 - x_1)^2 + (y_2 - y_1)^2}$$

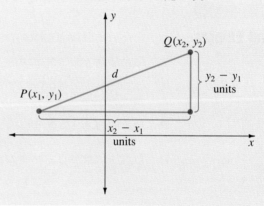

The distance formula holds regardless of the positions in the plane of the two points P and Q.

Example 1   Find the distance between $(4, -5)$ and $(-3, 1)$.

Solution   It makes no difference which point is called P and which is called Q. Setting $P = (4, -5)$ and $Q = (-3, 1)$,

$$d = \sqrt{(-3 - 4)^2 + [1 - (-5)]^2}$$
$$= \sqrt{(-7)^2 + (6)^2}$$
$$= \sqrt{49 + 36}$$
$$= \sqrt{85}$$

Example 2   Find x if the distance between $(1, 4)$ and $(x, -1)$ is $\sqrt{29}$.

Solution   Let d be the distance between $P = (1, 4)$ and $Q = (x, -1)$, then we are given that $d = \sqrt{29}$. Using the distance formula,

$$\sqrt{(x-1)^2 + (-1-4)^2} = \sqrt{29}$$

$(x-1)^2 + (-1-4)^2 = 29$ Square both sides.

$x^2 - 2x + 1 + 25 = 29$ Simplify.

$x^2 - 2x - 3 = 0$ Subtract 29 from both sides.

$(x-3)(x+1) = 0$ Factor the left side.

Either $x - 3 = 0$ or $x + 1 = 0$

Therefore, either $x = 3$ or $x = -1$, and so there are two points $(3, -1)$ and $(-1, -1)$ that are $\sqrt{29}$ units from $(1, 4)$. These two points can be checked by using the distance formula on $(3, -1)$ and $(1, 4)$ and on $(-1, -1)$ and $(1, 4)$ to see if the distances are both $\sqrt{29}$. ▲

We are now ready to derive the equation of a circle. First, we are given the definition of a circle.

D E F I N I T I O N

A **circle** is the set of all points in the plane the same distance from a given fixed point called the **center**. The **radius** of the circle is the length of any line segment determined by the center and any point on the circle.

   The following theorem gives the standard form for the equation of a circle.

T H E O R E M

The **equation of a circle** with center at (h, k) and radius r is given by

$$(x - h)^2 + (y - k)^2 = r^2$$

Proof   As shown in the figure, a circle of radius r with center at (h, k) is the set of all points (x, y) that are a distance r from (h, k).

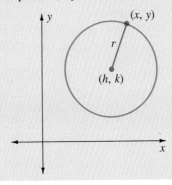

T H E O R E M (*continued*)

Using the distance formula,

$$\sqrt{(x - h)^2 + (y - k)^2} = r$$

Squaring both sides, we have the desired result:

$$(x - h)^2 + (y - k)^2 = r^2$$

Using the theorem for the equation of a circle, we can find the equation of a circle if we are given the center and the radius.

Example 3   Find the equation of a circle with center $(-4, 1)$ and radius 3.

Solution   Setting $h = -4$, $k = 1$, and $r = 3$ in the equation for a circle, we have

$$[x - (-4)]^2 + (y - 1)^2 = 3^2$$
$$(x + 4)^2 + (y - 1)^2 = 9$$ ▲

If the center of the circle of radius r is at the origin, then $(h, k) = (0, 0)$ and the equation becomes

$$(x - 0)^2 + (y - 0)^2 = r^2$$

or

$$x^2 + y^2 = r^2$$

Example 4   The equation of the circle whose center is the origin with a radius of 4 is $x^2 + y^2 = 16$.   ▲

Example 5   Find the equation of the circle whose center is $(-1, 3)$ and contains the point $(4, 1)$.

Solution   We start with the standard form

$$(x - h)^2 + (y - k)^2 = r^2$$

To find the equation of the circle, we must find h, k, and r. We are given that the center is $(-1, 3)$, so $h = -1$ and $k = 3$. Now, to find the radius r, recall that r is the length of any line segment determined by the center and any point on the circle. In particular, r is the distance between the center $(-1, 3)$ and the point $(4, 1)$.

$$r = \sqrt{(4 + 1)^2 + (1 - 3)^2} = \sqrt{25 + 4} = \sqrt{29}$$

The desired equation is

$$(x + 1)^2 + (y - 3)^2 = 29$$ ▲

We may sketch the graph of a circle, if we know the center and radius.

Example 6   Draw the graph of the circle whose equation is

$$(x - 4)^2 + (y + 5)^2 = 9$$

Solution   Comparing this equation to the standard form

$$(x - h)^2 + (y - k)^2 = r^2$$

we may rewrite our equation as

$$(x - 4)^2 + [y - (-5)]^2 = 3^2$$

Therefore, the center is $(4, -5)$ and the radius is 3.
This circle is drawn in the following figure.

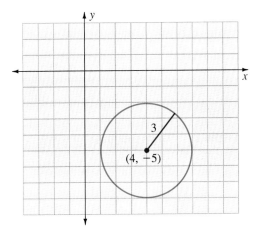

Sometimes the equation of a circle is not in standard form. By completing the square on the terms involving x as well as completing the square on the terms involving y, we can arrive at the standard form. The details are shown in the next example.

Example 7   Draw the graph of $x^2 + 8x + y^2 - 6y + 21 = 0$.

Solution   We can put this equation into the standard form $(x - h)^2 + (y - k)^2 = r^2$ by completing the square on x and on y.

$$x^2 + 8x + y^2 - 6y + 21 = 0$$

$$(x^2 + 8x + \quad) + (y^2 - 6y + \quad) = -21$$

$$(x^2 + 8x + 16) + (y^2 - 6y + 9) = -21 + 16 + 9$$

Add 16 to complete   Add 9 to complete   Add 16 and 9 to
the square on x.   the square on y.   balance the equation.

$$(x + 4)^2 + (y - 3)^2 = 4$$

In this form, we can "read off" the center and radius: the center is at $(-4, 3)$ and the radius is 2. The graph of this circle is shown in the following figure.

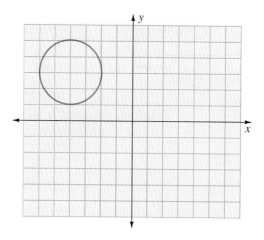

Example 8   Draw the graph of $y = -\sqrt{9 - x^2}$.

Solution   If we square both sides of this equation and then add x^2 to both sides, we have

$$y^2 = (-\sqrt{9 - x^2})^2$$
$$y^2 = 9 - x^2$$
$$x^2 + y^2 = 9$$

Therefore, the graph of $y = -\sqrt{9 - x^2}$ is part of the circle with center at the origin and a radius of 3. Now, since y is equal to a negative square root, $y < 0$. This means the graph lies below the x-axis. As shown in the figure, the graph is the lower half of the circle.

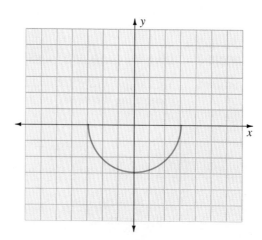

The final curve we consider in this section is the ellipse. Ellipses have applications in fields such as space science. For example, a communication satellite moves in an elliptical path around the earth.

DEFINITION

An **ellipse** is the set of all points in the plane such that the sum of the distances from two fixed points is constant. Each of the fixed points is called a **focus** (plural: **foci**).

The following figure shows an ellipse whose foci are $(c, 0)$ and $(-c, 0)$ with x-intercepts a and $-a$ and y-intercepts b and $-b$. The origin $(0, 0)$ is the **center** of the ellipse.

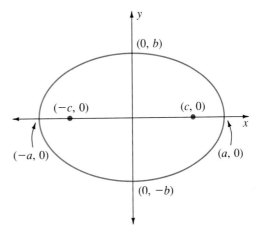

The following equation for an ellipse can be developed from the definition.

RULE

The **equation of an ellipse** with x-intercepts a and $-a$ and y-intercepts b and $-b$ has the form

$$\frac{x^2}{a^2} + \frac{y^2}{b^2} = 1$$

To graph an ellipse from its equation, first determine a and b, then use these intercepts to plot the four points $(a, 0)$, $(-a, 0)$, $(0, b)$, and $(0, -b)$.

Example 9   Graph the ellipse $\dfrac{x^2}{9} + \dfrac{y^2}{4} = 1$.

Solution   The x-intercepts are 3 and -3 and the y-intercepts are 2 and -2. We plot the points $(3, 0)$, $(-3, 0)$, $(0, 2)$, and $(0, -2)$ and draw the ellipse.

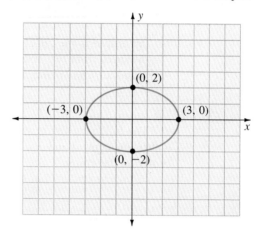

Just as the center of a circle can be at any point (h, k) in the plane, the center of an ellipse likewise can be located at any point. The standard form for the equation of an ellipse with center at (h, k) is given by

$$\frac{(x - h)^2}{a^2} + \frac{(y - k)^2}{b^2} = 1$$

Example 10   Graph the ellipse $\dfrac{(x - 3)^2}{16} + \dfrac{(y - 5)^2}{9} = 1$.

Solution   The center is $(3, 5)$, $a = 4$, and $b = 3$. Plot the center, then measure 4 units horizontally in each direction from the center and plot these two points. Next, measure vertically 3 units above the center and 3 units below the center and plot these two points. As a final step, draw the ellipse as shown in the figure.

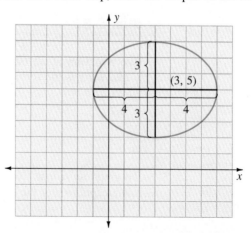

Just as in the case of a circle, the equation of an ellipse may not be in the standard form

$$\frac{(x - h)^2}{a^2} + \frac{(y - k)^2}{b^2} = 1$$

In the next example we show how to change such an equation to the standard form.

Example 11   Draw the graph of $9x^2 - 54x + 4y^2 + 8y + 49 = 0$.

Solution   We plan to complete the square on x and on y. Notice that the coefficient of x^2 is 9 and the coefficient of y^2 is 4. Therefore, our first step is to factor 9 from the two terms in x and factor 4 from the two terms in y before completing the square on x and on y.

$$9x^2 - 54x + 4y^2 + 8y + 49 = 0$$

$$9(x^2 - 6x + \quad) + 4(y^2 + 2y + \quad) = -49$$

$$9(x^2 - 6x + 9) + 4(y^2 + 2y + 1) = -49 + 9 \cdot 9 + 4 \cdot 1$$

Add 9 to complete   Add 1 to complete   Add $9 \cdot 9$ and $4 \cdot 1$ to
the square on x.   the square on y.   balance the equation.

$$9(x - 3)^2 + 4(y + 1)^2 = 36$$

As a final step, divide both sides of the equation by 36, since we want a 1 on the right side.

$$\frac{\overset{1}{\cancel{9}}(x - 3)^2}{\underset{4}{\cancel{36}}} + \frac{\overset{1}{\cancel{4}}(y + 1)^2}{\underset{9}{\cancel{36}}} = \frac{36}{36}$$

$$\frac{(x - 3)^2}{4} + \frac{(y + 1)^2}{9} = 1$$

The equation is now in standard form; its graph is shown in the following figure.

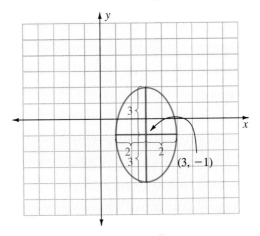

We summarize the techniques developed in the examples for graphing an ellipse.

S T R A T E G Y

Graphing an ellipse

Step 1   If necessary, put the equation in standard form

$$\frac{(x - h)^2}{a^2} + \frac{(y - k)^2}{b^2} = 1$$

It may be necessary to complete the square on x and on y.

Step 2   Determine the two numbers a and b by taking the (positive) square roots of the numbers in the denominators of the two fractions on the left side of the equation.

Step 3   Plot the center (h, k) and measure horizontally a units to the right and a units to the left from the center and plot the two resulting points. Next, measure vertically b units upward and b units downward from the center and plot the two resulting points.

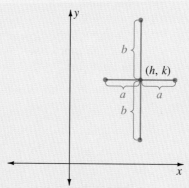

Step 4   Draw the ellipse determined by the four points found in Step 3.

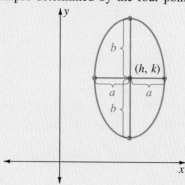

EXERCISE SET 9.3

For Exercises 1–8, find the distance between the two points.

1. $(2, 5)$ and $(7, 1)$

2. $(-1, 4)$ and $(3, 1)$

3. $(8, 2)$ and $(10, 6)$

4. $(8, -7)$ and $(-1, -4)$

5. $(0, 0)$ and $(0, -15)$

6. $(0, 0)$ and $(-11, 0)$

7. $(0, 5)$ and $(-2, 0)$

8. $(-3, 0)$ and $(0, -7)$

9. Find x so that the distance between $(x, -2)$ and $(-2, 2)$ is 5.

10. Find x so that the distance between $(x, 1)$ and $(0, 1 + 4\sqrt{3})$ is $5\sqrt{3}$.

11. Find y so that the distance between $(6, y)$ and $(1, 1)$ is 13.

12. Find y so that the distance between $(-3, y)$ and $(-2, \sqrt{3})$ is 2.

For Exercises 13–20 find the equation of the circle.

13. The center is $(4, 1)$ and the radius is 5.

14. The center is the origin and the radius is $\sqrt{5}$.

15. The center is the origin and the radius is $2\sqrt{3}$.

16. The center is $(-2, -7)$ and the radius is $4\sqrt{3}$.

17. The center is $(-3, -5)$ and the circle contains the point $(-4, -8)$.

18. The center is $(9, -4)$ and the circle contains the point $(13, 0)$.

19. The circle is centered at the origin and passes through the point $(-5, -3)$.

20. The circle is centered at $(0, -4)$ and passes through the point $(1, 1)$.

For Exercises 21–36, draw the graph of each equation.

21. $(x - 4)^2 + (y - 3)^2 = 9$

22. $(x - 2)^2 + (y - 2)^2 = 1$

23. $x^2 + y^2 = 49$

24. $x^2 + y^2 = 100$

25. $(x - 10)^2 + (y + 20)^2 = 100$

26. $(x + 5)^2 + (y - 10)^2 = 25$

27. $3x^2 + 3y^2 = 27$

28. $4x^2 + 4y^2 = 64$

29. $2(x + 1)^2 + 2y^2 = 8$

30. $3x^2 + 3(y + 2)^2 = 12$

31. $x^2 - 6x + y^2 - 4y + 4 = 0$

32. $x^2 + 2x + y^2 - 8y + 16 = 0$

33. $x^2 - 4x + y^2 = 0$

34. $x^2 + y^2 + 6y = 0$

35. $\dfrac{x^2}{2} - 4x + \dfrac{y^2}{2} - 5y = -8$

36. $\dfrac{x^2}{3} + 2x + \dfrac{y^2}{3} - 2y = -\dfrac{2}{3}$

For Exercises 37–40, draw the graph of each equation.

37. $y = \sqrt{4 - x^2}$

38. $y = \sqrt{16 - x^2}$

39. $y = -\sqrt{49 - x^2}$

40. $y = -\sqrt{25 - x^2}$

For Exercises 41–50, use the graphing strategy for the ellipse to draw the graph of each equation.

41. $\dfrac{x^2}{4} + \dfrac{y^2}{16} = 1$

42. $\dfrac{x^2}{4} + \dfrac{y^2}{9} = 1$

43. $\dfrac{x^2}{16} + \dfrac{y^2}{9} = 1$

44. $\dfrac{x^2}{25} + \dfrac{y^2}{16} = 1$

45. $(x-5)^2 + \dfrac{(y+3)^2}{4} = 1$

46. $\dfrac{(x+1)^2}{4} + \dfrac{(y-6)^2}{16} = 1$

47. $\dfrac{x^2}{9} + \dfrac{(y+4)^2}{25} = 1$

48. $\dfrac{(x+2)^2}{9} + \dfrac{(y+2)^2}{4} = 1$

49. $25x^2 + 50x + 4y^2 - 8y - 71 = 0$

50. $4x^2 - 24x + 9y^2 + 36y + 36 = 0$

REVIEW PROBLEMS

The following exercises review parts of Section 9.1. Doing these problems will help prepare you for the next section.

Write each equation in the form $y = a(x - h)^2 + k$.

51. $y = x^2 - 6x + 10$

52. $y = x^2 + 2x + 8$

53. $y = 3x^2 + 12x + 8$

54. $y = 4x^2 + 8x - 4$

55. $y = -2x^2 - 24x - 27$

56. $y = -3x^2 - 12x + 8$

57. $y = -x^2 + 6x - 12$

58. $y = -x^2 - 2x - 2$

E N R I C H M E N T   E X E R C I S E S

1. Find all points (x, y) so that the distance between (x, y) and $(1, 5)$ and the distance between (x, y) and $(2, 5)$ are both 2.

For Exercises 2–4, find an equation of the circle that satisfies the given conditions.

2. The circle is tangent to the y-axis and its center is $(3, -2)$.

3. The circle is tangent to the x-axis and its center is $(-1, -4)$.

4. The circle contains the points $(-3, 4)$ and $(7, 4)$ and its center lies on the line joining these two points.

5. What is the graph of $\dfrac{(x - h)^2}{a^2} + \dfrac{(y - k)^2}{b^2} = 1$ if $a = b$?

For Exercises 6 and 7, draw the graph of each equation.

6. $y = 2 + \sqrt{9 - (x - 1)^2}$

7. $y = \dfrac{5}{4}\sqrt{16 - (x - 4)^2}$

8. The city of Allentown is planning to construct a flower garden in Westend park. This garden has the shape of an ellipse as shown in the figure.

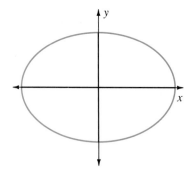

The equation for this ellipse is $324x^2 + 576y^2 = 186{,}624$, where x and y are measured in feet. How many linear feet of concrete are needed to make the walkways indicated in red?

9. Suppose that the foci of an ellipse are located at $(-3, 0)$ and $(3, 0)$ as shown in the following figure. Suppose the sum of the distances from a point (x, y) on the ellipse to the foci is 10 units. That is

$$d_1 + d_2 = 10$$

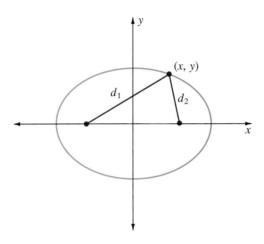

Show that the equation for this ellipse is

$$\frac{x^2}{25} + \frac{y^2}{16} = 1$$

9.4

Hyperbolas

▶ *To draw the graphs of hyperbolas*

▶ *To identify the graph of a general second degree equation*

In this section, we will draw the graphs of hyperbolas.

DEFINITION

A **hyperbola** is the set of all points in the plane such that the absolute value of the difference of the distances from two fixed points, called **foci,** is constant.

Suppose the two foci are the points F_1 and F_2 as shown in the following figure. Let (x, y) be an arbitrary point on the hyperbola and d_1 and d_2 the indicated distances. Then by the definition, we have that $|d_1 - d_2| =$ constant.

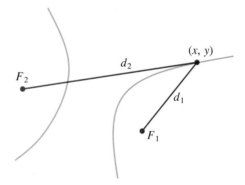

The following equations for hyperbolas can be developed from the definition.

RULE

Two standard forms for the hyperbola are the following:

1. The standard form of the equation of a hyperbola with center at the origin and vertices $(-a, 0)$ and $(a, 0)$ is given by

$$\frac{x^2}{a^2} - \frac{y^2}{b^2} = 1$$

2. The standard form for the equation of a hyperbola with center at the origin and vertices $(0, -b)$ and $(0, b)$ is given by

$$\frac{y^2}{b^2} - \frac{x^2}{a^2} = 1$$

If the hyperbola is given by Form 1, notice that if we set $y = 0$,

$$\frac{x^2}{a^2} - \frac{0^2}{b^2} = 1$$

$$\frac{x^2}{a^2} = 1$$

$$x^2 = a^2$$

$$x = \pm a$$

Therefore, the x-coordinates of the vertices are the x-intercepts of the hyperbola. To investigate the possibility of y-intercepts, we set $x = 0$. Then,

$$\frac{0^2}{a^2} - \frac{y^2}{b^2} = 1$$

$$-\frac{y^2}{b^2} = 1$$

$$y^2 = -b^2$$

From this last equation, we see that y has no real solutions, since y^2 cannot be negative. Therefore, the hyperbola, given by Form 1, does not have y-intercepts.

In a similar way, it can be shown that a hyperbola given by Form 2 has y-intercepts of $\pm b$ and no x-intercepts.

To graph a hyperbola given by either Form 1 or 2, we use two intersecting lines that the hyperbola approaches but never touches. These lines are called the **asymptotes.** They are given by $y = \pm\frac{b}{a}x$.

In the first example, we use the preceding discussion to graph a hyperbola.

Example 1   Graph the equation $\dfrac{x^2}{16} - \dfrac{y^2}{9} = 1$.

Solution   This equation is in standard Form 1, with $a^2 = 16$ and $b^2 = 9$, so $a = 4$ and $b = 3$. Since a is 4, the vertices are $(-4, 0)$ and $(4, 0)$ and the hyperbola opens to the left and to the right.

Next, we use a and b to draw the asymptotes. In addition to plotting the vertices $(-4, 0)$ and $(4, 0)$, also plot $(0, -b)$ and $(0, b)$, which for this hyperbola are $(0, -3)$ and $(0, 3)$. Now, draw a rectangle with the points just plotted being the midpoints of the sides as shown in Figure (a). Draw the two lines that pass through the origin (center) and the corners of the rectangle. These two lines are the asymptotes. Now, draw the hyperbola starting at the vertices and using the asymptotes as your guide as shown in Figure (b).

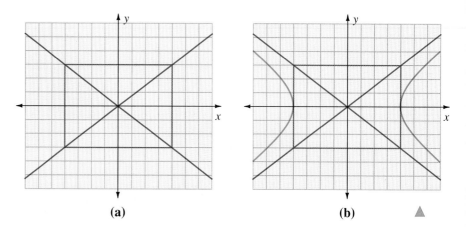

(a)   (b)   ▲

N O T E   *When graphing the hyperbola in Example 1, keep in mind that the rectangle and asymptotes are* not *part of the graph, but they are used as an aid to drawing the hyperbola.*

Example 2   Draw the graph of $\dfrac{y^2}{4} - \dfrac{x^2}{9} = 1$.

Solution   This equation is of Form 2 with $a = 3$ and $b = 2$. The graph has vertices at $(0, 2)$ and $(0, -2)$ and opens up and down. We draw the rectangle and the asymptotes. Using the vertices and asymptotes as a guide, we draw the hyperbola as shown in the following figure.

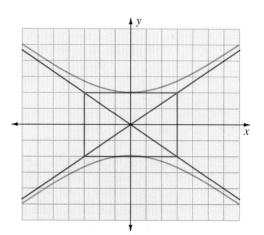

▲

Just as the center of an ellipse can be at any point (h, k) in the plane, the center of a hyperbola likewise can be located at any point. The standard forms for the equation of a hyperbola with center at (h, k) are given by these forms:

3. $\dfrac{(x - h)^2}{a^2} - \dfrac{(y - k)^2}{b^2} = 1$

or

4. $\dfrac{(y - k)^2}{b^2} - \dfrac{(x - h)^2}{a^2} = 1$

Example 3   Graph the hyperbola $\dfrac{(y - 4)^2}{16} - \dfrac{(x + 3)^2}{25} = 1.$

Solution   This equation is of Form 4. It has a center at $(-3, 4)$ with $a = 5$ and $b = 4$. Therefore, we plot the center, then determine the rectangle by measuring 5 units left and right and 4 units up and down. Draw the asymptotes and then draw the hyperbola using the vertices $(-3, 8)$ and $(-3, 0)$ and the asymptotes as guides. See the following figure.

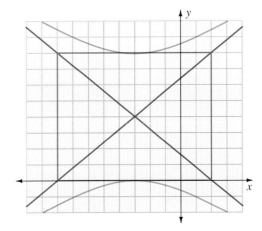

If the equation of a hyperbola is not in one of the standard forms, we must change it to a standard form before graphing. We are shown how to complete the square to obtain a standard form in the next example.

Example 4   Graph the equation $9x^2 + 18x - 4y^2 + 24y - 63 = 0.$

Solution   First factor 9 from the two terms in x and -4 from the two terms in y, then complete the square.

$$9x^2 + 18x - 4y^2 + 24y - 63 = 0$$

$$9(x^2 + 2x + \quad) - 4(y^2 - 6y + \quad) = 63$$

$$9(x^2 + 2x + 1) - 4(y^2 - 6y + 9) = 63 + 9 - 36$$

Add 1 to complete   Add 9 to complete   Add $9 \cdot 1 - 4 \cdot 9$ to
the square on x.   the square on y.   balance the equation.

$$\frac{9(x + 1)^2}{36} - \frac{4(y - 3)^2}{36} = \frac{36}{36}$$

$$\frac{(x + 1)^2}{4} - \frac{(y - 3)^2}{9} = 1$$

The equation is now in standard Form 3. Its graph is shown in the following figure.

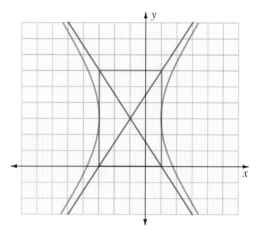

We summarize the graphing techniques for the hyperbola.

STRATEGY

Graphing a hyperbola

Step 1   If necessary, put the equation in standard form. It may be necessary to complete the square on x and on y.

Step 2   Determine the two numbers a and b. The number a is obtained by taking the (positive) square root of the number in the denominator of the fraction involving x. The number b is obtained by taking the (positive) square root of the number in the denominator of the fraction involving y.

Step 3   Plot the center (h, k) and draw the rectangle determined by the two numbers a and b. Draw the two lines that connect opposite corners of the rectangle. These lines are the asymptotes of the hyperbola.

Step 4   **(a)** If the hyperbola opens left and right, the vertices are the midpoints of the two vertical sides of the rectangle found in Step 3.

   (b) If the hyperbola opens up and down, the vertices are the midpoints of the two horizontal sides of the rectangle found in Step 3.

Draw the hyperbola using the vertices and the asymptotes as a guide.

There is another type of equation whose graph is a hyperbola. The general form is

$$xy = c, \qquad \text{where } c \neq 0$$

Examples of this type of equation are the following:

$$xy = 4, \qquad xy = -7, \qquad \text{and} \qquad y = \frac{1}{x}$$

For these hyperbolas, the asymptotes are the x- and y-axes.

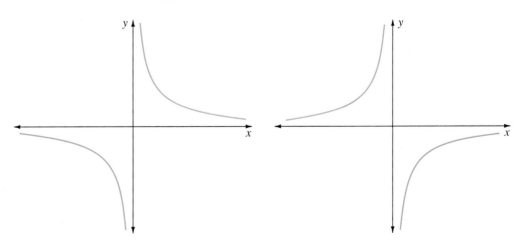

$xy = c$, where $c > 0$ $xy = c$, where $c < 0$

Example 5

Graph $xy = 3$.

Solution

First we solve for y, $y = \dfrac{3}{x}$, and use this equation to generate ordered pairs that belong to the graph. Since the y-axis is an asymptote, we choose values of x close to zero.

Choice of x	Determine y from $y = \dfrac{3}{x}$	(x, y)
$\dfrac{1}{3}$	$y = \dfrac{3}{\dfrac{1}{3}} = 9$	$\left(\dfrac{1}{3}, 9\right)$
$\dfrac{1}{2}$	$y = \dfrac{3}{\dfrac{1}{2}} = 6$	$\left(\dfrac{1}{2}, 6\right)$
1	$y = \dfrac{3}{1} = 3$	$(1, 3)$
2	$y = \dfrac{3}{2}$	$\left(2, \dfrac{3}{2}\right)$
3	$y = \dfrac{3}{3} = 1$	$(3, 1)$
4	$y = \dfrac{3}{4}$	$\left(4, \dfrac{3}{4}\right)$
6	$y = \dfrac{3}{6} = \dfrac{1}{2}$	$\left(6, \dfrac{1}{2}\right)$
9	$y = \dfrac{3}{9} = \dfrac{1}{3}$	$\left(9, \dfrac{1}{3}\right)$
$-\dfrac{1}{3}$	$y = \dfrac{3}{-\dfrac{1}{3}} = -9$	$\left(-\dfrac{1}{3}, -9\right)$
$-\dfrac{1}{2}$	$y = \dfrac{3}{-\dfrac{1}{2}} = -6$	$\left(-\dfrac{1}{2}, -6\right)$
-1	$y = \dfrac{3}{-1} = -3$	$(-1, -3)$
-2	$y = \dfrac{3}{-2} = -\dfrac{3}{2}$	$\left(-2, -\dfrac{3}{2}\right)$

$$-3 \qquad y = \frac{3}{-3} = -1 \qquad (-3, -1)$$

$$-4 \qquad y = \frac{3}{-4} = -\frac{3}{4} \qquad \left(-4, -\frac{3}{4}\right)$$

$$-6 \qquad y = \frac{3}{-6} = -\frac{1}{2} \qquad \left(-6, -\frac{1}{2}\right)$$

$$-9 \qquad y = \frac{3}{-9} = -\frac{1}{3} \qquad \left(-9, -\frac{1}{3}\right)$$

These points are plotted and a smooth curve drawn through them, keeping in mind that the coordinate axes are the asymptotes for this hyperbola.

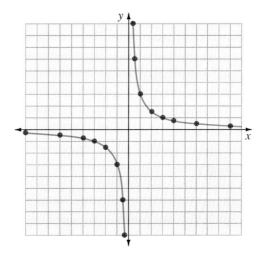

In our study of conic sections we have graphed **second degree equations.** These are equations of the form $Ax^2 + By^2 + Cx + Dy + E = 0$, where a and b are not both zero. In particular, we completed the square to convert this equation to one of the standard forms. However, not all second degree equations can be graphed. For example, the second degree equation $x^2 + y^2 = -1$ has no graph, since if x and y are real numbers, $x^2 + y^2$ is nonnegative and could not be equal to -1.

If the equation *can* be graphed, do we always get a conic section? For example, the equation $x^2 + y^2 = 0$ looks like the equation of a circle, except that the right side is zero, not a positive number. Upon close inspection, we see that the only pair of real numbers that satisfies this equation is $(0, 0)$. Therefore, the graph of $x^2 + y^2 = 0$ is the origin. This is an example of a **degenerate form** of a conic section. In general, we can make the following statement.

The degenerate form for a circle or ellipse is a point.

In this case, the second degree equation can be written either as

$$\text{degenerate circle:}\quad (x-h)^2 + (y-k)^2 = 0$$

or

$$\text{degenerate ellipse:}\quad \frac{(x-h)^2}{a^2} + \frac{(y-k)^2}{b^2} = 0$$

In either case, the graph is the point (h, k).

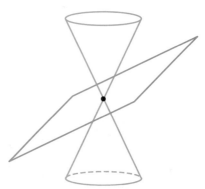

Degenerate ellipse

In the next example, we will see that the degenerate form for a hyperbola is two intersecting lines.

Example 6

Graph the equation, if possible.

$$x^2 - 2x - y^2 - 4y = 3$$

Solution

We complete the square on x and on y.

$$x^2 - 2x - y^2 - 4y = 3$$
$$(x^2 - 2x + \quad) - (y^2 + 4y + \quad) = 3$$
$$(x^2 - 2x + 1) - (y^2 + 4y + 4) = 3 + 1 - 4$$
$$(x - 1)^2 - (y + 2)^2 = 0$$

In this last form, the equation looks like the equation of a hyperbola, except that the right side is zero, not one. To see what the graph is, we add $(y + 2)^2$ to both sides, then take square roots.

$$(x - 1)^2 = (y + 2)^2$$
$$x - 1 = \pm(y + 2)$$

Therefore, either $x - 1 = y + 2$ or $x - 1 = -(y + 2)$. So, we obtain the equations of two lines:

$$x - y = 3 \quad \text{or} \quad x + y = -1$$

These lines are graphed in the following figure.

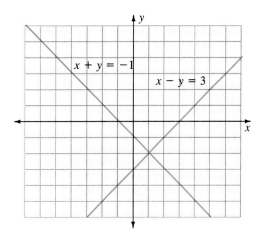

In general, we can make the following statement.

The degenerate form for a hyperbola is two intersecting lines.

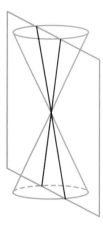

The general equation for a degenerate hyperbola is

$$\frac{(y-k)^2}{b^2} - \frac{(x-h)^2}{a^2} = 0$$

or the equivalent equation

$$\frac{(x-h)^2}{a^2} - \frac{(y-k)^2}{b^2} = 0$$

Either equation simplifies to the two linear equations:

$$y - k = \pm\frac{b}{a}(x-h)$$

EXERCISE SET 9.4

For Exercises 1–20, use the graphing strategy for the hyperbola to draw the graph of each equation.

1. $\dfrac{x^2}{4} - \dfrac{y^2}{16} = 1$

2. $\dfrac{x^2}{9} - \dfrac{y^2}{4} = 1$

3. $\dfrac{y^2}{36} - \dfrac{x^2}{25} = 1$

4. $\dfrac{y^2}{25} - \dfrac{x^2}{9} = 1$

5. $\dfrac{(y-5)^2}{16} - \dfrac{(x-6)^2}{4} = 1$

6. $\dfrac{(y+2)^2}{9} - \dfrac{(x-3)^2}{4} = 1$

7. $\dfrac{(x+6)^2}{25} - \dfrac{y^2}{9} = 1$

8. $\dfrac{x^2}{16} - \dfrac{(y-4)^2}{9} = 1$

9. $\dfrac{(x+2)^2}{4} - \dfrac{(y-3)^2}{4} = 1$

10. $\dfrac{(y+1)^2}{9} - \dfrac{(x-1)^2}{9} = 1$

11. $x^2 - y^2 = 1$

12. $y^2 - x^2 = 1$

13. $4y^2 - x^2 = 4$

14. $x^2 - 9y^2 = 9$

15. $4x^2 - 32x - 9y^2 + 90y - 197 = 0$

16. $9x^2 + 54x - 16y^2 + 96y - 207 = 0$

17. $4y^2 + 40y - 25x^2 = 0$

18. $-4x^2 + 32x + 9y^2 - 108y + 224 = 0$

19. $y^2 - 12y - x^2 - 8x + 11 = 0$

20. $x^2 - 2x - y^2 + 2y - 16 = 0$

21. Find the equations of the lines that are the asymptotes for the graph of $\dfrac{x^2}{16} - y^2 = 1.$

22. Find the equations of the lines that are the asymptotes of $x^2 - (y - 2)^2 = 1.$

For Exercises 23–30, draw the graph of each equation.

23. $xy = 1$

24. $xy = 4$

25. $y = \dfrac{6}{x}$

26. $y = \dfrac{5}{x}$

27. $xy = -3$

28. $xy = -4$

29. $y = -\dfrac{1}{x}$

30. $y = -\dfrac{5}{x}$

For Exercises 31–36, draw the graph of the following degenerate conic sections.

31. $x^2 + y^2 = 0$

32. $\dfrac{(x + 7)^2}{4} + \dfrac{(y - 8)^2}{9} = 0$

33. $\dfrac{(x - 4)^2}{4} - \dfrac{(y + 3)^2}{9} = 0$

34. $\dfrac{x^2}{25} - \dfrac{y^2}{4} = 0$

35. $(y - 3)^2 - x^2 = 0$

36. $y^2 - 9(x + 2)^2 = 0$

For Exercises 37–48, graph, when possible, each second degree equation. If the equation has no graph, write "no graph." (*Note:* Some of the graphs will be degenerate conic sections.)

37. $x^2 + y^2 - 4 = 0$

38. $16x^2 + 9y^2 - 32x - 18y - 119 = 0$

39. $\dfrac{1}{4}x^2 - \dfrac{1}{4}y^2 = 1$

40. $x^2 + y^2 - 4x - 6y + 13 = 0$

41. $x^2 + y^2 + 2x + 4y + 5 = 0$

42. $4x^2 + 9y^2 + 16x - 20 = 0$

43. $3x^2 + 2y^2 - 12x + 16 = 0$ **44.** $-4x^2 - y^2 + 24x - 6y - 41 = 0$

45. $2x^2 + 4x + y - 2 = 0$ **46.** $2x^2 - 8x - y + 5 = 0$

47. $5x^2 + 2y^2 - 40x + 4y + 102 = 0$ **48.** $x^2 - 9y^2 - 4x - 72y - 140 = 0$

REVIEW PROBLEMS

The following exercises review parts of Sections 6.3 and 6.4. Doing these problems will help prepare you for the next section.

Solve each equation.

49. $3x^2 - 4x + 7 = 3x^2 + 2x + 1$ **50.** $1 + 2x - 12x^2 = 4(1 - 3x^2)$

51. $x + 6 = x^2$ **52.** $x^2 + 2(x^2 - 1) = 6$

53. $y^2 + (2 - y)^2 = 2$

E N R I C H M E N T   E X E R C I S E S

For Exercises 1–4, graph each equation.

1. $x = \dfrac{5}{2}\sqrt{4 + y^2}$ **2.** $y = -2\sqrt{1 + x^2}$

3. $y = 2 - \sqrt{4 + (x - 3)^2}$

4. $x = -1 + \sqrt{9 + (y - 4)^2}$

psz 90d^Answers to Enrichment Exercises are on page A.62.

9.5

Nonlinear Systems of Equations

OBJECTIVES

▶ *To solve systems of equations in which exactly one equation is linear*

▶ *To solve systems with two nonlinear equations*

In Chapter 8, we solved linear systems of equations. In this section, we solve nonlinear systems of equations. For most nonlinear systems in our discussion, we will use the substitution method, although the elimination method can be used in special cases. The first kind of system we solve is the system in which one equation is linear and the other is a second degree equation. Geometrically, real number solutions to such systems will be any point of intersection of a straight line with a conic section.

Three possibilities exist: two solutions, one solution, or no solution. Examples of these situations are illustrated in the following figures.

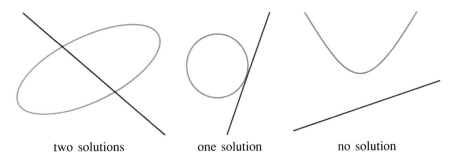

two solutions   one solution   no solution

While graphing a system is not practical for finding solutions, it would indicate to us how many solutions to expect when we algebraically solve the system. We are shown in the first example how to solve a system in which one of the equations is linear. In this case, we use the substitution method.

Example 1   Solve the system:

$$x^2 + y^2 = 13$$

$$2x + y = -1$$

Solution   We solve the linear equation for y, then substitute into the quadratic equation.

$$y = -2x - 1$$

$$x^2 + (-2x - 1)^2 = 13$$

Next, we solve the resulting equation in the single variable x.

$$x^2 + 4x^2 + 4x + 1 = 13$$

$$5x^2 + 4x - 12 = 0$$

$$(5x - 6)(x + 2) = 0$$

Therefore, either $x = \dfrac{6}{5}$ or $x = -2$.

To find the corresponding values for y, we use the linear equation $y = -2x - 1$.

When $x = \dfrac{6}{5}$, $y = -2\left(\dfrac{6}{5}\right) - 1 = -\dfrac{17}{5}$.

When $x = -2$, $y = -2(-2) - 1 = 3$.

There are two solutions: $\left(\dfrac{6}{5}, -\dfrac{17}{5}\right)$ and $(-2, 3)$. As a check, substitute each ordered pair into both equations. The following figure shows the graphs of this system and the two solutions.

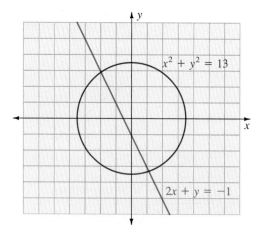

When a system contains two nonlinear equations whose graphs are conic sections, the system may have 4, 3, 2, 1, or no real solutions. Some examples of these possibilities are shown in the following figures.

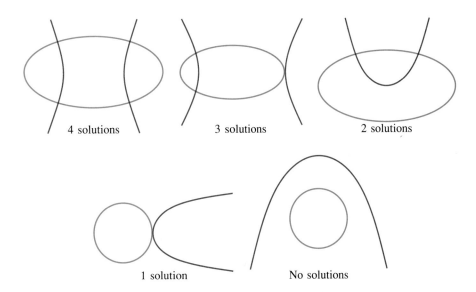

4 solutions   3 solutions   2 solutions

1 solution   No solutions

The next examples involve solving a system of two nonlinear equations. In this case, we can sometimes use the elimination method.

Example 2   Solve the system:

$$x^2 - y^2 = 1$$
$$x^2 + y^2 = 7$$

Solution   For solving this system, the elimination method works best. Add the two equations to eliminate y, then solve for x.

$$x^2 - y^2 = 1$$
$$\underline{x^2 + y^2 = 7} \qquad \text{Add.}$$
$$2x^2 \quad\;\; = 8$$
$$x^2 = 4$$

Therefore, $x = \pm 2$. Next, we substitute these two values of x into either original equation and solve for y.

When $x = 2$, $2^2 + y^2 = 7$
$$y^2 = 3$$
$$y = \pm\sqrt{3}$$

When $x = -2$, $(-2)^2 + y^2 = 7$
$$y^2 = 3$$
$$y = \pm\sqrt{3}$$

The solutions are $(2, \sqrt{3})$, $(2, -\sqrt{3})$, $(-2, \sqrt{3})$, and $(-2, -\sqrt{3})$. As a final step, these solutions should be checked in the original system. The graph of this system together with the four solutions is drawn in the following figure.

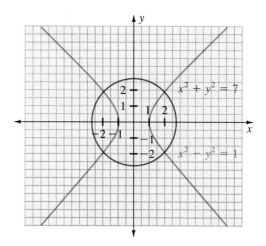

Example 3   Solve the system:

$$\frac{x^2}{4} + \frac{y^2}{9} = 1$$

$$x^2 + y^2 = 9$$

Solution   We use the elimination method and start by multiplying both sides of the first equation by 36 and the second equation by -4.

$$\frac{x^2}{4} + \frac{y^2}{9} = 1 \quad\text{—multiply by 36}\longrightarrow\quad 9x^2 + 4y^2 = 36$$

$$x^2 + y^2 = 9 \quad\text{—multiply by } -4 \longrightarrow\quad \underline{-4x^2 - 4y^2 = -36}$$
$$5x^2 \qquad\quad = 0$$

Therefore, $x^2 = 0$, which means that $x = 0$.

Replacing x by 0 in either of the original equations (we use the second equation),

$$0^2 + y^2 = 9$$

$$y^2 = 9$$

$$y = \pm 3 \qquad \text{Take square roots.}$$

The solutions to the system are $(0, 3)$ and $(0, -3)$. The two solutions are shown in the graph in the following figure.

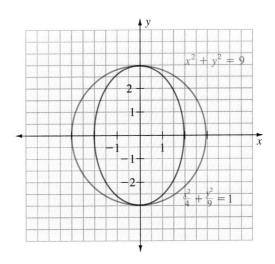

Example 4   Solve:

$$2x^2 - y^2 = 1$$

$$xy = -1$$

Solution   From the second equation, we have $y = -\dfrac{1}{x}$. We replace y by $-\dfrac{1}{x}$ in the first equation and solve for x.

$$2x^2 - y^2 = 1$$

$$2x^2 - \left(-\frac{1}{x}\right)^2 = 1$$

$$2x^2 - \frac{1}{x^2} = 1$$

$$2x^4 - 1 = x^2 \qquad \text{Multiply both sides by } x^2.$$

$$2x^4 - x^2 - 1 = 0$$

This last equation is quadratic in x^2, since $x^4 = (x^2)^2$.

$$2(x^2)^2 - x^2 - 1 = 0$$

$$(2x^2 + 1)(x^2 - 1) = 0 \qquad \text{Factor.}$$

Therefore, either $2x^2 + 1 = 0$ or $x^2 - 1 = 0$.
    Solving each equation, either

$$x^2 = -\frac{1}{2} \qquad \text{or} \qquad x^2 = 1$$

$$x = \pm\sqrt{-\frac{1}{2}} \qquad \text{or} \qquad x = \pm\sqrt{1} \qquad \text{Take square roots.}$$

$$x = \pm\sqrt{\frac{1}{2}}i \quad \text{or} \quad x = \pm 1$$

$$x = \pm\frac{\sqrt{2}}{2}i \quad \text{or} \quad x = \pm 1$$

We now take these four values for x and find the corresponding values for y using $y = -\frac{1}{x}$.

When $x = \frac{\sqrt{2}}{2}i$, $y = -\dfrac{1}{\dfrac{\sqrt{2}}{2}i} = i\sqrt{2}.$ Verify.

When $x = -\frac{\sqrt{2}}{2}i$, $y = -\dfrac{1}{-\dfrac{\sqrt{2}}{2}i} = -i\sqrt{2}.$ Verify.

When $x = 1$, then $y = -\dfrac{1}{1} = -1.$

When $x = -1$, then $y = -\dfrac{1}{-1} = 1.$

The four solutions are

$$\left(\frac{\sqrt{2}}{2}i, i\sqrt{2}\right), \quad \left(-\frac{\sqrt{2}}{2}i, -i\sqrt{2}\right), \quad (1, -1), \quad (-1, 1)$$

Although there are four solutions, notice that only two of them involve real numbers. This is reflected in the fact that the graphs intersect in two points. (See the following figure.)

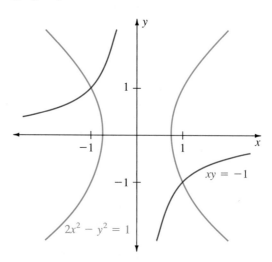

Sometimes a combination of elimination and substitution is used to solve a nonlinear system of equations. This procedure is illustrated in the next example.

Example 5   Solve:

$$x^2 + 2xy - 2y^2 = 6$$

$$-x^2 + 3xy + 2y^2 = -6$$

Solution   We can eliminate the squared terms by adding the two equations.

$$x^2 + 2xy - 2y^2 = 6$$

$$\underline{-x^2 + 3xy + 2y^2 = -6}$$

$$5xy \qquad\quad = 0$$

Therefore, either $x = 0$ or $y = 0$.

We now substitute $x = 0$ in either equation of the original system. From the first equation of the system,

$$0^2 + 2(0)y - 2y^2 = 6$$

$$-2y^2 = 6$$

$$y^2 = -3$$

$$y = \pm i\sqrt{3} \qquad \text{Take square roots.}$$

If $y = 0$,

$$x^2 + 2x(0) - 2(0)^2 = 6$$

$$x^2 = 6$$

$$x = \pm\sqrt{6} \qquad \text{Take square roots.}$$

The four solutions are $(0, i\sqrt{3})$, $(0, -i\sqrt{3})$, $(\sqrt{6}, 0)$, and $(-\sqrt{6}, 0)$. These should be checked in the original system.

The first equation has a hyperbola as its graph; however, it requires advanced techniques to draw it. Therefore, we shall omit the graphs of this system.   ▲

C A L C U L A T O R   C R O S S O V E R

Solve the following systems.

1. $xy = 4.92$ **2.** $x^2 + 3y^2 = 15.88$

$\quad x = 8.3y + 5.27$ $\quad x^2 - 2y^2 = 9.62$

Answers   **1.** $(9.5, 0.52)$, $(-4.3, -1.2)$ **2.** $(3.48, 1.12)$, $(3.48, -1.12)$, $(-3.48, 1.12)$, $(-3.48, -1.12)$

We finish this section with a word problem.

Example 6 A rectangle has an area of 8 square inches and a perimeter of 12 inches. Find the length and width of the rectangle.

Solution If the width is x and the length is y, then we obtain the system:

$$xy = 8 \qquad \text{The area is 8.}$$

$$2x + 2y = 12 \qquad \text{The perimeter is 12.}$$

To solve this system, we use the second equation to express y in terms of x, then substitute into the first equation.

$$2x + 2y = 12$$

$$x + y = 6 \qquad \text{Divide both sides by 2.}$$

$$y = 6 - x$$

Now, replace y by $6 - x$ in the first equation, then solve for x.

$$xy = 8$$

$$x(6 - x) = 8$$

$$6x - x^2 = 8$$

$$-x^2 + 6x - 8 = 0$$

$$x^2 - 6x + 8 = 0 \qquad \text{Multiply both sides by } -1.$$

$$(x - 4)(x - 2) = 0 \qquad \text{Factor.}$$

Therefore, either $x = 4$ or $x = 2$.

If $x = 4$, then $y = 6 - x = 6 - 4 = 2$.
If $x = 2$, then $y = 6 - x = 6 - 2 = 4$.

In either case, we obtain a rectangle with a width of 2 inches and a length of 4 inches.

We check our answer:

The area is $2 \cdot 4 = 8$ square inches.
The perimeter is $2(2) + 2(4) = 4 + 8 = 12$ inches. ▲

E X E R C I S E S E T 9.5

For Exercises 1–26, solve each system.

1. $x^2 + y^2 = 8$
    $y = x$

2. $x^2 + y^2 = 50$
    $x = -y$

3. $y = 3x^2 - 2x$
$\quad x = -2$

4. $2y^2 - x^2 = 1$
$\quad\quad y = 1$

5. $y = 3x^2 - 2x + 1$
$\quad 2x - y = 1$

6. $y = 2x^2 + 3x - 6$
$\quad y - 2x = 0$

7. $y = -2x^2 + x + 2$
$\quad 6x + y = -2$

8. $y = -x^2 - 5x + 4$
$\quad x - y = 3$

9. $y^2 - x^2 = 12$
$\quad x^2 + y = 0$

10. $x^2 + y^2 = 20$
$\quad\quad y^2 = x$

11. $\quad\quad x^2 + y^2 = 4$
$\quad x^2 + 3xy + y^2 = -2$

12. $x^2 + 2xy - y^2 = 17$
$\quad\quad x^2 - y^2 = 5$

13. $\quad\quad x(x - y) = 0$
$\quad x^2 + 3xy - y^2 = 8$

14. $x^2 - x + 2y^2 = 2$
$\quad\quad x(x - 2y) = 0$

15. $2y^2 - x^2 = 8$
$\quad\quad xy = 8$

16. $x^2 - 2xy + y^2 = 25$
$\quad\quad\quad xy = -6$

17. $(x + 1)^2 + (y - 3)^2 = 49$
$\quad\quad\quad\quad x + y = 9$

18. $(x - 4)^2 + (y + 4)^2 = 25$
$\quad\quad\quad\quad x + y = 7$

19. $y = 3x^2 - 4x + 5$
$\quad y = 3x^2 + 2x + 1$

20. $y = -2x^2 + 5x - 12$
$\quad y = -2x^2 + 3x - 4$

21. $y = (x - 2)^2 + 1$
$\quad y = x^2$

22. $y = -(x + 1)^2 - 2$
$\quad y = -x^2$

23. $\dfrac{x^2}{2} - \dfrac{y^2}{3} = 1$

$\dfrac{x^2}{3} + \dfrac{y^2}{2} = 1$

24. $x^2 + y^2 = 9$

$\dfrac{x^2}{25} + y^2 = 1$

25. $3xy - 4x^2 = 11$

$xy - 2x^2 = 1$

26. $3y^2 - 2xy = 8$

$y^2 + xy = 6$

27. The sum of the squares of two positive numbers is 73. The difference of the square of the larger number and three times the square of the smaller number is 37. Find the two numbers.

28. The sum of the squares of two numbers is $\dfrac{85}{9}$ and their product is 2. Find the two numbers.

29. The area of a right triangle is 16 square feet and its hypotenuse is $4\sqrt{5}$ feet. Find the lengths of the other two sides.

30. The area of a rectangle is 12 square cm and the length of a diagonal is 5 cm. Find the dimensions of the rectangle.

31. Find the dimensions of a rectangle whose area is 3 square meters and perimeter is 7 meters.

32. Find the dimensions of a rectangular rug that has an area of 32 square feet and a perimeter of 24 feet.

33. The supply equation for fresh bay scallops in the greater Lancaster area is given by

$$p = 2x^2 + 3$$

where p is the price per pound and x is the number of pounds, in thousands, of scallops. The demand equation is given by

$$p = -11x + 9$$

Find the equilibrium point.

34. The supply and demand equations for salad shrimp in the Lehigh Valley are given by

Supply: $p = \dfrac{1}{3}x + 2$

Demand: $p = -\dfrac{1}{6}x^2 + 10$

where p is the price per pound and x is the number of pounds, in thousands, of salad shrimp. Find the equilibrium point.

REVIEW PROBLEMS

The following exercises review parts of Section 8.5. Doing these problems will help prepare you for the next section.

Graph the system of inequalities.

35. $x + 2y < 4$

$2y > x$

36. $2y - x \geq -2$

$x \geq 0$ and $y \geq 0$

37. $3x + 2y \geq 6$

$x - y \leq 2$

E N R I C H M E N T   E X E R C I S E S

For Exercises 1–3, solve the system.

1. $y = 8x^3$
    $y = 2x$

2. $y = -x^3$
    $y = -x$

3. $(x - 2)^2 + (y - 2)^2 = 2$
    $xy = 1$

4. The product of two numbers is 6 and the sum of their reciprocals is $\dfrac{11}{12}$.

Find the two numbers.

5. Let A and P be the area and perimeter, respectively, of a rectangle. Express the length and width of the rectangle in terms of A and P.

Answers to Enrichment Exercises are on page A.63.

9.6

Nonlinear Systems of Inequalities

O B J E C T I V E

▶   *To graph nonlinear systems of inequalities*

The topic of this section is graphing nonlinear inequalities. Recall, in Section 7.4 we graphed linear inequalities. We first graphed the boundary line, then chose a test point not on the boundary to determine which half-plane the inequality described. This technique can also be applied to nonlinear inequalities.

Example 1   Graph $x^2 + y^2 \leq 9$.

Solution   The boundary is the circle $x^2 + y^2 = 9$ whose center is the origin with a radius of 3. The region in the plane that is the solution of $x^2 + y^2 < 9$ is either the region inside the circle or the region outside the circle. We use the origin $(0, 0)$ as a test point:

$$x^2 + y^2 \leq 9$$

$$0^2 + 0^2 \leq 9$$

$$0 \leq 9$$

Therefore, the test point satisfies the inequality, and the solution set is the set of points inside and on the circle as shown in the following figure.

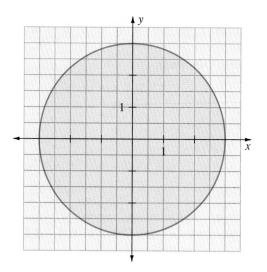

Example 2   Graph $\dfrac{x^2}{4} - \dfrac{y^2}{9} > 1$.

Solution   We graph the hyperbola $\dfrac{x^2}{4} - \dfrac{y^2}{9} = 1$ as shown in Figure (a). Notice that in this case, the boundary curve divides the plane into three parts. We select a test point from each part: $(-3, 0)$, $(0, 0)$, and $(3, 0)$.

Test point	$\dfrac{x^2}{4} - \dfrac{y^2}{9}$
$(-3, 0)$	$\dfrac{(-3)^2}{4} - \dfrac{0^2}{9} \overset{?}{>} 1$
	$\dfrac{9}{4} > 1$
$(0, 0)$	$\dfrac{0^2}{4} - \dfrac{0^2}{9} \overset{?}{>} 1$
	$0 \not> 1$
$(3, 0)$	$\dfrac{3^2}{4} - \dfrac{0^2}{9} \overset{?}{>} 1$
	$\dfrac{9}{4} > 1$

Therefore, the graph of our inequality is the two regions as indicated by test points $(-3, 0)$ and $(3, 0)$. The two regions that comprise the solution set are shown in Figure (b). Notice that the boundary curves are dashed, as they are not included in the solution set.

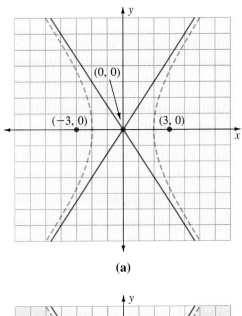

(a)

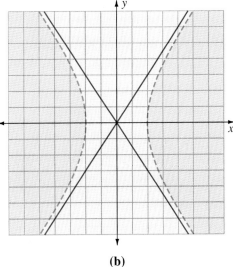

(b)

We now look at graphing nonlinear systems of inequalities. In Section 8.5, we graphed linear systems of inequalities. The idea remains the same, that is, we graph the solution sets of each inequality in the same coordinate plane, then find the intersection of these solution sets.

Example 3   Graph the system of inequalities:

$$y < -2x^2 + 8$$

$$y > 2x^2 - 8$$

Solution   The graph of $y = -2x^2 + 8$ is a parabola with vertex at $(0, 8)$ that opens downward and intersects the x-axis at $(2, 0)$ and $(-2, 0)$. The solution set of $y < -2x^2 + 8$ is the interior of the parabola. The graph of $y = 2x^2 - 8$ is a parabola with vertex $(0, -8)$ and opens upward. It also intersects the x-axis at $(2, 0)$ and $(-2, 0)$. The solution set of $y > 2x^2 - 8$ is the interior of this parabola. The solution set of the system is the shaded region that includes all points interior to both parabolas. Notice that neither boundary curve is included in the solution set as the inequalities are *strictly* less than and greater than. (See the following figure.)

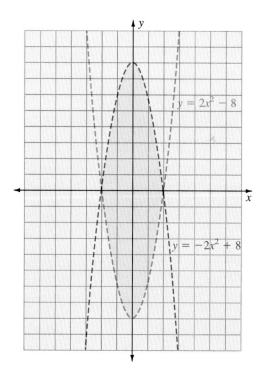

Example 4   Graph the system of inequalities.

$$\frac{x^2}{9} + \frac{y^2}{25} \geq 1$$

$$xy \geq 1$$

$$x + y < 10$$

Solution   We graph each region separately in the same coordinate plane, then take the intersection. The intersection is the region shaded in the figure below. Notice that part of the region is bounded by $x + y = 10$. This part of the boundary is a dashed line, since $x + y < 10$ contains a strict inequality.

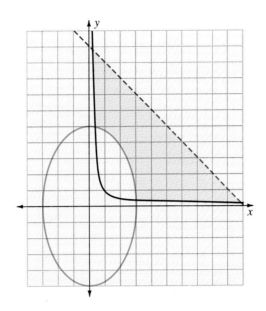

E X E R C I S E   S E T   9 . 6

For Exercises 1–10, graph the inequality.

1. $x^2 + y^2 > 16$

2. $x^2 + y^2 < 100$

3. $\dfrac{(x-3)^2}{4} + \dfrac{(y-5)^2}{9} \leq 1$

4. $\dfrac{(x+4)^2}{25} + \dfrac{(y+4)^2}{16} \geq 1$

5. $\dfrac{x^2}{9} - \dfrac{y^2}{4} \leq 1$

6. $\dfrac{y^2}{16} - \dfrac{x^2}{25} < 1$

7. $y^2 - x^2 > 1$

8. $x^2 - y^2 \geq 1$

9. $xy \geq -6$

10. $xy < 4$

For Exercises 11–24, graph the system of inequalities.

11. $y \leq -3x^2 + 3$
$y \geq 3x^2 - 3$

12. $\dfrac{x^2}{9} + \dfrac{y^2}{25} < 1$
$y + x^2 < 0$

13. $x^2 - y^2 > 1$
$x < 2$

14. $xy < 4$
$x + y < 5$

15. $x^2 - y^2 \geq 9$
$x^2 + y^2 \leq 25$

16. $y^2 - x^2 \geq 4$
$\dfrac{x^2}{4} + \dfrac{y^2}{25} \leq 1$

17. $xy > -4$
  $y \geq -x$

18. $y^2 - x^2 \geq 16$
  $x^2 + (y - 4)^2 < 16$

19. $y \geq (x - 2)^2 - 3$
  $x \geq 1$
  $4y - 3x \leq 9$

20. $x^2 - y^2 \geq 9$
  $(x - 3)^2 + y^2 \leq 9$
  $y \geq 0$

21. $\dfrac{x^2}{25} + \dfrac{y^2}{9} \leq 1$
  $-3x + 5y < 15$
  $3x + 5y < 15$

22. $y \geq (x - 5)^2 - 3$
  $(x - 5)^2 + (y + 3)^2 \leq 9$
  $x > 5$

23. $\dfrac{y^2}{16} - \dfrac{x^2}{9} \geq 1$
  $\dfrac{x^2}{25} + \dfrac{y^2}{36} \leq 1$
  $x \geq 0$ and $y \geq 0$

24. $x^2 + y^2 \leq 49$
  $x^2 - y^2 \geq 16$
  $x \leq 0$ and $y \geq 0$

REVIEW PROBLEMS

The following exercises review parts of Section 7.1. Doing these problems will help prepare you for the next section.

25. Draw a coordinate plane and plot the following points.
  (a) $A(-2, 0)$ **(b)** $B(4, 3)$ **(c)** $C(0, 1)$
  (d) $D(6, -4)$ **(e)** $E(-2, -5)$

26. Find the ordered pair for each point shown in the coordinate plane.

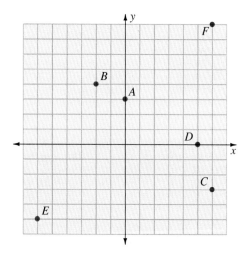

E N R I C H M E N T   E X E R C I S E S

Graph the inequality.

1. $|x| + |y| \le 5$

For Exercises 2–4, graph the system of inequalities.

2. $y \le -0.46x^2 + 4.2$
  $4.4x^2 + 20y^2 \le 88$

3. $x^2 + y^2 \le 12.25$
  $\dfrac{x^2}{6.76} + \dfrac{y^2}{30.25} \ge 1$

4. $\dfrac{y^2}{10.24} - \dfrac{x^2}{7.84} \ge 1$
  $x \ge (y - 5.5)^2$

Summary and review

Examples

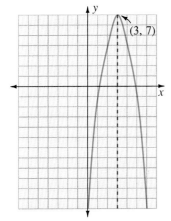

$y = -2(x - 3)^2 + 7$

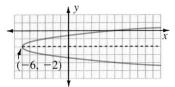

$x = 3(y + 2)^2 - 6$

Given $y = 3x^2 - 15x + 4$, since $a = 3$ is positive, y has a smallest value at $x = -\dfrac{-15}{2 \cdot 3} = \dfrac{5}{2}$ and this smallest value is $y = -\dfrac{59}{4}$.

Parabolas (9.1)

The two standard forms for a parabola are the following:

$$y = a(x - h)^2 + k$$
$$x = a(y - k)^2 + h$$

More Parabolas (9.2)

Given $y = ax^2 + bx + c$, where $a \neq 0$, then y has an extreme value that occurs when $x = -\dfrac{b}{2a}$. If $a > 0$, then y has a smallest value, and if $a < 0$, then y has a largest value.

Circles and ellipses (9.3)

The equation of a **circle** with center at (h, k) and radius r is

$$(x - h)^2 + (y - k)^2 = r^2$$

The equation of an **ellipse** is

$$\frac{(x - h)^2}{a^2} + \frac{(y - k)^2}{b^2} = 1$$

$\dfrac{(x + 2)^2}{16} + \dfrac{(y - 3)^2}{9} = 1$

Examples

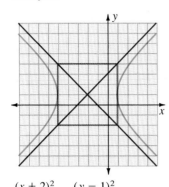

$$\frac{(x + 2)^2}{9} - \frac{(y - 1)^2}{9} = 1$$

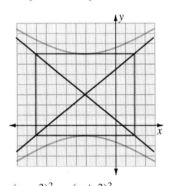

$$\frac{(y - 3)^2}{16} - \frac{(x + 3)^2}{25} = 1$$

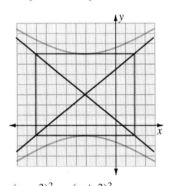

$$xy = -10$$

Solve:

$2x^2 - y^2 = 1$
$x^2 + y^2 = 5$

Hyperbolas (9.4)

The standard equations for **hyperbolas** are the following:

1. $\dfrac{(x - h)^2}{a^2} - \dfrac{(y - k)^2}{b^2} = 1$

2. $\dfrac{(y - k)^2}{b^2} - \dfrac{(x - h)^2}{a^2} = 1$

3. $xy = c$

Nonlinear systems of equations (9.5)

In this section, we solved nonlinear systems of equations.

Examples

We add the equations to eliminate y.

$$2x^2 - y^2 = 1$$
$$\underline{x^2 + y^2 = 5}$$
$$3x^2 = 6$$
$$x^2 = 2$$
$$x = \pm\sqrt{2}$$

To find y, replace x by $\pm\sqrt{2}$ in the second equation.

$$(\pm\sqrt{2})^2 + y^2 = 5$$
$$2 + y^2 = 5$$
$$y^2 = 3$$
$$y = \pm\sqrt{3}$$

The solutions are $(-\sqrt{2}, -\sqrt{3})$, $(-\sqrt{2}, \sqrt{3})$, $(\sqrt{2}, -\sqrt{3})$, $(\sqrt{2}, \sqrt{3})$

Nonlinear systems of inequalities (9.6)

Graphing a nonlinear system of inequalities is similar to graphing a linear system of inequalities.

Graph the system:

$$xy \geq 8$$
$$y \geq x^2 + 2$$

We graph each inequality, then shade the intersection as shown in the following figure.

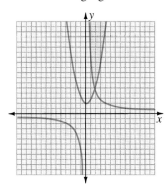

Section 9.1

For Exercises 1–6, graph each equation.

1. $y = x^2 - 4$

2. $y = -2x^2 + 10$

3. $y = -3(x + 2)^2 + 6$

4. $x = 2(y + 1)^2 - 4$

5. $y = 2x^2 - 4x + 5$

6. $y = -x^2 + 2x$

Section 9.2

7. Find the intercepts of the parabola given by

$$y = 2x^2 + 5x - 12$$

8. Given: $y = 4x^2 + 16x + 27$. **(a)** Determine if y has a largest or smallest value. **(b)** Find the value of x at which this extreme value occurs. **(c)** Find the extreme value.

9. A projectile is fired vertically upward. The height, h (in feet), that it is above the ground after t seconds is given by

$$h = 96t - 16t^2$$

When will the projectile reach its maximum height? What is the maximum height?

Section 9.3

For Exercises 10–12, find the distance between the two points.

10. $(3, 9)$ and $(6, 4)$ **11.** $(-2, 0)$ and $(4, 5)$ **12.** $(0, 0)$ and $\left(\dfrac{1}{2}, -\dfrac{3}{2} \right)$

For Exercises 13 and 14, find the equation of the circle with the given center and radius.

13. The center is $(3, -2)$ and the radius is 1.   **14.** The center is $(-4, -9)$ and the radius is 6.

For Exercises 15–18, draw the graph of the circle.

15. $(x + 2)^2 + (y - 6)^2 = 4$ **16.** $(x + 3)^2 + y^2 = 9$

17. $x^2 - 2x + y^2 + 2y - 23 = 0$ **18.** $x^2 + y^2 - 8y = 0$

For Exercises 19–22, draw the graph of each equation.

19. $\dfrac{x^2}{25} + \dfrac{y^2}{4} = 1$ **20.** $x^2 + 4y^2 = 4$

21. $\dfrac{(x + 4)^2}{36} + \dfrac{(y + 2)^2}{25} = 1$ **22.** $4x^2 - 24x + 9y^2 - 72y + 144 = 0$

For Exercises 23–29, draw the graph of each equation.

23. $x^2 - \dfrac{y^2}{4} = 1$

24. $\dfrac{y^2}{9} - \dfrac{x^2}{36} = 0$

25. $xy = -12$

26. $xy = 14$

27. $\dfrac{(x-2)^2}{9} - \dfrac{(y-5)^2}{16} = 1$ **28.** $4y^2 + 24y - x^2 + 12x - 4 = 0$ **29.** $4x^2 + 9(y-2)^2 = 0$

Section 9.5

For Exercises 30–35, solve each system.

30. $x^2 + y^2 = 20$
$\quad\quad y = 2x$

31. $3x^2 - 8y^2 = 1$
$\quad\quad\quad x = 5$

32. $2x^2 - 9y^2 = 1$
$\quad\quad\quad x - 3y = 0$

33. $x^2 - y^2 = 13$
$\quad\ x = y^2 + 1$

34. $2y^2 - x^2 = 26$
$\quad\ \ y^2 + x^2 = 1$

35. $x^2 + y^2 = 5$
$\quad\quad\ xy = -2$

36. The perimeter of a rectangular garden is 44 feet and its area is 120 square feet. Find the dimensions of the garden.

Section 9.6

For Exercises 37–40, graph the inequality.

37. $x^2 + 9y^2 \geq 9$

38. $x^2 - 4y^2 > 4$

39. $(x - 4)^2 + (y + 7)^2 < 25$

40. $xy \leq 6$

For Exercises 41–46, graph the system of inequalities.

41. $y \geq x^2 - 9$
  $y \leq -x^2 + 9$

42. $y^2 - x^2 \geq 1$
  $y \leq 3$
  $y \geq 0$

43. $y \leq -x^2 - 1$
  $xy < -4$
  $y \geq -8$

44. $\dfrac{x^2}{4} - \dfrac{y^2}{9} \geq 1$
  $\dfrac{x^2}{9} + \dfrac{y^2}{4} \leq 1$

45. $xy > 6$
  $2y - 3x \geq 0$

46. $x^2 - y^2 \geq 9$
  $x^2 + y^2 > 16$

CHAPTER 9 **TEST**

Section 9.1

For problems 1–4, graph each equation.

1. $y = 2x^2 - 10$

2. $y = -3x^2 + 14$

3. $y = 2(x - 5)^2 + 1$ **4.** $x = -(y + 4)^2 + 7$

Section 9.2

5. Find the intercepts of the parabola given by

$$y = x^2 - 2x - 5$$

6. Given: $y = -3x^2 + 12x - 8$. **(a)** Determine if y has a largest or smallest value. **(b)** Find the value of x at which this extreme value occurs. **(c)** Find the extreme value.

Section 9.3

7. Find the distance between the two points $(-3, 2)$ and $(1, 5)$.

8. Find the equation of the circle whose center is at the point $(-1, 6)$ and whose radius is 3.

9. Draw the graph of $x^2 - 8x + y^2 + 6y = -16$.

For problems 10–12, draw the graph of each equation.

10. $\dfrac{x^2}{49} + \dfrac{y^2}{25} = 1$ **11.** $\dfrac{(x - 6)^2}{36} + \dfrac{(y + 3)^2}{25} = 1$ **12.** $4x^2 + 9y^2 + 32x - 36y = -64$

Section 9.4

For problems 13–16, draw the graph of each equation.

13. $x^2 - \dfrac{y^2}{9} = 1$

14. $xy = -18$

15. $\dfrac{(y+3)^2}{9} - \dfrac{(x-4)^2}{16} = 1$

16. $9x^2 - 4y^2 = 0$

Section 9.5

For problems 17 and 18, solve the system.

17. $x^2 + y^2 = 13$
$3x + 2y = 0$

18. $2x^2 + 5y^2 = 38$
$xy = 6$

Section 9.6

For problems 19 and 20, graph the inequality.

19. $16x^2 + 9y^2 \leq 144$

20. $xy > -10$

For problems 21 and 22, graph the system of inequalities.

21. $y \leq -x^2 + 4$
$y \geq x^2 - 4$

22. $xy < 12$
$5x + 8y \leq 80$

Functions

Overview

There are many examples in the real world of dependence of one quantity upon another: the price of soybeans depends upon the supply; the level of blood sugar in the body after a meal depends upon the amount of insulin present; and the height of a rocket depends upon the time elapsed since ignition. Mathematics enters the picture if we can express the relationship between two quantities by an equation.

10.1

Relations and Functions

O B J E C T I V E S

▶ To understand the definition of relation and function

▶ To determine when a relation is also a function

▶ To determine the domain and the range of a function

In this section, we start with the concept of relation. Loosely speaking, a relation is a correspondence. For example, we can correspond the three people, Jay, Lyn, and Sam, to telephone numbers of phones that they use. We can display this correspondence in the following way:

$$
\begin{array}{ll}
\text{Jay} \longrightarrow & 821\text{-}0511 \\
\text{Lyn} \diagup\!\!\!\!\longrightarrow & 683\text{-}4111 \\
\text{Sam} \diagup &
\end{array}
$$

Jay and Lyn are brother and sister living at home, so both are assigned their common home phone number 821-0511. Lyn is also assigned her office phone number 683-4111. Lyn shares the office with Sam, so Sam is also assigned 683-4111. Sam's home phone has been disconnected.

We can think of this relation as an assignment between the two sets {Jay, Lyn, Sam} and {821-0511, 683-4111}. The first set is called the *domain* and the second set is called the *range* of the relation.

In general, the concept of relation can be defined in two different but equivalent ways. We give both definitions.

D E F I N I T I O N

Two definitions of relation

1. A **relation** is a correspondence between two sets, where each element of the first set is assigned one or more elements from the second set. The first set is called the **domain** and the second set is called the **range** of the relation.
2. A **relation** is a set of ordered pairs (x, y), where the x values come from a set called the **domain** and the y values come from a set called the **range** of the relation.

To see how these two definitions are the same, we first define a relation using the correspondence idea. We choose the first set (domain) to be $\{a, b, c\}$ and the second set (range) to be $\{1, 2, 3\}$. For this relation, we correspond the letter a to the number 1, the letter b to the number 2, and the letter c to the number 3. We can display this relation using the arrow notation:

$$\text{domain} \quad \text{range}$$

$a \longrightarrow 1$	means "a is assigned 1"
$b \longrightarrow 2$	means "b is assigned 2"
$c \longrightarrow 3$	means "c is assigned 3"

From definition 2, this relation can be displayed as a set of ordered pairs. The connection between the two definitions is as follows:

definition 1		definition 2
"a is assigned 1"	means	$(a, 1)$ is a member of the relation
"b is assigned 2"	means	$(b, 2)$ is a member of the relation
"c is assigned 3"	means	$(c, 3)$ is a member of the relation

Therefore, by definition 2, the relation is the set of ordered pairs $\{(a, 1), (b, 2), (c, 3)\}$.

Example 1   Consider the relation R that is given by

domain   range

$$0 \searrow 1$$
$$2$$

$$1 \longrightarrow 5$$

Express R as a set of ordered pairs. What is the domain and range of R?

Solution $R = \{(0, 1), (0, 2), (1, 5)\}$. The domain of R is $\{0, 1\}$ and the range of R is $\{1, 2, 5\}$.   ▲

Another way to specify a relation is to give a rule, or equation, that defines the correspondence. For example, the equation $y^2 = x - 2$ defines a relation of pairs (x, y) that satisfy the equation. For instance, when $x = 6$, $y^2 = 6 - 2 = 4$, so $y = \pm 2$ and the ordered pairs $(6, -2)$ and $(6, 2)$ are members of the relation.

What is the domain of a relation that is defined by an equation? We will use the following rule:

R U L E

Agreement on domain

If the domain is not specified, the **domain of a relation** is assumed to be the largest set of real numbers that can replace x to produce real numbers for y.

We can visualize a relation by drawing its graph.

D E F I N I T I O N

The **graph of a relation** is the graph of its ordered pairs.

The way to obtain information about a relation from its graph is illustrated in the next two examples.

Example 2   The graph of a relation R is shown in the following figure. Write R as a set of ordered pairs of numbers. What is the domain and the range of R?

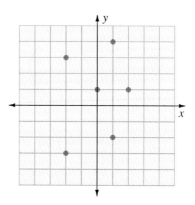

Solution   The graph of R consists of the six points shown in the figure. The relation is the set of coordinates of these points. Therefore,

$$R = \{(-2, -3), (-2, 3), (0, 1), (1, -2), (1, 4), (2, 1)\}$$

By looking at the elements of R, we can "read off" the domain and range.

domain $= \{-2, 0, 1, 2\}$ and   range $= \{-3, -2, 1, 3, 4\}$ ▲

Example 3   Draw the graph of the relation

$$\left\{ (x, y) \,\middle|\, \frac{x^2}{25} + \frac{y^2}{16} = 1 \right\}$$

From the graph, determine the domain and range.

Solution   The graph of this relation is the ellipse shown in the following figure. From the graph, the domain can be seen to be the red interval on the x-axis given by $[-5, 5] = \{x \mid -5 \leq x \leq 5\}$ and the range is the yellow interval on the y-axis given by $[-4, 4] = \{y \mid -4 \leq y \leq 4\}$.

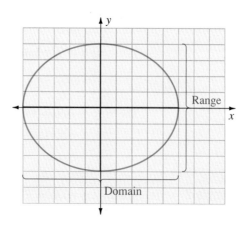

A very important concept in mathematics is that of a *function*. A function is a special kind of relation. Like the definition of relation, a function has two equivalent definitions:

DEFINITION

Two definitions of function

1. A **function** is a correspondence that assigns exactly one value from the range to each value from the domain.
2. A **function** is a set of ordered pairs in which no two different ordered pairs have the same first entry. That is, no first entry is repeated.

If the function is given by an equation such as $y = 4 - 3x$, we say that **y is a function of x.** The variable x takes on the numbers in the domain and is called the **independent variable.** The variable y takes on a value in the range when a domain value is chosen and is called the **dependent variable.** Note that y *depends* upon the value for x *independently* chosen from the domain.

N O T E   *Notice that all functions are relations, but we shall see that not all relations are functions.*

Example 4   Determine which of the following relations are functions.

(a) $a \longrightarrow 1$
$\quad b \nearrow$
$\quad c \longrightarrow 2$

(b) $a \searrow 1$
$\qquad\qquad 2$
$\quad b \longrightarrow 3$
$\quad d \nearrow$

(c) $\left\{ (-3, 0), (-2, 1), \left(1, \dfrac{1}{2}\right), \left(2, \dfrac{1}{4}\right) \right\}$

(d) $\left\{ \left(0, \dfrac{4}{3}\right), (2, 3), (-1, 1), (0, 1) \right\}$

(e) $\{ (2, 1), (3, 1), (4, -1) \}$

Solution   **(a)** This relation is a function, since each member of the domain is assigned exactly one member of the range. That is, a is assigned 1, b is assigned 1, and c is assigned 2. Notice that 1 is used twice, but it does not violate the definition of function.

(b) This relation is not a function, since the letter a corresponds to two different members of the range.

(c) This relation is a function, since no first entry is repeated.

(d) This relation is not a function, since 0 is the first entry in the two different ordered pairs $\left(0, \dfrac{4}{3}\right)$ and $(0, 1)$.

(e) This relation is a function, since no first entry is repeated. Note that a function may have the same *second* entry in two different ordered pairs. That is, the second entry may be repeated. In this example, the number 1 is used twice as the second entry.   ▲

In the next two examples, we again investigate the connection between a relation and its graph.

Example 5   Shown in the following figure is the graph of a relation R. Write R as a set of ordered pairs. Is this relation a function?

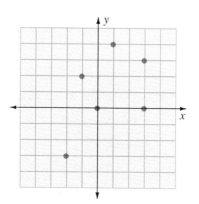

Solution   We can write R as a set of ordered pairs by listing the coordinates of the plotted points. Therefore,

$$R = \{(-2, -3), (-1, 2), (0, 0), (1, 4), (3, 0), (3, 3)\}$$

From this list, we see that R is not a function, since $(3, 0)$ and $(3, 3)$ are two members of R with the same first coordinate.   ▲

Example 6   Determine the domain and range for the relation S whose graph is shown in the following figure. Is the relation a function?

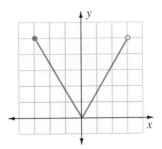

Solution   The domain is the restriction of the x-coordinate. If we project the graph down
             onto the x-axis, the x values span the interval from -3 to 3, including -3 but
             excluding 3. As shown in red on the x-axis in the following figure, the domain
             is $[-3, 3) = \{x \mid -3 \le x < 3\}$. To find the range, project the graph onto the y-
             axis. The range is $[0, 5] = \{y \mid 0 \le y \le 5\}$ as shown in yellow on the y-axis.

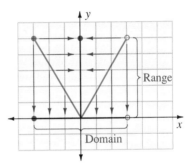

             Is S a function? Notice that for each x in the domain there is exactly one y in
             the range corresponding to x. This means that all ordered pairs (x, y) have dif-
             ferent first coordinates, so S is a function.   ▲

             N O T E   *Notice that the open circle in the graph of S (in Example 6) is
             used to indicate that (3, 5) is "missing" from the graph and that 3 is not in the
             domain of S.*

                 In Example 5, the relation R is not a function because $(3, 0)$ and $(3, 3)$ are
             both members of R. Notice that in the graph of R, these two points lie on the
             same vertical line $x = 3$.
                 In Example 6, the relation S is a function. Notice that no matter where a
             vertical line is placed in the graph of S, it meets the graph in precisely one
             point.
                 We generalize these observations from the last two examples and call it the
             vertical line test.

R U L E

Vertical line test

1. If any vertical line intersects the graph of a relation in more than one point, the relation is not a function.

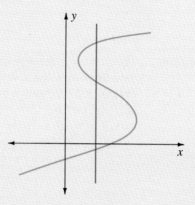

2. If any vertical line intersects the graph of a relation in at most one point, the relation is a function.

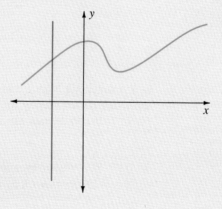

Example 7   Use the vertical line test to check if the relation $\{(x, y)\,|\,x^2 + y^2 = 36\}$ is a function.

Solution   The graph of this relation is shown in the following figure. From the graph, we see that a vertical line like the red one drawn in the figure will meet the circle in more than one point. Therefore, the relation is not a function.

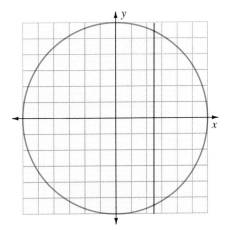

Example 8   Drawn in the figure below is the graph of a relation. Determine the domain and the range. Use the vertical line test to determine if the relation is a function.

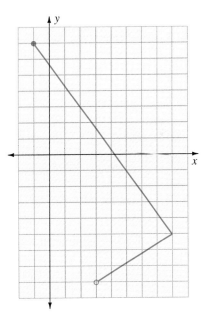

Solution   By projecting the graph onto the x-axis (see Figure (a) on the next page), we see that the domain is the interval $[-1, 8] = \{x \mid -1 \le x \le 8\}$ shaded in red.

To find the range, project the graph onto the y-axis as shown in Figure (b). The range is the interval $(-8, 7] = \{y \mid -8 < y \le 7\}$ shaded in yellow.

Finally, the relation is not a function, since we can draw a vertical line that meets the graph in more than one point. For example, the vertical red line in Figure (c) intersects the graph in two points.

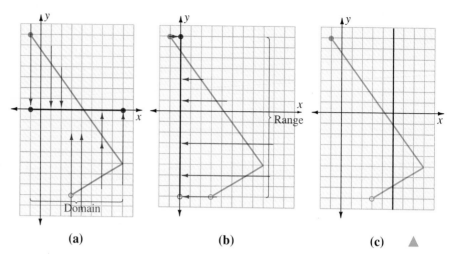

(a)   (b)   (c)   ▲

As we have seen so far, many relations and functions are given in equation form. It is easier to just write the equation rather than use the set-builder notation. For example, rather than write $\{(x, y) \mid y = x + 1\}$, we simply write $y = x + 1$. If the domain is not specified, it is assumed to be the largest set of replacements for x so that the equation is defined as a real number. For example, we cannot use any value of x that makes a denominator zero or requires taking the square root of a negative number.

Example 9   State the domain for $y = \dfrac{x}{3x - 9}$.

Solution   The domain is the set of all real numbers x such that the denominator, $3x - 9$, is not zero. That is, the domain is equal to $\{x \mid x \neq 3\}$.   ▲

Example 10   Find the domain and range for $y = (x - 1)^2 - 2$. Is this relation also a function?

Solution   To find the domain and range of this relation, we draw its graph.

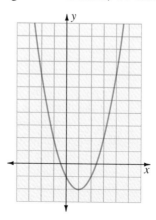

From the figure, we conclude that the domain is the set of all real numbers and the range is $\{y \mid y \geq -2\}$, which can be written in interval notation as $[-2, +\infty)$. This parabola is the graph of a function, since any vertical line crosses the graph in precisely one point.   ▲

E X E R C I S E   S E T   1 0 . 1

For Exercises 1 and 2, write the relation as a set of ordered pairs. State the domain and range.

1. $a \searrow 5$
   $b \searrow 3$
       2
   $c \nearrow$

2. $-1 \longrightarrow 1$
   $1 \nearrow$
   $9 \searrow 3$
      -3

For Exercises 3 and 4, from the graph, write the relation R as a set of ordered pairs. What is the domain and the range of R?

3.
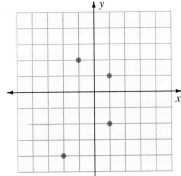

4.
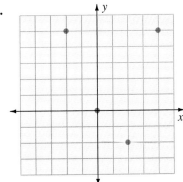

For Exercises 5 and 6, draw the graph of the relation. From the graph, determine the domain and the range.

5. $\left\{ (x, y) \,\middle|\, \dfrac{x^2}{9} + \dfrac{y^2}{4} = 1 \right\}$

6. $\{ (x, y) \mid x^2 + y^2 = 16 \}$

For Exercises 7–18, state the domain and range of each relation and determine if it is also a function.

7. $a \longrightarrow 4$
   $b \longrightarrow 5$
   $c \longrightarrow 6$

8. $m \longrightarrow -1$
   $n \longrightarrow -2$
   $p \longrightarrow 0$

9. 1 $\longrightarrow$ *a*

2 $\searrow$

3 $\longrightarrow$ *b*

10. 4 $\longrightarrow$ 2

-2

0 $\diagup$ 0

11. $\{(2, 3), (-3, 2), (-1, -2), (1, -1)\}$

12. $\{(6, -2), (2, -6), (0, 7), (1, 5)\}$

13. $\{(2, -2), (3, -5), (3, -6), (7, -1)\}$

14. $\{(3, -4), (-4, 6), (3, -5), (0, 0)\}$

15. $\{(-5, 1), (4, -1), (-3, 0), (6, -3)\}$

16. $\{(9, 8), (0, 1), (-4, 3), (-4, 0)\}$

17. $\{(-1, -3), (-3, 2), (-2, -2), (-3, 10)\}$

18. $\{(10, 9), (-9, 10), (-10, 9), (9, -10)\}$

For Exercises 19–22, from the graph, write the relation R as a set of ordered pairs. Determine the domain and range of R and whether this relation is also a function.

19.

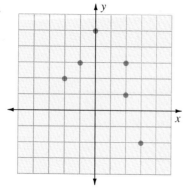

20.

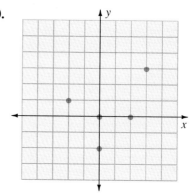

21.

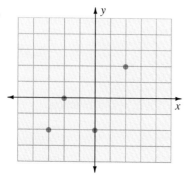

22.

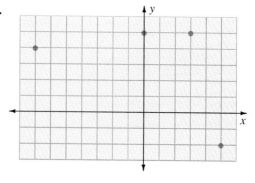

For Exercises 23–30, determine the domain and the range of each relation from its graph. Is the relation also a function?

23.

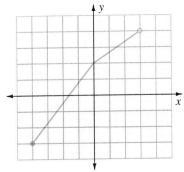

24.

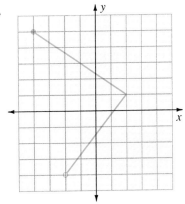

25.

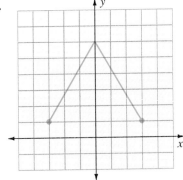

26.

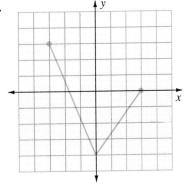

27.

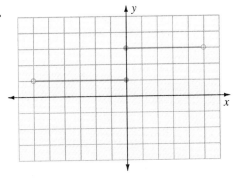

28.

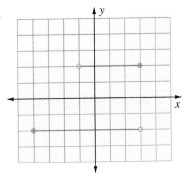

29.

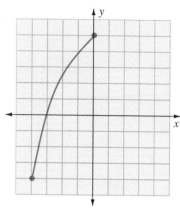

30.

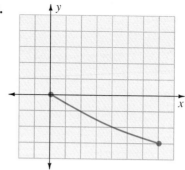

For Exercises 31–48, state the domain for each function.

31. $y = \dfrac{2}{x + 3}$

32. $y = -\dfrac{1}{x - 2}$

33. $y = \dfrac{x + 1}{x - 1}$

34. $y = \dfrac{2x - 3}{x + 5}$

35. $y = \sqrt{x - 10}$

36. $y = \sqrt{x + 2}$

37. $y = \sqrt{3x + 12}$

38. $y = \sqrt{4x - 8}$

39. $y = \dfrac{1}{\sqrt{2 - 10x}}$

40. $y = \dfrac{1}{\sqrt{2 - 3x}}$

41. $y = \dfrac{1}{(x + 1)(2x - 3)}$

42. $y = \dfrac{x}{(x - 1)(4x - 2)}$

43. $y = -\dfrac{3}{x^2 - 16}$

44. $y = \dfrac{2x}{1 - x^2}$

45. $y = \dfrac{x^2}{6x^2 - 7x + 2}$

46. $y = \dfrac{x}{4x^2 - 17x + 4}$

47. $y = \dfrac{1}{5 - 9x - 2x^2}$

48. $y = \dfrac{1}{3 - 4x - 4x^2}$

For Exercises 49–62, by graphing the relation, determine the domain and range. Is the relation also a function?

49. $x^2 + y^2 = 25$

50. $x^2 + y^2 = 1$

51. $y = x^2 + \dfrac{1}{2}$

52. $y = -x^2 + 1$

53. $y = x^2 - 2x - 3$

54. $y = x^2 + 4x + 1$

55. $(x - 1)^2 + (y + 2)^2 = 16$

56. $(x + 3)^2 + (y - 4)^2 = 25$

57. $\dfrac{x^2}{9} + \dfrac{y^2}{16} = 1$

58. $\dfrac{x^2}{25} + \dfrac{y^2}{49} = 1$

59. $4x^2 + 9y^2 = 16$

60. $16x^2 + 9y^2 = 81$

61. $\dfrac{x^2}{9} - \dfrac{y^2}{4} = 1$

62. $\dfrac{y^2}{16} - \dfrac{x^2}{9} = 1$

63. The blood sugar of a diabetic is regulated by a mixture of time-released insulin. The following graph shows the relationship between the amount of blood sugar y at time t, where t is the number of hours after the insulin was injected and y is measured in milligrams per deciliter (mg/dl).

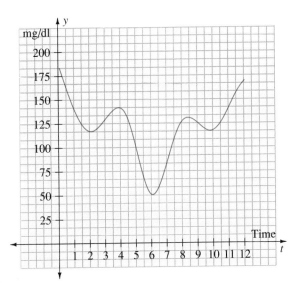

(a) Is y a function of t?
(b) What is the domain?
(c) At what time is the blood sugar the lowest?

(d) What is the lowest blood sugar?

64. An advertising company rates the influence a TV commercial has on a scale of 0 to 10. A particular commercial is aired in a large viewing area and a graph is constructed so that the horizontal axis is time t, in days, since the commercial was aired. The vertical axis measures the influence y of the commercial.

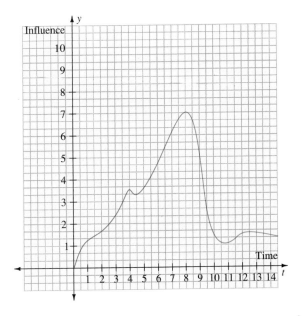

(a) Is y a function of t?
(b) What is the domain?
(c) When is the influence the greatest?

(d) What is the greatest influence?

REVIEW PROBLEMS

The following exercises review parts of Section 4.1. Doing these problems will help prepare you for the next section.

Simplify.

65. $\dfrac{(x-3)-(a-3)}{x-a}$

66. $\dfrac{(4x+1)-(4a+1)}{x-a}$

67. $\dfrac{(2x^2-5)-(2a^2-5)}{x-a}$

68. $\dfrac{(x^2+7)-(a^2+7)}{x-a}$

69. $\dfrac{4-2(x+h)-(4-2x)}{h}$

70. $\dfrac{5(x+h)-1-(5x-1)}{h}$

71. $\dfrac{(x+h)^2-x^2}{h}$

72. $\dfrac{2(x+h)^2-1-(2x^2-1)}{h}$

E N R I C H M E N T   E X E R C I S E S

Determine the domain and range of the relation whose graph is given. Is the relation also a function?

1.

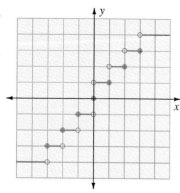

2.

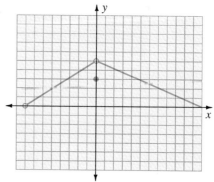

3.

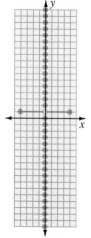

4.

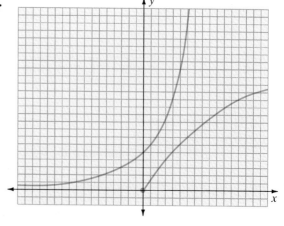

5. Consider the relation defined by

$$y = \begin{cases} -x^2, & \text{if } x \le 0 \\ 2, & \text{if } x = 0 \\ x^2, & \text{if } x > 0 \end{cases}$$

(a) Draw the graph of this relation.
(b) What is the domain and range?
(c) Is this relation also a function? Why?

6. Consider the relation defined by

$$y = \frac{|x - 1|}{x - 1}$$

(a) Draw the graph of this relation.
(b) What is the domain and range?
(c) Is this relation also a function? Why?

10.2

Function Notation and the Algebra of Functions

O B J E C T I V E S

▶ *To use the function notation*

▶ *To form the quotient* $\dfrac{f(x) - f(a)}{x - a}$ *and simplify*

▶ *To form the quotient* $\dfrac{f(x + h) - f(x)}{h}$ *and simplify*

▶ *To learn the algebra of functions*

Consider the function f defined by $y = 2x + 3$. We say that y is a function of x and write $y = f(x)$. The notation $f(x)$ is read "f of x," and can be used in place of y in the equation $y = 2x + 3$. Therefore, we can write

$$f(x) = 2x + 3$$

Function notation is convenient when dealing with functions. For example, when x is -1, $y = 2(-1) + 3 = 1$. Using the function notation,

$$f(x) = 2x + 3$$

$$f(-1) = 2(-1) + 3 \qquad \text{Replace } x \text{ by } -1.$$

$$= 1$$

N O T E   *Keep in mind that the function notation, f(x), does not mean "f times x." The symbolism f(x) is simply another name for y, that is, y = f(x).*

We can think of a function f as a machine that converts the input x into output $f(x)$. See the following figure.

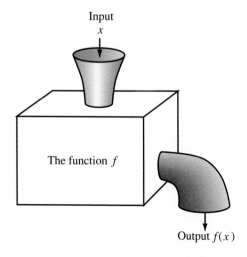

Input
x

The function f

Output $f(x)$

Example 1   If $f(x) = x^2 - x - 2$, find $f(-1)$, $f(0)$, and $f(3)$.

Solution   We replace x by -1, 0, and 3, respectively, in the formula for f.

$$f(x) = x^2 - x - 2$$
$$f(-1) = (-1)^2 - (-1) - 2 = 0$$
$$f(0) = 0^2 - 0 - 2 = -2$$
$$f(3) = 3^2 - 3 - 2 = 4$$ ▲

When we write $y = f(x)$, we say that $f(x)$ is the **value of f at x.** For example, if $f(x) = 4 - 3x^2$, then $f(2) = 4 - 3(2)^2 = -8$. That is, the value of f at 2 is -8.

Remember that f itself is a function and, therefore, is a set of ordered pairs of numbers. The connection between the function notation and f itself is the following.

If $b = f(a)$, then (a, b) is a member of f.

Conversely,

If (a, b) is a member of f, then $b = f(a)$.

Therefore, the second entry b of an ordered pair $(a, b) \in f$ is the value of the function at a.

Example 2   Let $g = \left\{ (-2, -1), (-1, -2), (0, 1), \left(11, \dfrac{4}{3} \right) \right\}$. Then,

$$g(-2) = -1, \ g(-1) = -2, \ g(0) = 1, \text{ and } g(11) = \frac{4}{3}$$ ▲

We now give more examples of using the function notation.

Example 3   If $f(x) = 2x^2 - 3x + 4$, then

$$f(-2) = 2(-2)^2 - 3(-2) + 4 = 18$$
$$f\left(-\frac{1}{2} \right) = 2\left(-\frac{1}{2} \right)^2 - 3\left(-\frac{1}{2} \right) + 4 = 6$$
$$f(0) = 2(0)^2 - 3(0) + 4 = 4$$
$$f(a) = 2a^2 - 3a + 4$$
$$f(x^2) = 2(x^2)^2 - 3(x^2) + 4 = 2x^4 - 3x^2 + 4$$
$$f(a + h) = 2(a + h)^2 - 3(a + h) + 4$$ ▲

Example 4   If $f(x) = 3x - 5$, form the quotient $\dfrac{f(x) - f(a)}{x - a}$ and simplify.

Solution

$$\frac{f(x) - f(a)}{x - a} = \frac{3x - 5 - (3a - 5)}{x - a}$$

$$= \frac{3x - 3a}{x - a}$$

$$= \frac{3(x - a)}{x - a} \qquad \text{Simplify.}$$

$$= 3$$ ▲

Example 5   If $f(x) = 1 - 3x^2$, form the quotient $\dfrac{f(x + h) - f(x)}{h}$ and simplify.

Solution   In the numerator, we have $f(x + h)$. To find $f(x + h)$, replace x by $x + h$ in the equation $f(x) = 1 - 3x^2$. Therefore,

$$f(x + h) = 1 - 3(x + h)^2$$

$$= 1 - 3(x^2 + 2xh + h^2)$$

$$= 1 - 3x^2 - 6xh - 3h^2$$

Now,

$$\frac{f(x + h) - f(x)}{h} = \frac{1 - 3x^2 - 6xh - 3h^2 - (1 - 3x^2)}{h}$$

$$= \frac{1 - 3x^2 - 6xh - 3h^2 - 1 + 3x^2}{h}$$

$$= \frac{-6xh - 3h^2}{h} \qquad \text{Simplify.}$$

$$= \frac{-3h(2x + h)}{h} \qquad \text{Factor.}$$

$$= \frac{-3\overset{1}{\cancel{h}}(2x + h)}{\underset{1}{\cancel{h}}}$$

$$= -3(2x + h)$$ ▲

The four basic operations—addition, subtraction, multiplication, and division—can be extended to functions according to the following definition.

DEFINITION

Let f and g be functions with a common domain. Then

1. $f + g$ is the function defined as

$$(f + g)(x) = f(x) + g(x)$$

2. $f - g$ is the function defined as

$$(f - g)(x) = f(x) - g(x)$$

3. fg is the function defined as

$$(fg)(x) = f(x)g(x)$$

4. $\dfrac{f}{g}$ is the function defined as

$$\left(\frac{f}{g}\right)(x) = \frac{f(x)}{g(x)}, \qquad \text{provided } g(x) \neq 0$$

The function notation is convenient to use in the algebra of functions.

Example 6   If $f(x) = x + 1$ and $g(x) = x^2 - 1$, find the following.

(a) $(f + g)(3)$ (b) $(f - g)(-1)$

(c) $(fg)(a)$ (d) $\left(\dfrac{g}{f}\right)(x)$

Solution   (a) $(f + g)(3) = f(3) + g(3)$. Now, $f(3) = 3 + 1 = 4$ and $g(3) = (3)^2 - 1 = 8$, therefore,

$$(f + g)(3) = 4 + 8 = 12$$

(b) $(f - g)(-1) = f(-1) - g(-1)$

$$= (-1 + 1) - [(-1)^2 - 1]$$

$$= 0 - 0 = 0$$

(c) $(fg)(a) = f(a)g(a) = (a + 1)(a^2 - 1)$

(d) $\left(\dfrac{g}{f}\right)(x) = \dfrac{g(x)}{f(x)} = \dfrac{x^2 - 1}{x + 1}$

$$= \frac{(x + 1)(x - 1)}{x + 1}$$

$$= x - 1, \text{ where } x \neq -1 \qquad \blacktriangle$$

N O T E   *The domain of the quotient function $\frac{g}{f}$ in Part (d) of the previous example is all x such that f(x) ≠ 0. Since f(−1) = 0, the domain of $\frac{g}{f}$ is $\{x \mid x \neq -1\}$. Notice that the equation for $\frac{g}{f}$ simplified to $\left(\frac{g}{f}\right)(x) = x - 1$ and this expression is defined when x = −1. However, the domain for $\frac{g}{f}$ is still $\{x \mid x \neq -1\}$.*

There are applications from business that involve the algebra of functions. We have dealt with these concepts in previous chapters, but we have not used the function notation.

Suppose a company makes and sells x units of a product a day (or week, month, etc.). Associated with the **level of production** of x units are the cost, revenue, and profit. We can think of each as a function of x.

$C(x)$ is the daily cost of making x units.

$R(x)$ is the daily revenue of selling x units.

$P(x)$ is the daily profit from selling x units.

The profit is equal to the difference of the revenue and cost. Using the function notation,

$$P(x) = R(x) - C(x)$$

Example 7   The Banner Tool Company makes and sells x tool kits each day. The daily cost function C, in dollars, is given by

$$C(x) = 18x + 245$$

and the daily revenue function R, in dollars, from selling x tool kits is given by

$$R(x) = -0.06x^2 + 38x$$

(a) Let $P(x)$ be the profit from making and selling x tool kits. Express P as a function of x.

(b) On one particular day, the level of production was 100 tool kits. What was the profit for that day?

Solution   **(a)** The profit function P is given by

$$P(x) = R(x) - C(x)$$
$$= -0.06x^2 + 38x - (18x + 245)$$
$$= -0.06x^2 + 20x - 245$$

(b) The profit when $x = 100$ is

$$P(100) = -0.06(100)^2 + 20(100) - 245$$

$$= -600 + 2000 - 245$$

$$= \$1,155$$

We finish this section with another way of combining functions. This operation on functions is not similar to one of the four basic operations on the set of real numbers. This operation is called *composition*.

D E F I N I T I O N

Let f and g be functions where the range of g is in the domain of f. The **composition of f by g** is defined by

$$(f \circ g)(x) = f[g(x)]$$

Using the function machine idea introduced earlier, we can illustrate the composition of two functions as shown in the following figure. Input x is fed into function machine g, and its output $g(x)$ is then entered into the function machine f. The final output is the quantity $(f \circ g)(x)$.

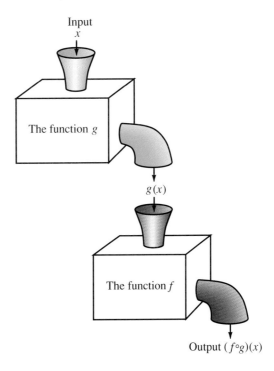

Example 8   If $f(x) = 3x - 4$ and $g(x) = 2x^2 + 1$, find the following.

(a) $(f \circ g)(4)$ (b) $(f \circ g)(-1)$

Solution   (a) Since $(f \circ g)(4) = f[g(4)]$, we first find $g(4)$.

$$g(4) = 2(4)^2 + 1 = 2(16) + 1 = 33$$

Therefore,

$$(f \circ g)(4) = f[g(4)]$$
$$= f[33]$$
$$= 3(33) - 4$$
$$= 99 - 4$$
$$= 95$$

(b) Since $(f \circ g)(-1) = f[g(-1)]$, we first find $g(-1)$.

$$g(-1) = 2(-1)^2 + 1 = 2(1) + 1 = 3$$

Therefore,

$$(f \circ g)(-1) = f[g(-1)]$$
$$= f[3]$$
$$= 3(3) - 4$$
$$= 9 - 4$$
$$= 5 \qquad \blacktriangle$$

Example 9   If $f(x) = x^2$ and $g(x) = 2x - 3$, write a formula for $f \circ g$.

Solution   There are two methods of applying the definition of $f \circ g$ to obtain the formula for the composition of f by g.

Method 1. $(f \circ g)(x) = f[g(x)]$
$$= f[2x - 3] \qquad \text{Use the formula for } g.$$
$$= (2x - 3)^2 \qquad \text{Use the formula for } f.$$

Method 2. $(f \circ g)(x) = f[g(x)]$
$$= [g(x)]^2 \qquad \text{Use the formula for } f.$$
$$= (2x - 3)^2 \qquad \text{Use the formula for } g. \qquad \blacktriangle$$

Example 10   If $f(x) = 4x + 3$ and $g(x) = \sqrt{x}$, write a formula for each of the following.

(a) $(f \circ g)(x)$ (b) $(g \circ f)(x)$

Solution   (a) $(f \circ g)(x) = f[g(x)]$
$$= f[\sqrt{x}] \qquad \text{Use the formula for } g.$$
$$= 4\sqrt{x} + 3 \qquad \text{Use the formula for } f.$$

(b) $(g \circ f)(x) = g[f(x)]$

$\qquad\qquad = g[4x + 3]$ Use the formula for f.

$\qquad\qquad = \sqrt{4x + 3}$ Use the formula for g.   ▲

In Example 10, notice that $(f \circ g)(x) \neq (g \circ f)(x)$. This means that the composition of functions is *not a commutative operation*. That is, it is not true that $f \circ g = g \circ f$ for all functions f and g.

E X E R C I S E   S E T   1 0 . 2

For Exercises 1–10, let $f(x) = 1 - 2x$ and $g(x) = x^2 - 3x + 5$. Find each of the following.

1. $f(1)$ **2.** $f(-2)$ **3.** $g(-4)$ **4.** $g(3)$

5. $f(-10)$ **6.** $f\left(-\dfrac{3}{2}\right)$ **7.** $f(2) + g(0)$ **8.** $g(2) + f(1)$

9. $f(4) - f(-1)$ **10.** $g(2) - g(-2)$

For Exercises 11–18, let

$$f = \{(-1, 1), (0, 0), (1, 1), (\sqrt{2}, 2), (2, -4)\}.$$

Find each of the following.

11. $f(-1)$ **12.** $f(0)$ **13.** $f(1)$ **14.** $f(\sqrt{2})$

15. $f(\sqrt{2}) + f(-1)$ **16.** $f(0) - f(1)$ **17.** $f[f(\sqrt{2})]$ **18.** $f[f(-1)]$

For Exercises 19–30, let $f(x) = 3x + 2$ and $g(x) = 1 - 2x$. Find each of the following.

19. $g(a)$ **20.** $f(z)$ **21.** $g(2 + h)$ **22.** $f(1 + h)$

23. $f(x + h)$ **24.** $g(a + h)$ **25.** $f[g(2)]$ **26.** $f[g(3)]$

27. $g[f(-1)]$ **28.** $g[f(5)]$ **29.** $f[g(x)]$ **30.** $g[f(x)]$

For Exercises 31–40, form the quotient $\dfrac{f(x) - f(a)}{x - a}$ and simplify.

31. $f(x) = 2x$ **32.** $f(x) = -5x$ **33.** $f(x) = 4x + 3$

34. $f(x) = x - 2$ **35.** $f(x) = 1 - x$ **36.** $f(x) = 2 - 3x$

37. $f(x) = x^2$ **38.** $f(x) = x^2 - 6$ **39.** $f(x) = -3x^2$

40. $f(x) = -2x^2$

For Exercises 41–50, form the quotient $\dfrac{f(x + h) - f(x)}{h}$ and simplify.

41. $f(x) = 3x$ **42.** $f(x) = x$ **43.** $f(x) = 2$

44. $f(x) = -4$ **45.** $f(x) = 3x + 5$ **46.** $f(x) = 2x - 6$

47. $f(x) = 4 - 9x$ **48.** $f(x) = 1 - 7x$ **49.** $f(x) = 2x^2 - 11$

50. $f(x) = 5x^2 + 4$

For Exercises 51–54, let $f(x) = 2x + 3$ and $g(x) = 6x^2 + 11x + 3$. Find each of the following.

51. $(f + g)(x)$ **52.** $(f - g)(x)$

53. $(fg)(x)$ **54.** $\left(\dfrac{f}{g}\right)(x)$

For Exercises 55–58, let $f(x) = 3x + 1$ and $g(x) = 9x^2 - 1$. Find each of the following.

55. $(f + g)(x)$ **56.** $(g - f)(x)$

57. $(fg)(x)$ **58.** $\left(\dfrac{g}{f}\right)(x)$

59. Tombstone Pizza Supplies makes and sells x cases of pizza trays a day. The daily cost function C, in dollars, is given by

$$C(x) = 225x + 450$$

and the daily revenue function R from selling x cases of trays is given by

$$R(x) = -0.03x^2 + 675x$$

(a) Let $P(x)$ be the profit from making and selling x cases of trays. Express P as a function of x.

(b) On one particular day, the level of production was 200 cases of trays. What was the profit for that day?

60. Wilderness Accessories makes and sells x 6-gallon portable air tanks each week. The cost $C(x)$, in dollars, of making x tanks is given by

$$C(x) = 65x + 320$$

and the revenue, $R(x)$, from selling x tanks is given by

$$R(x) = -0.15x^2 + 170x$$

(a) Let $P(x)$ be the profit from making and selling x tanks. Express P as a function of x.

(b) One week, the level of production was 400 tanks. What was the profit for that week?

For Exercises 61–64, let $f(x) = 3x - 5$ and $g(x) = 2 - 4x$. Find each of the following.

61. $(f \circ g)(-1)$ **62.** $(f \circ g)(0)$ **63.** $(g \circ f)(3)$ **64.** $(g \circ f)(-2)$

For Exercises 65–68, let $f(x) = -2x^2 + 3$ and $g(x) = x^2 - 4x - 2$. Find each of the following.

65. $(g \circ f)(\sqrt{3})$ **66.** $(f \circ g)(-\sqrt{2})$ **67.** $(g \circ g)(0)$ **68.** $(f \circ f)(-1)$

For Exercises 69–76, find $(f \circ g)(x)$ and $(g \circ f)(x)$.

69. $f(x) = 2x,\ g(x) = 4 - x$ **70.** $f(x) = -3x,\ g(x) = 9x - 1$

71. $f(x) = x^2 + 2,\ g(x) = 4x + 3$ **72.** $f(x) = 3x^2,\ g(x) = 6 - x$

73. $f(x) = \sqrt{x} - 10$, $g(x) = x + 3$ **74.** $f(x) = \sqrt{x} + 2$, $g(x) = 3x + 1$

75. $f(x) = \dfrac{2}{x}$, $g(x) = 4x - 3$ **76.** $f(x) = 2x + 2$, $g(x) = -\dfrac{1}{2x}$

For Exercises 77–80, show that $(f \circ g)(x) = x$ and $(g \circ f)(x) = x$.

77. $f(x) = 3x$, $g(x) = \dfrac{x}{3}$ **78.** $f(x) = \dfrac{5x}{2}$, $g(x) = \dfrac{2x}{5}$

79. $f(x) = 3x - 1$, $g(x) = \dfrac{1}{3}x + \dfrac{1}{3}$ **80.** $f(x) = 3 - 4x$, $g(x) = \dfrac{3}{4} - \dfrac{x}{4}$

REVIEW PROBLEMS

The following exercises review parts of Sections 7.1 and 9.1. Doing these problems will help prepare you for the next section.

Draw the graph of each equation.

Section 7.1

81. $y = -3x + 2$ **82.** $y = 2x - 5$ **83.** $y = -x$

84. $y = \dfrac{2}{3}x - 4$ **85.** $y = \dfrac{3}{2}$ **86.** $y = -\dfrac{5}{4}$

87. $y = 3(x + 3)^2 - 4$ **88.** $y = -2(x - 4)^2 + 3$ **89.** $y = -x^2 - 8x - 17$ **90.** $y = x^2 - 6x + 11$

E N R I C H M E N T   E X E R C I S E S

For Exercises 1–4, find functions $f(x)$ and $g(x)$ so that $F(x) = f[g(x)]$.

1. $F(x) = (3x + 10)^2$

2. $F(x) = \sqrt{1 + 2x^2}$

3. $F(x) = \dfrac{1}{(1 - 7x)^3}$

4. $F(x) = 2^{x-1}$

5. If $f(x) = 2x - 1$, find a function g such that

$$(f \circ g)(x) = x$$

6. If $f(x) = 4 - 3x$, find a function g such that

$$(f \circ g)(x) = x$$

Answers to Enrichment Exercises are on page A.73.

10.3

Some Special Functions

O B J E C T I V E S

▶ *To understand linear functions*

▶ *To understand quadratic functions*

▶ *To understand polynomial functions*

▶ *To understand exponential functions*

In Chapter 7, we have studied the graphs of linear equations. In this section, we will discover that nonvertical lines are the graphs of a special class of functions called *linear functions*.

D E F I N I T I O N

A function f of the form

$$f(x) = mx + b$$

where m and b are constants, is called a **linear function.**

The graph of any linear function is a straight line with slope m and y-intercept b. Some examples of linear functions are the following.

$$f(x) = 2x + 1, \qquad g(x) = -3x, \qquad f(x) = -5$$

When the slope $m = 0$, we obtain the **constant function** $f(x) = b$. For example, $f(x) = 2$ is a constant function. The value of the function f at *any* number x is the number 2. The graph of any constant function is a horizontal line.

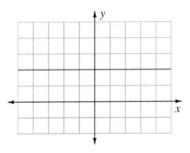

The graph of $f(x) = 2$.

Example 1   Draw the graph of the linear function

$$f(x) = 3x - 2$$

Solution   Recall from Section 7.1 that two points determine a line. Now, $f(0) = -2$ and $f(1) = 1$, so we plot $(0, -2)$ and $(1, 1)$ and draw the straight line between them as shown in the following figure.

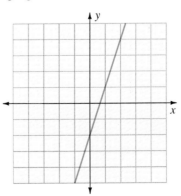

DEFINITION

A function f of the form

$$f(x) = ax^2 + bx + c$$

where a, b, and c are constants with $a \neq 0$, is called a **quadratic function.**

The graph of any quadratic function is a parabola. We practiced drawing parabolas in Section 9.1.

Example 2   Draw the graph of the quadratic function

$$f(x) = x^2 + x - 6$$

Solution   You might want to review the strategy for drawing parabolas on page 627. The y-intercept is -6. We set $f(x) = 0$ to find the x-intercepts.

$$f(x) = 0$$

$$x^2 + x - 6 = 0$$

$$(x + 3)(x - 2) = 0$$

The x-intercepts are -3 and 2.

The x-coordinate of the vertex is

$$x = -\frac{b}{2a} = -\frac{1}{2(1)} = -\frac{1}{2}$$

The y-coordinate of the vertex is

$$f\left(-\frac{1}{2}\right) = \left(-\frac{1}{2}\right)^2 - \frac{1}{2} - 6 = -6\frac{1}{4}$$

Using this information, we graph the parabola shown in the following figure.

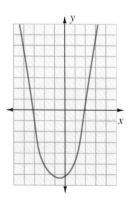

Recall that one result from Section 9.1 was that a parabola has a maximum value or a minimum value at its vertex depending on whether the coefficient of x^2 was negative or positive. Since the graph of a quadratic function is a parabola, we can restate this result in terms of the function concept. We say that the function $f(x) = ax^2 + bx + c$ has an *optimum* (maximum or minimum) *value* at $x = -\dfrac{b}{2a}$. This optimum value is equal to $f\left(-\dfrac{b}{2a}\right)$.

STRATEGY

Consider the quadratic function f given by

$$f(x) = ax^2 + bx + c$$

1. If $a > 0$, then f has a **minimum value** at $x = -\dfrac{b}{2a}$.

2. If $a < 0$, then f has a **maximum value** at $x = -\dfrac{b}{2a}$.

In either case, the **optimum value** of f is $f\left(-\dfrac{b}{2a}\right)$.

Example 3   Consider the function f defined by

$$f(x) = 3x^2 - 12x + 9$$

Determine if the optimum value of f is a minimum or maximum, then find this optimum value and where it occurs.

Solution   Since $a = 3 > 0$, the optimum value is a minimum. This minimum occurs at $x = -\dfrac{b}{2a} = -\dfrac{-12}{2(3)} = 2$. The minimum value of f is $f(2) = 3(2)^2 - 12(2) + 9 = -3$.   ▲

N O T E   *One important observation should be made about optimum values of a quadratic function. If the domain is the entire set of real numbers, then a quadratic function does have an optimum value. However, if the domain is restricted to some subset of the real numbers, then a quadratic function may or may not have an optimum value. These concepts are covered in calculus. In this book, a quadratic function will have an optimum value at its vertex.*

Linear and quadratic functions are both examples of *polynomial functions*.

Polynomial functions are functions of the form

$$f(x) = a_n x^n + a_{n-1} x^{n-1} + \cdots + a_2 x^2 + a_1 x + a_0$$

where $a_n \neq 0$ and n is a nonnegative integer.

To draw an accurate graph of a quadratic function such as $f(x) = x^3 + 3x^2 - 3x + 4$ requires techniques from calculus. However, by plotting enough points, we can obtain a very accurate sketch of the graph.

Example 4   Draw the graph of the function $f(x) = x^3 + 3x^2 - 3x + 4$.

Solution   Setting $y = f(x)$, we make the following table of points by first choosing the numbers x and then finding y from the formula.

x	-4	-3	-2	-1	0	1	2	3
y	0	13	14	9	4	5	18	49

The ordered pairs (x, y) are plotted, then a graph is drawn by "fitting" these points. See the following figure.

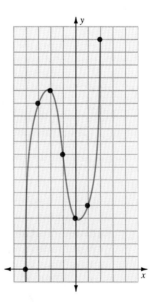

The graph of $f(x) = x^3 + 3x^2 - 3x + 4$.   ▲

Another type of function is called the *exponential function.* We will be studying certain kinds of exponential functions in Chapter 11. We give the general definition here.

DEFINITION

An **exponential function** is a function of the form

$$f(x) = b^x$$

where b is a positive constant, $b \neq 1$.

Two examples of exponential functions are the following:

$$f(x) = 3^x \qquad \text{and} \qquad f(x) = \left(\frac{1}{2}\right)^x$$

Example 5   Let $f(x) = 2^x$, then

$$f(0) = 2^0 = 1$$

$$f(-1) = 2^{-1} = \frac{1}{2} \qquad f(1) = 2^1 = 2$$

$$f(-2) = 2^{-2} = \frac{1}{2^2} = \frac{1}{4} \qquad f(2) = 2^2 = 4$$

$$f(-3) = 2^{-3} = \frac{1}{2^3} = \frac{1}{8} \qquad f(3) = 2^3 = 8 \qquad \blacktriangle$$

We show how to graph exponential functions in the next two examples. We will compute a few points, plot them, then draw the curve suggested by the pattern of these points.

Example 6   Graph the function given by $f(x) = 2^x$.

Solution   We set $y = 2^x$ and make a table of conveniently chosen x values and the corresponding y values.

x	-3	-2	-1	0	1	2	3
y	$\frac{1}{8}$	$\frac{1}{4}$	$\frac{1}{2}$	1	2	4	8

Next, we draw a coordinate plane, plot the points (x, y) from the table, and draw a smooth curve through them as shown in the following figure.

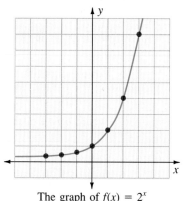

The graph of $f(x) = 2^x$

Example 7   Graph the function given by $f(x) = \left(\dfrac{1}{2}\right)^x$.

Solution   We set $y = \left(\dfrac{1}{2}\right)^x$ and make a table of conveniently chosen values of x and corresponding values of y.

x	-3	-2	-1	0	1	2	3
y	8	4	2	1	$\dfrac{1}{2}$	$\dfrac{1}{4}$	$\dfrac{1}{8}$

The points (x, y) are plotted and the curve is drawn in the following figure.

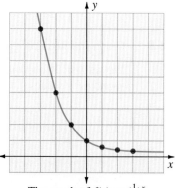

The graph of $f(x) = (\tfrac{1}{2})^x$

There are other functions that can be graphed with slight variations to the techniques we have developed in this section. The way to graph the absolute value function is shown in the next example.

Example 8 Graph the function $f(x) = |x|$.

Solution Recall, the absolute value is defined as

$$|x| = \begin{cases} x, & \text{if } x \geq 0 \\ -x, & \text{if } x < 0 \end{cases}$$

Therefore, since $f(x) = |x|$, we have

$$f(x) = \begin{cases} -x, & \text{if } x < 0 \\ x, & \text{if } x \geq 0 \end{cases}$$

To graph the function $y = f(x)$, we split the domain into two parts. If $x < 0$, we draw the half line $y = -x$.

If $x \geq 0$, we draw the half line $y = x$. The graph is "V" shaped as shown in the following figure.

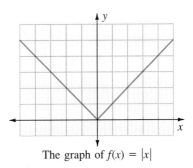

The graph of $f(x) = |x|$

Example 9 Draw the graph of $g(x) = |x - 4|$.

Solution Setting $y = g(x)$, we make a table of values.

x	0	1	2	3	4	5	6	7	8
y	4	3	2	1	0	1	2	3	4

The points (x, y) are plotted in the following figure. Notice that $g(4) = 0$. These points suggest the "V" shaped graph as shown in the figure.

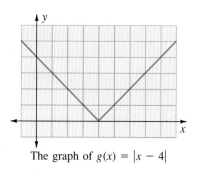

The graph of $g(x) = |x - 4|$ ▲

N O T E   *Notice that the graph of $g(x) = |x - 4|$ is the graph of $f(x) = |x|$ translated to the right by 4 units.*

E X E R C I S E   S E T   10.3

For Exercises 1–12, draw the graph of each function.

1. $f(x) = 3x - 1$

2. $f(x) = -2x + 3$

3. $f(x) = 4$

4. $f(x) = -3$

5. $g(x) = 2x^2 - 2$

6. $f(x) = -4x^2 + 1$

7. $g(x) = x^2 - 2x - 4$

8. $f(x) = x^2 + 4x$

9. $f(x) = -x^2 - 6x - 3$

10. $f(x) = -x^2 + 4x - 3$ **11.** $f(x) = 2x^3 + 4x^2 - 3x + 4$ **12.** $f(x) = 2x^3 + 3x^2 - 12x - 5$

For Exercises 13–22 determine if the optimum value of the quadratic function is a minimum or maximum, then find this optimum value and where it occurs. Do not graph the function.

13. $f(x) = 3x^2 - 6x + 1$

14. $f(x) = -4x^2 - 12x + 3$

15. $f(x) = -x^2 + 5x - 1$

16. $f(x) = x^2 + 6x + 10$

17. $f(x) = -2x^2 + 16x - 2$

18. $f(x) = 3x^2 + 18x - 8$

19. $f(x) = \dfrac{2}{3}x^2 - 4x + 1$

20. $f(x) = \dfrac{4}{3}x^2 + 16x + 2$

21. $f(t) = 640t - 16t^2$

22. $g(t) = 25 + 350t - 5t^2$

23. Tombstone Pizza Supplies makes and sells x cases of pizza trays a day. The daily profit function P is given by

$$P(x) = -0.03x^2 + 450x - 450$$

What level of production will maximize the profit?

24. Wilderness Accessories makes and sells x 6-gallon portable air tanks each week. The profit function P is given by

$$P(x) = -0.15x^2 + 105x - 320$$

What level of production will yield the maximum profit?

25. Let $f(x) = \left(\dfrac{3}{2}\right)^x$. Find

   (a) $f(0)$ **(b)** $f(1)$ **(c)** $f(2)$ **(d)** $f(-2)$

26. Let $g(x) = \left(\dfrac{1}{3}\right)^x$. Find

   (a) $g(0)$ **(b)** $g(1)$ **(c)** $g(-1)$ **(d)** $g(-2)$

For Exercises 27–40, graph each function.

27. $f(x) = 4^x$

28. $f(x) = 5^x$

29. $f(x) = \left(\dfrac{2}{3}\right)^x$

30. $f(x) = \left(\dfrac{1}{4}\right)^x$

31. $f(x) = 2^{-x}$

32. $f(x) = 3^{-x}$

33. $g(x) = 3 \cdot 2^{x/2}$

34. $g(x) = 3 \cdot 2^{x-1}$

35. $f(x) = |x| + 2$

36. $f(x) = |x| - 2$

37. $f(x) = |x - 1|$

38. $f(x) = |x + 2|$

39. $g(x) = |x + 2| - 1$

40. $g(x) = |x - 2| + 1$

41. If $f(x) = 4^x$ and $g(x) = 2^{2x}$, show that $f = g$.

42. If $f(x) = 9^{x/2}$ and $g(x) = 3^x$, show that $f = g$.

REVIEW PROBLEMS

The following exercises review parts of Section 2.4. Doing these problems will help prepare you for the next section.

Solve for y in terms of x.

43. $x = 2y - 4$

44. $x = -3y + 12$

45. $x = \dfrac{y}{4} - \dfrac{5}{4}$

46. $x = 5 - \dfrac{y}{3}$

47. $4y - 15 = 2x - 5$

48. $9y + 3x = -2$

49. $\dfrac{x}{3} - \dfrac{y}{4} = 1$

50. $-\dfrac{4y}{3} + 1 = \dfrac{x}{6} - 2$

E N R I C H M E N T   E X E R C I S E S

For Exercises 1–3, graph the function.

1. $f(x) = \dfrac{|x|}{x}$

2. $f(x) = \dfrac{x^2 + x - 2}{x - 1}$

3. $f(x) = \begin{cases} -2x - 3, & \text{if } x < 0 \\ x - 3, & \text{if } x > 0 \end{cases}$

4. In the same coordinate plane, draw the graphs of $f(x) = \left(\dfrac{1}{2}\right)^x$ and $g(x) = 2^x$. What is the geometrical connection between these two graphs?

5. The number $N(k)$ of bacteria present at the end of k periods of simple fission is

$$N(k) = N_o 2^k$$

where N_o is the number of bacteria at the start. Suppose there were 1,000 bacteria initially. Find
(a) $N(1)$ **(b)** $N(2)$ **(c)** $N(3)$

6. Refer to Exercise 5. Starting with N_o bacteria, how many periods of simple fissions are necessary to have 32 times the initial amount?

7. Refer to Exercise 5. If a simple fission occurs every 15 minutes, how long does it take to have 16 times the original amount?

Answers to Enrichment Exercises are on page A.75.

10.4

Inverse Relations and Functions

O B J E C T I V E S

▶ *To find the inverse of a relation*

▶ *To understand the definition of one-to-one function*

▶ *To draw the graph of a function and its inverse*

Consider the relation $S = \{(1, -1), (2, 3), (4, 0)\}$. If we reverse the order of the entries in each ordered pair, we obtain the *inverse* S^{-1} of S.

$$S^{-1} = \{(-1, 1), (3, 2), (0, 4)\}$$

D E F I N I T I O N

The **inverse** of a relation S is the relation S^{-1} obtained by reversing the order of the entries of each ordered pair belonging to S.

Together, S and S^{-1} are called inverse relations. Notice that the domains and ranges are interchanged; the domain of S^{-1} is the range of S and the range of S^{-1} is the domain of S.

Example 1   Let $T = \{(-1, 3), (2, -2), (4, 5), (6, 5)\}$. Then,

$$T^{-1} = \{(3, -1), (-2, 2), (5, 4), (5, 6)\} \qquad \blacktriangle$$

N O T E   *If S is a relation, then S^{-1} does not mean $\dfrac{1}{S}$. The superscript -1 means the inverse relation, not S raised to the -1 power.*

In Example 1, notice that T is a function since no two ordered pairs have the same first entry. However T^{-1} is not a function, since $(5, 4)$ and $(5, 6)$ are both members of T^{-1} having the same first entry.

There is a special group of functions that have inverses that are also functions. They are *one-to-one* functions.

D E F I N I T I O N

A function is **one-to-one,** provided that at most one x value is associated with a y value.

The function in Example 1 is not one-to-one, since two x values, 4 and 6, are associated with the single y value of 5.

One-to-one functions are functions and all functions are relations. These observations are shown pictorally in the following figure.

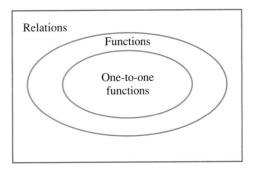

Example 2   Determine if the given relation is a function. If it is a function, is it one-to-one?

(a) $R = \{(-3, 4), (-2, -1), (-1, -4), (0, 3)\}$

(b) $y = x^2$

Solution   (a) R is a function, since no two ordered pairs have the same first coordinate. It is a one-to-one function, since no two ordered pairs have the same second coordinate.

(b) $y = x^2$ defines a quadratic function. This function is not one-to-one. For example, setting $y = 4$, then $x^2 = 4$ or $x = \pm 2$. Therefore, $(2, 4)$ and $(-2, 4)$ are two ordered pairs belonging to the function with the same second coordinate.   ▲

There is a graphical way to decide whether a function is one-to-one. We used a vertical line to tell if a relation is a function. We can use a horizontal line to tell if a function is one-to-one.

R U L E

The horizontal line test

If a horizontal line crosses the graph of a function at more than one point, then the function is *not* one-to-one.

Example 3   Use the horizontal line test to determine if the function $y = x^2$ is one-to-one.

Solution   The graph of this parabola is shown in the figure on the following page.

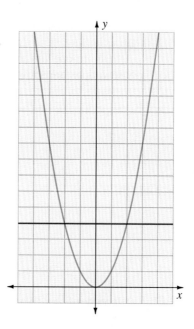

From the figure, we conclude that this function is not one-to-one. For example, the horizontal line $y = 4$ intersects the graph in more than one point.   ▲

A major significance of a function being one-to-one is that its inverse is also a function. We state this as a theorem without proof.

THEOREM

If a function f is one-to-one, then its inverse is also a function.

Example 4   The function given by $f(x) = x^3$ is one-to-one. Find each of the following.

(a) $f(2)$ **(b)** $f^{-1}(8)$

Solution   **(a)** Replacing x by 2 in the equation $f(x) = x^3$,

$$f(2) = 2^3 = 8$$

(b) By Part (a), $(2, 8)$ belongs to f. Therefore, $(8, 2)$ belongs to f^{-1} and we can write $f^{-1}(8) = 2$.   ▲

Suppose a function that is one-to-one is defined by the equation $y = f(x)$. By the theorem, it has an inverse that is also a function. The equation of the inverse can be found by using the following steps.

S T R A T E G Y

To find the inverse f^{-1} of a one-to-one function $y = f(x)$

Step 1   Start with $y = f(x)$.

Step 2   Interchange the role of x and y in this equation.

Step 3   Solve the equation in Step 2 for y.

Step 4 $f^{-1}(x)$ is given by the new equation found in Step 3.

Example 5   Find f^{-1}, if $f(x) = 3x - 4$.

Solution   Step 1 $y = 3x - 4$.

Step 2 $x = 3y - 4$ Interchange x and y.

Step 3   Solve for y.

$$x + 4 = 3y$$

$$\frac{x + 4}{3} = y \qquad \text{Divide both sides by 3.}$$

Step 4 f^{-1} is given by the equation

$$f^{-1}(x) = \frac{x + 4}{3} \qquad \text{or} \qquad \frac{1}{3}x + \frac{4}{3} \qquad \blacktriangle$$

Example 6   Find the equation for f^{-1} of the one-to-one function $f(x) = (x + 2)^3$.

Solution   We use the four-step method for finding the inverse.

Step 1 $y = (x + 2)^3$

Step 2 $x = (y + 2)^3$ Interchange x and y.

Step 3   To solve for y, start by taking the cube root of both sides.

$$x = (y + 2)^3$$

$$\sqrt[3]{x} = y + 2$$

$$\sqrt[3]{x} - 2 = y \qquad \text{Subtract 2 from both sides.}$$

Step 4   Therefore, the equation for f^{-1} is

$$f^{-1}(x) = \sqrt[3]{x} - 2 \qquad \blacktriangle$$

Example 7   Find the equation for f^{-1} of the one-to-one function $f(x) = \dfrac{2x + 1}{x - 3}$.

Solution   We use the four-step method.

Step 1 $y = \dfrac{2x + 1}{x - 3}$.

Step 2 $x = \dfrac{2y + 1}{y - 3}$.   Interchange x and y.

Step 3   Solve for y.

$$x = \frac{2y + 1}{y - 3}$$

$$x(y - 3) = 2y + 1 \qquad \text{Multiply both sides by } y - 3.$$

$$xy - 3x = 2y + 1 \qquad \text{Distributive property.}$$

$$xy - 2y = 3x + 1 \qquad \text{Add } 3x - 2y \text{ to both sides.}$$

$$y(x - 2) = 3x + 1 \qquad \text{Factor the left side.}$$

$$y = \frac{3x + 1}{x - 2} \qquad \text{Divide both sides by } x - 2.$$

Step 4   Therefore, the inverse function is given by

$$f^{-1}(x) = \frac{3x + 1}{x - 2}$$ ▲

There is a connection between the graph of a relation R and the graph of its inverse R^{-1}. Suppose $(a, b) \in R$, then $(b, a) \in R^{-1}$. In the following figure, the line segment joining (a, b) and (b, a) is perpendicular to the line $y = x$. Furthermore, the line $y = x$ divides the line segment into two equal parts. This means that (b, a) is the mirror image of (a, b) with respect to the line $y = x$.

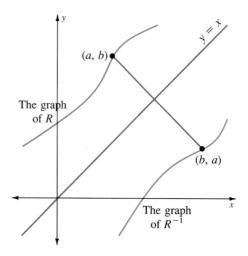

Therefore, given the graph of a function or relation, we can obtain the graph of its inverse by using this *symmetry property*.

P R O P E R T Y

Symmetry property

The graphs of a relation and its inverse are mirror images of each other with respect to the line $y = x$.

Example 8   The graph of a relation R is shown in the following figure. Use the symmetry property to draw the graph of the inverse of R.

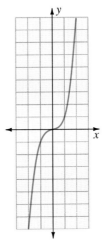

Solution   Using the line $y = x$ as a guide, we draw the mirror image of the given graph as shown in the following figure.

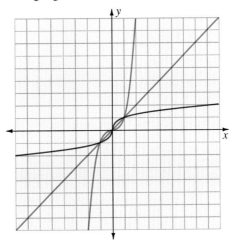

Example 9   Graph the function $y = x^2 - 4$. Then, use the symmetry property to graph the inverse.

Solution   First draw the parabola $y = x^2 - 4$. Then, draw the line $y = x$ as shown in Figure (a).

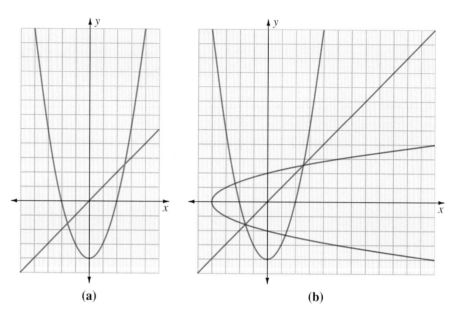

(a)   **(b)**

Next, use the symmetry property to draw the graph of the inverses. Notice that the vertex $(0, -4)$ is reflected to $(-4, 0)$, which is the vertex of the inverse parabola. The inverse graph will meet the original at two points on the line $y = x$. Now, sketch in the inverse parabola keeping in mind that the curve we are drawing is the mirror image, in the line $y = x$, of the original parabola. [See Figure (b).]   ▲

E X E R C I S E   S E T   10.4

1. Let $T = \{(-6, 1), (-6, -1), (6, 1), (6, -1)\}$. Find T^{-1}.

2. Let $S = \{(-3, 0), (-1, 1), (0, 2), (1, 1), (2, 0)\}$. Find S^{-1}.

For Exercises 3–8, determine if the relation is a function. If it is a function, is it one-to-one?

3. $S = \{(5, 6), (2, -7), (-1, 10), (3, -6)\}$

4. $T = \{(4, 0), (-4, 2), (-5, -1), (5, 3)\}$

5. $T = \{(-3, 0), (3, 9), (-2, 3), (3, 2)\}$

6. $S = \{(1, 9), (2, 8), (1, 7), (2, 6), (-9, -9)\}$

7. $R = \{(-3, 8), (-2, 7), (-1, 8), (0, 1)\}$

8. $R = \{(12, 10), (-4, -1), (4, -1), (3, 2), (-12, 0)\}$

For Exercises 9–16, determine if the given function is one-to-one.

9. $f(x) = 1 - x^2$

10. $f(x) = 4 + x^2$

11. $g(x) = -\dfrac{2}{3}x - 10$

12. $g(x) = 5x - 12$

13. $y = 2^x$

14. $y = \left(\dfrac{1}{3}\right)^x$

15. $y = |x|$

16. $y = |2x - 6|$

17. Consider the one-to-one function f given by $f(x) = x^3 - 2$. Find
  (a) $f(2)$ **(b)** $f^{-1}(6)$

18. Consider the one-to-one function f given by
  $f(x) = \dfrac{2x + 3}{4 - x}$. Find
  (a) $f(1)$ **(b)** $f^{-1}\left(\dfrac{5}{3}\right)$

For Exercises 19–36, each function is one-to-one. Use the four-step method to find the equation for f^{-1}. Use the $f^{-1}(x)$ notation in your final answer.

19. $f(x) = 5x - 15$

20. $f(x) = -3x + 12$

21. $f(x) = \dfrac{1}{3}x - 1$

22. $f(x) = \dfrac{1}{2}x + 10$

23. $f(x) = -\dfrac{2}{5}x + \dfrac{4}{5}$

24. $f(x) = -\dfrac{3}{4}x + \dfrac{9}{4}$

25. $f(x) = \dfrac{1}{x}$

26. $f(x) = \dfrac{3}{x}$

27. $f(x) = 1 - \dfrac{2}{x}$

28. $f(x) = 2 - \dfrac{1}{x}$

29. $y = x^3 - 1$

30. $y = (x - 5)^3$

31. $f(x) = 2(3x - 1)^3 + 4$

32. $f(x) = 6 - 3(5x + 2)^3$

33. $y = \dfrac{x + 1}{x - 1}$

34. $y = \dfrac{2x + 3}{x + 4}$

35. $f(x) = \dfrac{-3x + 1}{2 - 3x}$

36. $f(x) = \dfrac{4 - 6x}{-2x + 5}$

For Exercises 37–42, the graph of a relation is given. Use the symmetry property to draw the graph of the inverse.

37.

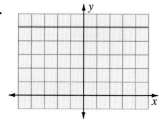

38.

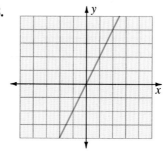

39.

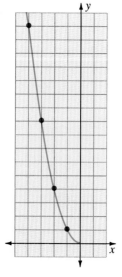

40.

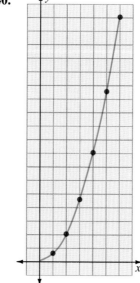

41.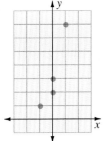

42.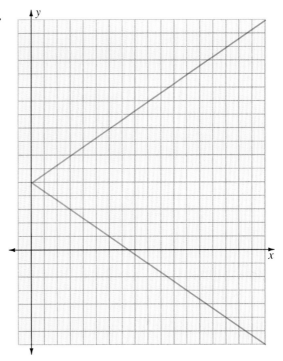

For Exercises 43–58, draw the graph of the relation or function. Then use the symmetry property to draw the graph of the inverse.

43. $f(x) = 2x - 1$

44. $f(x) = 3x - 2$

45. $x - 2y = 4$

46. $-x + 3y = -3$

47. $y = x^2 - 9$

48. $y = x^2 - 1$

49. $f(x) = 4 - x^2$

50. $g(x) = 9 - x^2$

51. $y = -3$

52. $x = 4$

53. $2x + 3y = 0$

54. $x + 4y = 0$

55. $y = 2^x$ **56.** $y = 3^x$ **57.** $y = \left(\dfrac{1}{2}\right)^x$

58. $y = \left(\dfrac{1}{3}\right)^x$

59. Let $f(x) = 5x - 4$.
   (a) Find $f^{-1}(x)$.
   (b) Show that $(f \circ f^{-1})(x) = x$.

60. Let $f(x) = -\sqrt[3]{x} + 7$.
   (a) Find $f^{-1}(x)$.
   (b) Show that $(f \circ f^{-1})(x) = x$.

   (c) Show that $(f^{-1} \circ f)(x) = x$.

(c) Show that $(f^{-1} \circ f)(x) = x$.

REVIEW PROBLEMS

The following exercises review parts of Section 10.3. Doing these problems will help prepare you for Section 11.1.

61. If $f(x) = 4^x$, find
   (a) $f(0)$ **(b)** $f(-1)$ **(c)** $f(1)$ **(d)** $f(3)$

62. If $g(x) = \left(\dfrac{1}{5}\right)^x$, find
   (a) $g(-2)$ **(b)** $g(0)$ **(c)** $g(1)$ **(d)** $g(2)$

ENRICHMENT EXERCISES

1. Draw the graph of the function f and its inverse.
$$f(x) = |x - 4|$$

2. Draw the graph of the relation and its inverse.
$$x^2 - y^2 = 1$$

3. Let $f(x) = \dfrac{|x|}{x}$. Draw the graph of f and its inverse.

4. Define f by
$$f(x) = \begin{cases} -x^2, & \text{if } x < 0 \\ 2, & \text{if } x = 0 \\ x^2, & \text{if } x > 0 \end{cases}$$

Draw the graph of f and its inverse.

5. An advertiser has found that the influence $f(t)$ that a TV commercial has in a particular viewing area t days after being aired is
$$f(t) = 10 - 2t$$
where $f(t)$ is a number on a scale of 1 to 10. How often should the commercial be aired if the influence should not drop below 4?

6. Let f be the function given by $y = mx + b$, where $m \neq 0$. Show that the graph of the inverse of f is a straight line with a slope of $\dfrac{1}{m}$.

Answers to Enrichment Exercises are on page A.77.

$\mathbf{S}$ummary and review

Examples

Relations and functions (10.1)

A **relation** S is a set of ordered pairs of numbers. The **domain** of S is the set of all first coordinates of the ordered pairs. The **range** of S is the set of all second coordinates of the ordered pairs.

Consider the relation $S =$ $\{(-2, 3), (-1, 4), (0, 0), (0, -1)\}$. The domain of S is $\{-2, -1, 0\}$; the range of S is $\{-1, 0, 3, 4\}$. The relation S is not a function.

A **function** is a relation in which no two different ordered pairs have the same first entry.

The **graph** of a relation is the set of points $P(x, y)$ in the coordinate plane whose coordinates belong to the relation.

To determine if a graph is the graph of a function, we use the **vertical line test.**

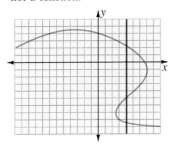

By the vertical line test, the graph in the above figure is not the graph of a function.

Function notation and the algebra of functions (10.2)

If $f(x) = 2x^2$ and $g(x) =$ $3x - 4$, then

$(f \circ g)(x) = 2(3x - 4)^2$

Consider a function such as $y = 3x - 1$. We say that y is a function of x and write $y = f(x)$. The connection between the function notation and f itself is the following.

> If $b = f(a)$, then $(a, b) \in f$.
> If $(a, b) \in f$, then $b = f(a)$.

The four basic operations of addition, subtraction, multiplication, and division are defined for functions. A fifth operation on functions is the following: Let f and g be functions with the range of g in the domain of f; the **composition of f by g** is defined by

$$(f \circ g)(x) = f[g(x)]$$

Some special functions (10.3)

In this section, we studied linear, quadratic, polynomial, and exponential functions.

Examples

Find the equation for the inverse of $f(x) = -3x + 6$.

Step 1   Set $y = -3x + 6$.

Step 2 $x = -3y + 6$.
Interchange x and y.

Step 3   Solve for y.

$$3y = -x + 6$$

$$y = -\frac{1}{3}x + 2$$

Step 4 $f^{-1}(x) = -\frac{1}{3}x + 2.$

Inverse relations and functions (10.4)

The **inverse** of a relation S is the relation S^{-1} obtained by reversing the order of the entries of each ordered pair belonging to S.

The inverse of a function may not be a function. To guarantee that the inverse is a function, we need the following condition of f: A function is **one-to-one**, provided that at most one x value is associated with a y value.

For a one-to-one function given by an equation, we can find the equation defining the inverse.

CHAPTER 10 REVIEW EXERCISE SET

Section 10.1

For Exercises 1 and 2, state the domain and range of each relation and determine if it is also a function.

1. $T = \{(8, -2), (2, -1), (6, -2), (-6, -1), (0, 2)\}$ **2.** $S = \{(-5, 5), (-2, 1), (-3, -1), (-2, -1), (0, 0)\}$

For Exercises 3–6, state the domain for each function.

3. $y = 3x - 9$

4. $y = \sqrt{x - 14}$

5. $y = \dfrac{x + 2}{2x - 7}$

6. $y = \dfrac{x - 3}{x^2 + x - 2}$

For Exercises 7–10, draw the graph of the relation. From the graph determine the domain and range. Is the relation also a function?

7. $x^2 + y^2 = 49$

8. $(x + 3)^2 + (y - 2)^2 = 25$

9. $y = -x^2 + 6$

10. $y^2 + 9x^2 = 9$

Section 10.2

For Exercises 11–16, let $f(x) = 4x + 12$ and $g(x) = 2x^2 + 4x - 6$. Find each of the following.

11. $f(-4)$

12. $g(-1)$

13. $f(2) + g(2)$

14. $-3g(-2)$

15. $(g + f)(x)$

16. $\left(\dfrac{g}{f}\right)(x)$

17. Given: $f(x) = 4x - 1$ and $g(x) = -3x + 2$. Find
  (a) $(f \circ g)(-1)$ **(b)** $(g \circ f)(0)$ **(c)** $(f \circ g)(a)$

For Exercises 18–20, find $(f \circ g)(x)$ and $(g \circ f)(x)$.

18. $f(x) = x^2$ and $g(x) = -x + 2$

19. $f(x) = \sqrt{x}$ and $g(x) = x - 5$

20. $f(x) = \dfrac{1}{3x + 7}$ and $g(x) = 1 - 2x$

For Exercises 21 and 22, form the quotient $\dfrac{f(x) - f(a)}{x - a}$ and simplify.

21. $f(x) = 4 - x$

22. $f(x) = 3x^2$

Section 10.3

For Exercises 23–26, draw the graph of each function.

23. $f(x) = \dfrac{3}{2}x - 5$

24. $f(x) = 2(x - 1)^2 - 1$

25. $f(x) = 3^{2x}$

26. $f(x) = \left(\dfrac{1}{4}\right)^x$

For Exercises 27–30, determine if the optimum value of the quadratic function is a minimum or maximum, then find this optimum value and where it occurs. Do not graph the function.

27. $f(x) = 5x^2 - 25x + 12$

28. $g(x) = \dfrac{x^2}{3} - 6x - 2$

29. $f(t) = -t^2 + t - 4$

30. $g(t) = -4t^2 + 8t - 9$

Section 10.4

31. Let $S = \{(0, 0), (-3, -2), (3, -3), (0, 2)\}$. Find S^{-1}.

For Exercises 32 and 33, determine if the relation is a function. If it is a function, is it one-to-one?

32. $T = \{(9, -1), (-9, 1), (9, 1), (-9, -1)\}$

33. $S = \{(4, -2), (5, 3), (2, -8), (1, 1), (0, -1)\}$

For Exercises 34–36, determine if the given function is one-to-one.

34. $f(x) = 3 - x^2$

35. $f(x) = -2x + 10$

36. $f(x) = |x|$

For Exercises 37–40, find the equation for f^{-1}. Use the $f^{-1}(x)$ notation in your final answer.

37. $f(x) = x + 11$

38. $f(x) = 3x + 9$

39. $f(x) = -6x - 2$

40. $f(x) = -\dfrac{2}{x}$

For Exercises 41–44, graph the function. Then, use the symmetry property to graph the inverse.

41. $f(x) = 3x + 1$

42. $g(x) = -x + 2$

43. $y = x^2 - 1$ **44.** $f(x) = -2^x$

C H A P T E R   1 0   **T E S T**

Section 10.1

1. Let $S = \{(-1, 2), (-2, 2), (0, -1), (-3, 0)\}$. State the domain and range of the relation S and determine if it is also a function.

For problems 2 and 3, state the domain of the given function.

2. $y = 4 - 5x$ **3.** $y = \dfrac{1}{\sqrt{x + 3}}$

For problems 4–6, draw the graph of the relation. From the graph determine the domain and range. Is the relation also a function?

4. $x^2 + y^2 = 100$ **5.** $y = -2x^2 + 8$

6. $\dfrac{(x - 4)^2}{16} + \dfrac{(y + 6)^2}{36} = 1$

Section 10.2

For problems 7–10, let $f(x) = 3x^2 - 2x - 5$ and $g(x) = -2x + 3$. Find each of the following.

7. $f(0)$

8. $g(-3)$

9. $4f(1)$

10. $(f + g)(x)$

11. Let $f(x) = 3x^2 - 1$ and $g(x) = 1 - 3x$. Find $(f \circ g)(x)$ and $(g \circ f)(x)$.

12. Given: $f(x) = 4x - 3$. Form the quotient $\dfrac{f(x) - f(a)}{x - a}$ and simplify.

Section 10.3

For problems 13–15, draw the graph of the given function.

13. $f(x) = -\dfrac{2}{3}x + 6$

14. $f(x) = \sqrt{16 - x^2}$

15. $f(x) = 3^{-x}$

16. Given: $f(x) = -3x^2 + 18x - 19$. Determine if the optimum value of the quadratic function f is a minimum or maximum, then find this optimum value and where it occurs. Do not graph the function.

Section 10.4

17. Given: $T = \{(-3, 1), (-2, 3), (-1, 2), (0, 0)\}$. Determine if T is a function. If it is a function, is it one-to-one?

18. For the function defined by $f(x) = -5x^2 + 3$, determine if f is one-to-one.

19. For the function given by $f(x) = 2x - 8$, find the equation for f^{-1}. Use the $f^{-1}(x)$ notation in your final answer.

20. Graph the function f given by $f(x) = \dfrac{4}{3}x - 3$. Then use the symmetric property to graph the inverse.

Exponential and Logarithmic Functions

CHAPTER

Overview

The introductions of the exponential function to the base e and logarithmic functions will enable us to solve an ever greater variety of problems. The applications are from different disciplines but have surprising similarities. For example, cell division in biology and the accumulation of interest from money in a savings account are both described using exponential functions.

In this chapter, we will have calculator solutions in the examples and not have Calculator Crossovers. You are encouraged to use a calculator, and not a table, when doing logarithmic computations. If you do not have a calculator, however, there is a logarithmic table in the Appendix.

11.1

The Exponential Functions

▶ *To define the exponential function $f(x) = e^x$*

▶ *To study applications of exponential functions*

In the last chapter, we introduced the exponential function. For convenience, we will repeat the definition.

D E F I N I T I O N

An **exponential function** is of the form

$$f(x) = b^x$$

where $b > 0$, $b \neq 1$.

The number b is called the **base.**

As we saw in Chapter 10, there are two basic graphs, depending upon the size of the base b. If $0 < b < 1$, the graph looks like the one in Figure (a). If $b > 1$, then the graph has the shape shown in Figure (b). In either case, the domain is the set of real numbers and the range is the set of positive real numbers.

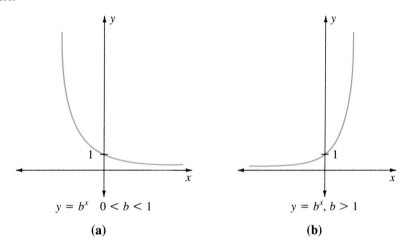

$y = b^x \quad 0 < b < 1$ $y = b^x, b > 1$

(a)   **(b)**

Suppose we wanted to solve an exponential equation such as $4^x = 8$. The following property can be used to solve certain exponential equations.

PROPERTY

If $b^x = b^y$, then $x = y$, where the base b is positive and $b \neq 1$.

Example 1   Solve the equation $4^x = 8$.

Solution   We can use the property stated above provided that we can express both sides of the equation as a power with the same base. Since $4 = 2^2$ and $8 = 2^3$, the given equation can be written as

$$(2^2)^x = 2^3$$

or

$$2^{2x} = 2^3$$

Using the above property,

$$2x = 3$$

so that

$$x = \frac{3}{2}$$

Next, we check our answer in the original equation.

Check $x = \dfrac{3}{2}$: $4^x = 8$

$$4^{3/2} \overset{?}{=} 8$$

$$(\sqrt{4})^3 \overset{?}{=} 8$$

$$2^3 \overset{?}{=} 8$$

$$8 = 8$$

▲

Exponential functions are used to describe growth or decay. In economics, for example, suppose the annual rate of inflation will be 3% a year for the next 5 years. This means that a $50 pair of shoes will cost $(50)(103\%) = (50)(1.03) = \$51.50$ next year. In 2 years, the pair of shoes will cost $(50)(1.03)(1.03) = (50)(1.03)^2 = \$53.05$. If A is the price of the pair of shoes t years from now, then

$$A = 50(1.03)^t, \ 0 \le t \le 5$$

For example, 5 years from now, the pair of shoes will cost

$$A = 50(1.03)^5 = \$57.96$$

The general form for **exponential growth** is

$$A = Pb^t$$

Example 2   It is predicted that the price of coffee will increase at an annual rate of 4% for the next 6 years. If a can of coffee costs $5.75, what is the predicted price in 3 years? In 6 years?

Solution   We start with the equation $A = Pb^t$. The base b is determined by the annual rate: $b = 1.04$. The number P is the present cost of the can of coffee. Therefore, in t years, the cost A of the can of coffee is given by

$$A = 5.75(1.04)^t, \qquad 0 \le t \le 6$$

To find the price in 3 years, set $t = 3$, and use a calculator to find A.

$$A = 5.75(1.04)^3 = \$6.47$$

To find the price in 6 years, set $t = 6$ and find A.

$$A = 5.75(1.04)^6 = \$7.28$$

Therefore, the $5.75 can of coffee will cost $6.47 in 3 years and $7.28 in 6 years.

▲

Example 3   Draw the graph of $f(x) = \dfrac{1}{3} \cdot 2^x$.

Solution   Setting $y = \dfrac{1}{3} \cdot 2^x$, we make the following table by first choosing convenient values for x, then finding the corresponding value for y.

x	-3	-2	-1	0	1	2	3	4
y	$\dfrac{1}{24}$	$\dfrac{1}{12}$	$\dfrac{1}{6}$	$\dfrac{1}{3}$	$\dfrac{2}{3}$	$\dfrac{4}{3}$	$\dfrac{8}{3}$	$\dfrac{16}{3}$

Next, we plot the pairs (x, y) and draw the curve suggested by these points as shown in the following figure.

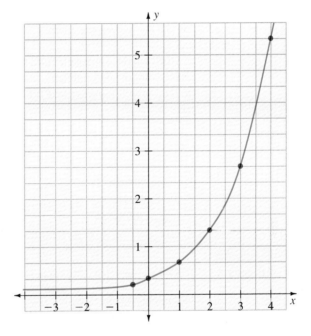

The value of certain objects like paintings or antiques may increase or appreciate in value as time passes. Consumer goods such as a car will generally decrease or depreciate with time.

Example 4   For tax purposes, a business depreciates a company car that costs $15,000 at an annual rate of 12% for the first 4 years.

(a) What is the value of the car after 1 year?

(b) If V is the value of the car after t years, find an equation that gives V as a function of t.

Solution   **(a)** Since the car depreciates 12% each year, after 1 year, the value of the car is $15{,}000(100\% - 12\%) = 15{,}000(88\%) = 15{,}000(0.88) = \$13{,}200$.

(b) Since the value of the car is 88% of the value the previous year,

$$V = 15{,}000(0.88)^t, \qquad 0 \le t \le 4 \qquad \blacktriangle$$

Another application of exponential functions is compound interest. **Compound interest** is interest paid on the interest previously earned as well as on the original investment. For example, suppose a **principal** of \$4,000 is invested at an interest rate of 6% compounded annually. The interest earned the first year is $4{,}000(0.06) = \$240$. The amount is added to the original principal, $\$4{,}000 + \$240 = \$4{,}240$, and this is the principal for the second year. The interest earned during the second year is $4{,}240(0.06) = \$254.40$. This amount is added to \$4,240, $\$4{,}240 + \$254.40 = \$4{,}494.40$, and this is the new principal for the third year. This pattern continues, where each year a new principal is formed by adding the interest earned from the previous year to the old principal.

In general, suppose P dollars is the original principal invested at an interest rate r compounded annually. At the end of 1 year, we have $P + Pr = P(1 + r)$ dollars, which is reinvested for the next year. At the end of the second year, we have

$$P(1 + r) + P(1 + r)r = P(1 + r) \cdot 1 + P(1 + r)r = P(1 + r)(1 + r)$$

$$= P(1 + r)^2 \text{ dollars}$$

We continue this pattern, so at the end of t years, the amount A in the account is

$$A = P(1 + r)^t$$

Example 5   If \$750 is invested in a savings account at 9% compounded annually, what is the amount in the account after $6\frac{1}{2}$ years?

Solution   Using the above formula with $r = 0.09$, the amount A in the account after t years is given by

$$A = 750(1 + 0.09)^t$$
$$= 750(1.09)^t$$

After $6\frac{1}{2}$ years, the amount in the account is

$$A = 750(1.09)^{6.5} = \$1{,}313.21 \qquad \blacktriangle$$

A principal P can be compounded more than just once a year. It can be compounded twice a year (semiannually), four times a year (quarterly), or in general, n times a year. The formula for the amount A after t years compounded n times a year at an interest rate r is

$$A = P\left(1 + \frac{r}{n}\right)^{nt}$$

Example 6 $800 is placed in a savings account at 6% interest compounded quarterly. What is the amount in the account after 3 years?

Solution   Using $P = 800$, $r = 0.06$, $n = 4$, and $t = 3$, we have

$$A = P\left(1 + \frac{r}{n}\right)^{nt}$$

$$= 800\left(1 + \frac{0.06}{4}\right)^{4 \cdot 3}$$

$$= 800(1.015)^{12}$$

$$= 956.49 \qquad\qquad \text{Use a calculator.}$$

Therefore, the $800 grew to $956.49 in this savings account.   ▲

Another form of compounding money in a savings account is called *compounding continuously*. If P dollars is **compounded continuously** at an interest rate r, then the amount A after t years is given by

$$A = Pe^{rt}$$

where e is an irrational number whose decimal expansion to seven decimal places is

$$e \approx 2.7182818$$

Just as the irrational number π is basic to formulas in geometry (the area of a circle is πr^2), the number e appears in the development of mathematics in areas such as statistics.

In fact, most calculators either have a button labeled $\boxed{e^x}$ or $\boxed{\ln x}$, the latter being closely related to e^x (see Section 11.2).

Example 7   Suppose $500 is invested at 10% compounded continuously. What is the amount of the investment after 8 years?

Solution   We let $P = 500$, $r = 0.1$, and $t = 8$ in the formula $A = Pe^{rt}$,

$$A = 500e^{(0.1)(8)}$$

$$= 500e^{0.8}$$

To evaluate $500e^{0.8}$ we use a calculator. If your calculator has an $\boxed{e^x}$ button, follow this sequence of steps:

Enter	Press	Press	Enter	Press	Display
0.8	e^x	$\times$	500	$=$	1112.7705

If your calculator does not have an e^x button, then it might have a button that is labeled either INV, IN, 2nd fcn, or shift along with $\ln x$. Follow the sequence of steps:

Enter	Press	Press	Press	Enter	Press	Display
0.8	INV	$\ln x$	$\times$	500	$=$	1112.7705

If you have any questions about using your calculator, be sure to refer to the booklet that accompanies it. Rounding off to the nearest hundredth, the answer to the problem is

$$A = 500e^{0.8} = \$1{,}112.77$$ ▲

Many applications in physics and social science rely on **growth** and **decay** equations of the form

$$Q = Q_0 e^{kt} \quad \text{Growth}$$

and

$$Q = Q_0 e^{-kt} \quad \text{Decay}$$

where k is a positive constant. The initial quantity present is Q_0, and Q is the quantity present at time t. Notice that Q_0 is the value of Q when $t = 0$, since for the growth equation

$$Q = Q_0 e^{k0} = Q_0 e^0 = Q_0 \cdot 1 = Q_0$$

and for the decay equation,

$$Q = Q_0 e^{-k0} = Q_0 e^0 = Q_0 \cdot 1 = Q_0$$

The graphs of the growth and decay equations are shown in the following figure. For the growth equation, as time t increases, the quantity Q becomes larger. For the decay equation, as time increases, Q becomes smaller.

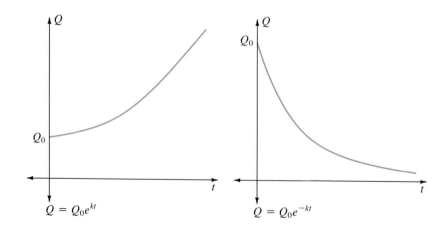

$$Q = Q_0 e^{kt} \qquad\qquad Q = Q_0 e^{-kt}$$

Let us look at an example of an application in growth and decay.

Example 8   A bacteria culture has an initial count of 400 and the bacteria count, Q, after t hours is given by

$$Q = 400e^{0.4t}$$

What is the count after 3 hours?

Solution   Set $t = 3$ and then find Q.

$$Q = 400e^{0.4(3)} = 400e^{1.2} = 1{,}328 \qquad \text{Rounded off to the nearest integer.}$$

Therefore, after 3 hours, the bacteria culture has a count of 1,328.   ▲

E X E R C I S E   S E T   11.1

For Exercises 1–12, solve the exponential equation.

1. $2^{x+1} = 8$

2. $\left(\dfrac{1}{4}\right)^x = 16$

3. $\dfrac{1}{27} = 9^x$

4. $\dfrac{1}{36} = 6^{3x}$

5. $2 \cdot 4^{-x} = 64$

6. $3 \cdot 5^{2x+1} = 75$

7. $9^{3t/2} = \left(\dfrac{2}{18}\right)^{2t}$

8. $8^{3t} = 16^{t-1}$

9. $10^{y-1} = 1$

10. $1 = 25^{4x+4}$

11. $7^{1-2x} = 49^x$

12. $16^{1-4x} = 32^{3-3x}$

For Exercises 13–20, graph the function.

13. $y = 3 \cdot \left(\dfrac{1}{2}\right)^x$

14. $y = 4 \cdot 3^{-x}$

15. $f(x) = -2 \cdot 3^x$

16. $f(x) = -4 \cdot \left(\dfrac{1}{2}\right)^x$

17. $y = 2^x - 2$ **18.** $y = \left(\dfrac{1}{3}\right)^x + 1$ **19.** $y = 4.7(0.8)^x$ **20.** $y = 3.5(2.4)^x$

For Exercises 21–26, find the price of each item t years from now for the given value of t, if the inflation rate is 3.5% a year for the next several years. Round your answer to the nearest hundredth.

21. A \$3.50 paperback book in 4 years.

22. A \$2.99 box of donuts in 5 years.

23. A \$1.59 roast beef sandwich in 2 years.

24. A \$13.89 toy in 3 years.

For Exercises 25 and 26, round to the nearest dollar.

25. A \$140,000 house in 4 years.

26. A \$25,500 car in 3 years.

27. A piece of earth-moving equipment that initially cost \$85,000 depreciates at 8% a year.
   (a) Write an equation that expresses the value V as a function of t, the number of years since purchase.
   (b) What is the value after 4 years?

28. A company car that cost \$14,500 new depreciates at a rate of 12% a year.
   (a) Write an equation that expresses the value of the car as a function of t, the number of years since purchase.
   (b) What is the value of the car after 3 years?

For Exercises 29–42, find the amount of the investment for the time indicated. Write your answer accurate to the nearest hundredth.

29. \$600 at 8% compounded annually for 3 years.

30. \$4,500 at 6% compounded annually for 5 years.

31. \$250 at 7% compounded semiannually for 2 years.

32. \$8,050 at 10% compounded semiannually for 4 years.

33. \$4,825 at 9% compounded quarterly for 5 years.

34. \$10,000 at 8% compounded quarterly for 7 years.

35. \$5,000 at 12% compounded monthly for 10 years.

36. \$6,700 at 6% compounded monthly for 8 years.

37. \$7,750 at $8\dfrac{1}{2}$% compounded semiannually for 3 years.

38. \$1,275 at $7\dfrac{1}{4}$% compounded annually for 2 years.

39. $3,400 at 12% compounded continuously for 1 year.

40. $9,950 at 10% compounded continuously for 2 years.

41. $12,500 at 8% compounded continuously for 6 years.

42. $38,450 at 13% compounded continuously for 12 years.

43. A bacteria culture initially has a count of 650 and the count Q after t hours is given by

$$Q = 650e^{0.3t}$$

What is the count after 4 hours?

44. A bacteria culture initially has a count of 750 and the count Q after t hours is given by

$$Q = 750e^{0.4t}$$

What is the count after $6\frac{1}{2}$ hours?

45. A culture of bacteria grows according to

$$Q = 1000e^{0.5t}$$

where Q is the number present at time t, in hours.
(a) How many are initially present?
(b) How many bacteria will be present in 3 hours?

46. The number Q of bacteria present at time t, in hours, is given by

$$Q = 7200e^{0.08t}$$

(a) How many are initially present?
(b) How many bacteria will be present in 2 hours?

47. Five hundred grams of a radioactive substance decays so that after t days the amount Q that is left is given by

$$Q = 500e^{-0.06t}$$

How many grams of the substance is left after 5 days?

48. Seven hundred grams of a radioactive substance decays so that after t days the amount Q that is left is given by

$$Q = 700e^{-0.09t}$$

How many grams of the substance is left after 6 days?

REVIEW PROBLEMS

The following exercises review parts of Section 10.4. Doing these problems will help prepare you for the next section.

Graph each function, then graph the inverse. Is the inverse also a function?

49. $y = 2x + 5$

50. $f(x) = -3x + 6$

51. $f(x) = x^2$

52. $f(x) = x^2 - 4$

53. $f(x) = 1$

54. $f(x) = -2$

ENRICHMENT EXERCISES

1. Recall that e is an irrational number whose decimal expansion to seven decimal places is 2.7182818. Complete the following table.

n	$\left(1 + \dfrac{1}{n}\right)^n$
1	
2	
10	
100	
1,000	
10,000	
100,000	

What do you think happens to the expression $\left(1 + \dfrac{1}{n}\right)^n$ when n becomes larger and larger?

2. How much money must be deposited now in an account that pays 8% compounded annually so that it grows to $10,000 in 12 years?

3. How much money must be deposited now in an account that pays $7\frac{1}{2}\%$ compounded continuously so that it grows to $40,000 in 20 years?

Answers to Enrichment Exercises are on page A.83.

11.2

Logarithmic Functions

O B J E C T I V E S

▶ To use the logarithmic notation

▶ To solve $y = \log_b x$ for y, x, or b

▶ To draw the graph of the logarithmic function

In Chapter 10, we graphed the exponential function $f(x) = b^x$, $b > 0$, $b \neq 1$. In this section, we investigate further properties of the exponential function. To graph an exponential function such as $f(x) = 2^x$, recall that we made a table of values by choosing x to be convenient numbers, then finding y from the equation $y = 2^x$.

x	-3	-2	-1	0	1	2	3
y	$\dfrac{1}{8}$	$\dfrac{1}{4}$	$\dfrac{1}{2}$	1	2	4	8

Next, we plotted the pairs (x, y) from this table in a coordinate system, then drew the curve suggested by these points.

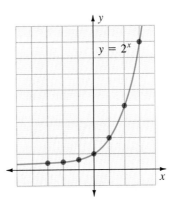

Notice that a horizontal line, *in any position,* intersects the graph in exactly one point, and, therefore, the function is one-to-one. This means that the inverse of $f(x) = 2^x$ is also a function. As we saw in Chapter 10, we can graph the inverse of $y = 2^x$ by reflecting the graph of $y = 2^x$ in the line $y = x$ as shown in the following figure.

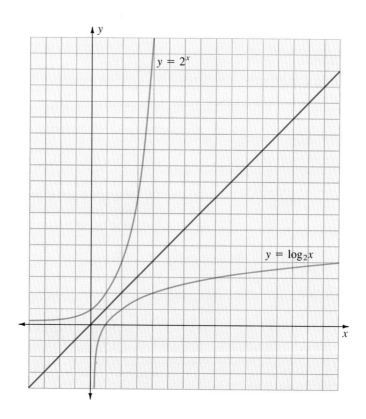

The four-step method for finding the equation of the inverse cannot be used on $f(x) = 2^x$ because we can only accomplish the first two steps. In particular,

Step 1   Set $y = 2^x$.

Step 2   Interchange the role of x and y, $x = 2^y$.

Step 3   Solve the equation in Step 2 for y.

We cannot do Step 3 (try it) even though we know that y is a function of x. However, this function is given a name by the following definition.

DEFINITION

If b is a positive number, $b \neq 1$, then

$$y = \log_b x \quad \text{means} \quad x = b^y$$

The symbolism $\log_b x$ is read "the logarithm of x to the base b." The function $y = \log_b x$ is called the **logarithmic function to the base b.**

Note that $\log_b x$ is the *exponent* to which b is raised in order to obtain x.

From the graph of $y = \log_2 x$ as shown on the previous page, we see that the domain is $\{x \mid x > 0\}$ and the range is the set of all real numbers. In general, we can make the following statement.

The domain of $y = \log_b x$ is the set of positive real numbers.

The range of $y = \log_b x$ is the set of all real numbers.

Example 1

(a) $8 = 2^3$ means the same as $\log_2 8 = 3$.

(b) $\left(\dfrac{1}{3}\right)^2 = \dfrac{1}{9}$ means the same as $\log_{1/3}\left(\dfrac{1}{9}\right) = 2$.

(c) $10^4 = 10{,}000$ means the same as $\log_{10} 10{,}000 = 4$.

(d) $\dfrac{1}{25} = 5^{-2}$ means the same as $\log_5\left(\dfrac{1}{25}\right) = -2$.

(e) $\log_2 16 = 4$ means the same as $16 = 2^4$.

(f) $\log_{1/4} 16 = -2$ means the same as $\left(\dfrac{1}{4}\right)^{-2} = 16$.

(g) $\log_a R = n$ means the same as $R = a^n$.

(h) $c^m = p$ means the same as $\log_c p = m$. ▲

N O T E   *Keep in mind that the base b of the logarithm $\log_b x$ is never negative. However, the logarithm $\log_b x$, itself, can be negative.*

For example, $\log_2\left(\dfrac{1}{8}\right) = -3$. The base is the positive number 2, while the logarithm of $\dfrac{1}{8}$ to this base is negative.

Example 2

Draw the graph of $y = \log_{1/3} x$.

Solution

By using the equivalent exponential form, $x = \left(\dfrac{1}{3}\right)^y$, we can construct a table of values by *first* choosing values for y, then finding the value of x from the equation $x = \left(\dfrac{1}{3}\right)^y$.

x	27	9	3	1	$\dfrac{1}{3}$	$\dfrac{1}{9}$	$\dfrac{1}{27}$
y	−3	−2	−1	0	1	2	3

We plot the points (x, y) and draw the curve suggested by them as shown in the following figure.

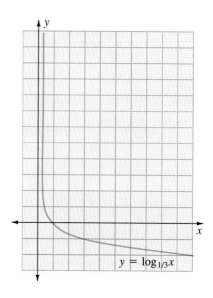

$$y = \log_{1/3} x$$

An important fact to keep in mind when working with logarithms is that

The logarithm of a number is an exponent.

Example 3   Find $\log_3 9$.

Solution   Setting $y = \log_3 9$, then $3^y = 9$. Since $9 = 3^2$, we have $3^y = 3^2$. Therefore, $y = 2$, that is, $\log_3 9 = 2$.   ▲

If b is a positive number, then $b^1 = b$ and $b^0 = 1$. The logarithmic forms for these two statements are stated in the following property:

P R O P E R T Y

If b is a positive real number, $b \neq 1$, then

$$\log_b b = 1 \qquad \text{and} \qquad \log_b 1 = 0$$

There are three variables in the equation $\log_b x = y$. If two of them are known, the remaining one can be determined. In the next example, we show how to find x when b and y are given.

Example 4   Solve for x: $\log_5 x = -2$.

Solution $\log_5 x = -2$ means $5^{-2} = x$. Therefore,

$$x = 5^{-2} = \frac{1}{5^2} = \frac{1}{25}$$

▲

Example 5   Solve for b: $\log_b 5 = \dfrac{1}{2}$.

Solution   The exponential form for this logarithmic statement is

$$b^{1/2} = 5$$

Therefore,

$$(b^{1/2})^2 = 5^2$$

$$b = 25$$

▲

Example 6   Solve for y: $\log_9 27 = y$.

Solution   We first write the equation in exponential form.

$$27 = 9^y$$

Both 27 and 9 can be written as powers of 3.

$$3^3 = (3^2)^y$$

$$3^3 = 3^{2y}$$

If two powers are the same, the exponents must be the same. Therefore, $3 = 2y$ or $y = \dfrac{3}{2}$. As a check, $\log_9 27 = \dfrac{3}{2}$ is the same as $27 = 9^{3/2}$. Now, $9^{3/2} = (\sqrt{9})^3 = 3^3 = 27$.

▲

There are two *bases* that are given special attention:

If $b = 10$, we call $\log_{10} x$ the **common logarithm** of x and write

$$\log_{10} x = \log x$$

Therefore, whenever you write $\log x$ with no base indicated, the base is understood to be 10.

The number e has such important applications in mathematics that the logarithm to the base e is given special treatment. If $b = e$, we call $\log_e x$ the **natural logarithm** of x and write

$$\log_e x = \ln x$$

Example 7   Find

(a) $\log 0.001$

(b) $\ln e^3$

Solution   (a) Set $y = \log 0.001$. Then $10^y = 0.001$. But, $0.001 = 10^{-3}$. Therefore, $10^y = 10^{-3}$, which means that $y = -3$. That is, $\log 0.001 = -3$.

(b) Set $y = \ln e^3$. Therefore, $e^y = e^3$, which means that $y = 3$. That is, $\ln e^3 = 3$.   ▲

In the next example we show how the logarithmic function of base 10 is used to measure earthquakes.

Example 8   Charles Richter, a seismologist, invented the **Richter scale** for measuring the magnitude M of an earthquake in terms of the total relative energy E. This scale is given by

$$M = \log E$$

The 1906 earthquake in San Francisco measured 8.4 on the Richter scale. The October, 1987, earthquake in Los Angeles measured 6.1 on the Richter scale. How many times more powerful, in terms of energy released, is the 1906 earthquake than the 1987 earthquake?

Solution   We find the relative energy E for each earthquake. Let E_1 be the relative energy of the 1906 earthquake and E_2 the relative energy of the 1987 earthquake. Then

$$8.4 = \log E_1, \quad \text{so } E_1 = 10^{8.4}$$

and

$$6.1 = \log E_2, \quad \text{so } E_2 = 10^{6.4}$$

To compare, we find the quotient $\dfrac{E_1}{E_2}$.

$$\frac{E_1}{E_2} = \frac{10^{8.4}}{10^{6.1}} = 10^{8.4-6.1} = 10^{2.3} \approx 200$$

So the 1906 earthquake was about 200 times more powerful than the 1987 earthquake.   ▲

E X E R C I S E   S E T   11.2

For Exercises 1–14, rewrite each exponential statement as a logarithmic statement and each logarithmic statement as an exponential statement.

1. $25 = 5^2$

2. $8 = 2^3$

3. $\dfrac{1}{9} = 3^{-2}$

4. $\dfrac{1}{8} = 2^{-3}$

5. $\left(\dfrac{1}{7}\right)^{-2} = 49$

6. $\left(\dfrac{1}{6}\right)^{-2} = 36$

7. $b^t = V$

8. $a^s = c$

9. $\log_4 8 = \dfrac{3}{2}$

10. $\log_9 81 = 2$

11. $\log 1{,}000 = 3$

12. $\log 0.1 = -1$

13. $\ln \sqrt{e} = \dfrac{1}{2}$

14. $\ln \sqrt[3]{e^2} = \dfrac{2}{3}$

For Exercises 15–22, draw the graph of each function.

15. $y = \log_4 x$

16. $y = \log_5 x$

17. $f(x) = \log x$

18. $g(x) = \log_{3/2} x$

19. $y = \log_{1/2} x$

20. $y = \log_{2/3} x$

21. $f(x) = \log_{1.8} x$

22. $g(x) = \log_{0.6} x$

For Exercises 23–34, simplify.

23. $\log_2 16$

24. $\log_3 \dfrac{1}{9}$

25. $\log_b b^2$

26. $\log_b b^3$

27. $\log_{1/3} \dfrac{1}{27}$

28. $\log_{1/2} 8$

29. $\log_{3/4} 1$

30. $\log_9 1$

31. $\log 0.001$

32. $\log 10{,}000$

33. $\ln \sqrt[4]{e^3}$

34. $\ln \sqrt[3]{e}$

For Exercises 35–50, solve each equation.

35. $\log_2 x = -3$

36. $\log_4 x = -2$

37. $\log_9 x = -\dfrac{1}{2}$

38. $\log_8 x = -\dfrac{1}{3}$

39. $\log_b 10 = 1$

40. $\log_b 36 = 2$

41. $\log_b 2 = -\dfrac{1}{3}$

42. $\log_b 3 = -\dfrac{1}{2}$

43. $\log_{16} 4 = y$

44. $\log_9 \dfrac{1}{27} = y$

45. $\log_8 16 = y$

46. $\log_{32} 8 = y$

47. $\log_b 1 = 0$

48. $\log_b 100{,}000 = 5$

49. $\log_2 x^2 = 3$

50. $\log_3 x^2 = 3$

51. An earthquake measured 5.1 on the Richter scale and an aftershock measured 3.2. How many times more powerful, in terms of energy released, is the earthquake than the aftershock?

52. One earthquake measured 4.8 and a second earthquake measured 3.9 on the Richter scale. How many times more powerful, in terms of energy released, is the first earthquake than the second?

REVIEW PROBLEMS

The following exercises review parts of Section 3.8. Doing these problems will help prepare you for the next section.

Solve each equation.

53. $x^2 + 4x - 12 = 0$

54. $x^2 - 2x - 3 = 0$

55. $4x^2 + 4x - 3 = 0$

56. $3x^2 - 5x + 2 = 0$

57. $8x^2 - 2x - 3 = 0$

58. $6x^2 - 17x + 12 = 0$

E N R I C H M E N T   E X E R C I S E S

For Exercises 1–5, evaluate the expression.

1. $\ln[\log_3(\log_2 8)]$

2. $\log_2[\log_5(\log_2 32)]$

3. $5^{\log_5 4 + \log_5 2}$

4. $2^{\log_2 3 - \log_2 6}$

5. $3^{-\log_3 4}$

6. Draw the graph of $f(x) = \log_2 |x|$.

7. Draw the graph of $f(x) = \log_{1/2} |x|$.

Answers to Enrichment Exercises are on page A.83.

11.3

Properties of Logarithms

▶ *To use the properties of logarithms to rewrite logarithmic expressions*

▶ *To solve a logarithmic equation using the properties of logarithms*

Recall from the last section that a logarithm is simply an exponent: $y = \log_b x$ means $x = b^y$. Therefore, it is reasonable to expect that logarithms have properties that are similar to properties of exponents. In particular, there are three basic properties of logarithms that we now state.

P R O P E R T I E S

Let u and v be positive numbers, $b > 0$, $b \neq 1$.

Product rule for logarithms

$$\log_b uv = \log_b u + \log_b v$$

That is, the log of a product is equal to the sum of the logs.

Quotient rule for logarithms

$$\log_b \frac{u}{v} = \log_b u - \log_b v$$

That is, the log of a quotient is equal to the difference of the logs.

Power rule for logarithms

$$\log_b u^r = r \log_b u, \qquad r \text{ is a rational number}$$

That is, the log of a power is the exponent times the log.

Proof of the product rule for logarithms

Recall the product rule *for exponents* that states that $b^m b^n = b^{m+n}$.

Setting $u = b^m$ and $v = b^n$, we can write these two equations as follow:

$$u = b^m \text{ is the same as } m = \log_b u$$

$$v = b^n \text{ is the same as } n = \log_b v$$

Now, $u \cdot v = b^m \cdot b^n = b^{m+n}$ and the statement

$$u \cdot v = b^{m+n} \text{ is the same as } m + n = \log_b uv$$

Therefore, $\log_b uv = m + n = \log_b u + \log_b v$ and the product rule is proved.

The other two properties are proved in a similar way and are left as exercises. These properties can be used to change the form of a logarithmic expression.

Example 1   Write $\log_3 4x$ as the sum of logarithms. Assume that x is positive.

Solution   By the product rule for logarithms,

$$\log_3 4x = \log_3 4 + \log_3 x \qquad \blacktriangle$$

Example 2   Write $\log_2 \dfrac{4x}{yz^3}$ in expanded form using the properties of logarithms.

Solution   We use the properties to rewrite the expression.

$$\log_2 \frac{4x}{yz^3} = \log_2 4x - \log_2 yz^3 \qquad \text{Quotient rule.}$$

$$= \log_2 4 + \log_2 x - (\log_2 y + \log_2 z^3) \qquad \text{Product rule.}$$

$$= 2 + \log_2 x - \log_2 y - 3\log_2 z \qquad \text{Power rule.} \quad \blacktriangle$$

The power rule for logarithms can be used to simplify expressions such as $\log_b \sqrt[3]{x}$. First write $\sqrt[3]{x}$ as $x^{1/3}$. The details follow:

$$\log_b \sqrt[3]{x} = \log_b x^{1/3} = \frac{1}{3}\log_b x$$

Example 3   Write $\log_b \sqrt[5]{\dfrac{x^3 y}{z^2}}$ in expanded form using properties of logarithms. Assume that all variables represent positive numbers.

Solution   We start by changing the fifth root to exponent form.

$$\log_b \sqrt[5]{\frac{x^3 y}{z^2}} = \log_b \left(\frac{x^3 y}{z^2}\right)^{1/5}$$

$$= \frac{1}{5}\log_b \frac{x^3 y}{z^2} \qquad \text{Power rule.}$$

$$= \frac{1}{5}[\log_b x^3 y - \log_b z^2] \qquad \text{Quotient rule.}$$

$$= \frac{1}{5}(\log_b x^3 + \log_b y - \log_b z^2) \qquad \text{Product rule.}$$

$$= \frac{1}{5}(3\log_b x + \log_b y - 2\log_b z) \qquad \text{Power rule.}$$

$$= \frac{3}{5}\log_b x + \frac{1}{5}\log_b y - \frac{2}{5}\log_b z \qquad \text{Distributive property.}$$

$$\blacktriangle$$

Example 4   Given that $\log_b 4 = 0.6931$ and $\log_b 9 = 1.0986$, find each of the following.

(a) $\log_b 36$ **(b)** $\log_b \dfrac{4}{9}$ **(c)** $\log_b 144$

Solution   We use properties of logarithms to express the given logarithm in terms of $\log_b 4$ and $\log_b 9$.

(a) $\log_b 36 = \log_b 4 \cdot 9$

$\qquad\qquad = \log_b 4 + \log_b 9$ $\qquad\qquad$ Product rule for logarithms.

$\qquad\qquad = 0.6931 + 1.0986$

$\qquad\qquad = 1.7917$

(b) $\log_b \dfrac{4}{9} = \log_b 4 - \log_b 9$ $\qquad\qquad$ Quotient rule for logarithms.

$\qquad\qquad = 0.6931 - 1.0986$

$\qquad\qquad = -0.4055$

(c) First, factor 144 as $16 \cdot 9 = 4^2 \cdot 9$. Therefore,

$\qquad \log_b 144 = \log_b 4^2 \cdot 9$

$\qquad\qquad\qquad = \log_b 4^2 + \log_b 9$ $\qquad$ Product rule.

$\qquad\qquad\qquad = 2 \log_b 4 + \log_b 9$ $\qquad$ Power rule.

$\qquad\qquad\qquad = 2(0.6931) + 1.0986$

$\qquad\qquad\qquad = 2.4848$ $\qquad\qquad\qquad\qquad$ ▲

Example 5   Write $\log_3 11 - \log_3 y + 2 \log_3 z - \log_3 5$ as one logarithm by using the properties of logarithms.

Solution   We start by regrouping the terms.

$\log_3 11 - \log_3 y + 2 \log_3 z - \log_3 5 = \log_3 11 + 2 \log_3 z - \log_3 y - \log_3 5$

Factor -1 from the last two terms. $\qquad = \log_3 11 + 2 \log_3 z - (\log_3 y + \log_3 5)$

Power rule. $\qquad = \log_3 11 + \log_3 z^2 - (\log_3 y + \log_3 5)$

Product rule (used twice). $\qquad = \log_3 11 z^2 - \log_3 5 y$

Quotient rule. $\qquad = \log_3 \dfrac{11 z^2}{5 y}$ $\qquad\qquad$ ▲

Two other properties of logarithms and exponents deal with the fact that the logarithmic and exponential functions are inverses of each other.

P R O P E R T I E S

If $b > 0$ and $b \neq 1$, then

$$b^{\log_b x} = x, \qquad \text{where } x > 0$$

and

$$\log_b b^x = x, \qquad \text{for any real number } x$$

Example 6   Find the value of each expression.

(a) $\log_7 7^3$ (b) $\log_2 16$ (c) $5^{\log_5 12}$

Solution   (a) Since $\log_b b^x = x$, $\log_7 7^3 = 3$.

(b) We can write 16 as a power of 2: $16 = 2^4$. Therefore,

$$\log_2 16 = \log_2 2^4 = 4 \qquad \log_b b^x = x.$$

(c) Since $b^{\log_b x} = x$, $5^{\log_5 12} = 12$.   ▲

Properties of logarithms can be used to solve certain logarithmic equations.

Example 7   Solve: $\log_3 x + \log_3(2x - 3) = 2$.

Solution   First, we use the product property to write the left side of the equation as a single logarithm.

$$\log_3 x + \log_3(2x - 3) = 2$$

$$\log_3 x(2x - 3) = 2$$

By the definition of logarithm, this last equation means

$$x(2x - 3) = 3^2$$

$$2x^2 - 3x - 9 = 0 \qquad \text{Simplify.}$$

$$(2x + 3)(x - 3) = 0 \qquad \text{Factor.}$$

Either $x = -\dfrac{3}{2}$ or $x = 3$. We check our answers in the original equation.

Check $x = -\dfrac{3}{2}$: $\log_3 x + \log_3(2x - 3) = 2$

$$\log_3\left(-\frac{3}{2}\right) + \log_3\left[2\left(-\frac{3}{2}\right) - 3\right] \stackrel{?}{=} 2$$

Notice that substituting $-\dfrac{3}{2}$ for x produces the expression $\log_3\left(-\dfrac{3}{2}\right)$. This logarithm is undefined since the domain of a logarithmic function is the set of *positive* real numbers. Therefore, $-\dfrac{3}{2}$ is an extraneous solution.

Check $x = 3$: $\log_3 x + \log_3(2x - 3) = 2$

$$\log_3 3 + \log_3(2 \cdot 3 - 3) \stackrel{?}{=} 2$$
$$1 + \log_3 3 \stackrel{?}{=} 2$$
$$1 + 1 = 2$$

The only solution is $x = 3$.   ▲

N O T E   *It is important to check your answers in the original equation. In the previous example, the equation obtained from the original equation,* $log_3 x(2x - 3) = 2$, *is satisfied by* $x = -\dfrac{3}{2}$.

The next property of logarithms will enable us to solve more equations.

P R O P E R T Y

Suppose x and y are positive numbers and $b > 0$, $b \neq 1$.

If $\log_b x = \log_b y$, then $x = y$.

Example 8

Solution

Solve: $\log_2 x + \log_2(x + 1) = \log_2 24 - 2\log_2 2$.

By using the properties of logarithms, we can write each side of the equation as a single logarithm.

$$\log_2 x + \log_2(x + 1) = \log_2 24 - 2\log_2 2$$
$$\log_2 x(x + 1) = \log_2 24 - \log_2 2^2$$
$$\log_2 x(x + 1) = \log_2 \frac{24}{4}$$
$$\log_2 x(x + 1) = \log_2 6$$

Next, we use the above property to obtain

$$x(x + 1) = 6$$
$$x^2 + x - 6 = 0$$
$$(x - 2)(x + 3) = 0$$

Therefore, either $x = 2$ or $x = -3$.

Check $x = 2$: $\log_2 x + \log_2(x + 1) = \log_2 24 - 2\log_2 2$

$$\log_2 2 + \log_2(2 + 1) \stackrel{?}{=} \log_2 24 - 2\log_2 2$$

$$\log_2 2 \cdot 3 \stackrel{?}{=} \log_2 \frac{24}{4}$$

$$\log_2 6 = \log_2 6$$

Check $x = -3$: $\log_2 x = \log_2(-3)$, and $\log_2(-3)$ is undefined since -3 is negative.

Therefore, 2 is the only solution.

E X E R C I S E   S E T   11.3

For Exercises 1–20, by using properties of logarithms, write each expression in expanded form. Assume that all variables represent positive real numbers.

1. $\log_3 7t$

2. $\log_6 2y$

3. $\log_{12} \dfrac{5}{z}$

4. $\log_9 \dfrac{R}{4}$

5. $\log_4 \dfrac{x^2 y}{z^5}$

6. $\log_3 \dfrac{2t^4}{r^2 s}$

7. $\log_b \dfrac{n^4 m^2}{pq}$

8. $\log_b \dfrac{u^3 v^2}{w}$

9. $\log_b b^5 c^3$

10. $\log_a av^2$

11. $\log \sqrt[3]{x^2 y}$

12. $\log \sqrt[4]{y^3 z}$

13. $\ln Pe^{rt}$

14. $\ln 2{,}000\, e^{10r}$

15. $\ln \dfrac{4}{e^3}$

16. $\ln \dfrac{e^3}{4}$

17. $\log\left(\dfrac{I}{I_0}\right)^{10}$

18. $\log\left(\dfrac{E}{E_0}\right)^{10}$

19. $\log_b \sqrt[3]{\dfrac{xy^4}{2z^3}}$

20. $\log_3 \sqrt[5]{\dfrac{4t^5}{r^4 s^3}}$

For Exercises 21–30, find the logarithm, given that $\log_5 2 = 0.4307$ and $\log_5 3 = 0.6825$.

21. $\log_5 6$

22. $\log_5 4$

23. $\log_5 12$

24. $\log_5 36$

25. $\log_5 \dfrac{2}{3}$

26. $\log_5 \dfrac{4}{9}$

27. $\log_5 0.25$

28. $\log_5 1.25$

29. $\log_5 30$

30. $\log_5 20$

For Exercises 31–38, write each expression as a single logarithm.

31. $\log_5 s - \log_5 2 + 3\log_5 t$

32. $\log_2 x - 3\log_2 z + \log_2 y$

33. $\log_b x + 4 \log_b y - \log_b z - \log_b t$

34. $\log_b w - 4 \log_b x - 2 \log_b y + 3 \log_b z$

35. $\dfrac{1}{2} \log_3 x - \dfrac{2}{3} \log_3 t$

36. $\dfrac{1}{3} \log_2 z + \dfrac{3}{2} \log_2 y$

37. $10 \log E_1 - 10 \log E_2$

38. $\ln A - rt \ln e$

For Exercises 39–64, solve each equation. Check for extraneous solutions.

39. $\log_3 4x + \log_3 2 = 1$

40. $\log_2 3 + \log_2 5x = 3$

41. $\log_4 y - \log_4 21 = \dfrac{1}{2}$

42. $\log_3 z - \log_3 10 = 3$

43. $2 \log_2 x = 3$

44. $2 \log_3 x = 1$

45. $\log_2 x^2 = 3$

46. $\log_3 x^2 = 1$

47. $\log(2x + 1) - \log(3x - 1) = 0$

48. $\ln(3t + 4) - \ln(4t - 1) = 0$

49. $\log x + \log(x + 3) = 1$

50. $\log_{12} x + \log_{12}(x + 1) = 1$

51. $\log_8 x + \log_8 x = \dfrac{2}{3}$

52. $\log_{16} y + \log_{16} y = \dfrac{3}{4}$

53. $\log_{12} x + \log_{12}(x - 2) = \log_{12} 1$

54. $\log_2 x + \log_2(3x + 2) = \log_2 1$

55. $\ln x + \ln(x - 1) = \ln 6$

56. $\log x + \log(2x - 3) = \log 2$

57. $\log_6 3 + \log_6 x = \log_6 6 - \log_6(x + 1)$

58. $\log_5(x + 1) + \log_5 4 = \log_5 15 - \log_5 x$

59. $\log_b(2x^2 - x) - \log_b(2x - 1) = \log_b 6$

60. $\log_b(3x - 6x^2) - \log_b x = \log_b 15$

61. $\log(x + 1) - \log(x - 2) = \log(x + 2) - \log x$

62. $\log_7(y - 4) - \log_7(y - 3) = \log_7(y + 1) - \log_7(y + 4)$

63. $2 \log_3 z + 2 \log_3 z = 2 + \log_3 z^2$

64. $\log_2(t - 3)^2 = \log_2(t - 3)$

REVIEW PROBLEMS

The following exercises review parts of Section 3.1. Doing these problems will help prepare you for the next section.

Simplify.

65. $b^2 \cdot b^4$

66. $x^{-4} \cdot x^8$

67. $\dfrac{c^{12}}{c^7}$

68. $(a^3)^2$

69. $(b^6)^{-2}$

70. $(b^{-3})^{-2}$

71. $(a^2 b^{-1})(a^{-1} b^4)^{-2}$

72. $(b^n)^{-2} b^{3n}$

Express each number in scientific notation.

73. 0.00023

74. 0.0000081

75. 152,000,000

76. 7,000,000,000

77. 59.3×10^{-4}

78. 628.5×10^{-7}

E N R I C H M E N T   E X E R C I S E S

Simplify.

1. $27^{-\log_3 2}$

2. $8^{\log_2 3}$

3. $18^{\log_4 2}$

A sound can be measured by its intensity I and its loudness L. The least intense sound that can be heard is denoted by I_0. The loudness L of a sound is a function of its intensity I given by

$$L = 10 \log I - 10 \log I_0$$

4. Show that $L = 10 \log \dfrac{I}{I_0}$.

5. Given two sounds of intensities I_1 and I_2, respectively, with corresponding loudness L_1 and L_2, show that

$$L_1 - L_2 = 10 \log \frac{I_1}{I_2}$$

Answers to Enrichment Exercises are on page A.84.

11.4

Solving Exponential Equations

O B J E C T I V E S

▶ *To use a calculator or the tables in the appendix to find logarithms and antilogarithms*

▶ *To solve exponential equations using a calculator or a table of logarithms*

▶ *To find a logarithm to a base other than 10 or e*

In this section, we solve equations involving logarithms where the solutions are not rational numbers. To solve these equations, we will use a calculator. Table II in the Appendix can also be used to evaluate common logarithms. We will first do examples by calculator, then explain how to use the table to evaluate logarithmic expressions.

Example 1   Use a calculator to find each logarithm to five decimal places.

(a) $\log 439$ **(b)** $\ln 0.12$ **(c)** $\log(-0.847)$

Solution

(a) *Enter*   *Press*   *Display*

439        $\boxed{\log}$ 2.64246

(b) *Enter*   *Press*   *Display*

0.12 $\boxed{\ln}$ -2.12026

(c) *Enter*   *Press*   *Display*

-0.847 $\boxed{\log}$ error

The reason that an error occurs in Part (c) is that -0.847 is not in the domain of the log function. Remember that you cannot take the logarithm of a negative number.   ▲

Example 2   Use a calculator to find the value of each accurate to three decimal places.

(a) $\dfrac{\log 4.5}{\log 8.2}$ **(b)** $\dfrac{\ln 7}{\ln 2.95}$

Solution

(a) This is the sequence of steps on a calculator.

Enter	*Press*	*Press*	*Enter*	*Press*	*Press*	*Display*
4.5	$\boxed{\log}$	$\boxed{\div}$	8.2	$\boxed{\log}$	$\boxed{=}$	0.7148201

Therefore, rounding to three decimal places,

$$\frac{\log 4.5}{\log 8.2} = 0.715$$

(b) Here is the sequence of steps on a calculator.

Enter	*Press*	*Press*	*Enter*	*Press*	*Press*	*Display*
7	$\boxed{\ln}$	$\boxed{\div}$	2.95	$\boxed{\ln}$	$\boxed{=}$	1.7987621

Therefore, rounding to three decimal places,

$$\frac{\ln 7}{\ln 2.95} = 1.799$$ ▲

COMMON   ERROR

When finding the value of a quotient such as $\dfrac{\log 4.5}{\log 8.2}$, note that

$$\frac{\log 4.5}{\log 8.2} \neq \log 4.5 - \log 8.2$$

From the example, accurate to three decimal places,

$$\frac{\log 4.5}{\log 8.2} = 0.715$$

Using a calculator, you will find that

$$\log 4.5 - \log 8.2 = -0.261$$

Suppose we are given the logarithm of some unknown number x, that is, we are given $\log x$ and we want to find x. The number x is called the **antilogarithm**, or **antilog**, of $\log x$. In Example 1, we found that $\log 439 = 2.64246$, so 439 is the antilog of 2.64246.

We can find the number by using the definition of logarithm. Recall,

$$\log x = y \text{ means the same as } x = 10^y$$

and

$$\ln x = y \text{ means the same as } x = e^y$$

Example 3   Find the antilog to the nearest thousandth.

(a) $\log x = 2.928$ (b) $\ln x = -6.021$

Solution   (a) $\log x = 2.928$ means the same as $x = 10^{2.928}$.
This is the sequence of steps to find $10^{2.928}$ with a calculator:

Enter	Press	Press	Display
2.928	INV	log	847.22741

Accurate to two decimal places,

$$x = 847.23$$

(b) $\ln x = -6.021$ means the same as

$$x = e^{-6.021}$$

The sequence of steps to find $e^{-6.021}$ with a calculator follows:

Enter	Press	Press	Press	Display
6.021	$+/-$	INV	ln	2.4272411^{-03}

Accurate to three decimal places,

$$x = 0.002$$

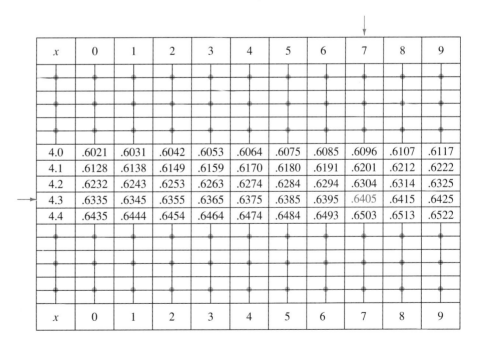

N O T E   *In Part (a) of Example 3, another way to evaluate $10^{2.928}$ using a calculator is the following sequence of steps:*

Enter	Press	Enter	Press	Display
10	y^x	2.928	$=$	847.22741

If a calculator is not available, Table II in the Appendix lists common logarithms of numbers between 1.00 and 9.99. For example, suppose we wanted to find the common logarithm of 4.37. We read down the left-hand column until we get to 4.3, then across until we are below the 7 in the top row or above the 7 in the bottom row.

x	0	1	2	3	4	5	6	7	8	9
4.0	.6021	.6031	.6042	.6053	.6064	.6075	.6085	.6096	.6107	.6117
4.1	.6128	.6138	.6149	.6159	.6170	.6180	.6191	.6201	.6212	.6222
4.2	.6232	.6243	.6253	.6263	.6274	.6284	.6294	.6304	.6314	.6325
4.3	.6335	.6345	.6355	.6365	.6375	.6385	.6395	.6405	.6415	.6425
4.4	.6435	.6444	.6454	.6464	.6474	.6484	.6493	.6503	.6513	.6522
x	0	1	2	3	4	5	6	7	8	9

Therefore, log 4.37 = 0.6405, which is accurate to four decimal places. More examples follow:

$$\log 6.05 = 0.7818, \qquad \log 3.60 = 0.5563, \qquad \log 8.98 = 0.9533$$

Recall that any decimal number can be written in scientific notation (Section 3.1). Converting a number to scientific notation and then using the product rule for logarithms enables us to use the table on numbers other than 1.00 to 9.99. For example,

$$\log 3{,}970 = \log(3.97 \times 10^3)$$
$$= \log 3.97 + \log 10^3$$
$$= \log 3.97 + 3$$

and

$$\log 0.000891 = \log(8.91 \times 10^{-4})$$
$$= \log 8.91 + \log 10^{-4}$$
$$= \log 8.91 - 4$$

In general, suppose a number is written in scientific notation, $x \times 10^n$, where x is between 1 and 10 and n is an integer, then

$$\log(x \times 10^n) = \log x + n$$

$\log x$ is called the **mantissa** and n is the **characteristic.**

Therefore, if x is in the table, we can find the logarithm of any number of the form $x \times 10^n$.

Example 4   Find each logarithm by using Table II of the Appendix.

(a) log 56,300   **(b)** log 0.00149

Solution   **(a)** $\log 56{,}300 = \log(5.63 \times 10^4)$

$$= \log 5.63 + \log 10^4$$
$$= 0.7505 + 4$$
$$= 4.7505$$

(b) $\log 0.00149 = \log(1.49 \times 10^{-3})$

$$= \log 1.49 - 3$$
$$= 0.1732 - 3$$
$$= -2.8268$$

▲

Example 5   Find each antilog by using Table II in the Appendix.

(a) $\log x = 0.6561$ **(b)** $\log x = -2.0953$

Solution   To find antilogarithms, we reverse the process that we used to find logarithms in the table.

(a) We look for 0.6561 in the heart of the table. It is located at the intersection of the Row 4.5 and Column 3. Therefore, $x = 4.53$.

(b) This antilog is negative and the table only has positive values. If we add 3 to the antilog, we will have a positive value. Therefore, we rewrite the antilog by adding *and subtracting* 3.

$$-2.0953 = 3 - 2.0953 - 3$$

$$= 0.9047 - 3$$

Next, we look up 0.9047 in the table. It is at the intersection of Row 8.0 and Column 3. The characteristic is -3, so the number is

$$x = 8.03 \times 10^{-3} = 0.00803$$ ▲

We can use logarithms to solve exponential equations. Previously, we could only solve equations of the form $2^x = 8$, where each side could be written as a power of the same base. The equation $3^x = 4$ cannot be solved by that method. We can solve it by using the following property.

PROPERTY

Suppose x and y are positive numbers.

If $x = y$, then $\log_b x = \log_b y$

Example 6   Solve: $3^x = 4$.

Solution   By the above property, we may take the logarithm of both sides and still maintain an equation.

$$3^x = 4$$

$$\log 3^x = \log 4$$

$$x \log 3 = \log 4 \qquad \text{Power rule: } \log_b x^r = r \log_b x.$$

$$x = \frac{\log 4}{\log 3}$$

$$= 1.26 \qquad \text{Accurate to two decimal places.} \quad ▲$$

Example 7   Solve: $e^{2x-1} = 2.6$.

Solution   Since the number e appears as a base, we use ln instead of log when taking logarithms of both sides.

$$e^{2x-1} = 2.6$$

$$\ln e^{2x-1} = \ln 2.6$$

$$(2x - 1)\ln e = \ln 2.6 \qquad \ln x^r = r \ln x.$$

$$(2x - 1)(1) = \ln 2.6 \qquad \ln e = 1.$$

$$2x - 1 = \ln 2.6$$

$$2x = \ln 2.6 + 1 \qquad \text{Add 1 to both sides.}$$

$$x = \frac{\ln 2.6 + 1}{2} \qquad \text{Divide both sides by 2.}$$

$$= 0.9778 \qquad \text{Accurate to four decimal places.} \quad \blacktriangle$$

In Section 11.1, we used the formula for the amount A after t years of P dollars compounded n times a year at an interest rate r:

$$A = P\left(1 + \frac{r}{n}\right)^{nt}$$

We use this formula in the next example.

Example 8   Two thousand dollars are invested at 8% compounded quarterly. How long does it take for the investment to double?

Solution   We set $P = 2{,}000$, $r = 0.08$, $n = 4$, and $A = 4{,}000$ in the formula $A = P\left(1 + \dfrac{r}{n}\right)^{nt}$, then solve for t.

$$4{,}000 = 2{,}000\left(1 + \frac{0.08}{4}\right)^{4t}$$

$$2 = (1.02)^{4t}$$

To solve this exponential equation, we take logarithms of both sides.

$$\log 2 = \log(1.02)^{4t}$$

$$\log 2 = 4t \log 1.02 \qquad \text{Power rule for logarithms.}$$

Therefore,

$$t = \frac{\log 2}{4 \log 1.02}$$

$$= 8.75$$

It takes about $8\dfrac{3}{4}$ years to double the investment. $\blacktriangle$

The last property of logarithms that we consider is a formula that expresses a logarithm in terms of logarithms to a different base.

P R O P E R T Y

Change of base property

If a and b are positive numbers other than 1, and $x > 0$, then

$$\log_a x = \frac{\log_b x}{\log_b a}$$

Proof By the definition of logarithm,

$$a^{\log_a x} = x$$

Taking the logarithms to the base b of both sides, we have

$$\log_b(a^{\log_a x}) = \log_b x$$

$$\log_a x \log_b a = \log_b x \qquad \text{Power rule for logarithms.}$$

Dividing both sides by $\log_b a$, we have the desired result:

$$\log_a x = \frac{\log_b x}{\log_b a}$$

This property allows us to find the logarithms to bases other than 10 or e.

Example 9 Find $\log_{12} 56.9$.

Solution We convert this logarithm in terms of logarithms to the base 10.

$$\log_{12} 56.9 = \frac{\log 56.9}{\log 12}$$

$$= 1.6263$$

Example 10 Find $\log_3 0.045$.

Solution Using the change of base property,

$$\log_3 0.045 = \frac{\log 0.045}{\log 3}$$

$$= -2.8227$$

E X E R C I S E   S E T   1 1 . 4

For Exercises 1–10, use a calculator to find each logarithm accurate to five decimal places.

1. log 35.8   **2.** log 923   **3.** log 0.0391   **4.** log 0.008

5. ln 62   **6.** ln 20.3   **7.** ln 5.501   **8.** ln 0.0239

9. log −3.40   **10.** ln −44.7

For Exercises 11–24, use a calculator to find the value of each, accurate to three decimal places.

11. $\dfrac{\log 83}{\log 0.592}$

12. $\dfrac{\log 0.00328}{\log 0.00086}$

13. $\dfrac{\ln 4.02}{\ln 6.81}$

14. $\dfrac{\ln 72.4}{\ln 0.506}$

15. $\dfrac{(\log 40.2)(\log 0.00491)}{\log 5.924}$

16. $\dfrac{\ln 82.4}{(\ln 0.091)(\ln 9.87)}$

17. $\dfrac{66 \ln 0.0311}{\ln 4.801}$

18. $\dfrac{4.91 \log 56.2}{\log 0.0998}$

19. $\dfrac{\log(9.45 + 0.783)}{3 \log 12.04}$

20. $\dfrac{\log(5.093 - 0.038)}{\log(8.3)^2}$

21. $\dfrac{(\log 5.4)^2}{\log(5.4)^2}$

22. $\dfrac{(\ln 13.8)^2}{\ln(13.8)^2}$

23. $\dfrac{\ln 34.1 - \ln 5.82}{1 + \ln 55.39}$

24. $\dfrac{\log 50.14 - 72.5}{42 - \log 66.2}$

For Exercises 25–36, use a calculator to find the antilogarithm to the nearest thousandth.

25. log x = 2.001   **26.** log x = 1.92   **27.** log x = 0.7281

28. log x = 0.0194   **29.** log x = −0.025   **30.** log x = −2.118

31. ln x = 4.81   **32.** ln x = 2.913   **33.** ln x = 0.008

34. ln x = 0.0069   **35.** ln x = −1.0034   **36.** ln x = −3.015

For Exercises 37–48, find the logarithm by using a calculator or Table II in the Appendix.

37. log 6.48   **38.** log 1.59   **39.** log 1.01

40. log 9.09   **41.** log 1,320   **42.** log 5,910

43. log 0.00815   **44.** log 0.0492   **45.** log 0.0058

46. log 0.00072   **47.** log 6,050,000   **48.** log 7,920,000

For Exercises 49–60, find each antilog by using a calculator or Table II in the Appendix.

49. log x = 0.9965   **50.** log x = 0.7435   **51.** log x = 0.0374

52. $\log x = 0.0170$

53. $\log x = 2.8169$

54. $\log x = 3.7364$

55. $\log x = 0.5977 - 2$

56. $\log x = 0.3502 - 3$

57. $\log x = -2.2557$

58. $\log x = -1.4750$

59. $\log x = -3.9318$

60. $\log x = -0.1024$

For Exercises 61–74, solve each equation. Write your answer accurate to two decimal places.

61. $4^x = 5.1$

62. $2^x = 7.09$

63. $e^{2x} = 3.71$

64. $e^{3x} = 7.32$

65. $(0.41)^x = 40.2$

66. $(0.68)^x = 9.24$

67. $10^{4x-3} = 60.3$

68. $10^{3x+4} = 0.85$

69. $(6.334)^x = 8.773$

70. $(5.917)^x = 2.204$

71. $e^{-0.4t} = 35$

72. $e^{-0.3t} = 9.81$

73. $1 - e^{-0.2t} = 0.5$

74. $8 - 8e^{-0.5t} = 3$

75. The tennis instructor estimates that the "learning curve" for one of her students is $y = 5(1 - e^{-0.04t})$, where t is the number of months of practice and y is the student's ability measured on a scale of 0 to 5.
(a) How long will it take the student to become a 2.5 player?
(b) In the instructor's estimation, will the student ever achieve a ranking of 5?

76. Sam is taking a typing course in high school. The teacher estimates that his learning curve is $y = 50(1 - e^{-0.16t})$, where y is Sam's typing speed in words per minute after t weeks of practice.
(a) How long will it take Sam to type 40 words per minute?
(b) In the teacher's estimation, will Sam ever type 50 words per minute?

77. The half-life of a radioactive substance is the amount of time required for it to decay to one half of the original amount. Carbon-14 has a half-life of 5,568 years. The amount A that is left after t years is given by

$$A = A_0 2^{-kt}$$

where A_0 is the initial amount of carbon-14. Find the value of k.

78. Refer to Exercise 77. All living wood contains the same concentration of carbon-14 that starts to decay when the tree is chopped down. A wooden artifact can be dated by determining the percentage of carbon-14 left. How old is a wooden artifact that contains 72% of the original amount of carbon-14?

For Exercises 79–86, use the change of base property and a calculator or Table II in the Appendix to find each logarithm. Write your answer accurate to four decimal places.

79. $\log_7 20$

80. $\log_4 5.2$

81. $\log_{15} 383,000$

82. $\log_{12} 28,700$

83. $\log_{0.5} 0.983$

84. $\log_{0.4} 0.796$

85. $\log_8 2,710,000$

86. $\log_5 895,000$

For Exercises 87–90, solve each equation. Write your answer accurate to four decimal places.

87. $\log_6 x = \log_5 27.1$

88. $\log_3 x = \log_2 x^2$

89. $\log_{12}(x + 1) = \log 657$

90. $\log_5 x = \log_6 (90.2)^2$

REVIEW PROBLEMS

The following exercises review parts of Section 10.2. Doing these problems will help prepare you for Section 12.1.

91. Let $f(x) = 3x - 5$. Find $f(1)$, $f(2)$, $f(3)$, and $f(4)$.

92. Let $f(x) = \dfrac{1}{x}$. Find $f(1)$, $f(2)$, $f(3)$, and $f(4)$.

93. Let $f(n) = n^2 + 1$. Find $f(2)$, $f(4)$, $f(6)$, and $f(8)$.

94. Let $f(n) = \dfrac{n + 1}{n^3}$. Find $f(3)$, $f(4)$, and $f(5)$.

95. Let $f(n) = 2n$. Find $f(1) + f(2) + f(3) + f(4)$.

96. Let $f(n) = 1 - \dfrac{1}{n}$. Find $f(1) + f(2) + f(3)$.

E N R I C H M E N T   E X E R C I S E S

In 1909, the Danish biochemist Sorensen invented the pH system for measuring the hydrogen ion concentration of a solution. The pH of a solution is given by

$$pH = -\log[H^+]$$

where $[H^+]$ is the concentration of hydrogen ions expressed in moles per liter. A pH level below 7 indicates that a solution is acidic, and a pH above 7 indicates that a solution is basic. A neutral solution, such as pure water, is neither acidic nor basic and has a pH of 7.

1. What is the hydrogen ion concentration $[H^+]$ of pure water?

2. If the hydrogen ion concentration of a solution is 0.00000000072 moles per liter. What is the pH, to the nearest tenth, of the solution? Is the solution acidic or basic?

3. A litmus pH indicator will change color at different pH levels. If the pH is below 4.5, the color change is red, and if the pH is above 8.3, the color change is blue. Between these values, it is a mixture of the two colors. A solution is tested and the color change is a mixture of the two colors. What can be said about the hydrogen ion concentration?

Answers to Enrichment Exercises are on page A.85.

Examples

$\mathbf{S}$ummary and review

The exponential functions (11.1)

An **exponential function** is of the form $f(x) = b^x$, where $b > 0$, $b \neq 1$. The number b is called the **base**.

Find the amount that $8,250 is worth at the end of 3 years, if it is invested at 8% compounded quarterly.

$$A = 8{,}250\left(1 + \frac{0.08}{4}\right)^{4 \cdot 3}$$
$$= \$10{,}463$$

One application of exponential functions is in **compound interest**. If P dollars is compounded n times a year at an annual interest rate r, then the amount A after t years is given by

$$A = P\left(1 + \frac{r}{n}\right)^{nt}$$

If P dollars is **compounded continuously** at an interest rate r, then the amount A after t years is given by

$$A = Pe^{rt}$$

The number e is irrational and is approximately equal to 2.7182818.

Applications in physics and social science rely on **growth** and **decay** equations of the form

$$Q = Q_0 e^{kt} \text{ (growth)} \qquad \text{and} \qquad Q = Q_0 e^{-kt} \text{ (decay)}$$

where k is a positive constant and Q_0 is the initial quantity present.

Logarithmic functions (11.2)

If b is a positive number, $b \neq 1$, then

$$y = \log_b x \text{ means the same as } x = b^y$$

$\log_2 16 = 4$ means $2^4 = 16$

The function $y = \log_b x$ is called the **logarithmic function to the base b.**

Keep in mind that the logarithm of a number is an exponent.

There are two bases that are given special attention.

$\log_{10} x = \log x$ is called the **common logarithm** of x.

$\log_e x = \ln x$ is called the **natural logarithm** of x.

Properties of logarithms (11.3) and (11.4)

We list several **properties of logarithms** that were introduced in Sections 11.3 and 11.4. Let x and y be positive numbers, $b > 0$, $b \neq 1$.

$\log_b \sqrt[7]{\dfrac{3x^4}{5z}} = \dfrac{1}{7}\log_b 3 +$

$\dfrac{4}{7}\log_b x - \dfrac{1}{7}\log_b 5 - \dfrac{1}{7}\log_b z$

Product rule: $\log_b xy = \log_b x + \log_b y$.

Quotient rule: $\log_b \dfrac{x}{y} = \log_b x - \log_b y$.

Power rule: $\log_b x^r = r \log_b x$, where r is a rational number.

Property: If $\log_b x = \log_b y$, then $x = y$.

Examples

Property: If $x = y$, then $\log_b x = \log_b y$.

Change of base property: $\log_a x = \dfrac{\log_b x}{\log_b a}$, where $a > 0$, $a \neq 1$.

Solving exponential and logarithmic equations (11.4)

Solve for x and write the answer accurate to the nearest thousandth.

$\log 3x = -1.89523$

Using a calculator,

$x = 0.004$

The use of a calculator or tables is necessary when solving exponential and logarithmic equations. Follow the instructions given in this section for their use. Check the manual that came with your calculator for additional help.

C H A P T E R 1 1 R E V I E W E X E R C I S E S E T

Section 11.1

For Exercises 1 and 2, solve the equation.

1. $2^x = 8\sqrt{2}$

2. $\left(\dfrac{1}{9}\right)^{x+1} = 27$

For Exercises 3–6, find the amount of the investment for the time indicated.

3. $550 at 6% compounded annually for 4 years.

4. $800 at 8% compounded quarterly for 2 years.

5. $1,000 at $4\dfrac{1}{2}$% compounded continuously for 3 years.

6. 70,000 at 24% compounded monthly for 1 year.

7. A bacteria culture initially has a count of 400 and the count Q after t hours is given by

$$Q = 400e^{0.45t}$$

What is the count after 5 hours?

8. Six hundred grams of a radioactive substance decays so that after t days, the amount Q that remains is given by

$$Q = 600e^{-0.08t}$$

How many grams of the substance is left after 3 days?

Section 11.2

For Exercises 9–16, rewrite each exponential statement as a logarithmic statement and each logarithmic statement as an exponential statement.

9. $121 = 11^2$

10. $7^{-2} = \dfrac{1}{49}$

11. $\log 0.01 = -2$

12. $\ln 532 = 6.2766$

13. $\log_3 9 = 2$

14. $\log_5 125 = 3$

15. $b^r = Q$

16. $\log_a P = s$

For Exercises 17–20, solve each equation.

17. $\log_4 x = -\dfrac{1}{2}$ **18.** $\log_b 32 = 4$ **19.** $\log_{11}\dfrac{1}{121} = y$ **20.** $\log_3 x^2 = 3$

Section 11.3

For Exercises 21–24, write each expression in expanded form.

21. $\log_2\dfrac{x^3 y^2}{z^4}$ **22.** $\log_5\dfrac{t^2}{25}$ **23.** $\ln Ae^{-rt}$ **24.** $\log \sqrt[3]{100z^3}$

For Exercises 25 and 26, write each expression as a single logarithm.

25. $\log_4 3x - \log_4 y$ **26.** $\ln P + rt \ln e$

For Exercises 27–33; solve each equation. Check for extraneous solutions.

27. $\log_2 7x - \log_2 3 = 2$ **28.** $\log_3 x^2 = -4$

29. $\log_b(x - 2) + \log_b\left(x - \dfrac{1}{2}\right) = 0$ **30.** $\log_{12}3 + \log_{12}x + \log_{12}(x - 1) = \log_{12}6$

31. $\log_a(x^2 - x) - \log_a x = \log_a 7$ **32.** $\log_2(3x + 1)^2 = [\log_2(3x + 1)]^2$

33. $\log_3 x^3 = (\log_3 x)^2$

Section 11.4

For Exercises 34–36, use a calculator to find the value of each, accurate to five decimal places.

34. $\log 5{,}829$ **35.** $\ln 0.09932$ **36.** $\dfrac{\log 39.32}{\log 84.21}$

For Exercises 37–40, use a calculator to find the antilogarithm to the nearest thousandth.

37. $\log x = 1.9032$ **38.** $\log x = -1.93728$ **39.** $\ln x = 4.8190$ **40.** $\ln x = -3.69184$

For Exercises 41–44, find each logarithm by using a calculator or Table II in the Appendix.

41. $\log 4.92$ **42.** $\log 0.0381$ **43.** $\log 47{,}300$ **44.** $\log(3.81)^2$

For Exercises 45–50, find each antilogarithm by using a calculator or Table II in the Appendix.

45. $\log x = 0.5441$ **46.** $\log x = 1.9547$ **47.** $\log x = 0.8215 - 3$

48. $\log x = -1.2403$ **49.** $\log x = -4.7144$ **50.** $\log x = -2.2526$

For Exercises 51–54, find each logarithm accurate to four decimal places.

51. $\log_5 39$ **52.** $\log_{30} 9.4$ **53.** $\log_{12} 40{,}900$ **54.** $\log_8 0.00492$

For Exercises 55–64, solve each equation. Write your answer to three decimal places.

55. $6^x = 48.9$

56. $(2.91)^x = 12.7$

57. $e^{-0.046t} = 32.8$

58. $2 - 2e^{-0.5t} = 1.5$

59. $\log_2 x = 9.81$

60. $\log_{15} x = 0.796$

61. $\log_3 4^{1/x} = x$

62. $\log_{12}(2.78)x = 4.23$

63. $\log_5 12 - \log_5 x = 0.6$ **64.** $\log_3 x = \log_4 20.7$

65. A mathematics professor estimates that the learning curve for her elementary algebra class is $y = 100(1 - e^{-0.3t})$, where t is the number of months into the semester and y is the class average measured in percent. By the end of the semester ($t = 4.5$), what is the class average?

66. Chicken consumption in the United States has increased 4% per year for the last 20 years. How long did it take the yearly consumption to double?

CHAPTER 11 **TEST**

Section 11.1

1. Solve the equation $9^x = 27$.

2. Suppose the rate of inflation is predicted to be 4% a year for the next 3 years. If a pair of basketball shoes costs $125 today, what is the predicted price of these shoes in 3 years?

3. $1000 is placed in a savings account at 7% interest, compounded quarterly. What is the amount in the account after 4 years?

4. A bacteria culture has an initial count of 600, and the bacteria count, Q, after t hours is given by

$$Q = 600e^{0.5t}$$

What is the count after 12 hours?

Section 11.2

For problems 5–8, rewrite each exponential statement as a logarithmic statement and each logarithmic statement as an exponential statement.

5. $5^{-3} = \dfrac{1}{125}$

6. $\log_3 27 = 3$

7. $a^r = T$

8. $\log_b M = n$

For problems 9 and 10, solve the equation.

9. $\log_3 x = 4$

10. $\log_b 2 = \dfrac{1}{2}$

Section 11.3

For problems 11 and 12, write the expression in expanded form.

11. $\log \dfrac{a\sqrt[3]{y}}{t^4}$

12. $\ln Q_0 e^{nt}$

For problems 13 and 14, write the expression as a single logarithm.

13. $\log_3 4 + \log_3 x - 2\log_3 H$

14. $\ln T + \dfrac{1}{2}\ln S$

For problems 15–17, solve the equation.

15. $\log_2 3x + \log_2 4 = 1$

16. $\log_5(x^2 + 1) = 2$

17. $\log_6 9 + \log_6 x - \log_6(x + 1) = \log_6(x + 4)$

For problems 18–25, solve for x.

18. $x = \log 35.79$

19. $x = \dfrac{\ln 8.3}{\ln 0.27}$

20. $\log x = 2.37125$

21. $\log x = -1.5706$

22. $\ln x = 0.1279$

23. $x = \log_{15} 25{,}281$

24. $7^x = 60.82$

25. $\log_3 x = 4.32$

Sequences and Series

Overview

In this chapter we study a special kind of function. There are many instances where an application is best described by a collection of discrete or separate values. For example, if Joe makes $20,500 a year and receives a $1,000 increase each year, then his salary over the next 6 years can be described by the six numbers: 21,500, 22,500, 23,500, 24,500, 25,500, 26,500. This is an example of a (finite) sequence of real numbers.

In Section 12.2, we investigate series of numbers and in Section 12.3, we define the arithmetic and geometric sequences. In the final section, we study the binomial theorem and the uses of the binomial coefficients.

12.1

Sequences

O B J E C T I V E S

▶ *To use the sequence notation*

▶ *To find the nth term of a sequence*

In this section, we introduce concepts that you will explore in more detail if you take further courses in algebra and calculus. We begin with the definition of *infinite sequence*.

D E F I N I T I O N

An **infinite sequence** of real numbers is a function whose domain is the set of positive integers

$$\{1, 2, 3, \ldots\}$$

If a is a sequence and n is a positive integer, we write a_n instead of the usual function notation of $a(n)$. The **first term** is a_1 and the **second term** is a_2, and in general, a_n is the **nth term** of the sequence. The sequence is frequently denoted by writing the terms of the sequence in order:

$$a_1, a_2, a_3, \ldots, a_n, \ldots$$

If the nth term a_n is given by a formula, then any term of the sequence can be found.

Example 1   Find the first five terms of the sequence whose nth term is given by the formula $a_n = 3n - 2$.

Solution   Replace n by 1, 2, 3, 4, and 5, respectively, in the formula for the general term.

General term: $a_n = 3n - 2$

$n = 1$: $a_1 = 3(1) - 2 = 1$

$n = 2$: $a_2 = 3(2) - 2 = 4$

$n = 3$: $a_3 = 3(3) - 2 = 7$

$n = 4$: $a_4 = 3(4) - 2 = 10$

$n = 5$: $a_5 = 3(5) - 2 = 13$

The first five terms of the sequence are 1, 4, 7, 10, and 13. We can write the sequence as

$$1, 4, 7, 10, 13, \ldots, 3n - 2, \ldots$$ ▲

Example 2   Find the first four terms of the sequence whose nth term is $a_n = \dfrac{1}{n}$.

Solution   Set $a_n = \dfrac{1}{n}$ and replace n, in turn, by 1, 2, 3, and 4.

$$a_1 = \frac{1}{1} = 1$$

$$a_2 = \frac{1}{2}$$

$$a_3 = \frac{1}{3}$$

$$a_4 = \frac{1}{4}$$

To show this pattern, the sequence can be written as

$$1, \frac{1}{2}, \frac{1}{3}, \frac{1}{4}, \ldots, \frac{1}{n} \ldots$$ ▲

In applications, many times we work with a *finite sequence*.

DEFINITION

If the domain of a function a is the set of positive integers $\{1, 2, 3, \cdots, n\}$ for some integer n, then a is called a **finite sequence.**

For example,

$$\frac{1}{2}, \frac{1}{4}, \frac{1}{8}, \frac{1}{16}$$

defines a finite sequence where $a_n = \dfrac{1}{2^n}$, $n \in \{1, 2, 3, 4\}$. Notice that a finite sequence ends after a fixed number of terms. The word "sequence" can refer to either an infinite or a finite sequence.

Example 3

A particular baseball card is predicted to increase 20% a year for the next 4 years. If this card is currently worth $40, write the value of the card for each of the next 4 years as a sequence.

Solution

Let a_n be the value of the card n years from now, where $n \in \{1, 2, 3, 4\}$. The value of the card next year will be $40 + 0.20(40) = \$48$. Therefore, $a_1 = \$48$. Next, we find a_2.

$$a_2 = 48 + 0.20(48) = 48 + 9.60 = \$57.60$$

To find a_3, add to $57.60 20% of $57.60. Therefore,

$$a_3 = \$57.60 + 0.20(57.60) = 57.60 + 11.52 = \$69.12$$

The final calculation is for a_4.

$$a_4 = 69.12 + 0.20(69.12) = \$82.94 \qquad \text{Rounding to the nearest hundredth.} \quad \blacktriangle$$

Sometimes the nth term of a sequence is given by a **recursive formula.** For example, consider the sequence where $a_1 = 2$ and $a_n = 3a_{n-1}$, $n \geq 2$. The nth term is three times the previous term. Since the first term is 2, we can find a_2 from the formula:

$$a_2 = 3a_{2-1} = 3a_1 = 3(2) = 6$$

Next, a_3 can be found by the formula and the fact that $a_2 = 6$:

$$a_3 = 3a_{3-1} = 3a_2 = 3(6) = 18$$

Next, a_4 can be found by the formula and the fact that $a_3 = 18$:

$$a_4 = 3a_{4-1} = 3a_3 = 3(18) = 54$$

and so on.

Example 4   Define a sequence as follows:

$$a_1 = 6 \quad \text{and} \quad a_n = \frac{a_{n-1}}{3}, \quad n \geq 2.$$

Write the first four terms of the sequence.

Solution   We start with $a_1 = 6$. Letting $n = 2$ in the formula,

$$a_2 = \frac{a_{2-1}}{3} = \frac{a_1}{3} = \frac{6}{3} = 2$$

For $n = 3$,

$$a_3 = \frac{a_{3-1}}{3} = \frac{a_2}{3} = \frac{2}{3}$$

For $n = 4$,

$$a_4 = \frac{a_{4-1}}{3} = \frac{a_3}{3} = \frac{\frac{2}{3}}{3} = \frac{2}{9}$$

The first four terms of the sequence are

$$6, 3, \frac{2}{3}, \frac{2}{9}$$

Example 5   Find the first four terms of a sequence defined by

$$a_1 = 12, \quad a_n = \frac{a_{n-1}}{n}, \quad n \geq 2$$

Solution   The first term a_1 is 12. To find the next three terms, let $n = 2, 3,$ and 4 in the formula.

$$n = 2: \quad a_2 = \frac{a_{2-1}}{2} = \frac{a_1}{2} = \frac{12}{2} = 6$$

$$n = 3: \quad a_3 = \frac{a_{3-1}}{3} = \frac{a_2}{3} = \frac{6}{3} = 2$$

$$n = 4: \quad a_4 = \frac{a_{4-1}}{4} = \frac{a_3}{4} = \frac{2}{4} = \frac{1}{2}$$

So, the first four terms of the sequence are

$$12, 6, 2, \frac{1}{2}.$$

In the previous examples, we generated the terms of the sequence from a formula for the nth term. In the next example, we are given the first four terms of the sequence, and from the pattern they form, we will write a formula for the nth term.

Example 6   Find a formula for the nth term of the sequence

$$1, 4, 9, 16, \ldots .$$

Solution   To find a pattern in the sequence of the first four terms, notice that each one is a perfect square.

$$a_1 = 1 = 1^2$$
$$a_2 = 4 = 2^2$$
$$a_3 = 9 = 3^2$$
$$a_4 = 16 = 4^2$$

Therefore, the formula for a_n is $a_n = n^2$.   ▲

Example 7   Find a formula for the nth term of the sequence

$$0, \frac{1}{4}, \frac{2}{9}, \frac{3}{16}, \ldots$$

Solution   Notice that except for the first term, the denominators are perfect squares and the numerators are one less than its position in the sequence. Let us see if this pattern is correct.

$$a_1 = 0 = \frac{1 - 1}{1^2}$$

$$a_2 = \frac{1}{4} = \frac{2 - 1}{2^2}$$

$$a_3 = \frac{2}{9} = \frac{3 - 1}{3^2}$$

$$a_4 = \frac{3}{16} = \frac{4 - 1}{4^2}$$

This pattern indicates that $a_n = \dfrac{n - 1}{n^2}$.   ▲

CALCULATOR CROSSOVER

Find the indicated term for each sequence.

1. $a_n = (34.2 - n)n^2$; a_7

2. $a_n = \dfrac{926.1}{n}$; a_{18}

3. $a_n = 4{,}302 - 53.01n$; a_{11}

4. $a_n = \dfrac{48.2}{\sqrt{n}}$; a_{56}

Answers   **1.** 1332.8   **2.** 51.45   **3.** 3718.89   **4.** 6.4410

EXERCISE   SET   12.1

For Exercises 1–20, find the first four terms of the sequence whose nth term is given.

1. $a_n = 4n - 5$

2. $a_n = 3 + 2n$

3. $a_n = 1 - n$

4. $a_n = 10 - 4n$

5. $a_n = \dfrac{2}{n}$

6. $a_n = -\dfrac{1}{n}$

7. $a_n = -\dfrac{3}{n^2}$

8. $a_n = \dfrac{4}{(n+1)^2}$

9. $a_n = \dfrac{n}{n+3}$

10. $a_n = \dfrac{2n-6}{n}$

11. $a_n = (-1)^n$

12. $a_n = (-1)^n n$

13. $a_n = (-2)^n$

14. $a_n = (-3)^n$

15. $a_n = \dfrac{(-1)^n}{n}$

16. $a_n = \dfrac{(-1)^n}{n^2}$

17. $a_n = 3\left(-\dfrac{1}{3}\right)^{n-1}$

18. $a_n = 6\left(-\dfrac{1}{2}\right)^{n+1}$

19. $a_n = n(n-1)$

20. $a_n = n(n-3)$

For Exercises 21–24, find the first four terms of the sequence from the given information.

21. $a_1 = 3,\ a_n = 4a_{n-1},\ n \ge 2$

22. $a_1 = 0,\ a_n = a_{n-1} + 2,\ n \ge 2$

23. $a_1 = 1,\ a_n = (-1)^n(a_{n-1} + 1),\ n \ge 2$

24. $a_1 = -1,\ a_n = (-1)^{n-1}(a_{n-1} - 2),\ n \ge 2$

From the pattern, write the next three terms of the sequence, then find a formula for the general term a_n.

25. $2, 4, 6, 8, \ldots$

26. $1, 3, 5, 7, \ldots$

27. $8, 9, 10, 11, \ldots$

28. $-3, -2, -1, 0, \ldots$

29. $-1, -1 + \sqrt{2}, -1 + 2\sqrt{2}, -1 + 3\sqrt{2}, \ldots$

30. $3, 3 - \sqrt{5}, 3 - 2\sqrt{5}, 3 - 3\sqrt{5}, \ldots$

31. $1, -2, 3, -4, \ldots$

32. $-1, 2, -3, 4, \ldots$

33. The value of a \$5,000 painting is predicted to increase 10% each year. Let a_n be the value of the painting n years from now. Find the first three terms of this sequence.

34. A colony of bacteria doubles each hour. If there are 1,000 bacteria now, let a_n be the number of bacteria in n hours. Find the first four terms of the sequence.

REVIEW PROBLEMS

The following exercises review parts of Section 1.4. Doing these problems will help prepare you for the next section.

Simplify the following numerical expressions.

35. $[2(1) - 3] + [2(2) - 3] + [2(3) - 3] + [2(4) - 3]$

36. $[5 - 3(1)] + [5 - 3(2)] + [5 - 3(3)]$

37. $2(1)^2 + 2(2)^2 + 2(3)^2 + 2(4)^2 + 2(5)^2$

38. $\dfrac{(-1)^1}{1} + \dfrac{(-1)^2}{2} + \dfrac{(-1)^3}{3}$

39. $\dfrac{1+1}{1+2} + \dfrac{2+1}{2+2} + \dfrac{3+1}{3+2}$

40. $(-2)^1 + (-2)^2 + (-2)^3 + (-2)^4$

E N R I C H M E N T   E X E R C I S E S

1. Find the first eight terms of the sequence defined by $a_1 = 1$, $a_2 = 1$, and $a_n = a_{n-1} + a_{n-2}$, $n \geq 3$.

2. Consider the sequence defined by $a_1 = 1$, $a_n = na_{n-1}$, $n \geq 2$.
   (a) Find the first six terms of the sequence.
   (b) Find a formula for a_n in terms of n only.

3. Consider the sequence defined by $a_1 = 3$, $a_2 = 4$, $a_n = \dfrac{a_{n-1}}{a_{n-2}}$, $n \geq 3$.

   (a) Find the first ten terms of the sequence.

   (b) Do you see a pattern formed by the terms in this sequence?

4. From the pattern, find the general term of $6, 12, 24, 48, \ldots$.

5. The deer population increases 6% each year in the Horicon Marsh national preserve. It is estimated that there are 5,000 deer in the preserve this year. Consider the sequence where the nth term is the predicted number of deer in the preserve in n years. Write a formula for the nth term.

Answers to Enrichment Exercises are on page A.86.

12.2

Series

O B J E C T I V E S

▶ *To use the sigma notation to write a series*

▶ *To expand a series written with the sigma notation*

Consider the sequence of five terms: $1, 3, 5, 7, 9$ and suppose we wanted to add them to form a series. This sum can be written as $1 + 3 + 5 + 7 + 9$. The sum of the terms of a sequence is called a *series*.

DEFINITION

A **series** is the sum of a sequence. If the sequence is finite, the corresponding series is a **finite series.** If the sequence is infinite, the corresponding series is an **infinite series.**

In this section, however, we study only finite series. Another way to indicate this sum is by using the **summation notation,** or **sigma** (Σ) **notation.** Notice that the sequence 1, 3, 5, 7, 9 has $a_i = 2i - 1$ as its general term, where $1 \leq i \leq 5$. Now, form the symbolism:

$$\sum_{i=1}^{5} (2i - 1)$$

The letter i is replaced, in turn, by each integer from 1 to 5. The symbols $i = 1$ written below Σ and 5 written above Σ indicate the values of i to be used for the first and last terms, respectively, of the sum. The letter i is called the **index of summation.**

$$\sum_{i=1}^{5} (2i - 1) \quad \text{is read "the sum from } i = 1 \text{ to 5 of } 2i - 1\text{"}$$

We write this sum in expanded form as

$$\sum_{i=1}^{5} (2i - 1) = [2(1) - 1] + [2(2) - 1] + [2(3) - 1] + [2(4) - 1] + [2(5) - 1]$$

$$= 1 + 3 + 5 + 7 + 9$$

$$= 25$$

N O T E   *The letter i used as the index of summation has no connection with its role in complex numbers. Other letters that are commonly used for summation indices are j, k, and n.*

Example 1   Find the value of the series $\displaystyle\sum_{i=1}^{4} (i^2 - i)$.

Solution   Replace i in $i^2 - i$ with all integers starting with 1 and ending with 4, then add.

$$\sum_{i=1}^{4} (i^2 - i) = (1^2 - 1) + (2^2 - 2) + (3^2 - 3) + (4^2 - 4)$$

$$= 0 + 2 + 6 + 12$$

$$= 20 \qquad \blacktriangle$$

Example 2   Evaluate the series $\displaystyle\sum_{k=1}^{3} \frac{k+1}{k+2}$.

Solution
$$\sum_{k=1}^{3} \frac{k+1}{k+2} = \frac{2}{3} + \frac{3}{4} + \frac{4}{5}$$
$$= \frac{40 + 45 + 48}{60}$$
$$= \frac{133}{60} \qquad\qquad \blacktriangle$$

Example 3   Find the value of $\displaystyle\sum_{j=1}^{4} (-3)^j$.

Solution
$$\sum_{j=1}^{4} (-3)^j = (-3)^1 + (-3)^2 + (-3)^3 + (-3)^4$$
$$= -3 + 9 - 27 + 81$$
$$= 60 \qquad\qquad \blacktriangle$$

A series may involve a variable. It is assumed that the variable takes on all real numbers for which each term of the series is defined.

Example 4   Write in expanded form: $\displaystyle\sum_{k=1}^{3} kx^k$.

Solution   Using the values of k from 1 to 3,

$$\sum_{k=1}^{3} kx^k = x + 2x^2 + 3x^3 \qquad\qquad \blacktriangle$$

In mathematics, we sometimes encounter a series of numbers that must be written in summation notation.

Example 5   Write the following series in summation notation:

$$2 + 4 + 6 + 8 + 10$$

Solution   First, find a formula for the general term of the sequence 2, 4, 6, 8, 10. In this case, $a_i = 2i$, $1 \leq i \leq 5$ describes this sequence. Here we are using the letter i instead of n that was used in the last section to write the general term. Now, we write the series in summation notation as

$$2 + 4 + 6 + 8 + 10 = \sum_{i=1}^{5} 2i \qquad\qquad \blacktriangle$$

Example 6   Write in summation notation:

$$1 + x + x^2 + x^3 + x^4 + x^5 + x^6$$

Solution   A formula for the general term is $a_i = x^i$, $0 \le i \le 6$. Therefore, the series can be written as

$$1 + x + x^2 + x^3 + x^4 + x^5 + x^6 = \sum_{i=0}^{6} x^i$$ ▲

E X E R C I S E   S E T   12.2

For Exercises 1–18, find the value of each series.

1. $\displaystyle\sum_{i=1}^{4} (i - 5)$ **2.** $\displaystyle\sum_{i=1}^{6} (i - 2)$ **3.** $\displaystyle\sum_{i=1}^{3} (i^2 - 2)$ **4.** $\displaystyle\sum_{i=1}^{3} 2i^2$

5. $\displaystyle\sum_{i=1}^{4} (i - 1)^2$ **6.** $\displaystyle\sum_{i=1}^{3} (i - 2)^2$ **7.** $\displaystyle\sum_{j=1}^{4} (2j - 6)$ **8.** $\displaystyle\sum_{k=1}^{3} (1 - 3k)$

9. $\displaystyle\sum_{k=1}^{8} (-1)^k$ **10.** $\displaystyle\sum_{k=1}^{5} (-1)^k$ **11.** $\displaystyle\sum_{i=1}^{5} \frac{i - 3}{i + 1}$ **12.** $\displaystyle\sum_{i=1}^{3} \frac{2i - 4}{i}$

13. $\displaystyle\sum_{n=1}^{4} \frac{1}{n}$ **14.** $\displaystyle\sum_{n=1}^{3} \frac{2}{n^2}$ **15.** $\displaystyle\sum_{k=1}^{5} 2^{k-3}$ **16.** $\displaystyle\sum_{j=1}^{3} 3^{2j-5}$

17. $\displaystyle\sum_{n=1}^{3} \frac{n^2 - 3n - 4}{n + 1}$ **18.** $\displaystyle\sum_{k=1}^{3} \frac{k^2 + k - 2}{k + 2}$

For Exercises 19–24, write in expanded form.

19. $\displaystyle\sum_{i=1}^{4} x^i$ **20.** $\displaystyle\sum_{i=1}^{5} x^i$ **21.** $\displaystyle\sum_{k=1}^{3} 2x^{k-1}$

22. $\displaystyle\sum_{k=1}^{2} 3x^{k-1}$ **23.** $\displaystyle\sum_{j=1}^{3} (-1)^j \frac{x^j}{j}$ **24.** $\displaystyle\sum_{k=1}^{2} \frac{y^k}{2^k}$

For Exercises 25–40, write each series in summation notation.

25. $1 + 2 + 3 + 4 + 5 + 6$ **26.** $1 + 2 + 3 + 4$

27. $2 + 3 + 4 + 5 + 6$ **28.** $3 + 4 + 5 + 6 + 7$

29. $-1 + 0 + 1 + 2 + 3$

30. $-2 - 1 + 0 + 1 + 2 + 3$

31. $1 + 3 + 5 + 7 + 9 + 11$

32. $2 + 4 + 6 + 8 + 10 + 12 + 14$

33. $1 - 2 + 3 - 4$

34. $3 - 4 + 5 - 6 + 7$

35. $x^3 + x^5 + x^7 + x^9 + x^{11}$

36. $x^6 + x^8 + x^{10} + x^{12}$

37. $\left(x - \dfrac{1}{2}\right) + \left(x - \dfrac{1}{2}\right)^2 + \left(x - \dfrac{1}{2}\right)^3 + \left(x - \dfrac{1}{2}\right)^4$

38. $(y + 5) + (y + 5)^2 + (y + 5)^3 + (y + 5)^4 + (y + 5)^5$

39. $1 - x^2 + x^4 - x^6 + x^8 - x^{10} + x^{12}$

40. $x - x^3 + x^5 - x^7 + x^9 - x^{11} + x^{13}$

41. A freely falling object dropped from a commercial airplane travels 16 feet the first second, 48 feet the second second, and 80 feet the third second. How far does it travel in 5 seconds?

42. A ball rolling down a hill travels 3 feet the first second, 5 feet the second second, and 7 feet the third second. How far does the ball roll in 6 seconds?

REVIEW PROBLEMS

The following exercises review parts of Section 1.5. Doing these problems will help prepare you for the next section.

43. Evaluate $a_1 + (n - 1)d$ for $a_1 = 2$, $n = 4$, and $d = -1$.

44. Evaluate $a_1 + (n - 1)d$ for $a_1 = -\dfrac{1}{2}$, $n = 8$, and $d = 6$.

45. Evaluate $\dfrac{a_1(1 - r^n)}{1 - r}$ for $a_1 = \dfrac{2}{5}$, $r = 3$, and $n = 2$.

46. Evaluate $\dfrac{a_1(1 - r^n)}{1 - r}$ for $a_1 = 4$, $r = \dfrac{1}{2}$, and $n = 3$.

E N R I C H M E N T   E X E R C I S E S

1. Consider a sequence $a_1, a_2, a_3, \ldots$ and a constant c,

(a) Show that $\displaystyle\sum_{i=1}^{3} ca_i = c\sum_{i=1}^{3} a_i$

(b) Show that $\displaystyle\sum_{i=1}^{4} ca_i = c\sum_{i=1}^{4} a_i$

(c) How do you think $\displaystyle\sum_{i=1}^{n} ca_i$ can be rewritten?

2. Let c be a constant.

(a) Show that $\displaystyle\sum_{i=1}^{3} c = 3c$.

(b) Show that $\displaystyle\sum_{i=1}^{4} c = 4c$.

(c) How do you think $\displaystyle\sum_{i=1}^{n} c$ can be rewritten?

3. Use the result of Exercise 2 to find the value of each sum.

(a) $\displaystyle\sum_{i=1}^{3} 4$ **(b)** $\displaystyle\sum_{k=1}^{10} (-2)$

4. Let $a_1, a_2, a_3, \ldots$ and $b_1, b_2, b_3, \ldots$ be two sequences.

(a) Show that $\displaystyle\sum_{i=1}^{3} (a_i + b_i) = \sum_{i=1}^{3} a_i + \sum_{i=1}^{3} b_i$.

(b) Show that $\displaystyle\sum_{i=1}^{4} (a_i + b_i) = \sum_{i=1}^{4} a_i + \sum_{i=1}^{4} b_i$.

(c) How do you think $\displaystyle\sum_{i=1}^{n} (a_i + b_i)$ can be rewritten?

A formula for the sum of the first n integers is the following:

$$1 + 2 + 3 + \cdots + n = \frac{n(n + 1)}{2}$$

A formula for the sum of the squares of the first n integers is the following:

$$1^2 + 2^2 + 3^2 + \cdots + n^2 = \frac{n(n + 1)(2n + 1)}{6}$$

Use these two formulas along with Exercises 1, 2, and 4 to find the value of each series.

5. $\displaystyle\sum_{i=1}^{6} (i + 2i^2)$

6. $\displaystyle\sum_{i=1}^{5} (3i^2 - i - 1)$

7. Express the value of this series in terms of n.

$$\sum_{i=1}^{n} (4i - 6i^2 + 3)$$

Answers to Enrichment Exercises are on page A.86.

12.3

Arithmetic and Geometric Sequences

O B J E C T I V E S

▶ *To find the general term of an arithmetic sequence*

▶ *To find the sum of the first n terms of an arithmetic series*

▶ *To find the general term of a geometric sequence*

▶ *To find the sum of the first n terms of a geometric series*

In this section we study two basic sequences: arithmetic and geometric sequences (also called *progressions*).

D E F I N I T I O N

An **arithmetic sequence** is a sequence in which there is a **common difference** d between successive terms. The ith term is equal to the previous term plus d,

$$a_i = a_{i-1} + d, \qquad i \geq 2$$

Given the first term a_1 and the common difference d, then all the terms of the arithmetic sequence are determined.

Example 1   Write the first four terms of the arithmetic sequence where $a_1 = 1$ and $d = 3$.

Solution   Using the recursive formula $a_i = a_{i-1} + d$, we generate the terms of the arithmetic sequence.

$$i = 2:\quad a_2 = a_{2-1} + d = a_1 + d = 1 + 3 = 4$$
$$i = 3:\quad a_3 = a_{3-1} + d = a_2 + d = 4 + 3 = 7$$
$$i = 4:\quad a_4 = a_{4-1} + d = a_3 + d = 7 + 3 = 10$$

Therefore, the first four terms of the sequence are

$$1, 4, 7, 10$$ ▲

N O T E   *In Example 1, notice that to find the next term of the sequence, we can simply add 3 to the previous term.*

Let us generate the terms of the general arithmetic sequence by using the recursive formula $a_i = a_{i-1} + d$.

First term: a_1
Second term: $a_2 = a_1 + d$
Third term: $a_3 = (a_1 + d) + d = a_1 + 2d$
Fourth term: $a_4 = (a_1 + 2d) + d = a_1 + 3d$
$\vdots$ $\vdots$ $\vdots$
nth term: $a_n = a_1 + (n - 1)d$

We therefore have the following result.

R U L E

The nth term of an arithmetic sequence

The general term a_n of an arithmetic sequence with the first term a_1 and common difference d is given by

$$a_n = a_1 + (n - 1)d$$

Example 2   Find the general term of the arithmetic sequence

$$5, 1, -3, -7, 11, \ldots$$

Solution   To find a_n, we need the first term a_1 and the common difference d. The first term is 5 and the common difference is $d = a_2 - a_1 = 1 - 5 = -4$. Therefore, the general term is

$$a_n = 5 + (n - 1)(-4)$$

which can be simplified to

$$a_n = 9 - 4n$$ ▲

Example 3   If the first and the twenty-first terms of an arithmetic sequence are 5 and 45, respectively, find the sixty-second term of the sequence.

Solution   We first find the common difference d. Using the formula for the general term with $a_1 = 5$,

$$a_n = a_1 + (n - 1)d$$
$$a_n = 5 + (n - 1)d$$

Since the twenty-first term is 45,

$$a_{21} = 5 + (21 - 1)d$$
$$45 = 5 + (21 - 1)d$$

Solving this last equation for d,

$$d = 2$$

Replacing d by 2 in the formula for a_n,

$$a_n = 5 + (n - 1)d$$
$$a_n = 5 + (n - 1)2$$
$$a_n = 2n + 3$$

To find the sixty-second term, replace n by 62 to obtain

$$a_{62} = 2(62) + 3$$
$$= 127 \qquad \blacktriangle$$

DEFINITION

Let $a_1, a_2, a_3, a_4, \ldots$ be a sequence. The **nth partial sum** S_n of the sequence is defined to be

$$S_n = a_1 + a_2 + a_3 + \cdots + a_n$$

That is, the nth partial sum is the series formed from the first n terms of an infinite sequence.

Example 4   Find S_1, S_2, S_3, and S_4 for the infinite sequence

$$2, 4, 6, 8, \ldots$$

Solution   Using the definition of nth partial sum,

$$S_1 = a_1 = 2$$
$$S_2 = a_1 + a_2 = 2 + 4 = 6$$
$$S_3 = 2 + 4 + 6 = 12$$
$$S_4 = a_1 + a_2 + a_3 + a_4 = 2 + 4 + 6 + 8 = 20 \qquad \blacktriangle$$

A formula for the nth partial sum of an arithmetic sequence is given in the next theorem.

THEOREM

The nth partial sum S_n of an arithmetic sequence with a first term a_1 and common difference d is given by

$$S_n = \frac{n}{2}[2a_1 + (n-1)d]$$

or

$$S_n = \frac{n}{2}(a_1 + a_n)$$

Proof   Let

$$S_n = a_1 + (a_1 + d) + \cdots + [a_1 + (n-2)d] + [a_1 + (n-1)d]$$

Reversing the order of the sum, we have

$$S_n = [a_1 + (n-1)d] + [a_1 + (n-2)d] + \cdots + (a_1 + d) + a_1$$

Adding the left sides and the corresponding terms of the right sides of these two equations,

$$2S_n = [2a_1 + (n-1)d] + [2a_1 + (n-1)d] + \cdots + [2a_1 + (n-1)d]$$

$$= n[2a_1 + (n-1)d] \qquad \text{The sum of the } n \text{ terms } 2a_1 + (n-1)d.$$

Dividing both sides of this last equation by 2, we obtain the first result.

$$S_n = \frac{n}{2}[2a_1 + (n-1)d]$$

Next, rewriting S_n, we have

$$S_n = \frac{n}{2}\{a_1 + [a_1 + (n-1)d]\}$$

$$= \frac{n}{2}(a_1 + a_n) \qquad\qquad a_n = a_1 + (n-1)d.$$

Example 5   Find the sum of the first eleven terms of the arithmetic sequence if the first term is -60 and the common difference is 10.

Solution   The sum of the first eleven terms is the eleventh partial sum S_{11}. We use the formula for the nth partial sum,

$$S_n = \frac{n}{2}[2a_1 + (n-1)d]$$

Next, replace n by 11, a_1 by -60, and d by 10.

$$S_{11} = \frac{11}{2}[2(-60) + (11 - 1)10]$$

$$= \frac{11}{2}[-120 + (10)10]$$

$$= \frac{11}{2}(-120 + 100)$$

$$= \frac{11}{2}(-20)$$

$$= \frac{11}{\underset{1}{\cancel{2}}}(\overset{-10}{\cancel{-20}})$$

$$= -110 \qquad\qquad \blacktriangle$$

The other basic sequence that we investigate in this section is the *geometric sequence* or *geometric progression*.

D E F I N I T I O N

A **geometric sequence** is a sequence in which each term after the first is obtained by multiplying the preceding term by the same number r, called the **common ratio.**

A geometric sequence is determined by the recursive formula

$$a_i = ra_{i-1}, \qquad i \geq 2$$

where a_1 is given.

Notice that r is the common ratio of successive terms,

$$r = \frac{a_i}{a_{i-1}}$$

Example 6   Write the first four terms of the geometric sequence where $a_1 = 3$ and $r = 2$.

Solution   Using the formula $a_i = ra_{i-1}$, we generate the terms of the geometric sequence.

$$i = 2: \quad a_2 = ra_1 = (2)(3) = 6$$
$$i = 3: \quad a_3 = ra_2 = (2)(6) = 12$$
$$i = 4: \quad a_4 = ra_3 = (2)(12) = 24$$

Therefore, the first four terms of the progression are

$$3, 6, 12, 24 \qquad\qquad \blacktriangle$$

If $a_1, a_2, a_3, \ldots$ is a geometric sequence with common ratio r, then we may generate the terms of the sequence through repeated use of the formula $a_i = ra_{i-1}$.

First term: a_1

Second term: $a_2 = ra_1$

Third term: $a_3 = ra_2 = r(ra_1) = r^2a_1$

Fourth term: $a_4 = ra_3 = r(r^2)a_1 = r^3a_1$

nth term: $a_n = r^{n-1}a_1$

Therefore, we have the following result.

R U L E

The nth term of a geometric sequence

The general term a_n of a geometric sequence with the first term a_1 and common ratio r is

$$a_n = a_1 r^{n-1}.$$

Example 7   Find the nth term, a_n, of the geometric sequence

$$-4, 2, -1, \frac{1}{2}, \ldots$$

Solution   The first term is $a_1 = -4$ and the common ratio is

$$r = \frac{a_2}{a_1} = \frac{2}{-4} = -\frac{1}{2}$$

Therefore, the nth term is

$$a_n = a_1 r^{n-1} = -4\left(-\frac{1}{2}\right)^{n-1} \qquad \blacktriangle$$

Example 8   Find the fifth term of the geometric sequence

$$4, 12, 36, \ldots$$

Solution   The first term is $a_1 = 4$ and the common ratio is $r = \dfrac{a_2}{a_1} = \dfrac{12}{4} = 3$. Therefore, the fifth term of the progression is

$$a_5 = a_1 r^4 = (4)3^4 = 324 \qquad \blacktriangle$$

In the next theorem, we give a formula for the sum of the first n terms of a geometric sequence. Recall, we use the notation S_n for the nth partial sum.

THEOREM

The sum S_n of the first n terms of a geometric sequence with first term a_1 and common ratio r is given by

$$S_n = \frac{a_1(1 - r^n)}{1 - r}, \qquad r \neq 1$$

Proof   We write the formula for S_n, then multiply both sides by r.

$$S_n = a_1 + a_1 r + a_1 r^2 + \cdots + a_1 r^{n-1}$$

$$rS_n = a_1 r + a_1 r^2 + a_1 r^3 + \cdots + a_1 r^n$$

If we subtract the second equation from the first, notice that the only two terms that remain are a_1 and $-a_1 r^n$. Therefore,

$$S_n - rS_n = a_1 - a_1 r^n$$

Next, we factor S_n from the two terms on the left side and a_1 from the two terms on the right side, to obtain

$$S_n(1 - r) = a_1(1 - r^n)$$

Dividing both sides by $1 - r$ yields the final result:

$$S_n = \frac{a_1(1 - r^n)}{1 - r}$$

Example 9   Find the sum of the first seven terms of the geometric sequence $4, 12, 36, \ldots$.

Solution   The first term a_1 is 4 and the common ratio r is $\dfrac{12}{4}$, or 3. Therefore, using the formula for S_n,

$$S_n = \frac{a_1(1 - r^n)}{1 - r}$$

$$S_7 = \frac{4(1 - 3^7)}{1 - 3}$$

$$= \frac{4(1 - 2{,}187)}{-2}$$

$$= 4{,}372 \qquad \blacktriangle$$

We now consider infinite geometric series such as

$$\sum_{i=1}^{\infty} 2\left(\frac{1}{2}\right)^{i-1} = 2 + 1 + \frac{1}{2} + \frac{1}{4} + \frac{1}{8} + \frac{1}{16} + \frac{1}{32} + \ldots$$

What is the value of this series? We cannot simply add infinitely many numbers. Instead, we look for a pattern in the sequence of nth partial sums.

$$S_1 = 2$$

$$S_2 = 2 + 1 = 3$$

$$S_3 = 2 + 1 + \frac{1}{2} = 3\frac{1}{2}$$

$$S_4 = 2 + 1 + \frac{1}{2} + \frac{1}{4} = 3\frac{3}{4}$$

$$S_5 = 2 + 1 + \frac{1}{2} + \frac{1}{4} + \frac{1}{8} = 3\frac{7}{8}$$

$$S_6 = 2 + 1 + \frac{1}{2} + \frac{1}{4} + \frac{1}{8} + \frac{1}{16} = 3\frac{15}{16}$$

$$S_7 = 2 + 1 + \frac{1}{2} + \frac{1}{4} + \frac{1}{8} + \frac{1}{16} + \frac{1}{32} = 3\frac{31}{32}$$

It appears that the numbers in this sequence

$$2, 3, 3\frac{1}{2}, 3\frac{3}{4}, 3\frac{7}{8}, 3\frac{15}{16}, 3\frac{31}{32}, \ldots$$

are approaching the number 4, and we write

$$2 + 1 + \frac{1}{2} + \frac{1}{4} + \frac{1}{8} + \frac{1}{16} + \ldots = 4$$

In general, consider an infinite geometric series

$$\sum_{i=1}^{\infty} a_1 r^{i-1} = a_1 + a_1 r + a_1 r^2 + \ldots$$

The nth partial sum is

$$S_n = \frac{a_1(1 - r^n)}{1 - r}$$

Now, for any fixed number r so that $|r| < 1$, then r^n approaches 0 as n increases without bound. Therefore, the nth partial sum S_n approaches $\frac{a_1(1 - 0)}{1 - r} = \frac{a_1}{1 - r}$ as n becomes larger and larger without bound. We say the geometric series has the sum of $\frac{a_1}{1 - r}$. We summarize this discussion in the following rule.

> **R U L E**
>
> **The value of an infinite geometric series**
>
> $$a_1 + a_1 r + a_1 r^2 + \ldots = \frac{a_1}{1-r}, \text{ provided that } |r| < 1$$

Using the sigma notation, the above rule may be stated as follows:

$$\sum_{i=1}^{\infty} a_1 r^{i-1} = \frac{a_1}{1-r}$$

Now we apply this formula to the geometric series given at the beginning of this discussion. For this series, $a_1 = 2$ and $r = \dfrac{1}{2}$. Therefore,

$$2 + 1 + \frac{1}{2} + \ldots = \frac{a_1}{1-r} = \frac{2}{1 - \dfrac{1}{2}} = \frac{2}{\dfrac{1}{2}} = 4$$

which agrees with our previous answer.

Example 10   Find the value of an infinite geometric series with

$$a_1 = 3 \quad \text{and} \quad r = -\frac{1}{2}$$

Solution   Using the rule stated above, the value is

$$\frac{a_1}{1-r} = \frac{3}{1 - \left(-\dfrac{1}{2}\right)} = \frac{3}{\dfrac{3}{2}} = 2 \qquad \blacktriangle$$

Example 11   Find the value of $\displaystyle\sum_{i=1}^{\infty} \left(\frac{2}{3}\right)^{i-1}$.

Solution   For this infinite geometric series, $a_1 = 1$ and $r = \dfrac{2}{3}$. Therefore, by the rule, the value of this series is

$$\frac{a_1}{1-r} = \frac{1}{1 - \dfrac{2}{3}} = \frac{1}{\dfrac{1}{3}} = 3 \qquad \blacktriangle$$

E X E R C I S E   S E T   1 2 . 3

For Exercises 1–12, write the first four terms of the arithmetic sequence from the given information.

1. $a_1 = 4, d = 1$

2. $a_1 = 2, d = 3$

3. $a_1 = -1, d = 5$

4. $a_1 = -3, d = 1$

5. $a_1 = -12, d = 10$

6. $a_1 = -13, d = 13$

7. $a_1 = 8, d = -3$

8. $a_1 = 4, d = -2$

9. $a_1 = -1, d = -1$

10. $a_1 = -3, d = -2$

11. $a_1 = \dfrac{1}{2}, d = \dfrac{1}{4}$

12. $a_1 = \dfrac{2}{3}, d = \dfrac{1}{6}$

For Exercises 13–18, find the general term, a_n, of the arithmetic sequence.

13. $4, 3, 2, 1, \ldots$

14. $-3, -1, 1, 3, \ldots$

15. $33, 11, -11, \ldots$

16. $24, 31, 38, \ldots$

17. $-2, -\dfrac{5}{2}, -3, \ldots$

18. $\dfrac{1}{3}, \dfrac{7}{6}, 2, \ldots$

19. If the first and eighteenth terms of an arithmetic sequence are 3 and 37, respectively, find the twenty-fifth term of the sequence.

20. If the first and fourteenth terms of an arithmetic sequence are 5 and -8, respectively, find the thirty-fourth term of the sequence.

21. If the first and thirty-third terms of an arithmetic sequence are 1 and -15, respectively, find the thirty-second term.

22. If the first and thirty-fourth terms of an arithmetic sequence are -2 and 9, respectively, find the tenth term of the sequence.

23. If the third and eleventh terms of an arithmetic sequence are 6 and 22, respectively, find the general term a_n.

24. If the eighth and twentieth terms of an arithmetic sequence are 12 and 36, respectively, find the general term a_n.

For Exercises 25–34, find the partial sums $S_1, S_2, S_3,$ and S_4 for the given sequence.

25. $1, 3, 7, 9, \ldots$

26. $4, 5, 6, 7, \ldots$

27. $1, 4, 9, 16, \ldots$

28. $1, 9, 25, 49, \ldots$

29. The nth term is $a_n = \dfrac{1}{n}$.

30. The nth term is $a_n = \dfrac{1}{n^2}$.

31. The nth term is $a_n = 2n - 3$.

32. The nth term is $a_n = -3n + 4$.

33. The nth term is $a_n = \dfrac{(-1)^n}{n^2}$.

34. The nth term is $a_n = \dfrac{(-1)^n}{n}$.

For Exercises 35–42, find the sum of the first n terms of the arithmetic sequence, where a_1, d, and n are given.

35. $a_1 = 4$, $d = 3$, and $n = 9$

36. $a_1 = 5$, $d = 2$, and $n = 7$

37. $a_1 = -1$, $d = 3$, and $n = 8$

38. $a_1 = -3$, $d = 2$, and $n = 12$

39. $a_1 = -1$, $d = -\dfrac{1}{2}$, and $n = 21$

40. $a_1 = -6$, $d = -\dfrac{2}{3}$, and $n = 6$

41. $a_1 = 100$, $d = -5$, and $n = 31$

42. $a_1 = 50$, $d = -10$, and $n = 43$

For Exercises 43–50, write the first four terms of the geometric sequence from the given information.

43. $a_1 = 2$, $r = 4$

44. $a_1 = 3$, $r = 2$

45. $a_1 = 4$, $r = \dfrac{1}{2}$

46. $a_1 = 18$, $r = \dfrac{1}{3}$

47. $a_1 = 2$, $r = -1$

48. $a_1 = 3$, $r = -2$

49. $a_1 = 27$, $r = -\dfrac{1}{3}$

50. $a_1 = 16$, $r = -\dfrac{1}{2}$

For Exercises 51–58, find the general term, a_n, of the geometric sequence.

51. $1, 2, 4, 8, \ldots$

52. $2, 16, 128, \ldots$

53. $3, 1, \dfrac{1}{3}, \dfrac{1}{9}, \ldots$

54. $4, 1, \dfrac{1}{4}, \dfrac{1}{16}, \ldots$

55. $3, -\dfrac{3}{2}, \dfrac{3}{4}, \ldots$

56. $2, -\dfrac{2}{3}, \dfrac{2}{9}, \ldots$

57. $-6, 2, -\dfrac{2}{3}, \ldots$

58. $-4, 2, -1, \ldots$

59. Find the fifth term of the geometric sequence
$$6, 12, 24, \ldots$$

60. Find the sixth term of the geometric sequence
$$2, 8, 32, \ldots$$

61. Find the fourth term of the geometric sequence
$$1, -\dfrac{1}{3}, \ldots$$

62. Find the fifth term of the geometric sequence
$$-1, \dfrac{1}{2}, \ldots$$

For Exercises 63–68, find the sum of the first five terms of the geometric sequence from the given information.

63. $a_1 = 4$, $r = 2$

64. $a_1 = 7$, $r = 2$

65. $a_1 = 3$, $r = \dfrac{1}{2}$

66. $a_1 = 2$, $r = \dfrac{1}{3}$

67. $a_1 = 4$, $r = -\dfrac{1}{2}$

68. $a_1 = 9$, $r = -\dfrac{1}{3}$

For Exercises 69–76, find the indicated sums.

69. $\displaystyle\sum_{i=1}^{4} 2(3)^{i-1}$

70. $\displaystyle\sum_{i=1}^{5} 3(2)^{i-1}$

71. $\displaystyle\sum_{j=1}^{3} (-2)^{j-1}$

72. $\displaystyle\sum_{k=1}^{3} (-3)^{k-1}$

73. $\displaystyle\sum_{i=1}^{4} (0.2)^{i-1}$

74. $\displaystyle\sum_{k=1}^{3} (0.3)^{k-1}$

75. $\displaystyle\sum_{k=1}^{3} -18\left(-\frac{1}{3}\right)^{k-1}$

76. $\displaystyle\sum_{i=1}^{3} -20\left(-\frac{1}{2}\right)^{i-1}$

For Exercises 77–82, find the sum of the first six terms of the geometric sequence.

77. $1, 4, 16, \ldots$

78. $1, 6, 36, \ldots$

79. $6, 3, \dfrac{3}{2}, \ldots$

80. $12, 4, \dfrac{4}{3}, \ldots$

81. $16, -4, 1, \ldots$

82. $27, -9, 3, \ldots$

83. The creator of the "energy ball" claims that the ball will rebound three-fourths the distance from which it fell. If the ball is dropped from a height of 64 feet, how far will it travel before it hits the ground the fifth time?

84. A steel ball, dropped from a height of 25 feet, rebounds on each bounce three-fifths the distance from which it fell. When the ball hits the ground the sixth time, how far has it traveled since it was first dropped?

For Exercises 85–88, find the sum of each series.

85. $3 + 5 + 7 + \cdots + 67$

86. $3 + 1 + (-1) + \cdots + (-197)$

87. $8 + 4 + 2 + \cdots + \dfrac{1}{8}$

88. $2 + 4 + 8 + \cdots + 4{,}096$

For Exercises 89–94, find the value of each infinite geometric series from the given information.

89. $a_1 = 5, r = \dfrac{3}{5}$

90. $a_1 = -8, r = \dfrac{1}{2}$

91. $a_1 = -15, r = -\dfrac{1}{4}$

92. $a_1 = -10, r = -\dfrac{2}{3}$

93. $a_1 = 28, r = -0.4$

94. $a_1 = 10, r = 0.2$

For Exercises 95–100, find the value of each infinite geometric series.

95. $\displaystyle\sum_{i=1}^{\infty} 2\left(\frac{1}{2}\right)^{i-1}$

96. $\displaystyle\sum_{i=1}^{\infty} 6\left(\frac{3}{4}\right)^{i-1}$

97. $\displaystyle\sum_{i=1}^{\infty} -8\left(\frac{3}{5}\right)^{i-1}$

98. $\displaystyle\sum_{i=1}^{\infty} -28\left(\frac{1}{5}\right)^{i-1}$

99. $\displaystyle\sum_{i=1}^{\infty} 87(-0.74)^{i-1}$

100. $\displaystyle\sum_{i=1}^{\infty} 13(0.35)^{i-1}$

REVIEW PROBLEMS

The following exercises review parts of Section 3.3. Doing these problems will help prepare you for the next section.

Multiply.

101. $(a + b)^3$

102. $(a - b)^3$

103. $(a + b)^4$

104. $(2x - y)^3$

105. $(x - 3y)^3$

106. $(3x + 2y)^3$

E N R I C H M E N T   E X E R C I S E S

Find the value of each infinite geometric series.

1. $6 + 2 + \dfrac{2}{3} + \ldots$

2. $1 + \dfrac{1}{4} + \dfrac{1}{16} + \ldots$

3. $8 - 4 + 2 - 1 + \ldots$

4. $1 - \dfrac{1}{3} + \dfrac{1}{9} - \dfrac{1}{27} + \ldots$

5. $0.999 \ldots = 0.9 + 0.09 + 0.009 + \ldots$

6. $0.24242424 \ldots = 0.24 + 0.0024 + 0.000024 + \ldots$

7. $\displaystyle\sum_{i=1}^{\infty} 3\left(\dfrac{1}{2}\right)^i$

8. $\displaystyle\sum_{i=1}^{\infty} 4\left(-\dfrac{3}{4}\right)^{i+1}$

9. $\displaystyle\sum_{i=2}^{\infty} (-6)(0.24)^i$

10. $\displaystyle\sum_{i=0}^{\infty} 5(0.83)^{i+1}$

11. For what values of x does this geometric series have a sum?

$$\frac{1}{x} + \frac{1}{x^2} + \frac{1}{x^3} + \frac{1}{x^4} + \ldots$$

Answers to Enrichment Exercises are on page A.87.

12.4

The Binomial Theorem

OBJECTIVES

▶ *To use Pascal's triangle to find the binomial coefficients*

▶ *To find n!*

▶ *To expand $(x + y)^n$ by the binomial theorem*

In Chapter 3, we developed a formula for squaring a binomial,

$$(a + b)^2 = a^2 + 2ab + b^2$$

In this section, we consider higher powers of a binomial. In general, we will derive a formula to expand $(a + b)^n$, where n is any positive integer. For small values of n, $(a + b)^n$ can be obtained by multiplying polynomials as explained in Section 3.3. For example,

$$(a + b)^1 = a + b$$
$$(a + b)^2 = a^2 + 2ab + b^2$$
$$(a + b)^3 = (a + b)^2(a + b) = a^3 + 3a^2b + 3ab^2 + b^3$$
$$(a + b)^4 = (a + b)^3(a + b) = a^4 + 4a^3b + 6a^2b^2 + 4ab^3 + b^4$$
$$(a + b)^5 = (a + b)^4(a + b) = a^5 + 5a^4b + 10a^3b^2 + 10a^2b^3 + 5ab^4 + b^5$$

and so forth.

Notice that the expansion of $(a + b)^n$ has $n + 1$ terms that contain powers of the variables a and b. The power of a starts at n and decreases by 1 for each term until it is zero in the last term. The power of b starts at zero in the first term and increases by 1 in each term until it is n in the last term. Furthermore, the sum of the exponents of the variables on any term is equal to n. The variables in the expansion of $(a + b)^n$ follow the pattern:

$$a^n, a^{n-1}b, a^{n-2}b^2, a^{n-3}b^3, \cdots, ab^{n-1}, b^n$$

Now, observe the coefficients of the terms on the far right. By arranging these expansions in a triangular shaped array, we obtain **Pascal's triangle.**

$$1 \cdot a + 1 \cdot b$$
$$1 \cdot a^2 + 2 \cdot ab + 1 \cdot b^2$$
$$1 \cdot a^3 + 3 \cdot a^2b + 3 \cdot ab^2 + 1 \cdot b^3$$
$$1 \cdot a^4 + 4 \cdot a^3b + 6 \cdot a^2b^2 + 4 \cdot ab^3 + 1 \cdot b^4$$
$$1 \cdot a^5 + 5 \cdot a^4b + 10 \cdot a^3b^2 + 10 \cdot a^2b^3 + 5 \cdot ab^4 + 1 \cdot b^5$$

$(a + b)^1$				1		1	
$(a + b)^2$			1		2		1
$(a + b)^3$		1		3		3		1
$(a + b)^4$	1		4		6		4		1
$(a + b)^5$	1		5		10		10		5		1

The nth row, counting from the top, contains the coefficients of the terms in the expanded form of $(a + b)^n$. Each row starts and ends with the number 1. Each entry, other than the first and last, in a given row can be obtained by adding the two numbers diagonally above it. For example, in the fourth row, 1 is the first entry; the second entry 4 is the sum of the two numbers above it, namely $1 + 3$; and so on.

Therefore, any row in Pascal's triangle can be generated from the previous row. For example, the row 6 would be

$$1 \quad 6 \quad 15 \quad 20 \quad 15 \quad 6 \quad 1$$

This tells us that the expansion of $(a + b)^6$ is

$$(a + b)^6 = a^6 + 6a^5b + 15a^4b^2 + 20a^3b^3 + 15a^2b^4 + 6ab^5 + b^6$$

From row 6 we could generate row 7, from row 7 we could generate row 8, and so on. Therefore, Pascal's triangle gives a method of expanding $(a + b)^n$ for any positive integer n.

There is another way to find the numerical coefficients of the terms in the expanded form of $(a + b)^n$. We start with a definition.

DEFINITION

If n is a positive integer, we define $n!$, called **n factorial,** by

$$n! = n(n - 1)(n - 2)(n - 3) \cdots 2 \cdot 1$$

We also define

$$0! = 1$$

Therefore, $n!$ is the product of n and all smaller consecutive integers down to one. For example,

$1! = 1$
$2! = 2 \cdot 1 = 2$
$3! = 3 \cdot 2 \cdot 1 = 6$
$4! = 4 \cdot 3 \cdot 2 \cdot 1 = 24$

and so on.

DEFINITION

If n and r are integers with $n \geq r \geq 0$, the symbol is called a **binomial coefficient** and is defined as

$$\binom{n}{r} = \frac{n!}{r!(n - r)!}$$

Example 1   Find the binomial coefficients.

(a) $\dbinom{4}{1}$ (b) $\dbinom{5}{2}$

Solution   (a) $\dbinom{4}{1} = \dfrac{4!}{1!(4-1)!}$

$$= \dfrac{4!}{1!\,3!}$$

$$= \dfrac{4 \cdot 3 \cdot 2 \cdot 1}{1(3 \cdot 2 \cdot 1)}$$

$$= \dfrac{4 \cdot \overset{1}{\cancel{3}} \cdot \overset{1}{\cancel{2}} \cdot 1}{1(\underset{1}{\cancel{3}} \cdot \underset{1}{\cancel{2}} \cdot 1)}$$

$$= \dfrac{4}{1}$$

$$= 4$$

(b) $\dbinom{5}{2} = \dfrac{5!}{2!(5-2)!}$

$$= \dfrac{5!}{2!\,3!}$$

$$= \dfrac{5 \cdot 4 \cdot 3 \cdot 2 \cdot 1}{(2 \cdot 1)(3 \cdot 2 \cdot 1)}$$

$$= \dfrac{5 \cdot \overset{2}{\cancel{4}} \cdot \overset{1}{\cancel{3}} \cdot \overset{1}{\cancel{2}} \cdot 1}{(\underset{1}{\cancel{2}} \cdot 1)(\underset{1}{\cancel{3}} \cdot \underset{1}{\cancel{2}} \cdot 1)}$$

$$= \dfrac{5 \cdot 2 \cdot 1 \cdot 1}{(1 \cdot 1)(1 \cdot 1 \cdot 1)}$$

$$= \dfrac{10}{1}$$

$$= 10 \qquad \blacktriangle$$

There is a connection between the binomial coefficients and Pascal's triangle. If we find the binomial coefficients in the following triangular array, we produce Pascal's triangle.

$$\binom{1}{0} \qquad \binom{1}{1}$$

$$\binom{2}{0} \qquad \binom{2}{1} \qquad \binom{2}{2}$$

$$\binom{3}{0} \qquad \binom{3}{1} \qquad \binom{3}{2} \qquad \binom{3}{3}$$

$$\binom{4}{0} \qquad \binom{4}{1} \qquad \binom{4}{2} \qquad \binom{4}{3} \qquad \binom{4}{4}$$

$$\binom{5}{0} \qquad \binom{5}{1} \qquad \binom{5}{2} \qquad \binom{5}{3} \qquad \binom{5}{4} \qquad \binom{5}{5}$$

Now it becomes clear why $\binom{n}{r}$ is called a binomial coefficient. For example, from row 3, we obtain

$$(a + b)^3 = \binom{3}{0}a^3 + \binom{3}{1}a^2b^1 + \binom{3}{2}a^1b^2 + \binom{3}{3}b^3$$

The general form is given in the next theorem.

THEOREM

The binomial theorem

If n is a positive integer, then

$(a + b)^n =$

$$\binom{n}{0}a^n + \binom{n}{1}a^{n-1}b + \binom{n}{2}a^{n-2}b^2 + \binom{n}{3}a^{n-3}b^3 + \cdots + \binom{n}{n}b^n$$

We can use either the binomial theorem or Pascal's triangle to find the numerical coefficients in the expansion of $(a + b)^n$. Both methods are illustrated in the next examples.

Example 2   Expand $(x + 2)^3$ and simplify.

Solution   Pascal's Triangle Method   We use row 3 in Pascal's triangle.

$$(x + 2)^3 = x^3 + 3x^2(2) + 3x(2)^2 + 2^3$$
$$= x^3 + 6x^2 + 12x + 8$$

Binomial Theorem Method   Using the binomial theorem with $n = 3$, we have

$$(x + 2)^3 = \binom{3}{0}x^3 + \binom{3}{1}x^2 2 + \binom{3}{2}x2^2 + \binom{3}{3}2^3$$

$$= \frac{3!}{0!(3 - 0)!}x^3 + \frac{3!}{1!(3 - 1)!}x^2 \cdot 2 + \frac{3!}{2!(3 - 2)!}x \cdot 4 + \frac{3!}{3!(3 - 3)!} \cdot 8$$

$$= x^3 + 3(2x^2) + 3(4x) + 8$$

$$= x^3 + 6x^2 + 12x + 8 \qquad \blacktriangle$$

Example 3   Expand $(x^2 - y)^5$ and simplify.

Solution   Think of x^2 as a and $-y$ as b. Using row 5 of Pascal's triangle,

$$(x^2 - y)^5 =$$
$$(x^2)^5 + 5(x^2)^4(-y) + 10(x^2)^3(-y)^2 + 10(x^2)^2(-y)^3 + 5x^2(-y)^4 + (-y)^5$$

$$= x^{10} - 5x^8y + 10x^6y^2 - 10x^4y^3 + 5x^2y^4 - y^5 \qquad \blacktriangle$$

Example 4   Write the sum of the first three terms in the expansion of $(x + y)^{12}$ and simplify.

Solution   By the binomial theorem, the sum of the first three terms in this expansion is

$$\binom{12}{0}x^{12} + \binom{12}{1}x^{11}y + \binom{12}{2}x^{10}y^2$$

This sum simplifies to

$$x^{12} + 12x^{11}y + 66x^{10}y^2 \qquad \blacktriangle$$

Example 5   Find the fifth term in the expansion of $(a + b)^7$ and simplify.

Solution   By the binomial theorem,

$$(a + b)^7 =$$
$$\binom{7}{0}a^7 + \binom{7}{1}a^6b + \binom{7}{2}a^5b^2 + \binom{7}{3}a^4b^3 + \binom{7}{4}a^3b^4 + \binom{7}{5}a^2b^5 + \binom{7}{6}ab^6 + \binom{7}{7}b^7$$

Now, $\binom{7}{0}a^7$ is the first term, $\binom{7}{1}a^6b$, the second term, and so on. Therefore, the fifth term of $(a + b)^7$ is

$$\binom{7}{4}a^3b^4 = \frac{7!}{4!(7 - 4)!}a^3b^4$$

$$= \frac{7!}{4!3!}a^3b^4$$

$$= \frac{7 \cdot 6 \cdot 5 \cdot \cancel{4} \cdot \cancel{3} \cdot \cancel{2} \cdot 1}{\cancel{4} \cdot \cancel{3} \cdot \cancel{2} \cdot 1 \cdot 3 \cdot 2 \cdot 1} a^3 b^4$$

$$= \frac{7 \cdot 6 \cdot 5}{3 \cdot 2 \cdot 1} a^3 b^4$$

$$= \frac{7 \cdot \overset{1}{\cancel{6}} \cdot 5}{\underset{1}{\cancel{3}} \cdot \underset{1}{\cancel{2}} \cdot 1} a^3 b^4$$

$$= 35 a^3 b^4 \qquad \blacktriangle$$

The form of the kth term in the expansion of $(a + b)^n$ is given in the next theorem.

THEOREM

Let n be a positive integer and let k be a positive integer so that $k \le n$. Then the kth term in the expansion of $(a + b)^n$ is the following:

$$\binom{n}{k-1} a^{n-(k-1)} b^{k-1}$$

Example 6   Find the eighth term in the expansion of $(x - y)^{10}$.

Solution   Setting $n = 10$ and $k = 8$ in the above theorem, the eighth term in the expansion of $(x - y)^{10}$ is

$$\binom{10}{7} x^{10-(8-1)} (-y)^{8-1}$$

This expression simplifies to

$$\frac{10!}{7!3!} x^3 (-y)^7 = -120 x^3 y^7 \qquad \blacktriangle$$

EXERCISE SET 12.4

For Exercises 1–16, use either Pascal's triangle or the binomial theorem to expand each of the following. Simplify.

1. $(a + 3)^4$

2. $(a + 2)^3$

3. $(x - 1)^3$

4. $(x - 3)^4$

5. $(2x + y)^5$

6. $(2x + b)^4$

7. $(3x - 2y)^3$

8. $(4x - 3y)^4$

9. $(x^2 + y^2)^6$

10. $(x^3 - y^2)^4$

11. $(x - 2z^2)^3$

12. $(2t^2 - s)^4$

13. $\left(\dfrac{a}{2} - 4\right)^4$

14. $\left(9 - \dfrac{t}{3}\right)^3$

15. $\left(2a^2 - \dfrac{b}{2}\right)^3$

16. $\left(3c^3 - \dfrac{d}{3}\right)^3$

For Exercises 17–24, write the sum of the first three terms of each expansion. Simplify.

17. $(x + y)^{11}$

18. $(a + b)^{10}$

19. $(x - y)^{10}$

20. $(a - b)^9$

21. $(x^2 + a)^{12}$

22. $(r^2 + t^2)^8$

23. $(s^3 - t^2)^{14}$

24. $(r - w^2)^{15}$

For Exercises 25–32, find the specified term of each expansion. Simplify.

25. Third term of $(a + b)^8$.

26. Fifth term of $(a + b)^6$.

27. Second term of $(x - y)^7$.

28. Fourth term of $(x - y)^5$.

29. Third term of $(2x - 3y)^5$.

30. Last term of $(t - 2y^2)^4$.

31. Middle term of $(3x + y^3)^4$.

32. Middle term of $\left(s^2 - \dfrac{3}{2}\right)^6$.

E N R I C H M E N T   E X E R C I S E S

The binomial coefficients $\dbinom{n}{r}$ have applications in counting techniques.

Consider the set of four elements $\{a, b, c, d\}$. The following table lists its subsets according to size: no elements, 1 element, 2 elements, 3 elements, or 4 elements.

Subsets with 0 elements: $\varnothing$
Subsets with 1 element: $\{a\}, \{b\}, \{c\}, \{d\}$
Subsets with 2 elements: $\{a, b\}, \{a, c\}, \{a, d\}, \{b, c\}, \{b, d\}, \{c, d\}$
Subsets with 3 elements: $\{a, b, c\}, \{a, b, d\}, \{a, c, d\}, \{b, c, d\}$
Subsets with 4 elements: $\{a, b, c, d\}$

From this information, we count the number of subsets of the five sizes.

0 elements	1 element	2 elements	3 elements	4 elements
1	4	6	4	1

Notice that these numbers form the fourth row of Pascal's triangle. Therefore, we have

0 elements	1 element	2 elements	3 elements	4 elements
$\binom{4}{0}$	$\binom{4}{1}$	$\binom{4}{2}$	$\binom{4}{3}$	$\binom{4}{4}$

Notice that the first number n in the binomial coefficient is the size of the original set, and the second number r is the size of the subsets that are being counted. Therefore,

$$\binom{n}{r} = \text{the number of possible subsets of size } r \text{ of a set of size } n$$

1. How many subsets of size 2 are possible from a set of size 5?

2. How many subsets of size 3 are possible from a set of size 6?

3. How many subsets of size 7 are possible from a set of size 8?

4. How many different committees of 4 people can be formed from 6 people?

5. How many different five-card poker hands are possible from a standard deck of 52 cards?

6. Show that a set of n elements has 2^n subsets. (*Hint:* Expand $(1 + 1)^n$ by the binomial theorem.)

Answers to Enrichment Exercises are on page A.87.

Summary and review

Examples

$1, \dfrac{1}{2}, \dfrac{1}{3}, \ldots$ is an infinite sequence with general term $a_n = \dfrac{1}{n}$.

$1, 3, 5, 7, \ldots$ is an arithmetic sequence with $a_1 = 1$ and $d = 2$.

Sequences (12.1 and 12.3)

In this section, we studied **infinite** and **finite sequences.**

An **arithmetic sequence** is a sequence with a general term of the form

$$a_n = a_1 + (n - 1)d$$

Examples

2, 4, 8, ... is a geometric series with $a_1 = 2$ and $r = 2$.

$$\sum_{i=1}^{3}(3i-1) = [3(1)-1]$$
$$+ [3(2)-1]$$
$$+ [3(3)-1]$$
$$= 2 + 5 + 8 = 15$$

For the sequence
$1, 3, 5, 7, 9, \ldots$,

$S_1 = 1, S_2 = 4, S_3 = 9, S_4 = 16$, and $S_5 = 25$

The seventh partial sum of the arithmetic sequence with $a_1 = 2$ and $a_n = 2 + (n-1)2$ is

$$S_7 = \frac{7}{2}[2 + 2 + (7-1)2]$$
$$= 56$$

The fourth partial sum of the geometric sequence with $a_1 = 12$ and $r = \dfrac{1}{2}$ is

$$S_4 = \frac{12\left[1-\left(\frac{1}{2}\right)^4\right]}{1-\frac{1}{2}}$$

$$= \frac{12\left[1-\frac{1}{16}\right]}{\frac{1}{2}}$$

$$= 24\left(\frac{15}{16}\right) = \frac{45}{2}$$

Using the third row, we can expand $(a + b)^3$:

$$(a+b)^3 = a^3 + 3a^2b$$
$$+3ab^2 + b^3$$

A **geometric sequence** has a general term of the form

$$a_n = a_1 r^{n-1}$$

Series (12.2)

The sum of the terms of a sequence is called a **series.** To indicate a sum, we use the **summation notation,** or **sigma notation.**

Let $a_1, a_2, a_3, \ldots$ be a sequence. The **nth partial sum** S_n of the sequence is defined to be

$$S_n = a_1 + a_2 + a_3 + \cdots + a_n$$

The nth partial sum of an arithmetic sequence is

$$S_n = \frac{n}{2}(a_1 + a_n) \qquad \text{or} \qquad S_n = \frac{n}{2}[2a_1 + (n-1)d]$$

The **nth partial sum** of a geometric sequence is

$$S_n = \frac{a_1(1-r^n)}{1-r}, \qquad r \neq 1$$

The **value of an infinite geometric series** is given by

$$\sum_{i=1}^{\infty} a_1 r^{i-1} = \frac{a_1}{1-r}, \qquad |r| < 1$$

The binomial theorem (12.4)

To expand binomials to powers higher than 2, it is convenient to use either Pascal's triangle or the binomial theorem. The numerical coefficients in the expansion of $(a + b)^n$ can be obtained from Pascal's triangle:

$$
\begin{array}{ccccccccc}
 & & & & 1 & & 1 & & \\
 & & & 1 & & 2 & & 1 & \\
 & & 1 & & 3 & & 3 & & 1 \\
 & 1 & & 4 & & 6 & & 4 & & 1
\end{array}
$$

and so on.

If n is a positive integer, $n!$, called **n factorial,** is defined as

$$n! = n(n-1)(n-2)(n-3) \cdots 2 \cdot 1$$

In addition, we define

$$0! = 1$$

Examples

$$\binom{3}{2} = \frac{3!}{2!(3-2)!} = \frac{3!}{2!1!}$$

$$= \frac{3 \cdot 2 \cdot 1}{2 \cdot 1 \cdot 1} = 3$$

If n and r are integers with $n \geq r \geq 0$, the symbol $\binom{n}{r}$ is called a **binomial coefficient** and is defined as

$$\binom{n}{r} = \frac{n!}{r!(n-r)!}$$

$$(a+b)^3 = \binom{3}{0}a^3 + \binom{3}{1}a^2b$$
$$+ \binom{3}{2}ab^2 + \binom{3}{3}b^3$$
$$= a^3 + 3a^2b + 3ab^2 + b^3$$

The binomial theorem states that the entries in Pascal's triangle are the binomial coefficients.

CHAPTER 12 REVIEW EXERCISE SET

Sections 12.1 and 12.3

For Exercises 1–7, find the first four terms of the given sequence.

1. $a_n = 3n - 9$

2. $a_n = (-1)^{n+1}n$

3. $a_n = \dfrac{n+1}{n+2}$

4. $a_1 = 2, a_n = -3a_{n-1}, n \geq 2$

5. $a_1 = -4, a_n = a_{n-1} + 5, n \geq 2$

6. $a_n = 2 + (n-1)3$

7. $a_n = 3\left(\dfrac{1}{2}\right)^{n-1}$

For Exercises 8–12, find the general term a_n.

8. $1, 3, 5, 7, \ldots$

9. $-1, 1, -1, 1, \ldots$

10. $1, -\dfrac{1}{2}, \dfrac{1}{3}, -\dfrac{1}{4}, \ldots$

11. $10, 5, \dfrac{5}{2}, \dfrac{5}{4}, \ldots$

12. $3, \dfrac{3}{4}, \dfrac{3}{16}, \dfrac{3}{64}, \ldots$

Sections 12.2 and 12.3

For Exercises 13–15, find the value of each series.

13. $\displaystyle\sum_{i=1}^{6}(3i-4)$

14. $\displaystyle\sum_{k=1}^{3}\left(\dfrac{1}{2}\right)^k$

15. $\displaystyle\sum_{i=1}^{4}\dfrac{4}{i^2}$

For Exercises 16–19, write each series in summation notation.

16. $2 + 5 + 8 + 11 + 14$

17. $\dfrac{1}{2} + \dfrac{3}{4} + \dfrac{5}{6} + \dfrac{7}{8} + \dfrac{9}{10}$

18. $1 - 2 + 4 - 8 + 16 - 32$

19. $\dfrac{x^3}{3} + \dfrac{x^5}{5} + \dfrac{x^7}{7} + \dfrac{x^9}{9}$

For Exercises 20–22, find the partial sums S_1, S_2, S_3, and S_4 for the given sequence.

20. $1, 5, 10, 16, \ldots$

21. The nth term is $a_n = \dfrac{2}{n}$.

22. The nth term is $a_n = (-1)^{n-1}n$.

For Exercises 23–26, find the sum of the first n terms of the arithmetic sequence, where a_1, d, and n are given.

23. $a_1 = -2$, $d = 3$, $n = 9$

24. $a_1 = 2$, $d = -4$, $n = 11$

25. $a_1 = \dfrac{1}{2}$, $d = 3$, $n = 8$

26. $a_1 = 0$, $d = -\dfrac{2}{3}$, $n = 16$

For Exercises 27–30, find the sum of the first n terms of the geometric sequence from the information.

27. $a_1 = \dfrac{1}{16}$, $r = 2$, $n = 5$

28. $a_1 = -\dfrac{1}{9}$, $r = -3$, $n = 3$

29. $a_1 = 32$, $r = -\dfrac{1}{2}$, $n = 4$

30. $a_1 = 6$, $r = \dfrac{2}{3}$, $n = 3$

For Exercises 31–34, find the value of each infinite geometric series.

31. $\displaystyle\sum_{i=1}^{\infty} 2\left(\dfrac{1}{3}\right)^{i-1}$

32. $\displaystyle\sum_{i=1}^{\infty} \left(-\dfrac{2}{5}\right)^{i-1}$

33. $\displaystyle\sum_{i=1}^{\infty} -6\left(-\dfrac{1}{3}\right)^{i-1}$

34. $\displaystyle\sum_{i=1}^{\infty} (0.3)^{i-1}$

Section 12.4

For Exercises 35–37, find the value of the binomial coefficient.

35. $\dbinom{6}{3}$

36. $\dbinom{12}{2}$

37. $\dbinom{8}{5}$

For Exercises 38–41, use either Pascal's triangle or the binomial theorem to expand each of the following. Simplify.

38. $(a + 4)^3$

39. $(x - 2)^4$

40. $(2x - 3y)^4$

41. $(x^2 + y^3)^5$

42. Find the sum of the first three terms in the expansion of $(x + 3y)^{10}$.

CHAPTER 12   TEST

Section 12.1

For problems 1–3, find the first four terms of the given sequence.

1. $a_n = 2n + 7$

2. $a_n = \dfrac{(-1)^n}{n^2}$

3. $a_1 = -3$, $a_n = a_{n-1} + 2$, $n \geq 2$

For problems 4–6, find the general term a_n from the given pattern.

4. $2, -4, 6, -8, \ldots$

5. $1, \dfrac{1}{3}, \dfrac{1}{5}, \dfrac{1}{7}, \ldots$

6. $2, \dfrac{2}{3}, \dfrac{2}{9}, \dfrac{2}{27}, \ldots$

Section 12.2

For problems 7 and 8, find the value of the series.

7. $\displaystyle\sum_{i=1}^{4} (1 - 2i)^2$

8. $\displaystyle\sum_{k=1}^{3} (-1)^{k-1} 2^k$

For problems 9 and 10, write the series in expanded form.

9. $\displaystyle\sum_{n=1}^{5} n x^{n+1}$

10. $\displaystyle\sum_{k=1}^{4} (-1)^{k-1} x^k$

For problems 11 and 12, express the series in summation notation.

11. $1 + 2 + 3 + 4 + 5 + 6 + 7$

12. $(z - 3) + (z - 3)^2 + (z - 3)^3$

Section 12.3

13. Write the first four terms of the arithmetic sequence in which $a_1 = -6$ and $d = 5$.

14. Find the general term, a_n, of the arithmetic sequence

$$8, 6, 4, 2, \ldots$$

15. If the first and eighteenth terms of an arithmetic sequence are 10 and -41, respectively, find the fifty-fourth term of the sequence.

16. Find the sum of the first twelve terms of the arithmetic sequence if the first term is -18 and the common difference is 6.

17. Write the first four terms of the geometric sequence in which $a_1 = 1$ and $r = -3$.

18. Find the sum of the first five terms of the geometric sequence

$$4, \dfrac{4}{3}, \dfrac{4}{9}, \ldots$$

Section 12.4

19. Find the binomial coefficients.

(a) $\dbinom{6}{2}$ (b) $\dbinom{5}{3}$

For problems 20–22, use either Pascal's triangle or the binomial theorem to expand each of the following. Simplify.

20. $(2a + 3)^3$

21. $(x - 3y)^4$

22. $(x^2 + 4t)^3$

23. Write the sum of the first three terms in the expansion of $(u - v)^{12}$.

24. Find the fourth term of $(2r - y)^6$.

Appendix

Table I Squares and square roots

n	n^2	$\sqrt{n}$	$\sqrt{10n}$	n	n^2	$\sqrt{n}$	$\sqrt{10n}$
1	1	1.000	3.162	51	2601	7.141	22.583
2	4	1.414	4.472	52	2704	7.211	22.804
3	9	1.732	5.477	53	2809	7.280	23.022
4	16	2.000	6.325	54	2916	7.348	23.238
5	25	2.236	7.071	55	3025	7.416	23.452
6	36	2.449	7.746	56	3136	7.483	23.664
7	49	2.646	8.367	57	3249	7.550	23.875
8	64	2.828	8.944	58	3364	7.616	24.083
9	81	3.000	9.487	59	3481	7.681	24.290
10	100	3.162	10.000	60	3600	7.746	24.495
11	121	3.317	10.488	61	3721	7.810	24.698
12	144	3.464	10.954	62	3844	7.874	24.900
13	169	3.606	11.402	63	3969	7.937	25.100
14	196	3.742	11.832	64	4096	8.000	25.298
15	225	3.873	12.247	65	4225	8.062	25.495
16	256	4.000	12.649	66	4356	8.124	25.690
17	289	4.123	13.038	67	4489	8.185	25.884
18	324	4.243	13.416	68	4624	8.246	26.077
19	361	4.359	13.784	69	4761	8.307	26.268
20	400	4.472	14.142	70	4900	8.367	26.458
21	441	4.583	14.491	71	5041	8.426	26.646
22	484	4.690	14.832	72	5184	8.485	26.833
23	529	4.796	15.166	73	5329	8.544	27.019
24	576	4.899	15.492	74	5476	8.602	27.203
25	625	5.000	15.811	75	5625	8.660	27.386
26	676	5.099	16.125	76	5776	8.718	27.568
27	729	5.196	16.432	77	5929	8.775	27.749
28	784	5.292	16.733	78	6084	8.832	27.928
29	841	5.385	17.029	79	6241	8.888	28.107
30	900	5.477	17.321	80	6400	8.944	28.284
31	961	5.568	17.607	81	6561	9.000	28.460
32	1024	5.657	17.889	82	6724	9.055	28.636
33	1089	5.745	18.166	83	6889	9.110	28.810
34	1156	5.831	18.439	84	7056	9.165	28.983
35	1225	5.916	18.708	85	7225	9.220	29.155
36	1296	6.000	18.974	86	7396	9.274	29.326
37	1369	6.083	19.235	87	7569	9.327	29.496
38	1444	6.164	19.494	88	7744	9.381	29.665
39	1521	6.245	19.748	89	7921	9.434	29.833
40	1600	6.325	20.000	90	8100	9.487	30.000
41	1681	6.403	20.248	91	8281	9.539	30.166
42	1764	6.481	20.494	92	8464	9.592	30.332
43	1849	6.557	20.736	93	8649	9.644	30.496
44	1936	6.633	20.976	94	8836	9.695	30.659
45	2025	6.708	21.213	95	9025	9.747	30.822
46	2116	6.782	21.448	96	9216	9.798	30.984
47	2209	6.856	21.679	97	9409	9.849	31.145
48	2304	6.928	21.909	98	9604	9.899	31.305
49	2401	7.000	22.136	99	9801	9.950	31.464
50	2500	7.071	22.361	100	10000	10.000	31.623

Table II Common logarithms

x	0	1	2	3	4	5	6	7	8	9
1.0	.0000	.0043	.0086	.0128	.0170	.0212	.0253	.0294	.0334	.0374
1.1	.0414	.0453	.0492	.0531	.0569	.0607	.0645	.0682	.0719	.0755
1.2	.0792	.0828	.0864	.0899	.0934	.0969	.1004	.1038	.1072	.1106
1.3	.1139	.1173	.1206	.1239	.1271	.1303	.1335	.1367	.1399	.1430
1.4	.1461	.1492	.1523	.1553	.1584	.1614	.1644	.1673	.1703	.1732
1.5	.1761	.1790	.1818	.1847	.1875	.1903	.1931	.1959	.1987	.2014
1.6	.2041	.2068	.2095	.2122	.2148	.2175	.2201	.2227	.2253	.2279
1.7	.2304	.2330	.2355	.2380	.2405	.2430	.2455	.2480	.2504	.2529
1.8	.2553	.2577	.2601	.2625	.2648	.2672	.2695	.2718	.2742	.2765
1.9	.2788	.2810	.2833	.2856	.2878	.2900	.2923	.2945	.2967	.2989
2.0	.3010	.3032	.3054	.3075	.3096	.3118	.3139	.3160	.3181	.3201
2.1	.3222	.3243	.3263	.3284	.3304	.3324	.3345	.3365	.3385	.3404
2.2	.3424	.3444	.3464	.3483	.3502	.3522	.3541	.3560	.3579	.3598
2.3	.3617	.3636	.3655	.3674	.3692	.3711	.3729	.3747	.3766	.3784
2.4	.3802	.3820	.3838	.3856	.3874	.3892	.3909	.3927	.3945	.3962
2.5	.3979	.3997	.4014	.4031	.4048	.4065	.4082	.4099	.4116	.4133
2.6	.4150	.4166	.4183	.4200	.4216	.4232	.4249	.4265	.4281	.4298
2.7	.4314	.4330	.4346	.4362	.4378	.4393	.4409	.4425	.4440	.4456
2.8	.4472	.4487	.4502	.4518	.4533	.4548	.4564	.4579	.4594	.4609
2.9	.4624	.4639	.4654	.4669	.4683	.4698	.4713	.4728	.4742	.4757
3.0	.4771	.4786	.4800	.4814	.4829	.4843	.4857	.4871	.4886	.4900
3.1	.4914	.4928	.4942	.4955	.4969	.4983	.4997	.5011	.5024	.5038
3.2	.5051	.5065	.5079	.5092	.5105	.5119	.5132	.5145	.5159	.5172
3.3	.5185	.5198	.5211	.5224	.5237	.5250	.5263	.5276	.5289	.5302
3.4	.5315	.5328	.5340	.5353	.5366	.5378	.5391	.5403	.5416	.5428
3.5	.5441	.5453	.5465	.5478	.5490	.5502	.5514	.5527	.5539	.5551
3.6	.5563	.5575	.5587	.5599	.5611	.5623	.5635	.5647	.5658	.5670
3.7	.5682	.5694	.5705	.5717	.5729	.5740	.5752	.5763	.5775	.5786
3.8	.5798	.5809	.5821	.5832	.5843	.5855	.5866	.5877	.5888	.5899
3.9	.5911	.5922	.5933	.5944	.5955	.5966	.5977	.5988	.5999	.6010
4.0	.6021	.6031	.6042	.6053	.6064	.6075	.6085	.6096	.6107	.6117
4.1	.6128	.6138	.6149	.6159	.6170	.6180	.6191	.6201	.6212	.6222
4.2	.6232	.6243	.6253	.6263	.6274	.6284	.6294	.6304	.6314	.6325
4.3	.6335	.6345	.6355	.6365	.6375	.6385	.6395	.6405	.6415	.6425
4.4	.6435	.6444	.6454	.6464	.6474	.6484	.6493	.6503	.6513	.6522
4.5	.6532	.6542	.6551	.6561	.6571	.6580	.6590	.6599	.6609	.6618
4.6	.6628	.6637	.6646	.6656	.6665	.6675	.6684	.6693	.6702	.6712
4.7	.6721	.6730	.6739	.6749	.6758	.6767	.6776	.6785	.6794	.6803
4.8	.6812	.6821	.6830	.6839	.6848	.6857	.6866	.6875	.6884	.6893
4.9	.6902	.6911	.6920	.6928	.6937	.6946	.6955	.6964	.6972	.6981
5.0	.6990	.6998	.7007	.7016	.7024	.7033	.7042	.7050	.7059	.7067
5.1	.7076	.7084	.7093	.7101	.7110	.7118	.7126	.7135	.7143	.7152
5.2	.7160	.7168	.7177	.7185	.7193	.7202	.7210	.7218	.7226	.7235
5.3	.7243	.7251	.7259	.7267	.7275	.7284	.7292	.7300	.7308	.7316
5.4	.7324	.7332	.7340	.7348	.7356	.7364	.7372	.7380	.7388	.7396
x	0	1	2	3	4	5	6	7	8	9

x	0	1	2	3	4	5	6	7	8	9
5.5	.7404	.7412	.7419	.7427	.7435	.7443	.7451	.7459	.7466	.7474
5.6	.7482	.7490	.7497	.7505	.7513	.7520	.7528	.7536	.7543	.7551
5.7	.7559	.7566	.7574	.7582	.7589	.7597	.7604	.7612	.7619	.7627
5.8	.7634	.7642	.7649	.7657	.7664	.7672	.7679	.7686	.7694	.7701
5.9	.7709	.7716	.7723	.7731	.7738	.7745	.7752	.7760	.7767	.7774
6.0	.7782	.7789	.7796	.7803	.7810	.7818	.7825	.7832	.7839	.7846
6.1	.7853	.7860	.7868	.7875	.7882	.7889	.7896	.7903	.7910	.7917
6.2	.7924	.7931	.7938	.7945	.7952	.7959	.7966	.7973	.7980	.7987
6.3	.7993	.8000	.8007	.8014	.8021	.8028	.8035	.8041	.8048	.8055
6.4	.8062	.8069	.8075	.8082	.8089	.8096	.8102	.8109	.8116	.8112
6.5	.8129	.8136	.8142	.8149	.8156	.8162	.8169	.8176	.8182	.8189
6.6	.8195	.8202	.8209	.8215	.8222	.8228	.8235	.8241	.8248	.8254
6.7	.8261	.8267	.8274	.8280	.8287	.8293	.8299	.8306	.8312	.8319
6.8	.8325	.8331	.8338	.8344	.8351	.8357	.8363	.8370	.8376	.8382
6.9	.8388	.8395	.8401	.8407	.8414	.8420	.8426	.8432	.8439	.8445
7.0	.8451	.8457	.8463	.8470	.8476	.8482	.8488	.8494	.8500	.8506
7.1	.8513	.8519	.8525	.8531	.8537	.8543	.8549	.8555	.8561	.8567
7.2	.8573	.8579	.8585	.8591	.8597	.8603	.8609	.8615	.8621	.8627
7.3	.8633	.8639	.8645	.8651	.8657	.8663	.8669	.8675	.8681	.8686
7.4	.8692	.8698	.8704	.8710	.8716	.8722	.8727	.8733	.8739	.8745
7.5	.8751	.8756	.8762	.8768	.8774	.8779	.8785	.8791	.8797	.8802
7.6	.8808	.8814	.8820	.8825	.8831	.8837	.8842	.8848	.8854	.8859
7.7	.8865	.8871	.8876	.8882	.8887	.8893	.8899	.8904	.8910	.8915
7.8	.8921	.8927	.8932	.8938	.8943	.8949	.8954	.8960	.8965	.8971
7.9	.8976	.8982	.8987	.8993	.8998	.9004	.9009	.9015	.9020	.9025
8.0	.9031	.9036	.9042	.9047	.9053	.9058	.9063	.9069	.9074	.9079
8.1	.9085	.9090	.9096	.9101	.9106	.9112	.9117	.9122	.9128	.9133
8.2	.9138	.9143	.9149	.9154	.9159	.9165	.9170	.9175	.9180	.9186
8.3	.9191	.9196	.9201	.9206	.9212	.9217	.9222	.9227	.9232	.9238
8.4	.9243	.9248	.9253	.9258	.9263	.9269	.9274	.9279	.9284	.9289
8.5	.9294	.9299	.9304	.9309	.9315	.9320	.9325	.9330	.9335	.9340
8.6	.9345	.9350	.9355	.9360	.9365	.9370	.9375	.9380	.9385	.9390
8.7	.9395	.9400	.9405	.9410	.9415	.9420	.9425	.9430	.9435	.9440
8.8	.9445	.9450	.9455	.9460	.9465	.9469	.9474	.9479	.9484	.9489
8.9	.9494	.9499	.9504	.9509	.9513	.9518	.9523	.9528	.9533	.9538
9.0	.9542	.9547	.9552	.9557	.9562	.9566	.9571	.9576	.9581	.9586
9.1	.9590	.9595	.9600	.9605	.9609	.9614	.9619	.9624	.9628	.9633
9.2	.9638	.9643	.9647	.9652	.9657	.9661	.9666	.9671	.9675	.9680
9.3	.9685	.9689	.9694	.9699	.9703	.9708	.9713	.9717	.9722	.9727
9.4	.9731	.9736	.9741	.9745	.9750	.9754	.9759	.9763	.9768	.9773
9.5	.9777	.9782	.9786	.9791	.9795	.9800	.9805	.9809	.9814	.9818
9.6	.9823	.9827	.9832	.9836	.9841	.9845	.9850	.9854	.9859	.9863
9.7	.9868	.9872	.9877	.9881	.9886	.9890	.9894	.9899	.9903	.9908
9.8	.9912	.9917	.9921	.9926	.9930	.9934	.9939	.9943	.9948	.9952
9.9	.9956	.9961	.9965	.9969	.9974	.9978	.9983	.9987	.9991	.9996
x	0	1	2	3	4	5	6	7	8	9

Answers to Selected Exercises

EXERCISE SET 1.1

1. $\dfrac{2}{3}$ **3.** $\dfrac{5}{6}$ **5.** $\dfrac{2}{7}$ **7.** $\dfrac{1}{7}$ **9.** $\dfrac{4}{5}$ **11.** $\dfrac{6}{35}$ **13.** $\dfrac{8}{75}$ **15.** $\dfrac{32}{105}$ **17.** $\dfrac{49}{5}$ **19.** $\dfrac{18}{5}$ **21.** 2 **23.** 1

25. $\dfrac{15}{2}$ **27.** 24 **29.** 6 **31.** $\dfrac{21}{10}$ **33.** $\dfrac{33}{10}$ **35.** 1 **37.** $\dfrac{2}{3}$ **39.** $\dfrac{85}{48}$ **41.** $\dfrac{5}{11}$ **43.** $\dfrac{9}{2}$ **45.** $\dfrac{4}{5}$

47. 1 **49.** $\dfrac{1}{3}$ **51.** $\dfrac{1}{5}$ **53.** $\dfrac{1}{2}$ **55.** $\dfrac{10}{9}$ **57.** 1 **59.** $\dfrac{4}{5}$ **61.** $\dfrac{7}{8}$ **63.** $7\dfrac{2}{3}$ feet **65.** $\dfrac{5}{8}$ of the pizza

67. 5 teaspoons **69.** Common factors, not common terms, can be divided out.

ENRICHMENT EXERCISES

1. $\dfrac{23}{36}$ **2.** $-\dfrac{1}{4}$ **3.** $1\dfrac{13}{14}$ **4.** $\dfrac{8}{3}$ **5.** $\dfrac{25}{9}$ **6.** $\dfrac{2}{7}$ **7.** $\dfrac{2}{15}$ **8.** $\dfrac{13}{14}$ **9.** $-\dfrac{5}{6}$ **10.** \$6.30

11. $\dfrac{a}{b} \div \dfrac{c}{d} = \dfrac{ad}{bc}$, since $\dfrac{ad}{bc} \cdot \dfrac{c}{d} = \dfrac{adc}{bcd} = \dfrac{ad\!\!\not{c}}{b\not{c}d} = \dfrac{a}{b}$

EXERCISE SET 1.2

1. $A = \{9, 11, 13, 15\}$ **3.** $B = \{x \mid x$ is a natural number between 3 and $11\}$ **5.** T **7.** T **9.** F **11.** T

13. $-1 + 2x$ **15.** $(4 + z)\sqrt{3}$ **17.** $\dfrac{20}{10}$ **19.** $-\dfrac{4}{2}$ **21.** $-\dfrac{4}{2}, \dfrac{5}{2}, \dfrac{7}{14}, 0, \dfrac{20}{10}, -0.333\ldots$ **23.** T **25.** F

27. **29.** **31.** The coordinate of A is -9. The coordinate of B

is -3. The coordinate of C is 2. The coordinate of D is 6. **33.** \$45 **35.** $-$\$10,450 **37.** -221 **39.** $>$ **41.** $<$

43. $>$ **45.** $>$ **47.** $\dfrac{7}{8}$ **49.** $-\dfrac{6}{5}$ **51.** $-1\dfrac{2}{7}$ **53.** 4.2971 **55.** -10.1289

A.1

57. (a) 1 and 2 **(b)** -5 and -4 **(c)** 3 and 4   **59.** $\{2, 4, 6, 8, 10\}$ **61.** $\{1, 2, 3\}$ **63.** $\varnothing$
65. $\{-2, -1, 0, 1, 2, 3, 4, \ldots\}$ **67.** $\{\ldots, -9, -6, -3, 0, 3, 6, 9, \ldots\}$ **69.** Drug B

E N R I C H M E N T   E X E R C I S E S

1. rational; $\dfrac{1}{4}$ **2.** rational; $\dfrac{410}{99}$ **3.** irrational   **4.** rational; $\dfrac{2}{3}$ **5.** rational; $\dfrac{9}{2}$

6. The number is 3.
7. 10   **8.** 10

E X E R C I S E   S E T   1 . 3

1. 4   **3.** $\dfrac{3}{4}$ **5.** 0   **7.** -8 **9.** $\dfrac{4}{11}$ **11.** 3.82   **13.** -32 **15.** 4.2   **17.** -90 **19.** 2   **21.** 50   **23.** 0

25. 1   **27.** $\dfrac{1}{2}$ **29.** 0   **31.** 18   **33.** $-1\dfrac{1}{2}$ **35.** 12   **37.** -72 **39.** $-\dfrac{3}{5}$ **41.** 0   **43.** -0.63 **45.** -2.3

47. (a) 2 **(b)** 8 **(c)** 8   **49. (a)** -8 **(b)** 8 **(c)** -2 **(d)** 16 **(e)** -8 **51.** $s \cdot s$ **53.** c **55.** $5a \cdot a \cdot a \cdot b \cdot b \cdot b \cdot b$

57. s^3 **59.** $2r^2t^3$ **61.** 9   **63.** $\dfrac{8}{27}$ **65.** 0.49   **67.** 100   **69.** -81 **71.** 2   **73.** 17   **75.** 0   **77.** -80

79. not defined   **81.** $\dfrac{5}{2}$ **83.** $-\dfrac{14}{5}$ **85.** not defined   **87.** 4   **89.** 2   **91.** 8

93. No. $|a + b| = |-4 + 3| = |-1| = 1$; $|a| + |b| = |-4| + |3| = 4 + 3 = 7$ **95.** $>$ **97.** $<$ **99.** \$4,500
101. 1,950 students   **103.** \$47   **105.** 840 bananas   **107.** \$32

E N R I C H M E N T   E X E R C I S E S

1. (d)   **2.** 19   **3.** 0.0877   **4.** 92,000   **5.** 252   **6.** yes; no   **7.** 24   **8.** 120   **9.** 6   **10.** 35

E X E R C I S E   S E T   1 . 4

1. $(33 + 2) + 51$ **3.** $(3 \cdot 5)2$ **5.** $c + 7$ **7.** $(-5)t$ **9.** $x(yz)$ **11.** $3(ab)$ **13.** $34 + [2d + (-r)]$ **15.** $2(rt)$
17. $4u$ **19.** $x + y$ **21.** $2r + 2v$ **23.** 0   **25.** commutative property of addition   **27.** multiplicative identity
29. commutative property of multiplication   **31.** commutative property of addition   **33.** additive inverse

35. closure (for addition)   **37.** $17 + 2w$ **39.** $12r$ **41.** $4z + 4$ **43.** r **45.** $\dfrac{7}{6}st$ **47.** $4s + 48$ **49.** $3r - 6$

51. $\dfrac{3}{2}s + 3$ **53.** $\dfrac{2}{3}c - 4$ **55.** $25 - 10a$ **57.** $1 + \dfrac{3}{5}b$ **59.** $\dfrac{6}{11}y - 1$ **61.** 21   **63.** -12 **65.** $-\dfrac{1}{10}$

67. $-\dfrac{7}{6}$ **69.** -4 **71.** 29   **73.** $\dfrac{6}{5}$ **75.** 3   **77.** 0   **79.** $-\dfrac{18}{7}$ **81.** $-\dfrac{3}{28}$

E N R I C H M E N T   E X E R C I S E S

1. (a) 2^7 **(b)** 2^8 **(c)** 2^9 **2.** x^5 **3.** x^{17} **4.** x^{n+m} **5.** x^6 **6.** x^6 **7.** x^{3n} **8.** x^{nm}

E X E R C I S E   S E T   1 . 5

1. $11x$ **3.** $-13(a + b)$ **5.** $-7x + 5$ **7.** $9c^2 - 19$ **9.** $-2s^2 + 1$ **11.** $10b - ay$ **13.** $-32a + 47b + 52$

15. $-5qp^2 + 3q^2p - 12$ **17.** $-\dfrac{1}{4}z^2 + \dfrac{7}{6}z^3$ **19.** $-150v - 25w$ **21.** $-7rsw + 3sw$ **23.** $y - 16$ **25.** $12 + 4s^2$

27. $2.2y - 0.6$ **29.** -1 **31.** $17r^3 - 25$ **33.** $15x^3y^2 - 36$ **35.** $-18 - 46r^2w$ **37.** $3x^2t + 12$ **39.** $-1 - 7z^2$

41. 28   **43.** -4 **45.** 1   **47.** 1   **49.** 0   **51.** $-\dfrac{4}{3}$ **53.** $3 + x$ **55.** $x - 20$ **57.** $\dfrac{x}{34}$ **59.** $12 - x$

61. $x + 92$ **63.** $2x + 48$ **65.** $0.065x$ **67.** $10(x - 3)$ **69.** $3x - \dfrac{4}{5}$ **71.** $30n$ dollars   **73.** $0.45t$ dollars

75. $\dfrac{y}{12}$ feet   **77.** Let $x =$ Debbie's age now, then $x - 12 =$ her age 12 years ago.   **79.** Let $x =$ one of the numbers, then $9 - x =$ the other number.   **81.** Let $x =$ the selling price of the lamp, then $0.30x - 15 = 30\%$ of the selling price decreased by 15 dollars.   **83.** Let $x =$ the number of defective items, then $45 - x =$ the number of nondefective items. **85.** Let $x =$ the enrollment at Salisbury High School, then $1.75x =$ the enrollment at Ridgemont High School.   **87.** Let $x =$ Mr. Kelly's income, then $67{,}000 - x =$ Mrs. Kelly's income.   **89.** Let $n =$ the first integer, then $n + 1$ is the next consecutive integer. The sum of these two consecutive integers is $n + (n + 1)$ or $2n + 1$ **91.** Let $n =$ the first odd integer, then $n + 2$ is the next consecutive odd integer. The square of the sum of these two consecutive odd integers is $[n + (n + 2)]^2$ or $(2n + 2)^2$ **93.** Let $n =$ the first even integer, then $n + 2$ and $n + 4$ are the next two consecutive even integers. The sum of their squares is $n^2 + (n + 2)^2 + (n + 4)^2$

ENRICHMENT EXERCISES

1. $-3 + 3x + 28y$ **2.** $-19x + 18x^2$ **3.** $2a^3b^2 - a^2b^3$ **4.** $4pq$ **5.** $12\dfrac{1}{2} - 3x$

CHAPTER 1 REVIEW EXERCISE SET

Section 1.1

1. $\dfrac{2}{3}$ **2.** $\dfrac{7}{9}$ **3.** $-\dfrac{1}{7}$ **4.** $-\dfrac{7}{2}$ **5.** $\dfrac{5}{2}$ **6.** $\dfrac{6}{11}$ **7.** 1   **8.** 6   **9.** $\dfrac{13}{25}$ **10.** 1

Section 1.2

11. $A = \{-1, 1, 3\}$ **12.** $-2 - 3x$ **13.** 2   **14.** $2, -\dfrac{9}{3}$ **15.** $-\sqrt{2}, 3\pi$ **16.** $-3\dfrac{1}{2}, \dfrac{4}{5}, 2, -\dfrac{9}{3}$ **17.** $-\sqrt{2}$

18.

19. $<$ **20.** $>$ **21.** $<$ **22.** $<$

Section 1.3

23. 2.7   **24.** 45   **25.** -1 **26.** -5 **27.** -4.7 **28.** 0   **29.** -5.3 **30.** not defined   **31.** $\dfrac{8}{3}$ **32.** $\dfrac{6}{5}$

33. 1   **34.** \$36.50   **35.** 1250

Section 1.4

36. $(24 + 3) + 2c$ **37.** $(2 \cdot 3)x$ **38.** $-2 + 4$ **39.** $3x$ **40.** $8t$ **41.** 0   **42.** $4s$ **43.** 1   **44.** $1 + 2v$

45. $5q - 7$ **46.** b **47.** $-\dfrac{3}{2}x$ **48.** $-\dfrac{3}{4}y$ **49.** $8x - 1$ **50.** $-10x + 7$ **51.** $7x - 7z + 18$ **52.** $4x - \dfrac{2}{3}$

53. -56 **54.** 1   **55.** $1\dfrac{1}{3}$ **56.** 11   **57.** -150 **58.** 2

Section 1.5

59. $14z$ **60.** $2y - 14$ **61.** $s^2 + 1$ **62.** $\dfrac{3}{4}x^2 + 2$ **63.** $7r^2 - 13r^3$ **64.** $25t - 6$ **65.** 0   **66.** $6 + 26t^2r^3$

67. $-2 + 2z^4$ **68.** -13 **69.** $x + 5$ **70.** $-7 - x$ **71.** $\frac{1}{3}x$ **72.** $3x + \frac{3}{2}$ **73.** $0.30x$ **74.** $2x - 43$

75. $\frac{1}{2 + x}$ **76.** Let x = Sam's age now, then $x - 10$ = his age 10 years ago.   **77.** Let n = the selling price of a VCR, then $0.35n - 150$ = 35% of the selling price decreased by \$150.   **78.** Let x = the number of defective items, then $84 - x$ = the number of nondefective items.   **79.** Let x = the length of one of the pieces, then $15 - x$ = the length of the other piece.   **80.** Let n = the first odd integer. Then, $n + 2$ is the next consecutive odd integer. The cube of their sum is $[n + (n + 2)]^3$ or $(2n + 2)^3$ **81.** Let n = the first even integer. Then, $n + 2$ is the next consecutive even integer. The sum of the cubes of these two integers is $n^3 + (n + 2)^3$

CHAPTER 1 TEST

Section 1.1

1. $\frac{4}{7}$ **2.** $\frac{14}{3}$ **3.** 1

Section 1.2

4. 3   **5.** $-5, 0, 3$ **6.** $-5, 0, \frac{1}{2}, 3$ **7.** $-\sqrt{2}$ **8.** $>$

Section 1.3

9. -3 **10.** 4.2   **11.** $\frac{1}{4}$

Section 1.4

12. $(3 + 2) + x$ **13.** $(4 \cdot 3)y$ **14.** $7 + w$ **15.** x **16.** $1 + 8a$ **17.** -6 **18.** -3

Section 1.5

19. $2x - 6$ **20.** $1 - x$ **21.** $x + 6$ **22.** $2x - 2$ **23.** $0.40x$ **24.** Let x = Tom's age now, then $x - 12$ = his age 12 years ago.   **25.** Let x = the number of defective items, then $20 - x$ = the number of nondefective items. **26.** Let x = the length of one piece, then $12 - x$ = the length of the other piece.

CHAPTER 2

EXERCISE SET 2.1

1. $\frac{7}{2}$ **3.** No member of $\{-1, 1, 4\}$ is a solution.   **5.** $x = 3$ **7.** $z = 2.5$ **9.** $h = -1,600$ **11.** $n = -\frac{2}{3}$

13. $b = -\frac{25}{6}$ **15.** $s = \frac{2}{3}$ **17.** $\{10\}$ **19.** $\left\{-\frac{3}{4}\right\}$ **21.** $\{3\}$ **23.** $t = -18$ **25.** $s = 3$ **27.** $m = -5$

29. $q = 49$ **31.** $R = -\frac{7}{3}$ **33.** $x = \frac{3}{5}$ **35.** $v = \frac{3}{4}$ **37.** $x = -2$ **39.** $w = 20$ **41.** $c = -\frac{14}{5}$ **43.** $z = \frac{2}{5}$

45. $b = -1$ **47.** $s = 1$ **49.** $\frac{x}{5} = \frac{2}{5}$; $x = 2$ **51.** $\frac{3x}{7} = -4$; $x = -\frac{28}{3}$ **53.** $2x + 10 = x - 8$; $x = -18$

55. $\dfrac{5x}{2} = 1.5$; $x = \dfrac{3}{5}$ **57.** $20 + 5x = 15 + 6x$; $x = 5$ **59.** $\dfrac{2}{3}x + \dfrac{5}{2} = 10 - \dfrac{1}{3}x$; $x = \dfrac{15}{2}$

Review Problems

61. $2x - 7$ **63.** $-12z + 6$ **65.** $11a - 8$ **67.** -23 **69.** $10 - x$ **71.** $0.45x$

ENRICHMENT EXERCISES

1. $x = \dfrac{12}{7}$ **2.** $y = -\dfrac{13}{6}$ **3.** $a = -\dfrac{1}{3}$ **4.** $\dfrac{x}{3x + 4} = -5$; $x = -\dfrac{5}{4}$ **5.** $4(5 + 2x) = 3(3x - 2)$; $x = 26$

6. $3\left(\dfrac{1}{x}\right) = -\dfrac{5}{7}$; $x = -\dfrac{21}{5}$ **7.** $\dfrac{1}{2x + 3} = \dfrac{1}{6}$; $x = \dfrac{3}{2}$

EXERCISE SET 2.2

1. $x = \dfrac{15}{4}$ **3.** $t = \dfrac{1}{6}$ **5.** $y = -\dfrac{1}{2}$ **7.** $w = -2$ **9.** $u = 5$ **11.** $z = 1$ **13.** $a = 6$ **15.** $x = -2$

17. $t = \dfrac{8}{3}$ **19.** $a = -1$ **21.** Contradiction; no solution.   **23.** $x = 1$ **25.** $T = \dfrac{1}{2}$ **27.** $c = -3$ **29.** $a = -5$

31. Identity; the solution set is the set of real numbers.   **33.** $x = \dfrac{7}{2}$ **35.** $x = \dfrac{17}{4}$ **37.** $p = \dfrac{2}{9}$ **39.** $R = \dfrac{7}{5}$

41. $w = 1$ **43.** $z = -\dfrac{31}{3}$ **45.** A conditional linear equation has exactly one solution; whereas, all real numbers are

solutions of an identity.   **47.** $4x + 5 = 3$; $x = -\dfrac{1}{2}$ **49.** $10x + 25 = -80$; $x = -10.5$

51. $2.6 + 1.3x = 0$; $x = -2$ **53.** $\dfrac{3}{4}x + 2 = -\dfrac{1}{4}$; $x = -3$

Review Problems

55. $10.99x$ dollars   **57.** $0.06z$ dollars   **59.** Let $x =$ one of the numbers, then $12 - x =$ the other number.   **61.** Let $x =$ the selling price of the sofa, then $0.25x + 15 = 25\%$ of the selling price increased by \$15.   **63.** Let $n =$ the first even integer, then $n + 2 =$ the next consecutive even integer, and $n + (n + 2) =$ the sum of these two integers.

ENRICHMENT EXERCISES

1. $x = \dfrac{d - b}{a}$ **2.** $x = \dfrac{d - b}{a - c}$ **3.** $x = \dfrac{d + 2a}{a}$ **4.** $x = \dfrac{1}{ac - 1}$ **5.** $a = -\dfrac{5}{2}$ **6.** $a = -14$ **7.** $c = -11$

EXERCISE SET 2.3

1. \$36,936   **3.** \$225   **5.** Let $x =$ Shawn's age now, then $x - 9 =$ Shawn's age 9 years ago.   **7.** Let $x =$ one of the numbers, then $40 - x =$ the other number.   **9.** Let $x =$ the amount to be invested at 6% interest, then $500 - x =$ the amount to be invested at 7% interest.   **11.** Let $x =$ the selling price of the desk, then $0.3x - 20 = \$20$ less than 30% of the selling price.   **13.** Let $x =$ Jeff's bowling score, then $\dfrac{3}{4}x =$ Tom's bowling score.   **15.** Let $x =$ the amount of money in one part, then $870 - x =$ the amount of money in the other part.   **17.** Let $x =$ the number of cats, then $47 - x =$ the number of dogs.   **19.** \$300   **21.** \$2.80   **23.** \$500   **25.** 9 and 36   **27.** 37.5 and 187.5   **29.** -12 and -10 **31.** -25 and -23 **33.** 28, 30, and 32   **35.** 9 feet and 12 feet   **37.** (a) $x + 3$ (b) $x - 4$

(c) $x - 1$ **(d)** Samantha is 10 years old; Tracy is 13 years old. **39.** Joe is 50 years old; his daughter is 20 years old.
41. 25 nickels and 19 dimes **43.** 140 pounds of the blend containing 15% butter; 210 pounds of the blend containing 65% butter. **45.** Forty-eight 50-cent comics; twenty-four 75-cent comics.

Review Problems

47. $-5x^3 + 7y$ **49.** $\frac{3}{2}h^2 + 3h$ **51.** 1 **53.** 210

ENRICHMENT EXERCISES

1. The first piece has a length of 8 feet. The second piece has a length of 12 feet. The third piece has a length of 11 feet.
2. $400,000

EXERCISE SET 2.4

1. 45 minutes **3.** The police car is traveling 70 mph; the sports car is traveling 80 mph. **5.** $1,700 **7.** $225
9. Use $5,000 for the certificate of deposit and put $10,000 into the savings account. **11. (a)** $A = 1$ sq yd; $P = 5$ yd
(b) $A = 2.34375$ sq cm; $P = 8.46$ cm **(c)** $A = 3.375$ sq in.; $P = 9.77$ in. **13.** $1,900 **15.** 33° and 99°

17. 90°, 30°, and 60° **19.** $r = \dfrac{C}{2\pi}$ **21.** $I = d^2 E$ **23.** $P = \dfrac{A}{rt + 1}$ **25.** $V_2 = \dfrac{p_1 V_1}{p_2}$ **27.** $T_1 = T_2(1 - E)$

29. $w = \dfrac{P - 2l}{2}$ **31.** $b = \dfrac{af}{a - f}$ **33.** $d = \dfrac{a_n - a}{n - 1}$ **35.** $x = \dfrac{y - 1}{3}$ **37.** $x = \dfrac{2y + 1}{4}$ **39.** $x = -\dfrac{2}{3}(t - 1)$

41. $x = \dfrac{y + 8}{2y - 3}$ **43.** $y = \dfrac{3}{4}x - 2$ **45.** $y = -\dfrac{A}{B}x - \dfrac{C}{B}$ **47.** $y = -\dfrac{10}{3}$ **49.** $x = 15$

51. 2,000 piano rolls per week **53. (a)** $F = 2.2t + 2.5$ **(b)** $57.50 **(c)** 10 miles **55.** $-40°$

Review Problems

Section 1.2

57. $<$ **59.** $>$ **61.** $>$

Section 1.5

63. $8 - 6x$ **65.** $11z - 24$

ENRICHMENT EXERCISES

1. The triangle has sides of lengths 5′, 6′, and 7′. The rectangle has sides of lengths 12′ and 2′. **2.** 16 feet by 24 feet
3. 252 **4.** $x = \dfrac{dy - b}{a - cy}$

EXERCISE SET 2.5

1. $x > 10$ **3.** $x \geq 9$ **5.** $x \leq 7$ **7.** 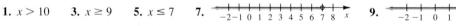 **9.**

11. **13.** **15.**

17. **19.** **21.**

23. **25.** **27.**

29. **31.** **33.**

35. $\{x \mid 0 < x \le 1\}$ **37.** $\left(-\dfrac{1}{3}, +\infty\right)$ **39.** $\{n \mid n < -2\}$ **41.** $(-\infty, 6)$ **43.** $(-\infty, 20]$ **45.** $[2, +\infty)$

47. $[-3, -2)$ **49.** $\left\{w \mid 5\dfrac{1}{2} < w < 8\right\}$ **51.** $(15, 25]$ **53.** $(1, +\infty)$ **55.** $(-\infty, 3]$ **57.** $\left\{t \mid t > -\dfrac{3}{4}\right\}$

59. $\left(-\dfrac{5}{3}, 3\right]$ **61.** $\left\{y \mid \dfrac{1}{2} \le y < 1\right\}$ **63.** $\left[\dfrac{3}{2}, +\infty\right)$ **65.** $x + 2 < 5$; $x < 3$ **67.** $x - 4 \le 1$; $x \le 5$

69. $3x + 4 \le 2x + 7$; $x \le 3$ **71.** $6 - 3x < -2x$; $x > 6$ **73.** $4 - x \le 5$; $x \ge -1$

75. $2x - \dfrac{5}{2} < 5x + \dfrac{7}{2}$; $x > -2$ **77.** $2 < 3x + 1 \le 3$; $\dfrac{1}{3} < x \le \dfrac{2}{3}$ **79.** She must bowl at least 185.

81. At least 200,000 boxes.   **83.** $(2, 5)$ **85.** $[-1, 5]$ **87.** the set of real numbers   **89.** $(-\infty, 2)$ **91.** $[2, 10)$

93. $(-\infty, 1) \cup (5, +\infty)$ **95.** $(-\infty, 2] \cup [5, +\infty)$ **97.** $\left(-\infty, -\dfrac{5}{2}\right)$

Review Problems

99. 41   **101.** 0   **103.** 1   **105.** $\dfrac{7}{2}$ **107.** within 4 units of the origin

E N R I C H M E N T   E X E R C I S E S

1. Since $c > 0$, $-c < 0$. Therefore, dividing by $-c$ changes the sense of the inequalities: $\dfrac{b}{-c} < \dfrac{-cx}{-c} < \dfrac{a}{-c}$. Therefore, $-\dfrac{b}{c} < x < -\dfrac{a}{c}$.   **2.** $(-2, 1)$ **3.** $\left(\dfrac{8}{3}, 4\right]$ **4.** $\left[-\dfrac{11}{2}, -\dfrac{9}{2}\right)$ **5.** Assume that $|x| < 2$. Case 1: $x \ge 0$. Then $|x| = x$, so $|x| < 2$ means that $x < 2$. That is, $0 \le x < 2$. Case 2: $x < 0$. Then $|x| = -x$, so $|x| < 2$ means that $-x < 2$ or $x > -2$. That is, $-2 < x < 0$. Conversely, if $-2 < x < 2$, then x is within 2 units of the origin. Therefore, $|x| < 2$.   **6.** Not true for all real numbers. For example, if $x = -2$ and $y = -3$, then $x > y$, but $(-2)^2 = 4$ and $(-3)^2 = 9$, so $(-2)^2$ is not greater than $(-3)^2$.

E X E R C I S E   S E T   2 . 6

1. either $x = 45$ or $x = -45$ **3.** either $z = \dfrac{5}{3}$ or $z = -1$ **5.** either $s = \dfrac{6}{5}$ or $s = \dfrac{2}{5}$ **7.** no solution   **9.** $x = -\dfrac{5}{4}$

11. either $x = \dfrac{3}{2}$ or $x = -\dfrac{3}{2}$ **13.** either $r = \dfrac{2}{5}$ or $r = -\dfrac{6}{5}$ **15.** either $x = 1$ or $x = -4$ **17.** either $y = 2$ or $y = -\dfrac{4}{3}$

19. either $x = 3$ or $x = \dfrac{1}{5}$ **21.** either $w = -\dfrac{1}{2}$ or $w = \dfrac{7}{8}$ **23.** $x = \dfrac{5}{2}$ **25.** all real numbers   **27.** $x = -\dfrac{1}{3}$

29. $(-2, 2)$ **31.** $[3, 5]$ **33.** $[-3, 0]$

35. $(-\infty, -2] \cup [1, +\infty)$ **37.** $\left[-\dfrac{1}{3}, 1\right]$

39. $(-\infty, 1) \cup (3, +\infty)$ **41.** $v = 3$ **43.** $\left(\dfrac{8}{7}, 2\right)$

45. $\left(-\infty, \dfrac{5}{6}\right] \cup \left[\dfrac{11}{6}, +\infty\right)$ **47.** $[-3, +\infty)$ **49.** $(-\infty, 6]$ **51.** $\left(-\infty, \dfrac{10}{3}\right) \cup \left(\dfrac{10}{3}, +\infty\right)$ **53.** $|x| \le 5$; $[-5, 5]$

55. $|4 + 2x| > 2$; $(-\infty, -3) \cup (-1, +\infty)$ **57.** $|x - 1| + (-6) > -2$; $(-\infty, -3) \cup (5, +\infty)$ **59.** $|x - 3| = |5x|$; either $x = -\dfrac{3}{4}$ or $x = \dfrac{1}{2}$

Review Problems

61. a^3 **63.** $3a^2b^3c$ **65.** $5x \cdot x \cdot x \cdot y \cdot y$ **67.** $\dfrac{2}{b \cdot b \cdot b}$ **69.** $\dfrac{4}{9}$ **71.** -16 **73.** 32   **75.** 8

E N R I C H M E N T   E X E R C I S E S

1. Let $a = -1$ and $b = 1$, then, $|a + b| = |1 + (-1)| = |0| = 0$. Whereas, $|a| + |b| = |1| + |-1| = 1 + 1 = 2$. Therefore, $|a + b| \neq |a| + |b|$.   **2.** $(-5, -3) \cup (3, 5)$ **3.** $(1, 2] \cup [4, 5)$ **4.** $-2 \leq x \leq 5$, and $x \neq \dfrac{3}{2}$ **5.** $\left(-\dfrac{1}{2}, \dfrac{1}{2}\right)$

6. $\left(-\infty, -\dfrac{7}{2}\right] \cup \left[-\dfrac{3}{2}, +\infty\right)$ **7.** $a - \delta < x < a + \delta$, and $x \neq a$

C H A P T E R   2   R E V I E W   E X E R C I S E   S E T

Sections 2.1 and 2.2

1. $x = -6$ **2.** $r = -5$ **3.** $t = 1$ **4.** $z = -\dfrac{12}{11}$ **5.** Identity. The solution set is the set of real numbers.

6. Contradiction. No solution.   **7.** $c = 6$ **8.** $a = -1$ **9.** $\dfrac{x}{3} = \dfrac{1}{5}$; $x = \dfrac{3}{5}$ **10.** $\dfrac{3x}{2} = -6$; $x = -4$

11. $3 + x = 12 + 4x$; $x = -3$ **12.** $3x + 1 = 6x - 4$; $x = \dfrac{5}{3}$ **13.** $3x + 10 = -2$; $x = -4$

14. $2x + \dfrac{3}{2} = -1$; $x = -\dfrac{5}{4}$

Sections 2.3 and 2.4

15. \$26,250   **16.** $-\dfrac{1}{4}$ and $\dfrac{3}{4}$ **17.** $-37, -35$, and -33 **18.** $3\dfrac{1}{4}$ ft and $4\dfrac{3}{4}$ ft   **19.** \$4,000 at 12% and \$1,000 at 8%

20. The second angle is 62°. The third angle is 71°.   **21.** $a = \dfrac{bf}{b - f}$ **22.** $P = \dfrac{I}{rt}$ **23.** $x = -\dfrac{1}{4}y + \dfrac{7}{4}$

24. $x = \dfrac{3y + 1}{2 - 4y}$

Section 2.5

25.   **26.**   **27.**

28.   **29.** $(-\infty, 3)$

30. $(2, +\infty)$ **31.** $[-2, +\infty)$

32. $\left(0, \dfrac{2}{3}\right]$ **33.** $x + (-5) \leq -2$; $(-\infty, 3]$ **34.** $3 - x \leq 7$; $[-4, +\infty)$

35. $-1 < 3x + 8 \leq 4$; $\left(-3, -\dfrac{4}{3}\right]$

Section 2.6

36. $x = 48$ or $x = -48$ **37.** $t = 3$ or $t = -\dfrac{9}{5}$ **38.** no solution   **39.** $w = -8$ or $w = -\dfrac{4}{7}$ **40.** $(-5, 5)$

41. $[-4, -2]$ **42.** $\left(-\infty, -\dfrac{4}{3}\right) \cup \left(-\dfrac{1}{3}, +\infty\right)$ **43.** $\left(-\infty, -\dfrac{3}{5}\right] \cup \left[\dfrac{11}{5}, +\infty\right)$ **44.** $z = -21$

CHAPTER 2 TEST

Sections 2.1 and 2.2

1. $x = -4$ **2.** $y = 7$ **3.** $t = 4$ **4.** Identity. The solution set is the set of real numbers.   **5.** $4 + 2x = 10;\ x = 3$

Sections 2.3 and 2.4

6. $37,500   **7.** $-4, -2, 0, 2$ **8.** The second angle is $70°$. The third angle is $35°$.   **9.** $r = \dfrac{I}{Pt}$ **10.** $x = \dfrac{1}{3}y + 4$

11. $x = -\dfrac{2 - y}{5y}$

Section 2.5

12.   **13.**   **14.**

15. $(1, +\infty)$ **16.** $(-\infty, -1]$

17. $[-2, 2)$ **18.** $(-\infty, 1)$

19. $x - 3 > -4;\ (-1, +\infty)$ **20.** $1 < 3x + (-4) \le 3;\ \left(\dfrac{5}{3}, \dfrac{7}{3}\right]$

Section 2.6

21. $x = 6$ or $x = -6$ **22.** $y = 1$ or $y = -2$ **23.** $[-9, 9]$ **24.** $(-\infty, 0) \cup (1, +\infty)$

CHAPTER 3

EXERCISE SET 3.1

1. x^9 **3.** $y^{16}z^3$ **5.** a^7x^7 **7.** 1 **9.** x^6 **11.** b^{16} **13.** $25a^2$ **15.** x^{11} **17.** $c^{40}z^{27}$ **19.** $-x^5$ **21.** $-a^4$
23. $27w^9$ **25.** $a^6b^6c^6$ **27.** $\dfrac{4}{9}a^6z^4$ **29.** $\dfrac{1}{a^{20}}$ **31.** $\dfrac{1}{uv}$ **33.** $\dfrac{1}{p^6q^3}$ **35.** $\dfrac{1}{2}a^2$ **37.** $\dfrac{1}{s}$ **39.** 3 **41.** m^6
43. $9x^3$ **45.** $\dfrac{1}{az}$ **47.** $\dfrac{49}{4}$ **49.** 4 **51.** $\dfrac{b^5}{a^5}$ **53.** $\dfrac{z^3}{8x^3}$ **55.** x^6 **57.** 125 **59.** $\dfrac{y^6x^4}{z^8}$ **61.** $\dfrac{t^5}{w^7}$
63. s^2t^2 **65.** 1 **67.** 3.91×10^{-4} **69.** 9.1×10^7 **71.** 9.3592×10^5 **73.** 3.8×10^3 **75.** $4,900$
77. 0.0004 **79.** 9.19263177×10^9 **81.** 6.2×10^{23} **83.** 2.822×10^{-4} **85.** 5.31×10^{-19}
87. 8.4×10^7; $84,000,000$ **89.** 2.4×10^4; $24,000$ **91.** 2×10^{-3}; 0.002 **93.** 2.3×10^{-2}; 0.023
95. 0.0009 **97.** $2,500,000,000$ **99.** 0.0006 **101.** $1,000$ **103.** 5.8692×10^{12} miles   **105.** 0.45 meter

Review Problems

107. $2 - 5x - 3x^2$ **109.** $-4R^2 + 10R + 3$ **111.** $-5ax^2 - 5a^2x$ **113.** $-2X^3 + X^2 - 4X + 13$
115. $-30 + 23x - 13x^2$

ENRICHMENT EXERCISES

1. x^{4n} **2.** x^{2s} **3.** x^{n+m} **4.** $\dfrac{1}{x^{m-n}}$ **5.** $\dfrac{1}{x^{m-n}}$ **6.** $n = 1$ **7.** $n = -3$

EXERCISE SET 3.2

1. 3   **3.** 6   **5.** 0   **7.** no degree   **9.** binomial   **11.** trinomial   **13.** monomial   **15.** $-2x^3 - 2x^2 + 19$; 3
17. $2z^2 + 12z$; 2   **19.** $4c + 4$; 1   **21.** $2w^2 - 12w$; 2   **23.** $-x^3 - 12x^2 + 7x$; 3   **25.** $P(1) = 3$; $P(-1) = -3$
27. $f(4) = 5$ **29.** 1830   **31.** $3x^4 + x^2 - x + 4$ **33.** $6a^4 - 6a^2 + a + 7$ **35.** $v^5 + 2v^4 - v^3 + 15v^2 - 7v - 9$
37. $t^4 + 7t^3 + 11t^2 - 5t + 5$ **39.** $-3u^2 + 3u - 5$ **41.** $9m^4 - m^3 + 6m - 2$ **43.** $t^3 + 10t - 39$
45. $3x^5 + 6x^3 - 4x + 15$ **47.** $9y^3 + 13y^2 - 2y - 4$ **49.** $2c^6 - 5c^5 + 4c^3 + 3c - 8$ **51.** $9a^2 - 3a + 11$
53. $r^5 - r^4 - r^3 + 10r^2 - 7$ **55.** $-8s^3 - 14s^2 + 8s - 2$ **57.** $-a^4 - 4$ **59.** $3n^3 - 11n^2$ **61.** $x^5 - 1$
63. $12x^2 - 5x + 1$ **65.** $4a^2 + ab$ **67.** $4rst - 9rs^2t - r^2st$ **69.** $13x + 9$ **71.** $5x^2 + 3x - 5$
73. $170 - 2a - 3a^2$ degrees

Review Problems

75. $\dfrac{1}{x^3}$ **77.** $25r^8t^2$ **79.** $\dfrac{z^{12}}{x^8}$ **81.** x^2y^{12}

ENRICHMENT EXERCISES

1. $-2a^3 + 6a^2 - 7a + 10$ **2.** $-8z^3 + 11z^2 - 2z + 9$ **3.** $a = 3, b = 5, c = 0, d = -\dfrac{1}{2}$

4. $a = \dfrac{3}{4}, b = -\dfrac{13}{3}, c = -2, d = 6$ **5.** $-54a^2b - 72ab^2 + 9$ **6.** $-75a^{10}$ **7.** $\dfrac{3}{2}, 2, \dfrac{5}{2}, 3$

EXERCISE SET 3.3

1. $14x^3y^5$ **3.** $-9u^8v^7$ **5.** $2xy$ **7.** $\dfrac{2s^2}{r}$ **9.** $-18x^3y^3$ **11.** $-3u^4v^5w^6$ **13.** $128m^7n^8$ **15.** $\dfrac{y}{8x^3}$

17. $-\dfrac{5z}{a}$ **19.** $10uv^2w^3$ **21.** $2x^2y^3z$ **23.** $z^5 - 2z^4 + 18z^3$ **25.** $2b^4 - 3b^9$ **27.** $-\dfrac{2}{3}t^5 + 3t^4 - \dfrac{2}{7}t^3$
29. $x^2 + x - 6$ **31.** $t^4 - 12t^2 + 35$ **33.** $6v^2 + 7v - 3$ **35.** $2b^4 - 5b^2 - 12$ **37.** $2x^4 - 19x^2 + 35$
39. $8a^2 - 2ab - 15b^2$ **41.** $x^4 - 4y^4$ **43.** $-8z^2 + 2z + 3$ **45.** $3x^3 + 4x^2 - 4x - 5$
47. $6z^5 + 15z^4 - 2z^3 - 3z^2 + 5z$ **49.** $3x^4 - x^3 - 5x^2 + 13x - 18$ **51.** $y^2 + 8y + 16$ **53.** $4x^2 + 40x + 100$
55. $a^6 + 2a^3c + c^2$ **57.** $9u^2 + 24uv^2 + 16v^4$ **59.** $\dfrac{4}{25}n^4 + \dfrac{4}{3}n^2m + \dfrac{25}{9}m^2$ **61.** $x^2 - 10x + 25$
63. $16z^2 - 8za + a^2$ **65.** $\dfrac{9}{4}h^2 - 3ha^2 + a^4$ **67.** $x^4 - 6x^2u^2 + 9u^4$ **69.** $\dfrac{4}{25}s^2 - 3s^2t^5 + \dfrac{225}{16}t^{10}$ **71.** $c^2 - 4$
73. $4x^2 - \dfrac{16}{25}$ **75.** $z^4 - \dfrac{1}{4}t^2$ **77.** $f^4 - 9g^2$ **79.** $r^2 + 2rt + t^2 - 9$ **81.** $25 - s^2 + 2st - t^2$
83. $16z^2 - y^2 + yr - \dfrac{1}{4}r^2$ **85.** $16 - 4s^2 + 12sv - 9v^2$ **87.** $-8az$ **89.** $z^3 + 3z^2b + 3zb^2 + b^3$
91. $4a^2x^2 + 9b^2y^4$ **93.** $16x^4 - 8x^2a^2 + a^4$ **95.** $A = 3x^2 - \dfrac{1}{2}x - 1$ **97.** \$1,800   **99.** 48 and 50
101. -4 and -2

Review Problems

103. $a^3 + 2a^2 - 5a + 9$ **105.** $-c^3 + 4c^2 - 20c + 5$ **107.** $z^2 - 4z - 3$

E N R I C H M E N T   E X E R C I S E S

1. (a) $(a + b)^3 = (a + b)(a + b)^2 = (a + b)(a^2 + 2ab + b^2) = a^3 + 2a^2b + ab^2 + a^2b + 2ab^2 + b^3 = a^3 + 3a^2b + 3ab^2 + b^3$
(b) $(a - b)^3 = (a - b)(a - b)^2 = (a - b)(a^2 - 2ab + b^2) = a^3 - 2a^2b + ab^2 - a^2b + 2ab^2 - b^3 = a^3 - 3a^2b + 3ab^2 - b^3$

2. $c^3 + \dfrac{3}{2}c^2 + \dfrac{3}{4}c + \dfrac{1}{8}$ **3.** $x^6 - 3x^4 + 3x^2 - 1$ **4.** $8v^3 + 12v^2a^2 + 6va^4 + a^6$ **5.** $(a + b)^4 = (a + b)^2(a + b)^2 =$
$(a^2 + 2ab + b^2)(a^2 + 2ab + b^2) = a^4 + 2a^3b + a^2b^2 + 2a^3b + 4a^2b^2 + 2ab^3 + a^2b^2 + 2ab^3 + b^4 =$
$a^4 + 4a^3b + 6a^2b^2 + 4ab^3 + b^4$ **6.** 9 terms   **7.** $4x^{6n} - 1$ **8.** $3x^{2r} - 2x^r y^s - 5y^{2s}$

E X E R C I S E   S E T   3.4

1. composite   **3.** prime   **5.** composite   **7.** $2 \cdot 3 \cdot 7$ **9.** $2^2 \cdot 3^2$ **11.** $3 \cdot 5^2 \cdot 7$ **13.** prime   **15.** $2 \cdot 5 \cdot 7$
17. $3 \cdot 7 \cdot 11$ **19.** $2^2 \cdot 5 \cdot 7$ **21.** prime   **23.** $2^2 \cdot 3^2 \cdot 7$ **25.** $2 \cdot 7^2$ **27.** $2 \cdot 3^3$ **29.** x^2 **31.** z^4 **33.** c^7
35. a **37.** n^{11} **39.** $2y^3$ **41.** $4s$ **43.** $3q^2$ **45.** $3C$ **47.** $-2x^{20}$ **49.** $4z^4r^2$ **51.** $9q_1$ **53.** $4v^8k^4m$
55. $5(5x^2 - x + 4)$ **57.** $x^2(3x^2 - 2x + 1)$ **59.** $2(x^2 + 4x + 2)$ **61.** $5(t^2 - 5t + 2)$ **63.** $4(3b^2 + 7b + 18)$
65. $15(3y^4 + 5y^2 - 2)$ **67.** $z^3(5z^3 - 3z^2 + 2)$ **69.** $c^2(c^4 - c - 1)$ **71.** $5s^5(s - 9)$ **73.** $5(7 - 3F^2)$
75. $36c^3d^3(c^3 - 3cd^2 - 7d)$ **77.** $5b^2z^2t^2(z - 2z^2t + b)$ **79.** $3m_1^3n_1^3(9n_1 - 3m_1 + 5m_1^2n_1^2)$ **81.** $(z + t)(2s + 3)$
83. $(h - t)(2r + 3b)$ **85.** $5(a - x)^2y^2[4(a - x) - 3y]$ **87.** $S = \dfrac{1}{2}n(n + 1)$ **89.** $(z^2 + y^2)(2 + t)$
91. $(3m - 2n)(s + z)$ **93.** $(2x - y)(7 - z)$ **95.** $(2N^2 - M^2)(2z - 3d)$ **97.** $(x^3 + 2y)(yz - a)$
99. $(y + z)(a - b)$ **101.** $(t - v^2)(u^2 - 2s)$ **103.** $A = P\left(1 + \dfrac{r}{2}\right) + P\left(1 + \dfrac{r}{2}\right)r\left(\dfrac{1}{2}\right) = P\left(1 + \dfrac{r}{2}\right)\left(1 + \dfrac{r}{2}\right)$
$\left(\text{Factor out } P\left(1 + \dfrac{r}{2}\right)\right) = P\left(1 + \dfrac{r}{2}\right)^2$ **105.** $A = 4(4 - \pi)r^2$ **107.** $A = \left(2 - \dfrac{1}{2}\pi\right)r^2$

Review Problems

109. $a^2 + 2a - 8$ **111.** $6w^2 + 7w + 2$ **113.** $5z^4 - 2z^2 - 7$

E N R I C H M E N T   E X E R C I S E S

1. $18t^2(3t^3 + 2t^2 - 4t + 7)$ **2.** $x^{3n}(x^2 + 1)$ **3.** $12y^{5n+7}(1 - 7y^2)$ **4.** $A = P\left(1 + \dfrac{r}{n}\right)^n$

E X E R C I S E   S E T   3.5

1. $(x + 2)(x + 1)$ **3.** $(t + 7)(t + 3)$ **5.** prime   **7.** $(r + 7)(r + 5)$ **9.** $(x + 3)(x - 2)$ **11.** $(t + 8)(t - 1)$
13. prime   **15.** $(a - 4)(a - 3)$ **17.** $(n - 3)(n - 8)$ **19.** $(x_1 + 8)(x_1 - 1)$ **21.** $(x + 2a)(x + 3a)$
23. $(a + 4v)(a - 3v)$ **25.** $(p - 8q)(p - 3q)$ **27.** $(C - 5R)(C - 6R)$ **29.** $y(y - 2)(y - 3)$ **31.** $5a^2(a + 2)(a - 1)$
33. $4z^5(z + 2)(z + 1)$ **35.** $7D^4(D + 1)(D - 3)$ **37.** $6x_1^5(x_1 - 8)(x_1 - 2)$ **39.** $9u(u^3 + u^2 + 2)$
41. $(2x + 3)(x + 1)$ **43.** $(5a + 3)(a + 2)$ **45.** $(2c + 3)(c - 7)$ **47.** $(3k - 2)(k - 2)$ **49.** $(p + 2)(5p - 7)$
51. $(2T - 1)(T - 6)$ **53.** $(3a_1 + 5)(2a_1 - 1)$ **55.** $(5x - 1)(2x - 5)$ **57.** $(y_2 + 1)(10y_2 - 7)$
59. $(11m - 10)(m - 1)$ **61.** $(2y + a)(y + 2a)$ **63.** $(n_1 + 5m_1)(4n_1 - m_1)$ **65.** $(x + h)(4x - 9h)$
67. $(2s + 5t)(2s + 3t)$ **69.** $2a^3(2a + 7)(a - 5)$ **71.** $5w^5(2w + 3)(3w + 1)$ **73.** $C^{10}(6C - 1)(7C - 5)$
75. $2u(5u - 6)(3u - 2)$ **77.** $(3x + 2)$ and $(x + 4)$ **79.** $B = -(n + 5)(2n - 11)$; 27 bushels per acre

Review Problems

81. $x^2 - 4$ **83.** $z^2 - \dfrac{1}{9}$ **85.** $s^4 - a^2$ **87.** $x^3 - 8$ **89.** $x^3 - y^3$

E N R I C H M E N T   E X E R C I S E S

1. $\dfrac{1}{2}(x + 5)(x - 3)$ **2.** $\dfrac{1}{2}(2z - 1)(z + 4)$ **3.** $0.1(a - 5)(a - 2)$ **4.** $\dfrac{1}{6}(2w + 1)(3w - 1)$ **5.** $\dfrac{1}{9}(6u + 5)(6u - 1)$

6. $a = \dfrac{9}{2}$ **7.** $(2x^n - 5)(3x^n + 1)$ **8.** $2a^2xy(3x - 1)(x - 4)$ **9.** $c(a + b)(a - b)$ **10.** $2a(3y + b)(2y - b)$

E X E R C I S E   S E T   3.6

1. $(x + 5)(x - 5)$ **3.** $(t + r)(t - r)$ **5.** $(n + p^2)(n - p^2)$ **7.** $(s + 3t)(s - 3t)$ **9.** $(2y + 1)(2y - 1)$

11. $(K + 5T)(K - 5T)$ **13.** $\left(v + \dfrac{1}{3}\right)\left(v - \dfrac{1}{3}\right)$ **15.** $\left(3m_1 + \dfrac{1}{7}\right)\left(3m_1 - \dfrac{1}{7}\right)$ **17.** prime   **19.** $(c + 9d)(c - 9d)$

21. prime   **23.** $(4s + 3y)(4s - 3y)$ **25.** $(q + 1 + 5p)(q + 1 - 5p)$ **27.** $\left(\dfrac{1}{10}t + \dfrac{1}{11}v\right)\left(\dfrac{1}{10}t - \dfrac{1}{11}v\right)$

29. $\left(\dfrac{2}{3}q + v^2\right)\left(\dfrac{2}{3}q - v^2\right)$ **31.** $8x(2x - 3)$ **33.** $-3(5k + 3)(k + 1)$ **35.** $h(h - 4)$ **37.** $(16r + 13)(-4r + 3)$

39. $(m + 4)(m - 4)(m + 1)$ **41.** $(x^2 + 4)(x + 1)(x - 1)$ **43.** $(w + 1)(w - 1)(w - 2)$ **45.** $(2b - 3)(2 + b)(2 - b)$
47. $(y + 3)^2$ **49.** $(z + 6)^2$ **51.** prime   **53.** $(x - 5)^2$ **55.** prime   **57.** $(v - 7)^2$ **59.** $(x + a)^2$
61. $(z + 3v)^2$ **63.** $(s - c)^2$ **65.** $(4k - z)^2$ **67.** prime   **69.** $(2d - 5)^2$ **71.** $64(p - 1)^2$

73. $4(m_1 + 1)^2(m_1 - 1)^2$ **75.** $(z + 3)(z^2 - 3z + 9)$ **77.** $\left(w + \dfrac{1}{2}\right)\left(w^2 - \dfrac{1}{2}w + \dfrac{1}{4}\right)$ **79.** $(4t + 1)(16t^2 - 4t + 1)$

81. $8(b - 5)(b^2 + 5b + 25)$ **83.** $(3d + 2v)(9d^2 - 6dv + 4v^2)$ **85.** $(ab - cd)(a^2b^2 + abcd + c^2d^2)$
87. $7(y + 3)(y - 3)$ **89.** $-2(w + 5)(w - 5)$ **91.** $x^3(x + 9)(x - 9)$ **93.** $10(k - 4)^2$ **95.** $6(m + 3)^2$
97. $4y^5(y + 2)(y - 2)$ **99.** $z(z - y)(z^2 + zy + y^2)$ **101.** $(x^2 + y^2)(x + y)(x - y)$ **103.** $2s^3r^2(r - 3s)^2$
105. $3z^2(z + 2)^2$ **107.** $25X^6(X - 1)^2$ **109.** $a(a - b)^2$ **111.** $2u(v + u)^2$ **113.** $hx(x - h)^2$

115. $y(3x + 1)^2$ **117.** $(3x + 2y + 4)(3x + 2y - 4)$ **119.** $k = -\dfrac{13}{2}$

121. $(a + b)(a^2 - ab + b^2) = a^3 - a^2b + ab^2 + a^2b - ab^2 + b^3 = a^3 + b^3$

Review Problems

123. $8x^3 - 6x^2 + 18x$ **125.** $6x^2 - 17x + 5$ **127.** $4b^2 - 20b + 25$ **129.** $t^3 + 8$ **131.** $16x^2 - 4a^2$
133. $4y^5 - 108y^2$

E N R I C H M E N T   E X E R C I S E S

1. $(2t + 1)(2t - 1)\left(t + \dfrac{1}{2}\right)$ **2.** $(x^2 + 2)(x^2 - 2)(x + 1)(x - 1)$ **3.** $(x + a)^2(x - a)^2$ **4.** $(2b)(3a^2 + b^2)$
5. $(x^n + y^n)(x^n - y^n)$ **6.** $(t^{2n} - 3)(t^{2n} - 1)$

E X E R C I S E   S E T   3.7

1. $x^2(x - 1)$ **3.** $(2y + 3)(y - 2)$ **5.** $5(r^2 + 25)$ **7.** $(y + 2)(y - 2)$ **9.** $2s^4(s + 1)(s^2 - s + 1)$
11. $(v - 2)(v^2 + 2v + 4)$ **13.** prime   **15.** $(4t - 1)(t - 2)$ **17.** $4(x_3 + 2)(x_3 - 2)$ **19.** $(D + 1)^2$
21. $6wy(3w + 2y + 5)$ **23.** $(q_1 - 3)^2$ **25.** $(a + d)(t + 1)$ **27.** $(d + 7)^2$ **29.** $(u + v)(u - v)(a + k)$
31. $(xy + 2z)(x^2y^2 - 2xyz + 4z^2)$ **33.** $(x + y)(s + t)(s - t)$ **35.** $uv(v + 1)(v - 1)$ **37.** $(z + t)(x + y)(x - y)$

39. $25b(3b + 5)(2b - 1)$ **41.** $(z + 9)(z - 8)$ **43.** $(H - 1)(H^2 + H + 1)$ **45.** $(z - h)^2$ **47.** $(p + q)^2$
49. $2y^2(y + 1)(y - 1)$ **51.** $(9x + 10r)(9x - 10r)$ **53.** $(2 - r)(6 + 5s)$ **55.** $(b + 2h)^2$
57. $mn(n + m)(n - m)(c - d)(c^2 + cd + d^2)$ **59.** $(3d + 4y)^2$ **61.** $2b^2(2b - 3)(4b^2 + 6b + 9)$
63. $(t + 10)(t - 10)(r + 3)(r - 3)$ **65.** $(f^2 + g)(f^4 - f^2g + g^2)$ **67.** $-(T + 5)(T - 3)$ **69.** $t(3w + 1)(w - 10)$
71. $(t + 1)^2(t - 1)^2$ **73.** $pq^2(p - 2q)^2$

Review Problems

75. $a = \dfrac{2}{3}$ **77.** $x = \dfrac{3}{2}$ **79.** $v = -15$ **81.** $y = -\dfrac{4}{3}$ **83.** $x = -\dfrac{2}{3}$ **85.** $x = -14$

E N R I C H M E N T   E X E R C I S E S

1. $(a + 2b + a^2 - b^2)(a + 2b - a^2 + b^2)$ **2.** $(2x^2 + 2x + 1)(2x^2 - 2x + 1)$ **3.** $(8t^2 + 4t + 1)(8t^2 - 4t + 1)$
4. $\left(\dfrac{9}{2}w^2 + 3w + 1\right)\left(\dfrac{9}{2}w^2 - 3w + 1\right)$

E X E R C I S E   S E T   3.8

1. $x = -3$ or $x = 2$ **3.** $a = \dfrac{1}{2}$ or $a = 7$ **5.** $z = \dfrac{1}{2}$ or $z = -10$ **7.** $x = 4$ or $x = -4$

9. $a = 6$ or $a = -6$ **11.** $u = \dfrac{1}{2}$ or $u = -\dfrac{1}{2}$ **13.** $s = \dfrac{2}{5}$ or $s = -\dfrac{2}{5}$ **15.** $x = 3$ **17.** $b = -2$

19. $b = 0$ or $b = 2$ **21.** $u = 0$ or $u = \dfrac{1}{3}$ **23.** $y = -7$ or $y = 3$ **25.** $y = -\dfrac{1}{3}$ or $y = 5$

27. $a = -\dfrac{3}{5}$ or $a = 4$ **29.** $c = 0$ or $c = \dfrac{5}{2}$ **31.** $s = 0$ or $s = 3$ **33.** $c = 2$ or $c = \dfrac{5}{2}$

35. $x = -3$ or $x = 2$ **37.** $k = -5$ or $k = 6$ **39.** $s = -4$ or $s = \dfrac{1}{2}$ **41.** $\left\{2, \dfrac{3}{2}, -4\right\}$ **43.** $\left\{0, \dfrac{1}{2}, -\dfrac{1}{3}\right\}$

45. $m = 0, m = \dfrac{3}{4}$, or $m = -2$ **47.** $\{2, -2, 3, -3\}$ **49.** $\{4, -4, -1\}$ **51.** $\left\{2, -2, \dfrac{1}{5}\right\}$ **53.** $a = -\dfrac{5}{2}$ or $a = 1$

55. $s = 1$ or $s = -1$ **57.** $a = -\dfrac{2}{3}$ or $a = \dfrac{4}{3}$ **59.** $x = -7$ or $x = \dfrac{4}{5}$ **61.** $\left\{2, -\dfrac{3}{2}\right\}$

63. $v = -\dfrac{2}{3}$ or $v = 1$ **65.** $u = \dfrac{1}{3}$ or $u = 2$ **67.** $x = \dfrac{1}{3}$ **69.** $x = \dfrac{3}{2}, x = -\dfrac{3}{2}, x = -\dfrac{1}{2}$, or $x = 1$

71. $\dfrac{2}{3}$ and $\dfrac{3}{2}$ **73.** -2 and 16 or -16 and 2

Review Problems

75. Let $x =$ Tom's age now, then $x - 15 =$ Tom's age 15 years ago.   **77.** Let $x =$ the length of one piece, then $10 - x =$ the length of the other piece.   **79.** $n + (n + 1)$, where $n =$ an integer   **81.** $n^2 + (n + 2)^2$, where $n =$ an odd integer

E N R I C H M E N T   E X E R C I S E S

1. $\left\{3, -3, \dfrac{1}{2}, -\dfrac{2}{3}\right\}$ **2.** $x = -2$ or $x = 1$ **3.** $x = 3, x = -3$, or $x = 1$ **4.** $k = 14; \dfrac{5}{7}$
5. $(x - 1)^2 + (y - 2)^2 = 9$ **6.** $(x - 3)^2 + (y - 4)^2 = 25$ **7.** $(x - 2)^2 + (y + 1)^2 = 7$

E X E R C I S E   S E T   3 . 9

1. -8 and 6 **3.** 9 and 11 **5.** 10 and 12 **7.** 8 and 9 **9.** -4 and -6 **11.** -2 and -10
13. 9 and -4 or 4 and -9 **15.** 3 and 4 **17.** 3 inches, 4 inches, and 5 inches
19. The hypotenuse is 13 cm, one leg is 5 cm, and the other leg is 12 cm.   **21.** 3 feet by 11 feet   **23.** 3 feet by 4 feet
25. $b = 6$ cm and $h = 4$ cm   **27.** $b = 4$ inches and $h = \dfrac{2}{3}$ inch   **29.** $b = 9$ feet and $h = 6$ feet   **31.** $1\dfrac{1}{4}$ miles
33. $b_1 = 4$ feet, $b_2 = 2$ feet, and $h = 3$ feet   **35.** 10 dozen   **37.** $1\dfrac{1}{2}$ seconds and $6\dfrac{1}{2}$ seconds   **39.** 4 trees

Review Problems

41. $\dfrac{3}{2}$ **43.** $\dfrac{5}{12}$ **45.** $\dfrac{4}{15}$ **47.** 0

E N R I C H M E N T   E X E R C I S E S

1. 3, 5, and 7 **2.** Cut the wire 5 inches from one end.   **3.** $x = 1$ or $x = -1$ **4.** 30 meters per second
5. 40 meters per second   **6.** 5 feet

C H A P T E R   3   R E V I E W   E X E R C I S E   S E T

Section 3.1

1. x^8 **2.** $-4r^8s^{18}$ **3.** a^2 **4.** $\dfrac{3c^2}{d}$ **5.** $\dfrac{z^9}{w^6}$ **6.** $\dfrac{1}{8u^2v^{10}}$ **7.** 5.9×10^6 **8.** 4.92×10^{-4} **9.** 2.14×10^{-5}
10. 2.316×10^5 **11.** 1.28×10^2; 128 **12.** 1.86; 1.86

Section 3.2

13. $5x^4 + 2x^3 - 3$; 4 **14.** $-3y^2 + 5y - 3$; 2 **15.** $4a^3 - 16a + 34$; 3 **16.** 820 **17.** $-2z^3 + 9z^2 - 9z + 8$
18. $210 - 9A$ degrees

Section 3.3

19. $-12x^7y^5$ **20.** $-9r^{12}t^{16}$ **21.** $\dfrac{yz^4}{4x}$ **22.** $9r^5 - 4r^4 + 12r^3$ **23.** $-\dfrac{4}{3}x^2 + \dfrac{3}{2}x^3 - \dfrac{2}{3}x^4$ **24.** $z^2 - 9z + 18$
25. $6c^2 + 13cd - 5d^2$ **26.** $s^2 - 16v^2$ **27.** $x^4 - 8x^2 + 16$ **28.** $4z^2 + 4z + 1$ **29.** $b^3 - 30b^2 + 300b - 1{,}000$
30. $9 - w^2 + 2wr - r^2$

Section 3.4

31. $5(9x^2 - 3x + 4)$ **32.** $2a^2(1 - 2a)$ **33.** $3b^2z(9b^3z - 5)$ **34.** $7(x + a)^2[4(x + a)^3 - 2(x + a) + 11]$
35. $(3x^2 - 4)(x + 2)$ **36.** $(2c - 5)(c^2 - 3)$ **37.** $(3z^2 + 4y)(z - 2y)$ **38.** $(v^2 + w^2)(2t + 3s^2)$

Section 3.5

39. $(2x - 3)(x + 5)$ **40.** $3(a - 4)(a - 1)$ **41.** prime   **42.** $(5u - 1)(3u + 2)$ **43.** $b(6b + 5)(b - 1)$
44. $u(2u + v)^2$ **45.** $2x^3(3x + 4)(x + 4)$ **46.** $3p(2p - 1)(p - 1)$

Section 3.6

47. $(D + 3)(D - 3)$ **48.** $(6w + 7)(6w - 7)$ **49.** $(m - 8)^2$ **50.** $(2s - 1)(4s^2 + 2s + 1)$ **51.** $4(3a + 2b)(3a - 2b)$
52. $(2h + 3k)^2$ **53.** $b(3b + y^2)(9b^2 - 3by^2 + y^4)$ **54.** $(p + 9)(p - 9)(2p - 3)$

Section 3.7

55. $15(2b + z)(2b - z)$ **56.** $(w - 11)^2$ **57.** prime   **58.** $3(t^2 + 6t + 16)$ **59.** $(7s^2 - 3r)^2$
60. $(2b - 1)(2a + 1)$ **61.** $(T + 4a)(T - 1)$ **62.** $3(k^2 + 4)(k + 2)(k - 2)$ **63.** $2n(4z + 3c)(16z^2 - 12zc + 9c^2)$
64. $(x - y)(x^2 + xy + y^2)(a - 2b)$

Section 3.8

65. $z = 0$, $z = \dfrac{2}{3}$, or $z = -\dfrac{8}{5}$ **66.** $x = 0$, $x = -3$, or $x = 3$ **67.** $c = -6$ **68.** $b = -7$ or $b = 3$

69. $x = -8$ or $x = 3$ **70.** $n = \dfrac{1}{2}$, $n = 2$, or $n = -2$ **71.** $x = 3$, $x = -3$, $x = 4$, or $x = -4$

72. $y = 2$ or $y = -1$

Section 3.9

73. 3 or 5   **74.** -8 and -12 **75.** width = 4 inches, length = 5 inches   **76.** -8 and -7 or 5 and 6   **77.** 9 seconds
78. 40 dozen frames

C H A P T E R   3   T E S T

Section 3.1

1. x^{11} **2.** $\dfrac{a^7}{8b^7}$ **3.** 1.06×10^{-1}; 0.106

Section 3.2

4. $x^4 + x^3 + 9x^2 + 2$; 4   **5.** $x^4 + 5x^3 + 3x^2 - 3x - 4$

Section 3.3

6. $12x^5 - 6x^3 + 24x^2$ **7.** $y^2 - 2y - 3$ **8.** $6a^2 - 11a + 3$ **9.** $16z^2 - 24z + 9$ **10.** $(2x^2 - 1)(2 + x)$
11. $(2x^2 + y)(3r + 2s)$

Section 3.5

12. $(3y + 1)(y - 2)$ **13.** $3(2x + 1)(x + 1)$ **14.** prime   **15.** $x^2y(x + y)(2x + y)$

Section 3.6

16. $(R + 7)(R - 7)$ **17.** $(p + 5)^2$ **18.** $2(2x + 3y)(2x - 3y)$ **19.** $(2a - t)(4a^2 + 2at + t^2)$

Section 3.7

20. $2r^2(r + 3)(r - 3)$ **21.** $(2z - 1)(3z - 4)$ **22.** prime   **23.** $(2t - b)(x + 3)$

Section 3.8

24. $y = -\dfrac{1}{2}$ **25.** $r = 3$ or $r = 5$ **26.** $x = \dfrac{1}{3}$ or $x = -\dfrac{3}{2}$

Section 3.9

27. 2 or 3   **28.** 6 seconds

CHAPTER 4

EXERCISE SET 4.1

1. $x = \dfrac{5}{3}$ **3.** $m = 2$ **5.** $p = 2$ **7.** The rational expression is defined for all real numbers.   **9.** $h = 0$

11. $k = -1$ or $k = \dfrac{3}{2}$ **13.** $t = 2$ **15.** $x = 1$, $x = -3$, or $x = 3$ **17.** $\dfrac{2a}{b}$ **19.** $-\dfrac{7}{3a + 5}$ **21.** $\dfrac{y - 1}{y - 3}$

23. $\dfrac{5}{4}$ **25.** $\dfrac{w - 1}{w + 2}$ **27.** 2   **29.** $\dfrac{5v + 2}{3v + 2}$ **31.** $\dfrac{1}{a - 2}$ **33.** $\dfrac{2(x + 5)}{2x - 7}$ **35.** $\dfrac{3(v - 1)}{5(v + 4)}$ **37.** $\dfrac{3}{d - 2}$

39. $\dfrac{2x - 1}{2x^3}$ **41.** $\dfrac{3x - 5}{(x + 1)(3x + 4)}$ **43.** $\dfrac{(a - 3)(a^2 + 2a + 4)}{2a}$ **45.** $-(x - 2)(x + 1)$ **47.** $-(2y - 3)(3y - 1)$

49. $-2h(2h + 1)(h + 2)$ **51.** $-5c^3(4c - 1)(c + 5)$ **53.** $-\dfrac{1}{5x + 1}$ **55.** $-\dfrac{1}{3a - 1}$ **57.** $-\dfrac{2s - 3}{3s - 1}$

59. $-\dfrac{2w + 1}{w + 5}$ **61.** -1 **63.** $\dfrac{k - 7}{7 + k}$ **65.** $-\dfrac{7}{6}$ **67.** $-\dfrac{1}{2}$ **69.** $\dfrac{3v - 4}{3v + 4}$ **71.** $-\dfrac{2n + 3}{2(4n - 1)}$ **73.** $\dfrac{3(k + 1)}{2(k + 2)}$

75. $\dfrac{a(a^2 + 9)}{a + 3}$ **77.** $-\dfrac{x^2 - 3}{x + 3}$ **79.** $-\dfrac{1}{2}$ **81.** $(a - b)(x + y)$ **83.** $\dfrac{5}{24}$ ohm   **85.** $\dfrac{x^n - 2}{x^n - 1}$ **87.** -1

89. $\dfrac{y^n - z^n}{y^n + z^n}$

Review Problems

91. $6x^3$ **93.** $\dfrac{2z}{x^2}$ **95.** $2x^3 - 8x^2 + 3x$ **97.** $-6x^4 + 7x^3 + x - 8$ **99.** $9t^3 + 8t + 7$

ENRICHMENT EXERCISES

1. $\dfrac{1}{x - 1}$ **2.** $0.6(z - 4)$ **3.** $\dfrac{1}{x^3}$ **4.** $\dfrac{2y^2}{3}$ **5.** $\dfrac{2a - 1}{a^2}$

EXERCISE SET 4.2

1. $7z - 2 - \dfrac{1}{z}$ **3.** $4u^2 - 3u + 1$ **5.** $-4 + \dfrac{2}{m} - \dfrac{1}{m^2}$ **7.** $2h^3 - \dfrac{5}{3h}$ **9.** $2z^5 - z + \dfrac{2}{z^2}$ **11.** $\dfrac{t}{5} - \dfrac{4}{5st} + \dfrac{3}{st^2}$

13. $\dfrac{2}{3r^2s} - \dfrac{8}{3rt^2} + \dfrac{s^3}{2}$ **15.** $2x + 3y$ **17.** $a + b$ **19.** $c + 2d$ **21.** $2a + 1$ **23.** $x - 2y$ **25.** $x - 2$

27. $6x^2 - x - 1$ **29.** $z + 5$ **31.** $x - 6 + \dfrac{-27}{x - 2}$ **33.** $y - 4$ **35.** $z^2 + 6z + 4 + \dfrac{20}{3z - 2}$

37. $3t^2 + 6t + 1 + \dfrac{-5}{t - 2}$ **39.** $\dfrac{3}{2}x^3 - 3x^2 + 6x + 12$; -48 **41.** $t^2 - 1$; 0 **43.** $2x^4 - x^3 - 2x^2 + x - \dfrac{1}{2}$; $-5\dfrac{1}{2}$

45. $3w^2 + w - 1$; 4 **47.** $x^2 - x + 1$; $2x - 1$ **49.** $x^4 - 4x^2 + 3$; 0 **51.** $x^3 + x$; 0 **53.** $a^4 - 2a^2 + 1$; 0

55. $(x - 1)(2x - 3)(x - 3)$ **57.** $(2x^2 - 1)(2x + 5)(3x - 1)$ **59.** $2x + 7$; 3 **61.** $x^3 - 3x^2 - x + 3$; -16

63. $P(1) = -3$ **65.** $P(-5) = -1$ **67.** $P\left(\dfrac{1}{2}\right) = -\dfrac{13}{4}$ **69.** $P\left(-\dfrac{1}{3}\right) = \dfrac{65}{27}$

Review Problems

71. $\dfrac{3}{2}$ **73.** $\dfrac{1}{8}$ **75.** 8   **77.** $-\dfrac{75}{2}$ **79.** $\dfrac{5}{3}$

E N R I C H M E N T   E X E R C I S E S

1. $a^{2n} - a^n - \dfrac{1}{a^n}$ **2.** $2x^{2n} + 5x^n + 3$ **3.** $2x^{2n} - x^n y^n - 3y^{2n}$ **4.** $c = -2$ **5.** $b = 1$ **6.** Suppose that when

$P(x)$ is divided by $x - c$, the remainder is zero. We know that there is a polynomial $Q(x)$ such that $\dfrac{P(x)}{x - c} = Q(x) + \dfrac{0}{x - c}$.

That is, $\dfrac{P(x)}{x - c} = Q(x)$. Therefore, $P(x) = (x - c)Q(x)$ and so $x - c$ is a factor of $P(x)$. Conversely, suppose $x - c$ is a factor

of $P(x)$. Then, there is a polynomial $Q(x)$ so that $P(x) = (x - c)Q(x)$. Therefore, $\dfrac{P(x)}{x - c} = Q(x)$. That is, $\dfrac{P(x)}{x - c} =$

$Q(x) + \dfrac{0}{x - c}$. Therefore, when $P(x)$ is divided by $x - c$, the remainder is zero.   **7. (a)** $x - 1$ is a factor.
(b) $x + 1$ is not a factor.   **(c)** $x - 2$ is a factor.

E X E R C I S E   S E T   4 . 3

1. $\dfrac{3}{40}$ **3.** $\dfrac{3x^3}{y^4}$ **5.** $\dfrac{a^2 b^2}{5}$ **7.** $\dfrac{9}{v^3}$ **9.** $-\dfrac{15}{4n}$ **11.** $\dfrac{z + 1}{5(z - 2)}$ **13.** 1   **15.** $\dfrac{2}{x}$ **17.** $\dfrac{s - 3}{s - 1}$ **19.** $2y + 3$

21. $\dfrac{t^2}{t + 1}$ **23.** $\dfrac{1}{3}$ **25.** $-\dfrac{4}{3k^7}$ **27.** 1   **29.** $\dfrac{s + 3}{s - 1}$ **31.** $\dfrac{3}{h^2}$ **33.** $2(v - 4)(v + 1)$ **35.** $s + 1$

37. $\dfrac{1}{2x + a}$ **39.** $\dfrac{1}{n - m}$ **41.** $\dfrac{x^2 + 2}{(2x + 3)(x + 3)}$ **43.** $\dfrac{z(z^2 - 4)}{4z - 1}$ **45.** $\dfrac{(x - 1)(3z + 1)}{(3x + 2)(2z + 1)}$ **47.** $\dfrac{z}{c}$ **49.** $\dfrac{3}{2}$

51. $\dfrac{16}{21}$ **53.** $\dfrac{3cd^2}{2}$ **55.** $\dfrac{(x - 2)(x - 3)}{(x + 2)(x + 1)}$ **57.** $\dfrac{(3u - v)(u - 2v)}{2(u^2 - uv + v^2)}$ **59.** $\dfrac{2}{2x + 1}$ **61.** 1   **63.** $-\dfrac{3}{2}$

65. $-\dfrac{2(2x + 3)}{x(x - 1)}$ **67.** $\dfrac{2y + 3}{y - 1}$ **69.** $\dfrac{x + 3}{x}$ **71.** $\dfrac{4a(2a + 1)}{a + 2}$ **73.** $2t$ **75.** $\dfrac{k - 4}{x + h}$ **77.** $\dfrac{(x - 2a)(x + a)}{2a}$

Review Problems

Section 3.4
79. $2^3 \cdot 3$ **81.** $2^4 \cdot 3^2$ **83.** prime

Section 3.7
85. $(x + 5)(x - 3)$ **87.** $x(x + 10)(x - 10)$ **89.** $(n + m)^2$

E N R I C H M E N T   E X E R C I S E S

1. $x^2 + y^2$ **2.** $\dfrac{b(a - b)}{a(a + b)}$ **3.** $\dfrac{R - T}{R^2 + T^2}$ **4.** $-\dfrac{x - y + z}{4a^2}$ **5.** $(x^3 + y^3)(x^2 + xy + y^2)(a^4 + b^4)(a - b)$

6.

Statement	Reason
$C\left(\dfrac{A}{B}\right) = \left(\dfrac{C}{1}\right)\left(\dfrac{A}{B}\right)$	$C = \dfrac{C}{1}$
$= \dfrac{C \cdot A}{1 \cdot B}$	Rule for multiplying two rational expressions
$= \dfrac{CA}{B}$	Multiplicative identity

E X E R C I S E   S E T   4.4

1. 6   **3.** 1   **5.** $\dfrac{17}{z}$ **7.** $-\dfrac{20}{z^2}$ **9.** $\dfrac{10}{b+1}$ **11.** $-\dfrac{q^2+6}{q^2}$ **13.** $\dfrac{x^2+7}{x^2+10}$ **15.** $\dfrac{2}{y-3}$ **17.** $\dfrac{1}{t+1}$

19. $\dfrac{1}{4p-1}$ **21.** $\dfrac{13}{10}$ **23.** $\dfrac{13}{36}$ **25.** $\dfrac{6}{5y}$ **27.** $\dfrac{(r-1)^2}{(r+2)(r-2)}$ **29.** $\dfrac{5m+1}{(m-2)(m-1)(m+1)}$

31. $-\dfrac{z(z-3)}{(2z-3)(z-1)(z+1)}$ **33.** $\dfrac{12x^2+52x-3}{(3x-1)(x+4)(x+5)}$ **35.** $\dfrac{2r^2+4r+3}{r(r+1)(r+2)}$ **37.** $\dfrac{2}{t-1}$ **39.** $-\dfrac{1}{y+5}$

41. $\dfrac{z^2+8z+17}{z+3}$ **43.** $\dfrac{3y}{10}$ **45.** $\dfrac{5c+14}{10c^2}$ **47.** $-\dfrac{5(1+3z)}{3z^2}$ **49.** $\dfrac{4y+3}{3y^2}$ **51.** $-\dfrac{2}{2x-1}$ **53.** $\dfrac{1}{x+2}$

55. $-\dfrac{1}{3y+1}$ **57.** $\dfrac{1}{x^2+x+1}$ **59.** $\dfrac{z+1}{z-3}$ **61.** $\dfrac{1}{x-y}$ **63.** $\dfrac{(u+2v)^2}{u(u-2v)^2}$ **65.** $-\dfrac{a^2+b^2}{ab(a+b)}$

67. $x+3\left(\dfrac{1}{x}\right)$; $\dfrac{x^2+3}{x}$ **69.** $\dfrac{1}{x+4}-2x$; $-\dfrac{2x^2+8x-1}{x+4}$ **71.** $\dfrac{1}{n}+\dfrac{1}{n+2}$; $\dfrac{2(n+1)}{n(n+2)}$ **73.** $R = 1\dfrac{7}{8}$ ohms

75. (a) $\dfrac{1}{R} = \dfrac{1}{R_1} + \dfrac{1}{R_2}$; $RR_1R_2\left(\dfrac{1}{R}\right) = RR_1R_2\left(\dfrac{1}{R_1} + \dfrac{1}{R_2}\right)$; $R_1R_2 = R(R_2 + R_1)$; $\dfrac{R_1R_2}{R_2 + R_1} = R$. Therefore, $R = \dfrac{R_1R_2}{R_1 + R_2}$.

(b) $R = 4\dfrac{4}{9}$ ohms

Review Problems

77. 14   **79.** $2x^3 - 3x^4$ **81.** $6x^2 - x$ **83.** $15a^3b^2 - 4a^2b^3$ **85.** $33st^3 + 4s^3t^3$

E N R I C H M E N T   E X E R C I S E S

1. $\dfrac{x+y}{xy(x^2+y^2)(x-y)}$ **2.** $\dfrac{x^n+3}{x^n}$ **3.** $\dfrac{3y^{2n}-4x^my^n+6x^{2m}}{30x^{2m}y^{2n}}$ **4.** $\dfrac{a^n+2}{a^n}$ **5.** 0   **6.** $\dfrac{2}{3(2x+1)}$

E X E R C I S E   S E T   4.5

1. $\dfrac{2}{3}$ **3.** $-\dfrac{4}{5}$ **5.** $-\dfrac{55}{28}$ **7.** $\dfrac{b}{2a}$ **9.** $\dfrac{5}{4s}$ **11.** $-\dfrac{x}{3}$ **13.** $\dfrac{7z^2}{4}$ **15.** $\dfrac{11}{16}$ **17.** $-\dfrac{13}{11}$ **19.** $\dfrac{10+3a}{2+a}$

21. $\dfrac{8c-42}{1+3c}$ **23.** $\dfrac{2+12w^2}{3w-2}$ **25.** $\dfrac{2x+3}{4x-x^2}$ **27.** $\dfrac{5s^3-3}{s+s^2}$ **29.** $\dfrac{12q^2+2}{3q-2}$ **31.** $\dfrac{p}{3}$ **33.** c **35.** $\dfrac{2}{a-3}$

37. $14-x$ **39.** $\dfrac{8-3n}{n-6}$ **41.** $\dfrac{2x-1}{1-3x}$ **43.** $\dfrac{3(17x^2+x-2)}{15x^2-2}$ **45.** $\dfrac{x^2+1}{2(2x+1)(x+1)}$ **47.** $\dfrac{3a+1}{a+4}$

49. $\dfrac{x^2+4}{2}$ **51.** $\dfrac{3z-1}{z+2}$ **53.** $\dfrac{x+1}{3x-1}$ **55.** $\dfrac{5-2n}{10n-1}$ **57.** $\dfrac{x-4}{x+1}$

Section 2.2

59. $x = -7$ **61.** $z = 3$ **63.** $x = -\dfrac{1}{7}$

Section 3.7

65. $x = 2$ or $x = -\dfrac{1}{2}$ **67.** $y = 0$, $y = \dfrac{1}{2}$, or $y = \dfrac{1}{3}$ **69.** $z = \dfrac{5}{6}$ or $z = \dfrac{1}{2}$

E N R I C H M E N T   E X E R C I S E S

1. -1 **2.** $\dfrac{1}{5}$ **3.** $\dfrac{1}{x(1-x)}$ **4.** $-\dfrac{1}{x(x+h)}$ **5.** $-\dfrac{2x+h}{x^2(x+h)^2}$ **6.** $-\dfrac{1}{(x-3)(a-3)}$

7. $-\dfrac{1}{(x+1)(x+\Delta x+1)}$

E X E R C I S E   S E T   4.6

1. $y = 2$ **3.** $c = -4$ **5.** $x = \dfrac{5}{6}$ **7.** $x = 4$ **9.** $t = \dfrac{5}{2}$ **11.** $s = -\dfrac{4}{5}$ **13.** $q = -1$ **15.** $x = -\dfrac{2}{5}$

17. $w = -2$ **19.** $r = \dfrac{1}{2}$ **21.** $v = 17$ **23.** $z = \dfrac{63}{23}$ **25.** $x = -46$ **27.** no solution   **29.** $n = -4$

31. $x = -10$ **33.** $z = 1$ or $z = -2$ **35.** $c = 2$ or $c = -3$ **37.** $u = 0$ or $u = -2$ **39.** $x = \dfrac{1}{2}$ or

$x = -2$ **41.** $z = -\dfrac{4}{3}$ **43.** $x = \dfrac{5}{3}$ **45.** $x = \dfrac{1}{3}$ **47.** $y = 2$ or $y = -\dfrac{3}{2}$ **49.** $x = -\dfrac{1}{2}$ or $x = -\dfrac{3}{2}$

51. $\dfrac{4}{27}$ and $\dfrac{4}{9}$ **53.** 2 and 6   **55.** $\dfrac{4}{23}$ **57.** (a) $q = \dfrac{fp}{p-f}$ (b) $q = 3$ cm   **59.** $q = \dfrac{Fr^2}{kp}$ **61.** $b = \dfrac{ak}{k-aW}$

63. $R = \dfrac{R_1 R_2}{R_1 + R_2}$ **65.** $t = \dfrac{z}{1-z}$ **67.** $V = \dfrac{62.4nT_k}{P}$ **69.** $y' = \dfrac{y - 2xy^2}{x}$

71. $x + \dfrac{3}{4}$ **73.** $\dfrac{3}{5}\left(x + \dfrac{10}{3}\right)$ **75.** $\dfrac{1}{x} - 3$ **77.** $\dfrac{1}{7-2x}$

E N R I C H M E N T   E X E R C I S E S

1. $n = -\dfrac{4}{3}$ or $n = \dfrac{5}{2}$ **2.** no solution   **3.** $v = 3$ or $v = -3$ **4.** $R_1 = 20$ ohms; $R_2 = 10$ ohms; $R_3 = 60$ ohms

5. $\dfrac{24}{5}$ **6.** 15 **7.** $-\dfrac{1}{4}$ and $\dfrac{7}{4}$

E X E R C I S E   S E T   4.7

1. 6 and 9   **3.** $-\dfrac{3}{4}$ and $\dfrac{1}{2}$ **5.** 5   **7.** 5 and -2 **9.** $\dfrac{1}{9}$; $\dfrac{4}{9}$; $\dfrac{1}{3}$ **11.** $1\dfrac{1}{3}$ hrs   **13.** 1.6 hrs   **15.** It would take

Allison 24 days. It would take John 8 days.   **17.** $5\dfrac{1}{7}$ hrs   **19.** $10\dfrac{1}{2}$ hrs   **21.** 4 hrs   **23.** 2 hrs   **25.** 35 mph

27. 10 mph   **29.** 4 km per hr

Review Problems

31. 16   **33.** 125   **35.** 1.69   **37.** 32   **39.** 81   **41.** 0.027   **43.** -32

ENRICHMENT EXERCISES

1. $2\frac{4}{7}$ hours   **2.** $3\frac{1}{3}$ hours   **3.** $3\frac{3}{7}$ hours   **4.** 75 mph

CHAPTER 4 REVIEW EXERCISE SET

Section 4.1

1. $\frac{9}{4}$ **2.** $y = 0, y = \pm 1$ **3.** The expression is defined for all real numbers.   **4.** $x = 2$ **5.** $\frac{4y}{x}$ **6.** $-3(3 - 4z)^2$

7. $\frac{a - 2}{a^2}$ **8.** -1 **9.** $-\frac{2(3x - 2)}{2x + 1}$ **10.** $\frac{z - a}{z^2 + az + a^2}$

Section 4.2

11. $2z - 3 - \frac{1}{z}$ **12.** $\frac{1}{3x} - 4y + 6x$ **13.** $2x + 1$ **14.** $x - 2a$ **15.** $3x - 5 + \frac{3}{2x + 3}$ **16.** $6y^3 - 2y^2b$

17. $5x + 9$; the remainder is 38.   **18.** $3x^3 - 3x^2 - 5x + 7$; the remainder is 8.   **19.** $(3x^2 + 4)(2x - 5)(x + 1)$

20. $(x^2 - x + 1)(x + 3)(x - 3)$

Section 4.3

21. $(x - 3)(2x - 5)$ **22.** -6 **23.** $\frac{p^2(p + 1)}{3p - 1}$ **24.** $\frac{4}{(x - 1)(3x - 1)}$ **25.** $\frac{z - 3}{z(z - 2)}$ **26.** $\frac{(x - 2y)(x + y)}{(x - y)^2}$

27. $-\frac{1}{14}$ **28.** $\frac{s^2}{3t^2}$ **29.** $\frac{x^2(x - 6)(x - 1)}{2(2x - 1)}$ **30.** $\frac{3z(2z - 3)}{z - 1}$

Section 4.4

31. $\frac{8}{x^2}$ **32.** $\frac{4}{z^2 + 1}$ **33.** $\frac{1}{a + 2}$ **34.** $-\frac{1}{t - 2}$ **35.** $-\frac{11}{24}$ **36.** $-\frac{2(b - 1)}{(2b + 1)(2b - 1)}$ **37.** $\frac{x + 1}{x - 2}$

38. $\frac{4}{(z - 1)(z + 1)}$ **39.** $\frac{2r + 3}{(r + 1)(r - 1)}$ **40.** $\frac{y - 19}{(y + 2)(y - 1)(3y - 1)}$

Section 4.5

41. $\frac{6}{7}$ **42.** $\frac{11xy^2}{6}$ **43.** $\frac{4T + 9}{5T - 6}$ **44.** $\frac{7y^2 - 2}{8y^2 + 3y}$ **45.** $\frac{a + 1}{a + 2}$ **46.** $\frac{x - 1}{x - 2}$ **47.** $\frac{2x + 3}{x - 1}$ **48.** $\frac{1}{2k + 1}$

Section 4.6

49. $x = \frac{7}{16}$ **50.** $y = \frac{3}{5}$ **51.** $r = -1$ **52.** $z = \frac{4}{3}$ **53.** $x = -3$ **54.** $x = \frac{3}{4}$ **55.** $h = \frac{A - 2\pi r^2}{2\pi r}$

56. $w = \frac{P - 2\ell}{2}$ **57.** $t = \frac{CT}{T - C}$ **58.** $d^2 = \frac{Gm_1m_2}{F}$ **59.** $x_1 = \frac{m}{y + 7} - h$ **60.** $y' = \frac{y}{2x}$

Section 4.7

61. $\frac{1}{8}$ and $\frac{1}{4}$ **62.** $-\frac{1}{3}$ **63.** (a) $\frac{1}{7}$ (b) $\frac{4}{7}$ (c) $\frac{3}{7}$ (d) $\frac{2}{7}t$ **64.** (a) $\frac{1}{4}$ (b) $\frac{2}{3}$ (c) $\frac{4}{9}$ (d) $\frac{1}{3}t$

65. 2 hours and 24 minutes   **66.** One pipe takes 10 hours; the other pipe takes 15 hours.   **67.** $3\frac{1}{3}$ hrs

68. 1 week and 5 days   **69.** Sam's speed is 64 mph; John's speed is 56 mph.   **70.** $6\frac{3}{5}$ mph

C H A P T E R   4   T E S T

1. $\dfrac{9x}{2a}$ **2.** -4 **3.** $\dfrac{x-2}{2(x-3)}$

4. $2x - 3 + \dfrac{-5}{3x-2}$; (quotient)(divisor) + remainder $= (2x-3)(3x-2) + (-5) = 6x^2 - 13x + 6 - 5 = 6x^2 - 13x + 1 =$ dividend   **5.** $2x^2 + x + 1$; 3

Section 4.3

6. $\dfrac{x-3}{3x+1}$ **7.** $t + 1$ **8.** $\dfrac{14}{xy^2z}$

Section 4.4

9. $\dfrac{5}{c}$ **10.** $\dfrac{1}{x-3}$ **11.** $-\dfrac{4}{y(y+2)(y-2)}$

Section 4.5

12. $-3xy$ **13.** $\dfrac{2}{t-2}$ **14.** $\dfrac{b^2+2b+4}{b+2}$

Section 4.6

15. $h = 12$ **16.** $x = -1$ or $x = 6$ **17.** $y = 2$ or $y = \dfrac{1}{3}$ **18.** $a = 5$ **19.** $x = 5$ or $x = 6$ **20.** $y = \dfrac{x}{1-3x}$

21. $y = \dfrac{2(x+1)}{1-2x}$

Section 4.7

22. $\dfrac{1}{4}$ and $\dfrac{3}{4}$ **23.** 20 miles per hour   **24.** $5\frac{1}{4}$ hours

C H A P T E R   5

E X E R C I S E   S E T   5.1

1. ± 3 **3.** ± 11 **5.** $\pm\dfrac{5}{3}$ **7.** ± 0.9 **9.** 9   **11.** $-\dfrac{2}{3}$ **13.** 0.18   **15.** 3   **17.** not a real number   **19.** 3

21. 2   **23.** -1 **25.** not a real number   **27.** 2   **29.** -2 **31.** $-\dfrac{1}{3}$ **33.** not a real number   **35.** -2

37. 9   **39.** 48   **41.** 1   **43.** -8 **45.** not a real number   **47.** $4s^2t$ **49.** $3x^2y^3$ **51.** $2a^2b$ **53.** $-2a^3cz^5$

55. $\dfrac{5}{8}$ meter   **57.** $\sqrt{3}$ meters   **59.** $\dfrac{3}{2}$ yards   **61.** $x \geq 5$ **63.** $r \geq -\dfrac{4}{5}$ **65.** $v \leq \dfrac{1}{3}$

67. $\dfrac{3\pi}{4}$ seconds, or approximately 2.355 seconds   **69.** 80 ft/sec

Review Problems

71. $\dfrac{1}{9}$ **73.** a^2 **75.** s^5t^5 **77.** $x^5 - x^6$ **79.** $\dfrac{s^8}{t^{16}v^8}$ **81.** $\dfrac{a^2}{b^4}$ **83.** x^6

ENRICHMENT EXERCISES

1. x^n **2.** a^n **3.** z^n **4.** x^n **5.** -1 **6.** 2 **7.** Let $b = \sqrt[n]{\sqrt[m]{a}}$. Then, $b^n = \sqrt[m]{a}$. So, $(b^n)^m = a$; that is, $b^{nm} = a$. Therefore, $b = \sqrt[nm]{a}$; that is, $\sqrt[n]{\sqrt[m]{a}} = \sqrt[nm]{a}$.   **8.** $\sqrt[12]{7}$ **9.** $\sqrt[20]{12}$.   **10.** $\sqrt[9]{-2}$ **11.** $\sqrt[12]{2}$.

EXERCISE SET 5.2

1. 10 **3.** -1 **5.** 36 **7.** 64 **9.** 4 **11.** $\dfrac{125}{343}$ **13.** $\dfrac{1}{9}$ **15.** not a real number   **17.** $\dfrac{1}{125}$ **19.** 2 **21.** $\dfrac{9}{16}$

23. 2 **25.** -16 **27.** $4^{5/6}$ **29.** $\dfrac{1}{t}$ **31.** 1 **33.** $a^{1/2}$ **35.** $s^{2/5}$ **37.** $x^{4/3}$ **39.** b^4 **41.** d **43.** $p^{91/30}$

45. $\dfrac{1}{y^{1/12}}$ **47.** $a^{2/3}$ **49.** $\dfrac{1}{z^{1/4}}$ **51.** $\dfrac{1}{x^{9/2}}$ **53.** $-2x^2y^{2/3}$ **55.** $\dfrac{1}{27z^{3/4}}$ **57.** $9st^{5/2}$ **59.** $\dfrac{1}{v^{9/2}w^6}$ **61.** m^2n^4

63. $\dfrac{x^{10}}{a^{12}}$ **65.** $\dfrac{x}{z^{1/2}}$ **67.** $\dfrac{s^{20/3}}{t^{20/3}}$ **69.** $\dfrac{a^3}{9x^5}$ **71.** $x^{5/3}$ **73.** $\dfrac{1}{b^{1/12}}$ **75.** $z^{5/6}$ **77.** $s^{(8/5)n+3}$ **79.** x^5t^3

81. $\dfrac{1}{a^3b^4}$ **83.** $x^{3/4} - 3x^{5/6}$ **85.** $4v^{3/2} - 8v^{3/10}$ **87.** $-4x^{3/4} + 2x^{1/4}$ **89.** $2 - \dfrac{10}{x^{3/2}}$ **91.** $x^{2/3}(x^{2/3} + 1)$

93. $3c^{2/5}(2 - 3c^{8/5})$ **95.** $q^{1/5}(q^{14/5} - 1)$ **97.** $z^{-5/3}(z^4 - 1)$ **99.** $y^{-2/3}(6 + 5y^{5/6})$

Review Problems

Section 3.3
101. $2x^2 + x - 10$ **103.** $5z^2 - 16z + 3$

Section 3.5
105. $(x - 2)(x - 3)$ **107.** $(2x - 1)(3x - 1)$

Section 4.2
109. $3a - 2b$

ENRICHMENT EXERCISES

1. x^5 **2.** $z^{3n-3/4}$ **3.** $a^{4+3/n}$ **4.** 0

EXERCISE SET 5.3

1. $5x^{12/5}y - 4x^{1/2}y^{3/2} + 3x^{8/5}y^{11/3}$ **3.** $2s^{1/7}t^{7/6} - 6s^{11/14}t^{8/3} + 8st^{5/6}$ **5.** $a^2 + 5a^{1/2} - a^{3/2} - 5$
7. $s - s^{3/4} + 6s^{1/4} - 6$ **9.** $x - \sqrt{x}\sqrt{y} - 2y$ **11.** $2r + \sqrt{r}\sqrt{s} - 6s$ **13.** $w^{4/3} - 1$ **15.** $4d - 4d^{1/2} + 1$
17. $q^3 + 2q^{3/2}a^{1/2} + a$ **19.** $x^5 - y^3$ **21.** $3x^{5/2} - x^2 - 2x^{3/2} - x^{1/2} + 1$ **23.** $z + c$ **25.** $s - t$ **27.** $w - 8$
29. $27 + d$ **31.** $x^3 - x^2 + 4x - 4$ **33.** $2s^4 + 2s^2 - 2s^{11/4} - 2s^{3/4}$ **35.** $a^{4/3}b^{1/2} - a^{7/6}b^{5/6} + 3a^{5/6}b^{7/6} - 3a^{2/3}b^{3/2}$
37. $(x^{1/3} + 2)(x^{1/3} - 2)$ **39.** $(z^{1/3} + 2)^2$ **41.** $(y^{1/5} + 4)(y^{1/5} - 1)$ **43.** $(3x^{1/7} + 4)(x^{1/7} - 1)$
45. $(3y^{1/3} + 2)(3y^{1/3} - 2)$ **47.** $(6r^{1/3} - 1)(r^{1/3} - 1)$ **49.** $(2y^{1/7} - 9)(y^{1/7} - 1)$ **51.** $7(x + 1)^{1/2}(3x + 4)$
53. $4(q^2 - 3)^{2/5}(10 - q^2)$ **55.** $(x^2 + 1)^{-1/2}(2x^2 + 1)$ **57.** $4a - 5a^2$ **59.** $9x^{3/2} + 11x^{17/6}$

61. $\dfrac{1}{3}w^{1/4}v^{7/6} - \dfrac{1}{2}v^{1/15}$ **63.** $\dfrac{2+3a}{\sqrt{a}}$ **65.** $\dfrac{12-11x}{x^{1/4}}$ **67.** $\dfrac{2}{(x^2+1)^{1/2}}$ **69.** $\dfrac{3\sqrt{1-t^2}-1}{\sqrt{1-t^2}}$

71. $-\dfrac{4}{(4+a^2)^{3/4}}$

Review Problems

73. $3a^3b$ **75.** $3xy^2z^3$ **77.** $2a^2b^7$

ENRICHMENT EXERCISES

1. $x^n y^{7m/4} + 3x^{(5n/3)-1}y^{17m/8}$ **2.** $\dfrac{b^{3s/14}}{a^{r/10}} - \dfrac{b^{s/42}}{a^{r/5}}$ **3.** $y^{4/3}(2y^{1/3}-3)(y^{1/3}-2)$

4. $a^{1/6}b^{1/3}(a^{1/3}+2b^{1/3})(a^{1/3}-2b^{1/3})$ **5.** $x^{1/3}y^{4/3}(x^{1/2}-y^{1/4})(x+x^{1/2}y^{1/4}+y^{1/2})$ **6.** $(z^2-3)^{1/4}(z^2-3)(z+1)(z-1)$
7. $-2(y+4)^{1/5}(y+4)(3y+10)$ **8.** $(3x^{1/n}-1)(x^{1/n}-3)$

EXERCISE SET 5.4

1. $\sqrt{14}$ **3.** 4   **5.** $\sqrt{3xy}$ **7.** $\sqrt[4]{6pq}$ **9.** $-\sqrt[3]{33yz}$ **11.** $\dfrac{\sqrt{3}}{5}$ **13.** $\dfrac{\sqrt{2}}{x^2}$ **15.** $\dfrac{\sqrt[3]{11}}{4}$ **17.** $\dfrac{\sqrt[3]{x}}{2}$ **19.** $4\sqrt{3}$
21. $2\sqrt{3}$ **23.** $3\sqrt[3]{10}$ **25.** $2\sqrt[3]{6c}$ **27.** $-4\sqrt{2az}$ **29.** 64   **31.** x^2 **33.** $2s^3\sqrt[3]{2s}$ **35.** u^2v **37.** $2s^4\sqrt[4]{2s}$
39. $2c\sqrt[5]{2b^4c}$ **41.** $r\sqrt[3]{r^2}$ **43.** $b\sqrt{b}$ **45.** $7u^5\sqrt{u}$ **47.** $s^3t\sqrt[4]{st}$ **49.** $10x\sqrt{xy}$ **51.** $-3a^3\sqrt{3a}$
53. $-2xz^4\sqrt{2xz}$ **55.** $x^5y^5\sqrt{y}$ **57.** $-z^2t\sqrt{2z}$ **59.** $7xy^3\sqrt{y}$ **61.** $\dfrac{3}{2}bc\sqrt{ac}$ **63.** $\dfrac{\sqrt[3]{10st}}{5t}$ **65.** $\dfrac{\sqrt[3]{2ac^2}}{2c}$
67. $\dfrac{p\sqrt{5pq}}{2q^3}$ **69.** $\dfrac{\sqrt[3]{12wr}}{3r}$ **71.** $\dfrac{\sqrt[3]{2t^2x^2}}{x^3}$ **73.** $\dfrac{s\sqrt[4]{t}}{2t}$ **75.** $\dfrac{2\sqrt{3}}{3}$ **77.** $\dfrac{\sqrt{5x}}{5}$ **79.** $\dfrac{c^2\sqrt{3c}}{3}$ **81.** $-\dfrac{5\sqrt{ab}}{ab}$
83. $\dfrac{\sqrt[3]{6}}{3}$ **85.** $\dfrac{\sqrt[4]{9}}{3}$ **87.** $\dfrac{7\sqrt[3]{y}}{y^2}$ **89.** $5\sqrt[4]{p}$ **91.** $\dfrac{4x\sqrt{21y}}{7y^2}$ **93.** $\dfrac{2x\sqrt{15y}}{3y}$ **95.** $\dfrac{ab\sqrt[3]{18a^2bc^2}}{2c^3}$ **97.** $\dfrac{\sqrt[4]{4u^3v^3w}}{uv}$
99. $\sqrt{29}$ **101.** 3   **103.** 5   **105.** $5\sqrt{2}$; 7.1   **107.** $7\sqrt{3}$; 12.1   **109.** $3\dfrac{1}{2}$ yards

Review Problems

Section 3.2

111. $8x^2$ **113.** $\dfrac{9}{4}x^2y$

Section 4.4

115. $-\dfrac{2}{x}$ **117.** $-\dfrac{3}{2x+1}$ **119.** $-\dfrac{5}{(y-3)(y+2)}$

ENRICHMENT EXERCISES

1. either $x \ge 3$ or $x \le -1$ **2.** either $x \ge 2$ or $x \le -4\dfrac{1}{2}$ **3.** all real numbers   **4.** $r-1$ **5.** $-z-4$

6. $y(y+3)$ **7.** 0.6   **8.** 0.3   **9.** 0.12   **10.** $\sqrt[km]{a^{kn}} = (a^{kn})^{1/km} = a^{kn/km} = a^{n/m} = (a^n)^{1/m} = \sqrt[m]{a^n}$ **11.** $\sqrt{x}$
12. $a\sqrt[3]{a}$ **13.** $z^2\sqrt{z}$ **14.** $b\sqrt[4]{a^3b}$ **15.** $yz^3\sqrt[3]{x^2y}$

EXERCISE SET 5.5

1. $-3\sqrt{3}$ **3.** $7\sqrt{2}$ **5.** $\sqrt{14}+\sqrt{7}$ **7.** $-12\sqrt[3]{3}$ **9.** $8\sqrt[4]{2}$ **11.** $7\sqrt{c}$ **13.** $-a^2\sqrt{a}+2a$

15. $10z^3\sqrt{z} + 7z^3$ **17.** $-2a^2\sqrt{a}$ **19.** $13\sqrt{3}$ **21.** 0 **23.** $3u^2v^2\sqrt[3]{2v^2}$ **25.** $12xy^2\sqrt[3]{2x} + 2xy^2\sqrt[3]{2y}$

27. $-\dfrac{4}{3}\sqrt{3}$ **29.** $\dfrac{11\sqrt{14}}{7}$ **31.** $\dfrac{(2\sqrt{2}-3)\sqrt{3}}{3}$ **33.** $-\dfrac{5x^2\sqrt{x}}{36}$ **35.** $\dfrac{\sqrt{6}+2}{2}$ **37.** $\dfrac{5\sqrt{6}}{3}$ **39.** $-\dfrac{17\sqrt{5}}{4}$

41. $\dfrac{3}{2x}$ **43.** $\dfrac{5\sqrt{6}}{6}$ **45.** $\dfrac{x^2-\sqrt{xy}}{3}$ **47.** $\dfrac{7\sqrt[3]{2}}{2}$ **49.** $-\dfrac{\sqrt[3]{3}}{3}$ **51.** $-\dfrac{3\sqrt[3]{a^2}}{a}$ **53.** $\dfrac{4\sqrt[4]{x}}{x}$ **55.** 1

57. $-2\sqrt{2}$ **59.** $\dfrac{1-3\sqrt{2}}{2}$ **61.** $\dfrac{3}{2}$

Review Problems

63. $6a^2bz - 8ab^2z$ **65.** $2x^2 + x - 3$ **67.** $4a^2 - 8ab + 4b^2$ **69.** $5p^2 - 9pq + 4q^2$ **71.** $8v^2 + 18vy^2 - 5y^4$

E N R I C H M E N T   E X E R C I S E S

1. 2 **2.** $-\dfrac{2}{9}$ **3.** -2 **4.** -1 **5.** $\dfrac{\sqrt{7}-1}{6}$ **6.** $2(\sqrt{3}+\sqrt{2})$ **7.** $\dfrac{2x(\sqrt{x}+1)}{x-1}$ **8.** $\sqrt{a}-\sqrt{b}$

9. $-2xyx^{1/n}y^{2/n}$ **10.** $3x^3y^2x^{2/n}y^{1/n}$

E X E R C I S E   S E T   5 . 6

1. $\sqrt{10}+\sqrt{22}$ **3.** $\sqrt{39}-2\sqrt{65}$ **5.** $2-3\sqrt{2}$ **7.** $3\sqrt{5}-\sqrt{3}$ **9.** $6+6\sqrt{2}$ **11.** $\sqrt{yz}-3z$ **13.** $x-2x^2$

15. $z-z^2$ **17.** $2ab^2 + ab\sqrt[4]{8ab^3}$ **19.** $56u$ **21.** $28-\sqrt{5}$ **23.** $2r+5\sqrt{r}+3$ **25.** $uv+6\sqrt{uv}+8$

27. $3x\sqrt[3]{x} - 7\sqrt[3]{x^2} + 4$ **29.** $5-2\sqrt{6}$ **31.** $x+4\sqrt{xz}+4z$ **33.** -2 **35.** $2\sqrt[3]{2}-1$ **37.** $x-1$

39. $3\sqrt[3]{3y^2}-1$ **41.** -26 **43.** 2 **45.** $\dfrac{5b}{4}$ **47.** -13 **49.** 39 **51.** $25x-4y$ **53.** $49xy-16z$

55. $-2(\sqrt{3}+2)$ **57.** $\dfrac{r(\sqrt{r}-1)}{r-1}$ **59.** $\dfrac{s\sqrt{t}-t\sqrt{s}}{s-t}$ **61.** $-\dfrac{4-\sqrt{6}}{5}$ **63.** $\dfrac{10\sqrt{2}-9}{7}$

65. $-\sqrt{3}+\sqrt{6}+6\sqrt{2}-6$ **67.** $\dfrac{s+\sqrt{s}-2}{s-1}$ **69.** $\dfrac{2\sqrt{x}-2\sqrt{y}-x\sqrt{y}+y\sqrt{x}}{x-y}$ **71.** $\dfrac{r-5\sqrt{rk}+4k}{r-16k}$

73. $\dfrac{\sqrt[3]{a^2}+\sqrt[3]{4a}+2\sqrt[3]{2}}{a-4}$ **75.** $\sqrt[3]{y^2}-\sqrt[3]{5y}+\sqrt[3]{25}$ **77.** $(s-1)(\sqrt[3]{4s^2}+\sqrt[3]{2s}+1)$ **79.** $3(2-\sqrt{5})$

81. $3x(x+5\sqrt{x})$ **83.** $a^2\sqrt{a}(a-3)$ **85.** $y^2\sqrt{y}(1-y)$ **87.** $r\sqrt[3]{r^2}(1-5r^2)$ **89.** $\dfrac{\sqrt{5}-5\sqrt{7}}{9}$

91. $\dfrac{9\sqrt{2}+2\sqrt{3}}{3}$ **93.** $\dfrac{4z-1}{2}$ **95.** $4+7x^2$ **97.** $\dfrac{1}{x}$ **99.** $\dfrac{1-3a}{4a^2}$ **101.** No, the conjugate of $2+\sqrt{3}$ is

$2-\sqrt{3}$, whereas, the opposite of $2+\sqrt{3}$ is $-(2+\sqrt{3})$ or $-2-\sqrt{3}$. Note that $2-\sqrt{3} \neq -2-\sqrt{3}$.

Review Problems

103. $x=0$ or $x=3$ **105.** $y=\dfrac{1}{3}$ or $y=1$ **107.** $a=1$ **109.** $z=0, z=\dfrac{2}{3}$, or $z=\dfrac{3}{2}$ **111.** $x=1$ or $x=-1$

E N R I C H M E N T   E X E R C I S E S

1. $-\dfrac{5}{2}$ **2.** $4-\sqrt{2}$ **3.** $-\dfrac{\sqrt{x}}{2}$ **4.** $\dfrac{1}{\sqrt{x+h}+\sqrt{x}}$ **5.** $\dfrac{1}{\sqrt{x}+\sqrt{a}}$ **6.** $\dfrac{1}{\sqrt[3]{(x+h)^2}+\sqrt[3]{x(x+h)}+\sqrt[3]{x^2}}$

7. $\dfrac{2x+h}{(x+h)\sqrt[3]{x+h}+\sqrt[3]{x^2(x+h)^2}+x\sqrt[3]{x}}$

8. (a) $x_1 + x_2 = \dfrac{-b + \sqrt{b^2 - 4ac}}{2a} + \dfrac{-b - \sqrt{b^2 - 4ac}}{2a} = \dfrac{-b + \sqrt{b^2 - 4ac} - b - \sqrt{b^2 - 4ac}}{2a} = \dfrac{-2b}{2a} = -\dfrac{b}{a}$

(b) $x_1 x_2 = \dfrac{-b + \sqrt{b^2 - 4ac}}{2a} \cdot \dfrac{-b - \sqrt{b^2 - 4ac}}{2a} = \dfrac{(-b)^2 - (\sqrt{b^2 - 4ac})^2}{4a^2} = \dfrac{b^2 - b^2 + 4ac}{4a^2} = \dfrac{4ac}{4a^2} = \dfrac{c}{a}$

E X E R C I S E   S E T   5 . 7

1. $x = 16$ **3.** no solution   **5.** $z = 10$ **7.** $t = \dfrac{5}{12}$ **9.** $s = 2$ **11.** $s = 9$ **13.** $p = 37$ **15.** $u = -\dfrac{4}{9}$

17. $c = \dfrac{6}{5}$ **19.** $x = -\dfrac{1}{8}$ **21.** $x = 1$ **23.** $y = 3$ **25.** $x = \dfrac{3}{5}$ **27.** no solution   **29.** $t = \dfrac{1}{2}$ **31.** no solution

33. $x = \dfrac{36}{7}$ **35.** $b = -1$ **37.** no solution   **39.** $x = \dfrac{2}{3}$ or $x = \dfrac{1}{3}$ **41.** $w = 0$ **43.** $b = 3$ or $b = 2$

45. $s = 2$ or $s = 5$ **47.** $a = 0$ **49.** $w = -2$ **51.** $x = 3$ **53.** $s = -\dfrac{35}{2}$ **55.** no solution   **57.** $x = \dfrac{15}{8}$

59. $b = 2$ or $b = 5$ **61.** $v = 2$ **63.** no solution   **65.** $x = \dfrac{1}{3}$ or $x = 3$ **67.** -1 **69.** 1 **71.** $-\dfrac{13}{2}$

Review Problems

Section 5.3
73. $\sqrt{6} - \sqrt{15}$ **75.** $3x - 7\sqrt{x} + 2$ **77.** $z - 2$

Section 5.5
79. $\dfrac{3x\sqrt{2} - \sqrt{2x}}{2x}$ **81.** $\dfrac{11 - 3\sqrt{15}}{7}$

E N R I C H M E N T   E X E R C I S E S

1. $x = \pm 1$ or $x = \pm 3$ **2.** no solution   **3.** $z = \pm 1$ or $z = \pm 2$ **4.** $x = 0$ or $x = 1$ **5.** $y = 5$ **6.** $x = -1$ or $x = -\dfrac{6}{5}$ **7.** The radicands must be nonnegative. Therefore, $x - 5 \geq 0$ or $x \geq 5$ and $4 - x \geq 0$ or $x \leq 4$. There are no numbers that satisfy $x \geq 5$ and $x \leq 4$.

E X E R C I S E   S E T   5 . 8

1. $3i$ **3.** $-10i$ **5.** $-i\sqrt{17}$ **7.** $2i\sqrt{7}$ **9.** $-5i\sqrt{2}$ **11.** $4 - 10i$ **13.** $2 + 2i\sqrt{2}$ **15.** $1 - 2\sqrt{5}$

17. $x = \dfrac{1}{4}, y = -\dfrac{5}{2}$ **19.** $x = \dfrac{3\sqrt{2}}{2}, y = -\dfrac{\sqrt{5}}{2}$ **21.** $10 + i$ **23.** -10 **25.** $10 + 6i$ **27.** $4 - 8i$ **29.** $12 - i$

31. $9 + 21i$ **33.** $5 - 12i$ **35.** $\dfrac{5}{16}$ **37.** $\dfrac{8}{5} + \dfrac{1}{5}i$ **39.** $1 - i$ **41.** $\dfrac{17}{13} + \dfrac{6}{13}i$ **43.** $3 - i$ **45.** $\dfrac{2}{3} + \dfrac{1}{3}i$

47. $8 - 12i$ **49.** $2 - 41i$ **51.** $-2 - 2i$ **53.** $8 - 5i$ **55.** $\dfrac{\sqrt{2}}{2}$ **57.** $\dfrac{1}{2} + \dfrac{\sqrt{7}}{4}i$ **59.** $-\dfrac{3}{17} + \dfrac{2\sqrt{2}}{17}i$ **61.** 1

63. i **65.** 1 **67.** $-i$ **69.** -1 **71.** $-18 - 3i$ **73.** $-\dfrac{1}{2} \pm \dfrac{\sqrt{3}}{2}i$ **75.** $\dfrac{1}{2} \pm \dfrac{1}{2}i$ **77.** $\pm 5i$

Review Problems

Section 2.2

79. $x = \dfrac{5}{2}$

Section 3.8

81. $t = \dfrac{1}{5}$ or $t = -\dfrac{7}{3}$ **83.** $x = 7$ or $x = -7$

ENRICHMENT EXERCISES

1. $(1 + i\sqrt{3})^3 = (1 + i\sqrt{3})^2(1 + i\sqrt{3}) = (-2 + 2i\sqrt{3})(1 + i\sqrt{3}) = -2 - 2i\sqrt{3} + 2i\sqrt{3} + 2i^2 \cdot 3 = -2 - 2 \cdot 3 = -8$

2. $(1 - i\sqrt{3})^3 = (1 - i\sqrt{3})^2(1 - i\sqrt{3}) = (-2 - 2i\sqrt{3})(1 - i\sqrt{3}) = -2 + 2i\sqrt{3} - 2i\sqrt{3} + 2i^2 \cdot 3 = -2 - 2 \cdot 3 = -8$

3. (a) $\overline{z + w} = \overline{(a + bi) + (c + di)} = \overline{(a + c) + (b + d)i} = (a + c) - (b + d)i = (a - bi) + (c - di) = \overline{z} + \overline{w}$

(b) $\overline{zw} = \overline{(a + bi)(c + di)} = \overline{(ac - bd) + (ad + bc)i} = (ac - bd) - (ad + bc)i = (a - bi)(c - di) = \overline{z}\,\overline{w}$

4. $\dfrac{1}{a + bi} = \dfrac{1 \cdot (a - bi)}{(a + bi)(a - bi)} = \dfrac{a - bi}{a^2 + b^2} = \dfrac{a}{a^2 + b^2} - \dfrac{b}{a^2 + b^2}i$ **5. (a)** $\dfrac{1}{2} - \dfrac{1}{2}i$ **(b)** $-i$ **(c)** $\dfrac{2}{7} + \dfrac{\sqrt{3}}{7}i$

CHAPTER 5 REVIEW EXERCISE SET

Section 5.1

1. 8   **2.** $-\dfrac{3}{4}$ **3.** -3 **4.** -2 **5.** $3x^4y^2$ **6.** $-2a^2b^3c^4$

Sections 5.2 and 5.3

7. $3^{3/2}$ **8.** $\dfrac{1}{z^{1/2}}$ **9.** $\dfrac{1}{x}$ **10.** s^8t^2 **11.** $w^{1/3}$ **12.** b^{20} **13.** $2x^{11/3}y^4 - 5x^{7/6}y^{7/2} + 7xy^{11/4}$

14. $s - \sqrt{sw} - 2w$ **15.** $c - b^3$ **16.** $a^5 - 4a^{5/2} + 4$ **17.** $z - h$ **18.** $3t^{5/3}(3t - 2)$

19. $3(a + 1)^{1/6}[6(a + 1)^{1/6} - 1]$ **20.** $\dfrac{1}{(t - 1)^{1/2}}$

Section 5.4

21. $2\sqrt{6}$ **22.** $-2\sqrt[3]{2}$ **23.** z^2t^3 **24.** $-\sqrt{3}s^4r^2$

Sections 5.4 and 5.6

25. $\dfrac{\sqrt{6}}{3}$ **26.** $\dfrac{3\sqrt{x}}{x}$ **27.** $2\sqrt{2s}$ **28.** $3\sqrt[3]{k}$ **29.** $\dfrac{4\sqrt[4]{v^3}}{3}$ **30.** $-3(2 + \sqrt{5})$ **31.** $-2(1 - \sqrt{3})$

32. $\dfrac{x - 2\sqrt{x} + 1}{x - 1}$ **33.** $\dfrac{2s\sqrt{t} + t\sqrt{s}}{4s - t}$ **34.** $c^{2/3} + (3c)^{1/3} + 9^{1/3}$

Section 5.5

35. $7\sqrt{3}$ **36.** $11\sqrt[4]{2}$ **37.** $-3a^2\sqrt{a}$ **38.** $-14\sqrt[3]{c^2}$ **39.** $-3uv\sqrt[3]{uv^2}$ **40.** $4x^2\sqrt{x}$ **41.** $4\sqrt{w}$

Section 5.6

42. $3 - 6\sqrt{2}$ **43.** $2s - 3s^2$ **44.** $x^2 - 2x^2\sqrt[3]{x^2}$ **45.** $a - 4\sqrt{ab} + 4b$ **46.** $3y\sqrt[3]{3y} - 4$

Section 5.7

47. $x = 64$ **48.** $t = 12$ **49.** $z = 1$ **50.** $z = \dfrac{9}{2}$ **51.** $x = \dfrac{3}{11}$ **52.** $y = 1$ **53.** no solution   **54.** $z = -2$

55. $s = -9$

Section 5.8

56. $10i$ **57.** $2 - 3i$ **58.** $4 + 2i\sqrt{10}$ **59.** $1 - 2\sqrt{5}$ **60.** $x = -\dfrac{5}{3},\ y = -5$ **61.** $10 - 6i$ **62.** $8 + 13i$

63. $2 + 2i$ **64.** $-7 - 17i$ **65.** $4 - i$ **66.** $\dfrac{9}{2} - 2i$ **67.** $\dfrac{6}{5} - \dfrac{8}{5}i$ **68.** $32 - 24i$ **69.** 1 **70.** i **71.** $2 + i$

72. $216i$

CHAPTER 5 TEST

Section 5.1

1. -10 **2.** 3 **3.** $4x^2y$

Sections 5.2 and 5.3

4. $\dfrac{1}{a^7}$ **5.** $x - 9y$ **6.** $b - 6\sqrt{b} + 9$ **7.** $-\dfrac{x}{(x + 2)^{1/2}}$

Section 5.4

8. $3\sqrt{6}$ **9.** $-3\sqrt[3]{2}$ **10.** $2\sqrt[4]{2}a^2b$

Sections 5.4 and 5.6

11. $\dfrac{3\sqrt{5}}{5}$ **12.** $\dfrac{2\sqrt{a}}{a}$ **13.** $-\dfrac{1 + \sqrt{3}}{2}$ **14.** $\dfrac{c + 2\sqrt{c}}{c - 4}$ **15.** $7\sqrt{2}$ **16.** $7a^2\sqrt{a}$ **17.** $4t - 9t^2\sqrt[3]{t}$

Section 5.7

17. $x = 16$ **18.** $y = \dfrac{1}{3}$ **19.** $x = \dfrac{1}{4}$ **20.** $t = -4$

Section 5.8

21. $-3 - 2\sqrt{6}i$ or $-3 - 2i\sqrt{b}$ **22.** $2 + 7i$ **23.** $-6 - 3i$ **24.** i **25.** $1 + 4i$

CHAPTER 6

EXERCISE SET 6.1

1. $z = \pm 3$ **3.** $w = \pm\dfrac{2}{3}$ **5.** $q = \pm 0.1$ **7.** $d = \pm 3\sqrt{2}$ **9.** $t = \pm 5\sqrt{5}$ **11.** $s = 2$ or $s = 10$

13. $z = -5$ or $z = 0$ **15.** $y = 1 \pm \dfrac{4\sqrt{2}}{3}$ **17.** $y = \pm\dfrac{\sqrt{6}}{6}$ **19.** $z = -3$ or $z = 6$ **21.** $r = \dfrac{-16 \pm 2\sqrt{3}}{5}$

23. $y = -\dfrac{3}{2} \pm 2\sqrt{2}$ **25.** $y = \dfrac{-3 \pm 4\sqrt{2}}{2}$ **27.** $3\sqrt{5}$ **29.** $\dfrac{3 \pm 3\sqrt{2}}{2}$ **31. (a)** $r = \sqrt{\dfrac{A}{\pi}}$

(b) $2\sqrt{5}$ inches ≈ 4.5 inches   **33. (a)** $r = \sqrt{\dfrac{3V}{\pi h}}$ **(b)** $6\sqrt{3}$ meters ≈ 10.392 meters   **35. (a)** $w = \sqrt{\dfrac{3V}{2h}}$

(b) width $= 5\sqrt{3}$ yards ≈ 8.66 yards; length $= 10\sqrt{3}$ yards ≈ 17.32 yards   **37.** $90\sqrt{2}$ feet ≈ 127 feet   **39.** $s = \dfrac{\sqrt{10}}{10}c$

41. $s = \dfrac{1}{5}c$ **43.** $s = \dfrac{2\sqrt{17}}{17}c$ **45.** The length of the hypotenuse is $\dfrac{10\sqrt{3}}{3}$ cm; the length of the side opposite $30°$ is

$\dfrac{5\sqrt{3}}{3}$ cm.   **47.** The length of the side opposite $60°$ is $\sqrt{30}$ ft; the length of the side opposite $30°$ is $\sqrt{10}$ ft.

49. The length of the hypotenuse is 8 in.; the length of the side opposite $30°$ is 4 in.

Review Problems

51. $x^2 + 6x + 9$ **53.** $y^2 - 4y + 4$ **55.** $x^2 - \dfrac{4}{3}x + \dfrac{4}{9}$ **57.** $9a^2 + 6a + 1$ **59.** $c^4 + 8c^2 + 16$

61. $x^4 + 2x^2y + y^2$

E N R I C H M E N T   E X E R C I S E S

1. $x = \dfrac{2}{3}$ or $x = 4$ **2.** $a = -\dfrac{8}{3}$ or $a = \dfrac{2}{11}$ **3.** all real numbers   **4.** $t = -\dfrac{5}{2}$ **5.** $x + \dfrac{b}{4} = \pm\sqrt{\dfrac{b^2 - 8c}{16}}$.

$x = -\dfrac{b}{4} \pm \sqrt{\dfrac{b^2 - 8c}{16}} = -\dfrac{b}{4} \pm \dfrac{\sqrt{b^2 - 8c}}{4} = \dfrac{-b \pm \sqrt{b^2 - 8c}}{4}$ **6.** $x + \dfrac{b}{2a} = \pm\sqrt{\dfrac{b^2 - 4ac}{4a^2}}$.

$x = -\dfrac{b}{2a} \pm \sqrt{\dfrac{b^2 - 4ac}{4a^2}} = -\dfrac{b}{2a} \pm \dfrac{\sqrt{b^2 - 4ac}}{2a} = \dfrac{-b \pm \sqrt{b^2 - 4ac}}{2a}$

E X E R C I S E   S E T   6 . 2

1. 16   **3.** $\dfrac{25}{36}$ **5.** $s = -11$ or $s = 3$ **7.** $t = 1$ or $t = 11$ **9.** $q = -8 \pm \sqrt{55}$ **11.** $s = -4$ or $s = 18$

13. $z = \dfrac{-5 \pm \sqrt{3}}{2}$ **15.** $c = \dfrac{7}{6}$ or $c = \dfrac{11}{6}$ **17.** $x = -\dfrac{3}{5}$ or $x = -\dfrac{1}{5}$ **19.** $w = -\dfrac{1}{3} \pm \sqrt{2}$ **21.** $v = \dfrac{3 \pm 2\sqrt{14}}{7}$

23. $r = -\dfrac{2}{5}$ or $r = -\dfrac{1}{5}$ **25.** $z = 14$ or $z = -8$ **27.** $b = -1$ or $b = \dfrac{7}{2}$ **29.** $x = -1 \pm \dfrac{\sqrt{6}}{3}$ **31.** $x = -1 \pm \sqrt{7}$

33. $s = -4$ or $s = -2$

Review Problems

35. $\dfrac{6}{5}$ **37.** -6 **39.** 3   **41.** $3, -1$ **43.** $\dfrac{-1 \pm \sqrt{2}}{2}$

E N R I C H M E N T   E X E R C I S E S

1. $x = \pm\sqrt{7}$ or $x = \pm 1$ **2.** $r = \pm 1$ **3.** $y = 16 \pm 8\sqrt{3}$ **4.** no real solution   **5.** $x = \dfrac{-b \pm \sqrt{b^2 - 4}}{2}$

6. $x = \dfrac{-b \pm \sqrt{b^2 - 4ac}}{2a}$ **7.** $x = 0$ or $x = 64$ **8.** $(x + 3)^2 + (y - 2)^2 = 4$

9. $\dfrac{(x - 1)^2}{4^2} + \dfrac{(y + 2)^2}{2^2} = 1$

EXERCISE SET 6.3

1. $a = 3, b = 4, c = -5$ **3.** $a = 1, b = -1, c = 1$ **5.** $a = 3, b = -12, c = 5$ **7.** $a = 3, b = -1, c = 0$

9. $a = 6, b = 7, c = -4$ **11.** $x = -9$ or $x = 3$ **13.** $s = 1$ or $s = 2$ **15.** $v = -4$ or $v = \dfrac{1}{2}$ **17.** $c = -\dfrac{3}{2}$ or

$c = -\dfrac{2}{3}$ **19.** no real solution   **21.** $m = \dfrac{-1 \pm \sqrt{13}}{6}$ **23.** $s = -1 \pm \sqrt{2}$ **25.** $v = 2 \pm \sqrt{6}$ **27.** $x = -3$

29. no real solution   **31.** $z = \dfrac{1 \pm 2\sqrt{2}}{2}$ **33.** $y = \dfrac{1 \pm \sqrt{6}}{10}$ **35.** $c = 0$ or $c = 3$ **37.** $s = \pm 2\sqrt{3}$

39. $x = \dfrac{1 \pm \sqrt{5}}{2}$ **41.** $x = \dfrac{2 \pm \sqrt{3}}{3}$ **43.** $x = \dfrac{1 \pm \sqrt{3}}{2}$; $x \approx 1.4$ or $x \approx -0.4$ **45.** $y = \dfrac{-5 \pm \sqrt{5}}{4}$; $y \approx -1.8$

or $y \approx -0.7$ **47.** $z = \dfrac{-3 \pm \sqrt{3}}{4}$; $z \approx -0.3$ or $z \approx -1.2$ **49.** $x = 0$ or $x = 1 \pm \sqrt{3}$ **51.** b is -3,

so $-b$ is 3. $x = \dfrac{3 \pm \sqrt{29}}{2}$.   **53. (1)** $x(2x + 3) = 2.\ 2x^2 + 3x - 2 = 0;\ (2x - 1)(x + 2) = 0.$ Either $x = \dfrac{1}{2}$ or $x = -2$.

(2) $x(2x + 3) = 2.\ 2x^2 + 3x = 2;\ x^2 + \dfrac{3}{2}x + \dfrac{9}{16} = 1 + \dfrac{9}{16};\ \left(x + \dfrac{3}{4}\right)^2 = \dfrac{25}{16};\ x + \dfrac{3}{4} = \pm\dfrac{5}{4};\ x = -\dfrac{3}{4} \pm \dfrac{5}{4}.$ Either

$x = -2$ or $x = \dfrac{1}{2}$.   **(3)** $x(2x + 3) = 2.\ 2x^2 + 3x - 2 = 0.\ x = \dfrac{-3 \pm \sqrt{9 - 4(2)(-2)}}{4} = \dfrac{-3 \pm \sqrt{9 + 16}}{4} = \dfrac{-3 \pm \sqrt{25}}{4} =$

$\dfrac{-3 \pm 5}{4}.$ Either $x = \dfrac{-3 + 5}{4} = \dfrac{1}{2}$ or $x = \dfrac{-3 - 5}{4} = -2.$

Review Problems

55. $\dfrac{3}{2}$ **57.** 3 in. by 7 in.

ENRICHMENT EXERCISES

1. $x = \pm 1$ or $x = \pm 3$ **2.** $x = -3 + 3\sqrt{3}$ or $x = -3 - \sqrt{3}$ **3.** $y = \dfrac{2 \pm \sqrt{10}}{3}$ **4.** $x = \sqrt{3}$ or $x = -3\sqrt{3}$

5. $r = 2\sqrt{2} \pm \sqrt{10}$ **6.** $x = \dfrac{b \pm \sqrt{(-b)^2 - 4(-a)(-c)}}{2(-a)} = \dfrac{b \pm \sqrt{b^2 - 4ac}}{-2a} = \dfrac{-b \pm \sqrt{b^2 - 4ac}}{2a}$

MISCELLANEOUS QUADRATIC EQUATIONS

1. $x = 2 \pm \sqrt{5}$ **3.** no real solution   **5.** $r = -\dfrac{3}{2}$ or $r = \dfrac{1}{3}$ **7.** $x = \dfrac{1 \pm \sqrt{6}}{2}$ **9.** $z = \pm 4$ **11.** $c = \dfrac{5}{2}$

13. $m = 0$ or $m = 4$ **15.** $t = -5$ or $t = 4$ **17.** $a = \dfrac{-11 \pm \sqrt{109}}{2}$ **19.** $b = -\dfrac{1}{2}$ or $b = -\dfrac{3}{5}$

21. $y = -\dfrac{7}{2}$ or $y = \dfrac{2}{3}$ **23.** $c = 0$ or $c = \sqrt{3}$ **25.** $d = -1 \pm \sqrt{5}$ **27.** $c = \dfrac{-4 \pm \sqrt{19}}{3}$ **29.** $s = 1$ or $s = \dfrac{5}{2}$

31. $s = 0$ or $s = \dfrac{\sqrt{10}}{2}$ **33.** $r = -1 \pm \sqrt{3}$ **35.** $k = 0$ or $k = 5$ **37.** no real solution   **39.** $z = \pm\sqrt{5}$

EXERCISE SET 6.4

1. (a) 15 feet; 18 feet **(b)** 2 seconds   **3.** 10 sides   **5.** 100 ft   **7.** $80\sqrt{5} \approx 178.9$ ft per sec

9. $-2\sqrt{2}$ and $-10\sqrt{2}$ or $2\sqrt{2}$ and $10\sqrt{2}$ **11.** width $= \dfrac{-3 + \sqrt{153}}{2} \approx 4.7$ cm; length $= \dfrac{3 + \sqrt{153}}{2} \approx 7.7$ cm

13. $-1 + 4\sqrt{3}$ in. by $1 + 4\sqrt{3}$ in., or about 5.9 in. by 7.9 in.   **15.** $4 + 3\sqrt{6}$ in. on each side, or about 11.3 in. on each side

17. 3 dozen tapes   **19.** either 3 or $\dfrac{1}{3}$ **21.** either $-\dfrac{3}{2}$ or $\dfrac{2}{3}$ **23.** 2 hours

Review Problems

Section 3.5

25. $(x - 3)(x - 1)$ **27.** $(2y + 3)(y - 2)$ **29.** $(8x - 3)(x - 1)$ **31.** $2(4t - 3)(t + 1)$

Section 4.4

33. $\dfrac{4 - x}{(x - 1)(x + 2)}$ **35.** $\dfrac{14t}{(2t - 1)(t + 3)}$ **37.** $\dfrac{2z^2 - 1}{z^2 - 1}$ **39.** $\dfrac{x^2}{x - 1}$

E N R I C H M E N T   E X E R C I S E S

1. about 0.3 second   **2.** about 7,547 meters per second   **3. (a)** The area of the triangle is $\dfrac{\sqrt{3}x^2}{36}$. The area of the

square is $\dfrac{(10 - x)^2}{16}$. Therefore, $A = \dfrac{(10 - x)^2}{16} + \dfrac{\sqrt{3}x^2}{36} = \dfrac{1}{4}\left[\dfrac{(10 - x)^2}{4} + \dfrac{\sqrt{3}}{9}x^2\right]$. **(b)** $1 + \sqrt{3}$ sq in.

E X E R C I S E   S E T   6 . 5

1. $-2 < x < 1$

3. $(-\infty, -3] \cup [4, +\infty)$

5. $x \le \dfrac{3}{2}$ or $x \ge 3$

7. $x < -\dfrac{3}{2}$ or $x > \dfrac{4}{3}$

9. all real numbers

11. $t < -2$ or $t > 2$

13. $(-3, 1) \cup (3, +\infty)$

15. $-1 < x < \dfrac{5}{2}$

17. $(-\infty, -1) \cup (0, 1)$

19. $x < 0$ or $0 < x < 3$

21. $x \le -1$ or $x \ge 3$

23. $[0, 2] \cup [4, +\infty)$

25. $-6 < x < 1$

27. $x > 4$

29. $(-\infty, -3) \cup [-2, +\infty)$

31. $x < 1$ or $x \ge 4$

33. $x < -3$ or $-\dfrac{4}{3} < x < 2$

35. $(-2, -1) \cup (2, 3)$

37. $-3 \le x \le 1$ or $x > 2$

39. $x < -1$, $2 < x < 4$, or $x > 6$

41. y is positive when $x < -5$ or $x > 5$. y is negative when $-5 < x < 5$.   **43.** y is positive when $x < -1$ or $x > \dfrac{8}{3}$.

y is negative when $-1 < x < \dfrac{8}{3}$.   **45.** y is positive when $x < -2$ or $x > 6$. y is negative when $-2 < x < 6$.

47. y is positive when $-10 < x < 0$ or $x > 4$. y is negative when $x < -10$ or $0 < x < 4$.   **49.** y is positive when $x > 2$.
y is negative when $x < 2$, $x \ne 1$.   **51.** $2 \le t \le 3$

Review Problems

53. $7i$ **55.** $2i\sqrt{14}$ **57** $-6 + 3i$ **59.** $\dfrac{1}{6} + 3i$ **61.** $8i\sqrt{2}$ **63.** $\dfrac{2 \pm i\sqrt{47}}{8}$

ENRICHMENT EXERCISES

1. $x < -2$, $x > 2$, or $-1 < x < 1$

2. $x < -3$, $x > 3$, or $-\sqrt{3} < x < \sqrt{3}$

3. $-2 \le x \le 1$ or $x \ge 2$

4. $-1 < x \le 4$ and $x \ne 2$ and $x \ne 0$

5. $-4 < x < 1$ or $x > 3$

EXERCISE SET 6.6

1. $x = \pm 7i$　　**3.** $r = -1 \pm 2i$　　**5.** $s = \dfrac{5 \pm 8\sqrt{2}}{3}$　　**7.** $v = \dfrac{3 \pm \sqrt{13}}{2}$　　**9.** $p = 1$ or $p = \dfrac{3}{2}$　　**11.** $a = -\dfrac{3}{2} \pm \dfrac{\sqrt{7}}{2}i$

13. $x = 2 \pm i$　　**15.** $z = 2 \pm \sqrt{3}$　　**17.** $n = 1 \pm \sqrt{22}$　　**19.** $u = \dfrac{9 \pm \sqrt{77}}{2}$　　**21.** $y = \dfrac{1}{4} \pm \dfrac{\sqrt{3}}{4}i$

23. $x = \dfrac{1}{4} \pm \dfrac{3\sqrt{2}}{8}i$　　**25.** $x = \sqrt{3} \pm 2i$　　**27.** $x = 3\sqrt{2} \pm i\sqrt{3}$　　**29.** 2 real roots　　**31.** 1 real root

33. 2 complex roots　　**35.** 2 real roots　　**37.** 2 complex roots　　**39.** $\dfrac{1}{2} \pm \dfrac{\sqrt{5}}{2}i$　　**41.** $x = -\dfrac{5}{3}$ or $x = -\dfrac{8}{3}$

43. $t = \dfrac{7 \pm i}{2}$　　**45.** $x = \dfrac{10}{3}$ or $x = 1$　　**47.** $x = \pm 2\sqrt{3}$ or $x = \pm i\sqrt{3}$　　**49.** $t = \pm 2i\sqrt{3}$ or $t = \pm i$

51. $x = \dfrac{27}{8}$ or $x = -\dfrac{1}{8}$　　**53.** $a = \dfrac{8}{125}$ or $a = 1$　　**55.** $x = \dfrac{1}{25}$ or $x = \dfrac{1}{9}$　　**57.** $t = \dfrac{1}{4}$ or $t = \dfrac{9}{4}$

59. $x = \dfrac{1}{32}$ or $x = 1$　　**61.** $x = 0$ or $x = 13$　　**63.** $x = -1$ or $x = 4$　　**65.** $t = \dfrac{10 \pm \sqrt{100 - h}}{4}$

67. $x = -2 \pm \sqrt{3 + y}$　　**69.** $t = \dfrac{v \pm \sqrt{v^2 - 20h}}{10}$

Review Problems

71. $\dfrac{3}{4} - 2x$　　**73.**

ENRICHMENT EXERCISES

1. $-(2i)^2 = -2^2 i^2 = -4(-1) = 4$　　**2.** $x^2 + 4 = 0$; $x^2 - (2i)^2 = 0$; $(x + 2i)(x - 2i) = 0$; therefore, $x = \pm 2i$

3. $x = \pm 5i$　　**4.** $r = \pm 10i$　　**5.** $t = \pm 2i\sqrt{2}$　　**6.** $z = \pm 3i\sqrt{3}$　　**7.** $x = 1$ or $x = 2 \pm i$　　**8.** $x = 3$ or $x = \dfrac{1 \pm i\sqrt{3}}{2}$

CHAPTER 6 REVIEW EXERCISES SET

Section 6.1

1. $x = \dfrac{11}{3}$ or $x = 1$　　**2.** $x = 4 \pm 2\sqrt{2}$

Section 6.2

3. $x = -1 \pm \sqrt{37}$　　**4.** $x = \dfrac{3}{2} \pm \dfrac{\sqrt{21}}{2}$

Section 6.3

5. $x = 1 \pm \sqrt{3}$ **6.** $r = 2 \pm \sqrt{3}$ **7.** $y = \dfrac{-1 \pm \sqrt{19}}{3}$ **8.** $s = 3$ or $s = -\dfrac{3}{2}$

Section 6.4

9. (a) 126.9 meters; 66 meters **(b)** approximately 11.3 seconds   **10.** Yes, the car was traveling 40 mph.
11. height $= -2.5 + 2.5\sqrt{17} \approx 7.8$ in. base $= 2.5 + 2.5\sqrt{17} \approx 12.8$ in.   **12.** $-2 + 4\sqrt{6} \approx 7.8$ feet by $2 + 4\sqrt{6} \approx 11.8$ feet
13. A square with common side $8 + 2\sqrt{14} \approx 15.5$ in.

Section 6.5

14. $1 < x < 5$ **15.** $x < -\dfrac{1}{2}$ or $x > \dfrac{1}{3}$

16. $(-5, -2] \cup (0, +\infty)$ **17.** $(-\infty, -1) \cup (2, 4]$

18. $(-\infty, -2) \cup (0, 2) \cup [4, +\infty)$

Section 6.6

19. $y = 3 \pm 4i$ **20.** $t = \dfrac{1}{2} \pm i$ **21.** $x = \dfrac{25}{9}$ or $x = 9$ **22.** $x = \pm 2i\sqrt{2}$ or $x = \pm i\sqrt{3}$ **23.** y is positive when

$x < -1$ or $x > 2$. y is negative when $-1 < x < 2$.   **24.** $t = \dfrac{25 \pm \sqrt{625 - 5h}}{5}$ **25.** $x = \dfrac{2 \pm \sqrt{7 + 3y}}{3}$

Sections 6.1–6.3 and 6.6

26. $x = \dfrac{2}{3} \pm i$ **27.** $x = -2$ or $x = \dfrac{1}{3}$ **28.** $t = 1 \pm \sqrt{5}$ **29.** $y = \dfrac{3 \pm i}{2}$ **30.** $x = \dfrac{-5 \pm 2i\sqrt{3}}{4}$ **31.** $x = 81$ or

$x = 1$ **32.** $x = \pm 2i$ or $x = \pm 2\sqrt{2}$ **33.** $t = 1$ or $t = -\dfrac{3}{5}$ **34.** $y = 3\sqrt{2}$ or $y = \sqrt{2}$ **35.** $t = \pm 2\sqrt{3}$ or $t = \pm 1$

36. $r = \dfrac{5}{2}$ or $r = \dfrac{2}{3}$ **37.** $x = \dfrac{7 \pm 4i\sqrt{3}}{6}$ **38.** $z = -\dfrac{1}{27}$ or $z = 8$ **39.** $a = \dfrac{5}{2}$ or $a = -\dfrac{3}{4}$ **40.** $x = \dfrac{1 \pm i\sqrt{7}}{4}$

41. $y = 4$ or $y = -\dfrac{1}{3}$ **42.** $k = \pm \sqrt{3}$ or $k = \pm 2i\sqrt{6}$ **43.** $x = \dfrac{7 \pm 4i}{3}$ **44.** $x = \pm \dfrac{1}{8}$ or $x = \pm 1$ **45.** $t = \pm 3\sqrt{3}$

CHAPTER 6 TEST

Section 6.1

1. $x = 0$ or $x = \dfrac{3}{2}$ **2.** $x = -2 \pm 2\sqrt{3}$

Section 6.2

3. $x = -2 \pm 3\sqrt{2}$

Section 6.3

4. $x = 2$ or $x = \dfrac{3}{2}$ **5.** $x = \dfrac{3 \pm \sqrt{21}}{2}$ **6.** $x = \dfrac{2 \pm \sqrt{2}}{2}$

Section 6.4

7. (a) 25 feet; 34 feet **(b)** 3 seconds   **8.** height $= -\dfrac{3}{2} + \dfrac{\sqrt{57}}{2} \approx 2.3$ inches, base $= \dfrac{3}{2} + \dfrac{\sqrt{57}}{2} \approx 5.3$ inches

Section 6.5

9. $(-\infty, -3) \cup (2, +\infty)$ **10.** $\left[-2, \dfrac{2}{3}\right]$ **11.** $[-3, 0) \cup (2, +\infty)$

Section 6.6

12. $x = 1 \pm i$ **13.** $x = -1 \pm i\sqrt{3}$ **14.** $x = \dfrac{-3 \pm 5i}{4}$

Sections 6.1, 6.3, and 6.6

15. $t = -2$ or $t = \dfrac{2}{3}$ **16.** $y = -2 \pm i$ **17.** $x = \dfrac{2 \pm \sqrt{10}}{2}$ **18.** $z = 1$ or $z = \dfrac{2}{3}$ **19.** $x = \pm i\sqrt{2}$
20. $x = -1$ or $x = \dfrac{4}{3}$

C H A P T E R   7

E X E R C I S E   S E T   7.1

1. linear   **3.** not linear   **5.** linear   **7. (a)** solution   **(b)** not a solution   **(c)** solution   **(d)** not a solution   **9.** $A(6, 0)$,
$B(3, -2)$, $C(2, 2.5)$, $D(-3.5, 0)$, $E(-3.5, -5)$, $F(0, 7)$ **11.** the first quadrant   **13.** the origin together with the positive
x-axis   **15.** the fourth quadrant   **17.** the x-axis   **19.** the vertical line one unit to the left of the y-axis   **21.** all
points in the first quadrant that are on the horizontal line 2 units above the x-axis   **23.** all points on the line $x = 2y$

25. 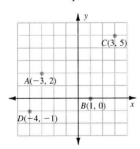   **27.** 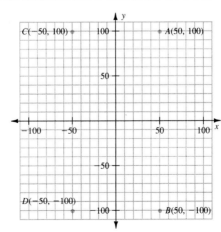   **29.** $(0, -4)$ **31.** $\left(2, \dfrac{4}{3}\right)$

33. $\left(-\dfrac{9}{4}, -10\right)$ **35.**

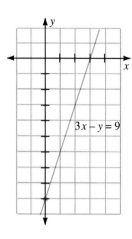

$3x - y = 9$

37.

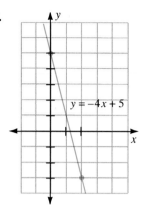

$y = -4x + 5$

39.

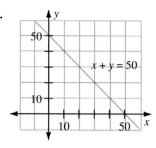

$x + y = 50$

41.

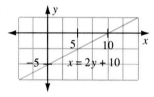

$x = 2y + 10$

43.

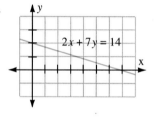

$2x + 7y = 14$

45.

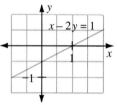

$x - 2y = 1$

47.

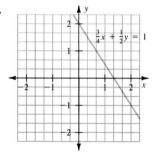

$\frac{3}{4}x + \frac{1}{2}y = 1$

49.

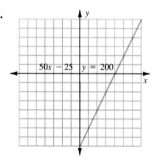

$50x - 25y = 200$

51.

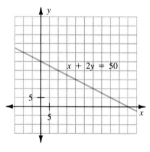

$x + 2y = 50$

53.

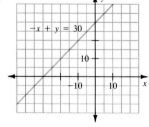

$-x + y = 30$

55.

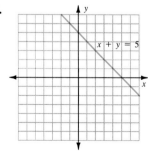

$x + y = 5$

57.

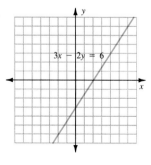

$3x - 2y = 6$

59.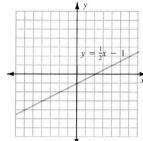

61.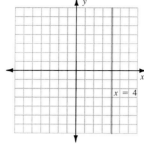

63.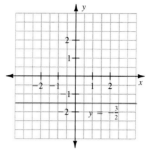

65.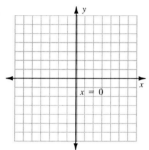

67.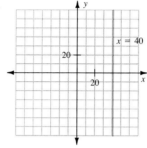

69.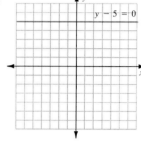

71.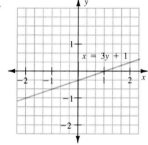

73.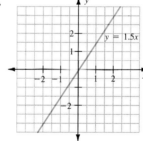

75.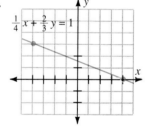

77. $y + 2x = -2$

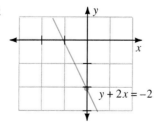

79. $y = 2(x - 5)$

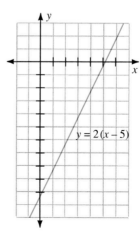

81.

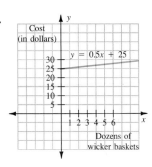

83. (a) $850 **(b)** 4 years old   **85. (a)** $(4, 8)$, $(6, 9)$

(b)

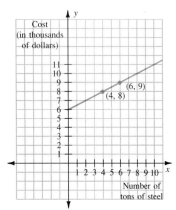

(d) $11,000   **87. (a)** $(0, 0.6)$, $(3, 1.2)$, $(4, 1.4)$

(b) yes

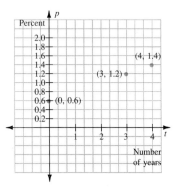

89. (x, y) is an ordered pair of real numbers, whereas $\{x, y\}$ is an unordered pair of real numbers.

Review Problems

91. $y = 5x - 9$ **93.** $y = \dfrac{4}{3}x + 2$ **95.** $x = -12t - 8$

E N R I C H M E N T   E X E R C I S E S

1.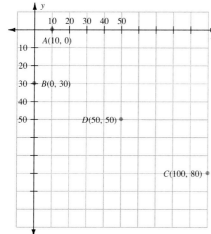

2. $m = \dfrac{5}{3}$ **3.** $b = -\dfrac{5}{14}$ **4.** $c = 2$ **5.** $d = -\dfrac{3}{2}$ **6.** $\left(\dfrac{3}{2}, 3\right)$

7. $(-21, -18)$ **8.** $\dfrac{y_2 - y_1}{x_2 - x_1} = \dfrac{(mx_2 + b) - (mx_1 + b)}{x_2 - x_1} = \dfrac{m(x_2 - x_1)}{x_2 - x_1} = m$

E X E R C I S E   S E T   7 . 2

1. $\dfrac{4}{7}$ **3.** 2   **5.** $-\dfrac{5}{3}$ **7.** undefined   **9.** 9   **11.** $-\dfrac{3}{2}$ **13.** $-\dfrac{5}{8}$ **15.** $\dfrac{9}{2}$ **17.** 0   **19.** undefined   **21.** 1

23. parallel   **25.** not parallel   **27.** not parallel   **29.** not parallel   **31.** parallel   **33.** colinear   **35.** colinear

37. colinear   **39.** colinear   **41.**   **43.** 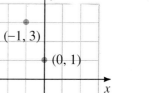   **45.** $1,180

47. no

Review Problems

49. $y = 4x - 5$ **51.** $x = -\dfrac{2}{5}t + \dfrac{1}{10}$ **53.** $y = mx + y_1 - mx_1$

E N R I C H M E N T   E X E R C I S E S

1. $k = \dfrac{1}{4}$ **2.** $k = \dfrac{1}{9}$ **3.** $k = -\dfrac{1}{3}$ or $k = 2$ **4.** line L_1 **5.** line L_1 **6.** 180 feet

E X E R C I S E   S E T   7.3

1. slope $= -1$; y-intercept $= -5$ **3.** slope $= 1$; y-intercept $= 2$ **5.** slope $= -\dfrac{3}{5}$; y-intercept $= 3$

7. slope $= \dfrac{1}{2}$; y-intercept $= -9$ **9.** slope $= 1$; y-intercept $= 0$ **11.** slope $= -1$; y-intercept $= 0$

13. slope $= \dfrac{3}{2}$; y-intercept $= \dfrac{1}{2}$ **15.** slope $= -\dfrac{5}{2}$; y-intercept $= 0$ **17.** slope $= 0$; y-intercept $= 0$ **19.** No, the slope is undefined. No, the line does not cross the y-axis.   **21.**

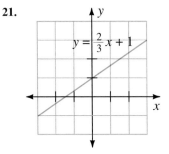

23.

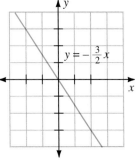

25.

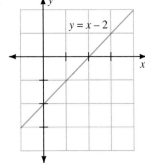

27.

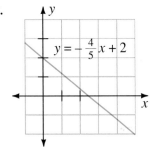

29.

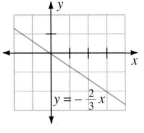

31.

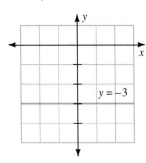

33.

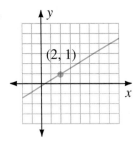

35.

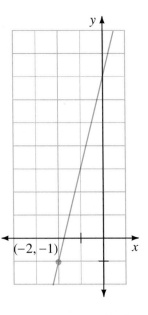

37.

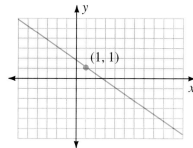

39.

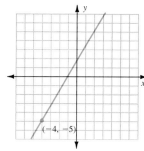

41.

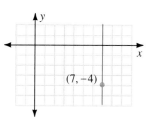

43. $-3x + y = 2$ **45.** $-2x + 5y = -21$ **47.** $32x + y = 2$ **49.** $x + 2y = 11$ **51.** $y = 7$ **53.** $-3x + y = -4$

55. $-2x + y = 0.5$ **57.** $y = \dfrac{10}{3}x$ **59.** $y = \dfrac{2}{5}x + \dfrac{13}{5}$ **61.** $y = -1$ **63.** $y = -x + 2$ **65.** $y = -1$

67. $y = -\dfrac{1}{3}x + \dfrac{3}{5}$ **69.** parallel   **71.** not parallel   **73.** parallel   **75.** $-3x + 8y = 16$ **77.** $x + 4y = -2$

79. $x = -50$ **81.** $y = \dfrac{10}{3}x - 4$ **83.** $y = x + 2$ **85.** $y = 8x + 4$ **87.** $y = \dfrac{4}{5}x - \dfrac{3}{5}$ **89.** $x = -1$

Review Problems

91. $x + 4 \le 3$; $x \le -1$ **93.** $3 - x > -4$; $x < 7$ **95.** $5x + 2 > -3$; $x > -1$

ENRICHMENT EXERCISES

1. $k = 13$ **2.** The line passes through $(0, b)$ and $(a, 0)$. Therefore, the slope is $m = \dfrac{0 - b}{a - 0} = -\dfrac{b}{a}$. Using the slope-intercept form, $y = -\dfrac{b}{a}x + b$, which simplifies to $\dfrac{x}{a} + \dfrac{y}{b} = 1$.   **3.** $-\dfrac{7x}{2} + \dfrac{2y}{7} = 1$ **4.** The slope of the line is

$\dfrac{y_2 - y_1}{x_2 - x_1}$. Using the point-slope form, $y - y_1 = \dfrac{y_2 - y_1}{x_2 - x_1}(x - x_1)$. Set $x = 0$, and solve for y: $y - y_1 = \dfrac{y_2 - y_1}{x_2 - x_1}(-x_1)$; $y =$

$\dfrac{-y_2 x_1 + y_1 x_1}{x_2 - x_1} + y_1 = \dfrac{-y_2 x_1 + y_1 x_1}{x_2 - x_1} + \dfrac{y_1(x_2 - x_1)}{x_2 - x_1} = \dfrac{-y_2 x_1 + y_1 x_1 + y_1 x_2 - y_1 x_1}{x_2 - x_1} = \dfrac{-y_2 x_1 + y_1 x_2}{x_2 - x_1} = \dfrac{y_1 x_2 - y_2 x_1}{x_2 - x_1}$

5. y-intercept $= -39$ **6.** The line has slope m and passes through $(0, b)$. Therefore, $y - b = m(x - 0)$, which simplifies to $y = mx + b$.

EXERCISE SET 7.4

1.

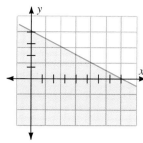

3.

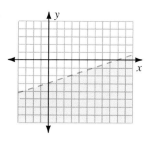

5.

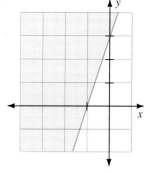

7.

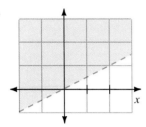

9.

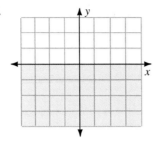

11.

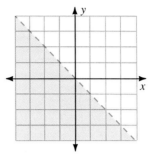

13.

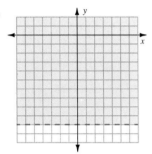

15.

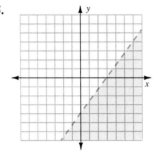

17.

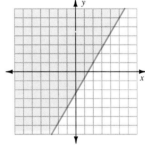

19.

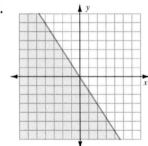

21.

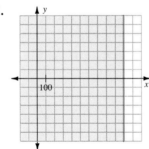

23. $x + y > 2$

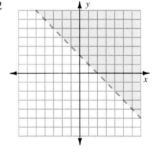

25. $2x - 3y < 18$

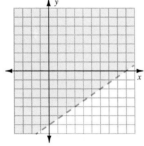

27. $y - 4x < 0$

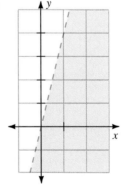

29. $2x \le 3y$

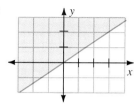

31. $x - 1 > \dfrac{y}{2}$

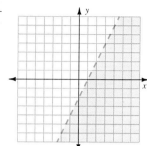

33. (a) $150x$ calories

(b) $30y$ calories **(c)** $150x + 30y \ge 450$ **(d)**

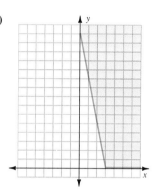

35. (a) $3x$ hours **(b)** $2y$ hours

(c) $3x + 2y \le 12$ **(d)**

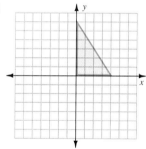

Review Problems

37. $30x$ dollars   **39.** $250x$ dollars

E N R I C H M E N T   E X E R C I S E S

1.

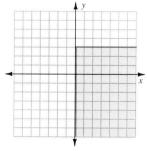

2.

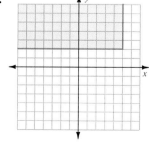

3.

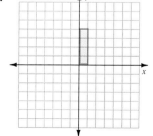

4. 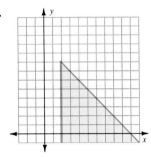   **5.**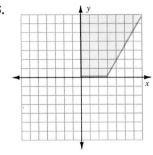

E X E R C I S E   S E T   7 . 5

1. (a) $V = -\dfrac{4}{7}t + 12$ **(b)**   **(c)** $6,300

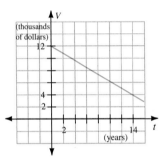   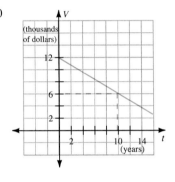

(d) $6,286   **3. (a)** $C = 5x + 40$ **(b)** $2,540   **5. (a)** $C = 525x + 55$ **(b)** $4,255   **7.** cost $= 0.20x + 175$

9. (a) $p = -0.00125x + 5.25$ **(b)** $3.69 per pound **(c)**   **11. (a)** $p = 0.004x - 0.2$

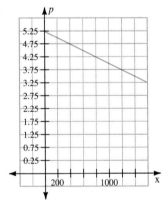

(b) $1.30 per head

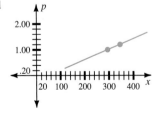

Review Problems

13. $k = \dfrac{5}{3}$ **15.** $k = \dfrac{1}{5}$ **17.** $k = 40$ **19.** $k = 4\sqrt{2}$ **21.** $k = 3$

E N R I C H M E N T   E X E R C I S E S

1. $L = 0.0000727t + 9.9982$ **2. (a)** $y = -\dfrac{275}{3}t + 300$ **(b)** $y = -\dfrac{125}{3}t + 375$ **(c)** $y = -\dfrac{350}{3}t + 1100$

3. (a) $C = \begin{cases} 50x + 200, & \text{if } 0 < x \le 15 \\ 35x + 425, & \text{if } x > 15 \end{cases}$ **(b)**

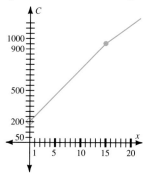

E X E R C I S E   S E T   7.6

1. $y = 27$ **3.** $A = 3$ **5.** $z = \dfrac{1}{12}$ **7.** $V = \dfrac{7}{27}$ **9.** $y = 18$ **11.** $z = \dfrac{32}{27}$ **13.** $y = 2\sqrt{3}$ **15.** $z = -4$

17. $w = 6\sqrt{6}$ **19.** y is reduced by a factor of $\dfrac{1}{3}$.   **21.** A is reduced by a factor of $\dfrac{1}{4}$.   **23.** 15%   **25.** $35.40

27. 312 miles   **29.** 16 bowls   **31.** 4,112 kg

33. The frequency of the second note is $\dfrac{5}{4}$ the frequency of the first note.

E N R I C H M E N T   E X E R C I S E S

1. If y varies directly as x, there is a constant k so that $y = kx$. From the given information, $y_1 = kx_1$ and $y_2 = kx_2$, where neither x_1 nor x_2 is zero. Therefore, $k = \dfrac{y_1}{x_1}$ and $k = \dfrac{y_2}{x_2}$ and so $\dfrac{y_1}{x_1} = \dfrac{y_2}{x_2}$.   **2.** $y = \dfrac{64}{15}$ **3.** Since y varies inversely as x, there is a constant k so that $y = \dfrac{k}{x}$. From the given information, $y_1 = \dfrac{k}{x_1}$ and $y_2 = \dfrac{k}{x_2}$. Therefore, $k = y_1x_1$ and $k = y_2x_2$ and so $y_1x_1 = y_2x_2$.   **4.** $y = -\dfrac{3}{2}$ **5.** 7.2 feet   **6.** 372 deer

C H A P T E R   7   R E V I E W   E X E R C I S E   S E T

Section 7.1

1. the second quadrant   **2.** any point below the x-axis
3. any point on the vertical line 2 units to the left of the y-axis   **4.** any point on the line $y = 3x$

5. 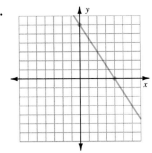   **6.** 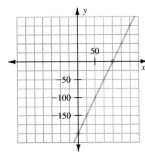   **7.**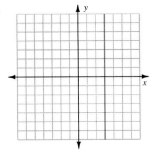

8. 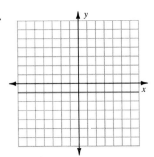   **9.** 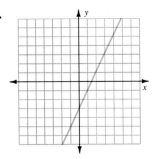   **10.**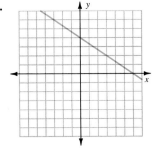

Section 7.2

11. $m = \dfrac{1}{3}$ **12.** $m = 0$ **13.** The slope is undefined.   **14.** $m = \dfrac{1}{8}$

Section 7.3

15. The slope is 4; the y-intercept is -5.   **16.** The slope is $\dfrac{4}{3}$; the y-intercept is $\dfrac{2}{3}$.   **17.**

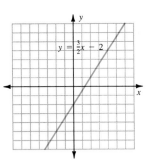

18. 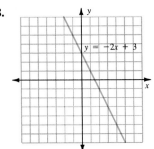   **19.** 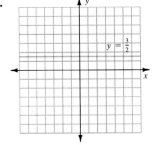   **20.**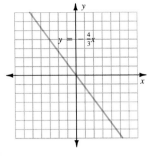

21. $3x + 4y = 10$ **22.** $-x + 2y = 0$ **23.** $y = -\dfrac{5}{6}x + \dfrac{31}{6}$ **24.** $y = -x + 2$ **25.** $y = x$ **26.** $y = -x - 1$

27. $y = \dfrac{3}{2}x + 7$ **28.** $y = -x - 3$

Section 7.4

29.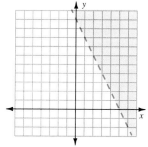

30.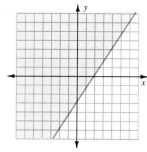

31.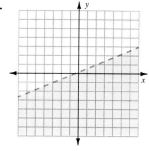

32.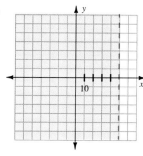

33.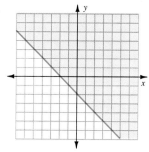

34.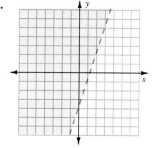

Section 7.5

35. (a) $V = -2t + 28$ **(b)**

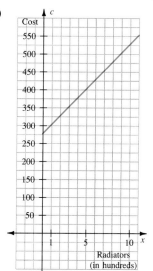

(c) \$16,000   **(d)** \$16,000   **36. (a)** $C = 25x + 275$ **(b)** \$400

(c)

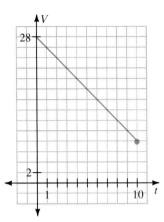

37. **(a)** $p = -8x + 5,600$ **(b)** $400 **(c)**

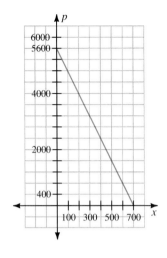

38. **(a)** $p = 0.25x + 0.5$ **(b)** $6 per pound **(c)**

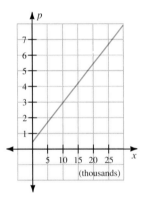

Section 7.6

39. $y = 100$ **40.** $y = -2$ **41.** $r = -108$ **42.** $w = -\sqrt{2}$

CHAPTER 7 TEST

Section 7.1

1.

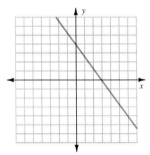

2.

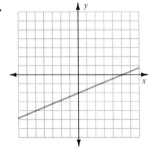

3.

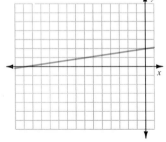

Section 7.2

4. $m = 2$ **5.** The slope is undefined.   **6.** $m = 0$

Section 7.3

7. The slope is $\dfrac{3}{2}$; the y-intercept is -6.   **8. (a)** 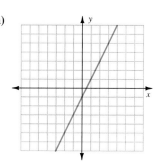 **(b)** $y = 2x - 1$

9. (a) 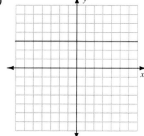 **(b)** $y = 3$ **10. (a)** 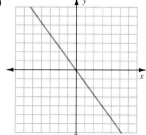 **(b)** $4x + 3y = 0$ **11.** $x + y = 3$

12. $y = \dfrac{1}{2}x - \dfrac{9}{2}$

Section 7.4

13. 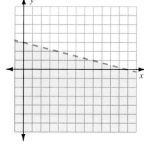   **14.** 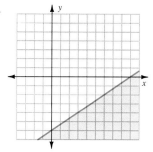   **15.**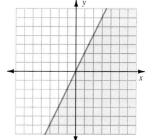

Section 7.5

16. (a) $C = 300x + 5{,}000$ **(b)** \$65,000   **17. (a)** $p = 0.5x$ **(b)** \$6.25 per pound **(c)**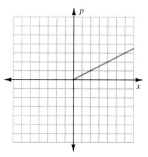

Section 7.6

18. $y = 15$ **19.** $s = -12$ **20.** $z = 18$

CHAPTER 8

EXERCISE SET 8.1

1. $(1, 1)$ is a solution. $(0, 2)$ is not a solution. $(5, -1)$ is not a solution.   **3.** $(1, 1)$ is not a solution. $(0, 5)$ is not a solution. $(0, 0)$ is a solution.   **5.** $x = 1$ and $y = 4$ **7.** $x = 0$ and $y = 4$ **9.** $x = 1$ and $y = 2$ **11.** no solution

13. infinitely many solutions   **15.** $x = -1$ and $y = 10$ **17.** $x = \dfrac{13}{2}$ and $y = -4$ **19.** $w = -4$ and $z = 7$

21. no solution   **23.** $r = \dfrac{1}{10}$ and $y = -\dfrac{1}{10}$ **25.** $m = 4$ and $n = 6$ **27.** no solution   **29.** $h = \dfrac{1}{3}$ and $k = -\dfrac{1}{4}$

31. $x = 2$ and $y = 0$ **33.** $x = 1$ and $y = -1$ **35.** infinitely many solutions   **37.** $x = 5$ and $y = 6$ **39.** $x = 3$ and $y = 6$ **41.** $a = 6$ and $b = 2$ **43.** $x = \dfrac{6}{7}$ and $y = \dfrac{6}{7}$ **45.** $x = 3$ and $y = 4$ **47.** $h = 4$ and $k = -2$

49. $x = 3$ and $y = 1$ **51.** $x = \dfrac{3}{2}$ and $t = \dfrac{1}{2}$ **53.** $x = -3.3$, $y = 2.2$ **55.** $A = 1.3$, $B = 1$ **57.** $X = 4$, $Y = -4$

59. $x_1 = -1$, $x_2 = 2$ **61.** $x = 7$, $y = 6$ **63.** 39 and -28 **65.** The first number is $\dfrac{1}{3}$. The second number is $\dfrac{1}{4}$.

67. -6 and -4 **69.** 42 degrees and 48 degrees   **71.** The length is 7 inches. The width is 5 inches.

73. (a)
$$-13x + 15y = 45$$
$$-13(15) + 15(16) \overset{?}{=} 45$$
$$-195 + 240 \overset{?}{=} 45$$
$$45 = 45; \ 39x - 45y = -135$$
$$39(15) - 45(16) \overset{?}{=} -135$$
$$585 - 720 \overset{?}{=} -135$$
$$-135 = -135$$
(b) No. It is a dependent system.

75. $a = \dfrac{1}{2}$, $b = \dfrac{1}{4}$

Review Problems

77. $0.1x + 0.25y$ dollars   **79.** $12.50x + 18y$ dollars

ENRICHMENT EXERCISES

1. one solution   **2.** no solution

3. (a) Multiply the first equation by b_2 and the second equation by $-b_1$, then add.
$$a_1b_2x + b_1b_2y = c_1b_2$$
$$-b_1a_2x - b_1b_2y = -b_1c_2 \ (+)$$
$$(a_1b_2 - b_1a_2)x = c_1b_2 - b_1c_2.$$
$$\text{Therefore, } x = \frac{c_1b_2 - b_1c_2}{a_1b_2 - b_1a_2}$$

(b) Multiply the first equation by $-a_2$ and the second equation by a_1, then add.
$$-a_1a_2x - b_1a_2y = -c_1a_2$$
$$a_1a_2x + a_1b_2y = a_1c_2 \ (+)$$
$$(a_1b_2 - b_1a_2)y = a_1c_2 - c_1a_2.$$
$$\text{Therefore, } y = \frac{a_1c_2 - c_1a_2}{a_1b_2 - b_1a_2}$$

(c) The system is either inconsistent or dependent.   **4.** $x = \dfrac{31}{13}$, $y = \dfrac{9}{13}$ **5.** $x = -\dfrac{17}{20}$, $y = \dfrac{93}{40}$ **6.** $x = 12$, $y = -12$

7. $x = \dfrac{30}{7}$, $y = -\dfrac{10}{7}$ **8. (a)** $x = \dfrac{1}{A + Bm}$, $y = \dfrac{m}{A + Bm}$ **(b)** $\varnothing$

EXERCISE SET 8.2

1. -9 and -3 **3.** $\dfrac{1}{2}$ and $\dfrac{3}{4}$ **5.** The numerator is -4. The denominator is -10.

7. The width is 2 cm. The length is 17 cm.   **9.** 6 slow terminals, 12 fast terminals

11. 343.75 mg of ingredient A and 156.25 mg of ingredient B   **13.** 20 ounces of ingredient A, 30 ounces of ingredient B
15. $r_b = 23.5$ km/hr, $r_c = 6.5$ km/hr   **17.** The speed of the wind is 75 mph. The speed of the plane is 150 mph.
19. 12 stoves per week   **21. (a)** 1,000 pounds of flounder at \$4 per pound   **(b)**

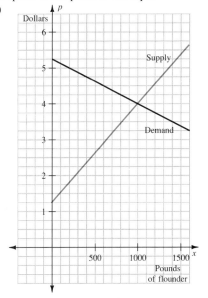

Review Problems

23. $-8y + 11z$ **25.** $-9x + 47z$ **27.** $-23y - 18z$

E N R I C H M E N T   E X E R C I S E S

(a) $R = 58.75x$ **(b)** 40 cases of figures per day   **(c)**

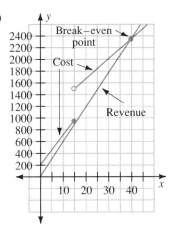

E X E R C I S E   S E T   8 . 3

1. $x = -6, y = -1, z = 0$ **3.** $x = 0, y = -1, z = 1$ **5.** $x = -\dfrac{13}{10}, y = -1, z = \dfrac{9}{10}$ **7.** $x = -3, y = -2, z = 2$

9. $x = \dfrac{5}{2}, y = -1, z = -3$ **11.** $x = -\dfrac{1}{3}, y = -\dfrac{3}{2}, z = \dfrac{1}{4}$ **13.** $x = \dfrac{1}{4}, y = -\dfrac{2}{3}, z = -\dfrac{1}{6}$ **15.** $x = 0, y = 0, z = 0$

17. $x = 0, y = 1, z = -1$ **19.** $x = -\dfrac{2}{3}, y = \dfrac{2}{7}, z = \dfrac{3}{2}$ **21.** $x = 3, y = -2, z = 1$ **23.** $x = \dfrac{1}{6}, y = \dfrac{1}{4}, z = \dfrac{1}{2}$

25. no solution   **27.** $x = -\dfrac{7}{6}, y = -\dfrac{1}{3}, z = -2$ **29.** $-2, 1$, and 3

31. Machine A can finish 12 bins, machine B can finish 8 bins, and machine C can finish 15 bins.

Review Problems

33. 5   **35.** 9   **37.** 7   **39.** 0

E N R I C H M E N T   E X E R C I S E S

1. $a = \dfrac{9}{2}, b = -3, c = 4$ **2.** $a = \pm i\sqrt{2}, b = \pm 1, c = \pm\dfrac{\sqrt{2}}{4}$ **3.** $x = \dfrac{1}{3}, y = 1, z = \dfrac{7}{6}, t = 1$

4. no

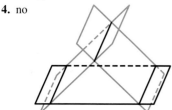

E X E R C I S E   S E T   8 . 4

1. -7 **3.** 6   **5.** -2 **7.** 5   **9.** 6   **11.** -3 **13.** -7 **15.** 15   **17.** $x = -1$ and $y = 3$ **19.** $x = 2$ and
$y = 1$ **21.** $x = -\dfrac{2}{3}$ and $y = \dfrac{4}{3}$ **23.** $x = -\dfrac{7}{2}$ and $y = \dfrac{2}{3}$ **25.** no solution   **27.** $x = -\dfrac{1}{4}, y = -8$ $z = -\dfrac{33}{4}$

29. $x = \dfrac{1}{2}, y = -\dfrac{1}{5}, z = \dfrac{1}{3}$ **31.** $x = -\dfrac{3}{5}, y = \dfrac{1}{5}, z = -1$ **33.** $x = -\dfrac{10}{7}, y = \dfrac{16}{7}, z = -\dfrac{24}{7}$ **35.** $x = 5, y = -1,$

$z = 0$ **37.** $\dfrac{1}{2}$ inch, $5\dfrac{1}{2}$ inches, and $7\dfrac{1}{2}$ inches.   **39.** 200 student tickets, 50 children tickets, 25 adult tickets

41. $a = \dfrac{3}{4}, b = \dfrac{1}{2}, c = \dfrac{1}{4}$

Review Problems

Section 8.1

43. $x = \dfrac{7}{2}, y = 3$

Section 8.3

45. $x = 4, y = 6, z = -1$ **47.** $x = -162, y = -26, z = -3$

E N R I C H M E N T   E X E R C I S E S

1. $x = -\dfrac{20}{23}, y = -\dfrac{15}{8}, z = \dfrac{10}{7}$ **2.** $x = \pm\dfrac{2}{3}i\sqrt{3}, y = \pm\dfrac{\sqrt{2}}{4}, z = \pm\dfrac{\sqrt{3}}{2}$ **3.** 0   **4.** 9 cases of tuning forks, 9 cases
of pitch pipes, 6 cases of mouth organs   **5.** 16 cases of hand excavators, 10 cases of drill bits, 12 cases of pinch valves

E X E R C I S E   S E T   8 . 5

1. $\begin{bmatrix} 1 & -2 & | & \frac{2}{3} \\ 2 & 4 & | & -7 \end{bmatrix}$ **3.** $\begin{bmatrix} 1 & -2 & | & 3 \\ 0 & 13 & | & -10 \end{bmatrix}$ **5.** $\begin{bmatrix} 1 & -4 & | & -5 \\ 0 & 1 & | & 0 \end{bmatrix}$ **7.** $\begin{bmatrix} 1 & -4 & 2 & | & 28 \\ 2 & 3 & -1 & | & 4 \\ 0 & -10 & -2 & | & -6 \end{bmatrix}$

9. $\begin{bmatrix} 1 & -4 & 3 & | & -1 \\ 0 & 5 & 4 & | & 2 \\ 0 & 7 & -7 & | & 7 \end{bmatrix}$ **11.** $\begin{bmatrix} 1 & -5 & \frac{3}{2} & | & -2 \\ 0 & 9 & -1 & | & 5 \\ 0 & 13 & -\frac{13}{4} & | & 2 \end{bmatrix}$ **13.** $x = 1, y = -1$ **15.** $x = 3, y = 6$ **17.** $x = \frac{1}{2}, y = \frac{3}{2}$

19. $\{(x, y) \mid x - 2y = -4\}$ **21.** no solution   **23.** $x = 17, y = 6, z = -1$ **25.** $x = 6, y = -3, z = -2$ **27.** $x = -4,$
$y = -11, z = 3$ **29.** $x = -5, y = 3, z = -4$ **31.** no solution   **33.** $x = 1, y = -1, z = 2$ **35.** $x = 3, y = -1, z = 2$
37. $x = -4, y = 0, z = 1$ **39.** $x = \frac{1}{2}, y = -\frac{1}{3}, z = 1$

Review Problems

41. 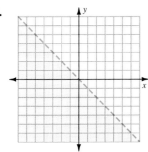   **43.** 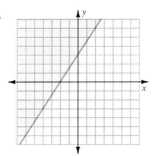   **45.**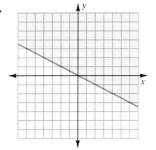

E N R I C H M E N T   E X E R C I S E S

1. $a = -\frac{1}{3}$ **2.** $x = \frac{3a + 2}{a^2 + 1}, y = \frac{-2a + 3}{a^2 + 1}$ **3.** $x = 1, y = -1, z = -1, t = 1$ **4.** $x = -2, y = -2, z = 3, t = -1$
5. $x = 1, y = 0, z = -1, s = 2, t = -2$ **6.** $x = -2, y = -1, z = 1, s = 2, t = 1$

E X E R C I S E   S E T   8 . 6

1. 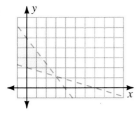   **3.** 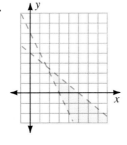   **5.**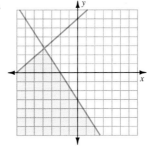

7.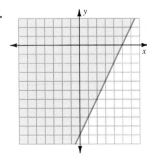

9.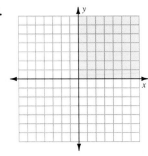

11.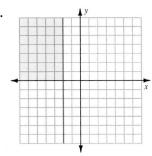

13.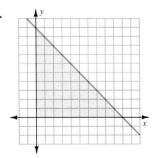

15.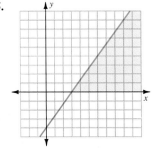

17.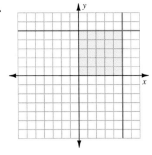

19.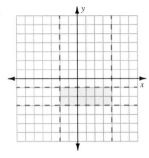

21.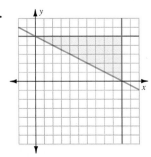

23. **(a)** $x + y \leq 20{,}000$; $y \leq 8{,}000$; $x \geq 0$ and $y \geq 0$

(b)

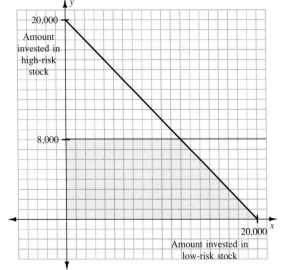

25. **(a)** $3x + 4y \leq 12$; $4x + 2y \leq 12$; $x \geq 0$ and $y \geq 0$

(b)

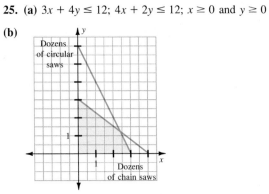

Review Problems

27. $x = 1$ or $x = -3$ **29.** $x = 2$ or $x = -\dfrac{1}{2}$ **31.** $x = 1$ or $x = 3$ **33.** $x = 1$ or $x = -5$ **35.** $x = \dfrac{3}{2}$ or $x = -\dfrac{2}{3}$

E N R I C H M E N T   E X E R C I S E S

1.

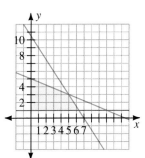

2.

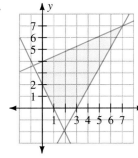

3.

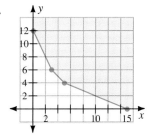

C H A P T E R   8   R E V I E W   E X E R C I S E   S E T

Section 8.1

1. $x = 0$, $y = 4$ **2.** no solution   **3.** $x = 2$, $y = 1$ **4.** $x = -2$, $y = -1$ **5.** $x = -5$, $y = 10$

6. $x = -\dfrac{1}{3}$, $y = -\dfrac{3}{2}$ **7.** $x = -\dfrac{9}{4}$, $y = \dfrac{5}{2}$ **8.** $x = \dfrac{1}{2}$, $y = \dfrac{1}{2}$

Section 8.2

9. The first number is $-\dfrac{14}{3}$; the second number is $-\dfrac{34}{3}$.   **10.** The width is 3 inches; the length is 12 inches.

11. The rate of the boat is 20 mph; the rate of the current is 4 mph.

12. 49 liters of the 20% iodine solution; 31 liters of the 60% iodine solution.   **13. (a)** $R = 350x$ **(b)** 10 beds per week

Section 8.3

14. $x = -\dfrac{1}{2}$, $y = \dfrac{1}{2}$, $z = -\dfrac{1}{2}$ **15.** $x = -\dfrac{3}{2}$, $y = -1$, $z = -\dfrac{2}{7}$ **16.** $x = -2$, $y = 0$, $z = 3$

17. $x = 1$, $y = -1$, $z = -1$

Section 8.4

18. -22 **19.** -3 **20.** 7   **21.** 91   **22.** 1   **23.** -10 **24.** $x = \dfrac{1}{12}$, $y = \dfrac{1}{3}$ **25.** $x = \dfrac{7}{6}$, $y = \dfrac{5}{2}$

26. infinitely many solutions   **27.** no solution   **28.** $x = 2$, $y = -1$, $z = -2$ **29.** $x = \dfrac{1}{2}$, $y = -\dfrac{1}{2}$, $z = 0$

30. $x = \dfrac{3}{4}$, $y = \dfrac{5}{4}$, $z = \dfrac{3}{4}$ **31.** $x = -\dfrac{1}{5}$, $y = \dfrac{2}{5}$, $z = \dfrac{1}{5}$ **32.** $x = 1$, $y = 1$, $z = 0$ **33.** $x = -\dfrac{5}{8}$, $y = -\dfrac{1}{8}$, $z = \dfrac{5}{8}$

Section 8.5

34. $x = -2$, $y = 1$ **35.** $x = \dfrac{1}{2}$, $y = \dfrac{1}{3}$ **36.** $x = -1$, $y = -1$, $z = 2$ **37.** $x = -2$, $y = -3$, $z = -4$

38. $x = 2$, $y = -3$, $z = -2$ **39.** $x = \dfrac{1}{2}$, $y = -\dfrac{1}{2}$, $z = -\dfrac{1}{3}$

Section 8.6

40.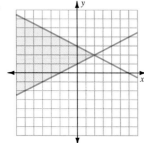

41.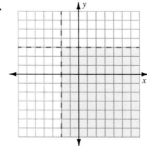

42.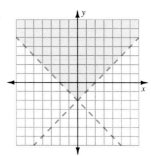

43.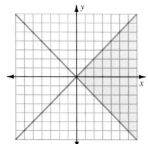

44.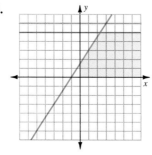

45.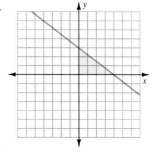

C H A P T E R   8   T E S T

Section 8.1

1. $x = 1$ and $y = -1$ **2.** $x = -\dfrac{1}{2}$ and $y = \dfrac{1}{3}$ **3.** $x = -4$ and $y = -1$ **4.** $x = \dfrac{2}{5}$ and $y = \dfrac{4}{3}$

Section 8.2

5. The first number is $-\dfrac{3}{2}$, the second number is $-\dfrac{7}{2}$.   **6.**
6 quarts of the 20% antifreeze; 10 quarts of the 12% antifreeze.

Section 8.3

7. $x = 1, y = -1, z = 2$ **8.** $x = 0, y = -1, z = -3$ **9.** $x = \dfrac{1}{2}, y = \dfrac{1}{3}, z = -\dfrac{1}{2}$

Section 8.4

10. -10 **11.** -12 **12.** -15 **13.** $x = 1, y = -1$ **14.** $x = \dfrac{1}{4}, y = \dfrac{1}{5}$ **15.** $x = -\dfrac{1}{2}, y = -\dfrac{3}{4}, z = \dfrac{1}{4}$

Section 8.5

16. $x = \dfrac{1}{2}, y = -\dfrac{1}{4}$ **17.** No solution   **18.** $x = 2, y = -1, z = -4$

Section 8.6

19.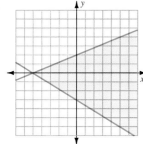

20.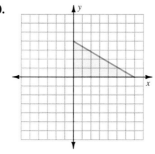

CHAPTER 9

EXERCISE SET 9.1

1.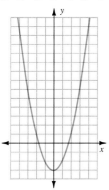

3.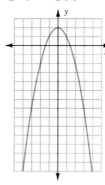

5.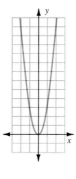

7.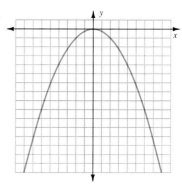

9.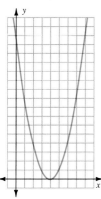

11.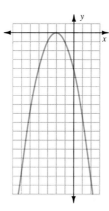

13.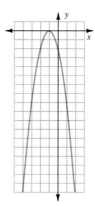

15.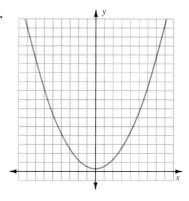

17.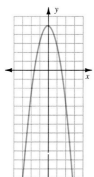

19.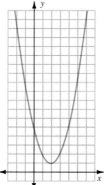

21.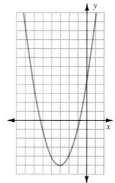

23.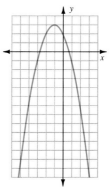

25.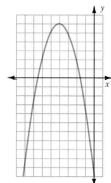

27.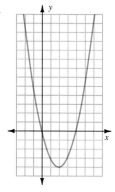

29.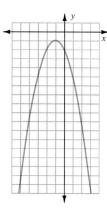

31.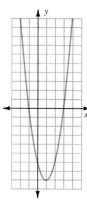

33.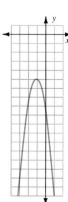

35.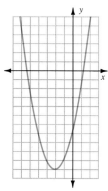

37.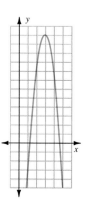

39.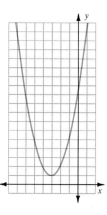

41.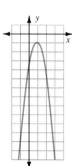

43.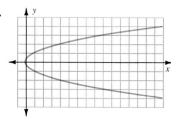

45.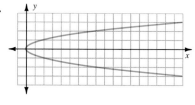

47.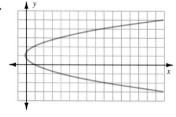

49.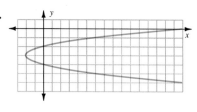

Review Problems

51. $x = -2$ or $x = -3$ **53.** $x = -1$ or $x = 3$ **55.** $x = -\dfrac{1}{2}$ or $x = \dfrac{2}{3}$ **57.** $x = 5 \pm 2\sqrt{2}$

ENRICHMENT EXERCISES

1.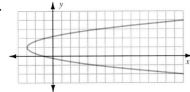
2.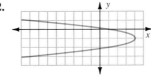
3.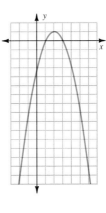

4.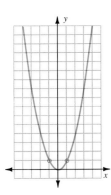
  5. $a = -\dfrac{1}{8}$ **6.** $k = 0$ **7.** $y = (x - 3)^2 + 6$

8. $y_0 = a(x_1 - h)^2 + k$; $y_0 = a(x_2 - h)^2 + k$. Therefore, $a(x_1 - h)^2 + k = a(x_2 - h)^2 + k$; $a(x_1 - h)^2 = a(x_2 - h)^2$; $(x_1 - h)^2 = (x_2 - h)^2$; $x_1 - h = \pm(x_2 - h)$. Divide both sides by a. Square root property. If $x_1 - h = x_2 - h$, then $x_1 = x_2$, contradicting that $x_1 \neq x_2$. Therefore, $x_1 - h = -(x_2 - h)$; $x_1 - h = -x_2 + h$; $x_1 + x_2 = 2h$; $\dfrac{x_1 + x_2}{2} = h$.

EXERCISE SET 9.2

1. y-intercept: -6; x-intercepts: 2 and -3 **3.** y-intercept: -4; x-intercepts: -4 and $\dfrac{1}{2}$ **5.** y-intercept: 0;

x-intercepts: 0 and 3   **7.** y-intercept: 1; x-intercept: $-\dfrac{1}{2}$ **9.** y-intercept: -2; x-intercepts: $\pm\sqrt{2}$ **11.** y-intercept: 1;

x-intercepts: $2 \pm \sqrt{3}$ **13.** y-intercepts: 2; no x-intercepts   **15.**

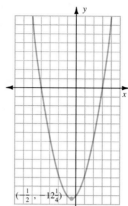

17.

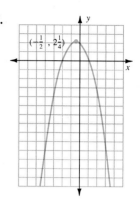

19.

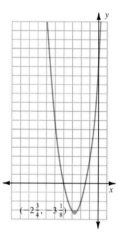

21. (a) y has a smallest value. (b) It occurs at $x = 12$. (c) The smallest value is $y = -9.8$.

23. (a) y has a largest value. (b) It occurs at $x = -\dfrac{1}{2}$. (c) The largest value is $y = 15$.

25. (a) y has a largest value. (b) It occurs at $x = -\dfrac{5}{8}$. (c) The largest value is $y = \dfrac{49}{8}$.

27. (a) y has a smallest value. (b) It occurs at $x = 3$. (c) The smallest value is $y = -\dfrac{37}{4}$.   **29.** 3 dozen dolls; $220

31. 4 seconds; 256 feet   **33.** 18 feet by 36 feet

Review Problems

35. $2\sqrt{15}$ **37.** $2\sqrt{2}$ **39.** $5\sqrt{2}$ **41.** $\dfrac{\sqrt{74}}{2}$

E N R I C H M E N T   E X E R C I S E S

1. 68.5 and 68.5   **2.** 14.5 and 14.5   **3.** 137 feet by 137 feet   **4.** Given $h = vt - 16t^2$, the time it takes to reach the maximum height is $-\dfrac{v}{2(-16)} = \dfrac{v}{32}$. Given $h = 2vt - 16t^2$, the time it takes to reach the maximum height is $-\dfrac{2v}{2(-16)} = 2\left(\dfrac{v}{32}\right)$.

EXERCISE SET 9.3

1. $\sqrt{41}$ **3.** $2\sqrt{5}$ **5.** 15 **7.** $\sqrt{29}$ **9.** $x = -5$ or $x = 1$ **11.** $y = -11$ or $y = 13$
13. $(x-4)^2 + (y-1)^2 = 25$ **15.** $x^2 + y^2 = 12$ **17.** $(x+3)^2 + (y+5)^2 = 10$ **19.** $x^2 + y^2 = 34$

21. 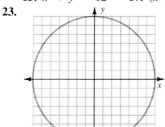 **23.** 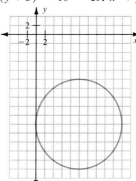 **25.**

27. 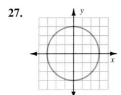 **29.** **31.** 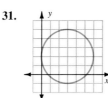 **33.**

35. 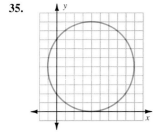 **37.** **39.** 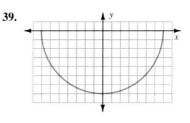 **41.**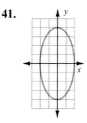

43. 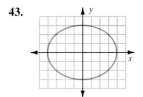 **45.** 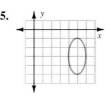 **47.** 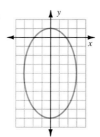 **49.**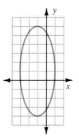

Review Problems

51. $y = (x-3)^2 + 1$ **53.** $y = 3(x+2)^2 - 4$ **55.** $y = -2(x+6)^2 + 45$ **57.** $y = -(x-3)^2 - 3$

ENRICHMENT EXERCISES

1. $\left(\frac{3}{2}, 5 + \frac{\sqrt{15}}{2}\right)$ and $\left(\frac{3}{2}, 5 - \frac{\sqrt{15}}{2}\right)$ **2.** $(x-3)^2 + (y+2)^2 = 9$ **3.** $(x+1)^2 + (y+4)^2 = 16$

4. $(x - 2)^2 + (y - 4)^2 = 25$ **5.** a circle   **6.** 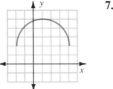   **7.** 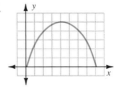   **8.** 84 feet

9.
$$d_1 + d_2 = 10$$
$$\sqrt{(x + 3)^2 + y^2} + \sqrt{(x - 3)^2 + y^2} = 10$$
$$\sqrt{(x + 3)^2 + y^2} = 10 - \sqrt{(x - 3)^2 + y^2}$$
$$[\sqrt{(x + 3)^2 + y^2}]^2 = [10 - \sqrt{(x - 3)^2 + y^2}]^2$$
$$(x + 3)^2 + y^2 = 100 - 20\sqrt{(x - 3)^2 + y^2} + (x - 3)^2 + y^2$$
$$x^2 + 6x + 9 + y^2 = 100 - 20\sqrt{(x - 3)^2 + y^2} + x^2 - 6x + 9 + y^2$$
$$12x - 100 = -20\sqrt{(x - 3)^2 + y^2}$$
$$3x - 25 = -5\sqrt{(x - 3)^2 + y^2}$$
$$(3x - 25)^2 = [-5\sqrt{(x - 3)^2 + y^2}]^2$$
$$9x^2 - 150x + 625 = 25[(x - 3)^2 + y^2]$$
$$9x^2 - 150x + 625 = 25(x^2 - 6x + 9 + y^2)$$
$$9x^2 - 150x + 625 = 25x^2 - 150x + 225 + 25y^2$$
$$9x^2 - 25x^2 - 25y^2 = 225 - 625$$
$$-16x^2 - 25y^2 = -400$$
$$\frac{-16x^2}{-400} - \frac{25y^2}{-400} = \frac{-400}{-400}$$
$$\frac{x^2}{25} + \frac{y^2}{16} = 1$$

E X E R C I S E   S E T   9 . 4

1. 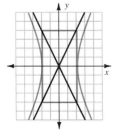   **3.** 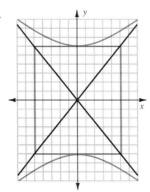   **5.** 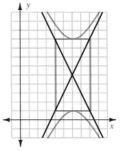   **7.**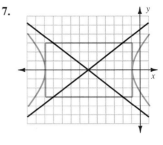

9. 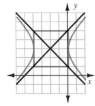   **11.**   **13.** 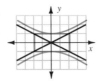   **15.**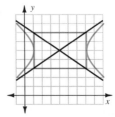

17.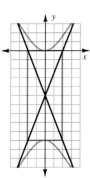

19.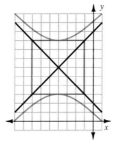

21. $y = \dfrac{1}{4}x$ and $y = -\dfrac{1}{4}x$

23.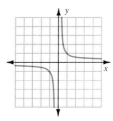

25.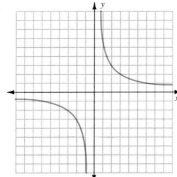

27.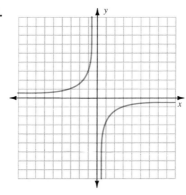

29.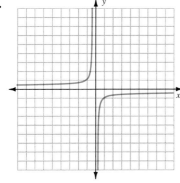

31.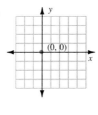

33.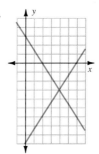

35.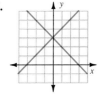

37.   **39.**   **41.**   **43.** no graph **45.**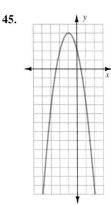

47. no graph

Review Problems

49. $x = 1$ **51.** $x = -2$ or $x = 3$ **53.** $y = 1$

E N R I C H M E N T   E X E R C I S E S

1. 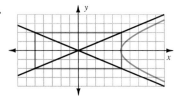   **2.** 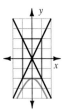   **3.** 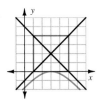   **4.**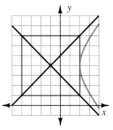

E X E R C I S E   S E T   9 . 5

1. $(2, 2), (-2, -2)$ **3.** $(-2, 16)$ **5.** $\left(\dfrac{2 + i\sqrt{2}}{3}, \dfrac{1 + 2i\sqrt{2}}{3}\right), \left(\dfrac{2 - i\sqrt{2}}{3}, \dfrac{1 - 2i\sqrt{2}}{3}\right)$ **7.** $\left(-\dfrac{1}{2}, 1\right), (4, -26)$

9. $(2, -4), (-2, -4), (i\sqrt{3}, 3), (-i\sqrt{3}, 3)$ **11.** $(\sqrt{2}, -\sqrt{2}), (-\sqrt{2}, \sqrt{2})$ **13.** $(0, 2i\sqrt{2}), (0, -2i\sqrt{2})$,

$\left(\dfrac{2}{3}\sqrt{6}, \dfrac{2}{3}\sqrt{6}\right), \left(-\dfrac{2}{3}\sqrt{6}, -\dfrac{2}{3}\sqrt{6}\right)$ **15.** $(2\sqrt{2}, 2\sqrt{2}), (-2\sqrt{2}, -2\sqrt{2}), (4i, -2i), (-4i, 2i)$ **17.** $(-1, 10), (6, 3)$

19. $\left(\dfrac{2}{3}, \dfrac{11}{3}\right)$ **21.** $\left(\dfrac{5}{4}, \dfrac{25}{16}\right)$ **23.** $\left(\dfrac{\sqrt{390}}{13}, \dfrac{\sqrt{78}}{13}\right), \left(\dfrac{\sqrt{390}}{13}, -\dfrac{\sqrt{78}}{13}\right); \left(-\dfrac{\sqrt{390}}{13}, \dfrac{\sqrt{78}}{13}\right), \left(-\dfrac{\sqrt{390}}{13}, -\dfrac{\sqrt{78}}{13}\right).$

25. $\left(2, \dfrac{9}{2}\right), \left(-2, -\dfrac{9}{2}\right)$ **27.** 3 and 8   **29.** 4 feet and 8 feet   **31.** $\dfrac{3}{2}$ meters by 2 meters

33. 500 pounds of scallops at \$3.50 per pound

Review Problems

35.   **37.**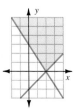

ENRICHMENT EXERCISES

1. $(0, 0)$, $\left(\frac{1}{2}, 1\right)$, $\left(-\frac{1}{2}, -1\right)$ **2.** $(0, 0)$, $(1, -1)$, $(-1, 1)$ **3.** $(1, 1)$ **4.** 4 and $\frac{3}{2}$

5. length $= \dfrac{P + \sqrt{P^2 - 16A}}{4}$; width $= \dfrac{P - \sqrt{P^2 - 16A}}{4}$

EXERCISE SET 9.6

1.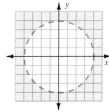

3.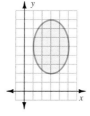

5.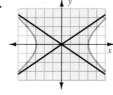

7.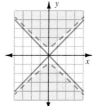

9.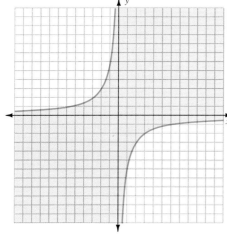

11.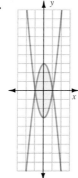

13.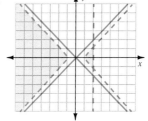

15.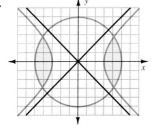

17.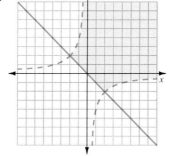

19.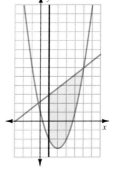

21.
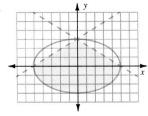

23.
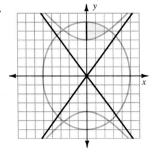

Review Problems

25.
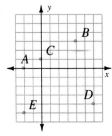

E N R I C H M E N T   E X E R C I S E S

1.
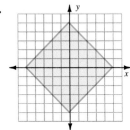

2.
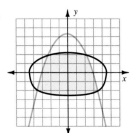

3.
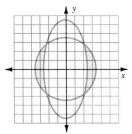

4.
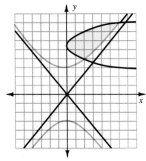

C H A P T E R   9   R E V I E W   E X E R C I S E   S E T

Section 9.1

1. 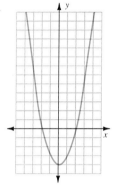   **2.** 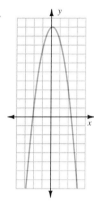   **3.** 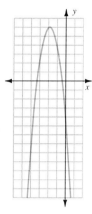   **4.**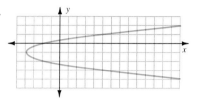

5. 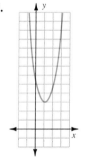   **6.**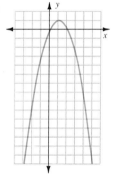

Section 9.2

7. y-intercept: -12; x-intercepts: -4 and $\dfrac{3}{2}$ **8. (a)** y has a smallest value.   **(b)** It occurs at $x = -2$.
(c) The smallest value is $y = 11$.   **9.** 3 seconds; 144 feet

Section 9.3

10. $\sqrt{34}$ **11.** $\sqrt{61}$ **12.** $\dfrac{\sqrt{10}}{2}$ **13.** $(x - 3)^2 + (y + 2)^2 = 1$ **14.** $(x + 4)^2 + (y + 9)^2 = 36$ **15.**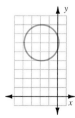

16.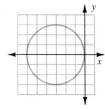

17.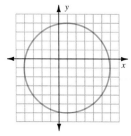

18.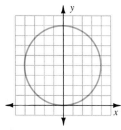

19.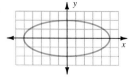

20.

21.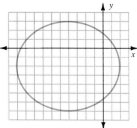

22.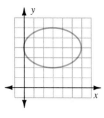

Section 9.4

23.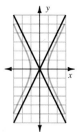

24.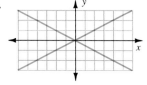

25.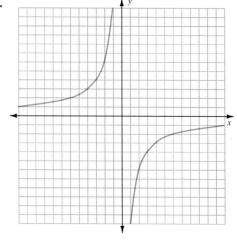

26.

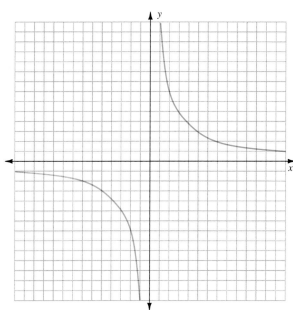

27.

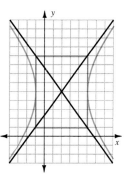

28.

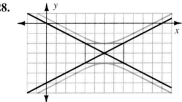

29.

Section 9.5

30. $(-2, -4)$ $(2, 4)$ **31.** $\left(5, -\dfrac{\sqrt{37}}{2}\right) \left(5, \dfrac{\sqrt{37}}{2}\right)$ **32.** $\left(-1, -\dfrac{1}{3}\right) \left(1, \dfrac{1}{3}\right)$

33. $(-3, -2i)$ $(-3, 2i)$ $(4, -\sqrt{3})$ $(4, \sqrt{3})$ **34.** $(-2i\sqrt{2}, -3)$ $(-2i\sqrt{2}, 3)$ $(2i\sqrt{2}, -3)$ $(2i\sqrt{2}, 3)$

35. $(2, -1)$ $(-2, 1)$ $(1, -2)$ $(-1, 2)$ **36.** 10 feet by 12 feet

Section 9.6

37.

38.

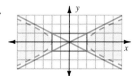

39.

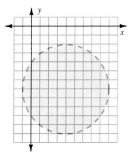

40.
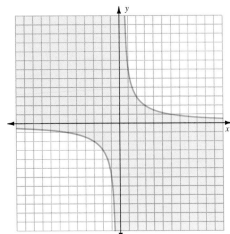

41.
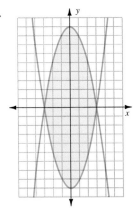

42.
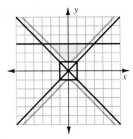

43.
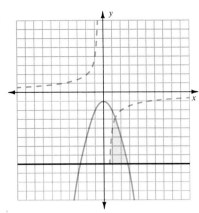

44.
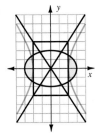

45.
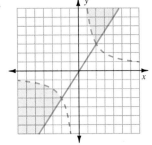

46.
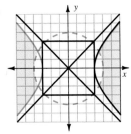

C H A P T E R   9   T E S T

Section 9.1

1.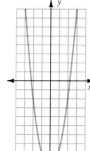

2.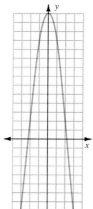

3.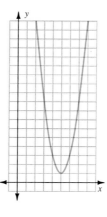

4.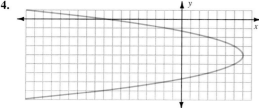

Section 9.2

5. y-intercept: -5, x-intercepts: $1 \pm \sqrt{6}$ **6. (a)** y has a largest value.   **(b)** It occurs at $x = 2$.
(c) The largest value is $y = 4$.

Section 9.3

7. 5   **8.** $(x + 1)^2 + (y - 6)^2 = 9$ **9.** 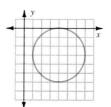   **10.**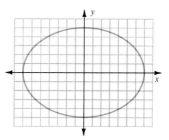

11. 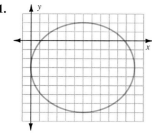   **12.**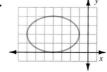

Section 9.4

13.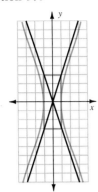

14.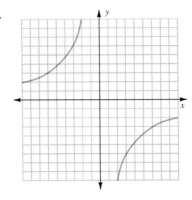

15.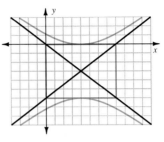

16.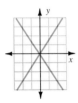

Section 9.5

17. $(2, -3)$ and $(-2, 3)$ **18.** $(3, 2)$, $(-3, -2)$, $\left(\sqrt{10}, \dfrac{3\sqrt{10}}{5}\right)$, $\left(-\sqrt{10}, -\dfrac{3\sqrt{10}}{5}\right)$

Section 9.6

19.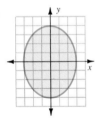

20.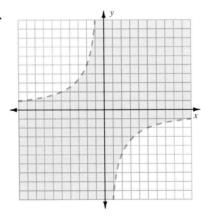

21.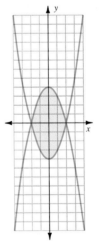

22.

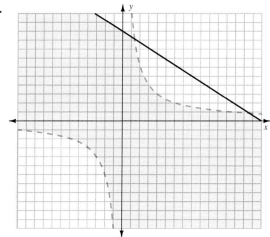

C H A P T E R   10

E X E R C I S E   S E T   10.1

1. $\{(a, 5), (a, 3), (b, 3), (b, 2), (c, 2)\}$; domain = $\{a, b, c\}$; range = $\{2, 3, 5\}$ **3.** $R = \{(-2, -4), (-1, 2), (1, -2), (1, 1)\}$; domain = $\{-2, -1, 1\}$; range = $\{-4, -2, 1, 2\}$ **5.** domain = $[-3, 3]$; range = $[-2, 2]$

7. domain = $\{a, b, c\}$; range = $\{4, 5, 6\}$; it is a function   **9.** domain = $\{1, 2, 3\}$; range = $\{a, b\}$; it is a function
11. domain = $\{-3, -1, 1, 2\}$; range = $\{-2, -1, 2, 3\}$; it is a function   **13.** domain = $\{2, 3, 7\}$; range = $\{-6, -5, -2, -1\}$; it is not a function   **15.** domain = $\{-5, -3, 4, 6\}$; range = $\{-3, -1, 0, 1\}$; it is a function   **17.** domain = $\{-3, -2, -1\}$; range = $\{-3, -2, 2, 10\}$; it is not a function   **19.** $R = \{(-2, 2), (-1, 3), (0, 5), (2, 1), (2, 3), (3, -2)\}$; domain = $\{-2, -1, 0, 2, 3\}$; range = $\{-2, 1, 2, 3, 5\}$; R is not a function   **21.** $R = \{(-3, -2), (-2, 0), (0, -2), (2, 2)\}$; domain = $\{-3, -2, 0, 2\}$; range = $\{-2, 0, 2\}$; R is a function   **23.** domain = $[-4, 3)$; range = $[-3, 4)$; it is a function
25. domain = $[-3, 3]$; range = $[1, 6]$; it is a function   **27.** domain = $(-6, 5)$; range = $\{1, 3\}$; it is not a function
29. domain = $[-4, 0]$; range = $[-4, 5]$; it is a function   **31.** $\{x \mid x \neq -3\}$ **33.** $\{x \mid x \neq 1\}$ **35.** $\{x \mid x \geq 10\}$
37. $\{x \mid x \geq -4\}$ **39.** $\left\{x \mid x < \dfrac{1}{5}\right\}$ **41.** $\left\{x \mid x \neq -1 \text{ and } x \neq \dfrac{3}{2}\right\}$ **43.** $\{x \mid x \neq \pm 4\}$ **45.** $\left\{x \mid x \neq \dfrac{2}{3} \text{ and } x \neq \dfrac{1}{2}\right\}$
47. $\left\{x \mid x \neq \dfrac{1}{2} \text{ and } x \neq -5\right\}$ **49.** domain = $[-5, 5]$; range = $[-5, 5]$; it is not a function   **51.** domain = the set of real numbers; range = $\left[\dfrac{1}{2}, +\infty\right)$; it is a function   **53.** domain = the set of real numbers; range = $[-4, +\infty)$; it is a function
55. domain = $[-3, 5]$; range = $[-6, 2]$; it is not a function   **57.** domain = $[-3, 3]$; range = $[-4, 4]$; it is not a function
59. domain = $[-2, 2]$; range = $\left[-\dfrac{4}{3}, \dfrac{4}{3}\right]$; it is not a function   **61.** domain = $\{x \mid \text{either } x \leq -3 \text{ or } x \geq 3\}$; range = the set of real numbers; it is not a function   **63. (a)** yes   **(b)** $\{t \mid 0 \leq t \leq 12\}$ **(c)** 6 hours after the insulin was injected
(d) 50 mg/dl

Review Problems

65. 1   **67.** $2(x + a)$ **69.** -2 **71.** $2x + h$

E N R I C H M E N T   E X E R C I S E S

1. domain = the set of real numbers; range = $\{-4, -3, -2, -1, 0, 1, 2, 3, 4\}$; it is a function   **2.** domain = $(-8, 12]$; range = $[0, 5)$; it is a function   **3.** domain = $[-4, 4]$; range = the set of integers; it is not a function
4. domain = the set of real numbers; range = $[0, +\infty)$; it is not a function   **5. (a)**   **(b)** domain =

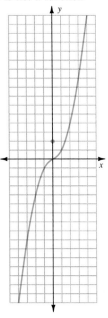

the set of real numbers; range = the set of real numbers.   **(c)** It is not a function, since $(0, 0)$ and $(0, 2)$ are both members of the relation.   **6. (a)**   **(b)** domain = $\{x \mid x \neq 1\}$; range = $\{-1, 1\}$ **(c)** yes, by the vertical line test

1. -1 **3.** 33   **5.** 21   **7.** 2   **9.** -10 **11.** 1   **13.** 1   **15.** 3   **17.** -4 **19.** $1 - 2a$ **21.** $-3 - 2h$
23. $3x + 3h + 2$ **25.** -7 **27.** 3   **29.** $-6x + 5$ **31.** 2   **33.** 4   **35.** -1 **37.** $x + a$ **39.** $-3(x + a)$
41. 3   **43.** 0   **45.** 3   **47.** -9 **49.** $2(2x + h)$ **51.** $(f + g)(x) = 6x^2 + 13x + 6$
53. $(fg)(x) = (2x + 3)(6x^2 + 11x + 3)$ **55.** $(f + g)(x) = 9x^2 + 3x$ **57.** $(fg)(x) = (3x + 1)(9x^2 - 1)$
59. (a) $P(x) = -0.03x^2 + 450x - 450$ **(b)** \$88,350   **61.** 13   **63.** -14 **65.** 19   **67.** 10   **69.** $(f \circ g)(x) = 8 - 2x$;
$(g \circ f)(x) = 4 - 2x$ **71.** $(f \circ g)(x) = 16x^2 + 24x + 11$; $(g \circ f)(x) = 4x^2 + 11$ **73.** $(f \circ g)(x) = \sqrt{x + 3} - 10$;
$(g \circ f)(x) = \sqrt{x} - 7$ **75.** $(f \circ g)(x) = \dfrac{2}{4x - 3}, x \neq \dfrac{3}{4}$; $(g \circ f)(x) = \dfrac{8}{x} - 3, x \neq 0$ **77.** $(f \circ g)(x) = 3\left(\dfrac{x}{3}\right) = x$;
$(g \circ f)(x) = \dfrac{3x}{3} = x$ **79.** $(f \circ g)(x) = 3\left(\dfrac{1}{3}x + \dfrac{1}{3}\right) - 1 = x + 1 - 1 = x$; $(g \circ f)(x) = \dfrac{1}{3}(3x - 1) + \dfrac{1}{3} = x - \dfrac{1}{3} + \dfrac{1}{3} = x$

Review Problems

Section 7.1

81.   **83.**   **85.**

Section 9.1

87. 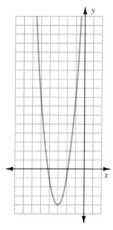   **89.**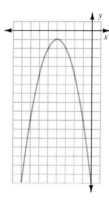

E N R I C H M E N T   E X E R C I S E S

1. $f(x) = x^2$; $g(x) = 3x + 10$ **2.** $f(x) = \sqrt{x}$; $g(x) = 1 + 2x^2$ **3.** $f(x) = \dfrac{1}{x^3}$; $g(x) = 1 - 7x$

4. $f(x) = 2^x$; $g(x) = x - 1$ **5.** $g(x) = \dfrac{1}{2}x + \dfrac{1}{2}$ **6.** $g(x) = -\dfrac{1}{3}x + \dfrac{4}{3}$

E X E R C I S E   S E T   1 0 . 3

1.   **3.**   **5.** 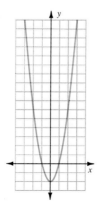   **7.** 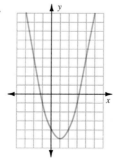   **9.**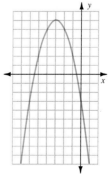

11. 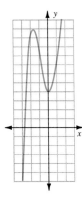　　　　**13.** Minimum; the minimum value is -2 at $x = 1$.

15. Maximum; the maximum value is $5\frac{1}{4}$ at $x = \frac{5}{2}$.　　　**17.** Maximum; the maximum value is 30 at $x = 4$.

19. Minimum; the minimum value is -5 at $x = 3$.　　　**21.** Maximum; the maximum value is 6,400 at $t = 20$.

23. 7,500 cases of trays　　**25. (a)** 1　**(b)** $\frac{3}{2}$　**(c)** $\frac{9}{4}$　**(d)** $\frac{4}{9}$　**27.** 　　**29.**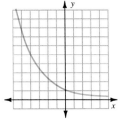

31. 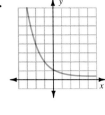　　**33.** 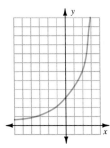　　**35.** 　　**37.**

39. 　　**41.** $f(x) = 4^x = (2^2)^x = 2^{2x} = g(x)$

Review Problems

43. $y = \frac{x}{2} + 2$　　**45.** $y = 4x + 5$　　**47.** $y = \frac{x}{2} + \frac{5}{2}$　　**49.** $y = \frac{4x}{3} - 4$

E N R I C H M E N T   E X E R C I S E S

1.   **2.**   **3.**   **4.** Each is the reflection of the other in the *y*-axis.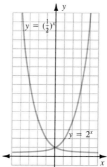

5. (a) 2,000 **(b)** 4,000 **(c)** 8,000   **6.** 5 periods   **7.** 1 hour

E X E R C I S E   S E T   10.4

1. $T^{-1} = \{(1, -6), (-1, -6), (1, 6), (-1, 6)\}$ **3.** *S* is a function; it is one-to-one.   **5.** *T* is not a function.
7. *R* is a function; it is not one-to-one.   **9.** *f* is not one-to-one.   **11.** *g* is one-to-one.   **13.** The function is one-to-one.

15. The function is not one-to-one.   **17. (a)** $f(2) = 6$ **(b)** $f^{-1}(6) = 2$ **19.** $f^{-1}(x) = \dfrac{1}{5}x + 3$ **21.** $f^{-1}(x) = 3x + 3$

23. $f^{-1}(x) = -\dfrac{5}{2}x + 2$ **25.** $f^{-1}(x) = \dfrac{1}{x}$ **27.** $f^{-1}(x) = \dfrac{2}{1 - x}$ **29.** $f^{-1}(x) = \sqrt[3]{x + 1}$

31. $f^{-1}(x) = \dfrac{1}{3}\sqrt[3]{\dfrac{x}{2} - 2} + \dfrac{1}{3}$ **33.** $f^{-1}(x) = \dfrac{x + 1}{x - 1}$ **35.** $f^{-1}(x) = \dfrac{2x - 1}{3x - 3}$ **37.**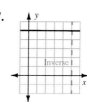

39. 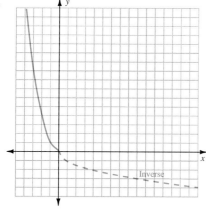   **41.** 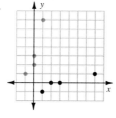   **43.**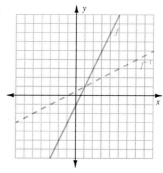

45.
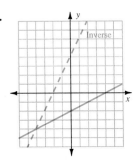

47.
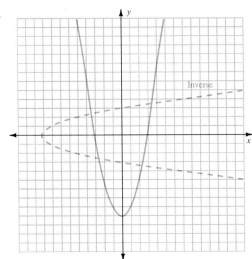

49.
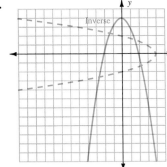

51.
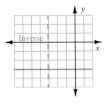

53.
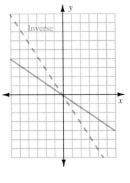

55.
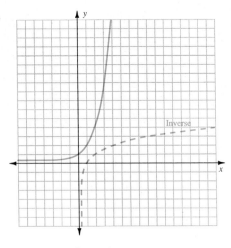

57.
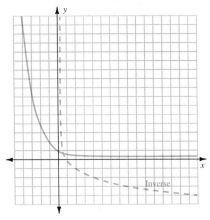

59. (a) $f^{-1}(x) = \dfrac{1}{5}x + \dfrac{4}{5}$ **(b)** $(f \circ f^{-1})(x) = f[f^{-1}(x)] = 5f^{-1}(x) - 4 = 5\left(\dfrac{1}{5}x + \dfrac{4}{5}\right) - 4 = x + 4 - 4 = x$

(c) $(f^{-1} \circ f)(x) = f^{-1}[f(x)] = \left(\dfrac{1}{5}\right)f(x) + \dfrac{4}{5} = \left(\dfrac{1}{5}\right)(5x - 4) + \dfrac{4}{5} = x - \dfrac{4}{5} + \dfrac{4}{5} = x$

Review Problems

61. (a) 1 **(b)** $\dfrac{1}{4}$ **(c)** 4 **(d)** 64

E N R I C H M E N T   E X E R C I S E S

1.

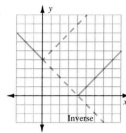

2.

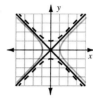

3.

4.

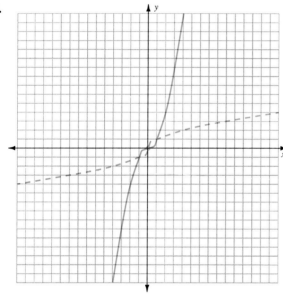

5. at least every 3 days

6. The inverse of f is given by $f^{-1}(x) = \dfrac{1}{m}x - \dfrac{b}{m}$. Therefore, its graph is a straight line with a slope of $\dfrac{1}{m}$.

C H A P T E R   10   R E V I E W   E X E R C I S E   S E T

Section 10.1

1. Domain $= \{-6, 0, 2, 6, 8\}$; range $= \{-2, -1, 2\}$; T is a function.   **2.** Domain $= \{-5, -3, -2, 0\}$; range $= \{-1, 0, 1, 5\}$;

S is not a function.   **3.** the set of real numbers   **4.** $\{x \mid x \ge 14\}$ **5.** $\left\{x \mid x \ne \dfrac{7}{2}\right\}$ **6.** $\{x \mid x \ne -2 \text{ and } x \ne 1\}$

7. Domain = $[-7, 7]$; range = $[-7, 7]$; the relation is not a function.

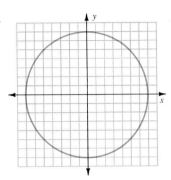

8. Domain = $[-8, 2]$; range = $[-3, 7]$; the relation is not a function.

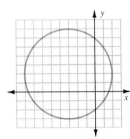

9. Domain = the set of real numbers; range = $(-\infty, 6]$; the relation is a function.

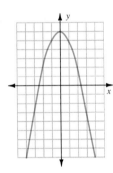

10. Domain = $[-1, 1]$; range = $[-3, 3]$; the relation is not a function.

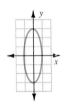

Section 10.2

11. -4 **12.** -8 **13.** 30   **14.** 18   **15.** $2x^2 + 8x + 6$ **16.** $\dfrac{x - 1}{2}$ **17. (a)** 19 **(b)** 5 **(c)** $-12a + 7$

18. $(f \circ g)(x) = (-x + 2)^2$; $(g \circ f)(x) = -x^2 + 2$ **19.** $(f \circ g)(x) = \sqrt{x - 5}$; $(g \circ f)(x) = \sqrt{x} - 5$

20. $(f \circ g)(x) = \dfrac{1}{10 - 6x}$; $(g \circ f)(x) = 1 - \dfrac{2}{3x + 7}$ **21.** -1 **22.** $3(x + a)$

Section 10.3

23.

24.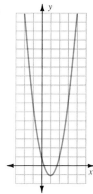

25.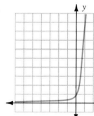

26.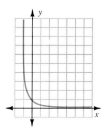

27. f has a minimum at $x = \dfrac{5}{2}$; this minimum value is $f\!\left(\dfrac{5}{2}\right) = -\dfrac{77}{4}$.   **28.** g has a minimum value at $x = 9$; this minimum value is $g(9) = -29$.   **29.** f has a maximum value at $t = \dfrac{1}{2}$; this maximum value is $f\!\left(\dfrac{1}{2}\right) = -\dfrac{15}{4}$.   **30.** g has a maximum value at $t = 1$; this maximum value is $g(1) = -5$.

Section 10.4

31. $S^{-1} = \{(0, 0), (-2, -3), (-3, 3), (2, 0)\}$ **32.** T is not a function.   **33.** S is a function; it is one-to-one.   **34.** It is not one-to-one.   **35.** It is one-to-one.   **36.** It is not one-to-one.   **37.** $f^{-1}(x) = x - 11$ **38.** $f^{-1}(x) = \dfrac{1}{3}x - 3$

39. $f^{-1}(x) = -\dfrac{1}{6}x - \dfrac{1}{3}$ **40.** $f^{-1}(x) = -\dfrac{2}{x}$ **41.** 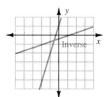   **42.**

43.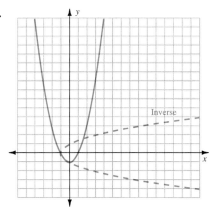

44.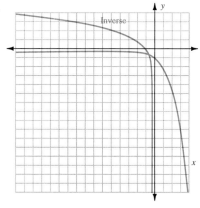

C H A P T E R   1 0   T E S T

Section 10.1

1. Domain = $\{-3, -2, -1, 0\}$; range = $\{-1, 0, 2\}$; S is a function.   **2.** the set of real numbers   **3.** $(-3, +\infty)$
4. Domain = $[-10, 10]$; range = $[-10, 10]$; the relation is not a function.

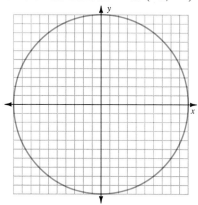

5. Domain = the set of real numbers; range = $(-\infty, 8]$; the relation is a function.

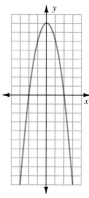

6. Domain = $[0, 8]$; range = $[-12, 0]$; the relation is not a function.

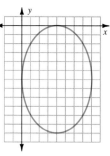

Section 10.2

7. -5 **8.** 9   **9.** -16 **10.** $3x^2 - 4x - 2$ **11.** $(f \circ g)(x) = 27x^2 - 18x + 2$; $(g \circ f)(x) = -9x^2 + 4$ **12.** 4

Section 10.3

13.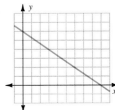

14.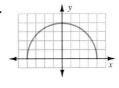

15.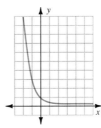

16. f has a maximum value at

$x = 3$; the maximum value is $f(3) = 8$.

Section 10.4

17. T is a function; it is one-to-one.   **18.** f is not one-to-one.   **19.** $f^{-1}(x) = \dfrac{1}{2}x + 4$

20.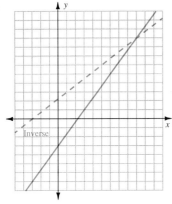

CHAPTER 11

EXERCISE SET 11.1

1. $x = 2$ **3.** $x = -\dfrac{3}{2}$ **5.** $x = -\dfrac{5}{2}$ **7.** $t = 0$ **9.** $y = 1$ **11.** $x = \dfrac{1}{4}$ **13.**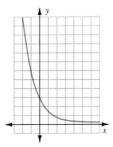

15.

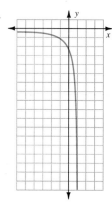

17.

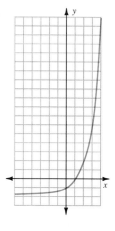

19.

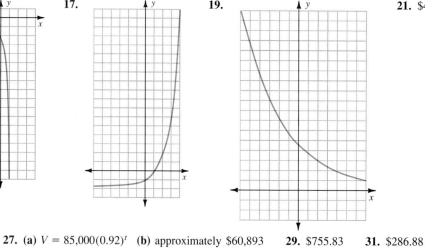

21. $4.02   **23.** $1.70

25. $160,653   **27. (a)** $V = 85,000(0.92)^t$ **(b)** approximately $60,893   **29.** $755.83   **31.** $286.88   **33.** $7529.46
35. $16,501.93   **37.** $9,948.51   **39.** $3,833.49   **41.** $20,200.93   **43.** approximately 2,158   **45. (a)** 1,000
(b) approximately 4,482   **47.** approximately 370.4 grams

Review Problems

49. The inverse is a function.

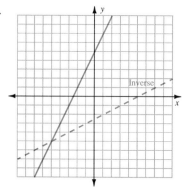

51. The inverse is not a function.

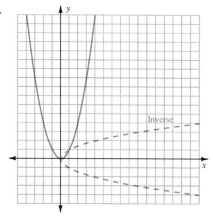

53. The inverse is not a function.

E N R I C H M E N T   E X E R C I S E S

1. 2; 2.25; 2.5937425; 2.7048138; 2.7169239; 2.7181459; 2.7182682. $\left(1 + \dfrac{1}{n}\right)^{n}$ becomes closer and closer to the number e.

2. approximately \$3,971.14   **3.** approximately \$8,925.21

E X E R C I S E   S E T   11.2

1. $\log_5 25 = 2$ **3.** $\log_3 \dfrac{1}{9} = -2$ **5.** $\log_{1/7} 49 = -2$ **7.** $\log_b V = t$ **9.** $8 = 4^{3/2}$ **11.** $1{,}000 = 10^3$

13. $\sqrt{e} = e^{1/2}$ **15.**

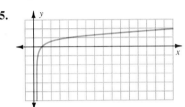

17.

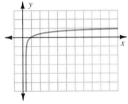

19.

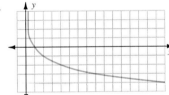

21.

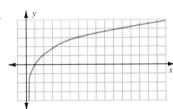

23. 4   **25.** 2   **27.** 3   **29.** 0

31. -3 **33.** $\dfrac{3}{4}$ **35.** $x = \dfrac{1}{8}$ **37.** $x = \dfrac{1}{3}$ **39.** $b = 10$ **41.** $b = \dfrac{1}{8}$ **43.** $y = \dfrac{1}{2}$ **45.** $y = \dfrac{4}{3}$
47. b is any positive real number.   **49.** $x = \pm 2\sqrt{2}$ **51.** about 79 times more powerful

Review Problems

53. $x = 2$ or $x = -6$ **55.** $x = -\dfrac{3}{2}$ or $x = \dfrac{1}{2}$ **57.** $x = -\dfrac{1}{2}$ or $x = \dfrac{3}{4}$

E N R I C H M E N T   E X E R C I S E S

1. e **2.** 0   **3.** 8   **4.** $\dfrac{1}{2}$ **5.** $\dfrac{1}{4}$ **6.**

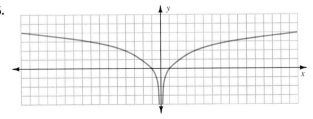

7.

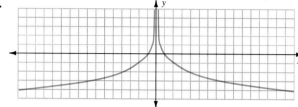

1. $\log_3 7 + \log_3 t$ **3.** $\log_{12} 5 - \log_{12} z$ **5.** $2 \log_4 x + \log_4 y - 5 \log_4 z$ **7.** $4 \log_b n + 2 \log_b m - \log_b p - \log_b q$

9. $5 + 3 \log_b c$ **11.** $\dfrac{2}{3} \log x + \dfrac{1}{3} \log y$ **13.** $\ln P + rt$ **15.** $\ln 4 - 3$ **17.** $10 \log I - 10 \log I_0$

19. $\dfrac{1}{3} \log_b x + \dfrac{4}{3} \log_b y - \dfrac{1}{3} \log_b 2 - \log_b z$ **21.** 1.1132 **23.** 1.5439 **25.** -0.2518 **27.** -0.8614 **29.** 2.1132

31. $\log_5 \dfrac{st^3}{2}$ **33.** $\log_b \dfrac{xy^4}{zt}$ **35.** $\log_3 \dfrac{\sqrt{x}}{\sqrt[3]{t^2}}$ **37.** $\log \left(\dfrac{E_1}{E_2} \right)^{10}$ **39.** $x = \dfrac{3}{8}$ **41.** $y = 42$ **43.** $x = 2\sqrt{2}$

45. $x = \pm 2\sqrt{2}$ **47.** $x = 2$ **49.** $x = 2$ **51.** $x = 2$ **53.** $x = 1 + \sqrt{2}$ **55.** $x = 3$ **57.** $x = 1$ **59.** $x = 6$
61. no solution **63.** $z = 3$

Review Problems

65. b^6 **67.** c^5 **69.** $\dfrac{1}{b^{12}}$ **71.** $\dfrac{a^4}{b^9}$ **73.** 2.3×10^{-4} **75.** 1.52×10^8 **77.** 5.93×10^{-3}

1. $\dfrac{1}{8}$ **2.** 27 **3.** $3\sqrt{2}$ **4.** $L = 10 \log I - 10 \log I_0 = 10(\log I - \log I_0) = 10 \log \dfrac{I}{I_0}$

5. By Exercise 4, $L_1 = 10 \log \dfrac{I_1}{I_0}$ and $L_2 = 10 \log \dfrac{I_2}{I_0}$. Therefore, $L_1 - L_2 = 10 \log \dfrac{I_1}{I_0} - 10 \log \dfrac{I_2}{I_0} = 10 \left(\log \dfrac{I_1}{I_0} - \log \dfrac{I_2}{I_0} \right) =$

$10 \log \dfrac{\dfrac{I_1}{I_0}}{\dfrac{I_2}{I_0}} = 10 \log \dfrac{I_1}{I_2}$

1. 1.55388 **3.** -1.40782 **5.** 4.12713 **7.** 1.70493 **9.** not a real number **11.** -8.429 **13.** 0.725
15. -4.794 **17.** -146.005 **19.** 0.312 **21.** 0.366 **23.** 0.353 **25.** 100.231 **27.** 5.347 **29.** 0.944
31. 122.732 **33.** 1.008 **35.** 0.367 **37.** 0.8116 **39.** 0.0043 **41.** 3.1206 **43.** -2.0888 or $0.9112 - 3$
45. -2.2366 or $0.7634 - 3$ **47.** 6.7818 **49.** 9.92 **51.** 1.09 **53.** 656 **55.** 0.0396 **57.** 0.00555
59. 0.000117 **61.** 1.18 **63.** 0.66 **65.** -4.14 **67.** 1.20 **69.** 1.18 **71.** -8.89 **73.** 3.47
75. (a) about 17 months **(b)** no **77.** approximately 0.0001796 **79.** 1.5395 **81.** 4.7472 **83.** 0.0247
85. 7.1233 **87.** 39.3822 **89.** $1,097.1552$

Review Problems

91. $f(1) = -2; f(2) = 1; f(3) = 4; f(4) = 7$ **93.** $f(2) = 5; f(4) = 17; f(6) = 37; f(8) = 65$ **95.** 20

E N R I C H M E N T E X E R C I S E

1. 10^{-7} moles per liter **2.** 9.1; basic **3.** The ion concentration is between 5.0×10^{-9} and 3.2×10^{-5}.

C H A P T E R   1 1   R E V I E W   E X E R C I S E   S E T

Section 11.1

1. $x = \dfrac{7}{2}$ **2.** $x = -\dfrac{5}{2}$ **3.** $694.36 **4.** $937.33 **5.** $1144.54 **6.** $88,776.93 **7.** about 3,795
8. about 471.98 grams

Section 11.2

9. $\log_{11} 121 = 2$ **10.** $\log_7 \dfrac{1}{49} = -2$ **11.** $10^{-2} = 0.01$ **12.** $e^{6.2766} = 532$ **13.** $3^2 = 9$ **14.** $5^3 = 125$

15. $\log_b Q = r$ **16.** $a^s = P$ **17.** $x = \dfrac{1}{2}$ **18.** $b = 2^{5/4}$ **19.** $y = -2$ **20.** $x = \pm 3\sqrt{3}$

Section 11.3

21. $3 \log_2 x + 2 \log_2 y - 4 \log_2 z$ **22.** $2 \log_5 t - 2$ **23.** $\ln A - rt$ **24.** $\dfrac{2}{3} + \log z$ **25.** $\log_4 \dfrac{3x}{y}$ **26.** $\ln P e^{rt}$

27. $x = \dfrac{12}{7}$ **28.** $x = \pm \dfrac{1}{9}$ **29.** $x = \dfrac{5}{2}$ **30.** $x = 2$ **31.** $x = 8$ **32.** $x = 0$ or $x = 1$ **33.** $x = 1$ or $x = 27$

Section 11.4

34. 3.76559 **35.** -2.30941 **36.** 0.82821 **37.** 80.020 **38.** 0.012 **39.** 123.841 **40.** 0.025 **41.** 0.6920
42. $0.5809 - 2$ **43.** 4.6749 **44.** 1.1619 using a calculator; 1.1618 using Table II. **45.** 3.5 **46.** 90.1
47. 0.00663 **48.** 0.0575 **49.** 0.0000193 **50.** 0.00559 **51.** 2.2763 **52.** 0.6588 **53.** 4.2734 **54.** -2.5557
55. 2.171 **56.** 2.379 **57.** -75.879 **58.** 2.773 **59.** 897.644 **60.** 8.633 **61.** ± 1.123 **62.** 13,209.667

63. 4.569 **64.** 11.038 **65.** 74% **66.** about $17\dfrac{2}{3}$ years

C H A P T E R   1 1   T E S T

Section 11.1

1. $x = \dfrac{3}{2}$ **2.** $140.61 **3.** $1,319.93 **4.** 242,057

Section 11.2

5. $\log_5 \dfrac{1}{125} = -3$ **6.** $3^3 = 27$ **7.** $\log_a T = r$ **8.** $b^n = M$ **9.** $x = 81$ **10.** $b = 4$

Section 11.3

11. $\log a + \dfrac{1}{3} \log y - 4 \log t$ **12.** $\ln Q_0 + nt$ **13.** $\log_3 \dfrac{4x}{H^2}$ **14.** $\ln T\sqrt{S}$ **15.** $x = \dfrac{1}{6}$ **16.** $x = \pm 2\sqrt{6}$
17. $x = 2$ **18.** $x = 1.5538$ **19.** $x = -1.6163$ **20.** $x = 235.0986$ **21.** $x = 0.0269$ **22.** $x = 1.1364$
23. $x = 3.7436$ **24.** $x = 2.1111$ **25.** $x = 115.1235$

CHAPTER 12

EXERCISE SET 12.1

1. $-1, 3, 7, 11$ **3.** $0, -1, -2, -3$ **5.** $2, 1, \dfrac{2}{3}, \dfrac{1}{2}$ **7.** $-3, -\dfrac{3}{4}, -\dfrac{1}{3}, -\dfrac{3}{16}$ **9.** $\dfrac{1}{4}, \dfrac{2}{5}, \dfrac{1}{2}, \dfrac{4}{7}$ **11.** $-1, 1, -1, 1$

13. $-2, 4, -8, 16$ **15.** $-1, \dfrac{1}{2}, -\dfrac{1}{3}, \dfrac{1}{4}$ **17.** $3, -1, \dfrac{1}{3}, -\dfrac{1}{9}$ **19.** $0, 2, 6, 12$ **21.** $3, 12, 48, 192$

23. $1, 2, -3, -2$ **25.** $10, 12, 14; a_n = 2n$ **27.** $12, 13, 14; a_n = n + 7$

29. $-1 + 4\sqrt{2}, -1 + 5\sqrt{2}, -1 + 6\sqrt{2}; a_n = -1 + (n - 1)\sqrt{2}$ **31.** $5, -6, 7; a_n = (-1)^{n+1}n$ **33.** $5{,}500, 6{,}050, 6{,}655$

Review Problems

35. 8 **37.** 110 **39.** $\dfrac{133}{60}$

ENRICHMENT EXERCISES

1. $1, 1, 2, 3, 5, 8, 13, 21$ **2. (a)** $1, 2, 6, 24, 120, 720$ **(b)** $a_n = n(n - 1)(n - 2)\cdots 2 \cdot 1$

3. (a) $3, 4, \dfrac{4}{3}, \dfrac{1}{3}, \dfrac{1}{4}, \dfrac{3}{4}, 3, 4, \dfrac{4}{3}, \dfrac{1}{3}$ **(b)** The sequence repeats itself after six terms.   **4.** $a_n = 2^{n-1}6$

5. $a_n = (1.06)^n 5{,}000$

EXERCISE SET 12.2

1. -10 **3.** 8 **5.** 14 **7.** -4 **9.** 0 **11.** $-\dfrac{4}{5}$ **13.** $\dfrac{25}{12}$ **15.** $\dfrac{31}{4}$ **17.** -6 **19.** $x + x^2 + x^3 + x^4$

21. $2 + 2x + 2x^2$ **23.** $-x + \dfrac{x^2}{2} - \dfrac{x^3}{3}$ **25.** $\displaystyle\sum_{i=1}^{6} i$ **27.** $\displaystyle\sum_{i=1}^{5} (i + 1)$ **29.** $\displaystyle\sum_{i=1}^{5} (i - 2)$ **31.** $\displaystyle\sum_{i=1}^{6} (2i - 1)$

33. $\displaystyle\sum_{i=1}^{4} (-1)^{i+1} i$ **35.** $\displaystyle\sum_{i=1}^{5} x^{2i+1}$ **37.** $\displaystyle\sum_{i=1}^{4} \left(x - \dfrac{1}{2}\right)^i$ **39.** $\displaystyle\sum_{i=1}^{7} (-1)^{i+1} x^{2i-2}$ **41.** 400 feet

Review Problems

43. -1 **45.** $\dfrac{8}{5}$

ENRICHMENT EXERCISES

1. (a) $\displaystyle\sum_{i=1}^{3} ca_i = ca_1 + ca_2 + ca_3 = c(a_1 + a_2 + a_3) = c\sum_{i=1}^{3} a_i$

(b) $\displaystyle\sum_{i=1}^{4} ca_i = ca_1 + ca_2 + ca_3 + ca_4 = c(a_1 + a_2 + a_3 + a_4) = c\sum_{i=1}^{4} a_i$ **(c)** $c\displaystyle\sum_{i=1}^{n} a_i$ **2. (a)** $\displaystyle\sum_{i=1}^{3} c = c + c + c = 3c$

(b) $\displaystyle\sum_{i=1}^{4} c = c + c + c + c = 4c$ **(c)** nc **3. (a)** 12 **(b)** -20

4. (a) $\displaystyle\sum_{i=1}^{3} (a_i + b_i) = (a_1 + b_1) + (a_2 + b_2) + (a_3 + b_3) = (a_1 + a_2 + a_3) + (b_1 + b_2 + b_3) = \sum_{i=1}^{3} a_i + \sum_{i=1}^{3} b_i$

(b) $\sum_{i=1}^{4}(a_i + b_i) = (a_1 + b_1) + (a_2 + b_2) + (a_3 + b_3) + (a_4 + b_4) = (a_1 + a_2 + a_3 + a_4) + (b_1 + b_2 + b_3 + b_4) = \sum_{i=1}^{4} a_i + \sum_{i=1}^{4} b_i$

(c) $\sum_{i=1}^{n} a_i + \sum_{i=1}^{n} b_i$ **5.** 203   **6.** 145   **7.** $-2n^3 - n^2 + 4n$

E X E R C I S E S E T 12.3

1. 4, 5, 6, 7   **3.** $-1, 4, 9, 14$ **5.** $-12, -2, 8, 18$ **7.** 8, 5, 2, -1 **9.** $-1, -2, -3, -4$ **11.** $\dfrac{1}{2}, \dfrac{3}{4}, 1, \dfrac{5}{4}$

13. $a_n = 5 - n$ **15.** $a_n = 55 - 22n$ **17.** $a_n = -\dfrac{3}{2} - \dfrac{1}{2}n$ **19.** 51   **21.** $-14\dfrac{1}{2}$ **23.** $a_n = 2n$

25. $S_1 = 1; S_2 = 4; S_3 = 11; S_4 = 20$ **27.** $S_1 = 1; S_2 = 5; S_3 = 14; S_4 = 30$ **29.** $S_1 = 1; S_2 = \dfrac{3}{2}; S_3 = \dfrac{11}{6}; S_4 = \dfrac{25}{12}$

31. $S_1 = -1; S_2 = 0; S_3 = 3; S_4 = 8$ **33.** $S_1 = -1; S_2 = -\dfrac{3}{4}; S_3 = -\dfrac{31}{36}; S_4 = -\dfrac{115}{144}$ **35.** $S_9 = 144$ **37.** $S_8 = 76$

39. $S_{21} = -126$ **41.** $S_{31} = 775$ **43.** 2, 8, 32, 128   **45.** $4, 2, 1, \dfrac{1}{2}$ **47.** $2, -2, 2, -2$ **49.** $27, -9, 3, -1$

51. $a_n = 2^{n-1}$ **53.** $a_n = 3\left(\dfrac{1}{3}\right)^{n-1}$ **55.** $a_n = 3\left(-\dfrac{1}{2}\right)^{n-1}$ **57.** $a_n = -6\left(-\dfrac{1}{3}\right)^{n-1}$ **59.** 96   **61.** $-\dfrac{1}{27}$

63. 124   **65.** $\dfrac{93}{16}$ **67.** $\dfrac{11}{4}$ **69.** 80   **71.** 3   **73.** 1.248   **75.** -14 **77.** 1,365   **79.** $\dfrac{189}{16}$ **81.** $\dfrac{819}{64}$

83. 326.5 feet   **85.** 1,155   **87.** $15\dfrac{7}{8}$ **89.** $\dfrac{25}{2}$ **91.** -12 **93.** 20   **95.** 4   **97.** -20 **99.** 50

Review Problems

101. $a_3 + 3a^2b + 3ab^2 + b^3$ **103.** $a^4 + 4a^3b + 6a^2b^2 + 4ab^3 + b^4$ **105.** $x^3 - 9x^2y + 27xy^2 - 27y^3$

E N R I C H M E N T E X E R C I S E S

1. 9   **2.** $\dfrac{4}{3}$ **3.** $\dfrac{16}{3}$ **4.** $\dfrac{3}{4}$ **5.** 1   **6.** $\dfrac{8}{33}$ **7.** 3   **8.** $\dfrac{9}{7}$ **9.** -0.4547 **10.** 24.4118
11. $(-\infty, -1) \cup (1, +\infty)$

E X E R C I S E S E T 12.4

1. $a^4 + 12a^3 + 54a^2 + 108a + 81$ **3.** $x^3 - 3x^2 + 3x - 1$ **5.** $32x^5 + 80x^4y + 80x^3y^2 + 40x^2y^3 + 10xy^4 + y^5$
7. $27x^3 - 54x^2y + 36xy^2 - 8y^3$ **9.** $x^{12} + 6x^{10}y^2 + 15x^8y^4 + 20x^6y^6 + 15x^4y^8 + 6x^2y^{10} + y^{12}$
11. $x^3 - 6x^2z^2 + 12xz^4 - 8z^6$ **13.** $\dfrac{a^4}{16} - 2a^3 + 24a^2 - 128a + 256$ **15.** $8a^6 - 6a^4b + \dfrac{3}{2}a^2b^2 - \dfrac{b^3}{8}$
17. $x^{11} + 11x^{10}y + 55x^9y^2$ **19.** $x^{10} - 10x^9y + 45x^8y^2$ **21.** $x^{24} + 12x^{22}a + 66x^{20}a^2$ **23.** $s^{42} - 14s^{39}t^2 + 91s^{36}t^4$
25. $28a^6b^2$ **27.** $-7x^6y$ **29.** $720x^3y^2$ **31.** $54x^2y^6$

E N R I C H M E N T E X E R C I S E S

1. 10   **2.** 20   **3.** 8   **4.** 15   **5.** 2,598,960   **6.** The number of subsets of a set of n elements is
$\dbinom{n}{0} + \dbinom{n}{1} + \dbinom{n}{2} + \dbinom{n}{3} + \cdots + \dbinom{n}{n} = (1 + 1)^n = 2^n$

CHAPTER   12   REVIEW   EXERCISE   SET

Sections 12.1 and 12.3

1. $-6, -3, 0, 3$ **2.** $1, -2, 3, -4$ **3.** $\dfrac{2}{3}, \dfrac{3}{4}, \dfrac{4}{5}, \dfrac{5}{6}$ **4.** $2, -6, 18, -54$ **5.** $-4, 1, 6, 11$ **6.** $2, 5, 8, 11$

7. $3, \dfrac{3}{2}, \dfrac{3}{4}, \dfrac{3}{8}$ **8.** $a_n = -1 + 2n$ **9.** $a_n = (-1)^n$ **10.** $a_n = \dfrac{(-1)^{n+1}}{n}$ **11.** $a_n = 10\left(\dfrac{1}{2}\right)^{n-1}$ **12.** $a_n = 3\left(\dfrac{1}{4}\right)^{n-1}$

Sections 12.2 and 12.3

13. 39 **14.** $\dfrac{7}{8}$ **15.** $\dfrac{205}{36}$ **16.** $\displaystyle\sum_{i=1}^{5}(-1 + 3i)$ **17.** $\displaystyle\sum_{i=1}^{5}\dfrac{2i-1}{2i}$ **18.** $\displaystyle\sum_{i=1}^{6}(-2)^{i-1}$ **19.** $\displaystyle\sum_{i=1}^{4}\dfrac{x^{2i+1}}{2i+1}$

20. $S_1 = 1; S_2 = 6; S_3 = 16; S_4 = 32$ **21.** $S_1 = 2; S_2 = 3; S_3 = \dfrac{11}{3}; S_4 = \dfrac{25}{6}$ **22.** $S_1 = 1; S_2 = -1; S_3 = 2; S_4 = -2$

23. 90 **24.** -198 **25.** 88 **26.** -80 **27.** $\dfrac{31}{16}$ **28.** $-\dfrac{7}{9}$ **29.** 20 **30.** $\dfrac{38}{3}$ **31.** 3 **32.** $\dfrac{5}{7}$ **33.** $-\dfrac{9}{2}$

34. $\dfrac{10}{7}$

Section 12.4

35. 20 **36.** 66 **37.** 56 **38.** $a^3 + 12a^2 + 48a + 64$ **39.** $x^4 - 8x^3 + 24x^2 - 32x + 16$
40. $16x^4 - 96x^3y + 216x^2y^2 - 216xy^3 + 81y^4$ **41.** $x^{10} + 5x^8y^3 + 10x^6y^6 + 10x^4y^9 + 5x^2y^{12} + y^{15}$
42. $x^{10} + 30x^9y + 405x^8y^2$

CHAPTER   12   TEST

Section 12.1

1. $9, 11, 13, 15$ **2.** $-1, \dfrac{1}{4}, -\dfrac{1}{9}, \dfrac{1}{16}$ **3.** $-3, -1, 1, 3$ **4.** $a_n = (-1)^{n+1}2n$ **5.** $a_n = \dfrac{1}{2n-1}$ **6.** $a_n = 2\left(\dfrac{1}{3}\right)^{n-1}$

Section 12.2

7. 84 **8.** 6 **9.** $x^2 + 2x^3 + 3x^4 + 4x^5 + 5x^6$ **10.** $x - x^2 + x^3 - x^4$ **11.** $\displaystyle\sum_{i=1}^{7} i$ **12.** $\displaystyle\sum_{i=1}^{3}(z-3)^i$

Section 12.3

13. $-6, -1, 4, 9$ **14.** $a_n = 10 - 2n$ **15.** -149 **16.** 180 **17.** $1, -3, 9, -27$ **18.** $\dfrac{484}{81}$

Section 12.4

19. 10 **20.** $8a^3 + 36a^2 + 54a + 27$ **21.** $x^4 - 12x^3y + 54x^2y^2 - 108xy^3 + 81y^4$ **22.** $x^6 + 12x^4t + 48x^2t^2 + 64t^3$
23. $u^{12} - 12u^{11}v + 66u^{10}v^2$ **24.** $-160r^3y^3$

How Text Meets State Requirements

The following table lists the Texas TASP skills and their location in *Intermediate Algebra*.

Description of TASP Skill	Chapter, Section
Use number concepts and computation skills	1.1, 1.2, 1.3, 1.4
Solve word problems involving integers, fractions, or decimals (including percents, ratios, and proportions)	1.1, 1.2, 1.3, 1.4
Interpret information from a graph, table, or chart	7.1, 7.2, 7.3, 7.5
Graph numbers or number relationships	7.1, 7.2, 7.4
Solve one- and two-variable equations	2.2, 8.1, 8.4, 8.5
Solve word problems involving one and two variables	2.4, 3.9, 4.7, 6.4, 7.5, 8.2
Understand operations with algebraic expressions	1.5, 2.1, 2.2, 2.3, 3.1, 3.2, 3.3, 3.4, 3.5, 3.6, 3.7, 3.8, 4.1, 4.2, 4.3, 4.4, 4.5, 5.1, 5.2, 5.3, 5.4, 5.5, 5.6
Solve problems involving quadratic equations	3.8, 3.9, 6.1, 6.2, 6.3, 6.4, 6.6
Solve problems involving geometric figures	
Apply reasoning skills	2.3, 2.4, 3.9, 4.7, 6.4, 7.5, 8.2
	All Enrichment Exercises

The following table lists the California ELM skills and the location of those skills in *Intermediate Algebra*.

Description of ELM Skill	Chapter, Section
Whole numbers and their operations	1.2, 1.3
Fractions and their operations	1.1, 1.3
Exponentiation and square roots	1.2, 5.1
Applications (percents and word problems)	2.3
Simplification of a polynomial by grouping—one and two variables	3.2
Evaluation of a polynomial—one and two variables	3.2
Addition and subtraction of polynomials	3.2
Multiplication of a monomial with a polynomial	3.3
Multiplication of two binomials	3.3
Squaring a binomial	3.3
Division of polynomials with a monomial divisor—no remainder	4.2
Division of polynomials with a linear binomial divisor—no remainder	4.2
Factoring polynomials by finding common factors	3.4
Factoring a trinomial	3.5
Factoring a difference of squares	3.6
Simplification of a rational expression by cancellation of common factors—one and two variables	4.1
Evaluation of a rational expression	4.1
Addition and subtraction of rational expressions	4.4
Multiplication and division of rational expressions	4.3
Simplification of a compound rational expression	4.5
Definition of exponentiation with positive exponents	3.1
Laws of exponents with positive exponents	3.1
Simplification of an expression with positive exponents	3.1
Definition of exponentiation with integral exponents	3.1

Description of ELM Skill	Chapter, Section
Laws of exponents with integral exponents	3.1
Simplification of an expression with integral exponents	3.1
Scientific notation	3.1
Definition of radical sign	5.1
Simplification of products under a single radical	5.4
Addition and subtraction of radical expressions	5.5
Multiplication of radical expressions	5.6
Solution of a simple radical equation	5.7
Solution of a linear equation in one unknown with numerical coefficients	2.2
Solution of a linear equation in one unknown with literal coefficients	2.4
Solution of a simple equation in one unknown which is reducible to a linear equation	2.2
Solution of a linear inequality in one unknown with numerical coefficients	2.5
Solution of two linear equations in two unknowns with numerical coefficients—by substitution	8.1
Solution of two linear equations in two unknowns with numerical coefficients—by elimination	8.1
Solution of a quadratic equation from factored form	3.8
Solution of a quadratic equation by factoring	3.8
Graphing a point on the number line	1.2
Graphing linear inequalities in one unknown	2.5, 2.6
Graphing a point in the coordinate plane	7.1
Graphing a simple linear equation: $y = mx$, $y = b$, $x = b$	7.1
Reading data from a graph	7.2
Measurement formulas for perimeter and area of triangles, squares, rectangles, and parallelograms	2.4
Measurement formulas for circumference and area of circles	2.4
The Pythagorean theorem	3.9

The following table lists the Florida CLAST skills and their location in *Intermediate Algebra*.

Skill Number	Description of CLAST Skill	Chapter, Section
I.A.1a	Add, subtract rational numbers	1.1, 1.3
1.A.1b	Multiply, divide rational numbers	1.1, 1.3
I.B.2a	Calculate distance	1.1
I.B.2b	Calculate areas	2.4
I.C.1a	Add, subtract real numbers	1.3
I.C.1b	Multiply, divide real numbers	1.3
I.C.2	Order of operations	1.4, 1.5
I.C.3	Scientific notation	3.1
I.C.4	Solve linear equations and inequalities	2.2, 2.5, 2.6
I.C.5	Use given formulas	2.4
I.C.6	Find function values	10.1, 10.2
I.C.7	Factor quadratic expressions	3.8
I.C.8	Find roots of quadratic equations	6.3, 6.6
II.A.1	Meaning of exponents	1.3, 5.2
II.A.3	Equivalent forms of rationals	1.2
II.A.4	Order relation	1.2
II.C.1	Properties of operations	1.4
II.C.2	Checking equations or inequalities	2.2, 2.5
II.C.3	Proportion and variation	7.6
II.C.4	Regions of coordinate plane	7.1
IV.B.1	Solve perimeter, area, volume problems	2.4
IV.B.2	Pythagorean property	3.9
IV.C.1	Solve real-world problems	2.3, 2.4, 3.9, 4.7, 6.4, 7.5, 8.2
IV.C.2	Solve problems involving structure and logic of algebra	2.1, 2.2

Index